PROBABILITY, RANDOM VARIABLES, AND STOCHASTIC PROCESSES

N Dessipris
Uni of Essex
1986

McGraw-Hill Series in Electrical Engineering

Consulting Editor
Stephen W. Director, Carnegie-Mellon University

Networks and Systems
Communications and Information Theory
Control Theory
Electronics and Electronic Circuits
Power and Energy
Electromagnetics
Computer Engineering
Introductory and Survey
Radio, Television, Radar, and Antennas

Previous Consulting Editors

Ronald M. Bracewell, Colin Cherry, James F. Gibbons, Willis W. Harman, Hubert Heffner, Edward W. Herold, John G. Linvill, Simon Ramo, Ronald A. Rohrer, Anthony E. Siegman, Charles Susskind, Frederick E. Terman, John G. Truxal, Ernst Weber, and John R. Whinnery

Communications and Information Theory

Consulting Editor
Stephen W. Director, Carnegie-Mellon University

Abramson: *Information Theory and Coding*
Antoniou: *Digital Filters: Analysis and Design*
Bennett: *Introduction to Signal Transmission*
Carlson: *Communications Systems*
Cherin: *An Introduction to Optical Fibers*
Davenport: *Probability and Random Processes: An Introduction for Applied Scientists and Engineers*
Drake: *Fundamentals of Applied Probability Theory*
Guiasu: *Information Theory with New Applications*
Keiser: *Optical Fiber Communications*
Melsa and Cohn: *Decision and Estimation Theory*
Papoulis: *Probability, Random Variables, and Stochastic Processes*
Papoulis: *Signal Analysis*
Papoulis: *The Fourier Integral and Its Applications*
Peebles: *Probability, Random Variables, and Random Signal Principles*
Proakis: *Digital Communications*
Schwartz: *Information Transmission, Modulation, and Noise*
Schwartz, Bennett, and Stein: *Communication Systems and Techniques*
Schwartz and Shaw: *Signal Processing*
Shooman: *Probabilistic Reliability: An Engineering Approach*
Taub and Schilling: *Principles of Communication Systems*
Viterbi: *Principles of Coherent Communication*
Viterbi and Omura: *Principles of Digital Communication and Coding*

PROBABILITY, RANDOM VARIABLES, AND STOCHASTIC PROCESSES

Second Edition

Athanasios Papoulis

Polytechnic Institute of New York

INTERNATIONAL STUDENT EDITION

McGRAW-HILL INTERNATIONAL BOOK COMPANY

Auckland Bogotá Guatemala Hamburg Johannesburg Lisbon London Madrid Mexico New Delhi Panama Paris San Juan São Paulo Singapore Sydney Tokyo

PROBABILITY, RANDOM VARIABLES, AND STOCHASTIC PROCESSES
INTERNATIONAL STUDENT EDITION

2nd printing 1985

This book was set in Times Roman.
The editor was T. Michael Slaughter;
The production supervisor was Diane Renda.
The cover was designed by Jerry Wilke.
Project supervision was done by Santype International Limited.

Library of Congress Cataloging in Publication Data
Papoulis, Athanasios, date
Probability, random variables, and stochastic processes.

(McGraw-Hill series in electrical engineering. Communications and information theory)
Includes index.
1. Probabilities. 2. Random variables.
3. Stochastic processes. I. Title. II. Series.
QA273. P2 1984 519.2 82-14927
ISBN 0-07-048468-6

When ordering this title use ISBN 0-07-Y66465-X

Printed in Singapore by
Chong Moh Offset Printing Pte. Ltd.

CONTENTS

PREFACE TO THE SECOND EDITION

This is an extensively revised edition reflecting the developments of the last two decades. Several new topics are added, important areas are strengthened, and sections of limited interest are eliminated. Most additions, however, deal with applications; the first ten chapters are essentially unchanged.

In the selection of the new material I have attempted to concentrate on subjects that not only are of current interest, but also contribute to a better understanding of the basic properties of stochastic processes. The new material includes the following:

Discrete time processes with applications in system theory
Innovations, factorization, spectral representation
Queueing theory, level crossings, spectra of FM signals, sampling theory
Mean square estimation, orthonormal expansions, Levinson's algorithm, Wold's decomposition, Wiener, lattice, and Kalman filters
Spectral estimation, windows, extrapolation, Burg's method, detection of line spectra

The book concludes with a self-contained chapter on entropy developed axiomatically from first principles. It is presented in the context of earlier chapters, and it includes the method of maximum entropy in parameter estimation and elements of coding theory.

As in the first edition, I made a special effort to stress the conceptual difference between mental constructs and physical reality. This difference is summarized in the following paragraph, taken from the first edition:

> *Scientific theories deal with concepts, not with reality. All theoretical results are derived from certain axioms by deductive logic. In physical sciences the theories are so formulated as to correspond in some useful sense to the real world, whatever that may mean. However, this correspondence is approximate, and the physical justification of all theoretical conclusions is based on some form of inductive reasoning.*

Responding to comments by a number of readers over the years, I would like to emphasize that this passage in no way questions the existence of natural laws (patterns). It is merely a reminder of the fundamental difference between concepts and reality.

During the preparation of the manuscript I had the benefit of lengthy discussions with a number of colleagues and friends. I thank in particular Hans Schreiber of Grumman, William Shanahan of Norden Systems, and my colleagues Frank Cassara and Basil Maglaris for their valuable suggestions. I wish also to express my appreciation to Mrs. Nina Adamo for her expert typing of the manuscript.

Athanasios Papoulis

PREFACE TO THE FIRST EDITION

Several years ago I reached the conclusion that the theory of probability should no longer be treated as adjunct to statistics or noise or any other terminal topic, but should be included in the basic training of all engineers and physicists as a separate course. I made then a number of observations concerning the teaching of such a course, and it occurs to me that the following excerpts from my early notes might give you some insight into the factors that guided me in the planning of this book:

"Most students, brought up with a deterministic outlook of physics, find the subject unreliable, vague, difficult. The difficulties persist because of inadequate definition of the first principles, resulting in a constant confusion between assumptions and logical conclusions. Conceptual ambiguities can be removed only if the theory is developed axiomatically. They say that this approach would require measure theory, would reduce the subject to a branch of mathematics, would force the student to doubt his intuition leaving him without convincing alternatives, but I don't think so. I believe that most concepts needed in the applications can be explained with simple mathematics, that probability, like any other theory, should be viewed as a conceptual structure and its conclusions should rely not on intuition but on logic. The various concepts must, of course, be related to the physical world, but such motivating sections should be separated from the deductive part of the theory. Intuition will thus be strengthened, but not at the expense of logical rigor.

"There is an obvious lack of continuity between the elements of probability as presented in introductory courses, and the sophisticated concepts needed in today's applications. How can the average student, equipped only with the probability of cards and dice, understand prediction theory or harmonic analysis? The applied books give at most a brief discussion of background material; their objective is not the use of the applications to strengthen the student's understanding of basic concepts, but rather a detailed discussion of special topics.

"Random variables, transformations, expected values, conditional densities, characteristic functions cannot be mastered with mere exposure. These concepts must be clearly defined and must be developed, one at a time, with sufficient elaboration. Special topics should be used to illustrate the theory, but they must be so presented as to minimize peripheral, descriptive material and to concentrate on probabilistic content. Only then the student can learn a variety of applications with economy and perspective."

I realized that to teach a convincing course, a course that is not a mere presentation of results but a connected theory, I would have to reexamine not only the development of special topics, but also the proofs of many results and the method of introducing the first principles.

"The theory must be mathematical (deductive) in form but without the generality or rigor of mathematics. The philosophical meaning of probability must somehow be discussed. This is necessary to remove the mystery associated with probability and to convince the student of the need for an axiomatic approach and a clear distinction between assumptions and logical conclusions. The axiomatic foundation should not be a mere appendix but should be recognized throughout the theory.

"Random variables must be defined as functions with domain an abstract set of experimental outcomes and not as points on the real line. Only then infinitely dimensional spaces are avoided and the extension to stochastic processes is simplified.

"The inadequacy of averages as definitions and the value of an underlying space is most obvious in the treatment of stochastic processes. Time averages must be introduced as stochastic integrals, and their relationship to the statistical parameters of the process must be established only in the form of ergodicity.

"The emphasis on second-order moments and spectra, utilizing the student's familiarity with systems and transform techniques, is justified by the current needs.

"Mean-square estimation (prediction and filtering), a topic of considerable importance, needs a basic reexamination. It is best understood if it is divorced from the details of integral equations or the calculus of variations, and is presented as an application of the orthogonality principle (linear regression), simply explained in terms of random variables.

"To preserve conceptual order, one must sacrifice continuity of special topics, introducing them as illustrations of the general theory."

These ideas formed the framework of a course that I taught at the Polytechnic Institute of Brooklyn. Encouraged by the students' reaction, I decided to make it into a book. I should point out that I did not view my task as an impersonal presentation of a complete theory, but rather as an effort to explain the essence of this theory to a particular group of students. The book is written neither for the handbook-oriented students nor for the sophisticated few who can learn the subject from advanced mathematical texts. It is written for the majority of engineers and physicists who have sufficient maturity to appreciate and follow a logical presentation, but, because of their limited mathematical background, would find a book such as Doob's too difficult for a beginning text.

Although I have included many useful results, some of them new, my hope is that the book will be judged not for completeness but for organization and clarity. In this context I would like to anticipate a criticism and explain my approach. Some readers will find the proofs of many important theorems lacking in rigor. I emphasize that it was not out of negligence, but after considerable thought, that I decided to give, in several instances, only plausibility arguments. I realize too well that "a proof is a proof or it is not." However, a rigorous proof must be preceded by a clarification of the new idea and by a plausible explanation of its validity. I felt that, for the purposes of this book, the emphasis should be placed on explanation, facility, and economy. I hope that this approach will give you not only a working knowledge, but also an incentive for a deeper study of this fascinating subject.

Although I have tried to develop a personal point of view in practically every topic, I recognize that I owe much to other authors. In particular, the books "Stochastic Processes" by J. L. Doob and "Théorie des Functions Aléatoires" by A. Blanc-Lapierre and R. Fortet influenced greatly my planning of the chapters on stochastic processes.

Finally, it is my pleasant duty to express my sincere gratitude to Mischa Schwartz for his encouragement and valuable comments, to Ray Pickholtz for his many ideas and constructive suggestions, and to all my colleagues and students who guided my efforts and shared my enthusiasm in this challenging project.

Athanasios Papoulis

PART ONE

PROBABILITY AND RANDOM VARIABLES

CHAPTER
ONE

THE MEANING OF PROBABILITY

1-1 INTRODUCTION

The theory of probability deals with averages of mass phenomena occurring sequentially or simultaneously: electron emission, telephone calls, radar detection, quality control, system failure, games of chance, statistical mechanics, turbulence, noise, birth and death rates, and queueing theory, among many others.

It has been *observed* that in these and other fields certain averages approach a constant value as the number of observations increases and this value remains the same if the averages are evaluated over any subsequence specified before the experiment is performed. In the coin experiment, for example, the percentage of heads approaches 0.5 or some other constant, and the same average is obtained if we consider every fourth, say, tossing (no betting system can beat the roulette).

The purpose of the theory is to describe and predict such averages in terms of probabilities of events. The probability of an event $\mathscr{A}$ is a number $P(\mathscr{A})$ assigned to this event. This number could be interpreted as follows:

If the experiment is performed n times and the event $\mathscr{A}$ occurs $n_{\mathscr{A}}$ times, then, *with a high degree of certainty*, the relative frequence $n_{\mathscr{A}}/n$ of the occurrence of $\mathscr{A}$ is *close* to $P(\mathscr{A})$:

$$P(\mathscr{A}) \simeq n_{\mathscr{A}}/n \tag{1-1}$$

provided that n is *sufficiently large*.

This interpretation is imprecise: the terms "with a high degree of certainty," "close," and "sufficiently large" have no clear meaning. However, this lack of precision cannot be avoided. If we attempt to define in probabilistic terms the "high degree of certainty" we shall only postpone the inevitable conclusion that probability, like any physical theory, is related to physical phenomena only in inexact terms. Nevertheless, the theory is an exact discipline developed logically from clearly defined axioms, and when it is applied to real problems, *it works*.

Observation, deduction, prediction In the applications of probability to real problems, the following steps must be clearly distinguished:

Step 1 (physical) We determine by an inexact process the probabilities $P(\mathscr{A}_i)$ of certain events $\mathscr{A}_i$.

This process could be based on the relationship (1-1) between probability and observation: The probabilistic data $P(\mathscr{A}_i)$ equal the observed ratios $n_{\mathscr{A}_i}/n$. It could also be based on "reasoning" making use of certain symmetries: If, out of a total of N outcomes, there are $N_{\mathscr{A}}$ outcomes favorable to the event $\mathscr{A}$, then $P(\mathscr{A}) \simeq N_{\mathscr{A}}/N$.

For example, if a loaded die is rolled 1000 times and *five* shows 203 times, then the probability of *five* equals about 0.2. If the die is fair, then, because of its symmetry, the probability of *five* equals 1/6.

Step 2 (conceptual) We assume that probabilities satisfy certain axioms, and by deductive reasoning we determine from the probabilities $P(\mathscr{A}_i)$ of certain events $\mathscr{A}_i$ the probabilities $P(\mathscr{B}_j)$ of other events $\mathscr{B}_j$.

For example, in the game with a fair die we deduce that the probability of the event *even* equals 3/6. Our reasoning is of the following form:

$$\text{If } P(1) = \cdots = P(6) = \frac{1}{6} \qquad \text{then} \qquad P(even) = \frac{3}{6}$$

Step 3 (physical) We make a physical *prediction* based on the numbers $P(\mathscr{B}_j)$ so obtained.

This step could rely on (1-1) applied in reverse: If we perform the experiment n times and an event $\mathscr{B}$ occurs $n_{\mathscr{B}}$ times, then $n_{\mathscr{B}} \simeq nP(\mathscr{B})$.

If, for example, we roll a fair die 1000 times, our prediction is that *even* will show about 500 times.

We could not emphasize too strongly the need for separating the above three steps in the solution of a problem. We must make a clear distinction between the data that are determined empirically and the results that are deduced logically.

Steps 1 and 3 are based on *inductive reasoning*. Suppose, for example, that we wish to determine the probability of *heads* of a given coin. Should we toss the coin 100 or 1000 times? If we toss it 1000 times and the average number of heads equals 0.48 what kind of prediction can we make on the basis of this observation? Can we deduce that at the next 1000 tossings the number of heads will be about 480? Such questions can be answered only inductively.

In this book, we consider only step 2, that is, from certain probabilities we derive *deductively* other probabilities. One might argue that such derivations are mere tautologies because the results are contained in the assumptions. This is true in the same sense that the intricate equations of motion of a satellite are included in Newton's laws.

To conclude, we repeat that the probability $P(\mathscr{A})$ of an event $\mathscr{A}$ will be interpreted as a number assigned to this event as mass is assigned to a body or resistance to a resistor. In the development of the theory, we will not be concerned about the "physical meaning" of this number. This is what is done in circuit analysis, in electromagnetic theory, in classical mechanics, or in any other scientific discipline. These theories are, of course, of no value to physics unless they help us solve real problems. We must assign specific, if only approximate, resistances to real resistors and probabilities to real events (step 1); we must also give physical meaning to all conclusions that are derived from the theory (step 3). But this link between concepts and observation must be separated from the purely logical structure of each theory (step 2).

As an illustration, we discuss in the next example the interpretation of the meaning of resistance in circuit theory.

Example 1-1 A resistor is commonly viewed as a two-terminal device whose voltage is proportional to the current

$$R = \frac{v(t)}{i(t)} \tag{1-2}$$

This, however, is only a convenient abstraction. A real resistor is a complex device with distributed inductance and capacitance having no clearly specified terminals. A relationship of the form (1-2) can, therefore, be claimed only within certain error, in certain frequency ranges, and with a variety of other qualifications. Nevertheless, in the development of circuit theory we ignore all these uncertainties. We assume that the resistance R is a precise number satisfying (1-2) and we develop a theory based on (1-2) and on Kirchhoff's laws. It would not be wise, we all agree, if at each stage of the development of the theory we were concerned with the *true* meaning of R.

1-2 THE DEFINITIONS

In this section, we discuss various definitions of probability and their role in our investigation.

Axiomatic Definition

We shall use the following concepts from set theory (for details see Chap. 2): The certain event $\mathscr{S}$ is the event that occurs in every trial. The union $\mathscr{A} + \mathscr{B}$ of two events $\mathscr{A}$ and $\mathscr{B}$ is the event that occurs when $\mathscr{A}$ or $\mathscr{B}$ or both occur. The intersection $\mathscr{A}\mathscr{B}$ of the events $\mathscr{A}$ and $\mathscr{B}$ is the event that occurs when both events $\mathscr{A}$ and $\mathscr{B}$ occur. The events $\mathscr{A}$ and $\mathscr{B}$ are mutually exclusive if the occurrence of one of them excludes the occurrence of the other.

We shall illustrate with the die experiment: The certain event is the event that occurs whenever any one of the six faces shows. The union of the events *even* and *less than 3* is the event *1 or 2 or 4 or 6* and their intersection is the event *2*. The events *even* and *odd* are mutually exclusive.

The axiomatic approach to probability is based on the following three postulates and on nothing else: The probability $P(\mathscr{A})$ of an event $\mathscr{A}$ is a positive number assigned to this event

$$P(\mathscr{A}) \geq 0 \tag{1-3}$$

The probability of the certain event equals 1

$$P(\mathscr{S}) = 1 \tag{1-4}$$

If the events $\mathscr{A}$ and $\mathscr{B}$ are mutually exclusive, then

$$P(\mathscr{A} + \mathscr{B}) = P(\mathscr{A}) + P(\mathscr{B}) \tag{1-5}$$

This approach to probability is relatively recent (A. Kolmogoroff†, 1933). However, in our view, it is the best way to introduce a probability even in elementary courses. It emphasizes the deductive character of the theory, it avoids conceptual ambiguities, it provides a solid preparation for sophisticated applications, and it offers at least a beginning for a deeper study of this important subject.

The axiomatic development of probability might appear overly mathematical. However, as we hope to show, this is not so. The elements of the theory can be adequately explained with basic calculus.

Relative Frequency Definition

The relative frequency approach is based on the following definition: The probability $P(\mathscr{A})$ of an event $\mathscr{A}$ is the limit

$$P(\mathscr{A}) = \lim_{n \to \infty} \frac{n_{\mathscr{A}}}{n} \tag{1-6}$$

where $n_{\mathscr{A}}$ is the number of occurrences of $\mathscr{A}$ and n is the number of trials.

This definition appears reasonable. Since probabilities are used to describe relative frequencies, it is natural to define them as limits of such frequencies. The problems associated with a priori definitions are eliminated, one might think, and the theory is founded on observation.

However, although the relative frequency concept is fundamental in the applications of probability (steps 1 and 3), its use as the basis of a deductive theory (step 2) must be challenged. Indeed, in a physical experiment, the numbers $n_{\mathscr{A}}$ and n might be large but they are only finite; their ratio cannot, therefore, be equated, even approximately, to a limit. If (1-6) is used to define $P(\mathscr{A})$, the limit

† A. Kolmogoroff: Grundbegriffe der Wahrscheinlichkeits Rechnung, *Ergeb. Math und ihrer Grensg.* vol. 2, 1933.

must be accepted as a *hypothesis*, not as a number that can be determined experimentally.

Early in the century, Von Mises† used (1-6) as the foundation for a new theory. At that time, the prevailing point of view was still the classical and his work offered a welcome alternative to the a priori concept of probability, challenging its metaphysical implications and demonstrating that it leads to useful conclusions mainly because it makes implicit use of relative frequencies based on our collective experience. The use of (1-6) as the basis for deductive theory has not, however, enjoyed wide acceptance even though (1-6) relates $P(\mathscr{A})$ to observed frequencies. It has generally been recognized that the axiomatic approach (Kolmogoroff) is superior.

We shall venture a comparison between the two approaches using as illustration the definition of the resistance R of an ideal resistor. We can define R as a limit

$$R = \lim_{n\to\infty} \frac{e(t)}{i_n(t)}$$

where $e(t)$ is a voltage source and $i_n(t)$ are the currents of a sequence of real resistors that tend in some sense to an ideal two-terminal element. This definition might show the relationship between real resistors and ideal elements but the resulting theory is complicated. An axiomatic definition of R based on Kirchhoff's laws is, of course, preferable.

Classical Definition

For several centuries, the theory of probability was based on the classical definition. This concept is used today to determine probabilistic data and as a working hypothesis. In the following, we explain its significance.

According to the classical definition, the probability $P(\mathscr{A})$ of an event $\mathscr{A}$ is determined a priori without actual experimentation: It is given by the ratio

$$P(\mathscr{A}) = \frac{N_{\mathscr{A}}}{N} \tag{1-7}$$

where N is the number of *possible* outcomes and $N_{\mathscr{A}}$ is the number of outcomes that are *favorable* to the event $\mathscr{A}$.

In the die experiment, the possible outcomes are six and the outcomes favorable to the event *even* are three, hence, $P(even) = 3/6$.

It is important to note, however, that the significance of the numbers N and $N_{\mathscr{A}}$ is not always clear. We shall demonstrate the underlying ambiguities with the following problem.

† Richard Von Mises: *Probability, Statistics and Truth*, English edition, H. Geiringer, ed., G. Allen and Unwin Ltd., London, 1957.

Example 1-2 We roll two dice and we want to find the probability p that the sum of the numbers that show equals 7.

To solve this problem using (1-7), we must determine the numbers N and $N_{\mathscr{A}}$. (*a*) We could consider as possible outcomes the 11 sums 2, 3, ..., 12. Of these, only one, namely the sum 7, is favorable, hence, $p = 1/11$. This result is of course wrong. (*b*) We could count as possible outcomes all pairs of numbers not distinguishing between the first and the second die. We have now 21 outcomes of which the pairs (3, 4), (5, 2), and (6, 1) are favorable. In this case, $N_{\mathscr{A}} = 3$ and $N = 21$, hence, $p = 3/21$. This result is also wrong. (*c*) We now reason that the above solutions are wrong because the outcomes in (*a*) and (*b*) are not *equally likely*. To solve the problem "correctly," we must count all pairs of numbers distinguishing between the first and the second die. The total number of outcomes is now 36 and the favorable outcomes are the six pairs (3, 4), (4, 3), (5, 2), (2, 5), (6, 1), and (1, 6), hence, $p = 6/36$.

The above example shows the need for refining definition (1-7). The improved version reads as follows:

The probability of an event equals the ratio of its favorable outcomes to the total number of outcomes provided that all outcomes are *equally likely*.

As we shall presently see, this refinement does not eliminate the problems associated with the classical definition.

Notes 1. The classical definition was introduced as a consequence of the *principle of insufficient reason*†: "In the absence of any prior knowledge, we *must* assume that the events $\mathscr{A}_i$ have equal probabilities." This conclusion is based on the subjective interpretation of probability as a *measure of our state of knowledge* about the events $\mathscr{A}_i$. Indeed, if it were not true that the events $\mathscr{A}_i$ have the same probability, then changing their indices we would obtain different probabilities without a change in the state of our knowledge.

2. As we explain in the last chapter, the principle of insufficient reason is equivalent to the *principle of maximum entropy*.

Critique The classical definition can be questioned on several grounds.

A. The term *equally likely* used in the improved version of (1-7) means, actually, *equally probable*. Thus, in the definition, use is made of the concept to be defined. As we have seen in Example 1-2, this leads often to difficulties in determining N and $N_{\mathscr{A}}$.

B. The definition can be applied only to a limited class of problems. In the die experiment, for example, it is applicable only if the six faces have the same probability. If the die is loaded and the probability of *four* equals 0.2, say, the number 0.2 cannot be derived from (1-7).

C. It appears from (1-7) that the classical definition is a consequence of logical imperatives divorced from experience. This, however, is not so. We accept

† H. Bernoulli, *Arts Conjectandi*, 1713.

certain alternatives as equally likely because of our collective experience. The probabilities of the outcomes of a fair die equal 1/6 not only because the die is symmetrical but also because it was observed in the long history of rolling dice that the ratio $n_{\mathscr{A}}/n$ in (1-1) is close to 1/6. The next illustration is, perhaps, more convincing:

We wish to determine the probability p that a newborn baby is a boy. It is generally assumed that $p = 1/2$; however, this is not the result of pure reasoning. In the first place, it is only approximately true that $p = 1/2$. Furthermore, without access to long records we would not know that the boy-girl alternatives are equally likely regardless of the sex history of the baby's family, the season or place of its birth, or other conceivable factors. It is only after long accumulation of records that such factors become irrelevant and the two alternatives are accepted as equally likely.

D. If the number of possible outcomes is infinite, then to apply the classical definition we must use length, area, or some other measure of infinity for determining the ratio $N_{\mathscr{A}}/N$ in (1-7). We illustrate the resulting difficulties with the following example known as *Bertrand paradox.*

Example 1-3 We are given a circle C of radius r and we wish to determine the probability p that the length l of a "randomly selected" cord AB is greater than the length $l\sqrt{3}$ of the inscribed equilateral triangle.

We shall show that this problem can be given at least three reasonable solutions.

I. If the center M of the cord AB lies inside the circle C_1 of radius $r/2$ shown in Fig. 1-1*a*, then $l > r\sqrt{3}$. It is reasonable, therefore, to consider as favorable outcomes all points inside the circle C_1 and as possible outcomes all points inside the circle C. Using as measure of their numbers the corresponding areas $\pi r^2/4$ and πr^2, we conclude that

$$p = \frac{\pi r^2/4}{\pi r^2} = \frac{1}{4}$$

II. We now assume that the end A of the cord AB is fixed. This reduces the number of possibilities but it has no effect on the value of p because the number of favorable locations of B is reduced proportionately. If B is on the 60° arc DBE of Fig. 1-1*b*, then $l > r\sqrt{3}$. The favorable outcomes are now the points on this arc

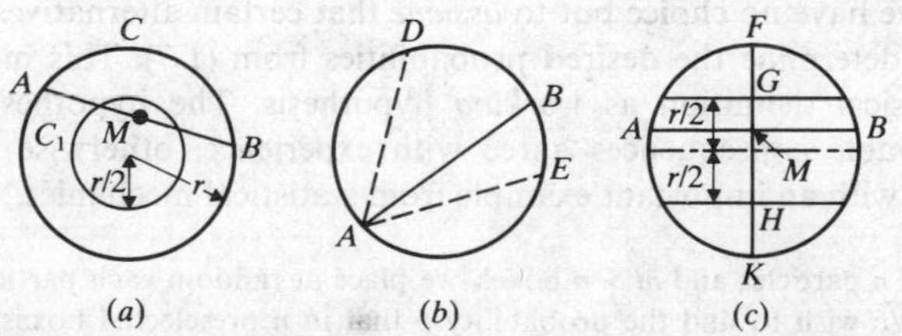

Figure 1-1

and the total outcomes all points on the circumference of the circle C. Using as their measurements the corresponding lengths $2\pi r/3$ and $2\pi r$, we obtain

$$p = \frac{2\pi r/3}{2\pi r} = \frac{1}{3}$$

III. We assume finally that the direction of AB is perpendicular to the line FK of Fig. 1-1*c*. As in II this restriction has no effect on the value of p. If the center M of AB is between G and H, then $l > r\sqrt{3}$. Favorable outcomes are now the points on GH and possible outcomes all points on FK. Using as their measures the respective lengths r and $2r$, we obtain

$$p = \frac{r}{2r} = \frac{1}{2}$$

We have thus found not one but three different solutions for the same problem! One might remark that these solutions correspond to three different experiments. This is true but not obvious and, in any case, it demonstrates the ambiguities associated with the classical definition, and the need for a clear specification of the outcomes of an experiment and the meaning of the terms "possible" and "favorable."

Validity We shall now discuss the value of the classical definition in the determination of probabilistic data and as a working hypothesis.

A. In many applications, the assumption that there are N equally likely alternatives is well established through long experience. Equation (1-7) is then accepted as self-evident. For example, "If a ball is selected at random from a box containing m black and n white balls, the probability that it is white equals $n/(m + n)$," or, "If a call occurs at random in the time interval $(0, T)$, the probability that it occurs in the interval (t_1, t_2) equals $(t_2 - t_1)/T$."

Such conclusions are of course, valid and useful; however, their validity rests on the meaning of the word *random*. The conclusion of the last example that "the unknown probability equals $(t_2 - t_1)/T$" is not a consequence of the "randomness" of the call. The two statements are merely equivalent and they follow not from a priori reasoning but from past records of telephone calls.

B. In a number of applications it is impossible to determine the probabilities of various events by repeating the underlying experiment a sufficient number of times. In such cases, we have no choice but to *assume* that certain alternatives are equally likely and to determine the desired probabilities from (1-7). This means that we use the classical definition as *working* hypothesis. The hypothesis is accepted if its observable consequences agree with experience, otherwise it is rejected. We illustrate with an important example from statistical mechanics.

Example 1-4 Given n particles and $m > n$ boxes, we place at random each particle in one of the boxes. We wish to find the probability p that in n preselected boxes, one and only one particle will be found.

Since we are interested only in the underlying assumptions, we shall only state the results (the proof is assigned as Prob. 3-14). We also verify the solution for $n = 2$ and $m = 6$. For this special case, the problem can be stated in terms of a pair of dice: the $m = 6$ faces correspond to the m boxes and the $n = 2$ dice to the n particles. We assume that the preselected faces (boxes) are 3 and 4.

The solution to this problem depends on the choice of possible and favorable outcomes. We shall consider the following three celebrated cases:

MAXWELL–BOLTZMANN STATISTICS If we accept as outcomes all possible ways of placing n particles in m boxes distinguishing the identity of each particle, then

$$p = \frac{n!}{m^n}$$

For $n = 2$ and $m = 6$ the above yields $p = 2/36$. This is the probability for getting 3, 4 in the game of two dice.

BOSE–EINSTEIN STATISTICS If we assume that the particles are not distinguishable, i.e., if all their permutations count as one, then

$$p = \frac{(m-1)!\,n!}{(n+m-1)!}$$

For $n = 2$ and $m = 6$ this yields $p = 1/21$. Indeed, if we do not distinguish between the two dice, then $N = 21$ and $N_{\mathscr{A}} = 1$ because the outcomes 3, 4 and 4, 3 are counted as one.

FERMI–DIRAC STATISTICS If we do not distinguish between the particles and also we assume that in each box we are allowed to place at most one particle, then

$$p = \frac{n!\,(m-n)!}{m!}$$

For $n = 2$ and $m = 6$ we obtain $p = 1/15$. This is the probability for 3, 4 if we do not distinguish between the dice and also we ignore the outcomes in which the two numbers that show are equal.

One might argue, as indeed it was in the early years of statistical mechanics, that only the first of these solutions is logical. The fact is that in the absence of direct or indirect experimental evidence this argument cannot be supported. The three models proposed are actually only *hypotheses* and the physicist accepts the one whose consequences agree with experience.

C. Suppose that we know the probability $P(\mathscr{A})$ of an event $\mathscr{A}$ in experiment 1 and the probability $P(\mathscr{B})$ of an event $\mathscr{B}$ in experiment 2. In general, from this information we cannot determine the probability $P(\mathscr{A}\mathscr{B})$ that both events $\mathscr{A}$ and $\mathscr{B}$ will occur. However, if we know that the two experiments are *independent*, then

$$P(\mathscr{A}\mathscr{B}) = P(\mathscr{A})P(\mathscr{B}) \tag{1-8}$$

In many cases, this independence can be established a priori by reasoning that the outcomes of experiment 1 have no effect on the outcomes of experiment 2.

For example, if in the coin experiment the probability of *heads* equals 1/2 and in the die experiment the probability of *even* equals 1/2, then, we conclude "logically" that if both experiments are performed, the probability that we get *heads* on the coin and *even* on the die equals $1/2 \times 1/2$. Thus, as in (1-7), we accept the validity of (1-8) as a logical necessity without recourse to (1-1) or to any other direct evidence.

D. The classical definition can be used as the basis of a deductive theory if we accept (1-7) as an *assumption*. In this theory, no other assumptions are used and postulates (1-3) to (1-5) become theorems. Indeed, the first two postulates are obvious and the third follows from (1-7) because, if the events $\mathscr{A}$ and $\mathscr{B}$ are mutually exclusive, then $N_{\mathscr{A}+\mathscr{B}} = N_{\mathscr{A}} + N_{\mathscr{B}}$, hence

$$P(\mathscr{A} + \mathscr{B}) = \frac{N_{\mathscr{A}+\mathscr{B}}}{N} = \frac{N_{\mathscr{A}}}{N} + \frac{N_{\mathscr{B}}}{N} = P(\mathscr{A}) + P(\mathscr{B})$$

As we show in (2-25), however, this is only a very special case of the axiomatic approach to probability.

1-3 PROBABILITY AND INDUCTION

In the applications of the theory of probability we are faced with the following question: Suppose that we know somehow from past observations the probability $P(\mathscr{A})$ of an event $\mathscr{A}$ in a given experiment. What conclusion can we draw about the occurrence of this event in a *single* future performance of this experiment?

We shall answer this question in two ways depending on the size of $P(\mathscr{A})$: We shall give one kind of an answer if $P(\mathscr{A})$ is a number distinctly different from 0 or 1, for example 0.6, and a different kind of an answer if $P(\mathscr{A})$ is close to 0 or 1, for example 0.999. Although the boundary between these two cases is not sharply defined, the corresponding answers are fundamentally different.

Case 1 Suppose that $P(\mathscr{A}) = 0.6$. In this case, the number 0.6 gives us only a "certain degree of confidence that the event $\mathscr{A}$ will occur." The known probability is thus used as a "measure of our belief" about the occurrence of $\mathscr{A}$ in a single trial. This interpretation of $P(\mathscr{A})$ is subjective in the sense that it cannot be verified experimentally. In a single trial, the event $\mathscr{A}$ will either occur or will not occur. If it does not, this will not be a reason for questioning the validity of the assumption that $P(\mathscr{A}) = 0.6$.

Case 2 Suppose, however, that $P(\mathscr{A}) = 0.999$. We can now state with practical certainty that at the next trial the event $\mathscr{A}$ will occur. This conclusion is objective in the sense that it can be verified experimentally. At the next trial the event $\mathscr{A}$ must occur. If it does not, we must seriously doubt, if not outright reject, the assumption that $P(\mathscr{A}) = 0.999$.

The boundary between these two cases, arbitrary though it is (0.9 or 0.999 99?), establishes in a sense the line separating "soft" from "hard" scientific conclusions. The theory of probability gives us the analytical tools (step 2) for transforming the "subjective" statements of case 1 to the "objective" statements of case 2. In the following, we explain briefly the underlying reasoning.

As we show in Chap. 3, the information that $P(\mathscr{A}) = 0.6$ leads to the conclusion that if the experiment is performed 1000 times, then "almost certainly" the number of times the event $\mathscr{A}$ will occur is between 550 and 650. This is shown by considering the repetition of the original experiment 1000 times as a *single* outcome of a new experiment. In this experiment the probability of the event

$$\mathscr{A}_1 = \{\text{the number of times } \mathscr{A} \text{ occurs is between 550 and 650}\}$$

equals 0.999 (see Prob. 3-5). We must, therefore, conclude that (case 2) the event $\mathscr{A}_1$ will occur with practical certainty.

We have thus succeeded, using the theory of probability, to transform the "subjective" conclusion about $\mathscr{A}$ based on the *given* information that $P(\mathscr{A}) = 0.6$, to the "objective" conclusion about $\mathscr{A}_1$ based on the *derived* conclusion that $P(\mathscr{A}_1) = 0.999$. We should emphasize, however, that both conclusions rely on inductive reasoning. Their difference, although significant, is only quantitative. As in case 1, the "objective" conclusion of case 2 is not a certainty but only an inference. This, however, should not surprise us; after all, no prediction about future events based on past experience can be accepted as logical certainty.

Our inability to make categorical statements about future events is not limited to probability but applies to all sciences. Consider, for example, the development of classical mechanics. It was *observed* that bodies fall according to certain patterns, and on this evidence Newton formulated the laws of mechanics and used them to *predict* future events. His predictions, however, are not logical certainties but only plausible inferences. To "prove" that the future will evolve in the predicted manner we must invoke methaphysical causes.

1-4 CAUSALITY VERSUS RANDOMNESS

We conclude with a brief comment on the apparent controversy between causality and randomness. There is no conflict between causality and randomness or between determinism and probability if we agree, as we must, that scientific theories are not *discoveries* of the laws of nature but rather *inventions* of the human mind. Their consequences are presented in deterministic form if we examine the results of a single trial; they are presented as probabilistic statements if we are interested in averages of many trials. In both cases, all statements are qualified. In the first case, the uncertainties are of the form "with certain errors and in certain ranges of the relevant parameters"; in the second, "with a high degree of certainty if the number of trials is large enough."

In the next example, we illustrate these two approaches.

Example 1-5 A rocket leaves the ground with an initial velocity v forming an angle θ with the horizontal axis (Fig. 1-2). We shall determine the distance $d = 0B$ from the origin to the reentry point B.

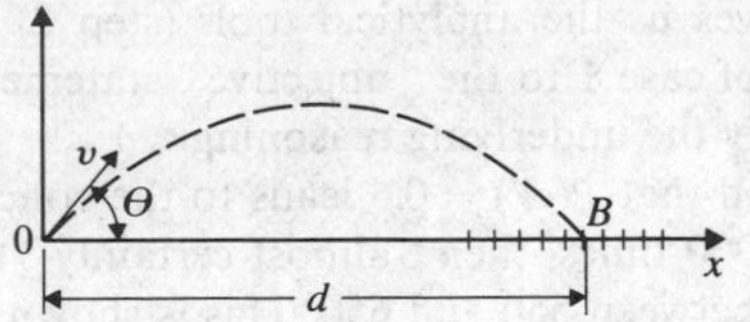

Figure 1-2

From Newton's law it follows that

$$d = \frac{v^2}{g} \sin 2\theta \tag{1-9}$$

The above seems to be an unqualified consequence of a causal law; however, this is not so. The result is approximate and it can be given a probabilistic interpretation.

Indeed, (1-9) is not the solution of a real problem but of an idealized model in which we have neglected air friction, air pressure, variation of g, and other uncertainties in the values of v and θ. We must, therefore, accept (1-9) only with qualifications. It holds within an error ε provided that the neglected factors are smaller than δ.

Suppose now that the reentry area consists of numbered holes and we want to find the reentry hole. Because of the uncertainties in v and θ we are in no position to give a deterministic answer to our problem. We can, however, ask a different question: If many rockets, nominally with the same velocity, are launched, what percentage will enter the nth hole? This question no longer has a causal answer; it can only be given a random interpretation.

Thus, the same physical problem can be subjected either to a deterministic or to a probabilistic analysis. One might argue that the problem is inherently deterministic because the rocket has a precise velocity even if we do not know it. If we did, we would know exactly the reentry hole. Probabilistic interpretations are, therefore, necessary because of our ignorance.

Such arguments can be answered with the statement that the physicists are not concerned with what *is true* but only with what *they can observe.*

CONCLUDING REMARKS

In this book, we present a deductive theory (step 2) based on the axiomatic definition of probability. Occasionally, we use the classical definition but only to determine probabilistic data (step 1).

To show the link between theory and applications (step 3), we give also a relative frequency interpretation of the important results. This part of the book, written in small print under the title *Frequency interpretation,* does not obey the rules of deductive reasoning on which the theory is based.

CHAPTER

TWO

THE AXIOMS OF PROBABILITY

2-1 SET THEORY

A *set* is a collection of objects called *elements*. For example, "car, apple, pencil" is a set whose elements are a car, an apple, and a pencil. The set "heads, tails" has two elements. The set "1, 2, 3, 5" has four elements. Most sets will be identified by script letters.

A *subset* $\mathscr{B}$ of a set $\mathscr{A}$ is another set whose elements are also elements of $\mathscr{A}$. All sets under consideration will be subsets of a set $\mathscr{S}$ which we shall call *space*.

The elements of a set will be identified mostly by the Greek letter ζ. Thus

$$\mathscr{A} = \{\zeta_1, \ldots, \zeta_n\} \tag{2-1}$$

will mean that the set $\mathscr{A}$ consists of the elements $\zeta_1, \ldots, \zeta_n$. We shall also identify sets by the properties of their elements. Thus,

$$\mathscr{A} = \{\text{all positive integers}\} \tag{2-2}$$

will mean the set whose elements are the numbers 1, 2, 3, ...

The notations

$$\zeta_i \in \mathscr{A} \qquad \zeta_i \notin \mathscr{A}$$

will mean that ζ_i is or is not an element of $\mathscr{A}$.

The *empty* or *null* set is by definition the set that contains no elements. This set will be denoted by $\{\emptyset\}$.

If a set consists of n elements, then the total number of its subsets equals 2^n.

Note In probability theory, we assign probabilities to the subsets (events) of $\mathscr{S}$ and we define various functions (random variables) whose domain consists of the elements of $\mathscr{S}$. We must be careful, therefore, to distinguish between the element ζ and the set $\{\zeta\}$ consisting of the single element ζ.

Example 2-1 We shall denote by f_i the faces of a die. These faces are the elements of the set $\mathscr{S} = \{f_1, \ldots, f_6\}$. In this case, $n = 6$, hence, $\mathscr{S}$ has $2^6 = 64$ subsets:

$$\{\emptyset\}, \{f_1\}, \ldots, \{f_1 f_2\}, \ldots, \{f_1 f_2 f_3\}, \ldots, \mathscr{S}$$

In general, the elements of a set are arbitrary objects. For example, the 64 subsets of the set $\mathscr{S}$ in the above example can be considered as the elements of another set. In Example 2-2, the elements of $\mathscr{S}$ are pairs of objects. In Example 2-3, $\mathscr{S}$ is the set of points in the square of Fig. 2-1.

Example 2-2 Suppose that a coin is tossed twice. The resulting outcomes are the four objects *hh*, *ht*, *th*, *tt* forming the set

$$\mathscr{S} = \{hh, ht, th, tt\}$$

where *hh* is an abbreviation for the element "heads–heads." The set $\mathscr{S}$ has $2^4 = 16$ subsets. For example,

$$\mathscr{A} = \{\text{heads at the first tossing}\} = \{hh, ht\}$$

$$\mathscr{B} = \{\text{only one head showed}\} = \{ht, th\}$$

$$\mathscr{C} = \{\text{heads shows at least once}\} = \{hh, ht, th\}$$

In the first equality, the sets $\mathscr{A}$, $\mathscr{B}$, and $\mathscr{C}$ are represented by their properties as in (2-2); in the second, in terms of their elements as in (2-1).

Example 2-3 In this example, $\mathscr{S}$ is the set of all points in the square of Fig. 2-1. Its elements are all ordered pairs of numbers (x, y) where

$$0 \le x \le T \qquad 0 \le y \le T$$

The shaded area is a subset $\mathscr{A}$ of $\mathscr{S}$ consisting of all points (x, y) such that $-b \le x - y \le a$. The notation

$$\mathscr{A} = \{-b \le x - y \le a\}$$

describes $\mathscr{A}$ in terms of the properties of x and y as in (2-2).

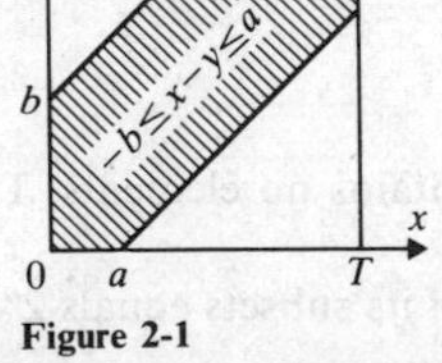

Figure 2-1

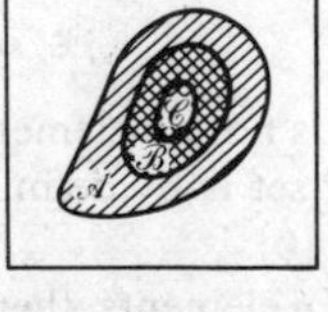

Figure 2-2

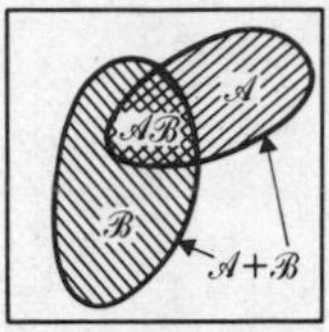

Figure 2-3

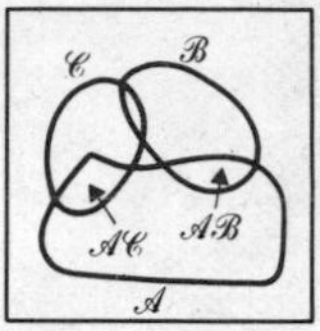

Figure 2-4

Set Operations

In the following, we shall represent a set $\mathscr{S}$ and its subsets by plane figures as in Fig. 2-2 (*Venn diagrams*).

The notation $\mathscr{B} \subset \mathscr{A}$ or $\mathscr{A} \supset \mathscr{B}$ will mean that $\mathscr{B}$ is a subset of $\mathscr{A}$, that is, that every element of $\mathscr{B}$ is an element of $\mathscr{A}$. Thus, for any $\mathscr{A}$,

$$\{\emptyset\} \subset \mathscr{A} \subset \mathscr{A} \subset \mathscr{S}$$

Transitivity If $\mathscr{C} \subset \mathscr{B}$ and $\mathscr{B} \subset \mathscr{A}$ then $\mathscr{C} \subset \mathscr{A}$
Equality $\mathscr{A} = \mathscr{B}$ iff† $\mathscr{A} \subset \mathscr{B}$ and $\mathscr{B} \subset \mathscr{A}$

Unions and intersections The *sum* or *union* of two sets $\mathscr{A}$ and $\mathscr{B}$ is a set whose elements are all elements of $\mathscr{A}$ or of $\mathscr{B}$ or of both (Fig. 2-3). This set will be written in the form

$$\mathscr{A} + \mathscr{B} \qquad \text{or} \qquad \mathscr{A} \cup \mathscr{B}$$

The above operation is commutative and associative:

$$\mathscr{A} + \mathscr{B} = \mathscr{B} + \mathscr{A} \qquad (\mathscr{A} + \mathscr{B}) + \mathscr{C} = \mathscr{A} + (\mathscr{B} + \mathscr{C})$$

We note that, if $\mathscr{B} \subset \mathscr{A}$, then $\mathscr{A} + \mathscr{B} = \mathscr{A}$. From this it follows that

$$\mathscr{A} + \mathscr{A} = \mathscr{A} \qquad \mathscr{A} + \{\emptyset\} = \mathscr{A} \qquad \mathscr{S} + \mathscr{A} = \mathscr{S}$$

The *product* or *intersection* of two sets $\mathscr{A}$ and $\mathscr{B}$ is a set consisting of all elements that are common to the sets $\mathscr{A}$ and $\mathscr{B}$ (Fig. 2-3). This set is written in the form

$$\mathscr{A}\mathscr{B} \qquad \text{or} \qquad \mathscr{A} \cap \mathscr{B}$$

The above operation is commutative, associative, and distributive (Fig. 2-4)

$$\mathscr{A}\mathscr{B} = \mathscr{B}\mathscr{A} \qquad (\mathscr{A}\mathscr{B})\mathscr{C} = \mathscr{A}(\mathscr{B}\mathscr{C}) \qquad \mathscr{A}(\mathscr{B} + \mathscr{C}) = \mathscr{A}\mathscr{B} + \mathscr{A}\mathscr{C}$$

We note that if $\mathscr{A} \subset \mathscr{B}$, then $\mathscr{A}\mathscr{B} = \mathscr{A}$. Hence

$$\mathscr{A}\mathscr{A} = \mathscr{A} \qquad \{\emptyset\}\mathscr{A} = \{\emptyset\} \qquad \mathscr{A}\mathscr{S} = \mathscr{A}$$

† Iff is an abbreviation for *if and only if.*

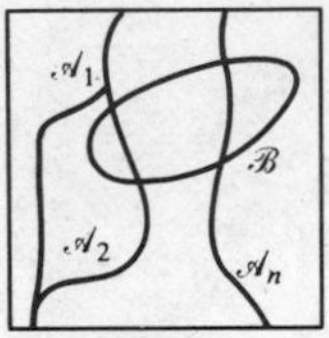

Figure 2-5

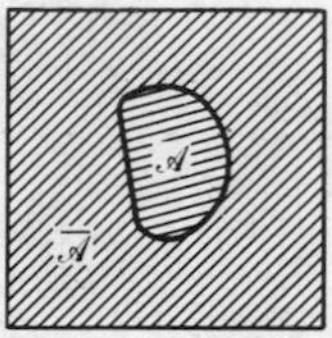

Figure 2-6

Note If two sets $\mathscr{A}$ and $\mathscr{B}$ are described by the properties of their elements as in (2-2), then their intersection $\mathscr{A}\mathscr{B}$ will be specified by including these properties in braces. For example, if

$$\mathscr{S} = \{1, 2, 3, 4, 5, 6\} \qquad \mathscr{A} = \{\text{even}\} \qquad \mathscr{B} = \{\text{less than } 5\}$$

then†

$$\mathscr{A}\mathscr{B} = \{\text{even, less than } 5\} = \{2, 4\} \tag{2-3}$$

Mutually exclusive sets Two sets, $\mathscr{A}$ and $\mathscr{B}$ are called *mutually exclusive* or *disjoint* if they have no common elements, i.e., if

$$\mathscr{A}\mathscr{B} = \{\emptyset\}$$

Several sets $\mathscr{A}_1, \mathscr{A}_2, \ldots$ are called mutually exclusive if

$$\mathscr{A}_i\mathscr{A}_j = \{\emptyset\} \qquad \text{for every } i \text{ and } j \neq i$$

Partitions A partition $\mathfrak{A}$ of a set $\mathscr{S}$ is a collection of mutually exclusive subsets $\mathfrak{A}_i$ of $\mathscr{S}$ whose union equals $\mathscr{S}$ (Fig. 2-5).

$$\mathscr{A}_1 + \cdots + \mathscr{A}_n = \mathscr{S} \qquad \mathscr{A}_i\mathscr{A}_j = \{\emptyset\} \qquad i \neq j \tag{2-4}$$

All partitions will be denoted by boldface German script (Fraktur) letters. Thus,

$$\mathfrak{A} = [\mathscr{A}_1, \ldots, \mathscr{A}_n]$$

Complements The complement $\bar{\mathscr{A}}$ of a set $\mathscr{A}$ is the set consisting of all elements of $\mathscr{S}$ that are not in $\mathscr{A}$ (Fig. 2-6). From the definition it follows that

$$\mathscr{A} + \bar{\mathscr{A}} = \mathscr{S} \qquad \mathscr{A}\bar{\mathscr{A}} = \{\emptyset\} \qquad \bar{\bar{\mathscr{A}}} = \mathscr{A} \qquad \bar{\mathscr{S}} = \{\emptyset\} \qquad \{\bar{\emptyset}\} = \mathscr{S}$$

If $\mathscr{B} \subset \mathscr{A}$ then $\bar{\mathscr{B}} \supset \bar{\mathscr{A}}$; if $\mathscr{A} = \mathscr{B}$, then $\bar{\mathscr{A}} = \bar{\mathscr{B}}$

† We should stress the difference in the meaning of commas in (2-1) and (2-3). In (2-1) the braces include all elements ζ_i and

$$\{\zeta_1, \ldots, \zeta_n\} = \{\zeta_1\} \cup \cdots \cup \{\zeta_n\}$$

is the union of the sets $\{\zeta_i\}$. In (2-3), the braces include the properties of the sets {even} and {less than 5}, and

$$\{\text{even, less than } 5\} = \{\text{even}\} \cap \{\text{less than } 5\}$$

is the intersection of the sets {even} and {less than 5}.

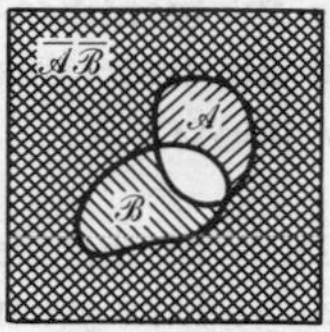

Figure 2-7

De Morgan law Clearly (see Fig. 2-7)

$$\overline{\mathscr{A} + \mathscr{B}} = \bar{\mathscr{A}}\bar{\mathscr{B}} \qquad \overline{\mathscr{A}\mathscr{B}} = \bar{\mathscr{A}} + \bar{\mathscr{B}} \tag{2-5}$$

Repeated application of (2-5) leads to the following:

If in a set identity we replace all sets by their complements, all unions by intersections, and all intersections by unions, the identity is preserved.

We shall demonstrate the above using as example the identity

$$\mathscr{A}(\mathscr{B} + \mathscr{C}) = \mathscr{A}\mathscr{B} + \mathscr{A}\mathscr{C} \tag{2-6}$$

From (2-5) it follows that

$$\overline{\mathscr{A}(\mathscr{B} + \mathscr{C})} = \bar{\mathscr{A}} + \overline{\mathscr{B} + \mathscr{C}} = \bar{\mathscr{A}} + \bar{\mathscr{B}}\bar{\mathscr{C}}$$

Similarly

$$\overline{\mathscr{A}\mathscr{B} + \mathscr{A}\mathscr{C}} = (\overline{\mathscr{A}\mathscr{B}})(\overline{\mathscr{A}\mathscr{C}}) = (\bar{\mathscr{A}} + \bar{\mathscr{B}})(\bar{\mathscr{A}} + \bar{\mathscr{C}})$$

and since the two sides of (2-6) are equal, their complements are also equal. Hence

$$\bar{\mathscr{A}} + \bar{\mathscr{B}}\bar{\mathscr{C}} = (\bar{\mathscr{A}} + \bar{\mathscr{B}})(\bar{\mathscr{A}} + \bar{\mathscr{C}}) \tag{2-7}$$

Duality principle As we know, $\bar{\mathscr{S}} = \{\emptyset\}$ and $\overline{\{\emptyset\}} = \mathscr{S}$. Furthermore, if in an identity like (2-7) all overbars are removed, the identity is preserved. This leads to the following version of De Morgan's law:

If in a set identity we replace all unions by intersections, all intersections by unions, and the sets $\mathscr{S}$ and $\{\emptyset\}$ by the sets $\{\emptyset\}$ and $\mathscr{S}$ respectively, the identity is preserved.

Applying the above to the identitities

$$\mathscr{A}(\mathscr{B} + \mathscr{C}) = \mathscr{A}\mathscr{B} + \mathscr{A}\mathscr{C} \qquad \mathscr{S} + \mathscr{A} = \mathscr{S}$$

we obtain the identities

$$\mathscr{A} + \mathscr{B}\mathscr{C} = (\mathscr{A} + \mathscr{B})(\mathscr{A} + \mathscr{C}) \qquad \{\emptyset\}\mathscr{A} = \{\emptyset\}$$

2-2 PROBABILITY SPACE

In probability theory, the following set terminology is used: The space $\mathscr{S}$ is called the *certain event*, its elements *experimental outcomes*, and its subsets *events*. The empty set $\{\emptyset\}$ is the *impossible event*, and the event $\{\zeta_i\}$ consisting of a single element ζ_i is an *elementary event*.

In the applications of probability theory to physical problems, the identification of experimental outcomes is not always unique. We shall illustrate this ambiguity with the die experiment as might be interpreted by players X, Y, and Z.

X says that the outcomes of this experiment are the six faces of the die forming the space $\mathscr{S} = \{f_1, \ldots, f_6\}$. This space has $2^6 = 64$ subsets and the event $\{\text{even}\}$ consists of the three outcomes f_2, f_4, and f_6.

Y wants to bet on *even* or *odd* only. He argues, therefore, that the experiment has only the two outcomes *even* and *odd* forming the space $\mathscr{S} = \{\text{even, odd}\}$. This space has only $2^2 = 4$ subsets and the event $\{\text{even}\}$ consists of a single outcome.

Z bets that *one* will show and the die will rest on the left side of the table. He maintains, therefore, that the experiment has infinitely many outcomes specified by the coordinates of its center and by the six faces. The event $\{\text{even}\}$ consists not of one or of three outcomes but of infinitely many.

In the following, when we talk about an experiment, we shall assume that its outcomes are clearly identified. In the die experiment, for example, $\mathscr{S}$ will be the set consisting of the six faces $f_1, \ldots, f_6$.

In the relative frequency interpretation of various results, we shall use the following terminology.

Trials A single performance of an experiment will be called a *trial*. At each trial we observe a single outcome ζ_i. We say that an event $\mathscr{A}$ *occurs* during this trial if it contains the element ζ_i. The certain event occurs at every trial and the impossible event never occurs. The event $\mathscr{A} + \mathscr{B}$ occurs when $\mathscr{A}$ or $\mathscr{B}$ or both occur. The event $\mathscr{A}\mathscr{B}$ occurs when both events $\mathscr{A}$ and $\mathscr{B}$ occur. If the events $\mathscr{A}$ and $\mathscr{B}$ are mutually exclusive and $\mathscr{A}$ occurs, then $\mathscr{B}$ does not occur. If $\mathscr{A} \subset \mathscr{B}$ and $\mathscr{A}$ occurs then $\mathscr{B}$ occurs. At each trial, either $\mathscr{A}$ or $\bar{\mathscr{A}}$ occurs.

If, for example, in the die experiment we observe the outcome f_5, then the event $\{f_5\}$, the event $\{\text{odd}\}$, and 30 other events occur.

The Axioms

We assign to each event $\mathscr{A}$ a number $P(\mathscr{A})$ which we call *the probability of the event* $\mathscr{A}$. This number is so chosen as to satisfy the following three conditions:

I $$P(\mathscr{A}) \geq 0 \tag{2-8}$$

II $$P(\mathscr{S}) = 1 \tag{2-9}$$

III If $\mathscr{A}\mathscr{B} = \{\emptyset\}$, then

$$P(\mathscr{A} + \mathscr{B}) = P(\mathscr{A}) + P(\mathscr{B}) \tag{2-10}$$

These conditions are the axioms of the theory of probability. In the development of the theory, all conclusions are based directly or indirectly on the axioms and only on the axioms. The following are simple consequences.

Properties The probability of the impossible event is zero

$$P\{\emptyset\} = 0 \tag{2-11}$$

Indeed, $\mathscr{A}\{\emptyset\} = \{\emptyset\}$ and $\mathscr{A} + \{\emptyset\} = \mathscr{A}$, therefore [see (2-10)]

$$P(\mathscr{A}) = P(\mathscr{A} + \emptyset) = P(\mathscr{A}) + P\{\emptyset\}$$

For any $\mathscr{A}$

$$P(\mathscr{A}) = 1 - P(\bar{\mathscr{A}}) \le 1 \tag{2-12}$$

because $\mathscr{A} + \bar{\mathscr{A}} = \mathscr{S}$ and $\mathscr{A}\bar{\mathscr{A}} = \{\emptyset\}$, hence

$$1 = P(\mathscr{S}) = P(\mathscr{A} + \bar{\mathscr{A}}) = P(\mathscr{A}) + P(\bar{\mathscr{A}})$$

For any $\mathscr{A}$ and $\mathscr{B}$

$$P(\mathscr{A} + \mathscr{B}) = P(\mathscr{A}) + P(\mathscr{B}) - P(\mathscr{A}\mathscr{B}) \le P(\mathscr{A}) + P(\mathscr{B}) \tag{2-13}$$

To prove the above, we write the events $\mathscr{A} + \mathscr{B}$ and $\mathscr{B}$ as unions of two mutually exclusive events:

$$\mathscr{A} + \mathscr{B} = \mathscr{A} + \bar{\mathscr{A}}\mathscr{B} \qquad \mathscr{B} = \mathscr{A}\mathscr{B} + \bar{\mathscr{A}}\mathscr{B}$$

Therefore [see (2-10)]

$$P(\mathscr{A} + \mathscr{B}) = P(\mathscr{A}) + P(\bar{\mathscr{A}}\mathscr{B}) \qquad P(\mathscr{B}) = P(\mathscr{A}\mathscr{B}) + P(\bar{\mathscr{A}}\mathscr{B})$$

Eliminating $P(\bar{\mathscr{A}}\mathscr{B})$, we obtain (2-13).

Finally, if $\mathscr{B} \subset \mathscr{A}$, then

$$P(\mathscr{A}) = P(\mathscr{B}) + P(\mathscr{A}\bar{\mathscr{B}}) \ge P(\mathscr{B}) \tag{2-14}$$

because $\mathscr{A} = \mathscr{B} + \mathscr{A}\bar{\mathscr{B}}$ and $\mathscr{B}(\mathscr{A}\bar{\mathscr{B}}) = \{\emptyset\}$.

Frequency interpretation The axioms of probability are so chosen that the resulting theory gives a satisfactory representation of the physical world. Probabilities as used in real problems must, therefore, be compatible with the axioms. Using the frequency interpretation

$$P(\mathscr{A}) \simeq \frac{n_{\mathscr{A}}}{n}$$

of probability, we shall show that they do.

I Clearly, $P(\mathscr{A}) \ge 0$ because $n_{\mathscr{A}} \ge 0$ and $n > 0$.

II $P(\mathscr{S}) = 1$ because $\mathscr{S}$ occurs at every trial, hence, $n_{\mathscr{S}} = n$.

III If $\mathscr{A}\mathscr{B} = 0$, then $n_{\mathscr{A}+\mathscr{B}} = n_{\mathscr{A}} + n_{\mathscr{B}}$ because if $\mathscr{A} + \mathscr{B}$ occurs then $\mathscr{A}$ or $\mathscr{B}$ occurs but not both. Hence

$$P(\mathscr{A} + \mathscr{B}) \simeq \frac{n_{\mathscr{A}+\mathscr{B}}}{n} = \frac{n_{\mathscr{A}}}{n} + \frac{n_{\mathscr{B}}}{n} \simeq P(\mathscr{A}) + P(\mathscr{B})$$

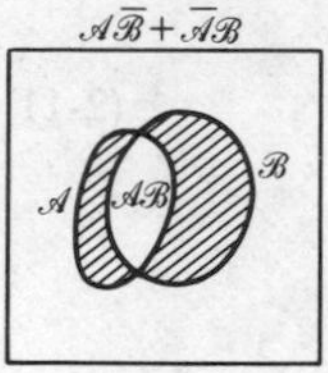

Figure 2-8

Equality of events Two events $\mathscr{A}$ and $\mathscr{B}$ are called *equal* if they consist of the same elements. They are called *equal with probability 1* if the set

$$(\mathscr{A} + \mathscr{B})(\overline{\mathscr{A}\mathscr{B}}) = \mathscr{A}\bar{\mathscr{B}} + \bar{\mathscr{A}}\mathscr{B}$$

consisting of all outcomes that are in $\mathscr{A}$ or in $\mathscr{B}$ but not in $\mathscr{A}\mathscr{B}$ (shaded area in Fig. 2-8) has zero probability.

From the definition it follows that (see Prob. 2-4) the events $\mathscr{A}$ and $\mathscr{B}$ are equal with probability 1 iff

$$P(\mathscr{A}) = P(\mathscr{B}) = P(\mathscr{A}\mathscr{B}) \tag{2-15}$$

If $P(\mathscr{A}) = P(\mathscr{B})$ then we say that $\mathscr{A}$ and $\mathscr{B}$ are *equal in probability.* In this case, no conclusion can be drawn about the probability of $\mathscr{A}\mathscr{B}$. If fact, the events $\mathscr{A}$ and $\mathscr{B}$ might be mutually exclusive.

From (2-15) it follows that, if an event $\mathscr{N}$ equals the impossible event with probability 1, then $P(\mathscr{N}) = 0$. This does not, of course, mean that $\mathscr{N} = \{\emptyset\}$.

The Class $\mathfrak{F}$ of Events

Events are subsets of $\mathscr{S}$ to which we have assigned probabilities. As we shall presently explain, we shall not consider as events all subsets of $\mathscr{S}$ but only a class $\mathfrak{F}$ of subsets.

One reason for this might be the nature of the application. In the die experiment, for example, we might want to bet only on *even* or *odd*. In this case, it suffices to consider as events only the four sets $\{\emptyset\}$, {even}, {odd}, and $\mathscr{S}$.

The main reason, however, for not including all subsets of $\mathscr{S}$ in the class $\mathfrak{F}$ of events is of a mathematical nature: In certain cases involving sets with infinitely many outcomes, it is impossible to assign probabilities to all subsets satisfying all the axioms including the generalized form (2-21) of axiom III.

The class $\mathfrak{F}$ of events will not be an arbitrary collection of subsets of $\mathscr{S}$. We shall assume that, if $\mathscr{A}$ and $\mathscr{B}$ are events, then $\mathscr{A} + \mathscr{B}$ and $\mathscr{A}\mathscr{B}$ are also events. We do so because we will want to know not only the probabilities of various events, but also the probabilities of their unions and intersections. This leads to the concept of a field.

Fields A field $\mathfrak{F}$ is a nonempty class of sets such that:

If $\qquad \mathscr{A} \in \mathfrak{F} \quad$ then $\quad \bar{\mathscr{A}} \in \mathfrak{F} \qquad$ (2-16)

If $\qquad \mathscr{A} \in \mathfrak{F} \quad$ and $\quad \mathscr{B} \in \mathfrak{F} \quad$ then $\quad \mathscr{A} + \mathscr{B} \in \mathfrak{F} \qquad$ (2-17)

These two properties give a minimum set of conditions for $\mathfrak{F}$ to be a field. All other properties follow:

$$\text{If} \qquad \mathscr{A} \in \mathfrak{F} \qquad \text{and} \qquad \mathscr{B} \in \mathfrak{F} \qquad \text{then} \qquad \mathscr{A}\mathscr{B} \in \mathfrak{F} \tag{2-18}$$

Indeed from (2-16) it follows that $\bar{\mathscr{A}} \in \mathfrak{F}$ and $\bar{\mathscr{B}} \in \mathfrak{F}$. Applying (2-17) and (2-16) to the sets $\bar{\mathscr{A}}$ and $\bar{\mathscr{B}}$, we conclude that

$$\bar{\mathscr{A}} + \bar{\mathscr{B}} \in \mathfrak{F} \qquad \overline{\bar{\mathscr{A}} + \bar{\mathscr{B}}} = \mathscr{A}\mathscr{B} \in \mathfrak{F}$$

A field contains the certain event and the impossible event:

$$\mathscr{S} \in \mathfrak{F} \qquad \{\emptyset\} \in \mathfrak{F} \tag{2-19}$$

Indeed, since $\mathfrak{F}$ is not empty, it contains at least one element $\mathscr{A}$, therefore [see (2-16)], it contains also $\bar{\mathscr{A}}$. Hence

$$\mathscr{A} + \bar{\mathscr{A}} = \mathscr{S} \in \mathfrak{F} \qquad \mathscr{A}\bar{\mathscr{A}} = \{\emptyset\} \in \mathfrak{F}$$

From the above it follows that all sets that can be written as unions or intersections of *finitely many* sets in $\mathfrak{F}$ are also in $\mathfrak{F}$. This is not, however, necessarily the case for infinitely many sets.

Borel fields Suppose that $\mathscr{A}_1, \ldots, \mathscr{A}_n, \ldots$ is an infinite sequence of sets in $\mathfrak{F}$. If the union and intersection of these sets belongs also to $\mathfrak{F}$, then $\mathfrak{F}$ is called a Borel field.

The class of all subjects of a set $\mathscr{S}$ is a Borel field. Suppose that $\mathfrak{C}$ is a class of subsets of $\mathscr{S}$ that is not a field. Attaching to it other subsets of $\mathscr{S}$, all subsets if necessary, we can form a field with $\mathfrak{C}$ as its subset. It can be shown that there exists a smallest Borel field containing all the elements of $\mathfrak{C}$.

Example 2-4 Suppose that $\mathscr{S}$ consists of the four elements a, b, c, d and $\mathfrak{C}$ consists of the sets $\{a\}$ and $\{b\}$. Attaching to $\mathfrak{C}$ the complements of $\{a\}$ and $\{b\}$ and their unions and intersections, we conclude that the smallest field containing $\{a\}$ and $\{b\}$ consists of the sets

$$\{\emptyset\} \qquad \{a\} \qquad \{b\} \qquad \{a, b\} \qquad \{c, d\} \qquad \{b, c, d\} \qquad \{a, c, d\} \qquad \mathscr{S}$$

Events In probability theory, events are certain subsets of $\mathscr{S}$ forming a Borel field. This permits us to assign probabilities not only to finite unions and intersections of events, but also to their limits.

For the determination of probabilities of sets that can be expressed as limits, the following extension of axiom III is necessary.

Repeated application of (2-10) leads to the conclusion that, if the events $\mathscr{A}_1, \ldots, \mathscr{A}_n$ are mutually exclusive, then

$$P(\mathscr{A}_1 + \cdots + \mathscr{A}_n) = P(\mathscr{A}_1) + \cdots + P(\mathscr{A}_n) \tag{2-20}$$

The extension of the above to infinitely many sets does not follow from (2-10). It is an additional condition known as the *axiom of infinite additivity*:

III*a* If the events $\mathscr{A}_1, \mathscr{A}_2, \ldots$ are mutually exclusive, then

$$P(\mathscr{A}_1 + \mathscr{A}_2 + \cdots) = P(\mathscr{A}_1) + P(\mathscr{A}_2) + \cdots \tag{2-21}$$

We shall assume that all probabilities satisfy axioms I, II, III, and III*a*.

Axiomatic Definition of an Experiment

In the theory of probability, an experiment is specified in terms of the following concepts:

1. The set $\mathscr{S}$ of all experimental outcomes.
2. The Borel field of all events of $\mathscr{S}$.
3. The probabilities of these events.

The letter $\mathscr{S}$ will be used to identify not only the certain event, but also the entire experiment.

We discuss next the determination of probabilities in experiments with finitely many and infinitely many elements.

Countable spaces If the space $\mathscr{S}$ consists of N outcomes and N is a finite number, then the probabilities of all events can be expressed in terms of the probabilities

$$P\{\zeta_i\} = p_i$$

of the elementary events $\{\zeta_i\}$. From the axioms it follows, of course, that the numbers p_i must be nonnegative and their sum must equal 1:

$$p_i \geq 0 \qquad p_1 + \cdots + p_N = 1 \tag{2-22}$$

Suppose that $\mathscr{A}$ is an event consisting of the r elements ζ_{k_i}. In this case, $\mathscr{A}$ can be written as the union of the elementary events $\{\zeta_{k_i}\}$. Hence, [see (2-20)]

$$P(\mathscr{A}) = P\{\zeta_{k_1}\} + \cdots + P\{\zeta_{k_r}\} = p_{k_1} + \cdots + p_{k_r} \tag{2-23}$$

The above is true even if $\mathscr{S}$ consists of an infinite but countable number of elements $\zeta_1, \zeta_2, \ldots$ [see (2-21)].

Classical definition If $\mathscr{S}$ consists of N outcomes and the probabilities p_i of the elementary events are all equal, then

$$p_i = \frac{1}{N} \tag{2-24}$$

In this case, the probability of an event $\mathscr{A}$ consisting of r elements equals r/N

$$P(\mathscr{A}) = \frac{r}{N} \tag{2-25}$$

This very special but important case is equivalent to the classical definition (1-7), with one important difference, however: In the classical definition, (2-25) is deduced as a logical necessity; in the axiomatic development of probability, (2-25) is a mere assumption.

Example 2-5(*a*) In the coin experiment, the space $\mathscr{S}$ consists of the outcomes h and t

$$\mathscr{S} = \{h, t\}$$

and its events are the four sets $\{\emptyset\}$, $\{t\}$, $\{h\}$, $\mathscr{S}$. If $P\{h\} = p$ and $P\{t\} = q$, then $p + q = 1$.

(*b*) We consider now the experiment of the tossing of a coin three times. The possible outcomes of this experiment are:

$$hhh,\ hht,\ hth,\ htt,\ thh,\ tht,\ tth,\ ttt$$

We shall assume that all elementary events have the same probability as in (2-24) (fair coin). In this case, the probability of each elementary event equals 1/8. Thus, the probability $P\{hhh\}$ that we get three heads equals 1/8. The event

$$\{\text{heads at the first two tossings}\} = \{hhh, hht\}$$

consists of the two outcomes hhh and hht, hence, its probability equals 2/8.

The real line If $\mathscr{S}$ consists of a noncountable infinity of elements, then its probabilities cannot be determined in terms of the probabilities of the elementary events. This is the case if $\mathscr{S}$ is the set of points in an n-dimensional space. In fact, most applications can be presented in terms of events in such a space. We shall discuss the determination of probabilities using as illustration the real line.

Suppose that $\mathscr{S}$ is the set of all real numbers. Its subsets can be considered as sets of points on the real line. It can be shown that it is impossible to define probabilities to all subjects of $\mathscr{S}$ so as to satisfy the axioms. To construct a probability space on the real line, we shall consider as events all intervals $x_1 \le x \le x_2$ and their countable unions and intersections. These events form a field $\mathfrak{F}$ that can be specified as follows:

It is the smallest Borel field that includes all half-lines $x \le x_i$ where x_i is any number.

This field contains all open and closed intervals, all points, and, in fact, every set of points on the real line that is of interest in the applications. One might wonder whether $\mathfrak{F}$ does not include *all* subsets of $\mathscr{S}$. Actually, it is possible to show that there exist sets of points on the real line that are not countable unions and intersections of intervals. Such sets, however, are of no interest in most applications. To complete the specification of $\mathscr{S}$, it suffices to assign probabilities to the events $\{x \le x_i\}$. All other probabilities can then be determined from the axioms.

Suppose that $\alpha(x)$ is a function such that (Fig. 2-9*a*)

$$\int_{-\infty}^{\infty} \alpha(x)\, dx = 1 \qquad \alpha(x) \ge 0 \tag{2-26}$$

We define the probability of the event $\{x \le x_i\}$ by the integral

$$P\{x \le x_i\} = \int_{-\infty}^{x_i} \alpha(x)\, dx \tag{2-27}$$

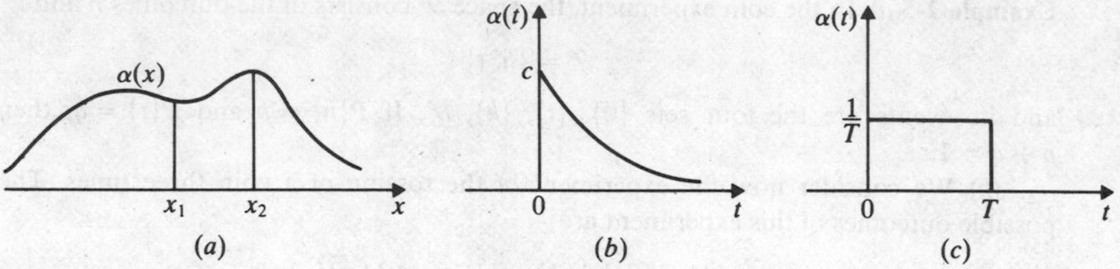

Figure 2-9

This specifies the probabilities of all events of $\mathscr{S}$. We maintain for example, that the probability of the event $\{x_1 < x \le x_2\}$ consisting of all points in the interval (x_1, x_2) is given by

$$P\{x_1 < x \le x_2\} = \int_{x_1}^{x_2} \alpha(x)\, dx \tag{2-28}$$

Indeed, the events $\{x \le x_1\}$ and $\{x_1 < x \le x_2\}$ are mutually exclusive and their union equals $\{x \le x_2\}$. Hence, [see (2-10)]

$$P\{x \le x_1\} + P\{x_1 < x \le x_2\} = P\{x \le x_2\}$$

and (2-28) follows from (2-27).

We note that, if the function $\alpha(x)$ is bounded, then the integral in (2-28) tends to zero as $x_1 \to x_2$. This leads to the conclusion that the probability of the event $\{x_2\}$ consisting of the single outcome x_2 is zero for every x_2. In this case, the probability of all elementary events of $\mathscr{S}$ equals zero, although the probability of their union equals 1. This is not in conflict with (2-21) because the total number of elements of $\mathscr{S}$ is not countable.

Example 2-6 A radioactive substance is selected at $t = 0$ and the time t of emission of a particle is observed. This process defines an experiment whose outcomes are all points on the positive t axis. This experiment can be considered as a special case of the real line experiment if we assume that $\mathscr{S}$ is the entire t axis and all events on the negative axis have zero probability.

Suppose then that the function $\alpha(t)$ in (2-26) is given by (Fig. 2-9b)

$$\alpha(t) = ce^{-ct}U(t) \qquad U(t) = \begin{cases} 1 & t \ge 0 \\ 0 & t < 0 \end{cases}$$

Inserting into (2-28), we conclude that the probability that a particle will be emitted in the time interval $(0, t_0)$ equals

$$\int_0^{t_0} e^{-ct}\, dt = 1 - e^{-ct_0}$$

Example 2-7 A telephone call occurs at *random* in the interval $(0, T)$. This means that the probability that it will occur in the interval $0 \le t \le t_0$ equals t_0/T. Thus, the outcomes of this experiment are all points in the interval (0, T) and the probability of the event {the call will occur in the interval (t_1, t_2)} equals

$$P\{t_1 \le t \le t_2\} = \frac{t_2 - t_1}{T}$$

This is again a special case of (2-28) with $\alpha(t) = 1/T$ for $0 \le t \le T$ and zero otherwise (Fig. 2-9c).

Probability masses The probability $P(\mathscr{A})$ of an event $\mathscr{A}$ can be interpreted as mass of the corresponding figure in its Venn diagram representation. Various identities have similar interpretations. Consider, for example, the identity $P(\mathscr{A} + \mathscr{B}) = P(\mathscr{A}) + P(\mathscr{B}) - P(\mathscr{A}\mathscr{B})$. The left side equals the mass of the event $\mathscr{A} + \mathscr{B}$. In the sum $P(\mathscr{A}) + P(\mathscr{B})$, the mass of $\mathscr{A}\mathscr{B}$ is counted twice (Fig. 2-3). To equate with $P(\mathscr{A} + \mathscr{B})$ we must, therefore, subtract $P(\mathscr{A}\mathscr{B})$.

2-3 CONDITIONAL PROBABILITY

The *conditional probability* of an event $\mathscr{A}$ assuming $\mathscr{M}$, denoted by $P(\mathscr{A} \mid \mathscr{M})$, is by definition the ratio

$$P(\mathscr{A} \mid \mathscr{M}) = \frac{P(\mathscr{A}\mathscr{M})}{P(\mathscr{M})} \tag{2-29}$$

where we assume that $P(\mathscr{M})$ is not zero.

The following properties follow readily from the definition:

$$\text{If} \quad \mathscr{M} \subset \mathscr{A} \quad \text{then} \quad P(\mathscr{A} \mid \mathscr{M}) = 1 \tag{2-30}$$

because then $\mathscr{A}\mathscr{M} = \mathscr{M}$. Similarly

$$\text{if} \quad \mathscr{A} \subset \mathscr{M} \quad \text{then} \quad P(\mathscr{A} \mid \mathscr{M}) = \frac{P(\mathscr{A})}{P(\mathscr{M})} \ge P(\mathscr{A}) \tag{2-31}$$

Frequency interpretation Denoting by $n_{\mathscr{A}}$, $n_{\mathscr{M}}$, and $n_{\mathscr{A}\mathscr{M}}$ the number of occurrences of the events $\mathscr{A}$, $\mathscr{M}$, and $\mathscr{A}\mathscr{M}$ respectively, we conclude from (1-1) that

$$P(\mathscr{A}) \simeq \frac{n_{\mathscr{A}}}{n} \qquad P(\mathscr{M}) \simeq \frac{n_{\mathscr{M}}}{n} \qquad P(\mathscr{A}\mathscr{M}) \simeq \frac{n_{\mathscr{A}\mathscr{M}}}{n}$$

Hence

$$P(\mathscr{A} \mid \mathscr{M}) = \frac{P(\mathscr{A}\mathscr{M})}{P(\mathscr{M})} \simeq \frac{n_{\mathscr{A}\mathscr{M}}/n}{n_{\mathscr{M}}/n} = \frac{n_{\mathscr{A}\mathscr{M}}}{n_{\mathscr{M}}} \tag{2-32}$$

This result can be phrased as follows: If we discard all trials in which the event $\mathscr{M}$ did not occur and we retain only the subsequence of trials in which $\mathscr{M}$ occurred, then $P(\mathscr{A} \mid \mathscr{M})$ equals the relative frequence of occurrence $n_{\mathscr{A}\mathscr{M}}/n_{\mathscr{M}}$ of the event $\mathscr{A}$ in that subsequence.

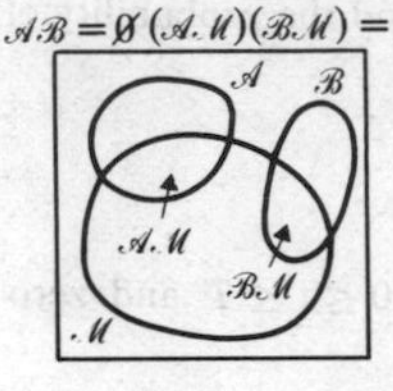

Figure 2-10

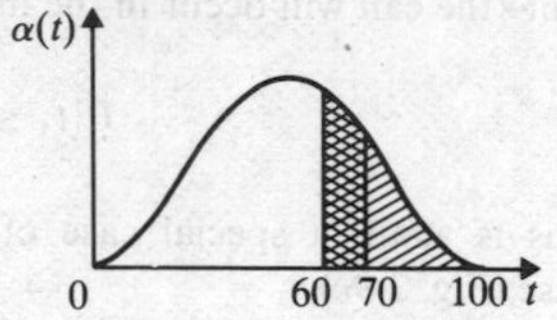

Figure 2-11

Fundamental remark We shall show that, for a specific $\mathscr{M}$, the conditional probabilities are indeed probabilities, i.e., they satisfy the axioms.

The first axiom is obviously satisfied because $P(\mathscr{A}\mathscr{M}) \geq 0$ and $P(\mathscr{M}) > 0$:

$$P(\mathscr{A} \mid \mathscr{M}) \geq 0 \tag{2-33}$$

The second follows from (2-30) because $\mathscr{M} \subset \mathscr{S}$:

$$P(\mathscr{S} \mid \mathscr{M}) = 1 \tag{2-34}$$

To prove the third, we observe that if the events $\mathscr{A}$ and $\mathscr{B}$ are mutually exclusive, then (Fig. 2-10) the events $\mathscr{A}\mathscr{M}$ and $\mathscr{B}\mathscr{M}$ are also mutually exclusive, hence,

$$P(\mathscr{A} + \mathscr{B} \mid \mathscr{M}) = \frac{P[(\mathscr{A} + \mathscr{B})\mathscr{M}]}{P(\mathscr{M})} = \frac{P(\mathscr{A}\mathscr{M}) + P(\mathscr{B}\mathscr{M})}{P(\mathscr{M})}$$

This yields the third axiom:

$$P(\mathscr{A} + \mathscr{B} \mid \mathscr{M}) = P(\mathscr{A} \mid \mathscr{M}) + P(\mathscr{B} \mid \mathscr{M}) \tag{2-35}$$

From the above it follows that all results involving probabilities hold also for conditional probabilities. The significance of this conclusion will be appreciated later.

Example 2-8 In the fair die experiment, we shall determine the conditional probability of the event $\{f_2\}$ assuming that the event *even* occurred. With

$$\mathscr{A} = \{f_2\} \qquad \mathscr{M} = \{\text{even}\} = \{f_2, f_4, f_6\}$$

we have $P(\mathscr{A}) = 1/6$ and $P(\mathscr{M}) = 3/6$. And since $\mathscr{A}\mathscr{M} = \mathscr{A}$, (2-29) yields

$$P\{f_2 \mid \text{even}\} = \frac{P\{f_2\}}{P\{\text{even}\}} = \frac{1}{3}$$

This equals the relative frequency of the occurrence of the event {two} in the subsequence whose outcomes are even numbers.

Example 2-9 We denote by t the age of a person when he dies. The probability that $t \leq t_o$ is given by

$$P\{t \leq t_o\} = \int_0^{t_o} \alpha(t)\, dt$$

where $\alpha(t)$ is a function determined from mortality records. We shall assume that

$$\alpha(t) = 3 \times 10^{-9} t^2 (1 - t)^2 \qquad 0 \le t \le 100 \text{ years}$$

and 0 otherwise (Fig. 2-11).

From (2-28) it follows that the probability that a person will die between the ages of 60 and 70 equals

$$P\{60 \le t \le 70\} = \int_{60}^{70} \alpha(t)\, dt = 0.154$$

This equals the number of people who die between the ages of 60 and 70 divided by the total population.

With

$$\mathscr{A} = \{60 \le t \le 70\} \qquad \mathscr{M} = \{t \ge 60\} \qquad \mathscr{A}\mathscr{M} = \mathscr{A}$$

it follows from (2-29) that the probability that a person will die between the ages of 60 and 70 assuming that he was alive at 60 equals

$$P\{60 \le t \le 70 \mid t \ge 60\} = \frac{\int_{60}^{70} \alpha(t)\, dt}{\int_{60}^{100} \alpha(t)\, dt} = 0.486$$

This equals the number of people who die between the ages 60 and 70 divided by the number of people that are alive at age 60.

Example 2-10 A box contains three white balls w_1, w_2, w_3 and two red balls r_1, r_2. We remove at random two balls in succession. What is the probability that the first removed ball is white and the second red?

We shall give two solutions to this problem. In the first, we apply (2-25); in the second, we use conditional probabilities.

FIRST SOLUTION The space of our experiment consists of all *ordered* pairs that we can form with the five balls:

$$w_1 w_2 \qquad w_1 w_3 \qquad w_1 r_1 \qquad w_1 r_2 \quad \cdots \quad r_2 w_1 \qquad r_2 w_2 \qquad r_2 w_3 \qquad r_2 r_1$$

The number of such pairs equals $5 \times 4 = 20$. The event {white first, red second} consists of the six outcomes

$$w_1 r_1 \qquad w_1 r_2 \qquad w_2 r_1 \qquad w_2 r_2 \qquad w_3 r_1 \qquad w_3 r_2$$

hence [see (2-25)] its probability equals 6/20.

SECOND SOLUTION Since the box contains three white and two red balls, the probability of the event $\mathscr{W}_1 = $ {white first} equals 3/5. If a white ball is removed, there remain two white and two red balls, hence, the conditional probability $P(\mathscr{R}_2 \mid \mathscr{W}_1)$ of the event $\mathscr{R}_2 = $ {red second} assuming {white first} equals 2/4. From this and (2-29) it follows that

$$P(\mathscr{W}_1 \mathscr{R}_2) = P(\mathscr{R}_2 \mid \mathscr{W}_1) P(\mathscr{W}_1) = \frac{2}{4} \times \frac{3}{5} = \frac{6}{20}$$

where $\mathscr{W}_1 \mathscr{R}_2$ is the event {white first, red second}.

Total Probability and Bayes' Theorem

If $\mathfrak{A} = [\mathscr{A}_1, \ldots, \mathscr{A}_n]$ is a partition of $\mathscr{S}$ and $\mathscr{B}$ is an arbitrary event (Fig. 2-5), then

$$P(\mathscr{B}) = P(\mathscr{B} \mid \mathscr{A}_1)P(\mathscr{A}_1) + \cdots + P(\mathscr{B} \mid \mathscr{A}_n)P(\mathscr{A}_n) \tag{2-36}$$

PROOF Clearly

$$\mathscr{B} = \mathscr{B}\mathscr{S} = \mathscr{B}(\mathscr{A}_1 + \cdots + \mathscr{A}_n) = \mathscr{B}\mathscr{A}_1 + \cdots + \mathscr{B}\mathscr{A}_n$$

But the events $\mathscr{B}\mathscr{A}_i$ and $\mathscr{B}\mathscr{A}_j$ are mutually exclusive because the events $\mathscr{A}_i$ and $\mathscr{A}_j$ are mutually exclusive [see (2-4)]. Hence

$$P(\mathscr{B}) = P(\mathscr{B}\mathscr{A}_1) + \cdots + P(\mathscr{B}\mathscr{A}_n)$$

and (2-36) follows because [see (2-29)]

$$P(\mathscr{B}\mathscr{A}_i) = P(\mathscr{B} \mid \mathscr{A}_i)P(\mathscr{A}_i) \tag{2-37}$$

This result is known as the *total probability theorem.*

Since $P(\mathscr{B}\mathscr{A}_i) = P(\mathscr{A}_i \mid \mathscr{B})P(\mathscr{B})$ we conclude with (2-37) that

$$P(\mathscr{A}_i \mid \mathscr{B}) = P(\mathscr{B} \mid \mathscr{A}_i)\,\frac{P(\mathscr{A}_i)}{P(\mathscr{B})} \tag{2-38}$$

Inserting (2-36) into (2-38), we obtain *Bayes' theorem*†:

$$P(\mathscr{A}_i \mid \mathscr{B}) = \frac{P(\mathscr{B} \mid \mathscr{A}_i)P(\mathscr{A}_i)}{P(\mathscr{B} \mid \mathscr{A}_1)P(\mathscr{A}_1) + \cdots + P(\mathscr{B} \mid \mathscr{A}_n)P(\mathscr{A}_n)} \tag{2-39}$$

Note The terms *a priori* and *a posteriori* are often used for the probabilities $P(\mathscr{A}_i)$ and $P(\mathscr{A}_i \mid \mathscr{B})$.

Example 2-11 We have four boxes. Box 1 contains 2000 components of which 5 percent are defective. Box 2 contains 500 components of which 40 percent are defective. Boxes 3 and 4 contain 1000 each with 10 percent defective. We select *at random* one of the boxes and we remove *at random* a single component.

(*a*) What is the probability that the selected component is defective?

The space of this experiment consists of 4000 good (*g*) components and 500 defective (*d*) components arranged as follows:

Box 1:	1900*g*, 100*d*	Box 2:	300*g*, 200*d*
Box 3:	900*g*, 100*d*	Box 4:	900*d*, 100*d*

We denote by $\mathscr{B}_i$ the event consisting of all components in the *i*th box and by $\mathscr{D}$ the event consisting of all defective components. Clearly

$$P(\mathscr{B}_1) = P(\mathscr{B}_2) = P(\mathscr{B}_3) = P(\mathscr{B}_4) = \frac{1}{4} \tag{2-40}$$

† The main idea of this theorem is due to Thomas Bayes (1763). However, its final form (2-39) was given by Laplace several years later.

because the boxes are selected at random. The probability that a component taken from a specific box is defective equals the ratio of the defective to the total number of components in that box. This means that

$$P(\mathscr{D} \mid \mathscr{B}_1) = \frac{100}{2000} = 0.05 \qquad P(\mathscr{D} \mid \mathscr{B}_2) = \frac{200}{500} = 0.4$$
$$P(\mathscr{D} \mid \mathscr{B}_3) = \frac{100}{1000} = 0.1 \qquad P(\mathscr{D} \mid \mathscr{B}_4) = \frac{100}{1000} = 0.1 \tag{2-41}$$

And since the events $\mathscr{B}_1, \mathscr{B}_2, \mathscr{B}_3, \mathscr{B}_4$ form a partition of $\mathscr{S}$, we conclude from (2-36) that

$$P(\mathscr{D}) = 0.05 \times \frac{1}{4} + 0.4 \times \frac{1}{4} + 0.1 \times \frac{1}{4} + 0.1 \times \frac{1}{4} = 0.1625$$

This is the probability that the selected component is defective.

(*b*) We examine the selected component and we find it defective. On the basis of this evidence, we want to determine the probability that it came from box 2.

We now want the conditional probability $P(\mathscr{B}_2 \mid \mathscr{D})$. Since

$$P(\mathscr{D}) = 0.1625 \qquad P(\mathscr{D} \mid \mathscr{B}_2) = 0.4 \qquad P(\mathscr{B}_2) = 0.25$$

(2-38) yields

$$P(\mathscr{B}_2 \mid \mathscr{D}) = 0.4 \times \frac{0.25}{0.1625} = 0.615$$

Thus, the a priori probability of selecting box 2 equals 0.25 and the a posteriori probability assuming that the selected component is defective equals 0.615. These probabilities have the following frequency interpretation: If the experiment is performed n times then box 2 is selected $0.25n$ times. If we consider only the $n_{\mathscr{D}}$ experiments in which the removed part is defective, then the number of times the part is taken from box 2 equals $0.615n_{\mathscr{D}}$.

We conclude with a comment on the distinction between assumptions and deductions: Equations (2-40) and (2-41) are not derived; they are merely reasonable *assumptions*. Based on these assumptions and on the axioms, we *deduce* that $P(\mathscr{D}) = 0.1625$ and $P(\mathscr{B}_2 \mid \mathscr{D}) = 0.615$.

Independence

Two events $\mathscr{A}$ and $\mathscr{B}$ are called *independent* if

$$P(\mathscr{A}\mathscr{B}) = P(\mathscr{A})P(\mathscr{B}) \tag{2-42}$$

The concept of independence is fundamental. In fact, it is this concept that justifies the mathematical development of probability, not merely as a topic in measure theory, but as a separate discipline. The significance of independence will be appreciated later in the context of repeated trials. We discuss here only various simple properties.

Frequency interpretation Denoting by $n_{\mathscr{A}}$, $n_{\mathscr{B}}$, and $n_{\mathscr{A}\mathscr{B}}$ the number of occurrences of the events $\mathscr{A}$, $\mathscr{B}$, and $\mathscr{A}\mathscr{B}$ respectively, we have

$$P(\mathscr{A}) \simeq \frac{n_{\mathscr{A}}}{n} \qquad P(\mathscr{B}) \simeq \frac{n_{\mathscr{B}}}{n} \qquad P(\mathscr{A}\mathscr{B}) \simeq \frac{n_{\mathscr{A}\mathscr{B}}}{n}$$

If the events $\mathscr{A}$ and $\mathscr{B}$ are independent, then

$$\frac{n_{\mathscr{A}}}{n} \simeq P(\mathscr{A}) = \frac{P(\mathscr{A}\mathscr{B})}{P(\mathscr{B})} \simeq \frac{n_{\mathscr{A}\mathscr{B}}/n}{n_{\mathscr{B}}/n} = \frac{n_{\mathscr{A}\mathscr{B}}}{n_{\mathscr{B}}}$$

Thus, if $\mathscr{A}$ and $\mathscr{B}$ are independent, then the relative frequence $n_{\mathscr{A}}/n$ of the occurrence of $\mathscr{A}$ in the original sequence of n trials equals the relative frequency $n_{\mathscr{A}\mathscr{B}}/n_{\mathscr{B}}$ of the occurrence of $\mathscr{A}$ in the subsequence in which $\mathscr{B}$ occurred.

We show next that if the events $\mathscr{A}$ and $\mathscr{B}$ are independent, then the events $\bar{\mathscr{A}}$ and $\mathscr{B}$ and the events $\bar{\mathscr{A}}$ and $\bar{\mathscr{B}}$ are also independent.

As we know, the events $\mathscr{A}\mathscr{B}$ and $\bar{\mathscr{A}}\mathscr{B}$ are mutually exclusive and

$$\mathscr{B} = \mathscr{A}\mathscr{B} + \bar{\mathscr{A}}\mathscr{B} \qquad P(\bar{\mathscr{A}}) = 1 - P(\mathscr{A})$$

From this and (2-42) it follows that

$$P(\bar{\mathscr{A}}\mathscr{B}) = P(\mathscr{B}) - P(\mathscr{A}\mathscr{B}) = [1 - P(\mathscr{A})]P(\mathscr{B}) = P(\bar{\mathscr{A}})P(\mathscr{B})$$

This establishes the independence of $\bar{\mathscr{A}}$ and $\mathscr{B}$. Repeating the argument, we conclude that $\bar{\mathscr{A}}$ and $\bar{\mathscr{B}}$ are also independent.

In the next two examples, we illustrate the concept of independence. In Example 2-12*a*, we start with a known experiment and we show that two of its events are independent. In Examples 2-12*b* and 2-13 we use the concept of independence to complete the specification of each experiment. This idea is developed further in the next chapter.

Example 2-12 If we toss a coin twice, we generate the four outcomes *hh*, *ht*, *th*, and *tt*.

(*a*) To construct an experiment with these outcomes, it suffices to assign probabilities to its elementary events. With a and b two positive numbers such that $a + b = 1$, we assume that

$$P\{hh\} = a^2 \qquad P\{ht\} = P\{th\} = ab \qquad P\{tt\} = b^2$$

These probabilities are consistent with the axioms because

$$a^2 + ab + ab + b^2 = (a + b)^2 = 1$$

In the experiment so constructed, the events

$$\mathscr{H}_1 = \{\text{heads at first tossing}\} = \{hh, ht\}$$

$$\mathscr{H}_2 = \{\text{heads at second tossing}\} = \{hh, th\}$$

consist of two elements each, and their probabilities are [see (2-23)]

$$P(\mathscr{H}_1) = P\{hh\} + P\{ht\} = a^2 + ab = a$$

$$P(\mathscr{H}_2) = P\{hh\} + P\{th\} = a^2 + ab = a$$

The intersection $\mathscr{H}_1\mathscr{H}_2$ of these two events consists of the single outcome $\{hh\}$. Hence

$$P(\mathscr{H}_1\mathscr{H}_2) = P\{hh\} = a^2 = P(\mathscr{H}_1)P(\mathscr{H}_2)$$

This shows that the events $\mathscr{H}_1$ and $\mathscr{H}_2$ are independent.

(b) The above experiment can be specified in terms of the probabilities $P(\mathscr{H}_1) = P(\mathscr{H}_2) = a$ of the events $\mathscr{H}_1$ and $\mathscr{H}_2$, and the information that these events are independent.

Indeed, as we have shown, the events $\bar{\mathscr{H}}_1$ and $\mathscr{H}_2$ and the events $\bar{\mathscr{H}}_1$ and $\bar{\mathscr{H}}_2$ are also independent. Furthermore

$$\mathscr{H}_1\mathscr{H}_2 = \{hh\} \qquad \mathscr{H}_1\bar{\mathscr{H}}_2 = \{ht\} \qquad \bar{\mathscr{H}}_1\mathscr{H}_2 = \{th\} \qquad \bar{\mathscr{H}}_1\bar{\mathscr{H}}_2 = \{tt\}$$

and $P(\bar{\mathscr{H}}_1) = 1 - P(\mathscr{H}_1) = 1 - a$, $P(\bar{\mathscr{H}}_2) = 1 - P(\mathscr{H}_2) = 1 - a$. Hence,

$$P\{hh\} = a^2 \qquad P\{ht\} = a(1-a) \qquad P\{th\} = (1-a)a \qquad P\{tt\} = (1-a)^2$$

Example 2-13 Trains X and Y arrive at a station at random between 8 A.M. and 8:20 A.M. Train X stops for four minutes and train Y stops for five minutes. Assuming that the trains arrive independently of each other, we shall determine various probabilities related to the times x and y of their respective arrivals. To do so we must first specify the underlying experiment.

The outcomes of this experiment are all points (x, y) in the square of Fig. 2-12. The event

$$\mathscr{A} = \{X \text{ arrives in the interval } (t_1, t_2)\} = \{t_1 \le x \le t_2\}$$

is a vertical strip as in Fig. 2-12*a* and its probability equals $(t_2 - t_1)/20$. This is our interpretation of the information that the train arrives at random. Similarly, the event

$$\mathscr{B} = \{Y \text{ arrives in the interval } (t_3, t_4)\} = \{t_3 \le y \le t_4\}$$

is a vertical strip and its probability equals $(t_4 - t_3)/20$.

Proceeding similarly, we can determine the probabilities of any horizontal or vertical sets of points. To complete the specification of the experiment, we must determine also the probabilities of their intersections. Translating the independence of the arrival times as independence of the events $\mathscr{A}$ and $\mathscr{B}$, we obtain

$$P(\mathscr{A}\mathscr{B}) = P(\mathscr{A})P(\mathscr{B}) = \frac{(t_2 - t_1)(t_4 - t_3)}{20 \times 20}$$

The event $\mathscr{A}\mathscr{B}$ is the rectangle shown in the figure. Since the coordinates of this rectangle are arbitrary, we conclude that the probability of any rectangle equals its

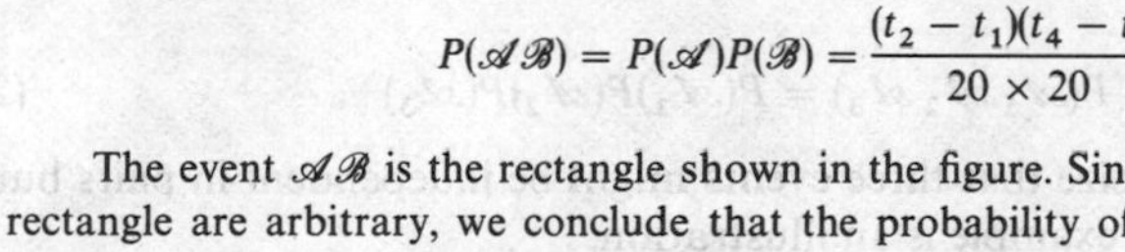

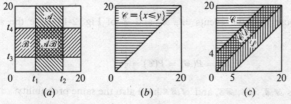

Figure 2-12

area divided by 400. In the plane, all events are unions and intersections of rectangles forming a Borel field. This shows that the probability that the point (x, y) will be in an arbitrary region R of the plane equals the area of R divided by 400. This completes the specification of the experiment.

(*a*) We shall determine the probability that train X arrives before train Y. This is the probability of the event

$$\mathscr{C} = \{x \le y\}$$

shown in Fig. 2-12*b*. This event is a triangle with area 200. Hence

$$P(\mathscr{C}) = \frac{200}{400}$$

(*b*) We shall determine the probability that the trains meet at the station. For the trains to meet, x must be less than $y + 5$ and y must be less than $x + 4$. This is the event

$$\mathscr{D} = \{-4 \le x - y \le 5\}$$

of Fig. 2-12*c*. As we see from the figure, the region $\mathscr{D}$ consists of two trapezoids with common base, and its area equals 159.5. Hence

$$P(\mathscr{D}) = \frac{159.5}{400}$$

(*c*) Assuming that the trains met, we shall determine the probability that train X arrived before train Y. We wish to find the conditional probability $P(\mathscr{C} \mid \mathscr{D})$. The event $\mathscr{C}\mathscr{D}$ is a trapezoid as shown and its area equals 72. Hence

$$P(\mathscr{C} \mid \mathscr{D}) = \frac{P(\mathscr{C}\mathscr{D})}{P(\mathscr{D})} = \frac{72}{159.5}$$

Independence of three events The events $\mathscr{A}_1, \mathscr{A}_2, \mathscr{A}_3$ are called (mutually) independent if they are independent in pairs:

$$P(\mathscr{A}_i \mathscr{A}_j) = P(\mathscr{A}_i)P(\mathscr{A}_j) \qquad i \ne j \tag{2-43}$$

and

$$P(\mathscr{A}_1 \mathscr{A}_2 \mathscr{A}_3) = P(\mathscr{A}_1)P(\mathscr{A}_2)P(\mathscr{A}_3) \tag{2-44}$$

We should emphasize that three events might be independent in pairs but not independent. The next example is an illustration.

Example 2-14 Suppose that the events $\mathscr{A}$, $\mathscr{B}$, and $\mathscr{C}$ of Fig. 2-13 have the same probability

$$P(\mathscr{A}) = P(\mathscr{B}) = P(\mathscr{C}) = \frac{1}{5}$$

and the intersections $\mathscr{A}\mathscr{B}$, $\mathscr{A}\mathscr{C}$, $\mathscr{B}\mathscr{C}$, and $\mathscr{A}\mathscr{B}\mathscr{C}$ have also the same probability

$$p = P(\mathscr{A}\mathscr{B}) = P(\mathscr{A}\mathscr{C}) = P(\mathscr{B}\mathscr{C}) = P(\mathscr{A}\mathscr{B}\mathscr{C})$$

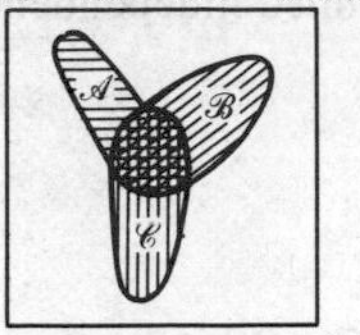

Figure 2-13

(*a*) If $p = 1/25$, then these events are independent in pairs but they are not independent because

$$P(\mathscr{A}\mathscr{B}\mathscr{C}) \neq P(\mathscr{A})P(\mathscr{B})P(\mathscr{C})$$

(*b*) If $p = 1/125$, then $P(\mathscr{A}\mathscr{B}\mathscr{C}) = P(\mathscr{A})P(\mathscr{B})P(\mathscr{C})$ but the events are not independent because

$$P(\mathscr{A}\mathscr{B}) \neq P(\mathscr{A})P(\mathscr{B})$$

From the independence of the events $\mathscr{A}$, $\mathscr{B}$, and $\mathscr{C}$ it follows that:

1. Any one of them is independent of the intersection of the other two.
   Indeed, from (2-43) and (2-44) it follows that

$$P(\mathscr{A}_1\mathscr{A}_2\mathscr{A}_3) = P(\mathscr{A}_1)P(\mathscr{A}_2)P(\mathscr{A}_3) = P(\mathscr{A}_1)P(\mathscr{A}_2\mathscr{A}_3) \tag{2-45}$$

   hence, the events $\mathscr{A}_1$ and $\mathscr{A}_2\mathscr{A}_3$ are independent.
2. If we replace one or more of these events with their complements, the resulting events are also independent.
   Indeed, since

$$\mathscr{A}_1\mathscr{A}_2 = \mathscr{A}_1\mathscr{A}_2\mathscr{A}_3 + \mathscr{A}_1\mathscr{A}_2\bar{\mathscr{A}}_3 \qquad P(\bar{\mathscr{A}}_3) = 1 - P(\mathscr{A}_3)$$

   we conclude with (2-45) that

$$P(\mathscr{A}_1\mathscr{A}_2\bar{\mathscr{A}}_3) = P(\mathscr{A}_1\mathscr{A}_2) - P(\mathscr{A}_1\mathscr{A}_2)P(\mathscr{A}_3) = P(\mathscr{A}_1)P(\mathscr{A}_2)P(\bar{\mathscr{A}}_3)$$

   Hence, the events $\mathscr{A}_1$, $\mathscr{A}_2$, and $\bar{\mathscr{A}}_3$ are independent because they satisfy (2-44) and, as we have shown earlier in the section, they are also independent in pairs.
3. Any one of them is independent of the union of the other two.
   To show that the events $\mathscr{A}_1$ and $\mathscr{A}_2 + \mathscr{A}_3$ are independent, it suffices to show that the events $\mathscr{A}_1$ and $\overline{\mathscr{A}_2 + \mathscr{A}_3} = \bar{\mathscr{A}}_2\bar{\mathscr{A}}_3$ are independent. This follows from 1 and 2.

Generalization The independence of n events can be defined inductively: Suppose that we have defined independence of k events for every $k < n$. We then say that the events $\mathscr{A}_1, \ldots, \mathscr{A}_n$ are independent if any $k < n$ of them are independent and

$$P(\mathscr{A}_1 \cdots \mathscr{A}_n) = P(\mathscr{A}_1) \cdots P(\mathscr{A}_n) \tag{2-46}$$

This completes the definition for any n because we have defined independence for $n = 2$.

PROBLEMS

2-1 Show that (*a*) $\overline{\overline{\mathscr{A}} + \overline{\mathscr{B}}} + \overline{\overline{\mathscr{A}} + \mathscr{B}} = \mathscr{A}$; (*b*) $(\mathscr{A} + \mathscr{B})\overline{(\mathscr{A}\mathscr{B})} = \mathscr{A}\overline{\mathscr{B}} + \mathscr{B}\overline{\mathscr{A}}$.

2-2 If $\mathscr{A} = \{2 \le x \le 5\}$ and $\mathscr{B} = \{3 \le x \le 6\}$, find $\mathscr{A} + \mathscr{B}$, $\mathscr{A}\mathscr{B}$, and $(\mathscr{A} + \mathscr{B})\overline{(\mathscr{A}\mathscr{B})}$.

2-3 Show that if $\mathscr{A}\mathscr{B} = \{\emptyset\}$, then $P(\mathscr{A}) \le P(\overline{\mathscr{B}})$.

2-4 Show that (*a*) if $P(\mathscr{A}) = P(\mathscr{B}) = P(\mathscr{A}\mathscr{B})$, then $P(\mathscr{A}\overline{\mathscr{B}} + \mathscr{B}\overline{\mathscr{A}}) = 0$; (*b*) if $P(\mathscr{A}) = P(\mathscr{B}) = 1$, then $P(\mathscr{A}\mathscr{B}) = 1$.

2-5 Prove and generalize the following identity

$$P(\mathscr{A} + \mathscr{B} + \mathscr{C}) = P(\mathscr{A}) + P(\mathscr{B}) + P(\mathscr{C}) - P(\mathscr{A}\mathscr{B}) - P(\mathscr{A}\mathscr{C}) - P(\mathscr{B}\mathscr{C}) + P(\mathscr{A}\mathscr{B}\mathscr{C})$$

2-6 Show that if $\mathscr{S}$ consists of a countable number of elements ζ_i and each subset $\{\zeta_i\}$ is an event, then all subsets of $\mathscr{S}$ are events.

2-7 If $\mathscr{S} = \{1, 2, 3, 4\}$, find the smallest field that contains the sets $\{1\}$, and $\{2, 3\}$.

2-8 If $\mathscr{A} \subset \mathscr{B}$, $P(\mathscr{A}) = 1/4$, and $P(\mathscr{B}) = 1/3$, find $P(\mathscr{A} \mid \mathscr{B})$ and $P(\mathscr{B} \mid \mathscr{A})$.

2-9 Show that $P(\mathscr{A}\mathscr{B} \mid \mathscr{C}) = P(\mathscr{A} \mid \mathscr{B}\mathscr{C})P(\mathscr{B} \mid \mathscr{C})$ and $P(\mathscr{A}\mathscr{B}\mathscr{C}) = P(\mathscr{A} \mid \mathscr{B}\mathscr{C})P(\mathscr{B} \mid \mathscr{C})P(\mathscr{C})$.

2-10 (*Chain rule*) Show that

$$P(\mathscr{A}_n \cdots \mathscr{A}_1) = P(\mathscr{A}_n \mid \mathscr{A}_{n-1}, \ldots, \mathscr{A}_1) \cdots P(\mathscr{A}_2 \mid \mathscr{A}_1)\mathscr{P}(\mathscr{A}_1)$$

2-11 We select at random m objects from a set $\mathscr{S}$ of n objects and we denote by $\mathscr{A}_m$ the set of the selected objects. Show that the probability p that a particular element ζ_o of $\mathscr{S}$ is in $\mathscr{A}_m$ equals m/n.

Hint: p equals the probability that a randomly selected point of $\mathscr{S}$ is in $\mathscr{A}_m$.

2-12 A call occurs at time t where t is a random point in the interval (0, 10). (*a*) Find $P\{6 \le t \le 8\}$; (*b*) Find $P\{6 \le t \le 8 \mid t > 5\}$.

2-13 The space $\mathscr{S}$ is the set of all positive numbers t. Show that if $P\{t_0 \le t \le t_0 + t_1 \mid t \ge t_0\} = P\{t \le t_1\}$ for every t_0 and t_1, then $P\{t \le t_1\} = 1 - e^{-ct_1}$ where c is a constant.

2-14 The events $\mathscr{A}$ and $\mathscr{B}$ are mutually exclusive. Can they be independent?

2-15 Show that if the events $\mathscr{A}_1, \ldots, \mathscr{A}_n$ are independent and $\mathscr{B}_i$ equals $\mathscr{A}_i$ or $\overline{\mathscr{A}}_i$ or $\mathscr{S}$, then the events $\mathscr{B}_1, \ldots, \mathscr{B}_n$ are also independent.

2-16 Show that $2^n - (n + 1)$ equations are needed to establish the independence of n events.

2-17 Box 1 contains 1 white and 999 red balls. Box 2 contains 1 red and 999 white balls. A ball is picked from a randomly selected box. If the ball is red what is the probability that it came from box 1?

2-18 Box 1 contains 1000 bulbs of which 10% are defective. Box 2 contains 2000 bulbs of which 5% are defective. Two bulbs are picked from a randomly selected box. (*a*) Find the probability that both bulbs are defective. (*b*) Assuming that both are defective, find the probability that they came from box 1.

2-19 A train and a bus arrive at the station at random between 9 A.M. and 10 A.M. The train stops for 10 minutes and the bus for x minutes. Find x so that the probability that the bus and the train will meet equals 0.5.

2-20 Show that a set $\mathscr{S}$ with n elements has

$$\frac{n(n-1)\cdots(n-k+1)}{1 \cdot 2 \cdots k} = \frac{n!}{k!(n-k)!}$$

k-element subsets.

CHAPTER

THREE

REPEATED TRIALS

3-1 COMBINED EXPERIMENTS

We are given two experiments: The first experiment is the rolling of a fair die

$$\mathscr{S}_1 = \{f_1, \ldots, f_6\} \qquad P_1\{f_i\} = \frac{1}{6}$$

The second experiment is the tossing of a fair coin

$$\mathscr{S}_2 = \{h, t\} \qquad P_2\{h\} = P_2\{t\} = \frac{1}{2}$$

We perform both experiments and we want to find the probability that we get "two" on the die and "heads" on the coin.

If we make the reasonable assumption that the outcomes of the first experiment are independent of the outcomes of the second, we conclude that the unknown probability equals $1/6 \times 1/2$.

The above conclusion is reasonable; however, the notion of independence used in its derivation does not agree with the definition given in (2-42). In that definition, the events $\mathscr{A}$ and $\mathscr{B}$ were subsets of the *same* space. In order to fit the above conclusion into our theory we must, therefore, construct a space $\mathscr{S}$ having as subsets the events "two" and "heads." This is done as follows:

The two experiments are viewed as a single experiment whose outcomes are pairs $\zeta_1\zeta_2$ where ζ_1 is one of the six faces of the die and ζ_2 is heads or tails†. The resulting space consists of the 12 elements

$$f_1h, \ldots, f_6 h, f_1t, \ldots, f_6 t$$

In this space, {two} is not an elementary event but a subset consisting of two elements

$$\{\text{two}\} = \{f_2 h, f_2 t\}$$

Similarly, {heads} is an event with six elements

$$\{\text{heads}\} = \{f_1h, \ldots, f_6 h\}$$

To complete the experiment, we must assign probabilities to all subsets of $\mathscr{S}$. Clearly, the event {two} occurs if the die shows "two" no matter what shows on the coin. Hence

$$P\{\text{two}\} = P_1\{f_2\} = \frac{1}{6}$$

Similarly

$$P\{\text{heads}\} = P_2\{h\} = \frac{1}{2}$$

The intersection of the events {two} and {heads} is the elementary event $\{f_2 h\}$. Assuming that the events {two} and {heads} are independent in the sense of (2-42), we conclude that $P\{f_2 h\} = 1/6 \times 1/2$ in agreement with our earlier conclusion.

Cartesian products Given two sets $\mathscr{S}_1$ and $\mathscr{S}_2$ with elements ζ_1 and ζ_2 respectively, we form all *ordered* pairs $\zeta_1\zeta_2$ where ζ_1 is any element of $\mathscr{S}_1$ and ζ_2 is any element of $\mathscr{S}_2$. The *cartesian product* of the sets $\mathscr{S}_1$ and $\mathscr{S}_2$ is a set $\mathscr{S}$ whose elements are all such pairs. This set is written in the form

$$\mathscr{S} = \mathscr{S}_1 \times \mathscr{S}_2$$

Example 3-1 The cartesian product of the sets

$$\mathscr{S}_1 = \{\text{car, apple, bird}\} \qquad \mathscr{S}_2 = \{h, t\}$$

has six elements

$$\mathscr{S}_1 \times \mathscr{S}_2 = \{\text{car-}h, \text{car-}t, \text{apple-}h, \text{apple-}t, \text{bird-}h, \text{bird-}t\}$$

Example 3-2 If $\mathscr{S}_1 = \{h, t\}$, $\mathscr{S}_2 = \{h, t\}$, then

$$\mathscr{S}_1 \times \mathscr{S}_2 = \{hh, ht, th, tt\}$$

† In the earlier discussion, the symbol ζ_i represented a *single* element of a set $\mathscr{S}$. In the following, ζ_i will represent also an arbitrary element of a set $\mathscr{S}_i$. It will be understood from the context whether ζ_i is one particular element or any element of $\mathscr{S}_i$.

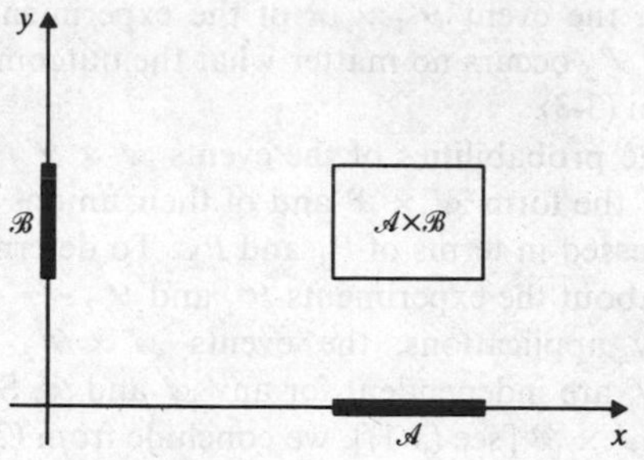

Figure 3-1

In this example, the sets $\mathscr{S}_1$ and $\mathscr{S}_2$ are identical. We note also that the element *ht* is different from the element *th*.

If $\mathscr{A}$ is a subset of $\mathscr{S}_1$ and $\mathscr{B}$ is a subset of $\mathscr{S}_2$, then the set

$$\mathscr{C} = \mathscr{A} \times \mathscr{B}$$

consisting of all pairs $\zeta_1\zeta_2$ where $\zeta_1 \in \mathscr{A}$ and $\zeta_2 \in \mathscr{B}$, is a subset of $\mathscr{S}$.

Forming similarly the sets $\mathscr{A} \times \mathscr{S}_2$ and $\mathscr{S}_1 \times \mathscr{B}$, we conclude that their intersection is the set $\mathscr{A} \times \mathscr{B}$

$$\mathscr{A} \times \mathscr{B} = (\mathscr{A} \times \mathscr{S}_2) \cap (\mathscr{S}_1 \times \mathscr{B}) \tag{3-1}$$

Note Suppose that $\mathscr{S}_1$ is the x axis, $\mathscr{S}_2$ is the y axis and $\mathscr{A}$ and $\mathscr{B}$ are two intervals:

$$\mathscr{A} = \{x_1 \le x \le x_2\} \qquad \mathscr{B} = \{y_1 \le y \le y_2\}$$

In this case, $\mathscr{A} \times \mathscr{B}$ is a rectangle, $\mathscr{A} \times \mathscr{S}_2$ is a vertical strip, and $\mathscr{S}_1 \times \mathscr{B}$ is a horizontal strip (Fig. 3-1).

We can thus interpret the cartesian product $\mathscr{A} \times \mathscr{B}$ of two arbitrary sets as a generalized rectangle.

Cartesian product of two experiments The cartesian product of two experiments $\mathscr{S}_1$ and $\mathscr{S}_2$ is a new experiment $\mathscr{S} = \mathscr{S}_1 \times \mathscr{S}_2$ whose events are all cartesian products of the form

$$\mathscr{A} \times \mathscr{B} \tag{3-2}$$

where $\mathscr{A}$ is an event of $\mathscr{S}_1$ and $\mathscr{B}$ is an event of $\mathscr{S}_2$, and their unions and intersections.

In this experiment, the probabilities of the events $\mathscr{A} \times \mathscr{S}_2$ and $\mathscr{S}_1 \times \mathscr{B}$ are such that

$$P(\mathscr{A} \times \mathscr{S}_2) = P_1(\mathscr{A}) \qquad P(\mathscr{S}_1 \times \mathscr{B}) = P_2(\mathscr{B}) \tag{3-3}$$

where $P_1(\mathscr{A})$ is the probability of the event $\mathscr{A}$ in the experiments $\mathscr{S}_1$ and $P_2(\mathscr{B})$ is the probability of the event $\mathscr{B}$ in the experiments $\mathscr{S}_2$. The above is motivated by the interpretation of $\mathscr{S}$ as a combined experiment. Indeed, the event $\mathscr{A} \times \mathscr{S}_2$ of the experiment $\mathscr{S}$ occurs if the event $\mathscr{A}$ of the experiment $\mathscr{S}_1$ occurs no matter what the outcome of $\mathscr{S}_2$ is. Similarly, the event $\mathscr{S}_1 \times \mathscr{B}$ of the experiment $\mathscr{S}$ occurs if the event $\mathscr{B}$ of the experiment $\mathscr{S}_2$ occurs no matter what the outcome of $\mathscr{S}_1$ is. This justifies the two equations in (3-3).

These equations determine only the probabilities of the events $\mathscr{A} \times \mathscr{S}_2$ and $\mathscr{S}_1 \times \mathscr{B}$. The probabilities of events of the form $\mathscr{A} \times \mathscr{B}$ and of their unions and intersections cannot in general be expressed in terms of P_1 and P_2. To determine them, we need additional information about the experiments $\mathscr{S}_1$ and $\mathscr{S}_2$.

Independent experiments In many applications, the events $\mathscr{A} \times \mathscr{S}_2$ and $\mathscr{S}_1 \times \mathscr{B}$ of the combined experiment $\mathscr{S}$ are independent for any $\mathscr{A}$ and $\mathscr{B}$. Since the intersection of these events equals $\mathscr{A} \times \mathscr{B}$ [see (3-1)], we conclude from (2-42) and (3-3) that

$$P(\mathscr{A} \times \mathscr{B}) = P(\mathscr{A} \times \mathscr{S}_2)P(\mathscr{S}_1 \times \mathscr{B}) = P_1(\mathscr{A})P_2(\mathscr{B}) \tag{3-4}$$

This completes the specification of the experiment $\mathscr{S}$ because all its events are unions and intersections of events of the form $\mathscr{A} \times \mathscr{B}$.

We note in particular that the elementary event $\{\zeta_1\zeta_2\}$ can be written as a cartesian product $\{\zeta_1\} \times \{\zeta_2\}$ of the elementary events $\{\zeta_1\}$ and $\{\zeta_2\}$ of $\mathscr{S}_1$ and $\mathscr{S}_2$. Hence

$$P\{\zeta_1\zeta_2\} = P_1\{\zeta_1\}P_2\{\zeta_2\} \tag{3-5}$$

Example 3-3 A box B_1 contains 10 white and 5 red balls and a box B_2 contains 20 white and 20 red balls. A ball is drawn from each box. What is the probability that the ball from B_1 will be white and the ball from B_2 red?

The above operation can be considered as a combined experiment. Experiment $\mathscr{S}_1$ is the drawing from B_1 and experiment $\mathscr{S}_2$ is the drawing from B_2. The space $\mathscr{S}_1$ has 15 elements: 10 white and 5 red balls. The event

$$\mathscr{W}_1 = \{\text{all white balls in } B_1\}$$

has 10 elements and its probability equals 10/15. The space $\mathscr{S}_2$ has 40 elements: 20 white and 20 red balls. The event

$$\mathscr{R}_2 = \{\text{all red balls in } B_2\}$$

has 20 elements and its probability equals 20/40. The space $\mathscr{S}_1 \times \mathscr{S}_2$ has 40 × 15 elements: all possible pairs that can be drawn.

We want the probability of the event

$$\mathscr{W}_1 \times \mathscr{R}_2 = \{\text{white from } B_1 \text{ and red from } B_2\}$$

Assuming independence of the two experiments, we conclude from (3-4) that

$$P(\mathscr{W}_1 \times \mathscr{R}_2) = P_1(\mathscr{W}_1)P_2(\mathscr{R}_2) = \frac{10}{15} \times \frac{20}{40}$$

Example 3-4 Consider the coin experiment where the probability of "heads" equals p and the probability of "tails" equals $q = 1 - p$. If we toss the coin twice, we obtain the space

$$\mathscr{S} = \mathscr{S}_1 \times \mathscr{S}_2 \qquad \mathscr{S}_1 = \mathscr{S}_2 = \{h, t\}$$

Thus, $\mathscr{S}$ consists of the four outcomes *hh*, *ht*, *th*, and *tt*. Assuming that the experiments $\mathscr{S}_1$ and $\mathscr{S}_2$ are independent we obtain

$$P\{hh\} = P_1\{h\}P_2\{h\} = p^2$$

Similarly

$$P\{ht\} = pq \qquad P\{th\} = qp \qquad P\{tt\} = q^2$$

We shall use the above to find the probability of the event

$$\mathscr{H}_1 = \{\text{heads at the first tossing}\} = \{hh, ht\}$$

Since $\mathscr{H}_1$ consists of the two outcomes *hh* and *ht*, (2-23) yields

$$P(\mathscr{H}_1) = P\{hh\} + P\{ht\} = p^2 + pq = p$$

This follows also from (3-4) because $\mathscr{H}_1 = \{h\} \times \mathscr{S}_2$.

Generalization Given n experiments $\mathscr{S}_1, \ldots, \mathscr{S}_n$, we define as their cartesian product

$$\mathscr{S} = \mathscr{S}_1 \times \cdots \times \mathscr{S}_n \tag{3-6}$$

the experiment whose elements are all ordered n tuplets $\zeta_1 \cdots \zeta_n$ where ζ_i is an element of the set $\mathscr{S}_i$. Events in this space are all sets of the form

$$\mathscr{A}_1 \times \cdots \times \mathscr{A}_n$$

where $\mathscr{A}_i \subset \mathscr{S}_i$, and their unions and intersections. If the experiments are independent and $P_i(\mathscr{A}_i)$ is the probability of the event $\mathscr{A}_i$ in the experiment $\mathscr{S}_i$, then

$$P(\mathscr{A}_1 \times \cdots \times \mathscr{A}_n) = P_1(\mathscr{A}_1) \cdots P_n(\mathscr{A}_n) \tag{3-7}$$

Example 3-5 If we toss the coin of Example 3-4 n times, we obtain the space $\mathscr{S} = \mathscr{S}_1 \times \cdots \times \mathscr{S}_n$ consisting of the 2^n elements $\zeta_1 \cdots \zeta_n$ where $\zeta_i = h$ or t. Clearly

$$P\{\zeta_1 \cdots \zeta_n\} = P_1\{\zeta_1\} \cdots P_n\{\zeta_n\} \qquad P_i\{\zeta_i\} = \begin{cases} p & \zeta_i = h \\ q & \zeta_i = t \end{cases} \tag{3-8}$$

If, in particular, $p = q = 1/2$, then

$$P\{\zeta_1 \cdots \zeta_n\} = \frac{1}{2^n} \tag{3-9}$$

From (3-8) it follows that, if the elementary event $\{\zeta_1 \cdots \zeta_n\}$ consists of k heads and $n - k$ tails (in a specific order), then

$$P\{\zeta_1 \cdots \zeta_n\} = p^k q^{n-k} \tag{3-10}$$

We note that the event $\mathscr{H}_1 = \{\text{heads at the first tossing}\}$ consists of 2^{n-1} outcomes $\zeta_1 \cdots \zeta_n$ where $\zeta_1 = h$ and $\zeta_i = t$ or h for $i > 1$. The event $\mathscr{H}_1$ can be written as a cartesian product

$$\mathscr{H}_1 = \{h\} \times \mathscr{S}_2 \times \cdots \times \mathscr{S}_n$$

Hence [see (3-7)]

$$P(\mathscr{H}_1) = P_1\{h\}P_2(\mathscr{S}_2) \cdots P_n(\mathscr{S}_n) = p$$

because $P_i(\mathscr{S}_i) = 1$. We can similarly show that if

$$\mathscr{H}_i = \{\text{heads at the } i\text{th tossing}\} \qquad \mathscr{T}_i = \{\text{tails at the } i\text{th tossing}\}$$

then

$$P(\mathscr{H}_i) = p \qquad P(\mathscr{T}_i) = q$$

Dual meaning of repeated trials In the theory of probability, the notion of repeated trials has two fundamentally different meanings. The first is the approximate relationship (1-1) between the probability $P(\mathscr{A})$ of an event $\mathscr{A}$ in an experiment $\mathscr{S}$ and the relative frequency of the occurrence of $\mathscr{A}$. The second is the creation of the experiment $\mathscr{S} \times \cdots \times \mathscr{S}$.

For example, the repeated tossings of a coin can be given the following two interpretations:

First interpretation (physical) Our experiment is the *single* tossing of a fair coin. Its space has two elements and the probability of each elementary event equals 1/2. A trial is the tossing of the coin *once.*

If we toss the coin n times and heads shows n_h times, then almost certainly $n_h/n \simeq 1/2$ provided that n is sufficiently large. Thus, the first interpretation of repeated trials is the above inprecise statement relating probabilities with observed frequencies.

Second interpretation (conceptual) Our experiment is now the tossing of the coin *n times* where n is any number large or small. Its space has 2^n elements and the probability of each elementary event equals $1/2^n$. A trial is the tossing of the coin *n times.* All statements concerning the number of heads are precise and in the form of probabilities.

We can, of course, give a relative frequency interpretation to these statements. However, to do so, we must repeat the *n tossing of the coin* a large number of times.

3-2 BERNOULLI TRIALS

It is well known from combinatorial analysis that, if a set has n elements, then the total number of its subsets consisting of k elements each equals

$$\binom{n}{k} = \frac{n(n-1)\cdots(n-k+1)}{1 \cdot 2 \cdots k} = \frac{n!}{k!\,(n-k)!} \tag{3-11}$$

For example, if $n = 4$ and $k = 2$, then

$$\binom{4}{2} = \frac{4 \cdot 3}{1 \cdot 2} = 6$$

Indeed, the two-element subsets of the four-element set *abcd* are

$$ab \qquad ac \qquad ad \qquad bc \qquad bd \qquad cd$$

The above result will be used to find the probability that an event occurs k times in n independent trials of an experiment $\mathscr{S}$. This problem is essentially the same as the problem of obtaining k heads in n tossings of a coin. We start, therefore, with the coin experiment.

Example 3-6 A coin with $P\{h\} = p$ is tossed n times. We maintain that the probability $p_n(k)$ that heads shows k times is given by

$$p_n(k) = \binom{n}{k} p^k q^{n-k} \qquad q = 1 - p \tag{3-12}$$

PROOF The experiment under consideration is the n-tossing of a coin. A single outcome is a particular sequence of heads and tails. The event $\{k \text{ heads in any order}\}$ consists of all sequences containing k heads and $n - k$ tails. The k heads of each such sequence form a k-element subset of a set of n heads. As we noted, there are $\binom{n}{k}$ such subsets. Hence, the event $\{k \text{ heads in any order}\}$ consists of $\binom{n}{k}$ elementary events containing k heads and $n - k$ tails in a specific order. Since the probability of each of these elementary events equals $p^k q^{n-k}$, we conclude that

$$P\{k \text{ heads in any order}\} = \binom{n}{k} p^k q^{n-k}$$

SPECIAL CASE If $n = 3$ and $k = 2$, then there are three ways of getting two heads, namely, *hht*, *hth*, and *thh*. Hence, $p_3(2) = 3p^2 q$ in agreement with (3-12).

Success or Failure of an Event $\mathscr{A}$ in n Independent Trials

We consider now our main problem. We are given an experiment $\mathscr{S}$ and an event $\mathscr{A}$ with

$$P(\mathscr{A}) = p \qquad P(\bar{\mathscr{A}}) = q \qquad p + q = 1$$

We repeat the experiment n times and the resulting product space we denote by $\mathscr{S}^n$. Thus

$$\mathscr{S}^n = \mathscr{S} \times \cdots \times \mathscr{S}$$

We shall determine the probability $p_n(k)$ that the event $\mathscr{A}$ occurs exactly k times.

Fundamental theorem

$$p_n(k) = P\{\mathscr{A} \text{ occurs } k \text{ times in any order}\} = \binom{n}{k} p^k q^{n-k} \tag{3-13}$$

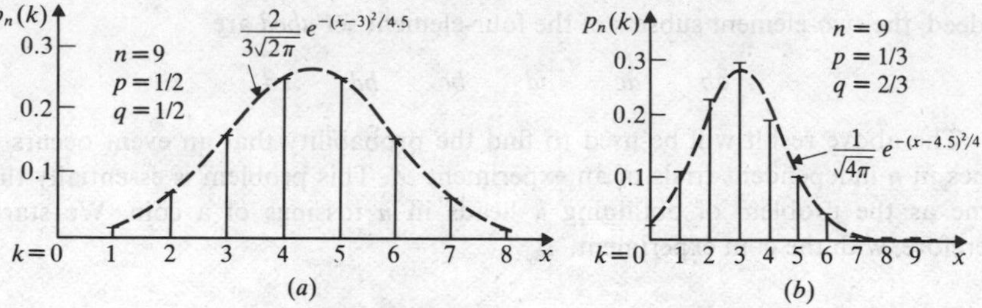

Figure 3-2

PROOF The event $\{\mathscr{A}$ occurs k times in a specific order$\}$ is a cartesian product $\mathscr{B}_1 \times \cdots \times \mathscr{B}_n$ where k of the events $\mathscr{B}_i$ equal $\mathscr{A}$ and the remaining $n-k$ equal $\bar{\mathscr{A}}$. As we know from (3-7), the probability of this event equals

$$P(\mathscr{B}_1) \cdots P(\mathscr{B}_n) = p^k q^{n-k}$$

because

$$P_i(\mathscr{B}_i) = \begin{cases} p & \text{if} \quad \mathscr{B}_i = \mathscr{A} \\ q & \text{if} \quad \mathscr{B}_i = \bar{\mathscr{A}} \end{cases}$$

In other words

$$P\{\mathscr{A} \text{ occurs } k \text{ times in a specific order}\} = p^k q^{n-k} \tag{3-14}$$

The event $\{\mathscr{A}$ occurs k times in any order$\}$ is the union of the $\binom{n}{k}$ events $\{\mathscr{A}$ occurs k times in a specific order$\}$ and since these events are mutually exclusive, we conclude from (2-20) that $p_n(k)$ is given by (3-13).

In Fig. 3-2, we plot $p_n(k)$ for $n = 9$. The meaning of the dashed curves will be explained later.

Example 3-7 A fair die is rolled five times. We shall find the probability $p_5(2)$ that "six" will show twice.

In the single rolling of a die, $\mathscr{A} = \{\text{six}\}$ is an event with probability 1/6. Setting

$$P(\mathscr{A}) = \frac{1}{6} \qquad P(\bar{\mathscr{A}}) = \frac{5}{6} \qquad n = 5 \qquad k = 2$$

in (3-13), we obtain

$$P_5(2) = \frac{5!}{2!\,3!}\left(\frac{1}{6}\right)^2\left(\frac{5}{6}\right)^3$$

Example 3-8 A pair of fair dice is rolled four times. We shall find the probability $p_4(0)$ that "seven" will not show at all.

The space of the single rolling of the two dice consists of the 36 elements $f_i f_j$. The event $\mathscr{A} = \{\text{seven}\}$ consists of the 6 elements

$$f_1 f_6, f_6 f_1, f_2 f_5, f_5 f_2, f_3 f_4, f_4 f_3$$

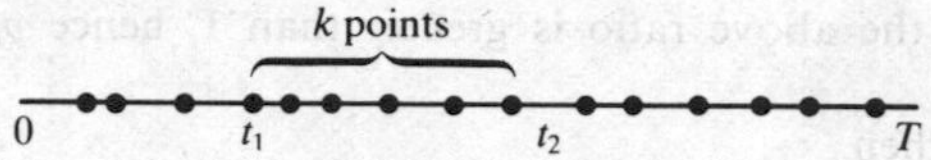

Figure 3-3

therefore, $P(\mathscr{A}) = 6/36$ and $P(\bar{\mathscr{A}}) = 5/6$. With $n = 4$ and $k = 0$, (3-13) yields

$$p_4(0) = \left(\frac{5}{6}\right)^4$$

Example 3-9 We place at random n points in the interval $(0, T)$. What is the probability that k of these points are in the interval (t_1, t_2) (Fig. 3-3)?

This example can be considered as a problem in repeated trials. The experiment $\mathscr{S}$ is the placing of a *single* point in the interval $(0, T)$. In this experiment, $\mathscr{A} = \{$the point is in the interval $(t_1, t_2)\}$ is an event with probability

$$P(\mathscr{A}) = p = \frac{t_2 - t_1}{T}$$

In the space $\mathscr{S}^n$, the event $\{\mathscr{A}$ occurs k times$\}$ means that k of the n points are in the interval (t_1, t_2). Hence, [see (3-13)]

$$P\{k \text{ points are in the interval } (t_1, t_2)\} = \binom{n}{k} p^k q^{n-k} \tag{3-15}$$

Example 3-10 A system containing n components is put into operation at $t = 0$. The probability that a particular component will fail in the interval $(0, t)$ equals

$$p = \int_0^t \alpha(\tau)\, d\tau \qquad \text{where} \qquad \alpha(t) \geq 0 \qquad \int_0^\infty \alpha(t)\, dt = 1 \tag{3-16}$$

What is the probability that k of these components will fail prior to time t?

This example can also be considered as a problem in repeated trials. Reasoning as above, we conclude that the unknown probability is given by (3-15).

Most likely number of successes We shall now examine the behavior of $p_n(k)$ as a function of k for a fixed n. We maintain that as k increases, $p_n(k)$ increases reaching a maximum for

$$k = k_{\max} = [(n + 1)p] \tag{3-17}$$

where the bracket means the largest integer that does not exceed $(n + 1)p$. If $(n + 1)p$ is an integer, then $p_n(k)$ is maximum for two consecutive values of k

$$k = k_1 = (n + 1)p \qquad \text{and} \qquad k = k_2 = k_1 - 1 = np - q$$

PROOF We form the ratio

$$\frac{p_n(k-1)}{p_n(k)} = \frac{kq}{(n - k + 1)p}$$

If this ratio is less than 1, that is, if $k < (n + 1)p$, then $p_n(k - 1)$ is less than $p_n(k)$. This shows that as k increases, $p_n(k)$ increases reaching its maximum for

$k = [(n+1)p]$. For $k > (n+1)p$, the above ratio is greater than 1, hence $p_n(k)$ decreases.

If $k_1 = (n+1)p$ is an integer, then

$$\frac{p_n(k_1 - 1)}{p_n(k_1)} = \frac{k_1 q}{(n - k_1 + 1)} = \frac{(n+1)pq}{[n - (n+1)p + 1]p} = 1$$

This shows that $p_n(k)$ is maximum for $k = k_1$ and $k = k_1 - 1$.

Example 3-11(*a*) If $n = 10$ and $p = 1/3$, then $(n+1)p = 11/3$, hence, $k_{\max} = [11/3] = 3$.

(*b*) If $n = 11$ and $p = 1/2$, then $(n+1)p = 6$, hence, $k_1 = 6$, $k_2 = 5$.

We shall, finally, find the probability

$$P\{k_1 \le k \le k_2\}$$

that the number k of occurrences of $\mathscr{A}$ is between k_1 and k_2. Clearly, the events $\{\mathscr{A}$ occurs k times$\}$, where k takes all values from k_1 to k_2, are mutually exclusive and their union is the event $\{k_1 \le k \le k_2\}$. Hence [see (3-13)]

$$P\{k_1 \le k \le k_2\} = \sum_{k=k_1}^{k_2} p_n(k) = \sum_{k=k_1}^{k_2} \binom{n}{k} p^k q^{n-k} \tag{3-18}$$

Example 3-12 An order of 10^4 parts is received. The probability that a part is defective equals 0.1. What is the probability that the total number of defective parts does not exceed 1100?

The experiment $\mathscr{S}$ is the selection of a single part. The probability of the event $\mathscr{A} = \{$the part is defective$\}$ equals 0.1. We want the probability that in 10^4 trials, $\mathscr{A}$ will occur at most 1100 times. With

$$p = 0.1 \qquad n = 10^4 \qquad k_1 = 0 \qquad k_2 = 1100$$

(3-18) yields

$$P\{0 \le k \le 1100\} = \sum_{k=0}^{1100} \binom{10^4}{k} (0.1)^k (0.9)^{10^4 - k} \tag{3-19}$$

3-3 ASYMPTOTIC THEOREMS

In the preceding section, we showed that if the probability $P(\mathscr{A})$ of an event $\mathscr{A}$ of a certain experiment equals p and the experiment is repeated n times, then the probability that $\mathscr{A}$ occurs k times in any order is given by (3-13) and the probability that k is between k_1 and k_2 by (3-18). In this section, we develop simple approximate formulas for evaluating these probabilities.

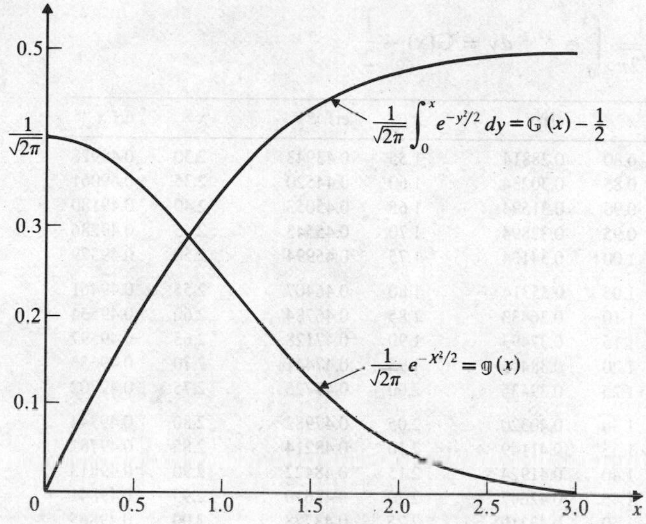

Figure 3-4

Gaussian functions In the following and throughout the book we use extensively the *normal* or *gaussian* function

$$\mathbb{g}(x) = \frac{1}{\sqrt{2\pi}} e^{-x^2/2} \tag{3-20}$$

and its integral (see Fig. 3-4 and Table 3-1).

$$\mathbb{G}(x) = \int_{-\infty}^{x} \mathbb{g}(y)\,dy = \frac{1}{\sqrt{2\pi}} \int_{-\infty}^{x} e^{-y^2/2}\,dy \tag{3-21}$$

As is well known

$$\int_{-\infty}^{\infty} e^{-\alpha x^2}\,dx = \sqrt{\frac{\pi}{\alpha}} \tag{3-22}$$

From this it follows that

$$\mathbb{G}(\infty) = \frac{1}{\sqrt{2\pi}} \int_{-\infty}^{\infty} e^{-x^2/2}\,dx = 1 \tag{3-23}$$

Since $\mathbb{g}(-x) = \mathbb{g}(x)$, we conclude that

$$\mathbb{G}(-x) = 1 - \mathbb{G}(x) \tag{3-24}$$

With a change of variables, (3-21) yields

$$\frac{1}{a\sqrt{2\pi}} \int_{x_1}^{x_2} e^{-(x-b)^2/2a^2}\,dx = \mathbb{G}\left(\frac{x_2 - b}{a}\right) - \mathbb{G}\left(\frac{x_1 - b}{a}\right) \tag{3-25}$$

for any a and b.

Table 3-1 $\operatorname{erf} x = \dfrac{1}{\sqrt{2\pi}} \displaystyle\int_0^x e^{-y^2/2}\, dy = \mathbb{G}(x) - \frac{1}{2}$

x	erf x	x	erf x	x	erf x	x	erf x
0.05	0.01994	0.80	0.28814	1.55	0.43943	2.30	0.48928
0.10	0.03983	0.85	0.30234	1.60	0.44520	2.35	0.49061
0.15	0.05962	0.90	0.31594	1.65	0.45053	2.40	0.49180
0.20	0.07926	0.95	0.32894	1.70	0.45543	2.45	0.49286
0.25	0.09871	1.00	0.34134	1.75	0.45994	2.50	0.49379
0.30	0.11791	1.05	0.35314	1.80	0.46407	2.55	0.49461
0.35	0.13683	1.10	0.36433	1.85	0.46784	2.60	0.49534
0.40	0.15542	1.15	0.37493	1.90	0.47128	2.65	0.49597
0.45	0.17364	1.20	0.38493	1.95	0.47441	2.70	0.49653
0.50	0.19146	1.25	0.39435	2.00	0.47726	2.75	0.49702
0.55	0.20884	1.30	0.40320	2.05	0.47982	2.80	0.49744
0.60	0.22575	1.35	0.41149	2.10	0.48214	2.85	0.49781
0.65	0.24215	1.40	0.41924	2.15	0.48422	2.90	0.49813
0.70	0.25804	1.45	0.42647	2.20	0.48610	2.95	0.49841
0.75	0.27337	1.50	0.43319	2.25	0.48778	3.00	0.49865

For large x, $\mathbb{G}(x)$ is given approximately by (see Prob. 3-8)

$$\mathbb{G}(x) \simeq 1 - \frac{1}{x}\, \mathbb{g}(x) \tag{3-26}$$

We note, finally, that $\mathbb{G}(x)$ is often expressed in terms of the *error function*

$$\operatorname{erf} x = \frac{1}{\sqrt{2\pi}} \int_0^x e^{-y^2/2}\, dy = \mathbb{G}(x) - \frac{1}{2}$$

DeMoivre–Laplace Theorem

It can be shown that, if $npq \gg 1$, then

$$\binom{n}{k} p^k q^{n-k} \simeq \frac{1}{\sqrt{2\pi npq}}\, e^{-(k-np)^2/2npq} \tag{3-27}$$

for k in the $\sqrt{npq}$ neighborhood of np. This important approximation, known as the DeMoivre–Laplace theorem, can be stated as an equality in the limit: The ratio of the two sides tends to 1 as $n \to \infty$. The proof is based on *Stirling's formula*

$$n! \simeq n^n e^{-n} \sqrt{2\pi n} \qquad n \to \infty \tag{3-28}$$

The details, however, will be omitted.†

† The proof can be found in F.2. (The notation F.2 means the second bibliographical listing of the references, beginning with the letter F.)

Thus, the evaluation of the probability of k successes in n trials, given exactly by (3-13), is reduced to the evaluation of the normal curve

$$\frac{1}{\sqrt{2\pi npq}}\, e^{-(x-np)^2/2npq} = \frac{1}{\sqrt{npq}}\, \mathbb{g}\left(\frac{x-np}{\sqrt{npq}}\right) \tag{3-29}$$

for $x = k$.

Example 3-13 A fair coin is tossed 1000 times. Find the probability p_a that heads will show 500 times and the probability p_b that heads will show 510 times.

In this example

$$p = q = 0.5 \qquad n = 1000 \qquad \sqrt{npq} = 5\sqrt{10}$$

(*a*) If $k = 500$ then $k - np = 0$ and (3-27) yields

$$p_a \simeq \frac{1}{\sqrt{2\pi npq}} = \frac{1}{10\sqrt{5\pi}} = 0.0252$$

(*b*) If $k = 510$ then $k - np = 10$ and (3-27) yields

$$p_b \simeq \frac{e^{-0.2}}{10\sqrt{5\pi}} = 0.0207$$

As the next example indicates, the approximation (3-27) is satisfactory even for moderate values of n.

Example 3-14 We shall determine $p_n(k)$ for $p = 0.5$, $n = 10$, and $k = 5$.

(*a*) Exactly from (3-13)

$$p_n(k) = \binom{n}{k} p^k q^{n-k} = \frac{10!}{5!\,5!}\,\frac{1}{2^{10}} = 0.246$$

(*b*) Approximately from (3-27)

$$p_n(k) \simeq \frac{1}{\sqrt{2\pi npq}}\, e^{-(k-np)^2/2npq} = \frac{1}{\sqrt{5\pi}} = 0.252$$

Approximate evaluation of $P\{k_1 \le k \le k_2\}$ Using the approximation (3-27), we shall show that

$$\sum_{k=k_1}^{k_2} \binom{n}{k} p^k q^{n-k} \simeq \mathbb{G}\left(\frac{k_2 - np}{\sqrt{npq}}\right) - \mathbb{G}\left(\frac{k_1 - np}{\sqrt{npq}}\right) \tag{3-30}$$

Thus, to find the probability that in n trials the number of occurrences of an event $\mathscr{A}$ is between k_1 and k_2, it suffices to evaluate the tabulated normal function $\mathbb{G}(x)$. The approximation is satisfactory if $npq \gg 1$ and the differences $k_1 - np$ and $k_2 - np$ are of the order of $\sqrt{npq}$.

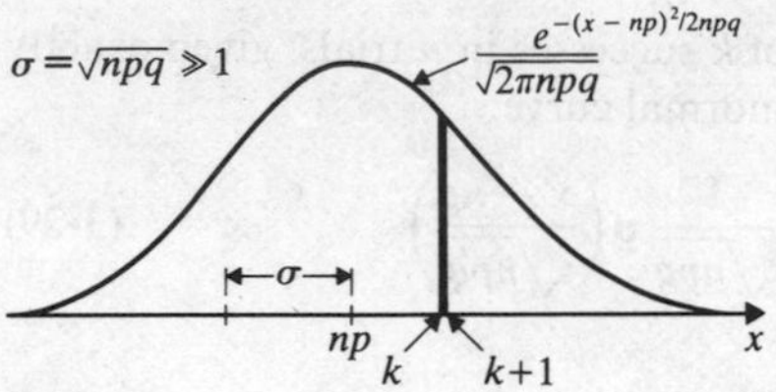

Figure 3-5

Proof Inserting (3-27) into (3-18), we obtain

$$P\{k_1 \le k \le k_2\} = \sum_{k=k_1}^{k_2} \binom{n}{k} p^k q^{n-k} \simeq \frac{1}{\sqrt{npq}} \sum_{k=k_1}^{k_2} \mathrm{g}\left(\frac{k-np}{\sqrt{npq}}\right) \tag{3-31}$$

Since $npq \gg 1$, the curve

$$\mathrm{g}\left(\frac{x-np}{\sqrt{npq}}\right) = \frac{1}{\sqrt{2\pi}} e^{-(x-np)^2/2npq}$$

is nearly constant in a unit interval (Fig. 3-5), hence, its area in that interval equals its ordinate. From this it follows that

$$\sum_{k=k_1}^{k_2} \mathrm{g}\left(\frac{k-np}{\sqrt{npq}}\right) \simeq \int_{k_1}^{k_2} \mathrm{g}\left(\frac{x-np}{\sqrt{npq}}\right) dx$$

Inserting into (3-31) and using (3-25), we obtain (3-30).

Note The approximations (3-27) and (3-30) improve if $\mathrm{g}(x)$ is expressed not as the derivative of $\mathrm{G}(x)$, but as its incremental slope

$$\mathrm{g}(x) \simeq \sqrt{npq}\left[\mathrm{G}\left(\frac{x+0.5}{\sqrt{npq}}\right) - \mathrm{G}\left(\frac{x-0.5}{\sqrt{npq}}\right)\right]$$

Substituting into (3-27) we obtain

$$\binom{n}{k} p^k q^{n-k} \simeq \mathrm{G}\left(\frac{k+0.5-np}{\sqrt{npq}}\right) - \mathrm{G}\left(\frac{k-0.5-np}{\sqrt{npq}}\right) \tag{3-32}$$

Hence

$$\sum_{k=k_1}^{k_2} \binom{n}{k} p^k q^{n-k} \simeq \mathrm{G}\left(\frac{k_2+0.5-np}{\sqrt{npq}}\right) - \mathrm{G}\left(\frac{k_1-0.5-np}{\sqrt{npq}}\right) \tag{3-33}$$

Example 3-15 A fair coin is tossed 10 000 times. What is the probability that the number of heads is between 4900 and 5100?

In this problem

$$n = 10\,000 \qquad p = q = 0.5 \qquad k_1 = 4900 \qquad k_2 = 5100$$

Since $(k_2 - np)/\sqrt{npq} = 100/50$ and $(k_1 - np)/\sqrt{npq} = -100/50$, we conclude from (3-30) that the unknown probability equals

$$\mathrm{G}(2) - \mathrm{G}(-2) = 2\mathrm{G}(2) - 1 = 0.9545$$

Example 3-16 Over a period of 12 hours 180 calls are made at random. What is the probability that in a four-hour interval the number of calls is between 50 and 70?

The above can be considered as a problem in repeated trials with $p = 4/12$ the probability that a particular call will occur in the four-hour interval. The probability that k calls will occur in this interval equals [see (3-27)]

$$\binom{180}{k}\left(\frac{1}{3}\right)^k\left(\frac{2}{3}\right)^{180-k} \simeq \frac{1}{4\sqrt{5\pi}}e^{-(k-60)^2/80}$$

and the probability that the number of calls is between 50 and 70 equals [see (3-30)]

$$\sum_{k=50}^{70}\binom{180}{k}\left(\frac{1}{3}\right)^k\left(\frac{2}{3}\right)^{180-k} \simeq \mathbb{G}(\sqrt{2.5}) - \mathbb{G}(-\sqrt{2.5}) \simeq 0.9876$$

Note It seems that we cannot use the approximation (3-30) if $k_1 = 0$ because the sum contains values of k that are not in the $\sqrt{npq}$ vicinity of np. However, the corresponding terms are small compared to the terms with k near np, hence the errors of their estimates are also small. Since

$$\mathbb{G}(-np/\sqrt{npq}) = \mathbb{G}(-\sqrt{np/q}) \simeq 0 \qquad \text{for} \qquad np/q \gg 1$$

we conclude that if not only $n \gg 1$ but also $np \gg 1$, then

$$\sum_{k=0}^{k_2}\binom{n}{k}p^k q^{n-k} \simeq \mathbb{G}\left(\frac{k_2 - np}{\sqrt{npq}}\right) \tag{3-34}$$

In the sum (3-19) of Example 3-12

$$np = 1000 \qquad npq = 900 \qquad \frac{k_2 - np}{\sqrt{npq}} = \frac{10}{3}$$

Using (3-34), we obtain

$$\sum_{k=0}^{1100}\binom{10^4}{k}(0.1)^k(0.9)^{10^4-k} \simeq \mathbb{G}\left(\frac{10}{3}\right) = 0.999\,36$$

We note that the sum of the terms of the above sum from 900 to 1100 equals $2\mathbb{G}(10/3) - 1 \simeq 0.998\,72$.

The Law of Large Numbers

According to the relative frequency interpretation of probability, if an event $\mathscr{A}$ with $P(\mathscr{A}) = p$ occurs k times in n trials, then $k \simeq np$. In the following, we rephrase this heuristic statement as a limit theorem.

We start with the observation that $k \simeq np$ does not mean that k will be close to np. In fact [(see (3-27)]

$$P\{k = np\} \simeq \frac{1}{\sqrt{2\pi npq}} \to 0 \qquad \text{as} \qquad n \to \infty \tag{3-35}$$

As we show in the next theorem, the approximation $k \simeq np$ means that the ratio k/n is close to p in the sense that, for any $\varepsilon > 0$, the probability that $|k/n - p| < \varepsilon$ tends to 1 as $n \to \infty$.

Theorem For any $\varepsilon > 0$

$$P\left\{\left|\frac{k}{n} - p\right| \le \varepsilon\right\} \to 1 \quad \text{as} \quad n \to \infty \tag{3-36}$$

PROOF The inequality $|k/n - p| < \varepsilon$ means that

$$n(p - \varepsilon) \le k \le n(p + \varepsilon)$$

With $k_1 = n(p - \varepsilon)$ and $k_2 = n(p + \varepsilon)$ we have

$$P\left\{\left|\frac{k}{n} - p\right| \le \varepsilon\right\} = P\{k_1 \le k \le k_2\} = \sum_{k=k_1}^{k_2} \binom{n}{k} p^k q^{n-k}$$

Inserting into (3-30), we obtain

$$P\{k_1 \le k \le k_2\} \simeq \mathbb{G}\left(\frac{k_2 - np}{\sqrt{npq}}\right) - \mathbb{G}\left(\frac{k_1 - np}{\sqrt{npq}}\right) = 2\mathbb{G}\left(\frac{n\varepsilon}{\sqrt{npq}}\right) - 1$$

But $\varepsilon\sqrt{n/pq} \to \infty$ as $n \to \infty$ for any ε. Hence

$$P\left\{\left|\frac{k}{n} - p\right| \le \varepsilon\right\} \simeq 2\mathbb{G}\left(\varepsilon\sqrt{\frac{n}{pq}}\right) - 1 \to 1 \quad \text{as} \quad n \to \infty \tag{3-37}$$

Example 3-17 Suppose that $p = q = 0.5$ and $\varepsilon = 0.05$. In this case

$$n(p - \varepsilon) = 0.45n \qquad n(p + \varepsilon) = 0.55n \qquad \varepsilon\sqrt{n/pq} = 0.1\sqrt{n}$$

In the table below we show the probability $2\mathbb{G}(0.1\sqrt{n}) - 1$ that k is between $0.45n$ and $0.55n$ for various values of n.

n	100	400	900
$0.1\sqrt{n}$	1	2	3
$2\mathbb{G}(0.1\sqrt{n}) - 1$	0.682	0.954	0.997

Example 3-18 We now assume that $p = 0.6$ and we wish to find n such that the probability that k is between $0.59n$ and $0.61n$ is at least 0.98.

In this case, $p = 0.6$, $q = 0.4$, and $\varepsilon = 0.01$. Hence

$$P\{0.59n \le k \le 0.61n\} \simeq 2\mathbb{G}(0.01\sqrt{n/0.24}) - 1$$

Thus, n must be such that

$$2\mathbb{G}(0.01\sqrt{n/0.24}) - 1 \ge 0.98$$

From Table 3-1 we see that $\mathbb{G}(x) > 0.49$ if $x > 2.35$. Hence, $0.01\sqrt{n/0.24} > 2.35$ yielding $n > 13\,254$.

Generalization of Bernoulli trials The experiment of repeated trials can be phrased in the following form: The events $\mathscr{A}_1 = \mathscr{A}$ and $\mathscr{A}_2 = \bar{\mathscr{A}}$ of the space $\mathscr{S}$

form a partition and their respective probabilities equal $p_1 = p$ and $p_2 = 1 - p$. In the space $\mathscr{S}^n$, the probability of the event $\{\mathscr{A}_1$ occurs $k_1 = k$ times and $\mathscr{A}_2$ occurs $k_2 = n - k$ times in any order$\}$ equals $p_n(k)$ as in (3-13). We shall now generalize.

Suppose that

$$\mathfrak{A} = [\mathscr{A}_1, \ldots, \mathscr{A}_r]$$

is a partition of $\mathscr{S}$ consisting of the r events $\mathscr{A}_i$ with

$$P(\mathscr{A}_i) = p_i \qquad p_1 + \cdots + p_r = 1$$

We repeat the experiment n times and we denote by $p_n(k_1, \ldots, k_r)$ the probability of the event $\{\mathscr{A}_1$ occurs k_1 times, ..., $\mathscr{A}_r$ occurs k_r times in any order$\}$ where

$$k_1 + \cdots + k_r = n$$

We maintain that

$$p_n(k_1, \ldots, k_r) = \frac{n!}{k_1! \cdots k_r!} p_1^{k_1} \cdots p_r^{k_r} \tag{3-38}$$

PROOF Repeated application of (3-11) leads to the conclusion that the number of events of the form $\{\mathscr{A}_1$ occurs k_1 times, ..., $\mathscr{A}_r$ occurs k_r times in a specific order$\}$ equals

$$\frac{n!}{k_1! \cdots k_r!}$$

Since the trials are independent, the probability of each such event equals

$$p_1^{k_1} \cdots p_r^{k_r}$$

and (3-38) results.

Example 3-19 A fair die is rolled 10 times. We shall determine the probability that f_1 shows 3 times, and "even" shows 6 times.

In this case

$$\mathscr{A}_1 = \{f_1\} \qquad \mathscr{A}_2 = \{f_2, f_4, f_6\} \qquad \mathscr{A}_3 = \{f_3, f_5\}$$

Clearly

$$p_1 = \frac{1}{6} \qquad p_2 = \frac{3}{6} \qquad p_3 = \frac{2}{6} \qquad k_1 = 3 \qquad k_2 = 6 \qquad k_3 = 1$$

and (3-38) yields

$$p_{10}(3, 6, 1) = \frac{10!}{3!\,6!\,1!}\left(\frac{1}{6}\right)^3 \left(\frac{1}{2}\right)^6 \frac{1}{3}$$

DeMoivre–Laplace theorem We can show as in (3-27) that, if k_i is in the $\sqrt{n}$ vicinity of np_i and n is sufficiently large, then

$$\frac{n!}{k_1! \cdots k_r!} p_1^{k_1} \cdots p_r^{k_r} \simeq \frac{\exp\left\{-\frac{1}{2}\left[\frac{(k_1 - np_1)^2}{np_1} + \cdots + \frac{(k_r - np_r)^2}{np_r}\right]\right\}}{\sqrt{(2\pi n)^{r-1} p_1 \cdots p_r}} \tag{3-39}$$

Equation (3-27) is a special case.

3-4 POISSON THEOREM AND RANDOM POINTS

We have shown in (3-13) that the probability that an event $\mathscr{A}$ occurs k times in n trials equals

$$\frac{n(n-1)\cdots(n-k+1)}{1 \cdot 2 \cdots k} p^k q^{n-k} \tag{3-40}$$

In the following, we obtain an approximate expression for this probability under the assumption that $p \ll 1$. If n is so large that $np \simeq npq \gg 1$, then we can use the DeMoivre–Laplace theorem (3-27). If, however, np is of the order of 1, (3-27) is no longer valid. In this case, the following approximation can be used: For k of the order of np

$$\frac{n!}{k!(n-k)!} p^k q^{n-k} \simeq e^{-np} \frac{(np)^k}{k!} \tag{3-41}$$

Indeed, if k is of the order of np, then $k \ll n$ and $kp \ll 1$. Hence

$$n(n-1)\cdots(n-k+1) \simeq n \cdot n \cdots n = n^k$$

$$q = 1 - p \simeq e^{-p} \qquad q^{n-k} \simeq e^{-(n-k)p} \simeq e^{-np}$$

Inserting into (3-40), we obtain (3-41).

The above approximation can be stated as a limit theorem (for the proof see F.2):

Poisson theorem If

$$n \to \infty \qquad p \to 0 \qquad np \to a$$

then

$$\frac{n!}{k!(n-k)!} p^k q^{n-k} \xrightarrow[n \to \infty]{} e^{-a} \frac{a^k}{k!} \tag{3-42}$$

Example 3-20 A system contains 1000 components. Each component fails independently of the others and the probability of its failure in one month equals 10^{-3}. We shall find the probability that the system will function (i.e., no component will fail) at the end of one month.

This can be considered as a problem in repeated trials with $p = 10^{-3}$, $n = 10^3$, and $k = 0$. Hence [see (3-15)]

$$P\{k = 0\} = q^n = 0.999^{1000}$$

Since $np = 1$, the approximation (3-41) yields

$$P\{k = 0\} \simeq e^{-np} = e^{-1} = 0.368$$

Applying (3-41) to the sum in (3-18), we obtain the following approximation for the probability that the number k of occurrence of $\mathscr{A}$ is between k_1 and k_2

$$P\{k_1 \le k \le k_2\} \simeq e^{-np} \sum_{k=k_1}^{k_2} \frac{(np)^k}{k!} \tag{3-43}$$

Example 3-21 An order of 3000 parts is received. The probability that a part is defective equals 10^{-3}. We wish to find the probability $P\{k > 5\}$ that there will be more than five defective parts in this order.

Clearly

$$P\{k > 5\} = 1 - P\{k \le 5\}$$

With $np = 3$, (3-43) yields

$$P\{k \le 5\} = e^{-3} \sum_{k=0}^{5} \frac{3^k}{k!} = 0.916$$

Hence

$$P\{k > 5\} = 0.084$$

Generalization of Poisson theorem Suppose that $\mathscr{A}_1, \ldots, \mathscr{A}_{m+1}$ are the $m + 1$ events of a partition with $P\{\mathscr{A}_i\} = p_i$. Reasoning as in (3-42), we can show that if $np_i \to a_i$ for $i \le m$, then

$$\frac{n!}{k_1! \cdots k_{m+1}!} p_1^{k_1} \cdots p_{m+1}^{k_{m+1}} \xrightarrow[n\to\infty]{} \frac{e^{-a_1}a_1^{k_1}}{k_1!} \cdots \frac{e^{-a_m}a_m^{k_m}}{k_m!} \tag{3-44}$$

Random Poisson Points

An important application of Poisson's theorem is the approximate evaluation of (3-15) as T and n tend to infinity. We repeat the problem: We place at random n points in the interval $(-T/2, T/2)$ and we denote by $P\{k \text{ in } t_a\}$ the probability that k of these points will lie in an interval (t_1, t_2) of length $t_2 - t_1 = t_a$. As we have shown in (3-15)

$$P\{k \text{ in } t_a\} = \binom{n}{k} p^k q^{n-k} \qquad \text{where} \qquad p = \frac{t_a}{T} \tag{3-45}$$

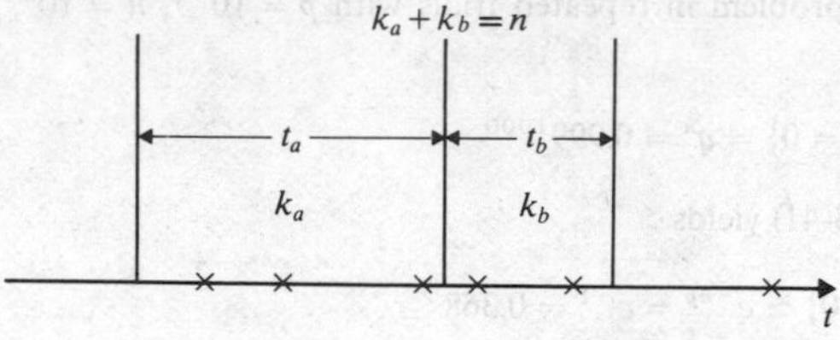

Figure 3-6

We now assume that $n \gg 1$ and $t_a \ll T$. Applying (3-41), we conclude that

$$P\{k \text{ in } t_a\} \simeq e^{-nt_a/T} \frac{(nt_a/T)^k}{k!} \tag{3-46}$$

for k of the order of nt_a/T.

Suppose, next, that n and T increase indefinitely but the ratio

$$\lambda = n/T$$

remains constant. The result is an infinite set of points covering the entire t axis from $-\infty$ to $+\infty$. As we see from (3-46), the probability that k of these points are in an interval of length t_a is given by

$$P\{k \text{ in } t_a\} = e^{-\lambda t_a} \frac{(\lambda t_a)^k}{k!} \tag{3-47}$$

Points in nonoverlapping intervals Returning for a moment to the original interval $(-T/2, T/2)$ containing n points, we consider two nonoverlapping subintervals t_a and t_b (Fig. 3-6).

We wish to determine the probability

$$P\{k_a \text{ in } t_a,\ k_b \text{ in } t_b\}$$

that k_a of the n points are in the interval t_a and k_b in the interval t_b. We maintain that

$$P\{k_a \text{ in } t_a,\ k_b \text{ in } t_b\} = \frac{n!}{k_a!\,k_b!\,k_3!}\left(\frac{t_a}{T}\right)^{k_a}\left(\frac{t_b}{T}\right)^{k_b}\left(1 - \frac{t_a}{T} - \frac{t_b}{T}\right)^{k_3} \tag{3-48}$$

where $k_3 = n - k_a - k_b$.

Proof The above can be considered as a generalized Bernoulli trial. The original experiment $\mathscr{S}$ is the random selection of a single point in the interval $(-T/2, T/2)$. In this experiment, the events $\mathscr{A}_1 = \{\text{the point is in } t_a\}$, $\mathscr{A}_2 = \{\text{the point is in } t_b\}$, and $\mathscr{A}_3 = \{\text{the point is outside the intervals } t_a \text{ and } t_b\}$ form a partition and

$$P(\mathscr{A}_1) = \frac{t_a}{T} \qquad P(\mathscr{A}_2) = \frac{t_b}{T} \qquad P(\mathscr{A}_3) = 1 - \frac{t_a}{T} - \frac{t_b}{T}$$

If the experiment $\mathscr{S}$ is performed n times, then the event $\{k_a \text{ in } t_a \text{ and } k_b \text{ in } t_b\}$ will equal the event $\{\mathscr{A}_1$ occurs $k_1 = k_a$ times, $\mathscr{A}_2$ occurs $k_2 = k_b$ times, and $\mathscr{A}_3$ occurs $k_3 = n - k_1 - k_2$ times$\}$. Hence, (3-48) follows from (3-38) with $r = 3$.

We note that the events $\{k_a \text{ in } t_a\}$ and $\{k_b \text{ in } t_b\}$ are *not* independent because the probability (3-48) of their intersection $\{k_a \text{ in } t_a, k_b \text{ in } t_b\}$ does not equal $P\{k_a \text{ in } t_a\}P\{k_b \text{ in } t_b\}$.

Suppose now that

$$\frac{n}{T} = \lambda \qquad n \to \infty \qquad T \to \infty$$

Since $nt_a/T = \lambda t_a$ and $nt_b/T = \lambda t_b$, we conclude from (3-48) and Prob. 3-15 that

$$P\{k_a \text{ in } t_a, k_b \text{ in } t_b\} = e^{-\lambda t_a} \frac{(\lambda t_a)^{k_a}}{k_a!} e^{-\lambda t_b} \frac{(\lambda t_b)^{k_b}}{k_b!} \tag{3-49}$$

From (3-47) and (3-49) it follows that

$$P\{k_a \text{ in } t_a, k_b \text{ in } t_b\} = P\{k_a \text{ in } t_a\}P\{k_b \text{ in } t_b\} \tag{3-50}$$

This shows that the events $\{k_a \text{ in } t_a\}$ and $\{k_b \text{ in } t_b\}$ are independent.

We have, thus, created an experiment whose outcomes are infinite sets of points on the t axis. These outcomes will be called *random Poisson points*. The experiment was formed by a limiting process; however, it is completely specified in terms of the following two properties:

1. The probability $P\{k_a \text{ in } t_a\}$ that the number of points in an interval (t_1, t_2) equals k_a is given by (3-47).
2. If two intervals (t_1, t_2) and (t_3, t_4) are nonoverlapping, then the events $\{k_a \text{ in } (t_1, t_2)\}$ and $\{k_b \text{ in } (t_3, t_4)\}$ are independent.

The experiment of random Poisson points is fundamental in the theory and the application of probability. As illustrations we mention electron emission, telephone calls, cars crossing a bridge, and shot noise, among many others.

Example 3-22 Consider two consecutive intervals (t_1, t_2) and (t_2, t_3) with respective lengths t_a and t_b. Clearly, (t_1, t_3) is an interval with length $t_c = t_a + t_b$. We denote by k_a, k_b, and $k_c = k_a + k_b$ the number of points in these intervals. We assume that the number of points k_c in the interval (t_1, t_3) is specified. We wish to find the probability that k_a of these points are in the interval (t_1, t_2). In other words, we wish to find the conditional probability

$$P\{k_a \text{ in } t_a \mid k_c \text{ in } t_c\}$$

With $k_b = k_c - k_a$, we observe that

$$\{k_a \text{ in } t_a, k_c \text{ in } t_c\} = \{k_a \text{ in } t_a, k_b \text{ in } t_b\}$$

Hence

$$P\{k_a \text{ in } t_a \mid k_c \text{ in } t_c\} = \frac{P\{k_a \text{ in } t_a, k_b \text{ in } t_b\}}{P\{k_c \text{ in } t_c\}}$$

From (3-47) and (3-49) it follows that the above fraction equals

$$\frac{e^{-\lambda t_a}[(\lambda t_a)^{k_a}/k_a!]e^{-\lambda t_b}[(\lambda t_b)^{k_b}/k_b!]}{e^{-\lambda t_c}[(\lambda t_c)^{k_c}/k_c!]}$$

Since $t_c = t_a + t_b$ and $k_c = k_a + k_b$, the above yields

$$P\{k_a \text{ in } t_a \,|\, k_c \text{ in } t_c\} = \frac{k_c!}{k_a!\,k_b!}\left(\frac{t_a}{t_c}\right)^{k_a}\left(\frac{t_b}{t_c}\right)^{k_b} \tag{3-51}$$

This result has the following useful interpretation: Suppose that we place at random k_c points in the interval (t_1, t_3). As we see from (3-15), the probability that k_a of these points are in the interval (t_1, t_2) equals the right side of (3-51).

Density of Poisson points The experiment of Poisson points is specified in terms of the parameter λ. We show next that this parameter can be interpreted as the density of the points. Indeed if the interval $\Delta t = t_2 - t_1$ is sufficiently small, then

$$\lambda\,\Delta t\, e^{-\lambda\,\Delta t} \simeq \lambda\,\Delta t$$

From this and (3-47) it follows

$$P\{\text{one point in } (t, t + \Delta t)\} \simeq \lambda\,\Delta t \tag{3-52}$$

Hence

$$\lambda = \lim_{\Delta t \to 0} \frac{P\{\text{one point in } (t, t + \Delta t)\}}{\Delta t} \tag{3-53}$$

Nonuniform density Using a nonlinear transformation of the t axis, we shall define an experiment whose outcomes are Poisson points specified by a minor modification of property 1.

Suppose that $\lambda(t)$ is a function such that $\lambda(t) \geq 0$ but otherwise arbitrary. We define the experiment of the nonuniform Poisson points as follows:

1. The probability that the number of points in the interval (t_1, t_2) equals k is given by

$$P\{k \text{ in } (t_1, t_2)\} = \exp\left[-\int_{t_1}^{t_2} \lambda(t)\,dt\right] \frac{\left[\int_{t_1}^{t_2} \lambda(t)\,dt\right]^k}{k!} \tag{3-54}$$

2. The same as in the uniform case.

The significance of $\lambda(t)$ as density remains the same. Indeed, with $t_2 - t_1 = \Delta t$ and $k = 1$, (3-54) yields

$$P\{\text{one point in } (t, t + \Delta t)\} \simeq \lambda(t)\,\Delta t \tag{3-55}$$

as in (3-52).

3-5 BAYES' THEOREM AND STATISTICS

As an application of repeated trials, we shall consider the important problem of assigning probabilities to various events of a real experiment. To be concrete, we assume that the experiment is the tossing of a given coin and our task is to find the probability $P\{h\} = p$ of the event "heads." We toss the coin 1000 times and we observe that "heads" shows 490 times. What conclusion can we draw from this observation?

Guided by the relative frequency interpretation of probability, we could conclude that p is a number close to 0.49. Such a conclusion is *reasonable* but it is based on inductive reasoning. In the following, we consider not the given problem but a "related" conceptual problem that can be solved deductively with the techniques of this chapter.

We have a pile of m coins. The probability of "heads" of the ith coin equals p_i. We select from this pile one coin and we toss it 1000 times. We observe that "heads" shows 490 times (in a specific order). On the basis of this observation, we shall find the probability X_r that we selected the rth coin.

We shall show that

$$X_r = \frac{p_r^k(1-p_r)^{n-k}}{p_1^k(1-p_1)^{n-k} + \cdots + p_m^k(1-p_m)^{n-k}} \tag{3-56}$$

where $k = 490$ and $n = 1000$.

PROOF We denote by $\mathscr{S}_c$ the space of the m coins and by $\mathscr{S}^n$ the space of the n-times tossings of a specific coin. The space $\mathscr{S}_c$ has m elements and the space $\mathscr{S}^n$ has 2^n elements. If we select a coin from $\mathscr{S}_c$ and toss it n times, we form the product space

$$\mathscr{S}_c \times \mathscr{S}^n$$

consisting of $m \times 2^n$ elements. A particular element is a sequence of heads and tails formed with a specific coin. In this space, the tossing of the ith coin is an event $\mathscr{C}_i$ consisting of 2^n elements, namely, all the sequences that we can obtain with the ith coin, and its probability equals $P(\mathscr{C}_i) = 1/m$ because the selection is random. The event

$$\mathscr{A} = \{k \text{ heads in a specific order}\}$$

has m elements and

$$P(\mathscr{A} \mid \mathscr{C}_i) = p_i^k(1-p_i)^{n-k} \tag{3-57}$$

This is the probability that we obtain k heads if we toss the ith coin n times. Our problem is to find the probability

$$X_r = P(\mathscr{C}_r \mid \mathscr{A})$$

that we have selected the rth coin assuming that k heads showed.

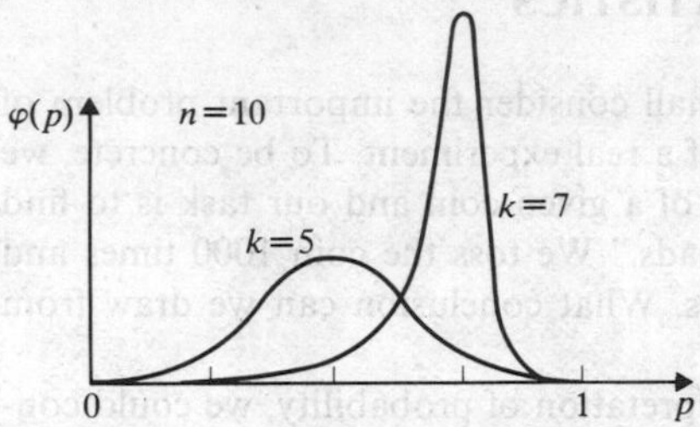

Figure 3-7

As we know (Bayes' theorem)

$$P(\mathscr{C}_r \mid \mathscr{A}) = \frac{P(\mathscr{A} \mid \mathscr{C}_r)P(\mathscr{C}_r)}{P(\mathscr{A} \mid \mathscr{C}_1)P(\mathscr{C}_1) + \cdots + P(\mathscr{A} \mid \mathscr{C}_m)P(\mathscr{C}_m)} \tag{3-58}$$

and (3-56) results.

The function

$$\varphi(p) = p^k(1-p)^{n-k}$$

is plotted in Fig. 3-7. It is maximum for $p = k/n$ and if n is large the maximum is sharp. This shows that X_r is maximum if r is closest to k/n.

PROBLEMS

3-1 A pair of fair dice is rolled 10 times. Find the probability that "seven" will show at least once.

Answer: $1 - (5/6)^{10}$.

3-2 A coin with $P\{h\} = p = 1 - q$ is tossed n times. Show that the probability that the number of heads is even equals $0.5[1 + (q - p)^n]$.

3-3 A fair coin is tossed 900 times. Find the probability that the number of heads is between 420 and 465.

Answer: $\mathbb{G}(2) + \mathbb{G}(1) - 1 \simeq 0.819$.

3-4 A fair coin is tossed n times. Find n such that the probability that the number of heads is between $0.45n$ and $0.52n$ is at least 0.9.

Answer: $\mathbb{G}(0.04\sqrt{n}) + \mathbb{G}(0.02\sqrt{n}) \geq 1.9$, hence, $n > 4556$.

3-5 If $P(\mathscr{A}) = 0.6$ and k is the number of successes of $\mathscr{A}$ in n trials (*a*) show that $P\{550 \leq k \leq 650\} = 0.999$, and (*b*) find n such that $P\{0.59n \leq k \leq 0.61n\} = 0.95$.

3-6 A system has 1000 components. The probability that a specific component will fail in the interval (a, b) equals $e^{-a/T} - e^{-b/T}$. Find the probability that in the interval $(0, T/4)$, no more than 100 components will fail.

3-7 A coin is tossed an infinite number of times. Show that the probability that k heads are observed at the nth tossing but not earlier equals $\binom{n-1}{k-1} p^k q^{n-k}$

3-8 Show that

$$\frac{1}{x}\left(1-\frac{1}{x^2}\right)\mathbb{g}(x) < 1 - \mathbb{G}(x) < \frac{1}{x}\,\mathbb{g}(x)$$

Hint: Prove the following inequalities and integrate from x to ∞

$$-\frac{d}{dx}\left(\frac{1}{x}e^{-x^2/2}\right) > e^{-x^2/2} \qquad -\frac{d}{dx}\left[\left(\frac{1}{x}-\frac{1}{x^3}\right)e^{-x^2/2}\right] < e^{-x^2/2}$$

3-9 Suppose that in n trials, the probability that an event $\mathscr{A}$ occurs at least once equals P_1. Show that, if $P(\mathscr{A}) = p$ and $pn \ll 1$, then $P_1 \simeq np$.

3-10 The probability that a driver will have an accident in 1 month equals 0.02. Find the probability that in 100 months he will have 3 accidents.

Answer: About $4e^{-2}/3$.

3-11 A fair die is rolled five times. Find the probability that *one* shows twice, *three* shows twice, and *six* shows once.

3-12 Show that (3-27) is a special case of (3-39) obtained with $r = 2$, $k_1 = k$, $k_2 = n - k$, $p_1 = p$, $p_2 = 1 - p$.

3-13 Players X and Y roll dice alternately starting with X. The player that rolls *eleven* wins. Show that the probability p that X wins equals 18/35.

Outline: Show that

$$P(\mathscr{A}) = P(\mathscr{A} \mid \mathscr{M})P(\mathscr{M}) + P(\mathscr{A} \mid \bar{\mathscr{M}})P(\bar{\mathscr{M}})$$

Set $\mathscr{A} = \{X \text{ wins}\}$, $\mathscr{M} = \{\textit{eleven} \text{ shows at first try}\}$. Note that $P(\mathscr{A}) = p$, $P(\mathscr{A} \mid \mathscr{M}) = 1$, $P(\mathscr{M}) = 2/36$, $P(\mathscr{A} \mid \bar{\mathscr{M}}) = 1 - p$.

3-14 We place at random n particles in $m > n$ boxes. Find the probability p that the particles will be found in n preselected boxes (one in each box). Consider the following cases: (*a*) M–B (Maxwell–Boltzmann) the particles are distinct; all alternatives are possible, (*b*) B–E (Bose–Einstein)—the particles cannot be distinguished; all alternatives are possible, (*c*) F–D (Fermi–Dirac)—The particles cannot be distinguished; at most one particle is allowed in a box.

Answer:

	M–B	B–E	F–D
$p =$	$\dfrac{n!}{m^n}$	$\dfrac{n!\,(m-1)!}{(m+n-1)!}$	$\dfrac{n!\,(m-m)!}{m!}$

Outline: (*a*) The number N of all alternatives equals m^n. The number $N_{\mathscr{A}}$ of favorable alternatives equals the $n!$ permutations of the particles in the preselected boxes. (*b*) Place the $m - 1$ walls separating the boxes in line ending with the n particles. This corresponds to one alternative where all particles are in the last box. All other possibilities are obtained by a permutation of the $n + m - 1$ objects consisting of the $m - 1$ walls and the n particles. All the $(m - 1)!$ permutations of the walls and the $n!$ permutations of the particles count as one alternative. Hence $N = (m + n - 1)!/(m - 1)!\,n!$ and $N_{\mathscr{A}} = 1$. (*c*) Since the particles are not distinguishable, N equals the number of ways of selecting n out of m objects: $N = \binom{m}{n}$ and $N_{\mathscr{A}} = 1$.

3-15 Reasoning as in (3-41), show that, if

$$k_1 + k_2 + k_3 = n \qquad p_1 + p_2 + p_3 = 1 \qquad k_1 p_1 \ll 1 \qquad k_2 p_2 \ll 1$$

then

$$\frac{n!}{k_1!\,k_2!\,k_3!} \simeq \frac{n^{k_1+k_2}}{k_1!\,k_2!} \qquad p_1^{k_1} p_2^{k_2} p_3^{k_3} \simeq e^{n(p_1+p_2)}$$

Use the above to justify (3-49).

CHAPTER

FOUR

THE CONCEPT OF A RANDOM VARIABLE

4-1 INTRODUCTION

A random variable (abbreviation: RV) is a number $\mathbf{x}(\zeta)$ assigned to every outcome ζ of an experiment. This number could be the gain in a game of chance, the voltage of a random source, the cost of a random component, or any other numerical quantity that is of interest in the performance of the experiment.

Example 4-1(*a*) In the die experiment, we assign to the six outcomes f_i the numbers $\mathbf{x}(f_i) = 10i$. Thus

$$\mathbf{x}(f_1) = 10, \ldots, \mathbf{x}(f_6) = 60$$

(*b*) In the same experiment, we assign the number 1 to every even outcome and the number 0 to every odd outcome. Thus

$$\mathbf{x}(f_1) = \mathbf{x}(f_3) = \mathbf{x}(f_5) = 0 \qquad \mathbf{x}(f_2) = \mathbf{x}(f_4) = \mathbf{x}(f_6) = 1$$

The meaning of a function An RV is a function whose domain is the set $\mathscr{S}$ of experimental outcomes. To clarify further this important concept, we review briefly the notion of a function. As we know, a function $x(t)$ is a rule of correspondence between values of t and x. The values of the independent variable t form a set $\mathscr{S}_t$ on the t axis called the *domain* of the function and the values of the dependent variable x form a set $\mathscr{S}_x$ on the x axis called the *range* of the function. The rule of correspondence between t and x could be a curve, a table, or a formula, for example, $x(t) = t^2$.

The notation $x(t)$ used to represent a function is ambiguous: it might mean either the particular number $x(t)$ corresponding to a specific t, or the function $x(t)$, namely, the rule of correspondence between any t in $\mathscr{S}_t$ and the corresponding x in $\mathscr{S}_x$. To distinguish between these two interpretations, we shall denote the latter by x, leaving its dependence on t understood.

The definition of a function can be phrased as follows: We are given two sets of numbers $\mathscr{S}_t$ and $\mathscr{S}_x$. To every $t \in \mathscr{S}_t$ we assign a number $x(t)$ belonging to the set $\mathscr{S}_x$. This leads to the following generalization: We are given two sets of objects $\mathscr{S}_\alpha$ and $\mathscr{S}_\beta$ consisting of the elements α and β respectively. We say that β is a function of α if to every element of the set $\mathscr{S}_\alpha$ we make correspond an element β of the set $\mathscr{S}_\beta$. The set $\mathscr{S}_\alpha$ is the domain of the function and the set $\mathscr{S}_\beta$ its range.

Suppose, for example, that $\mathscr{S}_\alpha$ is the set of children in a community and $\mathscr{S}_\beta$ the set of their fathers. The pairing of a child with his father is a function.

We note that to a given α there corresponds a single $\beta(\alpha)$. However, more than one element from $\mathscr{S}_\alpha$ might be paired with the same β (a child has only one father but a father might have more than one child). In Example 4-1*b*, the domain of the function consists of the six faces of the die. Its range, however, has only two elements, namely, the numbers 0 and 1.

The Random Variable

We are given an experiment specified by the space $\mathscr{S}$, the field of subsets of $\mathscr{S}$ called events and the probability assigned to these events. To every outcome ζ of this experiment we assign a number $\mathbf{x}(\zeta)$. We have thus created a function $\mathbf{x}$ with domain the set $\mathscr{S}$ and range a set of numbers. This function is called random variable if it satisfies certain mild conditions to be soon given.

All random variables will be written in boldface letters. The symbol $\mathbf{x}(\zeta)$ will indicate the number assigned to the specific outcome ζ and the symbol $\mathbf{x}$ will indicate the rule of correspondence between any element of $\mathscr{S}$ and the number assigned to it. In Example 4-1*a*, $\mathbf{x}$ is the table pairing the six faces of the die with the six numbers 10, ..., 60. The domain of this function is the set $\mathscr{S} = \{f_1, \ldots, f_6\}$ and its range is the set of the above six numbers. The expression $\mathbf{x}(f_2)$ is the number 20.

Events generated by random variables In the study of RVs, questions of the following form arise: What is the probability that the RV $\mathbf{x}$ is less than a given number x, or what is the probability that $\mathbf{x}$ is between the numbers x_1 and x_2. If, for example, the RV is the height of a person, we might want the probability that it will not exceed certain bounds. As we know, probabilities are assigned only to events; therefore, in order to answer such questions, we should be able to express the various conditions imposed on $\mathbf{x}$ as events.

We start with the meaning of the notation

$$\{\mathbf{x} \le x\}$$

This notation represents a subset of $\mathscr{S}$ consisting of all outcomes ζ such that $\mathbf{x}(\zeta) \le x$. We elaborate on its meaning: Suppose that the RV $\mathbf{x}$ is specified by a table. At the left column we list all elements ζ_i of $\mathscr{S}$ and at the right the corresponding values (numbers) $\mathbf{x}(\zeta_i)$ of $\mathbf{x}$. Given an arbitrary number x, we find all numbers $\mathbf{x}(\zeta_i)$ that do not exceed x. The corresponding elements ζ_i on the left column form the set $\{\mathbf{x} \le x\}$. Thus, $\{\mathbf{x} \le x\}$ is not a set of numbers but *a set of experimental outcomes.*

The meaning of

$$\{x_1 \le \mathbf{x} \le x_2\}$$

is similar. It represents a subset of $\mathscr{S}$ consisting of all outcomes ζ such that $x_1 \le \mathbf{x}(\zeta) \le x_2$ where x_1 and x_2 are two given numbers.

The notation

$$\{\mathbf{x} = x\}$$

is a subset of $\mathscr{S}$ consisting of all outcomes ζ such that $\mathbf{x}(\zeta) = x$.

Finally, if $\mathscr{I}$ is a set of numbers on the x axis, then

$$\{\mathbf{x} \in \mathscr{I}\}$$

represents the subset of $\mathscr{S}$ consisting of all outcomes ζ such that $\mathbf{x}(\zeta) \in \mathscr{I}$.

Example 4-2 We shall illustrate the above with the RV $\mathbf{x}(f_i) = 10i$ of the die experiment (Fig. 4-1).

The set $\{\mathbf{x} \le 35\}$ consists of the elements f_1, f_2, f_3 because $\mathbf{x}(f_i) \le 35$ only if $i = 1$, 2, or 3.

The set $\{\mathbf{x} \le 5\}$ is empty because there is no outcome such that $\mathbf{x}(f_i) \le 5$.

The set $\{20 \le \mathbf{x} \le 35\}$ consists of the elements f_2 and f_3 because $20 \le \mathbf{x}(f_i) \le 35$ only if $i = 2$ or 3.

The set $\{\mathbf{x} = 40\}$ consists of the element f_4 because $\mathbf{x}(f_i) = 40$ only if $i = 4$.

Finally, $\{\mathbf{x} = 35\}$ is the empty set because there is no experimental outcome such that $\mathbf{x}(f_i) = 35$.

Note In the applications, we are interested in the probability that an RV $\mathbf{x}$ takes values in a certain region $\mathscr{I}$ of the x axis. This requires that the set $\{\mathbf{x} \in \mathscr{I}\}$ be an event. As we noted in Sec. 2-2, that is not always possible. However, if $\{\mathbf{x} \le x\}$ is an event for every x and $\mathscr{I}$ is a countable union and intersection of intervals, then $\{\mathbf{x} \in \mathscr{I}\}$ is also an event. In

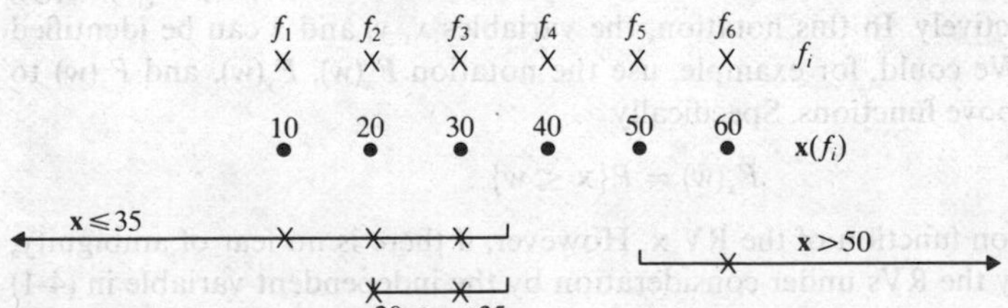

Figure 4-1

the definition of RVs we shall assume, therefore, that the set $\{\mathbf{x} \le x\}$ is an event. This mild restriction is mainly of mathematical interest.

We conclude with a formal definition of an RV.

Definition An RV $\mathbf{x}$ is a process of assigning a number $\mathbf{x}(\zeta)$ to every outcome ζ. The resulting function must satisfy the following two conditions but is otherwise arbitrary:

I. The set $\{\mathbf{x} \le x\}$ is an event for every x.
II. The probabilities of the events $\{\mathbf{x} = \infty\}$ and $\{\mathbf{x} = -\infty\}$ equal zero

$$P\{\mathbf{x} = \infty\} = 0 \qquad P\{\mathbf{x} = -\infty\} = 0$$

The second condition states that, although we allow $\mathbf{x}$ to be $+\infty$ or $-\infty$ for some outcomes, we demand that these outcomes form a set with zero probability.

A *complex* RV $\mathbf{z}$ is a sum

$$\mathbf{z} = \mathbf{x} + j\mathbf{y}$$

where $\mathbf{x}$ and $\mathbf{y}$ are real RVs. Unless otherwise stated, it will be assumed that all RVs are real.

4-2 DISTRIBUTION AND DENSITY FUNCTIONS

The elements of the set $\mathscr{S}$ that are contained in the event $\{\mathbf{x} \le x\}$ change as the number x takes various values. The probability $P\{\mathbf{x} \le x\}$ of the event $\{\mathbf{x} \le x\}$ is, therefore, a number that depends on x. This number is denoted by $F_x(x)$ and is called the *(cumulative) distribution function* of the RV $\mathbf{x}$.

Definition The distribution function of the RV $\mathbf{x}$ is the function

$$F_x(x) = P\{\mathbf{x} \le x\} \tag{4-1}$$

defined for every x from $-\infty$ to ∞.

The distribution functions of the RVs $\mathbf{x}$, $\mathbf{y}$, and $\mathbf{z}$ are denoted by $F_x(x)$, $F_y(y)$, and $F_z(z)$ respectively. In this notation, the variables x, y, and z can be identified by any letter. We could, for example, use the notation $F_x(w)$, $F_y(w)$, and $F_z(w)$ to represent the above functions. Specifically

$$F_x(w) = P\{\mathbf{x} \le w\}$$

is the distribution function of the RV $\mathbf{x}$. However, if there is no fear of ambiguity, we shall identify the RVs under consideration by the independent variable in (4-1) omitting the subscript. Thus, the distribution functions of the RVs $\mathbf{x}$, $\mathbf{y}$, and $\mathbf{z}$ will be denoted by $F(x)$, $F(y)$, and $F(z)$ respectively.

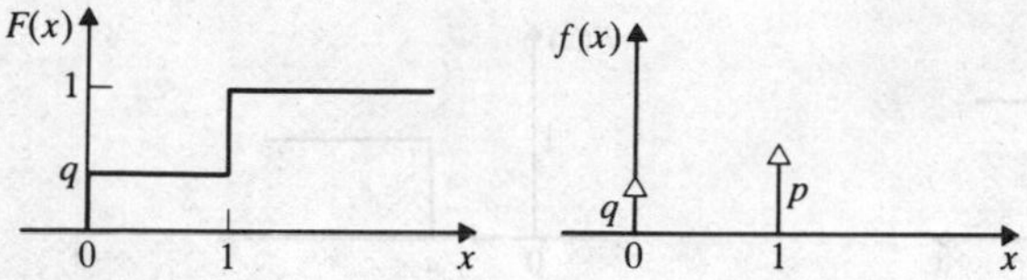

Figure 4-2

Example 4-3 In the coin-tossing experiment, the probability of heads equals p and the probability of tails equals q. We define the RV $\mathbf{x}$ such that

$$\mathbf{x}(h) = 1 \qquad \mathbf{x}(t) = 0$$

We shall find its distribution function $F(x)$ for every x from $-\infty$ to ∞.

If $x \geq 1$, then $\mathbf{x}(h) = 1 \leq x$ and $\mathbf{x}(t) = 0 \leq x$. Hence (Fig. 4-2)

$$F(x) = P\{\mathbf{x} \leq x\} = P\{h, t\} = 1 \qquad x \geq 1$$

If $0 \leq x < 1$, then $\mathbf{x}(h) = 1 > x$ and $\mathbf{x}(t) = 0 \leq x$. Hence

$$F(x) = P\{\mathbf{x} \leq x\} = P\{t\} = q \qquad 0 \leq x < 1$$

If $x < 0$, then $\mathbf{x}(h) = 1 > x$ and $\mathbf{x}(t) = 0 > x$. Hence

$$F(x) = P\{\mathbf{x} \leq x\} = P\{\emptyset\} = 0 \qquad x < 0$$

Example 4-4 In the die experiment of Example 4-2, the RV $\mathbf{x}$ is such that $\mathbf{x}(f_i) = 10i$. If the die is fair, then the distribution function of $\mathbf{x}$ is a staircase function as in Fig. 4-3.

We note, in particular, that

$$F(100) = P\{\mathbf{x} < 100\} = P(\mathscr{S}) = 1$$

$$F(35) = P\{\mathbf{x} \leq 35\} = P\{f_1, f_2, f_3\} = \frac{3}{6}$$

$$F(30.01) = P\{\mathbf{x} \leq 30.01\} = P\{f_1, f_2, f_3\} = \frac{3}{6}$$

$$F(30) = P\{\mathbf{x} \leq 30\} = P\{f_1, f_2, f_3\} = \frac{3}{6}$$

$$F(29.99) = P\{\mathbf{x} \leq 29.99\} = P\{f_1, f_2\} = \frac{2}{6}$$

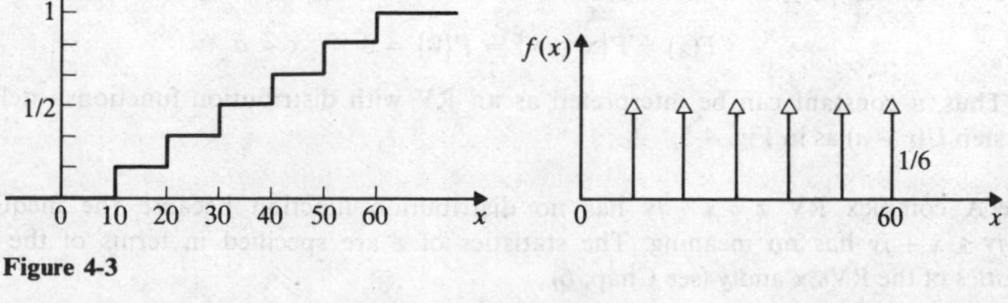

Figure 4-3

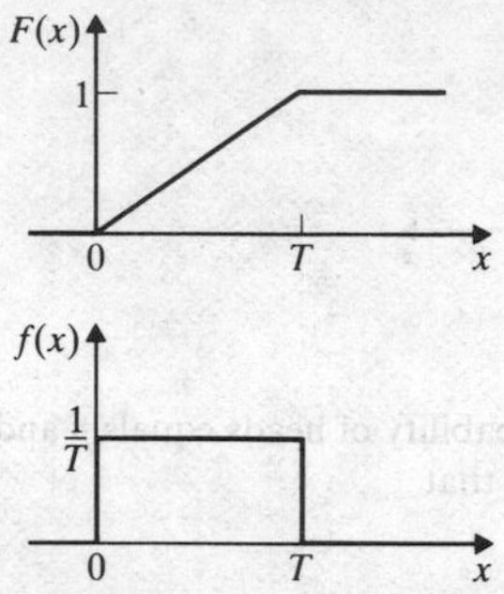

Figure 4-4

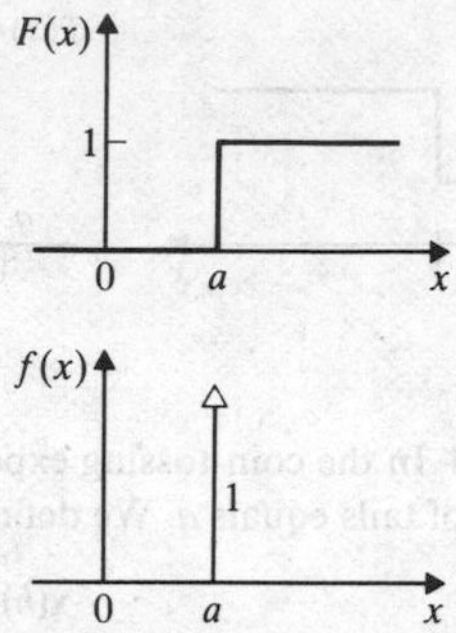

Figure 4-5

Example 4-5 A telephone call occurs at random in the interval (0, 1). In this experiment, the outcomes are time instances t between 0 and 1 and the probability that t is between t_1 and t_2 is given by

$$P\{t_1 \le t \le t_2\} = t_2 - t_1$$

We define the RV $\mathbf{x}$ such that

$$\mathbf{x}(t) = t \qquad 0 \le t \le 1$$

Thus, the variable t has a double meaning: It is the outcome of the experiment and the corresponding value $\mathbf{x}(t)$ of the RV $\mathbf{x}$. We shall show that the distribution function $F(x)$ of $\mathbf{x}$ is a ramp as in Fig. 4-4.

If $x > 1$, then $\mathbf{x}(t) \le x$ for every outcome. Hence

$$F(x) = P\{\mathbf{x} \le x\} = P\{0 \le t \le 1\} = P(\mathscr{S}) = 1 \qquad x > 1$$

If $0 \le x \le 1$, then $\mathbf{x}(t) \le x$ for every t in the interval $(0, x)$. Hence

$$F(x) = P\{\mathbf{x} \le x\} = P(0 \le t \le x\} \qquad 0 \le x \le 1$$

If $x < 0$, then $\{\mathbf{x} \le x\}$ is the impossible event because $\mathbf{x}(t) \ge 0$ for every t. Hence

$$F(x) = P\{\mathbf{x} \le x\} = P(\emptyset\} = 0 \qquad x < 0$$

Example 4-6 Suppose that an RV $\mathbf{x}$ is such that $\mathbf{x}(\zeta) = a$ for every ζ in $\mathscr{S}$. We shall find its distribution function.

If $x \ge a$, then $\mathbf{x}(\zeta) = a \le x$ for every ζ. Hence

$$F(x) = P\{\mathbf{x} \le x\} = P(\mathscr{S}) = 1 \qquad x \ge a$$

If $x < a$ then $\{\mathbf{x} \le x\}$ is the impossible event because $\mathbf{x}(\zeta) = a$. Hence

$$F(x) = P\{\mathbf{x} \le x\} = P\{\emptyset\} = 0 \qquad x < a$$

Thus, a constant can be interpreted as an RV with distribution function a delayed step $U(t - a)$ as in Fig. 4-5.

Note A complex RV $\mathbf{z} = \mathbf{x} + j\mathbf{y}$ has no distribution function because the inequality $\mathbf{x} + j\mathbf{y} \le x + jy$ has no meaning. The statistics of $\mathbf{z}$ are specified in terms of the *joint statistics* of the RVs $\mathbf{x}$ and $\mathbf{y}$ (see Chap. 6).

Frequency interpretation At a single trial, an outcome ζ occurs and the RV $\mathbf{x}$ takes the value $\mathbf{x}(\zeta)$. Suppose that the experiment is performed n times and the total number of trials such that $\mathbf{x}(\zeta) \leq x$ equals n_x where x is a given number. We then have [see (1-1) and (4-1)]

$$F(x) = P\{\mathbf{x} \leq x\} \simeq \frac{n_x}{n} \tag{4-2}$$

Properties of Distribution Functions

In the following, the expressions $F(x^+)$ and $F(x^-)$ will mean the limits

$$F(x^+) = \lim F(x + \varepsilon) \qquad F(x^-) = \lim F(x - \varepsilon) \qquad 0 < \varepsilon \to 0$$

The distribution function has the following properties

1.
$$F(+\infty) = 1 \qquad F(-\infty) = 0 \tag{4-3}$$

PROOF

$$F(+\infty) = P\{\mathbf{x} \leq \infty\} = P(\mathscr{S}) = 1 \qquad F(-\infty) = P\{\mathbf{x} = -\infty\} = 0$$

2. It is a nondecreasing function of x:

If
$$x_1 < x_2 \qquad \text{then} \qquad F(x_1) \leq F(x_2) \tag{4-4}$$

PROOF The event $\{\mathbf{x} \leq x_1\}$ is a subset of the event $\{\mathbf{x} \leq x_2\}$ because, if $\mathbf{x}(\zeta) \leq x_1$ for some ζ, then $\mathbf{x}(\zeta) \leq x_2$. Hence, [see (2-14)] $P\{\mathbf{x} \leq x_1\} \leq P\{\mathbf{x} \leq x_2\}$ and (4-4) results.

From (4-3) and (4-4) it follows that $F(x)$ increases from 0 to 1 as x increases from $-\infty$ to ∞. The value x_m of x such that $F(x_m) = 0.5$ is called the *median* of $\mathbf{x}$.

3. If
$$F(x_0) = 0 \qquad \text{then} \qquad F(x) = 0 \qquad \text{for every} \qquad x \leq x_0 \tag{4-5}$$

PROOF It follows from (4-4) because $F(-\infty) = 0$. The above leads to the following conclusion: Suppose that $\mathbf{x}(\zeta) \geq 0$ for every ζ. In this case, $F(0) = P\{\mathbf{x} \leq 0\} = 0$ because $\{\mathbf{x} \leq 0\}$ is the impossible event. Hence, $F(x) = 0$ for every $x \leq 0$.

4.
$$P\{\mathbf{x} > x\} = 1 - F(x) \tag{4-6}$$

PROOF The events $\{\mathbf{x} \leq x\}$ and $\{\mathbf{x} \geq x\}$ are mutually exclusive and

$$\{\mathbf{x} \leq x\} + \{\mathbf{x} > x\} = \mathscr{S}$$

Hence, $P\{\mathbf{x} \leq x\} + P\{\mathbf{x} > x\} = P(\mathscr{S}) = 1$ and (4-6) results.

5. The function $F(x)$ is continuous from the right:

$$F(x^+) = F(x) \tag{4-7}$$

Proof It suffices to show that $P\{\mathbf{x} \le x + \varepsilon\} \to F(x)$ as $\varepsilon \to 0$ because $P\{\mathbf{x} \le x + \varepsilon\} = F(x + \varepsilon)$ and $F(x + \varepsilon) \to F(x^+)$ by definition. To prove the above, we must show that the sets $\{\mathbf{x} \le x + \varepsilon\}$ tend to the set $\{\mathbf{x} \le x\}$ as $\varepsilon \to 0$ and to use the axiom III*a* of infinite additivity. We omit, however, the details of the proof because we have not introduced limits of sets.

6.
$$P\{x_1 < \mathbf{x} \le x_2\} = F(x_2) - F(x_1) \tag{4-8}$$

Proof The events $\{\mathbf{x} \le x_1\}$ and $\{x_1 < \mathbf{x} \le x_2\}$ are mutually exclusive because $\mathbf{x}(\zeta)$ cannot be less than x_1 and between x_1 and x_2. Furthermore

$$\{\mathbf{x} \le x_2\} = \{\mathbf{x} \le x_1\} + \{x_1 < \mathbf{x} \le x_2\}$$

Hence

$$P\{\mathbf{x} \le x_2\} = P\{\mathbf{x} \le x_1\} + P\{x_1 < \mathbf{x} \le x_2\}$$

and (4-8) results.

7.
$$P\{\mathbf{x} = x\} = F(x) - F(x^-) \tag{4-9}$$

Proof Setting $x_1 = x - \varepsilon$ and $x_2 = x$ in (4-8), we obtain

$$P\{x - \varepsilon < \mathbf{x} \le x\} = F(x) - F(x - \varepsilon)$$

and with $\varepsilon \to 0$, (4-9) results.

8.
$$P\{x_1 \le \mathbf{x} \le x_2\} = F(x_2) - F(x_1^-) \tag{4-10}$$

Proof It follows from (4-8) and (4-9) because

$$\{x_1 \le \mathbf{x} \le x_2\} = \{x_1 < \mathbf{x} \le x_2\} + \{\mathbf{x} = x_1\}$$

and the last two events are mutually exclusive.

Statistics We shall say that the statistics of an RV $\mathbf{x}$ are known if we can determine the probability $P\{\mathbf{x} \in \mathscr{I}\}$ that $\mathbf{x}$ is in a set $\mathscr{I}$ of the x axis consisting of countable unions or intersections of intervals. From (4-1) and the axioms it follows that the statistics of $\mathbf{x}$ are determined in terms of its distribution function. The preceding properties are special cases.

Continuous, discrete, and mixed types We shall say that an RV $\mathbf{x}$ is of *continuous type* if its distribution function $F(x)$ is continuous. In this case, $F(x^-) = F(x)$, hence

$$P\{\mathbf{x} = x\} = 0 \tag{4-11}$$

for every x.

We shall say that $\mathbf{x}$ is of *discrete type* if $F(x)$ is a staircase function as in Fig. 4-6. Denoting by x_i the discontinuity points of $F(x)$, we have

$$F(x_i) - F(x_i^-) = P\{\mathbf{x} = x_i\} = p_i \tag{4-12}$$

In this case, the statistics of $\mathbf{x}$ are determined in terms of x_i and p_i. If the points x_i are equidistant, i.e., if $x_i = a + bi$, then the RV $\mathbf{x}$ is of *lattice type*.

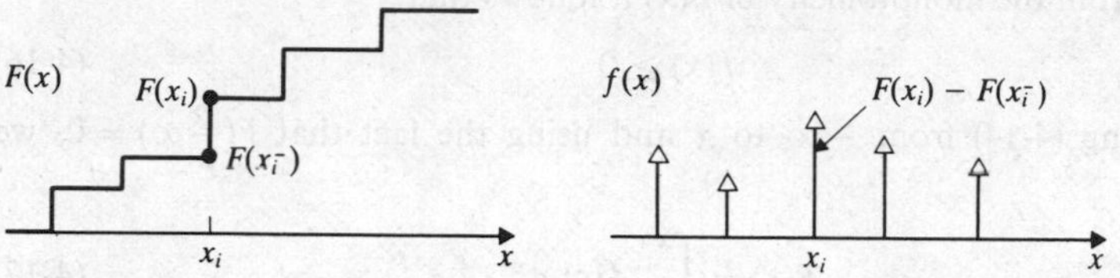

Figure 4-6

We shall say that $\mathbf{x}$ is of *mixed type* if $F(x)$ is discontinuous but not a staircase.

We note that if $\mathscr{S}$ has finitely many elements, then any RV defined on $\mathscr{S}$ is of discrete type. However, an RV $\mathbf{x}$ might be of discrete type even if $\mathscr{S}$ has infinitely many elements.

Example 4-7 If $\mathscr{A}$ is an arbitrary event of $\mathscr{S}$ and $\mathbf{x}_{\mathscr{A}}$ is an RV such that

$$\mathbf{x}_{\mathscr{A}}(\zeta) = \begin{cases} 1 & \zeta \in \mathscr{A} \\ 0 & \zeta \notin \mathscr{A} \end{cases} \tag{4-13}$$

then $\mathbf{x}_{\mathscr{A}}$ is called the *zero-one* RV associated with the event $\mathscr{A}$. Thus

$$\{\mathbf{x}_{\mathscr{A}} = 1\} = \mathscr{A} \qquad \{\mathbf{x}_{\mathscr{A}} = 0\} = \bar{\mathscr{A}}$$

Hence, $\mathbf{x}_{\mathscr{A}}$ is of discrete type taking only the two values 0 and 1 with

$$P\{\mathbf{x}_{\mathscr{A}} = 1\} = P(\mathscr{A}) \qquad P\{\mathbf{x}_{\mathscr{A}} = 0\} = 1 - P(\mathscr{A})$$

The space $\mathscr{S}$, however, might have infinitely many elements.

The Density Function

The derivative

$$f(x) = \frac{dF(x)}{dx} \tag{4-14}$$

of $F(x)$ is called the *density function* (known also as *frequency function*) of the RV $\mathbf{x}$.

If the RV $\mathbf{x}$ is of discrete type taking the values x_i with probabilities p_i, then

$$f(x) = \sum_i p_i\,\delta(x - x_i) \qquad p_i = P\{\mathbf{x} = x_i\} \tag{4-15}$$

where $\delta(x)$ is the impulse function (Fig. 4-6). The term $p_i\,\delta(x - x_i)$ is shown as a vertical arrow at $x = x_i$ with length equal to p_i.

In Example 4-2, the RV $\mathbf{x}$ is of discrete type taking the six values $x_1 = 10$, $\ldots, x_6 = 60$ with $p_i = 1/6$. Hence

$$f(x) = \frac{1}{6}\,[\delta(x - 10) + \delta(x - 20) + \cdots + \delta(x - 60)]$$

Properties From the monotonicity of $F(x)$ it follows that

$$f(x) \geq 0 \tag{4-16}$$

Integrating (4-14) from $-\infty$ to x and using the fact that $F(-\infty) = 0$, we obtain

$$F(x) = \int_{-\infty}^{x} f(\xi)\, d\xi \tag{4-17}$$

Since $F(\infty) = 1$, the above yields

$$\int_{-\infty}^{\infty} f(x)\, dx = 1 \tag{4-18}$$

From (4-17) it follows that

$$F(x_2) - F(x_1) = \int_{x_1}^{x_2} f(x)\, dx \tag{4-19}$$

Hence [see (4-8)]

$$P\{x_1 < \mathbf{x} \leq x_2\} = \int_{x_1}^{x_2} f(x)\, dx \tag{4-20}$$

If the RV $\mathbf{x}$ is of continuous type, then the set on the left might be replaced by the set $\{x_1 \leq \mathbf{x} \leq x_2\}$. However, if $F(x)$ is discontinuous at x_1 or x_2, then the integration must include the corresponding impulses of $f(x)$.

With $x_1 = x$ and $x_2 = x + \Delta x$ it follows from (4-20) that, if $\mathbf{x}$ is of continuous type, then

$$P\{x \leq \mathbf{x} \leq x + \Delta x\} \simeq f(x)\, \Delta x \tag{4-21}$$

provided that Δx is sufficiently small. This shows that $f(x)$ can be defined directly as a limit

$$f(x) = \lim_{\Delta x \to 0} \frac{P\{x \leq \mathbf{x} \leq x + \Delta x\}}{\Delta x} \tag{4-22}$$

Note As we see from (4-21), the probability that $\mathbf{x}$ is in a small interval of specified length Δx is proportional to $f(x)$ and it is maximum if that interval contains the point x_m where $f(x)$ is maximum. This point is called the *mode* or the *most likely value* of $\mathbf{x}$. An RV is called unimodal if it has a single mode.

Frequency interpretation We denote by Δn_x the number of trials such that

$$x \leq \mathbf{x}(\zeta) \leq x + \Delta x$$

From (1-1) and (4-21) it follows that

$$f(x)\, \Delta x \simeq \frac{\Delta n_x}{n} \tag{4-23}$$

4-3 SPECIAL CASES

In the preceding sections we defined RVs starting from known experiments. In this section and throughout the book we shall often consider RVs having specific distribution or density functions without any reference to a particular probability space.

Existence theorem To do so, we must show that given a function $f(x)$ or its integral

$$F(x) = \int_{-\infty}^{x} f(\xi)\, d\xi$$

we can construct an experiment and an RV $\mathbf{x}$ with distribution $F(x)$ or density $f(x)$. As we know, these functions must have the following properties:

The function $f(x)$ must be nonnegative and its area must be 1. The function $F(x)$ must be continuous from the right and, as x increases from $-\infty$ to ∞, it must increase monotonically from 0 to 1.

PROOF We consider as our space $\mathscr{S}$ the set of all real numbers, and as its events all intervals on the real line and their unions and intersections. We define the probability of the event $\{x \le x_1\}$ by

$$P\{x \le x_1\} = F(x_1) \tag{4-24}$$

where $F(x)$ is the given function. This specifies the experiment completely (see Sec. 2-2).

The outcomes of our experiment are the real numbers. To define an RV $\mathbf{x}$ on this experiment, we must know its value $\mathbf{x}(x)$ for every x. We define $\mathbf{x}$ such that

$$\mathbf{x}(x) = x \tag{4-25}$$

Thus, x is the outcome of the experiment and the corresponding value of the RV $\mathbf{x}$ (see also Example 4-5).

We maintain that the distribution function of $\mathbf{x}$ equals the given $F(x)$. Indeed, the event $\{\mathbf{x} \le x_1\}$ consists of all outcomes x such that $\mathbf{x}(x) \le x_1$. Hence

$$P\{\mathbf{x} \le x_1\} = P\{x \le x_1\} = F(x_1) \tag{4-26}$$

and since this is true for every x_1, the theorem is proved.

In the following, we discuss briefly a number of common densities.

Normal An RV $\mathbf{x}$ is called *normal* or *gaussian* if its density is the normal curve $\mathbb{g}(x)$ [see (3-20)], shifted and scaled

$$f(x) = \frac{1}{\sigma}\, \mathbb{g}\left(\frac{x-\eta}{\sigma}\right) = \frac{1}{\sigma\sqrt{2\pi}}\, e^{-(x-\eta)^2/2\sigma^2} \tag{4-27}$$

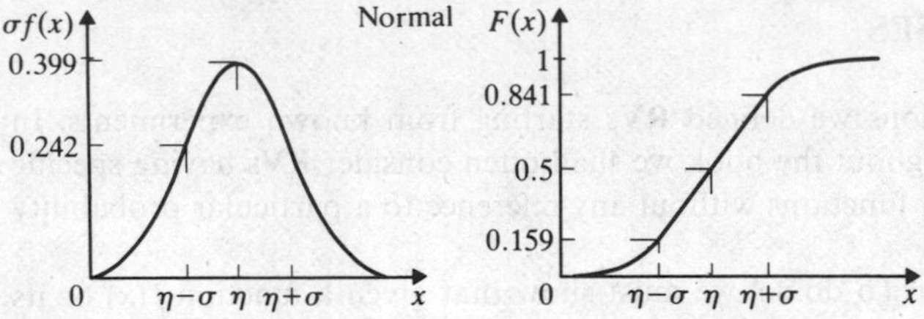

Figure 4-7

This is a bell-shaped curve, symmetrical about the line $x = \eta$ (Fig. 4-7) and its area equals 1 as it should [see (3-22)]. The corresponding distribution function is given by

$$F(x) = \mathbb{G}\left(\frac{x - \eta}{\sigma}\right) \tag{4-28}$$

where $\mathbb{G}(x)$ is the tabulated integral of $\mathfrak{g}(x)$ [see (3-21)].

We shall use the notation

$$N(\eta; \sigma)$$

to indicate that an RV **x** is normal as in (4-27). The significance of the constants η and σ will be given in Sec. 5-4 (η: mean, σ: standard deviation).

Example 4-8 An RV **x** is $N(1000; 50)$. We shall find the probability that **x** is between 900 and 1050. Clearly

$$P\{900 \le \mathbf{x} \le 1050\} = F(1050) - F(900) = \mathbb{G}(1) - \mathbb{G}(-2)$$

Since

$$\mathbb{G}(-x) = 1 - \mathbb{G}(x) \tag{4-29}$$

we conclude from Table 3-1 that

$$P\{900 \le \mathbf{x} \le 1050\} = \mathbb{G}(1) + \mathbb{G}(2) - 1 = 0.819$$

Uniform An RV **x** is called uniform between x_1 and x_2 if its density is constant in the interval (x_1, x_2) and zero elsewhere

$$f(x) = \begin{cases} \dfrac{1}{x_2 - x_1} & x_1 \le x \le x_2 \\ 0 & \text{otherwise} \end{cases} \tag{4-30}$$

The corresponding distribution function is a ramp as in Fig. 4-8.

Example 4-9 A resistor **r** is an RV uniform between 900 and 1100 Ω. We shall find the probability that **r** is between 950 and 1050 Ω.

Since $f(r) = 1/200$ in the interval (900, 1100), (4-20) yields

$$P\{950 \le \mathbf{r} \le 1050\} = \frac{1}{200}\int_{950}^{1050} dr = 0.5$$

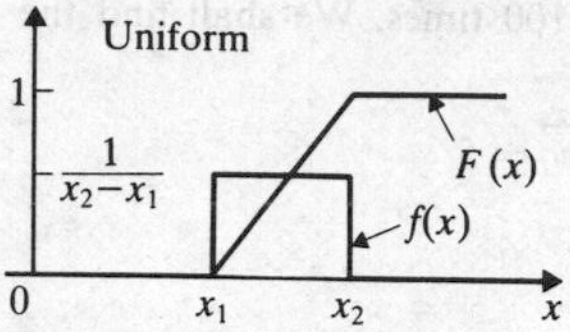

Figure 4-8

Binomial We say that an RV **x** has a *binomial* distribution of order n if it takes the values 0, 1, ..., n with

$$P\{\mathbf{x}=k\} = \binom{n}{k} p^k q^{n-k} \qquad p+q=1 \tag{4-31}$$

Thus, **x** is of lattice type and its density is a sum of impulses (Fig. 4-9a)

$$f(x) = \sum_{k=0}^{n} \binom{n}{k} p^k q^{n-k} \,\delta(n-k) \tag{4-32}$$

The corresponding distribution is a staircase function and in the interval $(0, n)$ it is given by

$$F(x) = \sum_{k=0}^{m} \binom{n}{k} p^k q^{n-k} \qquad m \le x < m+1 \tag{4-33}$$

We note that, if n is large, then [see (3-34)] $F(x)$ is close to an $N(np, \sqrt{npq})$ distribution. In other words

$$F(x) \simeq \mathbb{G}\left(\frac{x-np}{\sqrt{npq}}\right) \tag{4-34}$$

Example 4-10 (Bernoulli trials) In the experiment of the n tossings of a coin, an outcome is a sequence $\zeta_1 \cdots \zeta_n$ of k heads and $n-k$ tails where $k = 0, \ldots, n$. We define the RV **x** such that

$$\mathbf{x}(\zeta_1 \cdots \zeta_n) = k$$

Thus, **x** equals the number of heads. As we know [see (3-13)], the probability that $\mathbf{x} = k$ equals the right side of (4-31). Hence, **x** has a binomial distribution.

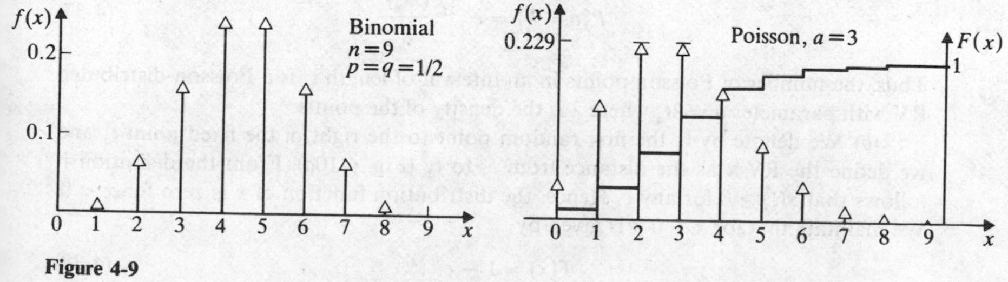

Figure 4-9

Suppose that the coin is fair and it is tossed $n = 100$ times. We shall find the probability that $\mathbf{x}$ is between 46 and 60. In this case

$$p = q = 0.5 \qquad np = 50 \qquad \sqrt{npq} = 5$$

and (4-34) yields

$$P\{40 \le \mathbf{x} \le 60\} = \mathbb{G}\left(\frac{60-50}{5}\right) - \mathbb{G}\left(\frac{40-50}{5}\right) = \mathbb{G}(2) - \mathbb{G}(-2) = 0.9545$$

Poisson An RV $\mathbf{x}$ is *Poisson* distributed with parameter a if it takes the values $0, 1, \ldots, n, \ldots$ with

$$P\{\mathbf{x} = k\} = e^{-a}\frac{a^k}{k!} \qquad k = 0, 1, \ldots \tag{4-35}$$

Thus, $\mathbf{x}$ is of lattice type with density

$$f(x) = e^{-a}\sum_{k=0}^{\infty}\frac{a^k}{k!}\,\delta(x-k) \tag{4-36}$$

The corresponding distribution is a staircase function as in Fig. 4-9*b*.

With $p_k = P\{\mathbf{x} = k\}$, it follows from (4-35) that

$$\frac{p_{k-1}}{p_k} = \frac{e^{-a}a^{k-1}/(k-1)!}{e^{-a}a^k/k!} = \frac{k}{a}$$

If the above ratio is less than 1, that is, if $k < a$, then $p_{k-1} < p_k$. This shows that, as k increases, p_k increases reaching its maximum for $k = [a]$. Hence

if $a < 1$, then p_k is maximum for $k = 0$;

if $a > 1$ but it is not an integer, then p_k increases as k increases, reaching its maximum for $k = [a]$;

if a is an integer, then p_k is maximum for $k = a - 1$ and $k = a$.

Example 4-11 (Poisson points) In the Poisson points experiment, an outcome ζ is a set of points $\mathbf{t}_i$ on the t axis.

(*a*) Given a constant t_o, we define the RV $\mathbf{n}$ such that its value $\mathbf{n}(\zeta)$ equals the number of points $\mathbf{t}_i$ in the interval $(0, t_o)$. Clearly, $\mathbf{n} = k$ means that the number of points in the interval $(0, t_o)$ equals k. Hence, [see (3-47)]

$$P\{\mathbf{n} = k\} = e^{-\lambda t_o}\frac{(\lambda t_o)^k}{k!} \tag{4-37}$$

Thus, the number of Poisson points in an interval of length t_o is a Poisson-distributed RV with parameter $a = \lambda t_o$ where λ is the density of the points.

(*b*) We denote by $\mathbf{t}_1$ the first random point to the right of the fixed point t_o and we define the RV $\mathbf{x}$ as the distance from t_o to $\mathbf{t}_1$ (Fig. 4-10*a*). From the definition it follows that $\mathbf{x}(\zeta) \ge 0$ for any ζ. Hence, the distribution function of $\mathbf{x}$ is zero for $x < 0$. We maintain that for $x > 0$ it is given by

$$F(x) = 1 - e^{-\lambda x} \tag{4-38}$$

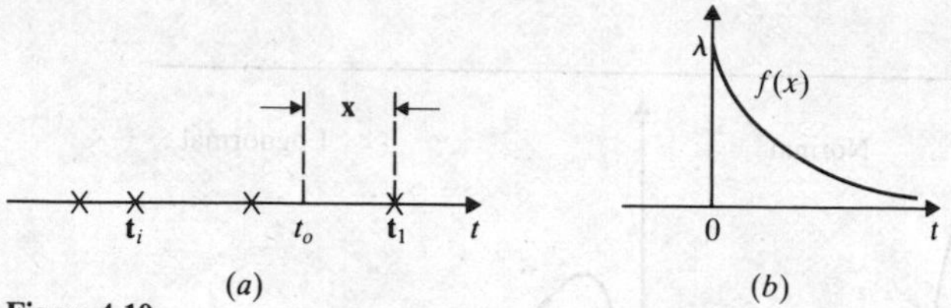

Figure 4-10

PROOF As we know, $F(x)$ equals the probability that $\mathbf{x} \leq x$ where x is a specific number. But $\mathbf{x} \leq x$ means that there is at least one point between t_o and $t_o + x$. Hence, $1 - F(x)$ equals the probability p_0 that there are no points in the interval $(t_o, t_o + x)$. And since the length of this interval equals x, (4-37) yields

$$p_0 = e^{-\lambda x} = 1 - F(x)$$

The corresponding density

$$f(x) = \lambda e^{-\lambda x} U(x) \tag{4-39}$$

is called *exponential* (Fig. 4-10*b*).

Gamma and Erlang densities The function

$$f(x) = Ax^b e^{-cx} U(x) \tag{4-40}$$

is known as *gamma density*. In the above, the numbers b and c are positive and A is such that

$$\int_0^\infty Ax^b e^{-cx}\, dx = 1$$

This yields

$$A = \frac{c^{b+1}}{\Gamma(b+1)} \qquad \text{where} \qquad \Gamma(b) = \int_0^\infty y^{b-1} e^{-y}\, dy$$

is the familiar *gamma function*. This function is called also "generalized factorial" because

$$\Gamma(b+1) = b\Gamma(b)$$

We note in particular that $\Gamma(1/2) = \sqrt{\pi}$.

If $b + 1 = n$ is an integer, then $\Gamma(n) = (n-1)!$ The resulting function

$$f(x) = \frac{c^n}{(n-1)!} x^{n-1} e^{-cx} U(x)$$

is called the *Erlang density*. A special case is the *exponential density*

$$f(x) = ce^{-cx} U(x)$$

In Table 4-1, we show a number of common densities. In the formulas of the various curves, a numerical factor is omitted. The omitted factor is determined from (4-18).

Table 4.1

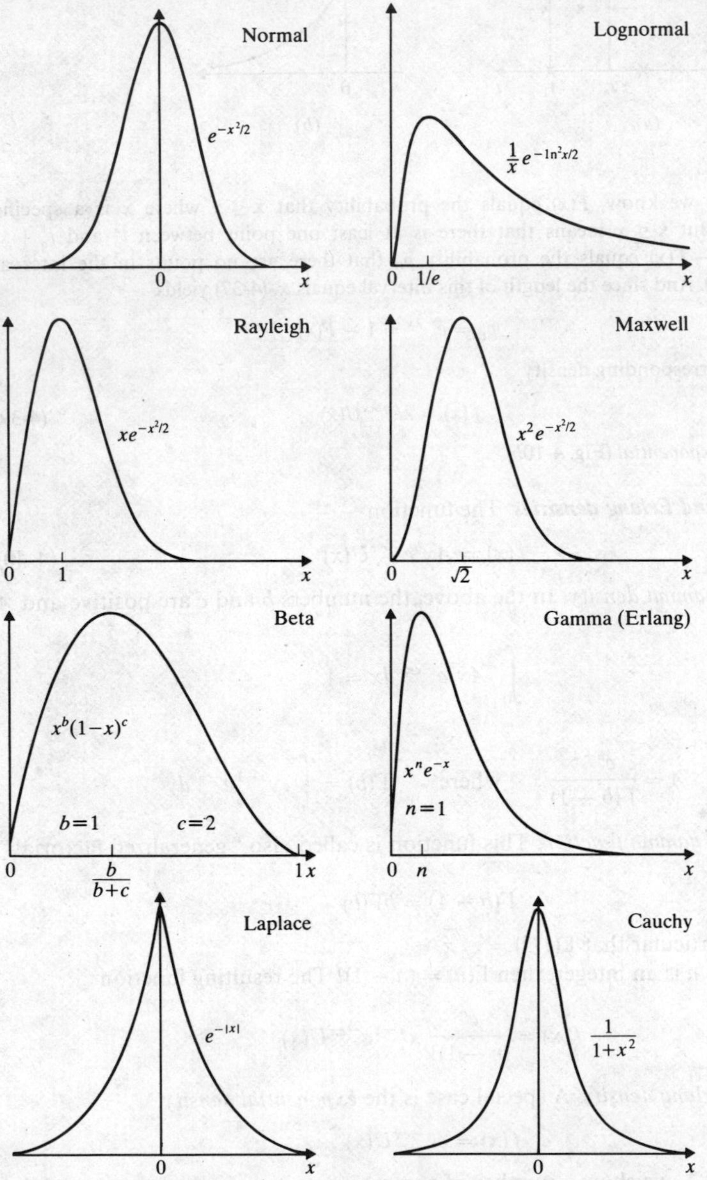
Normal
$e^{-x^2/2}$
0
x
Lognormal
$\frac{1}{x}e^{-\ln^2 x/2}$
0
$1/e$
x
Rayleigh
$xe^{-x^2/2}$
0
1
x
Maxwell
$x^2e^{-x^2/2}$
0
$\sqrt{2}$
x
Beta
$x^b(1-x)^c$
$b=1$
$c=2$
0
$\frac{b}{b+c}$
1
x
Gamma (Erlang)
$x^n e^{-x}$
$n=1$
0
n
x
Laplace
$e^{-|x|}$
0
x
Cauchy
$\frac{1}{1+x^2}$
0
x

4-4 CONDITIONAL DISTRIBUTIONS

We recall that the probability of an event $\mathscr{A}$ assuming $\mathscr{M}$ is given by

$$P(\mathscr{A} \mid \mathscr{M}) = \frac{P(\mathscr{A}\mathscr{M})}{P(\mathscr{M})} \qquad \text{where} \qquad P(\mathscr{M}) \neq 0$$

The *conditional distribution* $F(x \mid \mathscr{M})$ of an RV $\mathbf{x}$ assuming $\mathscr{M}$ is defined as the conditional probability of the event $\{\mathbf{x} \leq x\}$

$$F(x \mid \mathscr{M}) = P\{\mathbf{x} \leq x \mid \mathscr{M}\} = \frac{P\{\mathbf{x} \leq x, \mathscr{M}\}}{P(\mathscr{M})} \tag{4-41}$$

In the above, $\{\mathbf{x} \leq x, \mathscr{M}\}$ is the intersection of the events $\{\mathbf{x} \leq x\}$ and $\mathscr{M}$, that is, the event consisting of all outcomes ζ such that $\mathbf{x}(\zeta) \leq x$ and $\zeta \in \mathscr{M}$.

Thus, the definition of $F(x \mid \mathscr{M})$ is the same as the definition (4-1) of $F(x)$ provided that all probabilities are replaced by conditional probabilities. From this it follows (see Fundamental remark Sec. 2-3) that $F(x \mid \mathscr{M})$ has the same properties as $F(x)$. In particular [see (4-3) and (4-8)]

$$F(\infty \mid \mathscr{M}) = 1 \qquad F(-\infty \mid \mathscr{M}) = 0 \tag{4-42}$$

$$P\{x_1 < \mathbf{x} \leq x_2 \mid \mathscr{M}\} = F(x_2 \mid \mathscr{M}) - F(x_1 \mid \mathscr{M}) = \frac{P\{x_1 < \mathbf{x} \leq x_2, \mathscr{M}\}}{P(\mathscr{M})} \tag{4-43}$$

The *conditional density* $f(x \mid \mathscr{M})$ is the derivative of $F(x \mid \mathscr{M})$

$$f(x \mid \mathscr{M}) = \frac{dF(x \mid \mathscr{M})}{dx} = \lim_{\Delta x \to 0} \frac{P\{x \leq \mathbf{x} \leq x + \Delta x \mid \mathscr{M}\}}{\Delta x} \tag{4-44}$$

This function is nonnegative and its area equals 1.

Example 4-12 We shall determine the conditional distribution $F(x \mid \mathscr{M})$ of the RV $\mathbf{x}(f_i) = 10i$ of the fair-die experiment (Example 4-4), where $\mathscr{M} = \{f_2, f_4, f_6\}$ is the event "even."

If $x \geq 60$, then $\{\mathbf{x} \leq x\}$ is the certain event and $\{\mathbf{x} \leq x, \mathscr{M}\} = \mathscr{M}$. Hence (Fig. 4-11)

$$F(x \mid \mathscr{M}) = \frac{P(\mathscr{M})}{P(\mathscr{M})} = 1 \qquad x \geq 60$$

If $40 \leq x < 60$, then $\{\mathbf{x} \leq x, \mathscr{M}\} = \{f_2, f_4\}$. Hence

$$F(x \mid \mathscr{M}) = \frac{P\{f_2, f_4\}}{P(\mathscr{M})} = \frac{2/6}{3/6} \qquad 40 \leq x < 60$$

If $20 \leq x < 40$, then $\{\mathbf{x} \leq x, \mathscr{M}\} = \{f_2\}$. Hence

$$F(x \mid \mathscr{M}) = \frac{P\{f_2\}}{P(\mathscr{M})} = \frac{1/6}{3/6} \qquad 20 \leq x < 60$$

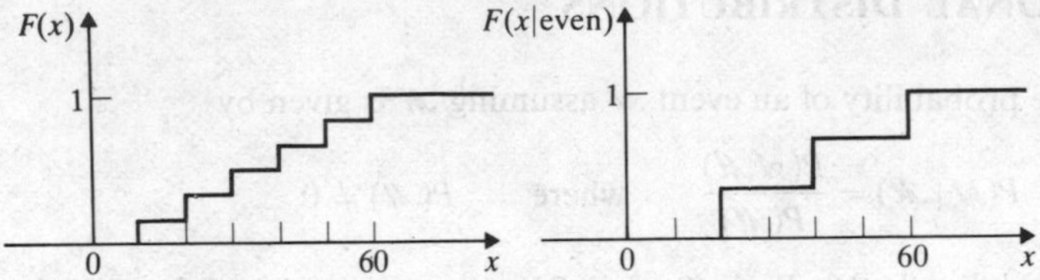

Figure 4-11

Finally, if $x < 20$, then $\{\mathbf{x} \le x, \mathcal{M}\} = \{\emptyset\}$. Hence,

$$F(x \mid \mathcal{M}) = 0 \qquad x < 20$$

To find $F(x \mid \mathcal{M})$ we must, in general, know the underlying experiment. However, if $\mathcal{M}$ is an event that can be expressed in terms of the RV $\mathbf{x}$, then, for the determination of $F(x \mid \mathcal{M})$, knowledge of $F(x)$ is sufficient. The following two cases are important illustrations.

I. We wish to find the conditional distribution of a RV $\mathbf{x}$ assuming that $\mathbf{x} \le a$ where a is number such that $F(a) \ne 0$. This is a special case of (4-41) with

$$\mathcal{M} = \{\mathbf{x} \le a\}$$

Thus, our problem is to find

$$F(x \mid \mathbf{x} \le a) = P\{\mathbf{x} \le x \mid \mathbf{x} \le a\} = \frac{P\{\mathbf{x} \le x, \mathbf{x} \le a\}}{P\{\mathbf{x} \le a\}}$$

If $x \ge a$, then $\{\mathbf{x} \le x, \mathbf{x} \le a\} = \{\mathbf{x} \le a\}$. Hence (Fig. 4-12)

$$F(x \mid \mathbf{x} \le a) = \frac{P\{\mathbf{x} \le a\}}{P\{\mathbf{x} \le a\}} = 1 \qquad x \ge a$$

If $x < a$, then $\{\mathbf{x} \le x, \mathbf{x} \le a\} = \{\mathbf{x} \le x\}$. Hence

$$F(x \mid \mathbf{x} \le a) = \frac{P\{\mathbf{x} \le x\}}{P\{\mathbf{x} \le a\}} = \frac{F(x)}{F(a)} \qquad x < a$$

Differentiating $F(x \mid \mathbf{x} \le a)$ with respect to x, we obtain the corresponding density: Since $F'(x) = f(x)$, the above yields

$$f(x \mid \mathbf{x} \le a) = \frac{f(x)}{F(a)} = \frac{f(x)}{\int_{-\infty}^{a} f(x)\,dx} \qquad \text{for} \qquad x < a \tag{4-45}$$

and it is zero for $x \ge a$.

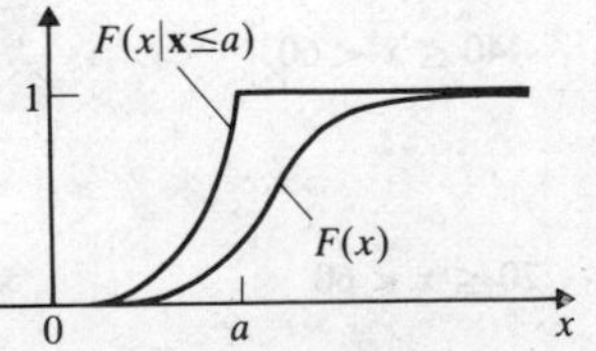

Figure 4-12

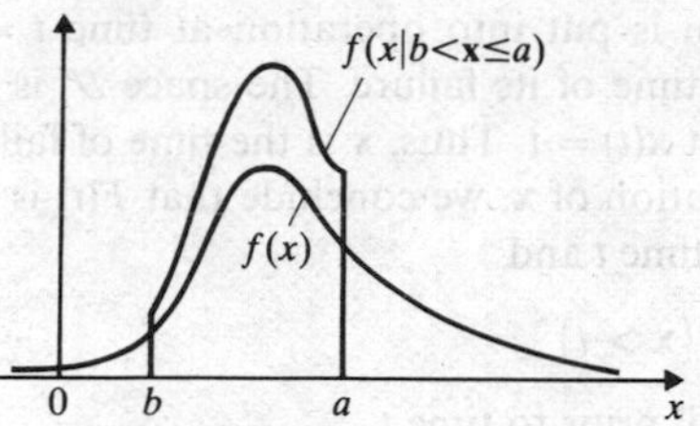

Figure 4-13

II. Suppose now that $\mathscr{M} = \{b < \mathbf{x} \le a\}$. In this case, (4-41) yields

$$F(x \mid b < \mathbf{x} \le a) = \frac{P\{\mathbf{x} \le x, b < \mathbf{x} \le a\}}{P\{b < \mathbf{x} \le a\}}$$

If $x \ge a$, then $\{\mathbf{x} \le x, b < \mathbf{x} \le a\} = \{b < \mathbf{x} \le a\}$. Hence

$$F(x \mid b < \mathbf{x} \le a) = \frac{F(a) - F(b)}{F(a) - F(b)} = 1 \qquad x \ge a$$

If $b \le x < a$, then $\{\mathbf{x} \le x, b < \mathbf{x} \le a\} = \{b < \mathbf{x} \le x\}$. Hence

$$F(x \mid b < \mathbf{x} \le a) = \frac{F(x) - F(b)}{F(a) - F(b)} \qquad b \le x < a$$

Finally, if $x < b$; then $\{\mathbf{x} \le x, b < \mathbf{x} \le a\} = \{\emptyset\}$. Hence

$$F(x \mid b < \mathbf{x} \le a) = 0 \qquad x < b$$

The corresponding density is given by

$$f(x \mid b < \mathbf{x} \le a) = \frac{f(x)}{F(a) - F(b)} \quad \text{for} \quad b \le x < a \tag{4-46}$$

and it is zero otherwise (Fig. 4-13).

Example 4-13 We shall determine the conditional density $f(x \mid |\mathbf{x} - \eta| \le k\sigma)$ of an $N(\eta; \sigma)$ RV. Since

$$P\{|\mathbf{x} - \eta| \le k\sigma\} = P\{\eta - k\sigma \le \mathbf{x} \le \eta + k\sigma\} = \mathbb{G}(k) - \mathbb{G}(-k) = 2\mathbb{G}(k) - 1$$

we conclude from (4-46) that

$$f(x \mid |\mathbf{x} - \eta| \le k\sigma) = \frac{1}{2\mathbb{G}(k) - 1} \frac{e^{-(x-\eta)^2/2\sigma^2}}{\sigma\sqrt{2\pi}}$$

for $\mathbf{x}$ between $\eta - k\sigma$ and $\eta + k\sigma$ and zero otherwise.

Frequency interpretation In a sequence of n trials, we reject all outcomes ζ such that $\mathbf{x}(\zeta) \le b$ or $\mathbf{x}(\zeta) > a$. In the subsequence of the remaining trials, $F(x \mid b < \mathbf{x} \le a)$ has the same frequency interpretation as $F(x)$ [see (4-2)].

System failure A randomly selected system is put into operation at time $t = 0$. We consider as experimental outcome the time of its failure. The space $\mathscr{S}$ is the positive t axis. We define an RV $\mathbf{x}$ such that $\mathbf{x}(t) = t$. Thus, $\mathbf{x}$ is the time of failure of the system. Denoting by $F(x)$ the distribution of $\mathbf{x}$, we conclude that $F(t)$ is the probability that the system will fail prior to time t and

$$1 - F(t) = P\{\mathbf{x} > t\}$$

is the probability that the system will not fail prior to time t.

With $\mathscr{M} = \{\mathbf{x} > t\}$, it follows from (4-41) that

$$F(x \mid \mathbf{x} > t) = \frac{P\{\mathbf{x} \le x, \mathbf{x} > t\}}{P\{\mathbf{x} > t\}} = \frac{F(x) - F(t)}{1 - F(t)} \tag{4-47}$$

for $x \ge t$ and zero otherwise.

Differentiating with respect to x, we conclude that the corresponding density is given by

$$f(x \mid \mathbf{x} > t) = \frac{F'(x)}{1 - F(t)} = \frac{f(x)}{\int_t^\infty f(x)\, dx} \qquad \text{for} \qquad x > t \tag{4-48}$$

and it is zero for $x < t$.

We note that the probability that the system will fail in the time interval $(x, x + dx)$ assuming that it did not fail prior to time t equals $f(x \mid \mathbf{x} > t)\, dx$.

Example 4-14 If

$$f(x) = ce^{-cx}U(x) \qquad F(x) = (1 - e^{-cx})U(x)$$

then

$$f(x \mid \mathbf{x} > t) = \frac{ce^{-cx}}{ce^{-ct}} = e^{-c(x-t)} = f(x - t) \qquad x > t \tag{4-49}$$

Conditional failure rate The conditional failure rate of a system is a function $\beta(t)$ defined as follows:

The product $\beta(t)\, dt$ equals the probability that the system fails in the interval $(t, t + dt)$ assuming that it did not fail prior to time t.

From the definition it follows that

$$\beta(t) = f(t \mid \mathbf{x} > t) \tag{4-50}$$

Hence

$$\beta(t) = \frac{F'(t)}{1 - F(t)} \tag{4-51}$$

Example 4-15(*a*) If $f(x) = ce^{-cx}U(x)$ is an exponential density, then

$$\beta(t) = \frac{ce^{-ct}}{1 - (1 - e^{-ct})} = c \tag{4-52}$$

(*b*) If $f(x)$ is the *gamma density*

$$f(x) = c^2 x e^{-cx} U(x) \qquad F(x) = (1 - cxe^{-cx} - e^{-cx})U(x) \tag{4-53}$$

then

$$\beta(t) = \frac{c^2 te^{-ct}}{cte^{-ct} + e^{-ct}} = \frac{c^2 t}{1 + ct}$$

Equation (4-51) can be used to express $F(x)$ in terms of $\beta(t)$. Since $F(0) = 0$, we conclude, integrating, that

$$-\ln\,[1 - F(x)] = \int_0^x \beta(t)\,dt$$

Hence

$$F(x) = 1 - \exp\left[-\int_0^x \beta(t)\,dt\right] \qquad f(x) = \beta(x)\exp\left[-\int_0^x \beta(t)\,dt\right] \tag{4-54}$$

We note that the function $\beta(t)$ equals the value of the conditional density $f(x \mid \mathbf{x} \geq t)$ for $x = t$. However, $\beta(t)$ *is not a density function* because its area does not equal 1. In fact, since

$$F(\infty) = 1 - \exp\left[-\int_0^\infty \beta(t)\,dt\right] = 1$$

we conclude that

$$\int_0^\infty \beta(t)\,dt = \infty \tag{4-55}$$

Conversely, if $\beta(t)$ is a nonnegative function satisfying (4-55), then the function $F(x)$ obtained from (4-54) is a distribution function because it increases as x increases and it equals 1 for $x = \infty$ [see (4-55)].

Example 4-16 Suppose that $f(x)$ is such that

$$f(x \mid \mathbf{x} \geq t) = f(x - t) \qquad x \geq t \tag{4-56}$$

In this case

$$\beta(t) = f(t \mid \mathbf{x} \geq t) = f(t - t) = f(0)$$

Thus, $\beta(t) = c = \text{constant}$, and (4-54) yields $f(x) = ce^{-cx}$.

We note that if $\beta(t)$ is constant, or, equivalently, if $\mathbf{x}$ has an exponential density, then the probability that the system fails in the interval $(t, t + dt)$, assuming that it did not fail prior to time t, is independent of t. In other words, the past has no effect on the statistics of the future. For this reason, (4-56) is called *memoryless property*.

4-5 TOTAL PROBABILITY AND BAYES' THEOREM

We shall now extend the results of Sec. 2-3 to random variables.

1. Setting $\mathscr{B} = \{\mathbf{x} \le x\}$ in (2-36), we obtain

$$P\{\mathbf{x} \le x\} = P\{\mathbf{x} \le x \mid \mathscr{A}_1\}P(\mathscr{A}_1) + \cdots + P\{\mathbf{x} \le x \mid \mathscr{A}_n\}P(\mathscr{A}_n)$$

Hence [see (4-41) and (4-44)]

$$F(x) = F(x \mid \mathscr{A}_1)P(\mathscr{A}_1) + \cdots + F(x \mid \mathscr{A}_n)P(\mathscr{A}_n) \tag{4-57}$$

$$f(x) = f(x \mid \mathscr{A}_1)P(\mathscr{A}_1) + \cdots + f(x \mid \mathscr{A}_n)P(\mathscr{A}_n) \tag{4-58}$$

In the above, the events $\mathscr{A}_1, \ldots, \mathscr{A}_n$ form a partition of $\mathscr{S}$.

Example 4-17 Suppose that the RV $\mathbf{x}$ is such that $f(x \mid \mathscr{M})$ is $N(\eta_1; \sigma_1)$ and $f(x \mid \bar{\mathscr{M}})$ is $N(\eta_2; \sigma_2)$ as in Fig. 4-14. Clearly, the events $\mathscr{M}$ and $\bar{\mathscr{M}}$ form a partition of $\mathscr{S}$. Setting $\mathscr{A}_1 = \mathscr{M}$ and $\mathscr{A}_2 = \bar{\mathscr{M}}$ in (4-58), we conclude that

$$f(x) = pf(x \mid \mathscr{M}) + (1-p)f(x \mid \bar{\mathscr{M}}) = \frac{p}{\sigma_1}\, \mathfrak{g}\left(\frac{x-\eta_1}{\sigma_1}\right) + \frac{1-p}{\sigma_2}\, \mathfrak{g}\left(\frac{x-\eta_2}{\sigma_2}\right)$$

where $p = P(\mathscr{M})$. This result is used in *hypothesis testing*.

2. From the identity

$$P(\mathscr{A} \mid \mathscr{B}) = \frac{P(\mathscr{B} \mid \mathscr{A})P(\mathscr{A})}{P(\mathscr{B})} \tag{4-59}$$

[see (2-38)] it follows that

$$P(\mathscr{A} \mid \mathbf{x} \le x) = \frac{P\{\mathbf{x} \le x \mid \mathscr{A}\}}{P\{\mathbf{x} \le x\}} P(\mathscr{A}) = \frac{F(x \mid \mathscr{A})}{F(x)} P(\mathscr{A}) \tag{4-60}$$

3. Setting $\mathscr{B} = \{x_1 < \mathbf{x} \le x_2\}$ in (4-59), we conclude with (4-43) that

$$P\{\mathscr{A} \mid x_1 < \mathbf{x} \le x_2\} = \frac{P\{x_1 < \mathbf{x} \le x_2 \mid \mathscr{A}\}}{P\{x_1 < \mathbf{x} \le x_2\}} P(\mathscr{A}) = \frac{F(x_2 \mid \mathscr{A}) - F(x_1 \mid \mathscr{A})}{F(x_2) - F(x_1)} P(\mathscr{A}) \tag{4-61}$$

4. The conditional probability $P(\mathscr{A} \mid \mathbf{x} = x)$ of the event $\mathscr{A}$ assuming $\mathbf{x} = x$ cannot be defined as in (2-29) because, in general, $P\{\mathbf{x} = x\} = 0$. We shall define

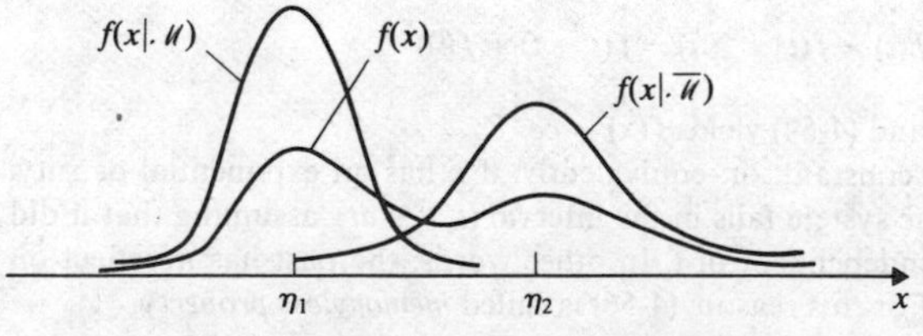

Figure 4-14

it as a limit

$$P\{\mathscr{A} \mid \mathbf{x} = x\} = \lim_{\Delta x \to 0} P\{\mathscr{A} \mid x < \mathbf{x} \le x + \Delta x\} \tag{4-62}$$

With $x_1 = x$, $x_2 = x + \Delta x$, we conclude from the above and (4-61) that

$$P\{\mathscr{A} \mid \mathbf{x} = x\} = \frac{f(x \mid \mathscr{A})}{f(x)} P(\mathscr{A}) \tag{4-63}$$

Total probability theorem As we know [see (4-42)]

$$F(\infty \mid \mathscr{A}) = \int_{-\infty}^{\infty} f(x \mid \mathscr{A})\, dx = 1$$

Multiplying (4-63) by $f(x)$ and integrating, we obtain

$$\int_{-\infty}^{\infty} P(\mathscr{A} \mid \mathbf{x} = x) f(x)\, dx = P(\mathscr{A}) \tag{4-64}$$

This is the continuous version of the total probability theorem (2-36).

Example 4-18 Suppose that the probability of heads in a coin-tossing experiment $\mathscr{S}$ is not a number, but an RV $\mathbf{p}$ with density $f(p)$ defined in some space $\mathscr{S}_c$. The experiment of the tossing of a randomly selected coin is a cartesian product $\mathscr{S}_c \times \mathscr{S}$. In this experiment, the event $\mathscr{H} = \{\text{head}\}$ consists of all pairs of the form $\zeta_c h$ where ζ_c is any element of $\mathscr{S}_c$ and h is the element heads of the space $\mathscr{S} = \{h, t\}$. We shall show that

$$P(\mathscr{H}) = \int_0^1 pf(p)\, dp \tag{4-65}$$

Proof The conditional probability of $\mathscr{H}$ assuming $\mathbf{p} = p$ is the probability of heads if the coin with $\mathbf{p} = p$ is tossed. In other words

$$P\{\mathscr{H} \mid \mathbf{p} = p\} = p \tag{4-66}$$

Inserting into (4-64), we obtain (4-65) because $f(p) = 0$ outside the interval (0, 1).

Bayes' theorem From (4-63) and (4-64) it follows that

$$f(x \mid \mathscr{A}) = \frac{P(\mathscr{A} \mid \mathbf{x} = x)}{P(\mathscr{A})} f(x) = \frac{P(\mathscr{A} \mid \mathbf{x} = x) f(x)}{\int_{-\infty}^{\infty} P(\mathscr{A} \mid \mathbf{x} = x) f(x)\, dx} \tag{4-67}$$

This is the continuous version of Bayes' theorem (2-39).

We discuss next an important application of (4-67) involving the use of past observations in the determination of unknown probabilities. This is a continuation of the topic introduced at the end of Sec. 3-4. As in that case, we state the problem in terms of the coin experiment. We are given the following information: A coin is tossed n times and "heads" shows k times. What conclusion can we draw about the probability that heads will show in subsequent tossings of the given coin?

This problem can be given two different interpretations.

1. The probability of heads is an unknown number p.

   As we noted earlier, any conclusions based on this interpretation are inductive, therefore, they cannot be explained within the framework of a deductive theory. We stress, however, that any method, inductive or deductive, must lead to the conclusion that, if n is large, then p is nearly equal the ratio k/n as in (1-1).

2. The probability of heads is an RV $\mathbf{p}$.

   Suppose that the density $f(p)$ of this RV is known somehow from past observations. We shall show that this knowledge can be updated on the basis of the given information.

   We assume as in Example 4-18, that $\mathbf{p}$ is defined in some space $\mathscr{S}_c$. The experiment of the n tossings of a randomly selected coin is the product $\mathscr{S}_c \times \mathscr{S}^n$ where $\mathscr{S}^n$ is the experiment of the n tossings of a specific coin. The elements of the space $\mathscr{S}_c \times \mathscr{S}^n$ are of the form $\zeta_c th \ldots t$, and $\mathbf{p}$ is an RV with density $f(p)$ if we assume that $\mathscr{S}_c$ and $\mathscr{S}^n$ are independent experiments. This means that the probability of future tossings with a specific coin is not affected by past outcomes and it leads to the conclusion that the conditional probability of the event

$$\mathscr{A} = \{k \text{ heads in a specific order}\}$$

assuming that the probability of the selected coin equals p, is given by

$$P(\mathscr{A} \mid \mathbf{p} = p) = p^k q^{n-k} \tag{4-68}$$

Inserting into (4-67), we obtain

$$f(p \mid \mathscr{A}) = \frac{p^k(1-p)^{n-k} f(p)}{\int_0^1 p^k(1-p)^{n-k} f(p)\, dp} \tag{4-69}$$

In the above, $f(p \mid \mathscr{A})$ is the *a posteriori* (updated) and $f(p)$ the *a priori* density of $\mathbf{p}$.

Large n For large n, the function

$$\varphi(p) = p^k(1-p)^{n-k}$$

has a sharp maximum at $p = k/n$. Therefore, if the a priori density $f(p)$ is smooth, then the product $f(p)\varphi(p)$ is highly concentrated near the point k/n as in Fig. 4-15*a*. However, if $f(p)$ has a sharp peak at $p = 0.5$, that is, if the coins are reasonably fair, then, for moderate values of n, the product $f(p)\varphi(p)$ has two maxima: one near k/n and the other near 0.5 as in Fig. 4-15*b*. As n increases, the sharpness of $\varphi(p)$ prevails and the resulting a posteriori density $f(p \mid \mathscr{A})$ is maximum near k/n as in Fig. 4-15*c*. As $n \to \infty$, $f(p \mid \mathscr{A})$ tends to $\delta(p - k/n)$.

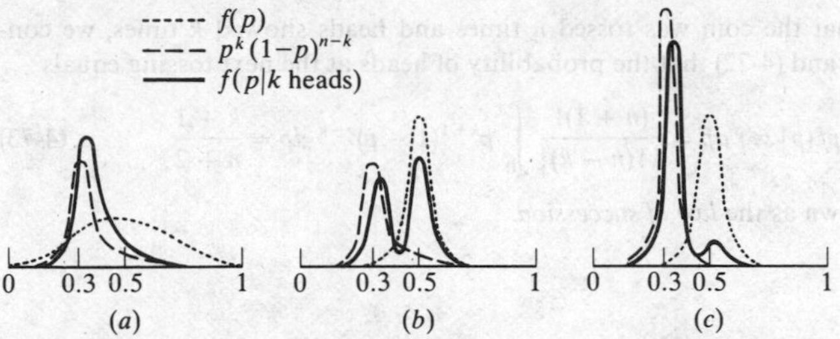

Figure 4-15

Summary If $\mathbf{p}$ is an RV with a priori density $f(p)$, then:

1*a*. The probability that $\mathbf{p}$ is between p_1 and p_2 equals

$$\int_{p_1}^{p_2} f(p)\, dp$$

1*b*. In a single tossing of the randomly selected coin the probability of heads equals [see (4-65)]

$$\int_0^1 pf(p)\, dp$$

We toss the selected coin n times and we observe k heads. On the basis of this evidence, we conclude that:

2*a*. The probability that $\mathbf{p}$ is between p_1 and p_2 equals

$$\int_{p_1}^{p_2} f(p \mid \mathscr{A})\, dp$$

2*b*. At the next tossing, the probability of heads equals

$$\int_0^1 pf(p \mid \mathscr{A})\, dp \tag{4-70}$$

where $f(p \mid \mathscr{A})$ is the a posteriori density (4-69). This follows from (4-65) if $f(p)$ is replaced by $f(p \mid \mathscr{A})$.

Example 4-19 Suppose that the RV $\mathbf{p}$ is uniform in the interval (0, 1). Setting $f(p) = 1$ in (4-69) and using the identity

$$\int_0^1 p^k(1-p)^{n-k}\, dp = \frac{k!\,(n-k)!}{(n+1)!} \tag{4-71}$$

we conclude that

$$f(p \mid \mathscr{A}) = \frac{(n+1)!}{k!\,(n-k)!}\, p^k(1-p)^{n-k} \tag{4-72}$$

This function is known as *beta density*.

Assuming that the coin was tossed n times and heads showed k times, we conclude from (4-70) and (4-72) that the probability of heads at the next tossing equals

$$\int_0^1 pf(p\,|\,\mathscr{A})\,dp = \frac{(n+1)!}{k!\,(n-k)!}\int_0^1 p^{k+1}(1-p)^{n-k}\,dp = \frac{k+1}{n+2} \tag{4-73}$$

This result is known as the *law of succession.*

PROBLEMS

4-1 The RV $\mathbf{x}$ is $N(10; 1)$. Find $f(x\,|\,(\mathbf{x}-10)^2 < 4)$.

4-2 Find $f(x)$ if $F(x) = (1 - e^{-\alpha x})U(x - c)$.

4-3 If $\mathbf{x}$ is $N(0, 2)$ find (*a*) $P\{1 \le \mathbf{x} \le 2\}$ and (*b*) $P\{1 \le \mathbf{x} \le 2\,|\,\mathbf{x} \ge 1\}$.

4-4 The space $\mathscr{S}$ consists of all points t_i in the interval (0, 1) and $P\{0 \le t_i \le y\} = y$ for every $y \le 1$. The function $G(x)$ is increasing from $G(-\infty) = 0$ to $G(\infty) = 1$, hence, it has an inverse $G^{(-1)}(y) = H(y)$. The RV $\mathbf{x}$ is such that $\mathbf{x}(t_i) = H(t_i)$. Show that $F_x(x) = G(x)$.

4-5 If $\mathbf{x}$ is $N(1000; 20)$ find (*a*) $P\{\mathbf{x} < 1024\}$, (*b*) $P\{\mathbf{x} < 1024\,|\,\mathbf{x} > 961\}$, and (*c*) $P\{31 < \sqrt{\mathbf{x}} \le 32\}$.

4-6 A fair coin is tossed three times and the RV $\mathbf{x}$ equals the total number of heads. Find and sketch $F_x(x)$ and $f_x(x)$.

4-7 A fair coin is tossed 900 times and the RV $\mathbf{x}$ equals the total number of heads. (*a*) Find $f_x(x)$: 1; exactly 2; approximately using (4-34). (*b*) Find $P\{435 \le \mathbf{x} \le 460\}$.

4-8 Show that, if $a \le \mathbf{x}(\zeta) \le b$ for every $\zeta \in \mathscr{S}$, then $F(x) = 1$ for $x > b$ and $F(x) = 0$ for $x < a$.

4-9 Show that if $\mathbf{x}(\zeta) \le \mathbf{y}(\zeta)$ for every $\zeta \in \mathscr{S}$, then $F_x(w) \ge F_y(w)$ for every w.

4-10 Show that if $\beta(t) = f(t\,|\,\mathbf{x} > t)$ is the conditional failure rate of the RV $\mathbf{x}$ and $\beta(t) = kt$, then $f(x)$ is a Rayleigh density.

4-11 Show that if $\beta_x(t) = f_x(t\,|\,\mathbf{x} > t)$, $\beta_y(t) = f_y(t\,|\,\mathbf{y} > t)$ and $\beta_x(t) = k\beta_y(t)$ then $1 - F_x(x) = [1 - F_y(x)]^k$.

4-12 Show that $P(\mathscr{A}) = P(\mathscr{A}\,|\,\mathbf{x} \le x)F(x) + P(\mathscr{A}\,|\,\mathbf{x} > x)[1 - F(x)]$.

4-13 Show that

$$F_x(x\,|\,\mathscr{A}) = \frac{P(\mathscr{A}\,|\,\mathbf{x} \le x)F_x(x)}{P(\mathscr{A})}$$

4-14 Show that if $P(\mathscr{A}\,|\,\mathbf{x} = x) = P(\mathscr{B}\,|\,\mathbf{x} = x)$ for every $x \le x_0$, then $P(\mathscr{A}\,|\,\mathbf{x} \le x_0) = P(\mathscr{B}\,|\,\mathbf{x} \le x_0)$.

Hint: Replace in (4-64) $P(\mathscr{A})$ and $f(x)$ by $P(\mathscr{A}\,|\,\mathbf{x} \le x_0)$ and $f(x\,|\,\mathbf{x} \le x_0)$.

4-15 The probability of *heads* of a random coin is an RV $\mathbf{p}$ uniform in the interval (0, 1). (*a*) Find $P\{0.3 \le \mathbf{p} \le 0.7\}$. (*b*) The coin is tossed 10 times and *heads* shows 6 times. Find the a posteriori probability that $\mathbf{p}$ is between 0.3 and 0.7.

4-16 The probability of *heads* of a random coin is an RV $\mathbf{p}$ uniform in the interval (0.4, 0.6). (*a*) Find the probability that at the next tossing of the coin *heads* will show. (*b*) The coin is tossed 100 times and *heads* shows 60 times. Find the probability that at the next tossing *heads* will show.

CHAPTER

FIVE

FUNCTIONS OF ONE RANDOM VARIABLE

5-1 THE RANDOM VARIABLE $g(\mathbf{x})$

Suppose that $\mathbf{x}$ is an RV and $g(x)$ is a function of the real variable x. The expression

$$\mathbf{y} = g(\mathbf{x})$$

is a new RV defined as follows: For a given ζ, $\mathbf{x}(\zeta)$ is a number and $g[\mathbf{x}(\zeta)]$ is another number specified in terms of $\mathbf{x}(\zeta)$ and $g(x)$. This number is the value $\mathbf{y}(\zeta) = g[\mathbf{x}(\zeta)]$ assigned to the RV $\mathbf{y}$.

The distribution function $F_y(y)$ of the RV so formed is the probability of the event $\{\mathbf{y} \le y\}$ consisting of all outcomes ζ such that $\mathbf{y}(\zeta) = g[\mathbf{x}(\zeta)] \le y$. Thus

$$F_y(y) = P\{\mathbf{y} \le y\} = P\{g(\mathbf{x}) \le y\} \tag{5-1}$$

For a specific y, the values of x such that $g(x) \le y$ form a set on the x axis denoted by $\mathscr{I}_y$. Clearly, $g[\mathbf{x}(\zeta)] \le y$ if $\mathbf{x}(\zeta)$ is a number in the set $\mathscr{I}_y$. Hence

$$F_y(y) = P\{\mathbf{x} \in \mathscr{I}_y\} \tag{5-2}$$

The above leads to the conclusion that for $g(\mathbf{x})$ to be an RV, the function $g(x)$ must have the following properties:

1. Its domain must include the range of the RV $\mathbf{x}$.
2. It must be a *Baire* function, i.e., for every y, the set $\mathscr{I}_y$ such that $g(x) \le y$ must consist of the union and intersection of a countable number of intervals. Only then $\{\mathbf{y} \le y\}$ is an event.
3. The events $\{g(\mathbf{x}) = \pm\infty\}$ must have zero probability.

5-2 THE DISTRIBUTION OF $g(\mathbf{x})$

We shall express the distribution function $F_y(y)$ of the RV $\mathbf{y} = g(\mathbf{x})$ in terms of the distribution function $F_x(x)$ of the RV $\mathbf{x}$ and the function $g(x)$. For this purpose, we must determine the set $\mathscr{I}_y$ of the x axis such that $g(x) \le y$, and the probability that $\mathbf{x}$ is in this set. The method will be illustrated with several examples. Unless otherwise stated, it will be assumed that $F_x(x)$ is continuous.

1. We start with the function $g(x)$ in Fig. 5-1. As we see from the figure, $g(x)$ is between a and b for any x. This leads to the conclusion that if $y \ge b$, then $g(x) \le y$ for every x, hence, $P\{\mathbf{y} \le y\} = 1$; if $y < a$ then there is no x such that $g(x) \le y$, hence, $P\{\mathbf{y} \le y\} = 0$. Thus

$$F_y(y) = \begin{cases} 1 & y \ge b \\ 0 & y < a \end{cases}$$

With x_1 and $y_1 = g(x_1)$ as shown, we observe that $g(x) \le y_1$ for $x \le x_1$. Hence

$$F_y(y_1) = P\{\mathbf{x} \le x_1\} = F_x(x_1)$$

We finally note that

$$g(x) \le y_2 \qquad \text{if} \qquad x \le x_2' \qquad \text{or if} \qquad x_2'' \le x \le x_2'''$$

Hence

$$F_y(y_2) = P\{\mathbf{x} \le x_2'\} + P\{x_2'' \le \mathbf{x} \le x_2'''\} = F_x(x_2') + F_x(x_2''') - F_x(x_2'')$$

because the events $\{\mathbf{x} \le x_2'\}$ and $\{x_2'' \le \mathbf{x} \le x_2'''\}$ are mutually exclusive.

Example 5-1

$$\mathbf{y} = a\mathbf{x} + b$$

To find $F_y(y)$, we must find the values of x such that $ax + b \le y$.

(*a*) If $a > 0$ then $ax + b \le y$ for $x \le (y - b)/a$ (Fig. 5-2*a*). Hence

$$F_y(y) = P\left\{\mathbf{x} \le \frac{y-b}{a}\right\} = F_x\left(\frac{y-b}{a}\right) \qquad a > 0$$

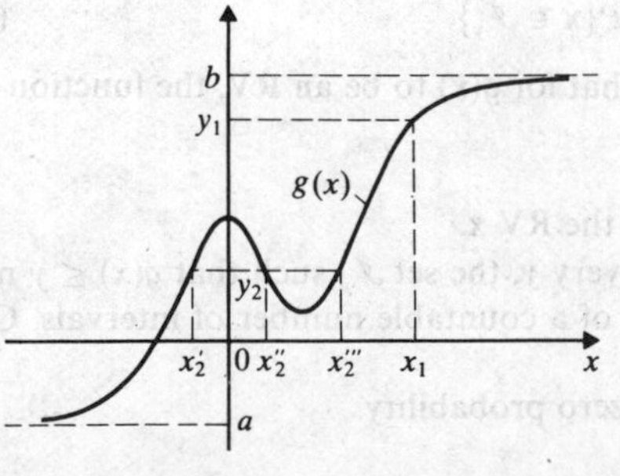

Figure 5-1

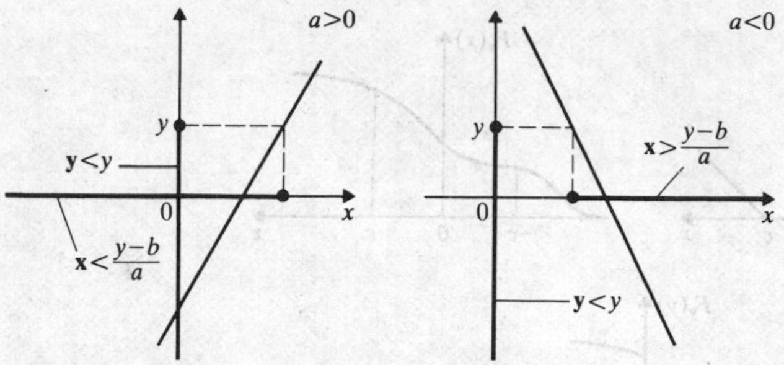

Figure 5-2

(*b*) If $a < 0$ then $ax + b \le y$ for $x > (y - b)/a$ (Fig. 5-2*b*). Hence

$$F_y(y) = P\left\{\mathbf{x} \ge \frac{y-b}{a}\right\} = 1 - F_x\left(\frac{y-b}{a}\right) \qquad a < 0$$

Example 5-2

$$\mathbf{y} = \mathbf{x}^2$$

If $y \ge 0$, then $x^2 \le y$ for $-\sqrt{y} \le x \le \sqrt{y}$ (Fig. 5-3*a*). Hence

$$F_y(y) = P\{-\sqrt{y} \le \mathbf{x} \le \sqrt{y}\} = F_x(\sqrt{y}) - F_x(-\sqrt{y}) \qquad y > 0$$

If $y < 0$, then there are no values of x such that $x^2 < y$. Hence

$$F_y(y) = P\{\emptyset\} = 0 \qquad y < 0$$

SPECIAL CASE If **x** is uniform in the interval $(-1, 1)$, then

$$F_x(x) = \frac{1}{2} + \frac{x}{2} \qquad |x| < \frac{1}{2}$$

(Fig. 5-3*b*). Hence

$$F_y(y) = \sqrt{y} \quad \text{for} \quad 0 \le y \le 1 \quad \text{and} \quad F_y(y) = \begin{cases} 1 & y > 1 \\ 0 & y < 0 \end{cases}$$

2. Suppose now that the function $g(x)$ is constant in an interval (x_0, x_1)

$$g(x) = y_1 \qquad x_0 < x \le x_1$$

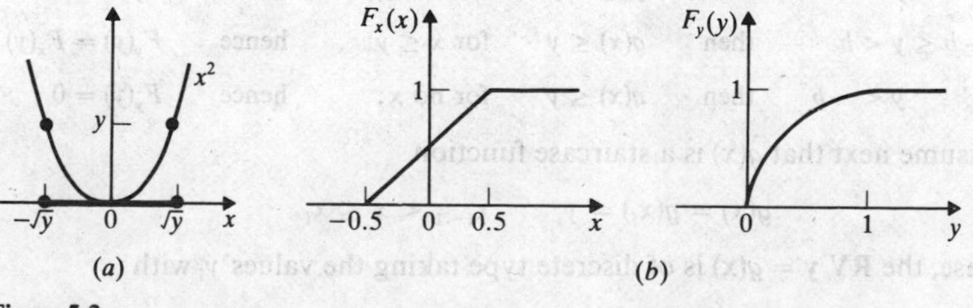

Figure 5-3

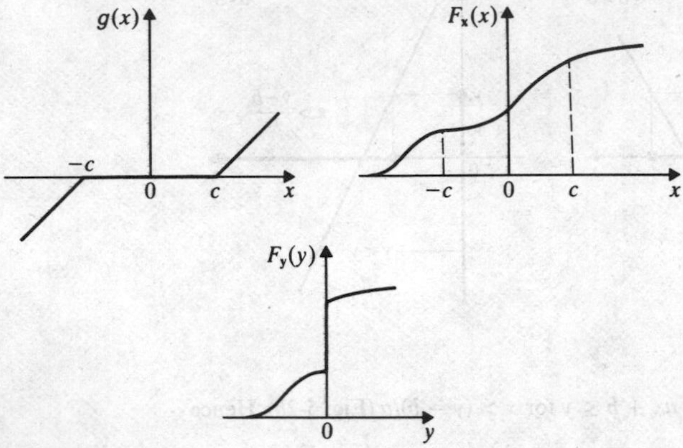

Figure 5-4

In this case

$$P\{\mathbf{y} = y_1\} = P\{x_0 < \mathbf{x} \le x_1\} = F_x(x_1) - F_x(x_0) \tag{5-3}$$

Hence, $F_y(y)$ is discontinuous at $\mathbf{y} = y_1$ and its discontinuity equals $F_x(x_1) - F_x(x_0)$.

Example 5-3 Consider the function (Fig. 5-4)

$$g(x) = 0 \qquad \text{for} \qquad -c \le x \le c \qquad \text{and} \qquad g(x) = \begin{cases} x - c & x > c \\ x + c & x < -c \end{cases}$$

In this case, $F_y(y)$ is discontinuous for $y = 0$ and its discontinuity equals $F_x(-c) - F_x(c)$. Furthermore

$$\text{If} \quad y \ge 0 \quad \text{then} \quad P\{\mathbf{y} \le y\} = P\{\mathbf{x} \le y + c\} = F_x(y + c)$$

$$\text{If} \quad y < 0 \quad \text{then} \quad P\{\mathbf{y} \le y\} = P\{\mathbf{x} \le y - c\} = F_x(y - c)$$

Example 5-4 (limiter) The curve $g(x)$ of Fig. 5-5 is constant for $x \le -b$ and $x \ge b$ and in the interval $(-b, b)$ it is a straight line. With $\mathbf{y} = g(\mathbf{x})$, it follows that $F_y(y)$ is discontinuous for $y = g(-b) = -b$ and $y = g(b) = b$ respectively. Furthermore

If	$y \ge b$	then	$g(x) \le y$	for every x;	hence	$F_y(y) = 1$
If	$-b \le y < b$	then	$g(x) \le y$	for $x \le y$;	hence	$F_y(y) = F_x(y)$
If	$y < -b$	then	$g(x) \le y$	for no x;	hence	$F_y(y) = 0$

3. We assume next that $g(x)$ is a staircase function

$$g(x) = g(x_i) = y_i \qquad x_{i-1} < x \le x_i$$

In this case, the RV $\mathbf{y} = g(\mathbf{x})$ is of discrete type taking the values y_i with

$$P\{\mathbf{y} = y_i\} = P\{x_{i-1} < \mathbf{x} \le x_i\} = F_x(x_i) - F_x(x_{i-1})$$

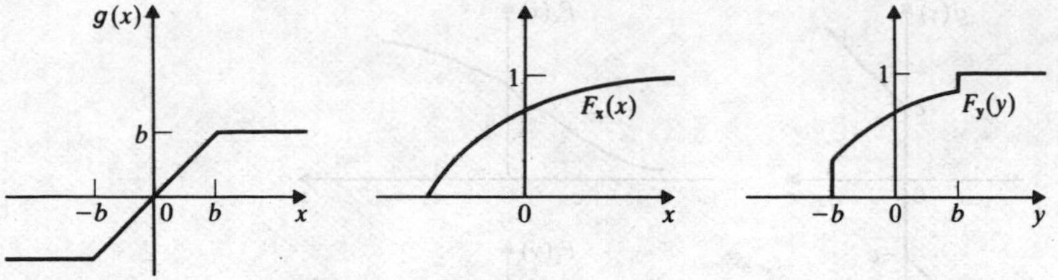

Figure 5-5

Example 5-5 (hard limiter) If

$$g(x) = \begin{cases} 1 & x > 0 \\ -1 & x \leq 0 \end{cases}$$

then $\mathbf{y}$ takes the values ± 1 with

$$P\{\mathbf{y} = -1\} = P\{\mathbf{x} \leq 0\} = F_x(0)$$
$$P\{\mathbf{y} = 1\} = P\{\mathbf{x} > 0\} = 1 - F_x(0)$$

Hence, $F_y(y)$ is a staircase function as in Fig. 5-6.

Example 5-6 (quantization) If

$$g(x) = ns \qquad (n-1)s < x \leq ns$$

then $\mathbf{y}$ takes the values $y_n = ns$ with

$$P\{\mathbf{y} = ns\} = P\{n-1)s < \mathbf{x} \leq ns\} = F_x(ns) - F_x(ns - s)$$

4. We assume, finally, that the function $g(x)$ is discontinuous at $x = x_0$ and such that

$$g(x) < g(x_0^-) \quad \text{for} \quad x < x_0 \qquad g(x) > g(x_0^+) \qquad \text{for} \qquad x > x_0$$

In this case, if y is between $g(x_0^-)$ and $g(x_0^+)$, then $g(x) < y$ for $x \leq x_0$. Hence

$$F_y(y) = P\{\mathbf{x} \leq x_0\} = F_x(x_0) \qquad g(x_0^-) \leq y \leq g(x_0^+)$$

Example 5-7 Suppose that

$$g(x) = \begin{cases} x + c & x \geq 0 \\ x - c & x < 0 \end{cases}$$

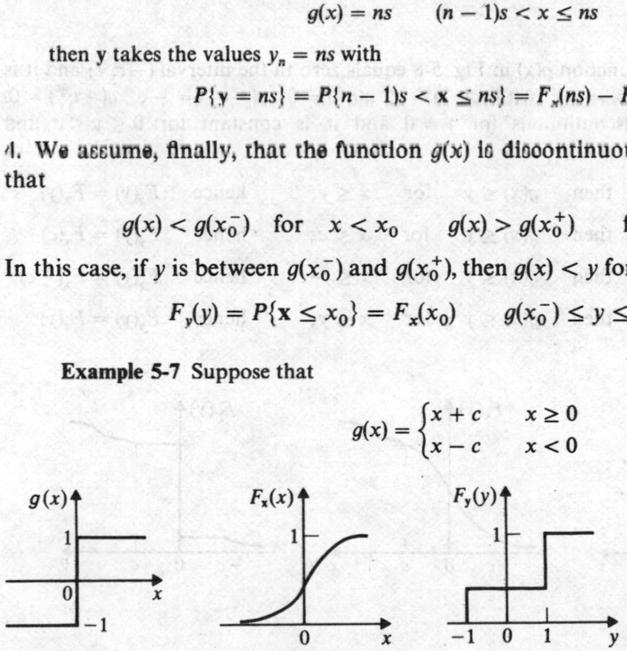

Figure 5-6

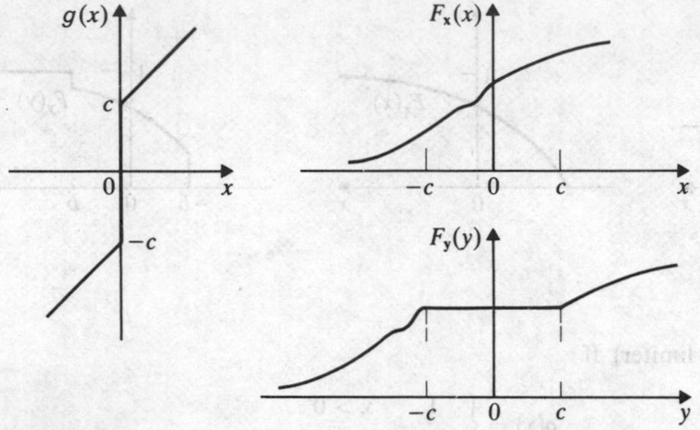

Figure 5-7

is discontinuous (Fig. 5-7). Thus, $g(x)$ is discontinuous for $x = 0$ with $g(0^-) = -c$ and $g(0^+) = c$. Hence, $F_y(y) = F_x(0)$ for $|y| \le c$. Furthermore

If $y \ge c$	then $g(x) \le y$	for $x \le y - c$;	hence	$F_y(y) = F_x(y - c)$
If $-c \le y \le c$	then $g(x) \le y$	for $x \le c$;	hence	$F_y(y) = F_x(0)$
If $y \le -c$	then $g(x) \le y$	for $x \le y + c$;	hence	$F_y(y) = F_x(y + c)$

Example 5-8 The function $g(x)$ in Fig. 5-8 equals zero in the interval $(-c, c)$ and it is discontinuous for $x = \pm c$ with $g(c^+) = c$, $g(c^-) = 0$, $g(-c^-) = -c$, $g(-c^+) = 0$. Hence, $F_y(y)$ is discontinuous for $y = 0$ and it is constant for $0 \le y \le c$ and $-c \le y \le 0$. Thus

If $y \ge c$	then $g(x) \le y$	for $x \le y$;	hence	$F_y(y) = F_x(y)$
If $0 \le y < c$	then $g(x) \le y$	for $x < c$;	hence	$F_y(y) = F_x(c)$
If $-c \le y < 0$	then $g(x) \le y$	for $x \le -c$;	hence	$F_y(y) = F_x(-c)$
If $y < -c$	then $g(x) \le y$	for $x \le y$;	hence	$F_y(y) = F_x(y)$

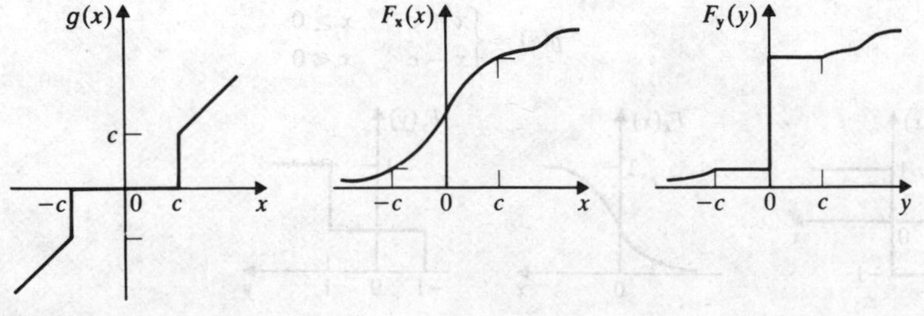

Figure 5-8

5. We now assume that the RV $\mathbf{x}$ is of discrete type taking the values x_k with probability p_k. In this case, the RV $\mathbf{y} = g(\mathbf{x})$ is also of discrete type taking the values $y_k = g(x_k)$.

If $y_k = g(x)$ for only one $x = x_k$, then

$$P\{\mathbf{y} = y_k\} = P\{\mathbf{x} = x_k\} = p_k$$

If, however, $y_k = g(x)$ for $x = x_k$ and $x = x_l$, then

$$P\{\mathbf{y} = y_k\} = P\{\mathbf{x} = x_k\} + P\{\mathbf{x} = x_l\} = p_k + p_l$$

Example 5-9

$$\mathbf{y} = \mathbf{x}^2$$

(*a*) If $\mathbf{x}$ takes the values 1, 2, ..., 6 with probability 1/6, then $\mathbf{y}$ takes the values $1^2, 2^2, \ldots, 6^2$ with probability 1/6.

(*b*) If, however, $\mathbf{x}$ takes the values $-2, -1, 0, 1, 2, 3$ with probability 1/6, then $\mathbf{y}$ takes the values 0, 1, 4, 9 with probabilities 1/6, 2/6, 2/6, 1/6 respectively.

Determination of $f_y(y)$

We wish to determine the density of $\mathbf{y} = g(\mathbf{x})$ in terms of the density of $\mathbf{x}$. Suppose, first, that the set $\mathscr{I}$ of the y axis is not in the range of the function $g(x)$, that is, that $g(x)$ is not a point of $\mathscr{I}$ for any x. In this case, the probability that $g(\mathbf{x})$ is in $\mathscr{I}$ equals zero. Hence, $f_y(y) = 0$ for $y \in \mathscr{I}$. It suffices therefore to consider the values of y such that for some x, $g(x) = y$.

Fundamental theorem To find $f_y(y)$ for a specific y, we solve the equation $y = g(x)$. Denoting its real roots by x_n

$$y = g(x_1) = \cdots = g(x_n) = \cdots \tag{5-4}$$

we shall show that

$$f_y(y) = \frac{f_x(x_1)}{|g'(x_1)|} + \cdots + \frac{f_x(x_n)}{|g'(x_n)|} + \cdots \tag{5-5}$$

where $g'(x)$ is the derivative of $g(x)$.

PROOF To avoid generalities, we assume that the equation $y = g(x)$ has three roots as in Fig. 5-9. As we know

$$f_y(y)\, dy = P\{y < \mathbf{y} \le y + dy\}$$

It suffices, therefore, to find the set of values x such that $y < g(x) \le y + dy$ and the probability that $\mathbf{x}$ is in this set. As we see from the figure, this set consists of the following three intervals

$$x_1 < x < x_1 + dx_1 \qquad x_2 + dx_2 < x < x_2 \qquad x_3 < x < x_3 + dx_3$$

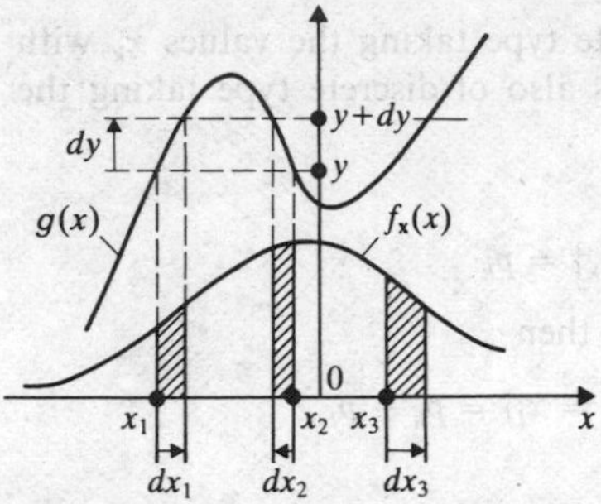

Figure 5-9

where $dx_1 > 0$, $dx_3 > 0$ but $dx_2 < 0$. From the above it follows that

$$P\{y < \mathbf{y} < y + dy\} = P\{x_1 < \mathbf{x} < x_1 + dx_1\} + P\{x_2 + dx_2 < \mathbf{x} < x_2\} + P\{x_3 < \mathbf{x} < x_3 + dx_3\}$$

The right side equals the shaded area in Fig. 5-9. Since

$$P\{x_1 < \mathbf{x} < x_1 + dx_1\} = f_x(x_1)\, dx_1 \qquad dx_1 = dy/g'(x_1)$$
$$P\{x_2 + dx_2 < \mathbf{x} < x_2\} = f_x(x_2)\,|dx_2| \qquad dx_2 = dy/g'(x_2)$$
$$P\{x_3 < \mathbf{x} < x_3 + dx_3\} = f_x(x_3)\, dx_3 \qquad dx_3 = dy/g'(x_3)$$

we conclude that

$$f_y(y)\, dy = \frac{f_x(x_1)}{g'(x_1)}\, dy + \frac{f_x(x_2)}{|g'(x_2)|}\, dy + \frac{f_x(x_3)}{g'(x_3)}\, dy$$

and (5-5) results.

We note, finally, that if $g(x) = y_1$ = constant for every x in the interval (x_0, x_1), then [see (5-3)] $F_y(y)$ is discontinuous for $y = y_1$, hence, $f_y(y)$ contains an impulse $\delta(y - y_1)$ of area $F_x(x_1) - F_x(x_0)$.

Conditional density The conditional density $f_y(y \mid \mathcal{M})$ of the RV $\mathbf{y} = g(\mathbf{x})$ assuming $\mathcal{M}$ is given by (5-5) if on the right side we replace the terms $f_x(x_i)$ by $f_x(x_i \mid \mathcal{M})$ (see, for example, Prob. 5-9).

Illustrations

We give next several applications of (5-2) and (5-5).

1. $$\mathbf{y} = a\mathbf{x} + b \qquad g'(x) = a$$

The equation $y = ax + b$ has a single solution $x = (y - b)/a$ for every y. Hence

$$f_y(y) = \frac{1}{|a|}\, f_x\!\left(\frac{y - b}{a}\right) \tag{5-6}$$

Special case If $\mathbf{x}$ is uniform in the interval (x_1, x_2), then $\mathbf{y}$ is uniform in the interval $(ax_1 + b, ax_2 + b)$.

Example 5-10 Suppose that the voltage $\mathbf{v}$ is an RV given by

$$\mathbf{v} = i(\mathbf{r} + r_0)$$

where $i = 0.01$ A and $r_0 = 1000\ \Omega$. If the resistance $\mathbf{r}$ is an RV uniform between 900 and 1100 Ω, then $\mathbf{v}$ is uniform between 19 and 21 V.

2. $$\mathbf{y} = \frac{1}{\mathbf{x}} \qquad g'(x) = -\frac{1}{x^2}$$

The equation $y = 1/x$ has a single solution $x = 1/y$. Hence

$$f_y(y) = \frac{1}{y^2} f_x\left(\frac{1}{y}\right) \tag{5-7}$$

Special case If $\mathbf{x}$ has a *Cauchy density* with parameter α

$$f_x(x) = \frac{\alpha/\pi}{x^2 + \alpha^2} \qquad \text{then} \qquad f_y(y) = \frac{1/\alpha\pi}{y^2 + 1/\alpha^2}$$

is also a Cauchy density with parameter $1/\alpha$.

Example 5-11 Suppose that the resistance $\mathbf{r}$ is uniform between 900 and 1100 Ω as in Fig. 5-10. We shall determine the density of the corresponding conductance

$$\mathbf{g} = 1/\mathbf{r}$$

Since $f_r(r) = 1/200$ S for r between 900 and 1100 it follows from (5-7) that

$$f_g(g) = \frac{1}{200g^2} \qquad \text{for} \qquad \frac{1}{1100} < g < \frac{1}{900}$$

and zero elsewhere.

3. $$\mathbf{y} = a\mathbf{x}^2 \qquad a > 0 \qquad g'(x) = 2ax$$

If $y < 0$, then the equation $y = ax^2$ has no real solutions, hence, $f_y(y) = 0$. If $y > 0$, then it has two solutions

$$x_1 = \sqrt{\frac{y}{a}} \qquad x_2 = -\sqrt{\frac{y}{a}}$$

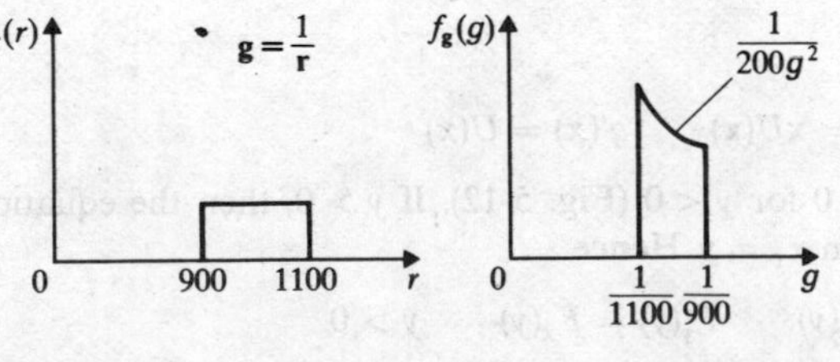

Figure 5-10

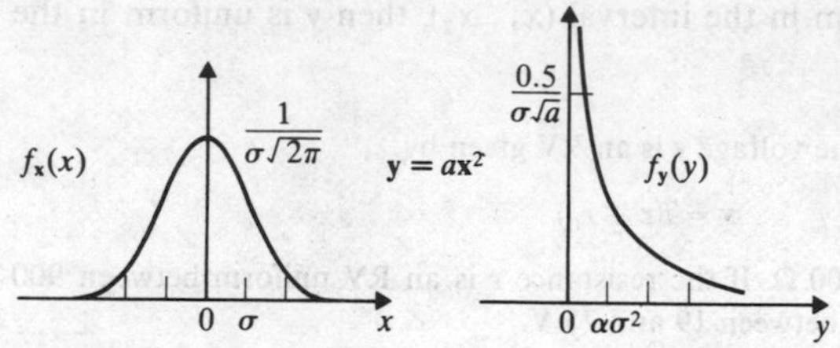

Figure 5-11

and (5-5) yields

$$f_y(y) = \frac{1}{2a\sqrt{y/a}}\left[f_x\left(\sqrt{\frac{y}{a}}\right) + f_x\left(-\sqrt{\frac{y}{a}}\right)\right] \qquad y > 0 \tag{5-8}$$

We note that $F_y(y) = 0$ for $y < 0$ and

$$F_y(y) = P\left\{-\sqrt{\frac{y}{a}} \le \mathbf{x} \le \sqrt{\frac{y}{a}}\right\} = F_x\left(\sqrt{\frac{y}{a}}\right) - F_x\left(-\sqrt{\frac{y}{a}}\right) \qquad y > 0$$

Special case Suppose that $\mathbf{x}$ is $N(0; \sigma)$

$$f_x(x) = \frac{1}{\sigma}\,\mathbb{g}\left(\frac{x}{\sigma}\right) \qquad F_x(x) = \mathbb{G}\left(\frac{x}{\sigma}\right)$$

(Fig. 5-11) where $\mathbb{g}(x)$ and $\mathbb{G}(x)$ are the normal curves introduced in Sec. 3-3. In this case, (5-8) yields

$$f_y(y) = \frac{1}{\sigma\sqrt{2\pi a y}}\, e^{-y/2a\sigma^2}\, U(y) \tag{5-9}$$

Example 5-12 The voltage across a resistor is an RV $\mathbf{e}$ uniform between 5 and 10 V. We shall determine the density of the power

$$\mathbf{w} = \frac{\mathbf{e}^2}{r} \qquad r = 1000\ \Omega$$

dissipated in r.

Since $f_e(e) = 1/5$ for e between 5 and 10 and zero elsewhere, we conclude from (5-8) with $a = 1/r$ that

$$f_w(w) = \sqrt{\frac{10}{w}} \qquad \frac{1}{40} < w < \frac{1}{10}$$

and zero elsewhere.

4.
$$\mathbf{y} = \mathbf{x}U(\mathbf{x}) \qquad g'(x) = U(x)$$

Clearly, $f_y(y) = 0$ and $F_y(y) = 0$ for $y < 0$ (Fig. 5-12). If $y > 0$, then the equation $y = xU(x)$ has a single solution $x_1 = y$. Hence

$$f_y(y) = f_x(y) \qquad F_y(y) = F_x(y) \qquad y > 0$$

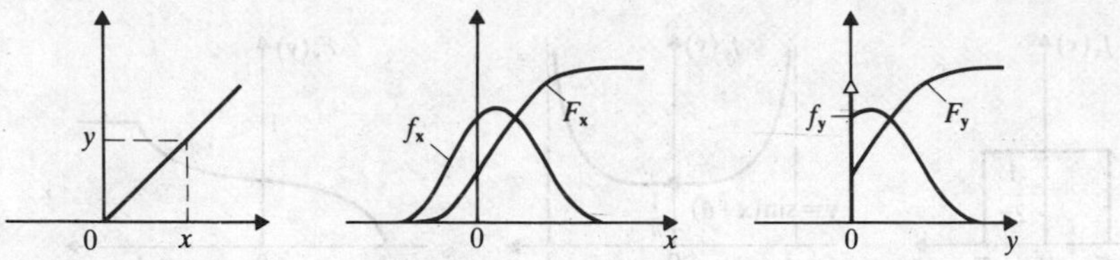

Figure 5-12

Thus, $F_y(y)$ is discontinuous at $y = 0$ with discontinuity $F_y(0^+) - F_y(0^-) = F_x(0)$. Hence

$$f_y(y) = f_x(y)U(y) + F_x(0)\,\delta(y)$$

5. $$\mathbf{y} = e^{\mathbf{x}} \qquad g'(x) = e^x$$

If $y > 0$, then the equation $y = e^x$ has the single solution $x = \ln y$. Hence

$$f_y(y) = \frac{1}{y} f_x(\ln y) \qquad y > 0$$

If $y < 0$, then $f_y(y) = 0$.

Special case If $\mathbf{x}$ is $N(\eta; \sigma)$, then

$$f_y(y) = \frac{1}{\sigma y \sqrt{2\pi}} e^{-(\ln y - \eta)^2/2\sigma^2} \tag{5-10}$$

This density is called *lognormal* (see Table 4-1).

6. $$\mathbf{y} = a \sin(\mathbf{x} + \theta) \qquad a > 0 \tag{5-11}$$

If $|y| > a$, then the equation $y = a \sin(x + \theta)$ has no solutions, hence, $f_y(y) = 0$. If $|y| < a$, then it has infinitely many solutions (Fig. 5-13*a*)

$$x_n = \arcsin \frac{y}{a} - \theta \qquad n = -\ldots, -1, 0, 1, \ldots$$

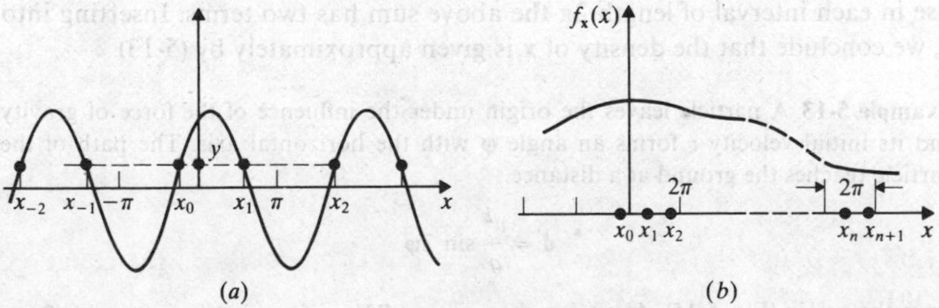

Figure 5-13

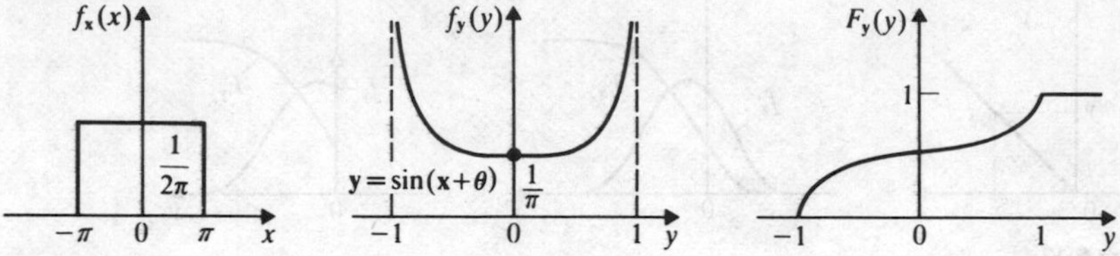

Figure 5-14

Since $g'(x_n) = a \cos (x_n + \theta) = \sqrt{a^2 - y^2}$, (5-5) yields

$$f_y(y) = \frac{1}{\sqrt{a^2 - y^2}} \sum_{n=-\infty}^{\infty} f_x(x_n) \qquad |y| < a \tag{5-12}$$

Special case Suppose that **x** is uniform in the interval $(-\pi, \pi)$. In this case, the equation $y = a \sin (x + \theta)$ has exactly two solutions in the interval $(-\pi, \pi)$ for any θ (Fig. 5-14). The function $f_x(x)$ equals $1/2\pi$ for these two values and it equals zero for any x_n outside the interval $(-\pi, \pi)$. Retaining the two nonzero terms in (5-12), we obtain

$$f_y(y) = \frac{2}{2\pi\sqrt{a^2 - y^2}} \qquad |y| < a \tag{5-13}$$

To find $F_y(y)$, we observe that $\mathbf{y} \le y$ if **x** is either between $-\pi$ and x_0 or between x_1 and π (Fig. 5-13*a*). Since the total length of the two intervals equals $\pi + 2x_0 + 2\theta$ we conclude, dividing by 2π, that

$$F_y(y) = \frac{1}{2} + \frac{1}{\pi} \arcsin \frac{y}{a} \qquad |y| < a \tag{5-14}$$

We note that although $f_y(\pm a) = \infty$, the probability that $\mathbf{x} = \pm a$ is zero.

Smooth phase If the density $f_x(x)$ of **x** is sufficiently smooth so that it can be approximated by a constant in any interval of length 2π (see Fig. 5-13*b*), then

$$\pi \sum_{n=-\infty}^{\infty} f_x(x_n) \simeq \int_{-\infty}^{\infty} f_x(x)\, dx = 1 \tag{5-15}$$

because in each interval of length 2π the above sum has two terms. Inserting into (5-12), we conclude that the density of **x** is given approximately by (5-13).

Example 5-13 A particle leaves the origin under the influence of the force of gravity and its initial velocity v forms an angle $\boldsymbol{\varphi}$ with the horizontal axis. The path of the particle reaches the ground at a distance

$$\mathbf{d} = \frac{v^2}{g} \sin 2\boldsymbol{\varphi}$$

from the origin (Fig. 5-15). Assuming that $\boldsymbol{\varphi}$ is an RV uniform between 0 and $\pi/2$, we shall determine: (*a*) the density of **d** (*b*) the probability that $\mathbf{d} \le d_0$.

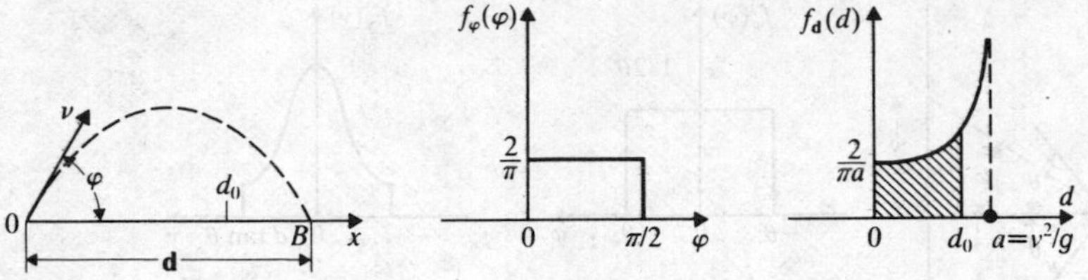

Figure 5-15

SOLUTION (*a*) Clearly

$$\mathbf{d} = a \sin \mathbf{x} \qquad a = v^2/g$$

where the RV $\mathbf{x} = 2\boldsymbol{\varphi}$ is uniform between 0 and π. If $0 < d < a$, then the equation $d = a \sin x$ has exactly two solutions in the interval $(0, \pi)$. Reasoning as in (5-13), we obtain

$$f_d(d) = \frac{2}{\pi\sqrt{a^2 - d^2}} \qquad 0 < d < a$$

and zero otherwise.

(*b*) The probability that $\mathbf{d} \leq d_0$ equals the shaded area in Fig. 5-15:

$$P\{\mathbf{d} \leq d_0\} = F_d(d_0) = \frac{2}{\pi} \arcsin \frac{d_0}{a}$$

7. $$\mathbf{y} = \tan \mathbf{x}$$

The equation $y = \tan x$ has infinitely many solutions for any y (Fig. 5-16*a*)

$$x_n = \arctan y \qquad n = \ldots, -1, 0, 1, \ldots$$

Since $g'(x) = 1/\cos^2 x = 1 + y^2$, (5-5) yields

$$f_y(y) = \frac{1}{1+y^2} \sum_{n=-\infty}^{\infty} f_x(x_n) \tag{5-16}$$

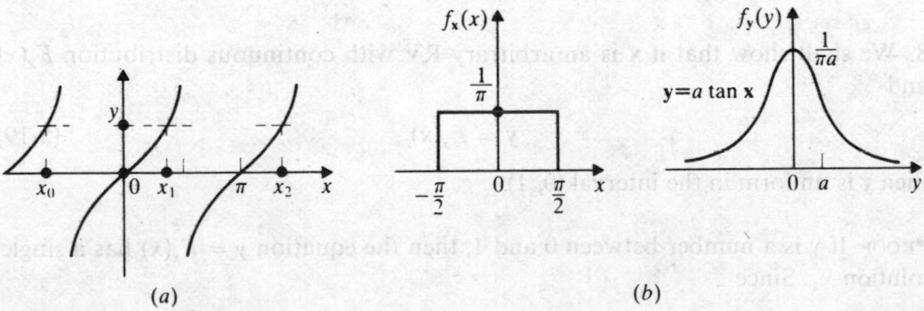

Figure 5-16

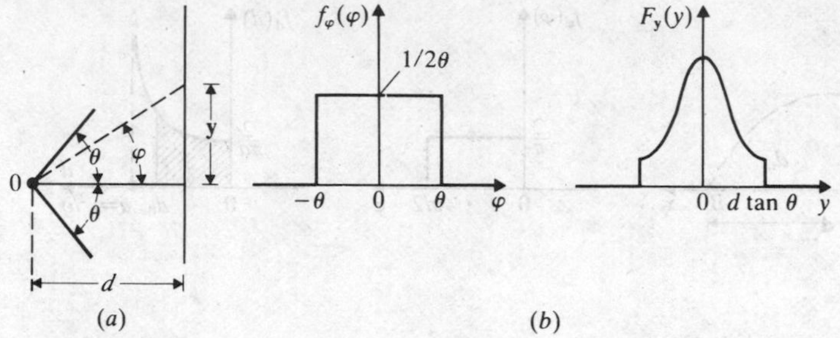

Figure 5-17

Special case If **x** is uniform in the interval $(-\pi/2, \pi/2)$ then the term $f_x(x_1)$ in (5-16) equals $1/\pi$ and all others are zero (Fig. 5-16*b*). Hence, **y** has a *Cauchy* density

$$f_y(y) = \frac{1/\pi}{1 + y^2} \tag{5-17}$$

As we see from the figure, $\mathbf{y} \le y$ if **x** is between $-\pi/2$ and x_1. Since the length of this interval equals $x_1 + \pi/2$, we conclude dividing by π that

$$F_y(y) = \frac{1}{\pi}\left(x_1 + \frac{\pi}{2}\right) = \frac{1}{2} + \frac{1}{\pi} \arctan y \tag{5-18}$$

Example 5-14 A particle leaves the origin in a free motion as in Fig. 5-17 crossing the vertical line $x = d$ at

$$\mathbf{y} = d \tan \boldsymbol{\varphi}$$

Assuming that the angle $\boldsymbol{\varphi}$ is uniform in the interval $(-\theta, \theta)$, we conclude as in (5-17) that

$$f_y(y) = \frac{d/2\theta}{d^2 + y^2} \qquad \text{for} \qquad |y| < d \tan \theta$$

and zero otherwise.

8. We shall show that if **x** is an arbitrary RV with continuous distribution $F_x(x)$ and

$$\mathbf{y} = F_x(\mathbf{x}) \tag{5-19}$$

then **y** is uniform in the interval (0, 1).

Proof If y is a number between 0 and 1, then the equation $y = F_x(x)$ has a single solution x_1. Since

$$g'(x) = F'_x(x) = f_x(x)$$

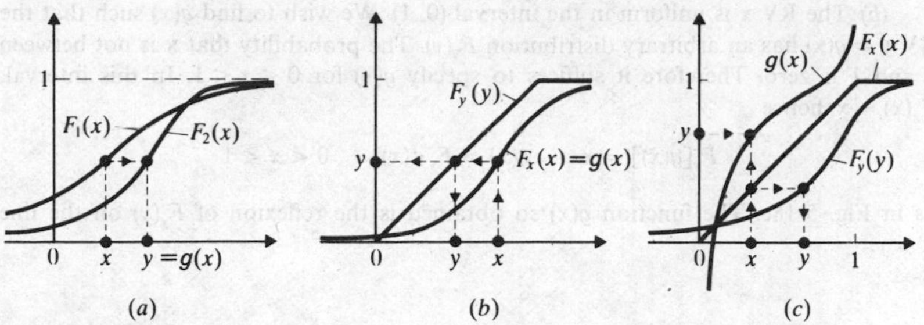

Figure 5-18

we conclude from (5-5) that

$$f_y(y) = \frac{f_x(x_1)}{g'(x_1)} = \frac{f_x(x_1)}{f_x(x_1)} = 1 \qquad 0 < y < 1$$

For $y < 0$ or $y > a$ the function $f_y(y)$ is zero because then the equation $y = F_x(x)$ has no real roots.

Synthesis In the preceding discussion, we were given the function $g(x)$ and the distribution of the RV $\mathbf{x}$ and we determined the density of the RV $\mathbf{y} = g(\mathbf{x})$. In the following, we consider the converse problem: We are given two distribution functions $F_1(x)$ and $F_2(y)$ and we wish to find a monotone increasing function $g(x)$ such that, if $\mathbf{y} = g(\mathbf{x})$ and

$$F_x(x) = F_1(x) \qquad \text{then} \qquad F_y(y) = F_2(y)$$

We maintain that the function $g(x)$ must be such that

$$F_2[g(x)] = F_1(x) \tag{5-20}$$

PROOF Suppose that x and y are such that $g(x) = y$. From the monotonicity of the unknown $g(x)$ it follows that if $g(\mathbf{x}) \le g(x)$ then $\mathbf{x} \le x$. Since $\mathbf{y} = g(\mathbf{x})$, we have

$$F_y(y) = P\{\mathbf{y} \le y\} = P\{g(\mathbf{x}) \le g(x)\} = P\{\mathbf{x} \le x\} = F_x(x)$$

and (5-20) results.

Equation (5-20) determines uniquely $g(x)$ in terms of $F_1(x)$ and $F_2(x)$. In Fig. 5-18*a*, we show the graphical solution of this equation.

Example 5-15(*a*) The RV $\mathbf{x}$ has an arbitrary distribution $F_x(x)$. We wish to find a function $g(x)$ such that the RV $\mathbf{y} = g(\mathbf{x})$ be uniform in (0, 1)

$$F_y(y) = y \qquad 0 \le y \le 1$$

Inserting into (5-20), we obtain

$$g(x) = F_x(x)$$

as in (5-19) (Fig. 5-18*b*).

(*b*) The RV **x** is uniform in the interval (0, 1). We wish to find $g(x)$ such that the RV $\mathbf{y} = g(\mathbf{x})$ has an arbitrary distribution $F_y(y)$. The probability that **x** is not between 0 and 1 is zero. Therefore it suffices to specify $g(x)$ for $0 \le x \le 1$. In this interval, $F_x(x) = x$, hence

$$F_y[g(x)] = x \qquad g(x) = F_y^{-1}(x) \qquad 0 < x < 1$$

as in Fig. 5-18*c*. The function $g(x)$ so obtained is the reflexion of $F_y(y)$ on the line $x = y$.

5-3 MEAN AND VARIANCE

The *expected value* or *mean* of an RV **x** is by definition the integral

$$E\{\mathbf{x}\} = \int_{-\infty}^{\infty} x f(x)\, dx \tag{5-21}$$

This number will also be denoted by η_x or η.

Example 5-16 If **x** is uniform in the interval (x_1, x_2), then $f(x) = 1/(x_2 - x_1)$ in this interval. Hence

$$E\{\mathbf{x}\} = \frac{1}{x_2 - x_1}\int_{x_1}^{x_2} x\, dx = \frac{x_1 + x_2}{2}$$

We note that, if the vertical line $x = a$ is an axis of symmetry of $f(x)$ then $E\{\mathbf{x}\} = a$; in particular, if $f(-x) = f(x)$, then $E\{\mathbf{x}\} = 0$. In the above example, $f(x)$ is symmetrical about the line $x = (x_1 + x_2)/2$.

Discrete type For discrete type RVs the integral in (5-21) can be written as a sum. Indeed, suppose that **x** takes the values x_i with probability p_i. In this case [see (4-15)]

$$f(x) = \sum_i p_i\, \delta(x - x_i)$$

Inserting into (5-21) and using the identity

$$\int_{-\infty}^{\infty} x\, \delta(x - x_i)\, dx = x_i$$

we obtain

$$E\{\mathbf{x}\} = \sum_i p_i x_i \qquad p_i = P\{\mathbf{x} = x_i\} \tag{5-22}$$

Example 5-17 If **x** takes the values 1, 2, ..., 6 with probability 1/6, then

$$E\{\mathbf{x}\} = \frac{1}{6}(1 + 2 + \cdots + 6) = 3.5$$

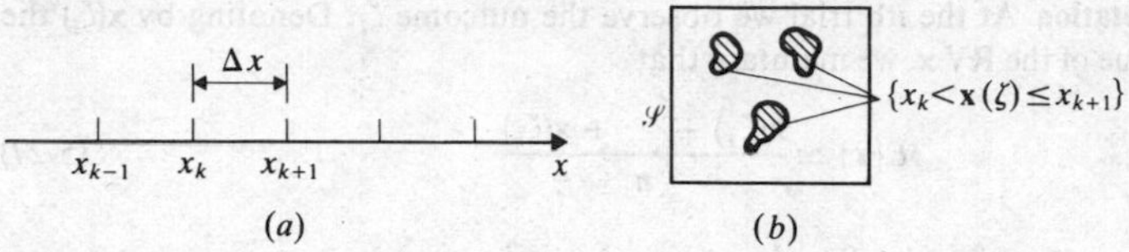

Figure 5-19

Conditional mean The conditional mean of an RV $\mathbf{x}$ assuming $\mathscr{M}$ is given by the integral in (5-21) if $f(x)$ is replaced by the conditional density $f(x \mid \mathscr{M})$

$$E\{\mathbf{x} \mid \mathscr{M}\} = \int_{-\infty}^{\infty} x f(x \mid \mathscr{M})\, dx \tag{5-23}$$

For discrete type RVs the above yields

$$E\{\mathbf{x} \mid \mathscr{M}\} = \sum_i x_i P\{\mathbf{x} = x_i \mid \mathscr{M}\} \tag{5-24}$$

Example 5-18 With $\mathscr{M} = \{\mathbf{x} \ge a\}$, it follows from (5-23) and (4-48) that

$$E\{\mathbf{x} \mid \mathbf{x} \ge a\} = \int_{-\infty}^{\infty} x f(x \mid \mathbf{x} \ge a)\, dx = \frac{\int_a^\infty x f(x)\, dx}{\int_a^\infty f(x)\, dx}$$

Lebesgue integral The mean of an RV can be interpreted as a Lebesgue integral. This interpretation is important in mathematics but it will not be used in our development. We make, therefore, only a passing reference:

We divide the x axis into intervals (x_k, x_{k+1}) of length Δx as in Fig. 5-19*a*. If Δx is small, then the Riemann integral in (5-21) can be approximated by a sum

$$\int_{-\infty}^{\infty} x f(x)\, dx \simeq \sum_{k=-\infty}^{\infty} x_k f(x_k)\, \Delta x \tag{5-25}$$

And since $f(x_k)\, \Delta x \simeq P\{x_k \le \mathbf{x} < x_k + \Delta x\}$, we conclude that

$$E\{\mathbf{x}\} \simeq \sum_{k=-\infty}^{\infty} x_k P\{x_k < \mathbf{x} < x_k + \Delta x\} \tag{5-26}$$

In the above, the sets $\{x_k < \mathbf{x} < x_k + \Delta x\}$ are differential events specified in terms of the RV $\mathbf{x}$, and their union is the space $\mathscr{S}$ (Fig. 5-19*b*). Hence, to find $E\{\mathbf{x}\}$, we multiply the probability of each differential event by the corresponding value of $\mathbf{x}$ and sum over all k. The resulting limit as $\Delta x \to 0$ is written in the form

$$E\{\mathbf{x}\} = \int_{\mathscr{S}} \mathbf{x}\, dP$$

and is called the *Lebesgue integral* of $\mathbf{x}$.

Frequency interpretation At the ith trial we observe the outcome ζ_i. Denoting by $\mathbf{x}(\zeta_i)$ the corresponding value of the RV $\mathbf{x}$, we maintain that

$$E\{\mathbf{x}\} \simeq \frac{\mathbf{x}(\zeta_1) + \cdots + \mathbf{x}(\zeta_n)}{n} \tag{5-27}$$

Proof Denoting by Δn_k the number of terms in (5-26) such that $\mathbf{x}(\zeta_k)$ is between x_k and $x_k + \Delta x$, we have

$$\mathbf{x}(\zeta_1) + \cdots + \mathbf{x}(\zeta_n) \simeq \sum x_k \, \Delta n_k$$

But $f(x_k)\,\Delta x \simeq \Delta n_k/n$ [see (4-23)]. Hence [see (5-25)]

$$\int_{-\infty}^{\infty} x f(x)\,dx \simeq \sum_{k=-\infty}^{\infty} x_k \frac{\Delta n_k}{n}$$

and (5-27) results.

The conditional mean $E\{\mathbf{x} \mid \mathscr{M}\}$ has a similar interpretation: We discard all outcomes in which the event $\mathscr{M}$ does not occur. In the remaining subsequence, $E\{\mathbf{x} \mid \mathscr{M}\}$ is given by average in (5-27).

***Mean of** $g(\mathbf{x})$* Given an RV $\mathbf{x}$ and a function $g(x)$, we form the RV $\mathbf{y} = g(\mathbf{x})$. As we see from (5-21), the mean of this RV is given by

$$E\{\mathbf{y}\} = \int_{-\infty}^{\infty} y f_y(y)\,dy \tag{5-28}$$

It appears, therefore, that to determine the mean of $\mathbf{y}$, we must find its density $f_y(y)$. This, however, is not necessary. As the next basic theorem shows, $E\{\mathbf{y}\}$ can be expressed directly in terms of the function $g(x)$ and the density $f_x(x)$ of $\mathbf{x}$.

Theorem

$$E\{g(\mathbf{x})\} = \int_{-\infty}^{\infty} g(x) f_x(x)\,dx \tag{5-29}$$

Proof We shall sketch a proof using the curve $g(x)$ of Fig. 5-20. With $y = g(x_1) = g(x_2) = g(x_3)$ as in the figure, we see that

$$f_y(y)\,dy = f_x(x_1)\,dx_1 + f_x(x_2)\,dx_2 + f_x(x_3)\,dx_3$$

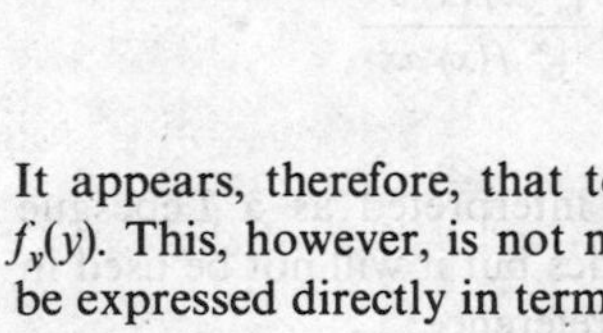

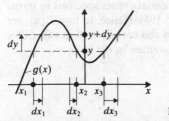

Figure 5-20

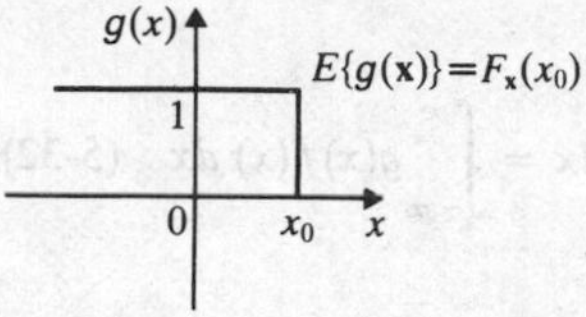

Figure 5-21

Multiplying by y, we obtain

$$yf_y(y)\,dy = g(x_1)f_x(x_1)\,dx_1 + g(x_2)f_x(x_2)\,dx_2 + g(x_3)f_x(x_3)\,dx_3$$

Thus, to each differential in (5-28) there correspond one or more differentials in (5-29). As dy covers the y axis, the corresponding dx's are nonoverlapping and they cover the entire x axis. Hence, the integrals in (5-28) and (5-29) are equal.

If $\mathbf{x}$ is of discrete type as in (5-22), then (5-29) yields

$$E\{g(\mathbf{x})\} = \sum_i g(x_i)P\{\mathbf{x} = x_i\} \tag{5-30}$$

Example 5-19 With x_0 an arbitrary number and $g(x)$ as in Fig. 5-21, (5-29) yields

$$E\{g(\mathbf{x})\} = \int_{-\infty}^{x_0} f_x(x)\,dx = F_x(x_0)$$

This shows that the distribution function of an RV can be expressed as expected value.

Example 5-20 In this example, we show that the probability of any event $\mathscr{A}$ can be expressed as expected value. For this purpose we form the zero-one RV $\mathbf{x}_{\mathscr{A}}$ associated with the event $\mathscr{A}$

$$\mathbf{x}_{\mathscr{A}}(\zeta) = \begin{cases} 1 & \zeta \in \mathscr{A} \\ 0 & \zeta \notin \mathscr{A} \end{cases}$$

Since this RV takes the values 1 and 0 with respective probabilities $P(\mathscr{A})$ and $P(\bar{\mathscr{A}})$, (5-22) yields

$$E\{\mathbf{x}_{\mathscr{A}}\} = 1 \cdot P(\mathscr{A}) + 0 \cdot P(\bar{\mathscr{A}}) = P(\mathscr{A})$$

Linearity From (5-29) it follows that

$$E\{a_1 g_1(\mathbf{x}) + \cdots + a_n g_n(\mathbf{x})\} = a_1 E\{g_1(\mathbf{x})\} + \cdots + a_n E\{g_n(\mathbf{x})\} \tag{5-31}$$

In particular, $E\{a\mathbf{x} + b\} = aE\{\mathbf{x}\} + b$

Complex RVs If $\mathbf{z} = \mathbf{x} + j\mathbf{y}$ is a complex RV then its expected value is by definition

$$E\{\mathbf{z}\} = E\{\mathbf{x}\} + jE\{\mathbf{y}\}$$

From this and (5-29) it follows that if

$$g(\mathbf{x}) = g_1(\mathbf{x}) + jg_2(\mathbf{x})$$

is a complex function of the real RV **x** then

$$E\{g(\mathbf{x})\} = \int_{-\infty}^{\infty} g_1(x)f(x)\,dx + j\int_{-\infty}^{\infty} g_2(x)f(x)\,dx = \int_{-\infty}^{\infty} g(x)f(x)\,dx \quad \text{(5-32)}$$

In other words, (5-29) holds even if $g(x)$ is complex.

Variance

The *variance* of an RV **x** is by definition the integral

$$\sigma^2 = \int_{-\infty}^{\infty} (x-\eta)^2 f(x)\,dx \quad \text{(5-33)}$$

where $\eta = E\{\mathbf{x}\}$. The positive constant σ, denoted also by σ_x, is called the *standard deviation* of **x**.

From the definition it follows that σ^2 is the mean of the RV $(\mathbf{x}-\eta)^2$. Thus

$$\sigma^2 = E\{(\mathbf{x}-\eta)^2\} = E\{\mathbf{x}^2 - 2\mathbf{x}\eta + \eta^2\} = E\{\mathbf{x}^2\} - 2\eta E\{\mathbf{x}\} + \eta^2$$

Hence

$$\sigma^2 = E\{\mathbf{x}^2\} - E^2\{\mathbf{x}\} \quad \text{(5-34)}$$

Example 5-21 If **x** is uniform in the interval $(-c, c)$, then $\eta = 0$ and

$$\sigma^2 = E\{\mathbf{x}^2\} = \frac{1}{2c}\int_{-c}^{c} x^2\,dx = \frac{c^2}{3}$$

Example 5-22 We have written the density of a normal RV in the form

$$f(x) = \frac{1}{\sigma\sqrt{2\pi}}\,e^{-(x-\eta)^2/2\sigma^2}$$

where up to now η and σ^2 were two arbitrary constants. We show next that η is indeed the mean of **x** and σ^2 its variance.

PROOF Clearly, $f(x)$ is symmetrical about the line $x = \eta$, hence, $E\{\mathbf{x}\} = \eta$. Furthermore

$$\int_{-\infty}^{\infty} e^{-(x-\eta)^2/2\sigma^2}\,dx = \sigma\sqrt{2\pi}$$

because the area of $f(x)$ equals 1. Differentiating with respect to σ, we obtain

$$\int_{-\infty}^{\infty} \frac{(x-\eta)^2}{\sigma^3}\,e^{-(x-\eta)^2/2\sigma^2}\,dx = \sqrt{2\pi}$$

Multiplying both sides by $\sigma^2/\sqrt{2\pi}$, we conclude that $E\{\mathbf{x}-\eta)^2\} = \sigma^2$ and the proof is complete.

Discrete type If the RV **x** is of discrete type, then

$$\sigma^2 = \sum_i p_i(x_i - \eta)^2 \qquad p_i = P\{\mathbf{x} = x_i\} \tag{5-35}$$

Example 5-23 The RV **x** takes the values 0 and 1 with probabilities p and $q = 1 - p$ respectively. In this case

$$E\{\mathbf{x}\} = 1 \cdot p + 0 \cdot q = p$$

$$E\{\mathbf{x}^2\} = 1^2 \cdot p + 0^2 \cdot q = p$$

Hence

$$\sigma^2 = E\{\mathbf{x}^2\} - E^2\{\mathbf{x}\} = p - p^2 = pq$$

Example 5-24 A Poisson distributed RV with parameter a takes the values 0, 1, ... with probabilities

$$P\{\mathbf{x} = k\} = e^{-a}\frac{a^k}{k!}$$

We shall show that its mean and variance equal a

$$E\{\mathbf{x}\} = a \qquad E\{\mathbf{x}^2\} = a^2 + a \qquad \sigma^2 = a \tag{5-36}$$

PROOF We differentiate twice the Taylor expansion of e^a

$$e^a = \sum_{k=0}^{\infty} \frac{a^k}{k!}$$

$$e^a = \sum_{k=0}^{\infty} k\frac{a^{k-1}}{k!} = \frac{1}{a}\sum_{k=1}^{\infty} k\frac{a^k}{k!}$$

$$e^a = \sum_{k=1}^{\infty} k(k-1)\frac{a^{k-2}}{k!} = \frac{1}{a^2}\sum_{k=1}^{\infty} k^2\frac{a^k}{k!} - \frac{1}{a^2}\sum_{k=1}^{\infty} k\frac{a^k}{k!}$$

Hence

$$E\{\mathbf{x}\} = e^{-a}\sum_{k=1}^{\infty} k\frac{a^k}{k!} = a \qquad E\{\mathbf{x}^2\} = e^{-a}\sum_{k=1}^{\infty} k^2\frac{a^k}{k!} = a^2 + a$$

and (5-36) results.

POISSON POINTS As we have shown in (3-47), the number **n** of Poisson points in an interval of length t_0 is a Poisson distributed RV with parameter $a = \lambda t_0$. From this it follows that

$$E\{\mathbf{n}\} = \lambda t_0 \qquad \sigma_n^2 = \lambda t_0 \tag{5-37}$$

This shows that the density λ of Poisson points equals the expected number of points per unit time.

5-4 MOMENTS

The following quantities are of interest in the study of RVs:

Moments

$$m_n = E\{\mathbf{x}^n\} = \int_{-\infty}^{\infty} x^n f(x)\, dx \tag{5-38}$$

Central moments

$$\mu_n = E\{(\mathbf{x} - \eta)^n\} = \int_{-\infty}^{\infty} (x - \eta)^n f(x)\, dx \tag{5-39}$$

Absolute moments

$$E\{|\mathbf{x}|^n\} \qquad E\{|\mathbf{x} - \eta|^n\} \tag{5-40}$$

Generalized moments

$$E\{(\mathbf{x} - a)^n\} \qquad E\{|\mathbf{x} - a|^n\} \tag{5-41}$$

We note that

$$\mu_n = E\{(\mathbf{x} - \eta)^n\} = E\left\{\sum_{k=0}^{n} \binom{n}{k} \mathbf{x}^k (-\eta)^{n-k}\right\}$$

Hence,

$$\mu_n = \sum_{k=0}^{n} \binom{n}{k} m_k (-\eta)^{n-k} \tag{5-42}$$

Similarly

$$m_n = E\{[(\mathbf{x} - \eta) + \eta]^n\} = E\left\{\sum_{k=0}^{n} \binom{n}{k} (\mathbf{x} - \eta)^k \eta^{n-k}\right\}$$

Hence

$$m_n = \sum_{k=0}^{n} \binom{n}{k} \mu_k \eta^{n-k} \tag{5-43}$$

In particular,

$$\mu_0 = m_0 = 1 \qquad m_1 = \eta \qquad \mu_1 = 0 \qquad \mu_2 = \sigma^2$$

and

$$\mu_3 = m_3 - 3\eta m_2 + 2\eta^3 \qquad m_3 = \mu_3 + 3\eta\sigma^2 + \eta^3$$

Notes 1. If the function $f(x)$ is interpreted as mass density on the x axis, then $E\{\mathbf{x}\}$ equals its center of gravity, $E\{\mathbf{x}^2\}$ equals the moment of inertia with respect to the origin, and σ^2 equals the central moment of inertia. The standard deviation σ is the radius of gyration.

2. The constants η and σ give only a limited characterization of $f(x)$. Knowledge of other moments provides additional information that can be used, for example, to distinguish between two densities with the same η and σ. In fact, if m_n is known for every n,

then, under certain conditions, $f(x)$ is determined uniquely [see also (5-69)]. The underlying theory is known in mathematics as the *moment problem.*

3. The moments of an RV are not arbitrary numbers but must satisfy various inequalities. For example [see (5-34)]

$$\sigma^2 = m_2 - m_1^2 \geq 0$$

Similarly, since the quadratic

$$E\{(\mathbf{x}^n - a)^2\} = m_{2n} - 2am_n + a^2$$

is nonnegative for any a, its discriminant cannot be positive. Hence

$$m_{2n} \geq m_n^2$$

Normal random variables We shall show that if

$$f(x) = \frac{1}{\sigma\sqrt{2\pi}} e^{-x^2/2\sigma^2}$$

then

$$E\{\mathbf{x}^n\} = \begin{cases} 0 & n = 2k+1 \\ 1 \cdot 3 \cdots (n-1)\sigma^n & n = 2k \end{cases} \tag{5-44}$$

$$E\{|\mathbf{x}|^n\} = \begin{cases} 1 \cdot 3 \cdots (n-1)\sigma^n & n = 2k \\ 2^k k!\, \sigma^{2k+1}\sqrt{2/\pi} & n = 2k+1 \end{cases} \tag{5-45}$$

The odd moments of $\mathbf{x}$ are zero because $f(-x) = -f(x)$. To prove the lower part of (5-44), we differentiate k times the identity

$$\int_{-\infty}^{\infty} e^{\alpha x^2}\, dx = \sqrt{\frac{\pi}{\alpha}}$$

This yields

$$\int_{-\infty}^{\infty} x^{2k} e^{-\alpha x^2}\, dx = \frac{1 \cdot 3 \cdots (2k-1)}{2^k} \sqrt{\frac{\pi}{\alpha^{2k+1}}}$$

and with $\alpha = 1/2\sigma^2$, (5-44) results.

Since $f(-x) = f(x)$, we have

$$E\{|\mathbf{x}|^{2k+1}\} = 2\int_0^{\infty} x^{2k+1} f(x)\, dx = \frac{2}{\sigma\sqrt{2\pi}} \int_0^{\infty} x^{2k+1} e^{-x^2/2\sigma^2}\, dx$$

With $y = x^2/2\sigma^2$, the above yields

$$\sqrt{\frac{2}{\pi}}\, \frac{(2\sigma^2)^{k+1}}{2\sigma} \int_0^{\infty} y^k e^{-y}\, dy$$

and (5-45) results because the last integral equals $k!$

We note in particular that

$$E\{\mathbf{x}^4\} = 3\sigma^4 = 3E^2\{\mathbf{x}^2\} \tag{5-46}$$

Example 5-25 If $\mathbf{x}$ has a *Rayleigh density*

$$f(x) = \frac{x}{\alpha^2} e^{-x^2/2\alpha^2} U(x)$$

then

$$E\{\mathbf{x}^n\} = \frac{1}{\alpha^2} \int_0^\infty x^{n+1} e^{-x^2/2\alpha^2}\, dx = \frac{1}{2\alpha^2} \int_{-\infty}^{\infty} |x|^{n+1} e^{-x^2/2\alpha^2}\, dx$$

From this and (5-45) it follows that

$$E\{\mathbf{x}^n\} = \begin{cases} 1 \cdot 3 \cdots n\alpha^n \sqrt{\pi/2} & n = 2k+1 \\ 2^k k!\, \alpha^{2k} & n = 2k \end{cases} \tag{5-47}$$

In particular

$$E\{\mathbf{x}\} = \alpha\sqrt{\pi/2} \qquad \sigma^2 = (2 - \pi/2)\alpha^2 \tag{5-48}$$

Example 5-26 If $\mathbf{x}$ has a *Maxwell density*

$$f(x) = \frac{\sqrt{2}}{\alpha^3 \sqrt{\pi}} x^2 e^{-x^2/2\alpha^2} U(x)$$

then

$$E\{\mathbf{x}^n\} = \frac{1}{\alpha^3\sqrt{2\pi}} \int_{-\infty}^{\infty} |x|^{n+2} e^{-x^2/2\alpha^2}\, dx \tag{5-49}$$

and (5-45) yields

$$E\{\mathbf{x}^n\} = \begin{cases} 1 \cdot 3 \cdots (n+1)\alpha^n & n = 2k \\ 2^k k!\, \alpha^{2k-1} \sqrt{\pi/2} & n = 2k-1 \end{cases} \tag{5-50}$$

In particular

$$E\{\mathbf{x}\} = 2\alpha\sqrt{2/\pi} \qquad E\{\mathbf{x}^2\} = 3\alpha^2 \tag{5-51}$$

Estimate of the mean of $g(\mathbf{x})$ The mean of the RV $\mathbf{y} = g(\mathbf{x})$ is given by

$$E\{g(\mathbf{x})\} = \int_{-\infty}^{\infty} g(x) f(x)\, dx \tag{5-52}$$

Hence, for its determination, knowledge of $f(x)$ is required. However, if $\mathbf{x}$ is concentrated near its mean, then $E\{g(\mathbf{x})\}$ can be expressed in terms of the moments μ_n of $\mathbf{x}$.

Suppose, first, that $f(x)$ is negligible outside an interval $(\eta - \varepsilon, \eta + \varepsilon)$ and in this interval, $g(x) \simeq g(\eta)$. In this case, (5-52) yields

$$E\{g(\mathbf{x})\} \simeq g(\eta) \int_{\eta-\varepsilon}^{\eta+\varepsilon} f(x)\, dx \simeq g(\eta) \tag{5-53}$$

This estimate can be improved if $g(x)$ is approximated by a polynomial

$$g(x) \simeq g(\eta) + g'(\eta)(x - a) + \cdots + g^{(n)}(\eta)\,\frac{(x-\eta)^n}{n!}$$

Inserting into (5-52), we obtain

$$E\{g(\mathbf{x})\} \simeq g(\eta) + g''(\eta)\,\frac{\sigma^2}{2} + \cdots + g^{(n)}(\eta)\,\frac{\mu_n}{n!} \tag{5-54}$$

In particular if $g(x)$ is approximated by a parabola, then

$$\eta_y = E\{g(\mathbf{x})\} \simeq g(\eta) + g''(\eta)\,\frac{\sigma^2}{2} \tag{5-55}$$

And if it is approximated by a straight line, then $\eta_y \simeq g(\eta)$. This shows that the slope of $g(x)$ has no effect on η_y; however, as we show next, it affects the variance σ_y^2 of $\mathbf{y}$.

Variance We maintain that the first-order estimate of σ_y^2 is given by

$$\sigma_y^2 \simeq |g'(\eta)|^2\sigma^2 \tag{5-56}$$

PROOF We apply (5-55) to the function $g^2(x)$. Since its second derivative equals $2(g')^2 + 2gg''$, we conclude that

$$\sigma_y^2 + \eta_y^2 = E\{g^2(\mathbf{x})\} \simeq g^2 + [(g')^2 + gg'']\sigma^2$$

Inserting the approximation (5-55) for η_y into the above and neglecting the σ^4 term, we obtain (5-56).

Example 5-27 A voltage $E = 120$ V is connected across a resistor whose resistance is an RV $\mathbf{r}$ uniform between 900 and 1100 Ω. Using (5-55) and (5-56) we shall estimate the mean and variance of the resulting current

$$\mathbf{i} = \frac{E}{\mathbf{r}}$$

Clearly, $E\{\mathbf{r}\} = \eta = 10^3$, $\sigma^2 = 100^2/3$. With $g(r) = E/r$, we have

$$g(\eta) = 0.12 \qquad g'(\eta) = -12 \times 10^{-5} \qquad g''(\eta) = 24 \times 10^{-8}$$

Hence

$$E\{\mathbf{i}\} \simeq 0.12 + 0.0004 \text{ A} \qquad \sigma_i^2 \simeq 48 \times 10^{-6} \text{ A}^2$$

Tchebycheff Inequality

A measure of the concentration of an RV near its mean η is its variance σ^2. In fact, as the following theorem shows, the probability that $\mathbf{x}$ is outside an arbitrary interval $(\eta - \varepsilon, \eta + \varepsilon)$ is negligible if the ratio σ/ε is sufficiently small. This result, known as *Tchebycheff inequality*, is fundamental.

Theorem For any $\varepsilon > 0$

$$P\{|\mathbf{x} - \eta| \geq \varepsilon\} \leq \frac{\sigma^2}{\varepsilon^2} \tag{5-57}$$

PROOF The proof is based on the fact that

$$P\{|\mathbf{x} - \eta| \geq \varepsilon\} = \int_{-\infty}^{-\eta-\varepsilon} f(x)\,dx + \int_{\eta+\varepsilon}^{\infty} f(x)\,dx = \int_{|x-\eta|\geq\varepsilon} f(x)\,dx$$

Indeed

$$\sigma^2 = \int_{-\infty}^{\infty} (x-\eta)^2 f(x)\,dx \geq \int_{|x-\eta|\geq\varepsilon} (x-\eta)^2 f(x)\,dx \geq \varepsilon^2 \int_{|x-\eta|\geq\varepsilon} f(x)\,dx$$

and (5-57) results because the last integral equals $P\{|\mathbf{x} - \eta| \geq \varepsilon\}$.

Notes 1. From (5-57) it follows that, if $\sigma = 0$, then the probability that $\mathbf{x}$ is outside the interval $(\eta - \varepsilon, \eta + \varepsilon)$ equals zero for any ε, hence, $\mathbf{x} = \eta$ with probability 1. Similarly, if

$$E\{\mathbf{x}^2\} = \eta^2 + \sigma^2 = 0 \qquad \text{then} \qquad \eta = 0 \qquad \sigma = 0$$

hence, $\mathbf{x} = 0$ with probability 1.

2. For specific densities, the bound in (5-57) is too high. Suppose, for example, that $\mathbf{x}$ is normal. In this case, $P\{|\mathbf{x} - \eta| \geq 3\sigma\} = 2\mathbb{G}(3) - 1 = 0.0027$. Inequality (5-57), however, yields $P\{|\mathbf{x} - \eta| \geq 3\sigma\} \leq 1/9$.

The significance of Tchebycheff's inequality is the fact that it holds for *any* $f(x)$ and can, therefore, be used even if $f(x)$ is not known.

3. The bound in (5-57) can be reduced if various assumptions are made about $f(x)$ [see *Chernoff bound* (Prob. 5-22)].

Generalization If $f(x) = 0$ for $x < 0$, then, for any $\alpha > 0$

$$P\{\mathbf{x} \geq \alpha\} \leq \frac{\eta}{\alpha} \tag{5-58}$$

PROOF

$$E\{\mathbf{x}\} = \int_0^{\infty} xf(x)\,dx \geq \int_{\alpha}^{\infty} xf(x)\,dx \geq \alpha \int_{\alpha}^{\infty} f(x)\,dx$$

and (5-58) results because the last integral equals $P\{\mathbf{x} \geq \alpha\}$.

Corollary Suppose that $\mathbf{x}$ is an arbitrary RV and a and n are two arbitrary numbers. Clearly, the RV $|\mathbf{x} - a|^n$ takes only positive values. Applying (5-58), with $\alpha = \varepsilon^n$, we conclude that

$$P\{|\mathbf{x} - a|^n \geq \varepsilon^n\} \leq \frac{E\{|\mathbf{x} - a|^n\}}{\varepsilon^n}$$

Hence,

$$P\{|\mathbf{x}-a| \geq \varepsilon\} \leq \frac{E\{|\mathbf{x}-a|^n\}}{\varepsilon^n} \tag{5-59}$$

This result is known as the *inequality of Bienaymé*. Tchebycheff's inequality is a special case obtained with $a = \eta$ and $n = 2$.

5-5 CHARACTERISTIC FUNCTIONS

The *characteristic function* of an RV is by definition the integral

$$\Phi(\omega) = \int_{-\infty}^{\infty} f(x)e^{j\omega x}\,dx \tag{5-60}$$

This function is maximum at the origin because $f(x) \geq 0$

$$|\Phi(\omega)| \leq \Phi(0) = 1 \tag{5-61}$$

If $j\omega$ is changed to s, the resulting integral

$$\mathbf{\Phi}(s) = \int_{-\infty}^{\infty} f(x)e^{sx}\,dx \qquad \mathbf{\Phi}(j\omega) = \Phi(\omega) \tag{5-62}$$

is the *moment* (*generating*) *function* of $\mathbf{x}$.

The function

$$\Psi(\omega) = \ln \Phi(\omega) = \mathbf{\Psi}(j\omega) \tag{5-63}$$

is the *second characteristic function* of $\mathbf{x}$.

Clearly [see (5-32)]

$$\Phi(\omega) = E\{e^{j\omega\mathbf{x}}\} \qquad \mathbf{\Phi}(s) = E\{e^{s\mathbf{x}}\}$$

This leads to the following:

If $\qquad \mathbf{y} = a\mathbf{x} + b \qquad$ then $\qquad \Phi_y(\omega) = e^{jb\omega}\Phi_x(a\omega) \qquad$ (5-64)

because

$$E\{e^{j\omega\mathbf{y}}\} = E\{e^{j\omega(a\mathbf{x}+b)}\} = e^{jb\omega}E\{e^{ja\omega\mathbf{x}}\}$$

Example 5-28 It is well known that

$$\frac{1}{\sigma\sqrt{2\pi}}\int_{-\infty}^{\infty} e^{-x^2/2\sigma^2}e^{j\omega x}\,dx = e^{-\sigma^2\omega^2/2}$$

From this and (5-64) it follows that the characteristic function of a *normal* RV with mean η and variance σ^2 equals

$$\Phi(\omega) = e^{j\eta\omega}e^{-\sigma^2\omega^2/2} \tag{5-65}$$

Inversion formula As we see from (5-60), $\Phi(-\omega)$ is the Fourier transform of $f(x)$. Hence, the properties of characteristic functions are essentially the same as the properties of Fourier transforms. We note, in particular, that $f(x)$ can be expressed in terms of $\Phi(\omega)$

$$f(x) = \frac{1}{2\pi}\int_{-\infty}^{\infty} \Phi(\omega)e^{-j\omega x}\, d\omega \tag{5-66}$$

Moment theorem Differentiating (5-62) n times, we obtain

$$\mathbf{\Phi}^{(n)}(s) = E\{\mathbf{x}^n e^{s\mathbf{x}}\}$$

Hence

$$\mathbf{\Phi}^{(n)}(0) = E\{\mathbf{x}^n\} = m_n \tag{5-67}$$

Thus, the derivatives of $\mathbf{\Phi}(s)$ at the origin equal the moments of $\mathbf{x}$. This justifies the name "moment function" given to $\mathbf{\Phi}(s)$.

In particular

$$\mathbf{\Phi}'(0) = m_1 = \eta \qquad \mathbf{\Phi}''(0) = m_2 = \eta^2 + \sigma^2 \tag{5-68}$$

Note Expanding $\mathbf{\Phi}(s)$ into a series near the origin and using (5-67), we obtain

$$\mathbf{\Phi}(s) = \sum_{n=0}^{\infty} \frac{m_n}{n!} s^n \tag{5-69}$$

This is valid only if all moments are finite and the series converges absolutely near $s = 0$. Since $f(x)$ can be determined in terms of $\mathbf{\Phi}(s)$, (5-69) shows that, under the stated conditions, the density of an RV is uniquely determined if all its moments are known.

Example 5-29(*a*) If $\mathbf{x}$ has an *exponential density*

$$f(x) = \lambda e^{-\lambda x} U(x)$$

then

$$\mathbf{\Phi}(s) = \int_0^{\infty} \lambda e^{-\lambda x} e^{sx}\, dx = \frac{\lambda}{\lambda - s} \tag{5-70}$$

$$\mathbf{\Phi}'(0) = \frac{1}{\lambda} \qquad \mathbf{\Phi}''(0) = \frac{2}{\lambda^2}$$

Hence

$$E\{\mathbf{x}\} = \frac{1}{\lambda} \qquad E\{\mathbf{x}^2\} = \frac{2}{\lambda^2} \qquad \sigma^2 = \frac{1}{\lambda^2}$$

(*b*) If $\mathbf{x}$ has an *Erlang density*

$$f(x) = \frac{\lambda^n}{(n-1)!} x^{n-1} e^{-\lambda x} U(x)$$

then

$$\Phi(s) = \frac{\lambda^n}{(n-1)!}\int_0^\infty x^{n-1}e^{-\lambda x}e^{sx}\,dx = \frac{\lambda^n}{(\lambda - s)^n} \tag{5-71}$$

$$\Phi'(0) = \frac{n}{\lambda} \qquad \Phi''(0) = \frac{n(n+1)}{\lambda^2}$$

Hence

$$E\{\mathbf{x}\} = \frac{n}{\lambda} \qquad E\{\mathbf{x}^2\} = \frac{n(n+1)}{\lambda^2} \qquad \sigma^2 = \frac{n}{\lambda^2}$$

Cumulants The cumulants λ_n of an RV $\mathbf{x}$ are by definition the derivatives

$$\frac{d^n\Psi(0)}{ds^n} = \lambda_n \tag{5-72}$$

of its second moment function $\Psi(s)$. Clearly [see (5-63)] $\Psi(0) = \lambda_0 = 0$, hence

$$\Psi(s) = \lambda_1 s + \frac{1}{2}\lambda_2 s^2 + \cdots + \frac{1}{n!}\lambda_n s^n + \cdots$$

We maintain that

$$\lambda_1 = \eta \qquad \lambda_2 = \sigma^2 \tag{5-73}$$

PROOF Since $\Phi = e^{\Psi}$, we conclude that

$$\Phi' = \Psi' e^{\Psi} \qquad \Phi'' = [\Psi'' + (\Psi')^2]e^{\Psi}$$

With $s = 0$, this yields

$$\Phi'(0) = \Psi'(0) = m_1 \qquad \Phi''(0) = \Psi''(0) + [\Psi'(0)]^2 = m_2$$

and (5-73) results.

Discrete Type

Suppose that $\mathbf{x}$ is a discrete type RV taking the values x_i with probability p_i. In this case, (5-60) yields

$$\Phi(\omega) = \sum_i p_i e^{j\omega x_i} \tag{5-74}$$

Thus, $\Phi(\omega)$ is a sum of exponentials. The moment function of $\mathbf{x}$ can be defined as in (5-62). However, if $\mathbf{x}$ takes only integer values, then a definition in terms of z transforms is preferable.

Lattice type If $\mathbf{n}$ is a lattice type RV taking integer values, then its moment function is by definition the sum

$$\Gamma(z) = E\{z^{\mathbf{n}}\} = \sum_{n=-\infty}^{\infty} p_n z^n \tag{5-75}$$

Thus, $\mathbf{\Gamma}(1/z)$ is the z transform of the sequence $p_n = P\{\mathbf{n} = n\}$. With $\Phi(\omega)$ as in (5-74), the above yields

$$\Phi(\omega) = \mathbf{\Gamma}(e^{j\omega}) = \sum_{n=-\infty}^{\infty} p_n e^{jn\omega}$$

Thus, $\Phi(\omega)$ is the *discrete Fourier transform* (DFT) of p_n and

$$\mathbf{\Psi}(s) = \ln \mathbf{\Gamma}(e^s) \tag{5-76}$$

Moment theorem Differentiating (5-75) k times, we obtain

$$\mathbf{\Gamma}^{(k)}(z) = E\{\mathbf{n}(\mathbf{n}-1)\cdots(\mathbf{n}-k+1)z^{\mathbf{n}-k}\}$$

With $z = 1$, this yields

$$\mathbf{\Gamma}^{(k)}(1) = E\{\mathbf{n}(\mathbf{n}-1)\cdots(\mathbf{n}-k+1)\} \tag{5-77}$$

We note, in particular, that $\mathbf{\Gamma}(1) = 1$ and

$$\mathbf{\Gamma}'(1) = E\{\mathbf{n}\} \qquad \mathbf{\Gamma}''(1) = E\{\mathbf{n}^2\} - E\{\mathbf{n}\} \tag{5-78}$$

Example 5-30(*a*) If $\mathbf{n}$ takes the values 0 and 1 with $E\{\mathbf{n} = 1\} = p$ and $E\{\mathbf{n} = 0\} = q$, then

$$\mathbf{\Gamma}(z) = pz + q$$

$$\mathbf{\Gamma}'(1) = E\{\mathbf{n}\} = p \qquad \mathbf{\Gamma}''(1) = E\{\mathbf{n}^2\} - E\{\mathbf{n}\} = 0$$

(*b*) If $\mathbf{n}$ has the binomial distribution

$$p_n = P\{\mathbf{n} = n\} = \binom{m}{n} p^n q^{m-n} \qquad 0 \le n \le m$$

then

$$\mathbf{\Gamma}(z) = \sum_{n=0}^{m} \binom{m}{n} p^n q^{m-n} z^n = (pz+q)^m$$

$$\mathbf{\Gamma}'(1) = mp \qquad \mathbf{\Gamma}''(1) = m(m-1)p^2$$

Hence

$$E\{\mathbf{n}\} = mp \qquad \sigma^2 = mpq$$

Example 5-31 If $\mathbf{n}$ is Poisson distributed with parameter a

$$P\{\mathbf{n} = n\} = e^{-a}\frac{a^n}{n!} \qquad n = 0, 1, \ldots$$

then

$$\mathbf{\Gamma}(z) = e^{-a}\sum_{n=0}^{\infty} a^n \frac{z^n}{n!} = e^{a(z-1)} \tag{5-79}$$

In this case [see (5-76)]

$$\boldsymbol{\Psi}(s) = a(e^s - 1) \qquad \boldsymbol{\Psi}'(0) = a \qquad \boldsymbol{\Psi}''(0) = a$$

and (5-73) yields $E\{\mathbf{n}\} = a$, $\sigma^2 = a$ in agreement with (5-36).

Determination of the density of $g(\mathbf{x})$ We show next that characteristic functions can be used to determine the density $f_y(y)$ of the RV $\mathbf{y} = g(\mathbf{x})$ in terms of the density $f_x(x)$ of $\mathbf{x}$.

From (5-32) it follows that the characteristic function

$$\Phi_y(\omega) = \int_{-\infty}^{\infty} e^{j\omega y} f_y(y)\, dy$$

of the RV $\mathbf{y} = g(\mathbf{x})$ equals

$$\Phi_y(\omega) = E\{e^{j\omega g(\mathbf{x})}\} = \int_{-\infty}^{\infty} e^{j\omega g(x)} f_x(x)\, dx \tag{5-80}$$

If, therefore, the above integral can be written in the form

$$\int_{-\infty}^{\infty} e^{j\omega y} h(y)\, dy$$

it will follow that (uniqueness theorem)

$$f_y(y) = h(y)$$

As the following examples show, this method leads to simple results if the transformation $y = g(x)$ is one-to-one.

Example 5-32 Suppose that $\mathbf{x}$ is $N(0; \sigma)$ and $\mathbf{y} = a\mathbf{x}^2$. Inserting into (5-80), and using the evenness of the integrand, we obtain

$$\Phi_y(\omega) = \int_{-\infty}^{\infty} e^{j\omega a x^2} f(x)\, dx = \frac{2}{\sigma\sqrt{2\pi}} \int_0^{\infty} e^{ja\omega x^2} e^{-x^2/2\sigma^2}\, dx$$

As x increases from 0 to ∞, the transformation $y = ax^2$ is one-to-one. Since

$$dy = 2ax\, dx = 2\sqrt{ay}\, dx$$

the above yields

$$\Phi_y(\omega) = \frac{2}{\sigma\sqrt{2\pi}} \int_0^{\infty} e^{j\omega y} e^{-y/2a\sigma^2} \frac{dy}{2\sqrt{ay}}$$

Hence

$$f_y(y) = \frac{e^{-y/2a\sigma^2}}{\sigma\sqrt{2\pi a y}} U(y)$$

in agreement with (5-9).

Example 5-33 We assume finally that **x** is uniform in the interval $(-\pi/2, \pi/2)$ and $\mathbf{y} = \sin \mathbf{x}$. In this case

$$\Phi_y(\omega) = \int_{-\infty}^{\infty} e^{j\omega \sin x} f(x)\, dx = \frac{1}{\pi} \int_{-\pi/2}^{\pi/2} e^{j\omega \sin x}\, dx$$

As x increases from $-\pi/2$ to $\pi/2$, the function $y = \sin x$ increases from -1 to 1 and

$$dy = \cos x\, dx = \sqrt{1 - y^2}\, dx$$

Hence

$$\Phi_y(\omega) = \frac{1}{\pi} \int_{-1}^{1} e^{j\omega y} \frac{dy}{\sqrt{1 - y^2}}$$

This leads to the conclusion that

$$f_y(y) = \frac{1}{\pi\sqrt{1 - y^2}} \qquad \text{for} \qquad |y| < 1$$

and zero otherwise, in agreement with (5-13).

PROBLEMS

5-1 Express the density $f_y(y)$ of the RV $\mathbf{y} = g(\mathbf{x})$ in terms of $f_x(x)$ if (*a*) $g(x) = |x|$; (*b*) $g(x) = e^{-x}U(x)$.

5-2 Find $F_y(y)$ and $f_y(y)$ if $F_x(x) = (1 - e^{-2x})U(x)$ and (*a*) $\mathbf{y} = (\mathbf{x} - 1)U(\mathbf{x} - 1)$; (*b*) $\mathbf{y} = \mathbf{x}^2$.

5-3 Show that, if the RV **x** has a Cauchy density and $\mathbf{y} = \arctan \mathbf{x}$, then **x** is uniform in the interval $(-\pi/2, \pi/2)$.

5-4 The RV **x** is uniform in the interval $(-2\pi, 2\pi)$. Find $f_y(y)$ if (*a*) $\mathbf{y} = \mathbf{x}^3$, (*b*) $\mathbf{y} = \mathbf{x}^4$, and (*c*) $\mathbf{y} = 2 \sin (3\mathbf{x} + 40°)$.

5-5 The RV **x** is uniform in the interval $(-1, 1)$. Find $g(x)$ such that if $\mathbf{y} = g(\mathbf{x})$ then $f_y(y) = 2e^{-2y}U(y)$.

5-6 Given an RV **x**, we form the RV $\mathbf{y} = g(\mathbf{x})$. (*a*) Find $f_y(y)$ if $g(x) = 2F_x(x) + 4$. (*b*) Find $g(x)$ such that **y** is uniform in the interval (8, 10).

5-7 A fair coin is tossed 10 times and **x** equals the number of heads. (*a*) Find $F_x(x)$. (*b*) Find $F_y(y)$ if $\mathbf{y} = (\mathbf{x} - 3)^2$.

5-8 If **t** is an RV of continuous type and $\mathbf{y} = a \sin \omega\mathbf{t}$, show that

$$f_y(y) \xrightarrow[\omega\to\infty]{} \begin{cases} 1/\pi\sqrt{a^2 - y^2} & |y| < a \\ 0 & |y| > a \end{cases}$$

5-9 Show that if $\mathbf{y} = \mathbf{x}^2$, then

$$f_y(y \mid \mathbf{x} \geq 0) = \frac{U(y)}{1 - F_x(0)} \frac{f_x(\sqrt{y})}{2\sqrt{y}}$$

5-10 (*a*) Show that if $\mathbf{y} = a\mathbf{x} + b$, then $\sigma_y = a\sigma_x$. (*b*) Find η_y and σ_y if $\mathbf{y} = (\mathbf{x} - \eta_x)/\sigma_x$.

5-11 Show that if **x** has a Rayleigh density with parameter α and $\mathbf{y} = b + c\mathbf{x}^2$, then $\sigma_y^2 = 4c^2\alpha^2$.

5-12 (*a*) Show that if m is the median of $\mathbf{x}$, then

$$E\{|\mathbf{x}-a|\} = E\{|\mathbf{x}-m|\} + 2\int_a^m (x-a)f(x)\,dx$$

for any a. (*b*) Find c such that $E\{|\mathbf{x}-c|\}$ is minimum.

5-13 Show that if the RV $\mathbf{x}$ is $N(\eta; \sigma)$, then

$$E\{|\mathbf{x}|\} = \sigma\sqrt{\frac{2}{\pi}}\,e^{-\eta^2/2\sigma^2} + 2\eta\mathbb{G}\left(\frac{\eta}{\sigma}\right) - \eta$$

5-14 If $\mathbf{x}$ is $N(0, 2)$ and $\mathbf{y} = 3\mathbf{x}^2$, find η_y, σ_y, and $f_y(y)$.

5-15 Show that

$$E\{\mathbf{x}\} = \int_{-\infty}^{0} F_x(x)\,dx + \int_0^{\infty} [1 - F_x(x)]\,dx$$

5-16 Show that if $\mathfrak{A} = [\mathscr{A}_1, \ldots, \mathscr{A}_n]$ is a partition of $\mathscr{S}$, then

$$E\{\mathbf{x}\} = E\{\mathbf{x}\,|\,\mathscr{A}_1\}P(\mathscr{A}_1) + \cdots + E\{\mathbf{x}\,|\,\mathscr{A}_n\}P(\mathscr{A}_n).$$

5-17 Show that if $\mathbf{x} \geq 0$ and $E\{\mathbf{x}\} = \eta$, then $P\{\mathbf{x} \geq \sqrt{\eta}\} \leq \sqrt{\eta}$.

5-18 Using (5-55), find $E\{\mathbf{x}^3\}$ if $\eta_x = 10$ and $\sigma_x = 2$.

5-19 If $\mathbf{x}$ is uniform in the interval (10, 12) and $\mathbf{y} = \mathbf{x}^3$ (*a*) find $f_y(y)$; (*b*) find $E\{\mathbf{y}\}$: 1, exactly; 2, using (5-55).

5-20 We are given an even concave function $g(x)$ and an RV $\mathbf{x}$ whose density $f(x)$ is symmetrical as in Fig. P5-20 with a single maximum at $x = \eta$. Show that the mean $E\{g(\mathbf{x}-a)\}$ of the RV $g(\mathbf{x}-a)$ is minimum if $a = \eta$.

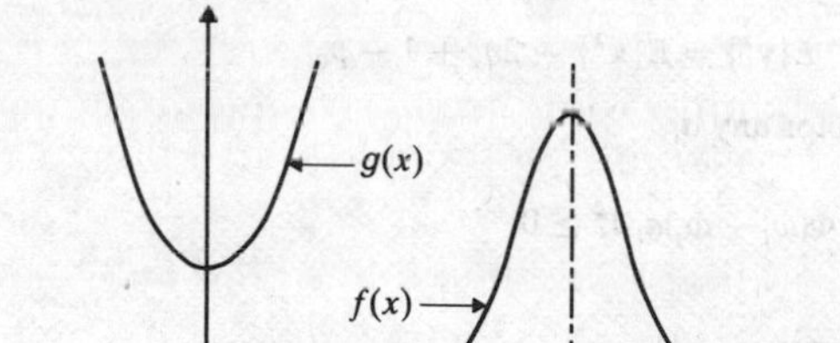

5-21 Show that if $\mathbf{x}$ and $\mathbf{y}$ are two RVs with densities $f_x(x)$ and $f_y(y)$ respectively, then

$$E\{\log f_x(\mathbf{x})\} > E\{\log f_y(\mathbf{x})\}$$

5-22 (*Chernoff bound*). (*a*) Show that for any $\alpha > 0$ and for any real s

$$P\{e^{s\mathbf{x}} \geq \alpha\} \leq \frac{\mathbf{\Phi}(s)}{\alpha} \qquad \text{where} \qquad \mathbf{\Phi}(s) = E\{e^{s\mathbf{x}}\} \tag{i}$$

Hint: Apply (5-57) to the RV $\mathbf{y} = e^{s\mathbf{x}}$. (*b*) For any A

$$P\{\mathbf{x} \geq A\} \leq e^{-sA}\mathbf{\Phi}(s) \qquad s > 0$$

$$P\{\mathbf{x} \leq A\} \leq e^{-sA}\mathbf{\Phi}(s) \qquad s < 0$$

Hint: Set $\alpha = e^{sA}$ in (i).

5-23 Show that (*a*) if $f(x)$ is a Cauchy density, then $\Phi(\omega) = e^{-\alpha|\omega|}$; (*b*) if $f(x)$ is a Laplace density, then $\Phi(\omega) = \alpha^2/(\alpha^2 + \omega^2)$.

5-24 Show that if $E\{\mathbf{x}\} = \eta$, then

$$E\{e^{s\mathbf{x}}\} = e^{s\eta} \sum_{n=0}^{\infty} \mu_n \frac{s^n}{n!} \qquad \mu_n = E\{(\mathbf{x} - \eta)^n\}$$

5-25 Show that if $\Phi_x(\omega_1) = 1$ for some $\omega_1 \neq 0$, then the RV $\mathbf{x}$ is of lattice type taking the values $x_n = 2\pi n/\omega_1$.

Hint:

$$0 = 1 - \Phi_x(\omega_1) = \int_{-\infty}^{\infty} (1 - e^{j\omega_1 x}) f_x(x)\, dx$$

5-26 The RV $\mathbf{x}$ has zero mean, central moments μ_n, and cumulants λ_n. Show that (*a*) $\lambda_3 = \mu_3$, $\lambda_4 = \mu_4 - 3\mu_2$; (*b*) if $\mathbf{y}$ is $N(0; \sigma_y)$ and $\sigma_y = \sigma_x$, then $E\{\mathbf{x}^4\} = E\{\mathbf{y}^4\} + \lambda_4$.

5-27 An RV $\mathbf{x}$ has a *geometric* distribution if

$$P\{\mathbf{x} = k\} = pq^k \qquad k = 0, 1, \ldots \qquad p + q = 1$$

Find $\Gamma(z)$ and show that $\eta_x = q/p$, $\sigma_x^2 = 2q/p^2$.

5-28 An RV $\mathbf{x}$ has a *Pascal* (or *negative binomial*) distribution if

$$P\{\mathbf{x} = k\} = \binom{-n}{k} p^n (-q)^k = \binom{n+k-1}{k} p^n q^k \qquad k = 0, 1, \ldots$$

Find $\Gamma(z)$ and show that $\eta_x = nq/p$, $\sigma_x^2 = nq/p^2$

5-29 The RV $\mathbf{x}$ takes the values 0, 1, ... with $P\{\mathbf{x} = k\} = p_k$. Show that if

$$\mathbf{y} = (\mathbf{x} - 1)U(\mathbf{x} - 1) \qquad \text{then} \qquad \Gamma_y(z) = p_0 + [z\Gamma_x(z) - p_0]$$

$$\eta_y = \eta_x - 1 + p_0 \qquad E\{\mathbf{y}^2\} = E\{\mathbf{x}^2\} - 2\eta_x + 1 - p_0$$

5-30 Show that, if $\Phi(\omega) = E\{e^{j\omega\mathbf{x}}\}$, then for any a_i

$$\sum_{i=1}^{n} \sum_{j=1}^{n} \Phi(\omega_i - \omega_j) a_i a_j^* \geq 0$$

Hint:

$$E\left\{\left|\sum_{i=1}^{n} a_i e^{j\omega_i \mathbf{x}}\right|^2\right\} \geq 0$$

5-31 The RV $\mathbf{x}$ is $N(0; \sigma)$. (*a*) Using characteristic functions, show that if $g(x)$ is a function such that $g(x)e^{-x} \to 0$ as $|x| \to \infty$, then

$$\frac{dE\{g(\mathbf{x})\}}{dv} = \frac{1}{2} E\left\{\frac{d^2 g(\mathbf{x})}{d\mathbf{x}^2}\right\} \qquad v = \sigma^2 \tag{i}$$

(*b*) The moments μ_n of $\mathbf{x}$ are functions of v. Using (i) show that

$$\mu_n(v) = \frac{n(n-1)}{2} \int_0^v \mu_{n-2}(\beta)\, d\beta$$

5-32 Show that, if $\mathbf{n}$ is an integer-valued RV with moment function $\Gamma(z)$ as in (5-75), then

$$P\{\mathbf{n} = k\} = \frac{1}{2\pi} \int_{-\pi}^{\pi} \Gamma(e^{j\omega}) e^{-jk\omega}\, d\omega$$

CHAPTER

SIX

TWO RANDOM VARIABLES

6-1 JOINT STATISTICS

We are given two RVs **x** and **y**, defined as in Sec. 4-1, and we wish to determine their joint statistics, i.e., the probability that the point (**x**, **y**) is in a specified region† D in the xy plane. The distribution functions $F_x(x)$ and $F_y(y)$ of the given RVs determine their separate (marginal) statistics but not their joint statistics. In particular, the probability of the event

$$\{\mathbf{x} \le x\} + \{\mathbf{y} \le y\} = \{\mathbf{x} \le x, \mathbf{y} \le y\}$$

cannot be expressed in terms of $F_x(x)$ and $F_y(y)$. In the following, we show that the joint statistics of the RVs **x** and **y** are completely determined if the probability of this event is known for every x and y.

Joint Distribution and Density

The joint distribution $F_{xy}(x, y)$ or, simply, $F(x, y)$ of two RVs **x** and **y** is the probability of the event

$$\{\mathbf{x} \le x, \mathbf{y} \le y\} = \{(\mathbf{x}, \mathbf{y}) \in D_1\}$$

where x and y are two arbitrary real numbers and D_1 is the quadrant shown in Fig. 6-1*a*

$$F(x, y) = P\{\mathbf{x} \le x, \mathbf{y} \le y\} \qquad (6\text{-}1)$$

† The region D is arbitrary subject only to the mild condition that it can be expressed as a countable union or intersection of rectangles.

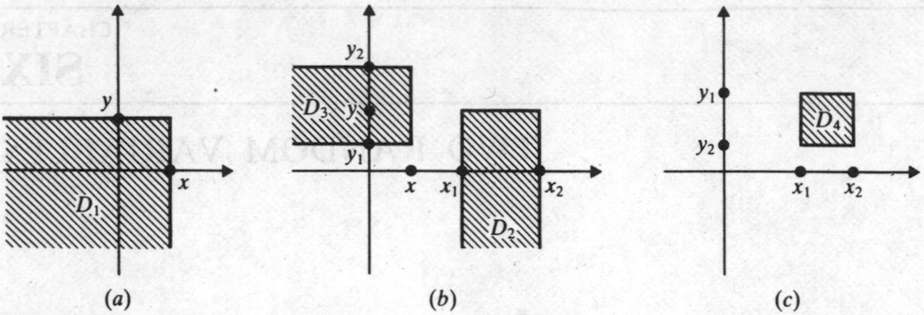

Figure 6-1

Properties 1. The function $F(x, y)$ is such that

$$F(-\infty, y) = 0 \qquad F(x, -\infty) = 0 \qquad F(\infty, \infty) = 1$$

PROOF As we know, $P\{\mathbf{x} = -\infty\} = P\{\mathbf{y} = -\infty\} = 0$. And since

$$\{\mathbf{x} = -\infty, \mathbf{y} \le y\} \subset \{\mathbf{x} = -\infty\} \qquad \{\mathbf{x} \le x, \mathbf{y} = -\infty\} \subset \{\mathbf{y} = -\infty\}$$

the first two equations follow. The last is a consequence of the identities

$$\{\mathbf{x} \le \infty, \mathbf{y} \le \infty\} = \mathscr{S} \qquad P(\mathscr{S}) = 1$$

2. The event $\{x_1 < \mathbf{x} \le x_2, \mathbf{y} \le y\}$ consists of all points $(\mathbf{x}, \mathbf{y})$ in the vertical half-strip D_2 and the event $\{\mathbf{x} \le x, y_1 < \mathbf{y} \le y_2\}$ consists of all point $(\mathbf{x}, \mathbf{y})$ in the horizontal half-strip D_3 of Fig. 6-1*b*. We maintain that

$$P\{x_1 < \mathbf{x} < x_2, \mathbf{y} \le y\} = F(x_2, y) - F(x_1, y) \tag{6-2}$$

$$P\{\mathbf{x} < x, y_1 < \mathbf{y} \le y_2\} = F(x, y_2) - F(x, y_1) \tag{6-3}$$

PROOF Clearly

$$\{\mathbf{x} \le x_2, \mathbf{y} \le y\} = \{\mathbf{x} \le x_1, \mathbf{y} \le y\} + \{x_1 < \mathbf{x} \le x_2, \mathbf{y} \le y\}$$

The last two events are mutually exclusive, hence [see (2-10)]

$$P\{\mathbf{x} \le x_2, \mathbf{y} \le y\} = P\{\mathbf{x} \le x_1, \mathbf{y} \le y\} + P\{x_1 < \mathbf{x} \le x_2, \mathbf{y} \le y\}$$

and (6-2) results. The proof of (6-3) is similar.

3. $\quad P\{x_1 < \mathbf{x} \le x_2, y_1 < \mathbf{y} \le y_2\}$

$$= F(x_2, y_2) - F(x_1, y_2) - F(x_2, y_1) + F(x_1, y_1) \tag{6-4}$$

This is the probability that $(\mathbf{x}, \mathbf{y})$ is in the rectangle D_4 of Fig. 6-1*c*.

PROOF It follows from (6-2) and (6-3) because

$$\{x_1 < \mathbf{x} \le x_2, \mathbf{y} \le y_2\} = \{x_1 < \mathbf{x} \le x_2, \mathbf{y} \le y_1\} + \{x_1 < \mathbf{x} \le x_2, y_1 < \mathbf{y} \le y_2\}$$

and the last two events are mutually exclusive.

Joint density The joint density of **x** and **y** is by definition the function

$$f(x, y) = \frac{\partial^2 F(x, y)}{\partial x\, \partial y} \tag{6-5}$$

From this and property 1 it follows that

$$F(x, y) = \int_{-\infty}^{x} \int_{-\infty}^{y} f(\alpha, \beta)\, d\alpha\, d\beta \tag{6-6}$$

Joint statistics We shall now show that the probability that the point (**x**, **y**) is in a region D of the xy plane equals the integral of $f(x, y)$ in D. In other words

$$P\{(\mathbf{x}, \mathbf{y}) \in D\} = \int_D \int f(x, y)\, dx\, dy \tag{6-7}$$

where $\{(\mathbf{x}, \mathbf{y}) \in D\}$ is the event consisting of all outcomes ζ such that the point $[\mathbf{x}(\zeta), \mathbf{y}(\zeta)]$ is in D.

PROOF As we know, the ratio

$$\frac{F(x + \Delta x, y + \Delta y) - F(x, y + \Delta y) - F(x + \Delta x, y) + F(x, y)}{\Delta x\, \Delta y}$$

tends to $\partial^2 F(x, y)/\partial x\, \partial y$ as $\Delta x \to 0$ and $\Delta y \to 0$. Hence, [see (6-4) and (6-5)]

$$P\{x < \mathbf{x} \le x + \Delta x, y < \mathbf{y} \le y + \Delta y) \simeq f(x, y)\, \Delta x\, \Delta y \tag{6-8}$$

We have, thus, shown that the probability that (**x**, **y**) is in a differential rectangle equals $f(x, y)$ times the area $\Delta x\, \Delta y$ of the rectangle. This proves (6-7) because the region D can be written as the limit of the union of such rectangles.

Marginal statistics In the study of several RVs, the statistics of each are called marginal. Thus, $F_x(x)$ is the *marginal distribution* and $f_x(x)$ the *marginal density* of **x**. In the following, we express the marginal statistics of **x** and **y** in terms of their joint statistics $F(x, y)$ and $f(x, y)$.

We maintain that

$$F_x(x) = F(x, \infty) \qquad F_y(y) = F(\infty, y) \tag{6-9}$$

$$f_x(x) = \int_{-\infty}^{\infty} f(x, y)\, dy \qquad f_y(y) = \int_{-\infty}^{\infty} f(x, y)\, dx \tag{6-10}$$

PROOF Clearly $\{\mathbf{x} \le \infty\} = \{\mathbf{y} \le \infty\} = \mathscr{S}$, hence

$$\{\mathbf{x} \le x\} = \{\mathbf{x} \le x, \mathbf{y} \le \infty\} \qquad \{\mathbf{y} \le y\} = \{\mathbf{x} \le \infty, \mathbf{y} \le y\}$$

The probabilities of the two sides above yield (6-9).

Differentiating (6-6), we obtain

$$\frac{\partial F(x, y)}{\partial x} = \int_{-\infty}^{y} f(x, \beta)\, d\beta \qquad \frac{\partial F(x, y)}{\partial y} = \int_{-\infty}^{x} f(\alpha, y)\, d\alpha \tag{6-11}$$

Setting $y = \infty$ in the first and $x = \infty$ in the second equation, we obtain (6-10) because [see (6-9)]

$$f_x(x) = \frac{\partial F(x, \infty)}{\partial x} \qquad f_y(y) = \frac{\partial F(\infty, y)}{\partial y}$$

Existence theorem From properties 1 and 3 it follows that

$$F(-\infty, y) = 0 \qquad F(x, -\infty) = 0 \qquad F(\infty, \infty) = 1 \tag{6-12}$$

and

$$F(x_2, y_2) - F(x_1, y_2) - F(x_2, y_1) + F(x_1, y_1) \geq 0 \tag{6-13}$$

for every $x_1 < x_2$, $y_1 < y_2$. Hence [see (6-6) and (6-8)]

$$\int_{-\infty}^{\infty}\int_{-\infty}^{\infty} f(x, y)\, dx\, dy = 1 \qquad f(x, y) \geq 0 \tag{6-14}$$

Conversely, given $F(x, y)$ or $f(x, y)$ as above, we can find two RVs **x** and **y**, defined in some space $\mathscr{S}$, with distribution $F(x, y)$ or density $f(x, y)$. This can be done by extending the existence theorem of Sec. 4-3 to joint statistics.

Joint normality We shall say that the RVs **x** and **y** are *jointly normal* if their joint density is given by

$$f(x, y) = A \exp\left\{-\frac{1}{2(1 - r^2)}\left[\frac{(x - \eta_1)^2}{\sigma_1^2} - 2r\,\frac{(x - \eta_1)(y - \eta_2)}{\sigma_1\sigma_2} + \frac{(y - \eta_2)^2}{\sigma_2^2}\right]\right\} \tag{6-15}$$

This function is positive and its integral equals 1 if

$$A = \frac{1}{2\pi\sigma_1\sigma_2\sqrt{1 - r^2}} \qquad |r| < 1 \tag{6-16}$$

Thus, $f(x, y)$ is an exponential and its exponent is a negative quadratic because $|r| < 1$. The function $f(x, y)$ will be denoted by

$$N(\eta_1, \eta_2; \sigma_1, \sigma_2; r)$$

As we shall presently see, η_1 and η_2 are the expected values of **x** and **y**, and σ_1^2 and σ_2^2 their variances. The significance of r will be given later (correlation coefficient).

We maintain that the marginal densities of **x** and **y** are given by

$$f_x(x) = \frac{1}{\sigma_1\sqrt{2\pi}}\, e^{-(x-\eta_1)^2/2\sigma_1{}^2} \qquad f_y(y) = \frac{1}{\sigma_2\sqrt{2\pi}}\, e^{-(y-\eta_2)^2/2\sigma_2{}^2} \tag{6-17}$$

PROOF To prove the above, we must show that if (6-15) is inserted into (6-10), the result is (6-17). The bracket in (6-15) can be written in the form

$$\left[\cdots\right] = \left[\frac{x-\eta_1}{\sigma_1} - r\,\frac{y-\eta_2}{\sigma_2}\right]^2 + (1-r^2)\,\frac{(y-\eta_2)^2}{\sigma_2^2}$$

Hence

$$\int_{-\infty}^{\infty} f(x, y)\,dx = A \exp\left[-\frac{(y-\eta_2)^2}{2\sigma_2^2}\right]$$
$$\times \int_{-\infty}^{\infty} \exp\left\{-\frac{1}{2(1-r^2)}\left[\frac{x-\eta_1}{\sigma_1} - r\,\frac{y-\eta_2}{\sigma_2}\right]^2\right\} dx$$

The last integral is a constant B (independent of x and y). Therefore [see (6-10)]

$$f_y(y) = ABe^{-(y-\eta_2)^2/2\sigma_2^2}$$

And since $f_y(y)$ is a density, its area must equal 1. This yields $AB = 1/\sigma_2\sqrt{2\pi}$ and the second equation in (6-17) results. The proof of the first is similar.

Notes 1. From (6-17) it follows that if two RVs are jointly normal, they are also marginally normal. However, as the next example shows, *the converse is not true.*

2. Joint normality can be defined as follows: Two RVs **x** and **y** are jointly normal if the sum $a\mathbf{x} + b\mathbf{y}$ is normal for every a and b [see (8-56)].

Example 6-1 We shall construct two RVs $\mathbf{x}_1$ and $\mathbf{y}_1$ that are marginally but not jointly normal. We start with two jointly normal RVs **x** and **y** with density $f(x, y)$ as in (6-15). Adding and subtracting small masses in the region D of Fig. 6-2 consisting of four circles as shown, we obtain a new function $f_1(x, y)$ such that $f_1(x, y) = f(x, y) \pm \varepsilon$ in D and $f_1(x, y) = f(x, y)$ everywhere else. The function $f_1(x, y)$ so formed is a density, hence, it defines two new RVs $\mathbf{x}_1$ and $\mathbf{y}_1$. These RVs are not jointly normal because $f_1(x, y)$ is not of the form (6-15). We maintain, however, that they are marginally normal. Indeed, the densities of $\mathbf{x}_1$ and $\mathbf{x}_2$ are determined by the masses in the vertical strip $x_1 < x < x_1 + dx$ and the horizontal strip $y_1 < y < y_1 + dy$. As we see from the figure, the masses in these strips have not changed. This shows that $\mathbf{x}_1$ and $\mathbf{y}_1$ are normal because **x** and **y** are normal.

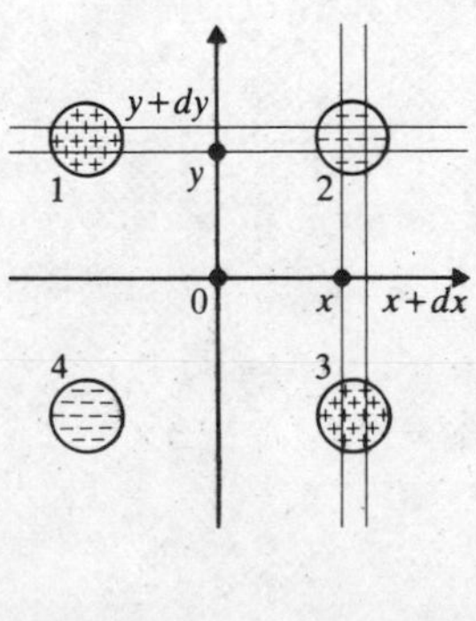

Figure 6-2

Discrete type Suppose that the RVs **x** and **y** are of discrete type taking the values x_i and y_k with respective probabilities

$$P\{\mathbf{x} = x_i\} = p_i \qquad P\{\mathbf{y} = y_k\} = q_k \tag{6-18}$$

Their joint statistics are determined in terms of the *joint probabilities*

$$P\{\mathbf{x} = x_i, \mathbf{y} = y_k\} = p_{ik} \tag{6-19}$$

Clearly

$$\sum_{i,k} p_{ik} = 1 \tag{6-20}$$

because, as i and k take all possible values, the events $\{\mathbf{x} = x_i, \mathbf{y} = y_k\}$ are mutually exclusive and their union equals the certain event.

We maintain that the *marginal probabilities* p_i and q_k can be expressed in terms of the joint probabilities p_{ik}

$$p_i = \sum_k p_{ik} \qquad q_k = \sum_i p_{ik} \tag{6-21}$$

This is the discrete version of (6-10).

PROOF The events $\{\mathbf{y} = y_k\}$ form a partition of $\mathscr{S}$. Hence, as k ranges over all possible values, the events $\{\mathbf{x} = x_i, \mathbf{y} = y_k\}$ are mutually exclusive and their union equals $\{\mathbf{x} = x_i\}$. This yields the first equation in (6-21) [see (2-36)]. The proof of the second is similar.

Probability Masses

The probability that the point (**x**, **y**) is in a region D of the plane can be interpreted as probability mass in this region. Thus, the mass in the entire plane equals 1. The mass in the half-plane $\mathbf{x} \le x$ to the left of the line L_x of Fig. 6-3 equals $F_x(x)$. The mass in the half-plane $\mathbf{y} \le y$ below the line L_y equals $F_y(y)$. The mass in the cross-hatched quadrant $(\mathbf{x} \le x, \mathbf{y} \le y)$ equals $F(x, y)$. Finally, the mass in the clear quadrant $(\mathbf{x} > x, \mathbf{y} > y)$ equals

$$P\{\mathbf{x} > x, \mathbf{y} > y\} = 1 - F_x(x) - F_y(y) + F(x, y) \tag{6-22}$$

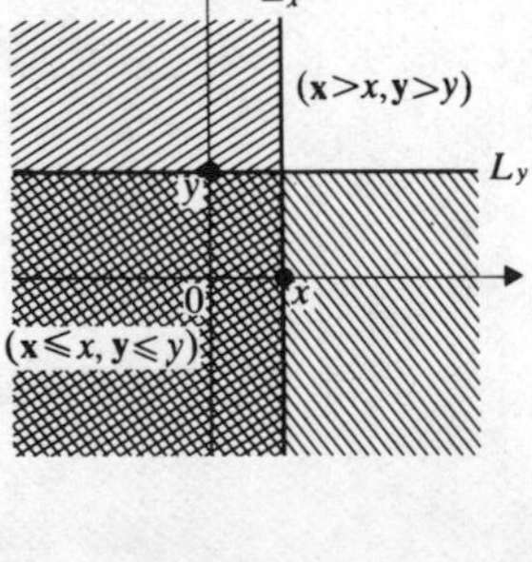

Figure 6-3

The probability mass in a region D equals the integral [see (6-7)]

$$\int_D \int f(x, y)\, dx\, dy$$

If, therefore, $f(x, y)$ is a bounded function, it can be interpreted as surface mass density.

Example 6-2 Suppose that

$$f(x, y) = \frac{1}{2\pi\sigma^2} e^{-(x^2+y^2)/2\sigma^2} \tag{6-23}$$

We shall find the mass m in the circle $x^2 + y^2 \le a^2$. Inserting (6-23) into (6-7) and using the transformation

$$x = r \cos \theta \qquad y = r \sin \theta$$

we obtain

$$m = \frac{1}{2\pi\sigma^2} \int_0^a \int_{-\pi}^{\pi} e^{-r^2/2\sigma^2} r\, dr\, d\theta = 1 - e^{-a^2/2\sigma^2} \tag{6-24}$$

Point masses If the RVs **x** and **y** are of discrete type taking the values x_i and y_k, then the probability masses are zero everywhere except at the points (x_i, y_k). We have, thus, only point masses and the mass at each point equals p_{ik} [see (6-19)]. The probability $p_i = P\{\mathbf{x} = x_i\}$ equals the sum of all mass p_{ik} on the line $x = x_i$ in agreement with (6-21).

If $i = 1, \ldots, M$ and $k = 1, \ldots, N$, then the number of possible point masses on the plane equals MN. However, as the next example shows, some of these masses might be zero.

Example 6-3 (*a*) In the fair-die experiment, **x** equals the number of faces shown and **y** equals twice this number. Thus

$$\mathbf{x}(f_i) = i \qquad \mathbf{y}(f_i) = 2i \qquad i = 1, \ldots, 6$$

In other words, $x_i = i$, $y_k = 2k$ and

$$p_{ik} = P\{\mathbf{x} = i, \mathbf{y} = 2k\} = \begin{cases} \frac{1}{6} & i = k \\ 0 & i \ne k \end{cases}$$

Thus, there are masses only on the six points $(i, 2i)$ and the mass of each point equals 1/6 (Fig. 6-4*a*).

(*b*) We toss the die twice obtaining the 36 outcomes $f_i f_k$ and we define **x** and **y** such that **x** equals the first number that shows and **y** the second

$$\mathbf{x}(f_i f_k) = i \qquad \mathbf{y}(f_i f_k) = k \qquad i, k = 1, \ldots, 6$$

Thus, $x_i = i$, $y_k = k$, and $p_{ik} = 1/36$. We have, therefore, 36 point masses (Fig. 6-4*b*) and the mass of each point equals 1/36. On the line $x = i$ there are 6 points with total mass 1/6.

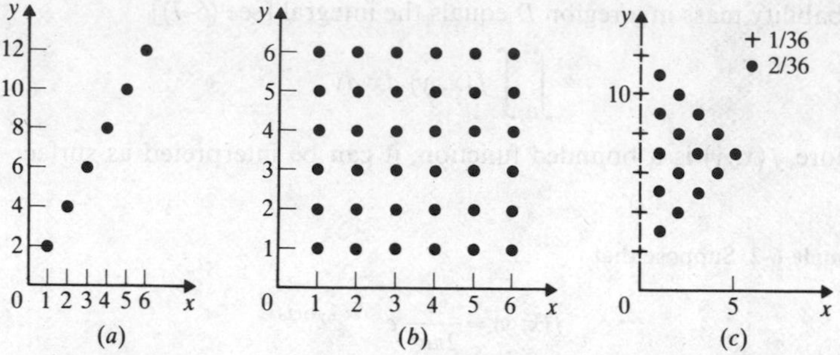

Figure 6-4

(c) Again the die is tossed twice but now

$$\mathbf{x}(f_i f_k) = |i - k| \qquad \mathbf{y}(f_i f_k) = i + k$$

In this case, $\mathbf{x}$ takes the values 0, 1, ..., 5 and $\mathbf{y}$ the values 2, 3, ..., 12. The number of possible points equals $5 \times 11 = 55$; however, only 21 have positive masses (Fig. 6-4c). Specifically, if $\mathbf{x} = 0$ then $\mathbf{y} = 2$, or 4, ..., or 12 because if $\mathbf{x} = 0$, then $i = k$ and $\mathbf{y} = 2i$. There are, therefore, 6 mass points on this line and the mass of each point equals 1/36. If $\mathbf{x} = 1$, then $\mathbf{y} = 3$, or 5, ..., or 11. There are, therefore, 5 mass points on the line $x = 1$ and the mass of each point equals 2/36. For example, if $\mathbf{x} = 1$ and $\mathbf{y} = 7$, then $i = 3, k = 4$, or $i = 4, k = 3$, hence, $P\{\mathbf{x} = 1, \mathbf{y} = 7\} = 2/36$.

Line masses The following cases lead to line masses:

1. If $\mathbf{x}$ is of discrete type taking the values x_i and $\mathbf{y}$ is of continuous type, then all probability masses are on the vertical lines $x = x_i$ (Fig. 6-5a). In particular, the mass between the points y_1 and y_2 on the line $x = x_i$ equals the probability of the event

$$\{\mathbf{x} = x_i, y_1 \le \mathbf{y} \le y_2\}$$

2. If $\mathbf{y} = g(\mathbf{x})$, then all masses are on the curve $y = g(x)$. In this case, $F(x, y)$ can be expressed in terms of $F_x(x)$. For example, with x and y as in Fig. 6-5b, $F(x, y)$ equals the masses on the curve $y = g(x)$ to the left of the point A and

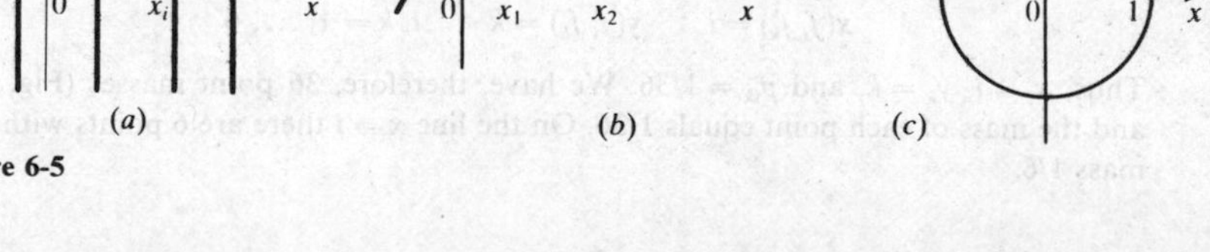

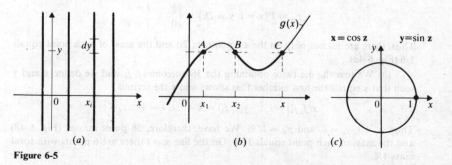

Figure 6-5

between B and C (heavy line). The masses to the left of A equal $F_x(x_1)$. The masses between B and C equal $F_x(x_3) - F_x(x_2)$. Hence

$$F(x, y) = F_x(x_1) + F_x(x_3) - F_x(x_2) \qquad y = g(x_1) = g(x_2) = g(x_3)$$

3. If $\mathbf{x} = g(\mathbf{z})$ and $\mathbf{y} = h(\mathbf{z})$, then all probability masses are on the curve $x = g(z)$, $y = h(z)$ specified parametrically. For example, if $g(z) = \cos z$, $h(z) = \sin z$, then the curve is a circle (Fig. 6-5c). In this case, the joint statistics of **x** and **y** can be expressed in terms of $F_z(z)$.

Independence

Two RVs **x** and **y** are called (*statistically*) *independent* if the events $\{\mathbf{x} \in A\}$ and $\{\mathbf{y} \in B\}$ are independent [see (2-40)], i.e., if

$$P\{\mathbf{x} \in A, \mathbf{y} \in B\} = P\{\mathbf{x} \in A\}P\{\mathbf{y} \in B\} \tag{6-25}$$

where A and B are two arbitrary sets on the x and y axes respectively.

Applying the above to the events $\{\mathbf{x} \le x\}$ and $\{\mathbf{y} \le y\}$, we conclude that, if the RVs **x** and **y** are independent, then

$$F(x, y) = F_x(x)F_y(y) \tag{6-26}$$

Hence

$$f(x, y) = f_x(x)f_y(y) \tag{6-27}$$

It can be shown that, if (6-26) or (6-27) is true, then (6-25) is also true, i.e., the RVs **x** and **y** are independent [see (6-7)].

If the RVs **x** and **y** are of discrete type as in (6-19) and independent, then

$$p_{ik} = p_i p_k \tag{6-28}$$

This follows if we apply (6-25) to the events $\{\mathbf{x} = x_i\}$ and $\{\mathbf{y} = y_k\}$.

Example 6-4 (Buffon's needle) A fine needle of length $2a$ is dropped at random on a board covered with parallel lines distance $2b$ apart as in Fig. 6-6a. We shall show that the probability p that the needle intersects one of the lines equals $2a/\pi b$.

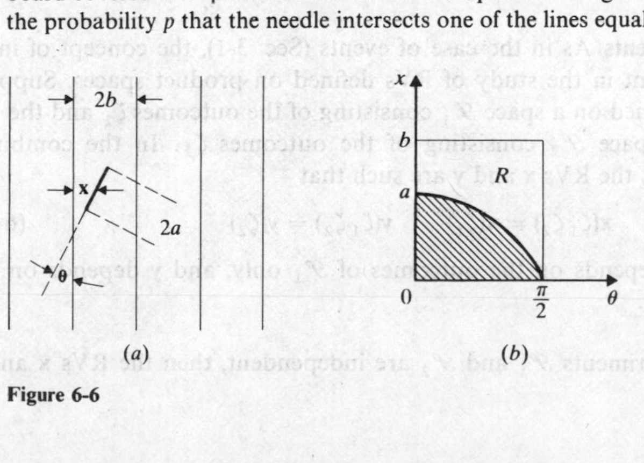

Figure 6-6

In terms of RVs the above experiment can be phrased as follows: We denote by **x** the distance from the center of the needle to the nearest line and by **θ** the angle between the needle and the direction of the lines. We assume that the RVs **x** and **θ** are independent, **x** is uniform in the interval (0, *b*), and **θ** is uniform in the interval (0, $\pi/2$). From this it follows that

$$f(x, \theta) = f_x(x) f_\theta(\theta) = \frac{1}{b}\frac{2}{\pi} \qquad 0 \le x \le b \qquad 0 \le \theta \le \frac{\pi}{2}$$

and zero elsewhere. Hence, the probability that the point $(\mathbf{x}, \boldsymbol{\theta})$ is in a region D included in the rectangle R of Fig. 6-6*b* equals the area of D times $2/\pi b$.

The needle intersects the lines if $x < b \cos \theta$. Hence, p equals the shaded area of Fig. 6-6*b* times $2/\pi b$

$$p = P\{\mathbf{x} < a \cos \boldsymbol{\theta}\} = \frac{2}{\pi b} \int_0^a a \cos \theta \, d\theta = \frac{2a}{\pi b}$$

The above can be used to determine experimentally the number π using the relative frequency interpretation of p: If the needle is dropped n times and it intersects the lines n_i times, then

$$\frac{n_i}{n} \simeq p = \frac{2a}{\pi b} \qquad \text{hence} \qquad \pi \simeq \frac{2an}{bn_i}$$

Theorem If the RVs **x** and **y** are independent then the RVs

$$\mathbf{z} = g(\mathbf{x}) \qquad \mathbf{w} = h(\mathbf{y})$$

are also independent.

Proof We denote by A_z the set of points on the x axis such that $g(x) \le z$ and by B_w the set of points on the y axis such that $h(y) \le w$. Clearly

$$\{\mathbf{z} \le z\} = \{\mathbf{x} \in A_z\} \qquad \{\mathbf{w} \le w\} = \{\mathbf{y} \in B_w\} \tag{6-29}$$

Therefore, the events $\{\mathbf{z} \le z\}$ and $\{\mathbf{w} \le w\}$ are independent because the events $\{\mathbf{x} \in A_z\}$ and $\{\mathbf{y} \in B_w\}$ are independent.

Independent experiments As in the case of events (Sec. 3-1), the concept of independence is important in the study of RVs defined on product spaces. Suppose that the RV **x** is defined on a space $\mathscr{S}_1$ consisting of the outcomes ζ_1 and the RV **y** is defined on a space $\mathscr{S}_2$ consisting of the outcomes ζ_2. In the combined experiment $\mathscr{S}_1 \times \mathscr{S}_2$ the RVs **x** and **y** are such that

$$\mathbf{x}(\zeta_1 \zeta_2) = \mathbf{x}(\zeta_1) \qquad \mathbf{y}(\zeta_1 \zeta_2) = \mathbf{y}(\zeta_2) \tag{6-30}$$

In other words, **x** depends on the outcomes of $\mathscr{S}_1$ only, and **y** depends on the outcomes of $\mathscr{S}_2$ only.

Theorem If the experiments $\mathscr{S}_1$ and $\mathscr{S}_2$ are independent, then the RVs **x** and **y** are independent.

PROOF We denote by $\mathscr{A}_x$ the set $\{\mathbf{x} \le x\}$ in $\mathscr{S}_1$ and by $\mathscr{B}_y$ the set $\{\mathbf{y} \le y\}$ in $\mathscr{S}_2$. In the space $\mathscr{S}_1 \times \mathscr{S}_2$

$$\{\mathbf{x} \le x\} = \mathscr{A}_x \times \mathscr{S}_2 \qquad \{\mathbf{y} \le y\} = \mathscr{S}_1 \times \mathscr{B}_y$$

From the independence of the two experiments it follows that [see (3-4)] the events $\mathscr{A}_x \times \mathscr{S}_2$ and $\mathscr{S}_1 \times \mathscr{B}_y$ are independent. Hence, the events $\{\mathbf{x} \le x\}$ and $\{\mathbf{y} \le y\}$ are also independent.

Example 6-5 A die with $P\{f_i\} = p_i$ is tossed twice and the RVs **x** and **y** are such that

$$\mathbf{x}(f_i f_k) = i \qquad \mathbf{y}(f_i f_k) - k$$

Thus, **x** equals the first number that shows and **y** equals the second, hence, the RVs **x** and **y** are independent. This leads to the conclusion that

$$p_{ik} = P\{\mathbf{x} = i, \mathbf{y} = k\} = p_i p_k$$

Circular Symmetry

We say that the statistics of two RVs **x** and **y** are circularly symmetrical if their joint density depends only on the distance from the origin, i.e., if

$$f(x, y) = g(r) \qquad r = \sqrt{x^2 + y^2} \tag{6-31}$$

Theorem If the RVs **x** and **y** are circularly symmetrical and independent, then they are normal with zero mean and equal variance.

PROOF From (6-31) and (6-27) it follows that

$$g(\sqrt{x^2 + y^2}) = f_x(x) f_y(y) \tag{6-32}$$

Since

$$\frac{\partial g(r)}{\partial x} = \frac{dg(r)}{dr}\frac{\partial r}{\partial x} \qquad \text{and} \qquad \frac{\partial r}{\partial x} = \frac{x}{r}$$

we conclude differentiating (6-32) with respect to x that

$$\frac{x}{r} g'(r) = f'_x(x) f_y(y)$$

Dividing both sides by $xg(r) = x f_x(x) f_y(y)$, we obtain

$$\frac{1}{r}\frac{g'(r)}{g(r)} = \frac{1}{x}\frac{f'_x(x)}{f_x(x)} \tag{6-33}$$

The right side above is independent of y and the left side is a function of $r = \sqrt{x^2 + y^2}$. This shows that both sides are independent of x and y. Hence

$$\frac{1}{r}\frac{g'(r)}{g(r)} = \alpha = \text{constant}$$

From this it follows that

$$\frac{d \ln g(r)}{dr} = \alpha r \qquad g(r) = Ae^{\alpha r^2/2}$$

and (6-31) yields

$$f(x, y) = g(\sqrt{x^2 + y^2}) = Ae^{\alpha(x^2+y^2)/2} \tag{6-34}$$

Thus, the RVs **x** and **y** are normal with zero mean and variance $\sigma^2 = -1/\alpha$.

6-2 ONE FUNCTION OF TWO RANDOM VARIABLES

Given two RVs **x** and **y** and a function $g(x, y)$, we form the RV

$$\mathbf{z} = g(\mathbf{x}, \mathbf{y})$$

We shall express the statistics of **z** in terms of the function $g(x, y)$ and the joint statistics of **x** and **y**.

With z a given number, we denote by D_z the region of the xy plane such that $g(x, y) \le z$. This region might not be simply connected (Fig. 6-7). Clearly

$$\{\mathbf{z} \le z\} = \{g(\mathbf{x}, \mathbf{y}) \le z\} = \{(\mathbf{x}, \mathbf{y}) \in D_z\}$$

Hence [see (6-7)]

$$F_z(z) = P\{\mathbf{z} \le z\} = P\{(\mathbf{x}, \mathbf{y}) \in D_z\} = \int\int_{D_z} f(x, y)\, dx\, dy \tag{6-35}$$

Thus, to determine $F_z(z)$ it suffices to find the region D_z for every z and to evaluate the above integral.

The density of **z** can be determined similarly. With ΔD_z the region of the xy plane such that $z < g(x, y) < z + dz$, we have

$$\{z < \mathbf{z} \le z + dz\} = \{(\mathbf{x}, \mathbf{y}) \in \Delta D_z\}$$

Hence

$$f_z(z)\, dz = P\{z < \mathbf{z} \le z + dz\} = \iint_{\Delta D_z} f(x, y)\, dx\, dy \tag{6-36}$$

Illustrations

In the following, we use (6-35) and (6-36) to find the statistics of various functions of **x** and **y**.

1. $\mathbf{z} = \mathbf{x} + \mathbf{y}$

The region D_z of the xy plane such that $x + y \le z$ is the shaded part of

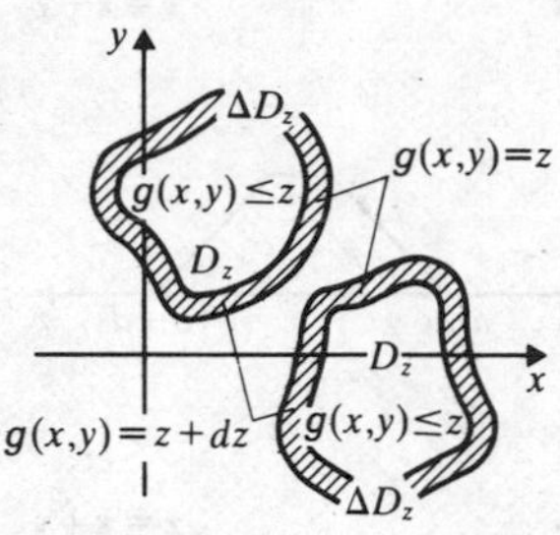

Figure 6-7

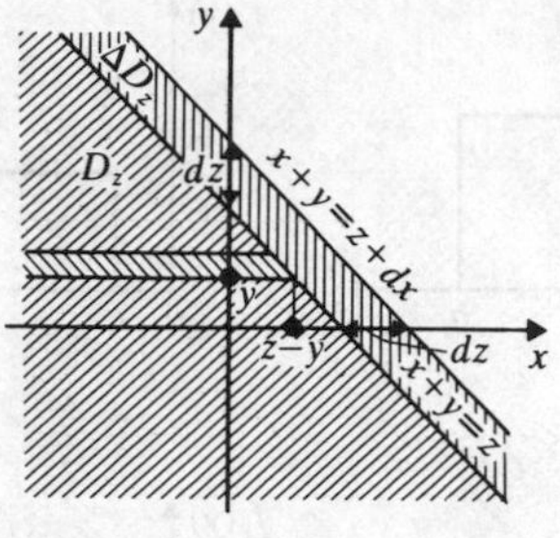

Figure 6-8

Fig. 6-8 to the left of the line $x + y = z$. Integrating over suitable strips, we obtain

$$F_z(z) = \int_{-\infty}^{\infty} \int_{-\infty}^{z-y} f(x, y)\, dx\, dy \tag{6-37}$$

We can find $f_z(z)$ either by differentiating $F_z(z)$ or directly from (6-36). The region ΔD_z such that $z < x + y < z + dz$ is a diagonal strip bounded by the lines $x + y = z$ and $x + y = z + dz$. The coordinates of a point of this region are $z - y$, y and the area of a differential equals $dy\, dz$. Hence

$$f_z(z)\, dz = \int_{-\infty}^{\infty} f(z - y, y)\, dy\, dz \tag{6-38}$$

Independence and convolution If the RVs **x** and **y** are independent, then

$$f(x, y) = f_x(x) f_y(y)$$

Inserting into (6-38), we obtain

$$f_z(z) = \int_{-\infty}^{\infty} f_x(z - y) f_y(y)\, dy \tag{6-39}$$

The above integral is the convolution of the functions $f_x(x)$ and $f_y(y)$. We thus reach the following fundamental conclusion:

If two RVs are independent, then the density of their sum equals the convolution of their densities.

We note that, if $f_x(x) = 0$ for $x < 0$ and $f_y(y) = 0$ for $y < 0$, then $f_z(0) = 0$ for $z < 0$ and

$$f_z(z) = \int_0^z f_x(z - y) f_y(y)\, dy \qquad z > 0 \tag{6-40}$$

Example 6-6 It follows from (6-39) that the convolution of two rectangles is a trapezoid. Hence, if the RVs **x** and **y** are uniform in the intervals (a, b) and (c, d) respectively, then the density of their sum $\mathbf{z} = \mathbf{x} + \mathbf{y}$ is a trapezoid as in Fig. 6-9*a*. If, in particular, $b - a = d - c$, then $f_z(z)$ is a triangle as in Fig. 6-9*b*.

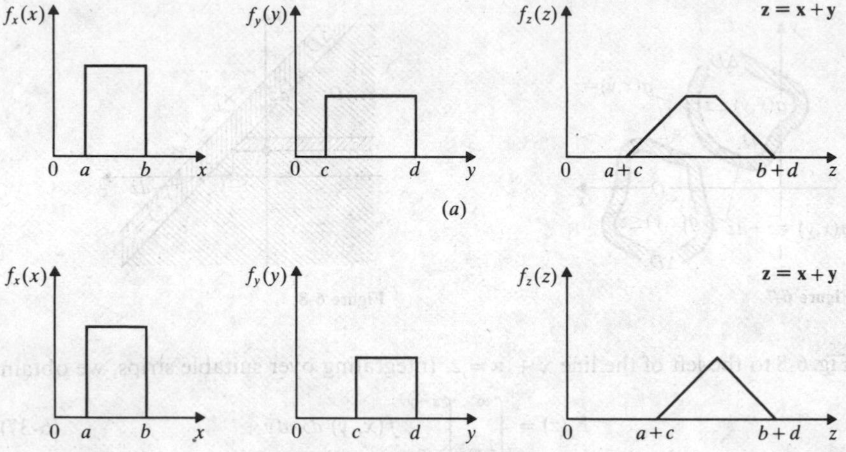

Figure 6-9 (b)

Suppose, for example, that resistors $\mathbf{r}_1$ and $\mathbf{r}_2$ are two independent RVs uniform between 900 and 1100 Ω. From the above it follows that, if they are connected in series, the density of the resulting resistor $\mathbf{r} = \mathbf{r}_1 + \mathbf{r}_2$ is a triangle between 1800 and 2200 Ω. In particular, the probability that $\mathbf{r}$ is between 1900 and 2100 Ω equals 0.75.

Example 6-7 If the RVs $\mathbf{x}$ and $\mathbf{y}$ are independent and

$$f_x(x) = \alpha e^{-\alpha x} U(x) \qquad f_y(y) = \beta e^{-\beta y} U(y)$$

(Fig. 6-10) then for $z > 0$,

$$f_z(z) = \alpha\beta \int_0^z e^{-\alpha(z-y)} e^{-\beta y} dy = \begin{cases} \dfrac{\alpha\beta}{\beta - \alpha}(e^{-\alpha z} - e^{-\beta z}) & \beta \neq \alpha \\ \alpha^2 z e^{-\alpha z} & \beta = \alpha \end{cases} \tag{6-41}$$

2. $\mathbf{z = x/y}$

The region D_z of the xy plane such that $x/y \le z$ is the shaded part of Fig. 6-11. Integrating over suitable strips, we obtain

$$F_z(z) = \int_0^\infty \int_{-\infty}^{yz} f(x, y)\, dx\, dy + \int_{-\infty}^0 \int_{yz}^\infty f(x, y)\, dx\, dy \tag{6-42}$$

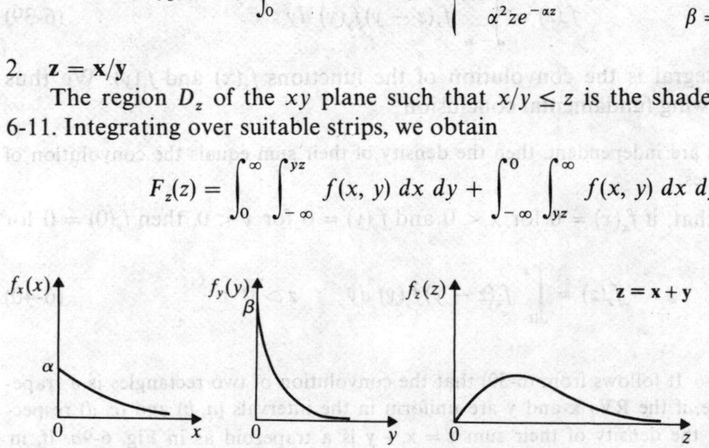

Figure 6-10

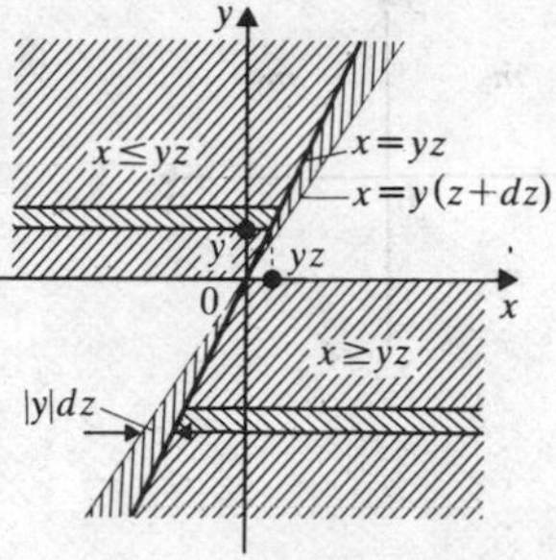

Figure 6-11

The region ΔD_z such that $z < x/y < z + dz$ is a triangle sector bounded by the lines $x = yz$ and $x = y(z + dz)$. The coordinates of a point in this region are zy, y and the area of a differential equals $|y|\,dy\,dz$. Inserting into (6-36) and cancelling dz, we obtain

$$f_z(z) = \int_{-\infty}^{\infty} |y|\, f(zy, y)\, dy \tag{6-43}$$

Normal densities We maintain that, if the RVs **x** and **y** are jointly normal with zero mean

$$f(x, y) = \frac{1}{2\pi\sigma_1\sigma_2\sqrt{1-r^2}} \exp\left[-\frac{1}{2(1-r^2)}\left(\frac{x^2}{\sigma_1^2} - 2r\frac{xy}{\sigma_1\sigma_2} + \frac{y^2}{\sigma_2^2}\right)\right] \tag{6-44}$$

then their ratio $\mathbf{z} = \mathbf{x}/\mathbf{y}$ has a *Cauchy* density centered at $r\sigma_1/\sigma_2$

$$f_z(z) = \frac{\sigma_1\sigma_2\sqrt{1-r^2}/\pi}{\sigma_2^2(z - r\sigma_1/\sigma_2)^2 + \sigma_1^2(1-r^2)} \tag{6-45}$$

PROOF Inserting (6-44) into (6-43) and using the fact that $f(-x, -y) = f(x, y)$, we obtain

$$f_z(z) = \frac{2}{2\pi\sigma_1\sigma_2\sqrt{1-r^2}} \int_0^{\infty} y \exp\left\{-\frac{y^2}{2(1-r^2)}\left[\frac{z^2}{\sigma_1^2} - 2r\frac{z}{\sigma_1\sigma_2} + \frac{1}{\sigma_1^2}\right]\right\} dy$$

and (6-45) results because the above integral equals $(1 - r^2)$ divided by the quantity in brackets.

Integrating (6-45) from $-\infty$ to z, we obtain the corresponding distribution function

$$F_z(z) = \frac{1}{2} + \frac{1}{\pi}\arctan\frac{\sigma_2 z - r\sigma_1}{\sigma_1\sqrt{1-r^2}} \tag{6-46}$$

Quadrant masses Using (6-46), we shall show that the probability masses m_1, m_2, m_3, m_4 in the four quadrants of the xy plane are given by

$$m_1 = m_3 = \frac{1}{4} + \frac{\alpha}{2\pi} \qquad m_2 = m_4 = \frac{1}{4} - \frac{\alpha}{2\pi} \tag{6-47}$$

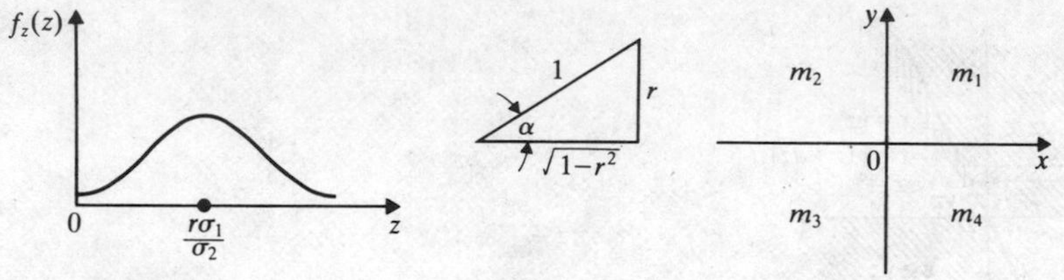

Figure 6-12

where (Fig. 6-12)

$$\alpha = \arcsin r = \arctan r/\sqrt{1-r^2} \qquad -\pi/2 < \alpha \le \pi/2$$

PROOF The second and fourth quadrant is the region of the plane such that $x/y \le 0$. The probability that the point $(\mathbf{x}, \mathbf{y})$ is in this region equals, therefore, the probability that the RV $\mathbf{z} = \mathbf{x}/\mathbf{y}$ is negative. Hence

$$m_2 + m_4 = P\{\mathbf{z} \le 0\} = F_z(0) = \frac{1}{2} - \arctan\frac{r}{\sqrt{1-r^2}}$$

and (6-47) results because

$$m_1 = m_3 \qquad m_2 = m_4 \qquad m_1 + m_2 + m_3 + m_4 = 1$$

This useful result could have been obtained by integrating $f(x, y)$ in each quadrant; the above method is, however, simpler.

3. $\mathbf{z} = \sqrt{\mathbf{x}^2 + \mathbf{y}^2}$

The region D_z is the circle $x^2 + y^2 \le z^2$ and $F_z(z)$ equals the probability masses in this circle. If $f(x, y) = g(r)$ is circularly symmetrical, then

$$F_z(z) = 2\pi \int_0^z rg(r)\,dr \qquad z > 0$$

Normal densities (*a*) If

$$f(x, y) = \frac{1}{2\pi\sigma^2}\, e^{-(x^2+y^2)/2\sigma^2} \tag{6-48}$$

then

$$F_z(z) = \frac{1}{\sigma^2}\int_0^z re^{-r^2/2\sigma^2}\,dr = 1 - e^{-z^2/2\sigma^2} \qquad z > 0 \tag{6-49}$$

Hence

$$f_z(z) = \frac{z}{\sigma^2}\, e^{-z^2/2\sigma^2} U(z) \tag{6-50}$$

Thus, if the RVs **x** and **y** are normal, independent with zero mean and equal variance, then the RV $\mathbf{z} = \sqrt{\mathbf{x}^2 + \mathbf{y}^2}$ has a Rayleigh density.

(*b*) Suppose now that

$$f(x, y) = \frac{1}{2\pi\sigma^2} e^{-[(x-\eta)^2 + y^2]/2\sigma^2} \tag{6-51}$$

In this case, the region ΔD_z of the xy plane such that $z < \sqrt{x^2 + y^2} < z + dz$ is a circular ring with inner radius z and thickness dz. With

$$x = z \cos \theta \qquad y = z \sin \theta \qquad dx\, dy = z\, dz\, d\theta$$

it follows that

$$f_z(z)\, dz = \iint_{\Delta D_z} f(x, y)\, dx\, dy = \frac{1}{2\pi\sigma^2} \int_0^{2\pi} e^{-[(z\cos\theta - \eta)^2 + (z \sin\theta)^2]/2\sigma^2} z\, dz\, d\theta$$

Hence

$$f_z(z) = \frac{z}{2\pi\sigma^2} e^{-(z^2+\eta^2)/2\sigma^2} \int_0^{2\pi} e^{z\eta\cos\theta/\sigma^2}\, d\theta$$

This yields

$$f_z(z) = \frac{z}{\sigma^2} I_0\left(\frac{z\eta}{\sigma^2}\right) e^{-(z^2+\eta^2)/2\sigma^2} \qquad z > 0 \tag{6-52}$$

where

$$I_0(x) = \frac{1}{2\pi} \int_0^{2\pi} e^{x\cos\theta}\, d\theta \tag{6-53}$$

is the modified Bessel function.

Example 6-8 Consider the sine wave

$$\mathbf{x} \cos \omega t + \mathbf{y} \sin \omega t = \mathbf{r} \cos (\omega t + \boldsymbol{\theta})$$

Since $\mathbf{r} = \sqrt{\mathbf{x}^2 + \mathbf{y}^2}$, it follows from the above that, if the RVs **x** and **y** are normal as in (6-49), then the density of **r** is Rayleigh as in (6-50).

4. $\mathbf{z} = \max(\mathbf{x}, \mathbf{y}) \qquad \mathbf{w} = \min(\mathbf{x}, \mathbf{y})$

(*a*) The region D_z of the xy plane such that $\max(x, y) \le z$ is the set of points such that $x \le z$ *and* $y \le z$ (shaded in Fig. 6-13*a*). Hence

$$F_z(z) = F_{xy}(z, z) \tag{6-54}$$

If the RVs **x** and **y** are *independent*, then

$$F_z(z) = F_x(z)F_y(z) \qquad f_z(z) = f_x(z)F_y(z) + f_y(z)F_x(z) \tag{6-55}$$

(*b*) The region D_w of the xy plane such that $\min(x, y) \le w$ is the set of points such that $x \le w$ *or* $y \le w$ (shaded in Fig. 6-13*b*). Hence

$$F_w(z) = F_x(w) + F_y(w) - F_{xy}(w, w) \tag{6-56}$$

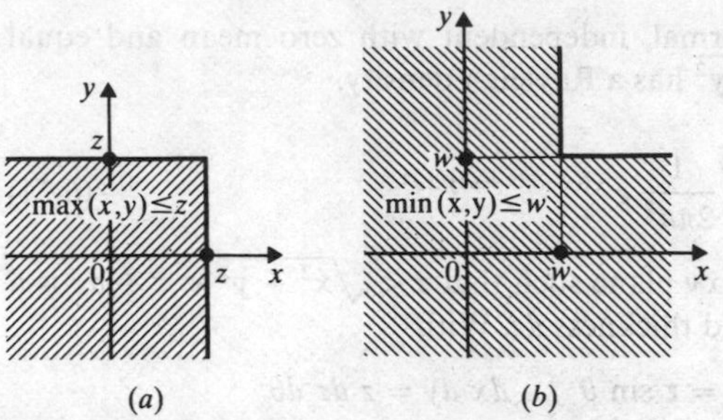

Figure 6-13

If the RVs **x** and **y** are *independent* then it is simpler to express the result in terms of the *reliability function*

$$R_x(x) = P\{\mathbf{x} > x\} = 1 - F_x(x) \tag{6-57}$$

Defining $R_y(y)$ and $R_w(w)$ similarly, we conclude from (6-56) that

$$R_w(w) = R_x(w)R_y(w) \qquad f_w(w) = f_x(w)R_y(w) + f_y(w)R_x(w) \tag{6-58}$$

Example 6-9 The times of failure of two random systems S_1 and S_2 are two RVs **x** and **y**. An outcome of the underlying experiment is the selection of two particular systems. The corresponding values $\mathbf{x}(\zeta)$ and $\mathbf{y}(\zeta)$ of the RVs **x** and **y** are the time intervals from the moment the systems are put into operation until they fail. Thus, if S_1 is put into operation at $t = 0$, then $F_x(t)$ is the probability that it will fail prior to time t and $f_x(t)\,dt$ is the probability that it will fail in the interval $(t, t + dt)$.

We connect the two systems in series or in parallel or in standby forming the combined system S as in Fig. 6-14.

We shall determine the distribution function $F_z(z)$ of the time of failure **z** of S.

(*a*) We say that the systems S_1 and S_2 are connected in *series* if S fails when at least one of the two systems fails. Hence

$$\mathbf{z} = \min(\mathbf{x}, \mathbf{y})$$

(*b*) We say that the systems S_1 and S_2 are connected in *parallel* if S fails when both systems fail. Hence,

$$\mathbf{z} = \max(\mathbf{x}, \mathbf{y})$$

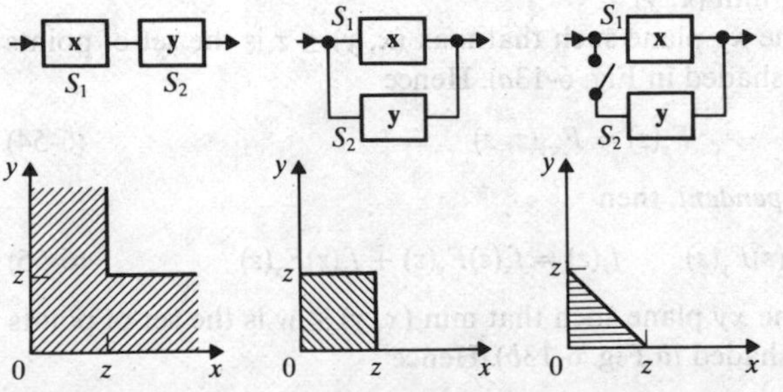

Figure 6-14

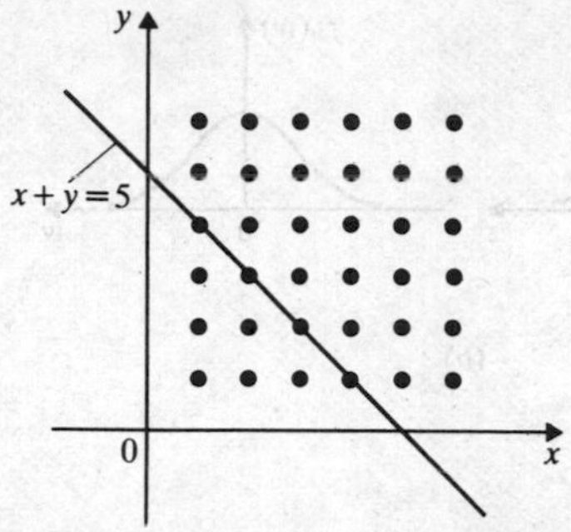

Figure 6-15

(*c*) In a *standby* connection, S_1 is put into operation while S_2 is idle. The moment S_1 fails, S_2 is put into operation and S fails when S_2 fails. Hence

$$\mathbf{z} = \mathbf{x} + \mathbf{y}$$

From (6-56), (6-54), and (6-37) it follows that $F_z(z)$ equals the mass in the shaded regions of Fig. 6-14.

Discrete type If the RVs **x** and **y** are of discrete type taking the values x_i and y_k, then the RV $\mathbf{z} = g(\mathbf{x}, \mathbf{y})$ is also of discrete type taking the values $z_r = g(x_i, y_k)$. The probability that $\mathbf{z} = z_r$ equals the sum of the point masses on the curve $g(x, y) = z_r$.

Example 6-10 A fair die is tossed twice and the RVs **x** and **y** are such that

$$\mathbf{x}(f_i f_k) = i \qquad \mathbf{x}(f_i f_k) = k$$

The *xy* plane has 36 equal point masses as in Fig. 6-15. The RV $\mathbf{z} = \mathbf{x} + \mathbf{y}$ takes the values $z_r = x_i + y_k$ with probabilities $p_r = m/36$ where m is the number of points on the line $x + y = z_r$. As we see from the figure

$$z_r = 2 \quad 3 \quad 4 \quad 5 \quad 6 \quad 7 \quad 8 \quad 9 \quad 10 \quad 11 \quad 12$$

$$p_r = \frac{1}{36} \; \frac{2}{36} \; \frac{3}{36} \; \frac{4}{36} \; \frac{5}{36} \; \frac{6}{36} \; \frac{5}{36} \; \frac{4}{36} \; \frac{3}{36} \; \frac{2}{36} \; \frac{1}{36}$$

For example, there are four mass points on the line $x + y = 5$, hence, $p_5 = 4/36$.

6-3 TWO FUNCTIONS OF TWO RANDOM VARIABLES

Given two RVs **x** and **y** and two functions $g(x, y)$ and $h(x, y)$, we form the RVs

$$\mathbf{z} = g(\mathbf{x}, \mathbf{y}) \qquad \mathbf{w} = h(\mathbf{x}, \mathbf{y}) \tag{6-59}$$

We shall express the joint statistics of **z** and **w** in terms of the functions $g(x, y)$ and $h(x, y)$ and the joint statistics of **x** and **y**.

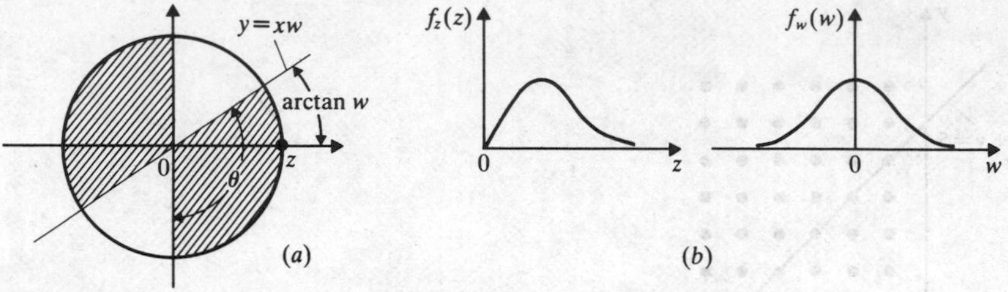

Figure 6-16

With z and w two given numbers, we denote by D_{zw} the region of the xy plane such that $g(x, y) \le z$ and $h(x, y) \le w$. Clearly

$$\{\mathbf{z} \le z, \mathbf{w} \le w\} = \{(\mathbf{x}, \mathbf{y}) \in D_{zw}\}$$

Hence [see (6-7)]

$$F_{zw}(z, w) = P\{(\mathbf{x}, \mathbf{y}) \in D_{zw}\} = \iint_{D_{zw}} f_{xy}(x, y)\, dx\, dy \tag{6-60}$$

Suppose, for example, that

$$\mathbf{z} = \sqrt{\mathbf{x}^2 + \mathbf{y}^2} \qquad \mathbf{w} = \mathbf{y}/\mathbf{x} \tag{6-61}$$

In this case, the set D_{zw} such that

$$\sqrt{x^2 + y^2} \le z \qquad y/x \le w$$

is the shaded region of Fig. 6-16*a*, and $F_{zw}(z, w)$ equals the mass in this region.

Example 6-11 If

$$f_{xy}(x, y) = \frac{1}{2\pi\sigma^2} e^{-(x^2+y^2)/2\sigma^2} \qquad \mathbf{z} = \sqrt{\mathbf{x}^2 + \mathbf{y}^2} \qquad \mathbf{w} = \mathbf{y}/\mathbf{x}$$

then [see (6-49)] the mass in the circle $x^2 + y^2 \le z^2$ equals $1 - e^{-z^2/2\sigma^2}$. Since $f_{xy}(x, y)$ has circular symmetry we conclude that for $z > 0$

$$F_{zw}(z, w) = \frac{2\theta}{2\pi}(1 - e^{-z^2/2\sigma^2}) \qquad \theta = \frac{\pi}{2} + \arctan w$$

and $F_{zw}(z, w) = 0$ for $z < 0$. This is a product of a function of z times a function of w. Hence, the RV $\mathbf{z}$ and $\mathbf{w}$ are *independent* with

$$F_z(z) = (1 - e^{-z^2/2\sigma})U(z) \qquad F_w(w) = \frac{1}{2} + \frac{1}{\pi} \arctan w$$

In other words, $\mathbf{z}$ has a *Rayleigh* density and $\mathbf{w}$ has a *Cauchy* density [see (5-18)] as in Fig. 6-16*b*.

Joint Density

We shall determine the joint density of the RVs

$$\mathbf{z} = g(\mathbf{x}, \mathbf{y}) \qquad \mathbf{w} = h(\mathbf{x}, \mathbf{y})$$

in terms of the joint density of **x** and **y**.

Fundamental theorem To find $f_{zw}(z, w)$ we solve the system

$$g(x, y) = z \qquad h(x, y) = w \tag{6-62}$$

Denoting by (x_n, y_n) its real roots

$$g(x_n, y_n) = z \qquad h(x_n, y_n) = w$$

we maintain that

$$f_{zw}(z, w) = \frac{f_{xy}(x_1, y_1)}{|J(x_1, y_1)|} + \cdots + \frac{f_{xy}(x_n, y_n)}{|J(x_n, y_n)|} + \cdots \tag{6-63}$$

where

$$J(x, y) = \begin{vmatrix} \dfrac{\partial z}{\partial x} & \dfrac{\partial z}{\partial y} \\[2ex] \dfrac{\partial w}{\partial x} & \dfrac{\partial w}{\partial y} \end{vmatrix} = \begin{vmatrix} \dfrac{\partial x}{\partial z} & \dfrac{\partial x}{\partial w} \\[2ex] \dfrac{\partial y}{\partial x} & \dfrac{\partial y}{\partial w} \end{vmatrix}^{-1} \tag{6-64}$$

is the *jacobian* of the transformation (6-62).

PROOF We denote by ΔD_{zw} the region in the xy plane such that

$$z < g(x, y) < z + dz \qquad w < h(x, y) < w + dw$$

This region consists of differential parallelograms, one for each (x_n, y_n) as in Fig. 6-17. The area of each parallelogram equals $dz\, dw/|J(x_n, y_n)|$ and its mass equals

$$f_{xy}(x_n, y_n)\, dz\, dw/|J(x_n, y_n)|$$

Since $f_{zw}(z, w)\, dz\, dw$ equals the mass in ΔD_{zw}, we conclude summing the masses in all parallelograms, that

$$f_{zw}(z, w)\, dz\, dw = \frac{f_{xy}(x_1, y_1)\, dz\, dw}{|J(x_1, y_1)|} + \cdots + \frac{f_{xy}(x_n, y_n)\, dz\, dw}{|J(x_n, y_n)|} + \cdots$$

and (6-63) results.

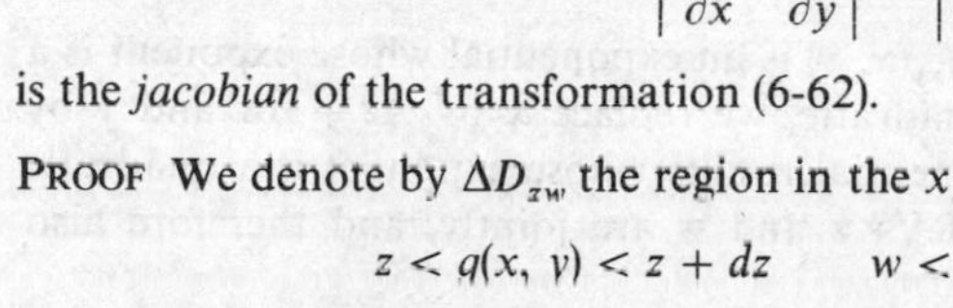

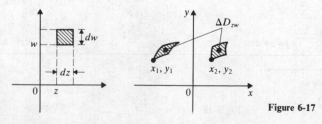

Figure 6-17

If the system (6-62) has no solutions in some region of the zw plane, then $f_{zw}(z, w) = 0$ in that region.

We shall illustrate the above theorem with two special cases.

Linear transformation

$$\mathbf{z} = a\mathbf{x} + b\mathbf{y} \qquad \mathbf{w} = c\mathbf{x} + d\mathbf{y} \tag{6-65}$$

If $ad - bc \neq 0$, then the system $ax + by = z$, $cx + dy = w$ has one and only one solution

$$x = Az + Bw \qquad y = Cz + Dw$$

Since $J(x, y) = ad - bc$, (6-63) yields

$$f_{zw}(z, w) = \frac{1}{|ad - bc|} f_{xy}(Az + Bw, Cz + Dw) \tag{6-66}$$

Joint normality From (6-66) it follows that if the RVs $\mathbf{x}$ and $\mathbf{y}$ are jointly normal and

$$\mathbf{z} = a\mathbf{x} + b\mathbf{y} \qquad \mathbf{w} = c\mathbf{x} + d\mathbf{y}$$

then $\mathbf{z}$ and $\mathbf{w}$ are also jointly normal.

PROOF Joint normality means that $f_{xy}(x, y)$ is an exponential whose exponent is a quadratic in x and y. If, in this quadratic, we replace x by $Az + Bw$ and y by $Cz + Dw$ as in (6-66), then an exponential results whose exponent is a quadratic in z and w. This shows that the RVs $\mathbf{z}$ and $\mathbf{w}$ are jointly, and therefore also marginally, normal.

From the above it follows that, if $\mathbf{x}$ and $\mathbf{y}$ are jointly normal and $\mathbf{z} = \mathbf{x} + \mathbf{y}$ then $\mathbf{z}$ is normal. We should emphasize, however, that if $\mathbf{x}$ and $\mathbf{y}$ are marginally but not jointly normal, then $\mathbf{z}$ is not, in general, normal. We give next a counter example.

Example 6-12 We shall construct two marginally normal RVs $\mathbf{x}_1$ and $\mathbf{y}_1$ such that their sum $\mathbf{z}_1 = \mathbf{x}_1 + \mathbf{y}_1$ is not normal: We start with two jointly normal RVs $\mathbf{x}$ and $\mathbf{y}$ and add and subtract masses on the four circles of Fig. 6-18. The resulting mass distribution specifies the joint density of the RVs $\mathbf{x}_1$ and $\mathbf{y}_1$. As we have shown in Example 6-1, these RVs are marginally normal. However, their sum $\mathbf{z}_1$ is not normal.

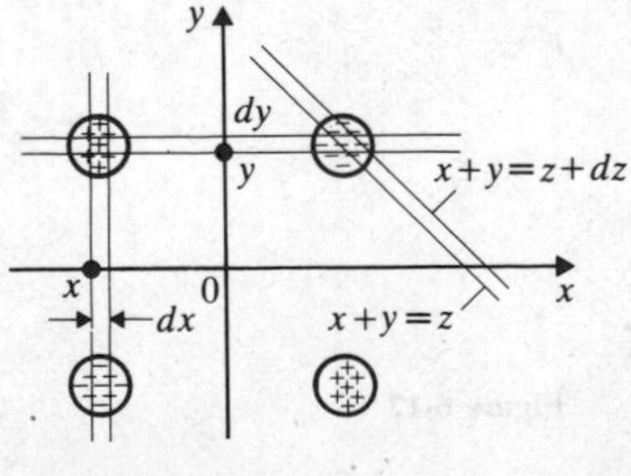

Fig. 6-18

Rotation A special case of (6-65) is the transformation

$$\mathbf{z} = \mathbf{x} \cos \varphi + \mathbf{y} \sin \varphi \qquad \mathbf{w} = -\mathbf{x} \sin \varphi + \mathbf{y} \cos \varphi \tag{6-67}$$

In this case $a = d = \cos \varphi$, $b = -c = \sin \varphi$, and $ad - bc = 1$. Hence

$$\mathbf{x} = \mathbf{z} \cos \varphi - \mathbf{w} \sin \varphi \qquad \mathbf{y} = \mathbf{z} \sin \varphi + \mathbf{w} \cos \varphi$$

and (6-66) yields

$$f_{zw}(z, w) = f_{xy}(z \cos \varphi - w \sin \varphi, z \sin \varphi + w \cos \varphi) \tag{6-68}$$

Thus, if two RVs are rotated by an angle φ, their probability masses are rotated in the opposite direction by the same angle.

Circular symmetry If $f_{xy}(x, y)$ is circularly symmetrical as in (6-31), then

$$f_{xy}(x, y) = f_{xy}(x \cos \varphi - y \sin \varphi, x \sin \varphi + y \cos \varphi) \tag{6-69}$$

because

$$(x \cos \varphi - y \sin \varphi)^2 + (x \sin \varphi + y \cos \varphi)^2 = x^2 + y^2$$

Hence [see (6-68)]

$$f_{zw}(z, w) = f_{xy}(z, w) = g(\sqrt{z^2 + w^2}) \tag{6-70}$$

Conversely, if the RVs **x**, **y** and **z**, **w** have the same statistics for every φ, then their joint density is circularly symmetrical. From (6-34) it follows that if **x** and **y** are also independent, then they are normal.

Polar coordinates Consider the RVs

$$\mathbf{r} = \sqrt{\mathbf{x}^2 + \mathbf{y}^2} \qquad \boldsymbol{\varphi} = \arctan \mathbf{y}/\mathbf{x} \tag{6-71}$$

where we assume that $\mathbf{r} \geq 0$ and $-\pi < \boldsymbol{\varphi} \leq \pi$. With this assumption, the system $\sqrt{x^2 + y^2} = r$, $\arctan y/x = \varphi$ has a single solution

$$x = r \cos \varphi \qquad y = r \sin \varphi \qquad \text{for} \quad r > 0$$

Since [see (6-65)]

$$J(x, y) = \begin{vmatrix} \cos \varphi & \sin \varphi \\ -r \sin \varphi & r \cos \varphi \end{vmatrix}^{-1} = \frac{1}{r}$$

we conclude from (6-63) that

$$f_{r\varphi}(r, \varphi) = r f_{xy}(r \cos \varphi, r \sin \varphi) \qquad r > 0 \tag{6-72}$$

and zero for $r < 0$.

Example 6-13 We shall show that if

$$\mathbf{x} \cos \omega t + \mathbf{y} \sin \omega t = \mathbf{r} \cos (\omega t - \boldsymbol{\varphi}) \qquad |\varphi| < \pi$$

and the RVs **x** and **y** are $N(0, \sigma)$ and independent, then the RVs **r** and $\boldsymbol{\varphi}$ are independent, $\boldsymbol{\varphi}$ is uniform in the interval $(-\pi, \pi)$, and **r** has a Rayleigh distribution.

PROOF Since $x = r\cos\varphi$, $y = r\sin\varphi$, and

$$f_{xy}(x, y) = \frac{1}{2\pi\sigma^2} e^{-(x^2+y^2)/2\sigma^2}$$

(6-72) yields

$$f_{r\varphi}(r, \varphi) = \frac{r}{2\pi\sigma^2} e^{-r^2/2\sigma^2} \qquad r > 0 \qquad |\varphi| < \pi$$

and zero otherwise. This is a product of a function of r times a function of φ. Hence, the RVs $\mathbf{r}$ and $\boldsymbol{\varphi}$ are independent with

$$f_r(r) = \frac{r}{\sigma^2} e^{-r^2/2\sigma^2} \qquad f_\varphi(\varphi) = \frac{1}{2\pi}$$

for $r > 0$, $-\pi < \varphi \le \pi$ and zero otherwise. The proportionality factors are so chosen as to make the area of each term equal to 1.

Auxiliary variables The determination of the density of *one* function $\mathbf{z} = g(\mathbf{x}, \mathbf{y})$ of two RVs can be determined from (6-63) where $\mathbf{w}$ is a conveniently chosen auxiliary variable, for example $\mathbf{w} = \mathbf{x}$ or $\mathbf{w} = \mathbf{y}$. The density of $\mathbf{z}$ is then found by integrating the function $f_{zw}(z, w)$ so obtained.

Example 6-14 We shall find the density of the RV

$$\mathbf{z} = a\mathbf{x} + b\mathbf{y}$$

using as auxiliary variable the function $\mathbf{w} = \mathbf{y}$.

The system $z = ax + by$, $w = y$ has a single solution: $x = (z - bw)/a$, $y = w$. Since

$$J(x, y) = \begin{vmatrix} a & b \\ 0 & 1 \end{vmatrix} = a$$

it follows from (6-63) that

$$f_{zw}(z, w) = \frac{1}{|a|} f_{xy}\left(\frac{z - by}{a}, y\right)$$

Hence

$$f_z(z) = \frac{1}{|a|} \int_{-\infty}^{\infty} f_{xy}\left(\frac{z - by}{a}, y\right) dy$$

Example 6-15 With

$$\mathbf{z} = \mathbf{xy} \qquad \mathbf{w} = \mathbf{x}$$

the system $xy = z$, $x = w$ has a single solution: $x = w$, $y = z/w$. In this case, $J = -w$ and (6-63) yields

$$f_{zw}(z, w) = \frac{1}{|w|} f_{xy}\left(w, \frac{z}{w}\right)$$

Hence, the density of the RV $\mathbf{z} = \mathbf{xy}$ is given by

$$f_z(z) = \int_{-\infty}^{\infty} \frac{1}{|w|} f_{xy}\left(w, \frac{z}{w}\right) dw$$

PROBLEMS

6-1 If x and y are the zero-one RVs associated with the events $\mathscr{A}$ and $\mathscr{B}$ respectively, (*a*) find the probability masses in the *x*-*y* plane and (*b*) show that the RVs x and y are independent iff the events $\mathscr{A}$ and $\mathscr{B}$ are independent.

6-2 The RVs x and y are independent and z = x + y. Find $f_y(y)$ if

$$f_x(x) = ce^{-cx}U(x) \qquad f_z(z) = c^2ze^{-cz}U(z)$$

6-3 The RVs x and y are independent and y is uniform in the interval (0, 1). Show that, if z = x + y, then

$$f_z(z) = F_x(z) - F_x(z-1)$$

6-4 (*a*) The function $g(x)$ is monotone increasing and $y = g(x)$. Show that

$$F_{xy}(x, y) = \begin{cases} F_x(x) & \text{if} \quad y > g(x) \\ F_y(y) & \text{if} \quad y < g(x) \end{cases}$$

(*b*) Find $F_{xy}(x, y)$ if $g(x)$ is monotone decreasing.

6-5 Express $F_{zw}(z, w)$ in terms of $f_{xy}(x, y)$ if z = max (x, y), w = min (x, y).

6-6 The RVs x and y are $N(0, 2)$ and independent. Find $f_z(z)$ and $F_z(z)$ if (*a*) z = 2x + 3y, and (*b*) z = x/y.

6-7 The RVs x and y are independent with

$$f_x(x) = \frac{x}{\alpha^2}e^{-x^2/2\alpha^2}U(x) \qquad f_y(y) = \begin{cases} 1/\pi\sqrt{1-y^2} & |y| < 1 \\ 0 & |y| > 1 \end{cases}$$

Show that the RV z = xy is $N(0, \alpha)$.

6-8 The RVs x and y are independent with Rayleigh densities

$$f_x(x) = \frac{x}{\alpha^2}e^{-x^2/2\alpha^2}U(x) \qquad f_y(y) = \frac{y}{\beta^2}e^{-y^2/2\beta^2}U(y)$$

(*a*) Show that if z = x/y, then

$$f_z(z) = \frac{2\alpha^2}{\beta^2}\frac{z}{(z^2 + \alpha^2/\beta^2)^2}U(z) \tag{i}$$

(*b*) Using (i), show that for any $k > 0$

$$P\{\mathbf{x} \le k\mathbf{y}\} = \frac{k^2}{k^2 + \alpha^2/\beta^2}$$

6-9 The RVs x and y are independent with exponential densities

$$f_x(x) = \alpha e^{-\alpha x}U(x) \qquad f_y(y) = \beta e^{-\beta y}U(y)$$

Find the densities of the following RVs:

1. 2x + y 2. x − y 3. $\dfrac{\mathbf{x}}{\mathbf{y}}$ 4. max (x, y) 5. min (x, y)

6-10 The RVs x and y are independent and each is uniform in the interval (0, *a*). Find the density of the RV z = |x − y|.

6-11 Show that (*a*) the convolution of two normal densities is a normal density, and (*b*) the convolution of two Cauchy densities is a Cauchy density.

6-12 The RVs **x** and **θ** are independent and **θ** is uniform in the interval $(-\pi, \pi)$. Show that if $\mathbf{z} = \mathbf{x}\cos(wt + \boldsymbol{\theta})$, then

$$f_z(z) = \frac{1}{\pi}\int_{-\infty}^{-|z|} \frac{f_x(y)}{\sqrt{y^2 - z^2}}\,dy + \frac{1}{\pi}\int_{|z|}^{\infty} \frac{f_x(y)}{\sqrt{y^2 - z^2}}\,dy$$

6-13 The RVs **x** and **y** are independent, **x** is $N(0, \sigma)$, and **y** is uniform in the interval $(0, \pi)$. Show that if $\mathbf{z} = \mathbf{x} + a\cos\mathbf{y}$ then

$$f_z(z) = \frac{1}{\pi\sigma\sqrt{2\pi}}\int_0^{\pi} e^{-(x - a\cos y)^2/2\sigma^2}\,dy$$

6-14 The RVs **x** and **y** are of discrete type, independent, with $P\{\mathbf{x} = n\} = a_n$, $P\{\mathbf{y} = n\} = b_n$, $n = 0, 1, \ldots$. Show that, if $\mathbf{z} = \mathbf{x} + \mathbf{y}$, then

$$P\{\mathbf{z} = n\} = \sum_{k=0}^{n} a_k b_{n-k}$$

6-15 The RV **x** is of discrete type taking the values x_n with $P\{\mathbf{x} = x_n\} = p_n$ and the RV **y** is of continuous type and independent of **x**. Show that if $\mathbf{z} = \mathbf{x} + \mathbf{y}$ and $\mathbf{w} = \mathbf{xy}$, then

$$f_z(z) = \sum_n f_y(z - x_n)p_n \qquad f_w(w) = \sum_n \frac{1}{|x_n|} f_y\left(\frac{z}{x_n}\right)p_n$$

6-16 The RVs **x** and **y** are normal, independent, with the same variance. Show that, if $\mathbf{z} = \sqrt{\mathbf{x}^2 + \mathbf{y}^2}$, then $f_z(z)$ is given by (6-52) where $\eta = \sqrt{\eta_x^2 + \eta_y^2}$.

6-17 The RVs $\mathbf{x}_1$ and $\mathbf{x}_2$ are jointly normal with zero mean. Show that their density can be written in the form

$$f(x_1, x_2) = \frac{1}{2\pi\sqrt{\Delta}} \exp\left\{-\frac{1}{2} XC^{-1}X^t\right\} \qquad C = \begin{bmatrix} \mu_{11} & \mu_{12} \\ \mu_{21} & \mu_{22} \end{bmatrix}$$

where $X\colon [x_1, x_2]$, $\mu_{ij} = E\{\mathbf{x}_i \mathbf{x}_j\}$, and $\Delta = \mu_{11}\mu_{22} - \mu_{12}^2$.

6-18 Show that if the RVs **x** and **y** are normal and independent, then

$$P\{\mathbf{xy} < 0\} = \mathbb{G}\left(\frac{\eta_x}{\sigma_x}\right) + \mathbb{G}\left(\frac{\eta_y}{\sigma_y}\right) - 2\mathbb{G}\left(\frac{\eta_x}{\sigma_x}\right)\mathbb{G}\left(\frac{\eta_y}{\sigma_y}\right)$$

CHAPTER

SEVEN

MOMENTS AND CONDITIONAL STATISTICS

7-1 JOINT MOMENTS

Given two RVs **x** and **y** and a function $g(x, y)$, we form the RV $\mathbf{z} = g(\mathbf{x}, \mathbf{y})$. The expected value of this RV is given by

$$E\{\mathbf{z}\} = \int_{-\infty}^{\infty} z f_z(z)\, dz \tag{7-1}$$

However, as the next theorem shows, $E\{\mathbf{z}\}$ can be expressed directly in terms of the function $g(x, y)$ and the joint density $f(x, y)$ of **x** and **y**.

Theorem

$$E\{g(\mathbf{x}, \mathbf{y})\} = \int_{-\infty}^{\infty}\int_{-\infty}^{\infty} g(x, y) f(x, y)\, dx\, dy \tag{7-2}$$

PROOF The proof is similar to the proof of (5-29). We denote by ΔD_z the region of the xy plane such that $z < g(x, y) < z + dz$. Thus, to each differential in (7-1) there corresponds a region ΔD_z in the xy plane. As dz covers the z axis, the regions ΔD_z are not overlapping and they cover the entire xy plane. Hence, the integrals in (7-1) and (7-2) are equal.

We note that the expected value of $g(\mathbf{x})$ can be determined either from (7-2) or from (5-29) as a single integral

$$E\{g(\mathbf{x})\} = \int_{-\infty}^{\infty}\int_{-\infty}^{\infty} g(x) f(x, y)\, dx\, dy = \int_{-\infty}^{\infty} g(x) f_x(x)\, dx$$

This is consistent with the relationship (6-10) between marginal and joint densities.

If the RVs **x** and **y** are of discrete type taking the values x_i and y_k with probability p_{ik} as in (6-19), then

$$E\{g(\mathbf{x}, \mathbf{y})\} = \sum_{i,k} g(x_i, y_k)p_{ik} \tag{7-3}$$

Linearly From (7-2) it follows that

$$E\left\{\sum_{k=1}^{n} a_k\, g_k(\mathbf{x}, \mathbf{y})\right\} = \sum_{k=1}^{n} a_k\, E\{g_k(\mathbf{x}, \mathbf{y})\} \tag{7-4}$$

This fundamental result will be used extensively.

We note in particular that

$$E\{\mathbf{x} + \mathbf{y}\} = E\{\mathbf{x}\} + E\{\mathbf{y}\} \tag{7-5}$$

Thus, the expected value of the sum of two RVs equals the sum of their expected values. We should stress, however, that in general

$$E\{\mathbf{xy}\} \neq E\{\mathbf{x}\}E\{\mathbf{y}\}$$

Frequency interpretation As in (5-27)

$$E\{\mathbf{x} + \mathbf{y}\} \simeq \frac{\mathbf{x}(\zeta_1) + \mathbf{y}(\zeta_1) + \cdots + \mathbf{x}(\zeta_n) + \mathbf{y}(\zeta_n)}{n}$$

$$= \frac{\mathbf{x}(\zeta_1) + \cdots + \mathbf{x}(\zeta_n)}{n} + \frac{\mathbf{y}(\zeta_1) + \cdots + \mathbf{y}(\zeta_n)}{n} \simeq E\{\mathbf{x}\} + E\{\mathbf{y}\}$$

However, in general

$$E\{\mathbf{xy}\} \simeq \frac{\mathbf{x}(\zeta_1)\mathbf{y}(\zeta_1) + \cdots + \mathbf{x}(\zeta_n)\mathbf{y}(\zeta_n)}{n}$$

$$\neq \frac{\mathbf{x}(\zeta_1) + \cdots + \mathbf{x}(\zeta_n)}{n} \times \frac{\mathbf{y}(\zeta_1) + \cdots + \mathbf{y}(\zeta_n)}{n} \simeq E\{\mathbf{x}\}E\{\mathbf{y}\}$$

Covariance The covariance C or C_{xy} of two RVs **x** and **y** is by definition the number

$$C = E\{(\mathbf{x} - \eta_x)(\mathbf{y} - \eta_y)\} \tag{7-6}$$

where $E\{\mathbf{x}\} = \eta_x$ and $E\{\mathbf{y}\} = \eta_y$. Expanding the product in (7-6) and using (7-4), we obtain

$$C = E\{\mathbf{xy}\} - E\{\mathbf{x}\}E\{\mathbf{y}\} \tag{7-7}$$

Correlation coefficient The correlation coefficient r or r_{xy} of the RVs **x** and **y** is by definition the ratio

$$r = \frac{C}{\sigma_x \sigma_y} \tag{7-8}$$

We maintain that

$$|r| \le 1 \qquad |C| \le \sigma_x \sigma_y \tag{7-9}$$

PROOF Clearly

$$E\{[a(\mathbf{x} - \eta_x) + (\mathbf{y} - \eta_y)]^2\} = a^2\sigma_x^2 + 2aC + \sigma_y^2 \tag{7-10}$$

The above is a positive quadratic for any a, hence, its discriminant is negative. In other words

$$C^2 - \sigma_x^2\sigma_y^2 \le 0 \tag{7-11}$$

and (7-9) results.

We note that the RVs $\mathbf{x}$, $\mathbf{y}$ and $\mathbf{x} - \eta_x$, $\mathbf{y} - \eta_y$ have the same covariance and correlation coefficient.

Example 7-1 We shall show that the correlation coefficient of two jointly normal RVs is the parameter r in (6-15). It suffices to assume that $\eta_x = \eta_y = 0$ and to show that $E\{\mathbf{xy}\} - r\sigma_1\sigma_2$.

Since

$$\frac{x^2}{\sigma_1^2} - 2r\frac{xy}{\sigma_1\sigma_2} + \frac{y^2}{\sigma_2^2} = \left(\frac{x}{\sigma_1} - r\frac{y}{\sigma_2}\right)^2 + (1 - r^2)\frac{y^2}{\sigma_2^2}$$

we conclude with (6-44) that

$$E\{\mathbf{xy}\} = \frac{1}{\sigma_2\sqrt{2\pi}}\int_{-\infty}^{\infty} ye^{-y^2/2\sigma_2^2}\int_{-\infty}^{\infty}\frac{x}{\sigma_1\sqrt{2\pi(1-r^2)}}\exp\left[-\frac{(x - ry\sigma_1/\sigma_2)^2}{2\sigma_1^2(1-r^2)}\right]dx\,dy$$

The inner integral is a normal density with mean $ry\sigma_1/\sigma_2$ multiplied by x, hence, it equals $ry\sigma_1/\sigma_2$. This yields

$$E\{\mathbf{xy}\} = \frac{r\sigma_1/\sigma_2}{\sigma_2\sqrt{2\pi}}\int_{-\infty}^{\infty} y^2e^{-y^2/2\sigma_2^2}\,dy = r\sigma_1\sigma_2$$

Uncorrelatedness Two RVs are called uncorrelated if their covariance is zero. This can be phrased in the following equivalent forms

$$C = 0 \qquad r = 0 \qquad E\{\mathbf{xy}\} = E\{\mathbf{x}\}E\{\mathbf{y}\}$$

Orthogonality Two RVs are called orthogonal if

$$E\{\mathbf{xy}\} = 0$$

We shall use the notation

$$\mathbf{x} \perp \mathbf{y}$$

to indicate that $\mathbf{x}$ and $\mathbf{y}$ are orthogonal.

Note (*a*) If $\mathbf{x}$ and $\mathbf{y}$ are uncorrelated, then $\mathbf{x} - \eta_x \perp \mathbf{y} - \eta_y$. (*b*) If $\mathbf{x}$ and $\mathbf{y}$ are uncorrelated and $\eta_x = 0$ or $\eta_y = 0$ then $\mathbf{x} \perp \mathbf{y}$.

Vector space of random variables We shall find it convenient to interpret RVs as vectors in an abstract space. In this space, the second moment

$$E\{\mathbf{xy}\}$$

of the RVs **x** and **y** is by definition their *inner product* and $E\{\mathbf{x}^2\}$ and $E\{\mathbf{y}^2\}$ are the squares of their lengths. The ratio

$$\frac{E\{\mathbf{xy}\}}{\sqrt{E\{\mathbf{x}^2\}E\{\mathbf{y}^2\}}}$$

is the cosine of their angle.

We maintain that

$$E^2\{\mathbf{xy}\} \le E\{\mathbf{x}^2\}E\{\mathbf{y}^2\} \tag{7-12}$$

This is the *cosine inequality* and its proof is similar to the proof of (7-11): The quadratic

$$E\{(a\mathbf{x} - \mathbf{y})^2\} = a^2E\{\mathbf{x}^2\} - 2aE\{\mathbf{xy}\} + E\{\mathbf{y}^2\}$$

is positive for every a, hence, its discriminant is negative and (7-12) results. If (7-12) is an equality, then the above quadratic is zero for some $a = a_0$, hence, $\mathbf{y} = a_0\mathbf{x}$. This agrees with the geometric interpretation of RV because, if (7-12) is an equality, then the vectors **x** and **y** are on the same line.

The following illustration is an example of the correspondence between vectors and RVs: Consider two RVs **x** and **y** such that $E\{\mathbf{x}^2\} = E\{\mathbf{y}^2\}$. Geometrically, this means that the vectors **x** and **y** have the same length. If, therefore, we construct a parallelogram with sides **x** and **y**, it will be a rhombus with diagonals $\mathbf{x} + \mathbf{y}$ and $\mathbf{x} - \mathbf{y}$ (Fig. 7-1). These diagonals are perpendicular because

$$E\{(\mathbf{x} + \mathbf{y})(\mathbf{x} - \mathbf{y})\} = E\{\mathbf{x}^2 - \mathbf{y}^2\} = 0$$

Theorem If two RVs are independent, i.e., if

$$f(x, y) = f_x(x)f_y(y) \tag{7-13}$$

then they are uncorrelated.

PROOF It suffices to show that

$$E\{\mathbf{xy}\} = E\{\mathbf{x}\}E\{\mathbf{y}\} \tag{7-14}$$

From (7-2) and (7-13) it follows that

$$E\{\mathbf{xy}\} = \int_{-\infty}^{\infty}\int_{-\infty}^{\infty} xyf_x(x)f_y(y)\,dx\,dy = \int_{-\infty}^{\infty} xf_x(x)\,dx \int_{-\infty}^{\infty} yf_y(y)\,dy$$

and (7-14) results.

If the RVs **x** and **y** are independent, then the RVs $g(\mathbf{x})$ and $h(\mathbf{y})$ are also independent [see (6-29)]. Hence

$$E\{g(\mathbf{x})h(\mathbf{y})\} = E\{g(\mathbf{x})\}E\{h(\mathbf{y})\} \tag{7-15}$$

This is not, in general, true if **x** and **y** are merely uncorrelated.

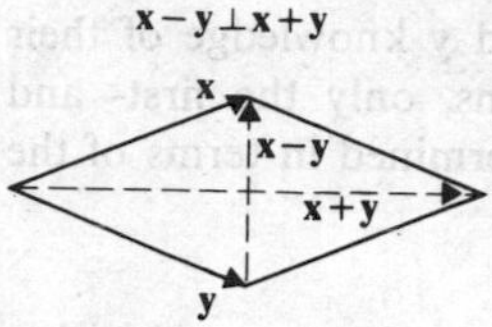

Figure 7-1

We note, finally, that if two RVs are uncorrelated they are not necessarily independent. However, for normal RVs uncorrelatedness is equivalent to independence. Indeed, if the RVs **x** and **y** are jointly normal and $r = 0$, then [see (6-15)], $f(x, y) = f_x(x) f_y(y)$.

Variance of the sum of two RVs If $\mathbf{z} = \mathbf{x} + \mathbf{y}$, then $\eta_z = \eta_x + \eta_y$, hence

$$\sigma_z^2 = E\{(\mathbf{z} - \eta_z)^2\} = E\{[(\mathbf{x} - \eta_x) + (\mathbf{y} - \eta_y)]^2\}$$

From this and (7-10) it follows that

$$\sigma_z^2 = \sigma_x^2 + 2r\sigma_x\sigma_y + \sigma_y^2 \tag{7-16}$$

The above leads to the conclusion that if $r = 0$ then

$$\sigma_z^2 = \sigma_x^2 + \sigma_y^2 \tag{7-17}$$

Thus, if two RVs are uncorrelated, then the variance of their sum equals the sum of their variances.

It follows from (7-14) that this is true also if **x** and **y** are independent.

Moments

The mean

$$m_{kr} = E\{\mathbf{x}^k\mathbf{y}^r\} = \int_{-\infty}^{\infty}\int_{-\infty}^{\infty} x^k y^r f(x, y)\, dx\, dy \tag{7-18}$$

of the product $\mathbf{x}^k\mathbf{y}^r$ is by definition the joint moment of the RVs **x** and **y** of order $k + r = n$.

Thus, $m_{10} = \eta_x$, $m_{01} = \eta_y$ are the first-order moments and

$$m_{20} = E\{\mathbf{x}^2\} \qquad m_{11} = E\{\mathbf{xy}\} \qquad m_{02} = E\{\mathbf{y}^2\}$$

are the second-order moments.

The joint central moments of **x** and **y** are the moments of $\mathbf{x} - \eta_x$ and $\mathbf{y} - \eta_y$

$$\mu_{kr} = E\{(\mathbf{x} - \eta_x)^k(\mathbf{y} - \eta_y)^r\} = \int_{-\infty}^{\infty}\int_{-\infty}^{\infty} (x - \eta_x)^k(y - \eta_y)^r f(x, y)\, dx\, dy \tag{7-19}$$

Clearly, $\mu_{10} = \mu_{01} = 0$ and

$$\mu_{11} = C \qquad \mu_{20} = \sigma_x^2 \qquad \mu_{02} = \sigma_y^2$$

Absolute and generalized moments are defined similarly [see (5-40) and (5-41)].

For the determination of the joint statistics of **x** and **y** knowledge of their joint density is required. However, in many applications, only the first- and second-order moments are used. These moments are determined in terms of the five parameters

$$\eta_x \quad \eta_y \quad \sigma_x \quad \sigma_y \quad r_{xy}$$

If **x** and **y** are jointly normal, then [see (6-15)] the above parameters determine uniquely $f(x, y)$.

Example 7-2 The RVs **x** and **y** are jointly normal with

$$\eta_x = 10 \qquad \eta_y = 0 \qquad \sigma_x = 2 \qquad \sigma_y = 1 \qquad r_{xy} = 0.5$$

We shall find the joint density of the RVs

$$\mathbf{z} = \mathbf{x} + \mathbf{y} \qquad \mathbf{w} = \mathbf{x} - \mathbf{y}$$

Clearly

$$\eta_z = \eta_x + \eta_y = 10 \qquad \eta_w = \eta_x - \eta_y = 10$$

$$\sigma_z^2 = \sigma_x^2 + \sigma_y^2 + 2r_{xy}\sigma_x\sigma_y = 7 \qquad \sigma_w^2 = \sigma_x^2 + \sigma_y^2 - 2r_{xy}\sigma_x\sigma_y = 3$$

$$E\{\mathbf{zw}\} = E\{\mathbf{x}^2 - \mathbf{y}^2\} = (100 + 4) - 1 = 103$$

$$r_{zw}^2 = \frac{E\{\mathbf{zw}\} - E\{\mathbf{z}\}E\{\mathbf{w}\}}{\sigma_z^2\sigma_w^2} = \frac{3}{7 \times 3}$$

As we know [see (6-66)], the RVs **z** and **w** are jointly normal because they are linearly dependent on **x** and **y**. Hence, their joint density is

$$N(10, 10; \sqrt{7}, \sqrt{3}; 1/\sqrt{7})$$

Estimate of the mean of $g(\mathbf{x}, \mathbf{y})$ If the function $g(x, y)$ is sufficiently smooth near the point (η_x, η_y), then the mean η_g and variance σ_g^2 of $g(\mathbf{x}, \mathbf{y})$ can be estimated in terms of the mean, variance, and covariance of **x** and **y**

$$\eta_g \simeq g + \frac{1}{2}\left(\frac{\partial^2 g}{\partial x^2}\sigma_x^2 + 2\frac{\partial^2 g}{\partial x\, \partial y} r\sigma_x\sigma_y + \frac{\partial^2 g}{\partial y^2}\sigma_y^2\right) \tag{7-20}$$

$$\sigma_g^2 \simeq \left(\frac{\partial g}{\partial x}\right)^2 \sigma_x^2 + 2\left(\frac{\partial g}{\partial x}\right)\left(\frac{\partial g}{\partial y}\right) r\sigma_x\sigma_y + \left(\frac{\partial g}{\partial y}\right)^2 \sigma_y^2 \tag{7-21}$$

where the function $g(x, y)$ and its derivatives are evaluated at $x = \eta_x$ and $y = \eta_y$.

PROOF We expand $g(x, y)$ into a series about the point (η_x, η_y)

$$g(x, y) = g(\eta_x, \eta_y) + (x - \eta_x)\frac{\partial g}{\partial x} + (y - \eta_y)\frac{\partial g}{\partial y} + \cdots \tag{7-22}$$

Inserting the above into (7-2), we obtain the moment expansion of $E\{g(\mathbf{x}, \mathbf{y})\}$ in terms of the derivatives of $g(x, y)$ at (η_x, η_y) and the joint moments μ_{kr} of **x** and **y**.

Using only the first five terms in (7-22), we obtain (7-20). Equation (7-21) follows if we apply (7-20) to the function $[g(x, y) - \eta_g]^2$ and neglect moments of order higher than 2.

7-2 JOINT CHARACTERISTIC FUNCTIONS

The *joint characteristic function* of the RVs **x** and **y** is by definition the integral

$$\Phi(\omega_1, \omega_2) = \int_{-\infty}^{\infty}\int_{-\infty}^{\infty} f(x, y)e^{j(\omega_1 x + \omega_2 y)}\, dx\, dy \tag{7-23}$$

From the above and the two-dimensional *inversion formula* for Fourier transforms it follows that

$$f(x, y) = \frac{1}{4\pi^2}\int_{-\infty}^{\infty}\int_{-\infty}^{\infty} \Phi(\omega_1, \omega_2)e^{-j(\omega_1 x + \omega_2 y)}\, d\omega_1\, d\omega_2 \tag{7-24}$$

Clearly

$$\Phi(\omega_1, \omega_2) = E\{e^{j(\omega_1 \mathbf{x} + \omega_2 \mathbf{y})}\} \tag{7-25}$$

The logarithm

$$\Psi(\omega_1, \omega_2) = \ln \Phi(\omega_1, \omega_2) \tag{7-26}$$

of $\Phi(\omega_1, \omega_2)$ is the joint second characteristic function of **x** and **y**.

The *marginal* characteristic functions

$$\Phi_x(\omega) = E\{e^{j\omega \mathbf{x}}\} \qquad \Phi_y(\omega) = E\{e^{j\omega \mathbf{y}}\} \tag{7-27}$$

of **x** and **y** can be expressed in terms of their joint characteristic function $\Phi(\omega_1, \omega_2)$. From (7-25) and (7-27) it follows that

$$\Phi_x(\omega) = \Phi(\omega, 0) \qquad \Phi_y(\omega) = \Phi(0, \omega) \tag{7-28}$$

We note that, if $\mathbf{z} = a\mathbf{x} + b\mathbf{y}$, then

$$\Phi_z(\omega) = E\{e^{j(a\mathbf{x} + b\mathbf{y})\omega}\} = \Phi(a\omega, b\omega) \tag{7-29}$$

Hence

$$\Phi_z(1) = \Phi(a, b)$$

Cramér–Wold theorem The above shows that if $\Phi_z(\omega)$ is known for every a and b, then $\Phi(\omega_1, \omega_2)$ is uniquely determined. In other words, if the density of $a\mathbf{x} + b\mathbf{y}$ is known for every a and b, then the joint density $f(x, y)$ of **x** and **y** is uniquely determined.

Independence and convolution If the RVs **x** and **y** are independent, then [see (7-15)]

$$E\{e^{j(\omega_1 \mathbf{x} + \omega_2 \mathbf{y})}\} = E\{e^{j\omega_1 \mathbf{x}}\}E\{e^{j\omega_2 \mathbf{y}}\}$$

From this it follows that

$$\Phi(\omega_1, \omega_2) = \Phi_x(\omega_1)\Phi_y(\omega_2) \tag{7-30}$$

Conversely, if (7-30) is true, then the RVs **x** and **y** are independent. Indeed, inserting (7-30) into the inversion formula (7-24) and using (5-66) we conclude that $f(x, y) = f_x(x) f_y(y)$.

Convolution theorem If the RVs **x** and **y** are independent and $\mathbf{z} = \mathbf{x} + \mathbf{y}$, then

$$E\{e^{j\omega\mathbf{z}}\} = E\{e^{j\omega(\mathbf{x}+\mathbf{y})}\} = E\{e^{j\omega\mathbf{x}}\}E\{e^{j\omega\mathbf{y}}\}$$

Hence

$$\Phi_z(\omega) = \Phi_x(\omega)\Phi_y(\omega) \qquad \Psi_z(\omega) = \Psi_x(\omega) + \Psi_y(\omega) \tag{7-31}$$

As we know [see (6-39)], the density of **z** equals the convolution of $f_x(x)$ and $f_y(y)$. From this and (7-31) it follows that the characteristic function of the convolution of two densities equals the product of their characteristic functions.

Example 7-3 We shall show that if the RVs **x** and **y** are *independent* and Poisson distributed with parameters a and b respectively, then their sum $\mathbf{z} = \mathbf{x} + \mathbf{y}$ is also Poisson distributed with parameter $a + b$.

PROOF As we know (see Example 5-31)

$$\Psi_x(\omega) = a(e^{j\omega} - 1) \qquad \Psi_y(\omega) = b(e^{j\omega} - 1)$$

Hence

$$\Psi_z(\omega) = \Psi_x(\omega) + \Psi_y(\omega) = (a + b)(e^{j\omega} - 1)$$

It can be shown that the converse is also true: If the RVs **x** and **y** are *independent* and their sum is Poisson distributed, then **x** and **y** are also Poisson distributed. The proof of this difficult theorem will not be given.

Example 7-4 It was shown in Sec. 6-3 that if the RVs **x** and **y** are jointly normal, then the sum $a\mathbf{x} + b\mathbf{y}$ is also normal. In the following we reestablish a special case of the above using (7-30): If **x** and **y** are *independent* and normal, then their sum $\mathbf{z} = \mathbf{x} + \mathbf{y}$ is also normal.

PROOF In this case [see (5-65)]

$$\Psi_x(\omega) = j\eta_x \omega - \tfrac{1}{2}\sigma_x^2\omega^2 \qquad \Psi_y(\omega) = j\eta_y \omega - \tfrac{1}{2}\sigma_y^2 \omega^2$$

Hence

$$\Psi_z(\omega) = j(\eta_x + \eta_y)\omega - \tfrac{1}{2}(\sigma_x^2 + \sigma_y^2)\omega^2$$

It can be shown that the converse is also true (Cramér theorem): If the RVs **x** and **y** are *independent* and their sum is normal, then they are also normal. The proof of this difficult theorem will not be given.†

† E. Lukacs: *Characteristic Functions*, Hafner Publishing Co., New York, 1960.

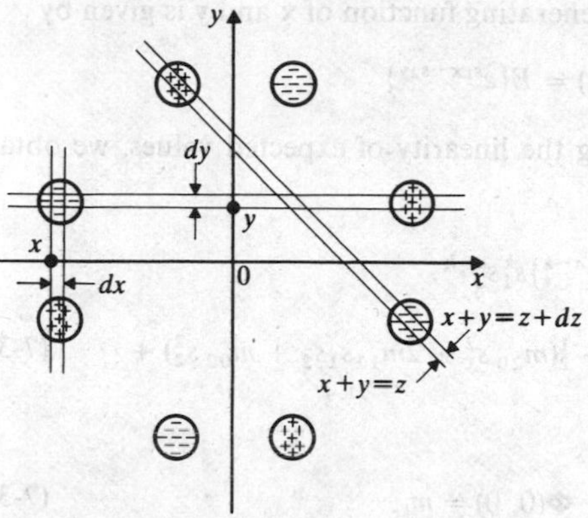

Figure 7-2

Normal RVs We shall show that the joint characteristic function of two jointly normal RVs is given by

$$\Phi(\omega_1, \omega_2) = e^{j(\eta_1\omega_1 + \eta_2\omega_2)}e^{-\frac{1}{2}(\omega_1^2\sigma_1^2 + 2r\sigma_1\sigma_2\omega_1\omega_2 + \omega_2^2\sigma_2^2)} \tag{7-32}$$

PROOF This can be derived by inserting $f(x, y)$ into (7-23). The following simpler proof is based on the fact that the RV $\mathbf{z} = \omega_1\mathbf{x} + \omega_2\mathbf{y}$ is normal and

$$\Psi_z(\omega) = j\eta_z\omega - \tfrac{1}{2}\sigma_z^2\omega^2 \tag{7-33}$$

Since

$$\eta_z = \omega_1\eta_1 + \omega_2\eta_2 \qquad \sigma_z^2 = \omega_1^2\sigma_1^2 + 2r\omega_1\omega_2\sigma_1\sigma_2 + \omega_2^2\sigma_2^2$$

and $\Phi_z(\omega) = \Phi(\omega_1\omega, \omega_2\omega)$, (7-32) follows from (7-33) with $\omega = 1$.

The above proof is based on the fact that the RV $\mathbf{z} = \omega_1\mathbf{x} + \omega_2\mathbf{y}$ is normal for any ω_1 and ω_2; this leads to the following conclusion: If it is known that the sum $a\mathbf{x} + b\mathbf{y}$ is normal for every a and b, then the RVs $\mathbf{x}$ and $\mathbf{y}$ are jointly normal. We should stress, however, that this is not true if $a\mathbf{x} + b\mathbf{y}$ is normal for only a finite set of values of a and b. A counter example can be formed by a simple extension of the construction in Fig. 7-2.

Example 7-5 We shall construct two RVs $\mathbf{x}_1$ and $\mathbf{x}_2$ with the following properties: $\mathbf{x}_1$, $\mathbf{x}_2$, and $\mathbf{x}_1 + \mathbf{x}_2$ are normal but $\mathbf{x}_1$ and $\mathbf{x}_2$ are not jointly normal.

Suppose that $\mathbf{x}$ and $\mathbf{y}$ are two jointly normal RVs with mass density $f(x, y)$. Adding and subtracting small masses in the region D of Fig. 7-2 consisting of eight circles as shown, we obtain a new function $f_1(x, y)$ such that $f_1(x, y) = f(x, y) \pm \varepsilon$ in D and $f_1(x, y) = f(x, y)$ everywhere else. The function $f_1(x, y)$ is a density, hence it defines two new RVs $\mathbf{x}_1$ and $\mathbf{y}_1$. These RVs are obviously not jointly normal. However, they are marginally normal because $\mathbf{x}$ and $\mathbf{y}$ are marginally normal and the masses in any vertical or horizontal strip have not changed. Furthermore, the RV $\mathbf{z}_1 = \mathbf{x}_1 + \mathbf{y}_1$ is also normal because $\mathbf{z} = \mathbf{x} + \mathbf{y}$ is normal and the masses in any diagonal strip of the form $z \le x + y \le z + dz$ have not changed.

Moment theorem The moment generating function of **x** and **y** is given by

$$\mathbf{\Phi}(s_1, s_2) = E\{e^{s_1\mathbf{x}+s_2\mathbf{y}}\}$$

Expanding the exponential and using the linearity of expected values, we obtain the series

$$\mathbf{\Phi}(s_1, s_2) = \sum_{n=0}^{\infty} \frac{1}{n!} \sum_{k=0}^{n} \binom{n}{k} E\{\mathbf{x}^k\mathbf{y}^{n-k}\} s_1^k s_2^{n-k}$$

$$= 1 + m_{10}s_1 + m_{01}s_2 + \tfrac{1}{2}(m_{20}s_1^2 + 2m_{11}s_1s_2 + m_{02}s_2^2) + \cdots \tag{7-34}$$

From this it follows that

$$\frac{\partial^k\, \partial^r}{\partial s_1^k\, \partial s_2^r} \mathbf{\Phi}(0, 0) = m_{kr} \tag{7-35}$$

The derivatives of the function $\mathbf{\Psi}(s_1, s_2) = \ln \mathbf{\Phi}(s_1, s_2)$ are by definition the joint cumulants λ_{kr} of **x** and **y**. It can be shown that

$$\lambda_{10} = m_{10} \qquad \lambda_{01} = m_{01} \qquad \lambda_{20} = \mu_{20} \qquad \lambda_{02} = \mu_{02} \qquad \lambda_{11} = \mu_{11}$$

Hence

$$\mathbf{\Psi}(s_1, s_2) = \eta_1 s_1 + \eta_2 s_2 + \tfrac{1}{2}(\sigma_1^2 s_1^2 + 2r\sigma_1\sigma_2 s_1 s_2 + \sigma_2^2 s_2^2) + \cdots \tag{7-36}$$

Example 7-6 Using (7-34), we shall show that if the RVs **x** and **y** are jointly normal with zero mean, then

$$E\{\mathbf{x}^2\mathbf{y}^2\} = E\{\mathbf{x}^2\}E\{\mathbf{y}^2\} + 2E^2\{\mathbf{xy}\} \tag{7-37}$$

PROOF As we see from (7-32)

$$\mathbf{\Phi}(s_1, s_2) = e^{-A} \qquad A = \tfrac{1}{2}(\sigma_1^2 s_1^2 + 2Cs_1s_2 + \sigma_2^2 s_2^2)$$

where $C = E\{\mathbf{xy}\} = r\sigma_1\sigma_2$. To prove (7-37), we shall equate the coefficient

$$\frac{1}{4!}\binom{4}{2} E\{\mathbf{x}^2\mathbf{y}^2\}$$

of $s_1^2 s_2^2$ in (7-34) with the corresponding coefficient of the expansion of e^{-A}. In this expansion, the factors $s_1^2 s_2^2$ appear only in the terms

$$\frac{A^2}{2} = \frac{1}{8}(\sigma_1^2 s_1^2 + 2Cs_1s_2 + \sigma_2^2 s_2^2)^2$$

Hence

$$\frac{1}{4!}\binom{4}{2} E\{\mathbf{x}^2\mathbf{y}^2\} = \frac{1}{8}(2\sigma_1^2\sigma_2^2 + 4C^2)$$

and (7-37) results.

7-3 CONDITIONAL DISTRIBUTIONS

As we have noted, conditional distributions can be expressed as conditional probabilities

$$F_z(z \mid \mathcal{M}) = P\{\mathbf{z} \le z \mid \mathcal{M}\} = \frac{P\{\mathbf{z} < z, \mathcal{M}\}}{P(\mathcal{M})}$$

$$F_{zw}(z, w \mid \mathcal{M}) = P\{\mathbf{z} \le z, \mathbf{w} \le w \mid \mathcal{M}\} = \frac{P\{\mathbf{z} \le z, \mathbf{w} \le w, \mathcal{M}\}}{P(\mathcal{M})} \tag{7-38}$$

The corresponding densities are obtained by appropriate differentiations.

In this section, we evaluate these functions for various special cases.

Example 7-7 We shall determine the conditional distribution $F_y(y \mid \mathbf{x} \le x)$ and density $f_y(y \mid \mathbf{x} \le x)$.

With $\mathcal{M} = \{\mathbf{x} \le x\}$, (7-38) yields

$$F_y(y \mid \mathbf{x} \le x) = \frac{P\{\mathbf{x} \le x, \mathbf{y} \le y\}}{P\{\mathbf{x} \le x\}} = \frac{F(x, y)}{F_x(x)}$$

$$f_y(y \mid \mathbf{x} \le x\} = \frac{\partial F(x, y)/\partial y}{F_x(x)}$$

Example 7-8 We shall next determine the joint distribution $F(x, y \mid \mathcal{M})$ for $\mathcal{M} = \{x_1 < \mathbf{x} \le x_2\}$. In this case, $F(x, y \mid \mathcal{M})$ is given by

$$F(x, y \mid x_1 < \mathbf{x} \le x_2) = \frac{P\{\mathbf{x} \le x, \mathbf{y} \le y, x_1 < \mathbf{x} \le x_2\}}{P(x_1 < \mathbf{x} \le x_2)}$$

$$= \begin{cases} \dfrac{F(x_2, y) - F(x_1, y)}{F_x(x_2) - F_x(x_1)} & x > x_2 \\[2ex] \dfrac{F(x, y) - F(x_1, y)}{F_x(x_2) - F_x(x_1)} & x_1 < x \le x_2 \end{cases}$$

and it equals zero for $x < x_1$.

Since $f = \partial^2 F/\partial x\, \partial y$, the above yields

$$f(x, y \mid x_1 < \mathbf{x} \le x_2) = \frac{f(x, y)}{F_x(x_2) - F_x(x_1)} \qquad x_1 < x \le x_2 \tag{7-39}$$

and zero otherwise.

The determination of the conditional density of $\mathbf{y}$ assuming $\mathbf{x} = x$ is of particular interest. This density cannot be derived directly from (7-38) because, in general, the event $\{\mathbf{x} = x\}$ has zero probability. It can, however, be defined as a limit.

Suppose first that

$$\mathcal{M} = \{x_1 < \mathbf{x} \le x_2\}$$

In this case, (7-38) yields

$$F_y(y \mid x_1 < \mathbf{x} \le x_2) = \frac{P\{x_1 < \mathbf{x} \le x_2, \mathbf{y} \le y\}}{P\{x_1 < \mathbf{x} \le x_2\}} = \frac{F(x_2, y) - F(x_1, y)}{F_x(x_2) - F_x(x_1)}$$

Differentiating with respect to y, we obtain

$$f_y(y \mid x_1 < \mathbf{x} \le x_2) = \frac{\int_{x_1}^{x_2} f(x, y)\, dx}{F_x(x_2) - F_x(x_1)} \tag{7-40}$$

because [see (6-6)]

$$\frac{\partial F(x, y)}{\partial y} = \int_{-\infty}^{x} f(\alpha, y)\, d\alpha$$

To determine $f_y(y \mid \mathbf{x} = x)$, we set $x_1 = x$ and $x_2 = x + \Delta x$ in (7-40). This yields

$$f_y(y \mid x < \mathbf{x} \le x + \Delta x) = \frac{\int_{x}^{x+\Delta x} f(\alpha, y)\, d\alpha}{F_x(x + \Delta x) - F_x(x)} \simeq \frac{f(x, y)\, \Delta x}{f_x(x)\, \Delta x}$$

Hence

$$f_y(y \mid \mathbf{x} = x) = \lim_{\Delta x \to 0} f_y(y \mid x < \mathbf{x} \le x + \Delta x) = \frac{f(x, y)}{f_x(x)}$$

If there is no fear of ambiguity, the function $f_y(y \mid \mathbf{x} = x)$ will be written in the form $f(y \mid x)$. Defining $f(x \mid y)$ similarly, we obtain

$$f(y \mid x) = \frac{f(x, y)}{f(x)} \qquad f(x \mid y) = \frac{f(x, y)}{f(y)} \tag{7-41}$$

If the RVs $\mathbf{x}$ and $\mathbf{y}$ are independent, then

$$f(x, y) = f(x)f(y) \qquad f(y \mid x) = f(y) \qquad f(x \mid y) = f(x)$$

Notes 1. For a specific x, the function $f(x, y)$ is a *profile* of $f(x, y)$, that is, it equals the intersection of the surface $f(x, y)$ by the plane $x =$ constant. The conditional density $f(y \mid x)$ is the equation of this curve normalized by the factor $1/f(x)$ so as to make its area 1. The function $f(x \mid y)$ has a similar interpretation: It is the normalized equation of the intersection of the surface $f(x, y)$ by the plane $y =$ constant.

2. As we know, the product $f(y)\, dy$ equals the probability of the event $\{y \le \mathbf{y} \le y + dy\}$. Extending this to conditional probabilities, we obtain

$$f_y(y \mid x_1 < \mathbf{x} \le x_2)\, dy = \frac{P\{x_1 < \mathbf{x} \le x_2, y < \mathbf{y} \le y + dy\}}{P\{x_1 < \mathbf{x} \le x_2\}}$$

This equals the mass in the rectangle of Fig. 7-3*a* divided by the mass in the vertical strip $x_1 < \mathbf{x} \le x_2$. Similarly, the product $f(y \mid x)\, dy$ equals the ratio of the mass in the differential rectangle $dx\, dy$ of Fig. 7-3*b* over the mass in the vertical strip $(x, x + dx)$.

3. The joint statistics of $\mathbf{x}$ and $\mathbf{y}$ are determined in terms of their joint density $f(x, y)$. Since

$$f(x, y) = f(y \mid x) f(x)$$

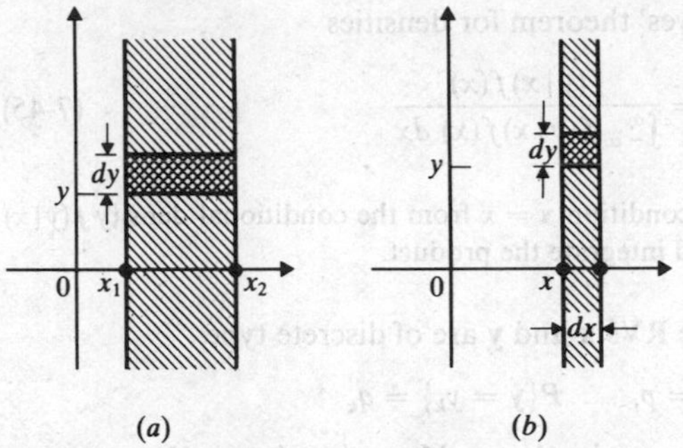

Figure 7-3

we conclude that they are also determined in terms of the marginal density $f(x)$ and the conditional density $f(y \mid x)$.

Example 7-9 We shall show that, if the RVs **x** and **y** are jointly normal with zero mean as in (6-44), then

$$f(y \mid x) = \frac{1}{\sigma_2\sqrt{2\pi(1-r^2)}} \exp\left[-\frac{(y - r\sigma_2 x/\sigma_1)^2}{2\sigma_2^2(1-r^2)}\right] \tag{7-42}$$

PROOF The exponent in (6-44) equals

$$\frac{(y - r\sigma_2 x/\sigma_1)^2}{2\sigma_2^2(1-r^2)} - \frac{x^2}{2\sigma_1^2}$$

Division by $f(x)$ removes the term $-x^2/2\sigma_1^2$ and (7-42) results.

The same reasoning leads to the conclusion that if **x** and **y** are jointly normal with $E\{\mathbf{x}\} = \eta_1$ and $E\{\mathbf{y}\} = \eta_2$, then $f(y \mid x)$ is given by (7-42) if y and x are replaced by $y - \eta_2$ and $x - \eta_1$ respectively. In other words, for a given x, $f(y \mid x)$ is a normal density with mean $\eta_2 + r\sigma_2(x - \eta_1)/\sigma_1$ and variance $\sigma_2^2(1 - r^2)$.

Bayes' theorem and total probability From (7-41) it follows that

$$f(x \mid y) = \frac{f(y \mid x) f(x)}{f(y)} \tag{7-43}$$

This is the density version of (2-38).

The denominator $f(y)$ can be expressed in terms of $f(y \mid x)$ and $f(x)$. Since

$$f(y) = \int_{-\infty}^{\infty} f(x, y)\,dx \quad \text{and} \quad f(x, y) = f(y \mid x) f(x)$$

we conclude that (total probability)

$$f(y) = \int_{-\infty}^{\infty} f(y \mid x) f(x)\,dx \tag{7-44}$$

Inserting into (7-43), we obtain Bayes' theorem for densities

$$f(x \mid y) = \frac{f(y \mid x) f(x)}{\int_{-\infty}^{\infty} f(y \mid x) f(x)\, dx} \tag{7-45}$$

Note As (7-44) shows, to remove the condition $\mathbf{x} = x$ from the conditional density $f(y \mid x)$ we multiply by the density $f(x)$ of $\mathbf{x}$ and integrate the product.

Discrete type Suppose that the RVs $\mathbf{x}$ and $\mathbf{y}$ are of discrete type

$$P\{\mathbf{x} = x_i\} = p_i \qquad P\{\mathbf{y} = y_k\} = q_k$$

$$P\{\mathbf{x} = x_i, \mathbf{y} = y_k\} = p_{ik} \qquad i = 1, \ldots, M \qquad k = 1, \ldots, N$$

where [see (6-21)]

$$p_i = \sum_k p_{ik} \qquad q_k = \sum_i p_{ik}$$

From the above and (2-29) it follows that

$$P\{\mathbf{y} = y_k \mid \mathbf{x} = x_i\} = \frac{P\{\mathbf{x} = x_i, \mathbf{y} = y_k\}}{P\{\mathbf{x} = x_i\}} = \frac{p_{ik}}{p_i}$$

Markoff matrix We denote by π_{ik} the above conditional probabilities

$$P\{\mathbf{y} = y_k \mid \mathbf{x} = x_i\} = \pi_{ik}$$

and by Π the $M \times N$ matrix whose elements are π_{ik}. Clearly

$$\pi_{ik} = \frac{p_{ik}}{p_i} \tag{7-46}$$

Hence

$$\pi_{ik} \geq 0 \qquad \sum_k \pi_{ik} = 1 \tag{7-47}$$

Thus, the elements of the matrix Π are positive and the sum on each row equals 1. Such a matrix is called *Markoff*. The conditional probabilities

$$P\{\mathbf{x} = x_i \mid \mathbf{y} = y_k\} = \pi^{ki} = \frac{p_{ik}}{q_k}$$

are the elements of an $N \times M$ Markoff matrix.

If the RVs $\mathbf{x}$ and $\mathbf{y}$ are independent, then

$$p_{ik} = p_i q_k \qquad \pi_{ik} = q_k \qquad \pi^{ki} = p_i$$

We note that

$$\pi^{ki} = \pi_{ik} \frac{p_i}{q_k} \qquad q_k = \sum_i \pi_{ik} p_i \tag{7-48}$$

These equations are the discrete versions of Eqs. (7-43) and (7-44).

7-4 CONDITIONAL EXPECTED VALUES

Applying theorem (5-29) to conditional densities, we obtain the conditional mean of $g(\mathbf{y})$

$$E\{g(\mathbf{y}) \mid \mathscr{M}\} = \int_{-\infty}^{\infty} g(y) f(y \mid \mathscr{M})\, dy \tag{7-49}$$

This can be used to define the conditional moments of $\mathbf{y}$.

Using a limit argument as in (7-41), we can define also the conditional mean $E\{g(\mathbf{y}) \mid x\}$. In particular

$$\eta_{y|x} = E\{\mathbf{y} \mid x\} = \int_{-\infty}^{\infty} y f(y \mid x)\, dy \tag{7-50}$$

is the *conditional mean* of $\mathbf{y}$ assuming $\mathbf{x} = x$, and

$$\sigma_{y|x}^2 = E\{(\mathbf{y} - \eta_{y|x})^2 \mid x\} = \int_{-\infty}^{\infty} (y - \eta_{y|x})^2 f(y \mid x)\, dy \tag{7-51}$$

is its *conditional variance*.

For a given x, the integral in (7-50) is the center of gravity of the masses in the vertical strip $(x, x + dx)$. The locus of these points as x varies from $-\infty$ to ∞, is the function

$$\varphi(x) = \int_{-\infty}^{\infty} y f(y \mid x)\, dy \tag{7-52}$$

known as *regression line* (Fig. 7-4).

Note If the RVs $\mathbf{x}$ and $\mathbf{y}$ are functionally related, i.e., if $\mathbf{y} = g(\mathbf{x})$, then the probability masses on the xy plane are on the line $y = g(x)$ (see Fig. 6-4*b*), hence, $E\{\mathbf{y} \mid x\} = g(x)$.

Example 7-10 If the RVs $\mathbf{x}$ and $\mathbf{y}$ are normal as in Example 7-9, then

$$E\{\mathbf{y} \mid x\} = \eta_2 + r\sigma_2 \frac{x - \eta_1}{\sigma_1} \tag{7-53}$$

is a straight line with slope $r\sigma_2/\sigma_1$ passing through the point (η_1, η_2). Since for normal RVs the conditional mean $E\{\mathbf{y} \mid x\}$ coincides with the maximum of $f(y \mid x)$, we conclude that the locus of the maxima of all profiles of $f(x, y)$ is the straight line (7-53).

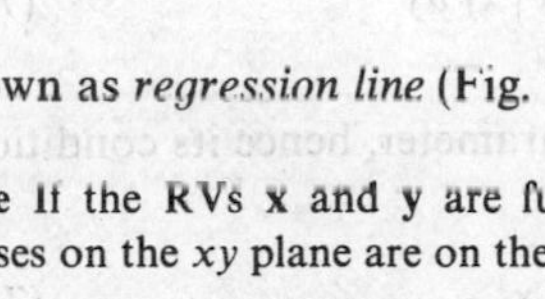

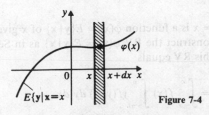

Figure 7-4

From theorem (7-2) it follows that

$$E\{g(\mathbf{x}, \mathbf{y}) \mid \mathcal{M}\} = \int_{-\infty}^{\infty}\int_{-\infty}^{\infty} g(x, y) f(x, y \mid \mathcal{M})\, dx\, dy \tag{7-54}$$

This expression can be used to determine $E\{g(\mathbf{x}, \mathbf{y}) \mid x\}$, however, the conditional density $f(x, y \mid x)$ consists of line masses on the line x-constant. To avoid dealing with line masses, we shall define $E\{g(\mathbf{x}, \mathbf{y}) \mid x\}$ as a limit:

As we have shown in Example 7-8, the conditional density $f(x, y \mid x < \mathbf{x} < x + \Delta x)$ is zero outside the strip $(x, x + \Delta x)$ and in this strip it is given by (7-39) where $x_1 = x$ and $x_2 = x + \Delta x$. It follows therefore, from (7-54) with $\mathcal{M} = \{x < \mathbf{x} \le x + \Delta x\}$, that

$$E\{g(\mathbf{x}, \mathbf{y}) \mid x < \mathbf{x} \le x + \Delta x\} = \int_{-\infty}^{\infty}\int_{x}^{x+\Delta x} g(\alpha, y) \frac{f(\alpha, y)\, d\alpha}{F_x(x + \Delta x) - F_x(x)}\, dy$$

As $\Delta x \to 0$, the inner integral tends to $g(x, y) f(x, y)/f(x)$. Defining $E\{g(\mathbf{x}, \mathbf{y}) \mid x\}$ as the limit of the above, we obtain

$$E\{g(\mathbf{x}, \mathbf{y}) \mid x\} = \int_{-\infty}^{\infty} g(x, y) f(y \mid x)\, dy \tag{7-55}$$

We also note that

$$E\{g(x, \mathbf{y}) \mid x\} = \int_{-\infty}^{\infty} g(x, y) f(y \mid x)\, dy \tag{7-56}$$

because $g(x, \mathbf{y})$ is a function of the RV $\mathbf{y}$, with x a parameter, hence its conditional expected value is given by (7-49). Thus

$$E\{g(\mathbf{x}, \mathbf{y}) \mid x\} = E\{g(x, \mathbf{y}) \mid x\} \tag{7-57}$$

One might be tempted from the above to conclude that (7-57) follows directly from (7-49); however, this is not so. The functions $g(\mathbf{x}, \mathbf{y})$ and $g(x, \mathbf{y})$ have the same expected value assuming $\mathbf{x} = x$ but they are not equal. The first is a function $g(\mathbf{x}, \mathbf{y})$ of the RVs $\mathbf{x}$ and $\mathbf{y}$ and for a specific ζ it takes the value $g[\mathbf{x}(\zeta), \mathbf{y}(\zeta)]$. The second is a function $g(x, \mathbf{y})$ of the real variable x and the RV $\mathbf{y}$, and for a specific ζ it takes the value $g[x, \mathbf{y}(\zeta)]$ where x is an arbitrary number.

Conditional Expected Values as RVs

The conditional mean of $\mathbf{y}$ assuming $\mathbf{x} = x$ is a function $\varphi(x) = E\{\mathbf{y} \mid x\}$ of x given by (7-52). Using this function, we can construct the RV $\varphi(\mathbf{x}) = E\{\mathbf{y} \mid \mathbf{x}\}$ as in Sec. 5-1. As we see from (5-29), the mean of this RV equals

$$E\{\varphi(\mathbf{x})\} = \int_{-\infty}^{\infty} \varphi(x) f(x)\, dx = \int_{-\infty}^{\infty} f(x) \int_{-\infty}^{\infty} y f(y \mid x)\, dy\, dx$$

Since $f(x, y) = f(x)f(y \mid x)$, the above yields

$$E\{E\{\mathbf{y} \mid \mathbf{x}\}\} = \int_{-\infty}^{\infty}\int_{-\infty}^{\infty} yf(x, y)\, dx\, dy = E\{\mathbf{y}\} \tag{7-58}$$

This basic result can be generalized: The conditional mean $E\{g(\mathbf{x}, \mathbf{y}) \mid x\}$ of $g(\mathbf{x}, \mathbf{y})$ assuming $\mathbf{x} = x$ is a function of the real variable x. It defines, therefore, the function $E\{g(\mathbf{x}, \mathbf{y}) \mid \mathbf{x}\}$ of the RV $\mathbf{x}$. As we see from (7-2) and (7-54), the mean of $E\{g(\mathbf{x}, \mathbf{y}) \mid \mathbf{x}\}$ equals

$$\int_{-\infty}^{\infty} f(x) \int_{-\infty}^{\infty} g(x, y) f(y \mid x)\, dy\, dx = \int_{-\infty}^{\infty}\int_{-\infty}^{\infty} g(x, y) f(x, y)\, dx\, dy$$

But the last integral equals $E\{g(\mathbf{x}, \mathbf{y})\}$, hence

$$E\{E\{g(\mathbf{x}, \mathbf{y}) \mid \mathbf{x}\}\} = E\{g(\mathbf{x}, \mathbf{y})\} \tag{7-59}$$

We note, finally, that

$$\begin{aligned} E\{g_1(\mathbf{x})g_2(\mathbf{y}) \mid x\} &= E\{g_1(x)g_2(\mathbf{y}) \mid x\} = g_1(x)E\{g_2(\mathbf{y}) \mid x\} \\ E\{g_1(\mathbf{x})g_2(\mathbf{y})\} &= E\{E\{g_1(\mathbf{x})g_2(\mathbf{y}) \mid \mathbf{x}\}\} = E\{g_1(\mathbf{x})E\{g_2(\mathbf{y}) \mid \mathbf{x}\}\} \end{aligned} \tag{7-60}$$

Example 7-11 Suppose that the RVs $\mathbf{x}$ and $\mathbf{y}$ are $N(0, 0; \sigma_1, \sigma_2; r)$. As we know

$$E\{\mathbf{x}^2\} = \sigma_1^2 \qquad E\{\mathbf{x}^4\} = 3\sigma_1^4$$

Furthermore, $f(y \mid x)$ is a normal density with mean $r\sigma_2 x/\sigma_1$ and variance $\sigma_2\sqrt{1 - r^2}$. Hence

$$E\{\mathbf{y}^2 \mid x\} = \eta_{y|x}^2 + \sigma_{y|x}^2 = \left(\frac{r\sigma_2 x}{\sigma_1}\right)^2 + \sigma_2^2(1 - r^2)$$

Using (7-60), we shall show that

$$E\{\mathbf{xy}\} = r\sigma_1\sigma_2$$

$$E\{\mathbf{x}^2\mathbf{y}^2\} = E\{\mathbf{x}^2\}E\{\mathbf{y}^2\} + 2E^2\{\mathbf{xy}\}$$

PROOF

$$E\{\mathbf{xy}\} = E\{\mathbf{x}E\{\mathbf{y} \mid \mathbf{x}\}\} = E\left\{r\sigma_2\frac{\mathbf{x}^2}{\sigma_1}\right\} = r\sigma_2\frac{\sigma_1^2}{\sigma_1}$$

$$E\{\mathbf{x}^2\mathbf{y}^2\} = E\{\mathbf{x}^2E\{\mathbf{y}^2 \mid \mathbf{x}\}\} = E\left\{\mathbf{x}^2\left[r^2\sigma_2^2\frac{\mathbf{x}^2}{\sigma_1^2} + \sigma_2^2(1 - r^2)\right]\right\}$$

$$= 3\sigma_1^4 r^2\frac{\sigma_2^2}{\sigma_1^2} + \sigma_1^2\sigma_2^2(1 - r^2) = \sigma_1^2\sigma_2^2 + 2r^2\sigma_1^2\sigma_2^2$$

and the proof is complete [see also (7-37)].

7-5 MEAN SQUARE ESTIMATION

The estimation problem is fundamental in the applications of probability and it will be discussed in detail later (Chap. 13). In this section, we introduce the main ideas using as illustration the estimation of an RV **y** in terms of another RV **x**. Throughout this analysis, the optimality criterion will be the minimization of the mean square value (abbreviation: MS) of the estimation error.

We start with a brief explanation of the underlying concepts in the context of repeated trials, considering first the problem of estimating the RV **y** by a constant.

Frequency interpretation As we know, the distribution function $F(y)$ of the RV **y** determines completely its statistics. This does not, of course, mean that if we know $F(y)$ we can predict the value $\mathbf{y}(\zeta)$ of **y** at some future trial. Suppose, however, that we wish to estimate the unknown $\mathbf{y}(\zeta)$ by some number c. As we shall presently see, knowledge of $F(y)$ can guide us in the selection of c.

If **y** is estimated by a constant c, then, at a particular trial, the error $\mathbf{y}(\zeta) - c$ results and our problem is to select c so as to minimize this error in some sense. A reasonable criterion for selecting c might be the condition that, in a long series of trials, the error is close to zero

$$\frac{\mathbf{y}(\zeta_1) - c + \cdots + \mathbf{y}(\zeta_n) - c}{n} \simeq 0$$

As we see from (5-27), this would lead to the conclusion that c should equal the *mean* of **y**.

Another criterion for selecting a might be the minimization of the average of $|\mathbf{y}(\zeta) - c|$. In this case, the optimum c is the *median* of **y** (see Prob. 5-12).

In our analysis, we consider only MS estimates. This means that c should be such as to minimize the average of $|\mathbf{y}(\zeta) - c|^2$. This criterion is in general useful but it is selected mainly because it leads to simple results. As we shall soon see, the best c is again the *mean* of **y**.

Suppose now that at each trial we observe the value $\mathbf{x}(\zeta)$ of the RV **x**. On the basis of this observation it might be best to use as the estimate of **y** not the same number c at each trial, but a number that depends on the observed $\mathbf{x}(\zeta)$. In other words, we might use as the estimate of **y** a function $c(\mathbf{x})$ of the RV **x**. The resulting problem is the optimum determination of this function.

It might be argued that, if at a certain trial we observe $\mathbf{x}(\zeta)$, then we can determine the outcome ζ of this trial, and hence also the corresponding value $\mathbf{y}(\zeta)$ of **y**. This, however, is not so. The same number $\mathbf{x}(\zeta) = x$ is observed for every ζ in the set $\{\mathbf{x} = x\}$. If, therefore, this set has many elements and the values of **y** are different for the various elements of this set, then the observed $\mathbf{x}(\zeta)$ does not determine uniquely $\mathbf{y}(\zeta)$. However, we know now that ζ is an element of the subset $\{\mathbf{x} = x\}$. This information reduces the uncertainty about the value of **y**. In the subset $\{\mathbf{x} = x\}$, the RV **x** equals x and the problem of determining $c(x)$ is reduced to the problem of determining the constant $c(x)$. As we noted, if the optimality criterion is the minimization of the MS error, then $c(x)$ must be the average of **y** in this set. In other words, $c(x)$ must equal the conditional mean of **y** assuming that $\mathbf{x} = x$.

We shall illustrate with an example. Suppose that the space $\mathscr{S}$ is the set of all children in a community and the RV **y** is the height of each child. A particular outcome ζ is a specific child and $\mathbf{y}(\zeta)$ is the height of this child. From the preceding discussion it follows that if we wish to estimate **y** by a number, this number must equal the mean of **y**. We now assume that each selected child is weighed. On the basis of this observation, the estimate of the height of the child can be improved. The weight is an RV **x**, hence the optimum estimate of **y** is now the conditional mean $E\{\mathbf{y} \mid x\}$ of **y** assuming $\mathbf{x} = x$ where x is the observed weight.

In the context of probability theory, the MS estimation of the RV **y** by a constant c can be phrased as follows: Find c such that the second moment (MS error)

$$e = E\{(\mathbf{y} - c)^2\} = \int_{-\infty}^{\infty} (y - c)^2 f(y)\, dy \qquad (7\text{-}61)$$

of the difference (error) $\mathbf{y} - c$ is minimum. Clearly, e depends on c and it is minimum if

$$\frac{de}{dc} = \int_{-\infty}^{\infty} 2(y - c) f(y)\, dy = 0 \qquad \text{that is, if} \qquad c = \int_{-\infty}^{\infty} y f(y)\, dy$$

Thus

$$c = E\{\mathbf{y}\} = \int_{-\infty}^{\infty} y f(y)\, dy \qquad (7\text{-}62)$$

This result is well known from mechanics: The moment of inertia with respect to a point c is minimum if c is the center of gravity of the masses.

Nonlinear MS estimation We wish to estimate **y** not by a constant but by a function $c(\mathbf{x})$ of the RV **x**. Our problem now is to find the function $c(x)$ such that the MS error

$$e = E\{[\mathbf{y} - c(\mathbf{x})]^2\} = \int_{-\infty}^{\infty} \int_{-\infty}^{\infty} [y - c(x)]^2 f(x, y)\, dx\, dy \qquad (7\text{-}63)$$

is minimum.

We maintain that

$$c(x) = E\{\mathbf{y} \mid x\} = \int_{-\infty}^{\infty} y f(y \mid x)\, dy \qquad (7\text{-}64)$$

PROOF Since $f(x, y) = f(y \mid x) f(x)$, (7-63) yields

$$e = \int_{-\infty}^{\infty} f(x) \int_{-\infty}^{\infty} [y - c(x)]^2 f(y \mid x)\, dy\, dx$$

The integrands above are positive. Hence, e is minimum if the inner integral is minimum for every x. This integral is of the form (7-61) if c is changed to $c(x)$ and $f(y)$ is changed to $f(y \mid x)$. Hence, it is minimum if $c(x)$ equals the integral in (7-62) provided that $f(y)$ is changed to $f(y \mid x)$. The result is (7-64).

Thus, the optimum $c(x)$ is the regression line $\varphi(x)$ of Fig. 7-4.

As we noted in the beginning of the section, if $\mathbf{y} = g(\mathbf{x})$, then $E\{\mathbf{y} \mid x\} = g(x)$, hence, $c(x) = g(x)$ and the resulting MS error is zero. This is not surprising because, if $\mathbf{x}$ is observed and $\mathbf{y} = g(\mathbf{x})$, then $\mathbf{y}$ is determined uniquely.

If the RVs $\mathbf{x}$ and $\mathbf{y}$ are independent, then $E\{\mathbf{y} \mid x\} = E\{\mathbf{y}\} =$ constant. In this case, knowledge of $\mathbf{x}$ has no effect on the estimate of $\mathbf{y}$.

Linear MS Estimation

The solution of the nonlinear MS estimation problem is based on knowledge of the function $\varphi(x)$. An easier problem, using only second-order moments, is the linear MS estimation of $\mathbf{y}$ in terms of $\mathbf{x}$. The resulting estimate is not as good as the nonlinear estimate; however, it is used in many applications because of the simplicity of the solution.

The linear estimation problem is the estimation of the RV $\mathbf{y}$ in terms of a linear function $A\mathbf{x} + B$ of $\mathbf{x}$. The problem now is to find the constants A and B so as to minimize the MS error

$$e = E\{[\mathbf{y} - (A\mathbf{x} + B)]^2\} \tag{7-65}$$

We maintain that $e = e_m$ is minimum if

$$A = \frac{\mu_{11}}{\mu_{20}} = \frac{r\sigma_y}{\sigma_x} \qquad B = \eta_y - A\eta_x \tag{7-66}$$

and

$$e_m = \mu_{02} - \frac{\mu_{11}^2}{\mu_{02}} = \sigma_y^2(1 - r^2) \tag{7-67}$$

PROOF For a given A, e is the MS error of the estimation of $\mathbf{y} - A\mathbf{x}$ by the constant B. Hence, e is minimum if $B = E\{\mathbf{y} - A\mathbf{x}\}$ as in (7-62). With B so determined, (7-65) yields

$$e = E\{[(\mathbf{y} - \eta_y) - A(\mathbf{x} - \eta_x)]^2\} = \sigma_y^2 - 2Ar\sigma_x\sigma_y + A^2\sigma_x^2$$

This is minimum if $A = r\sigma_y/\sigma_x$ and (7-66) results. Inserting into the above quadratic, we obtain (7-67).

Terminology In the above, the sum $A\mathbf{x} + B$ is the *nonhomogeneous* linear estimate of $\mathbf{y}$ in terms of $\mathbf{x}$. If $\mathbf{y}$ is estimated by a straight line $a\mathbf{x}$ passing through the origin, the estimate is called *homogeneous.*

The RV $\mathbf{x}$ is the data of the estimation, the RV $\boldsymbol{\varepsilon} = \mathbf{y} - (A\mathbf{x} + B)$ is the *error* of the estimation, and the number $e = E\{\boldsymbol{\varepsilon}^2\}$ is the MS error.

Fundamental note In general, the nonlinear estimate $\varphi(x) = E\{\mathbf{y} \mid x\}$ of $\mathbf{y}$ in terms of $\mathbf{x}$ is not a straight line and the resulting MS error $E\{[\mathbf{y} - \varphi(\mathbf{x})]^2\}$ is smaller than the MS error e_m of the linear estimate $A\mathbf{x} + B$. However, if the RVs $\mathbf{x}$ and $\mathbf{y}$ are jointly normal, then [see (7-53)]

$$\varphi(x) = \frac{r\sigma_y x}{\sigma_x} + \eta_y - \frac{r\sigma_y \eta_x}{\sigma_x}$$

is a straight line as in (7-66). In other words:

For normal RVs, nonlinear and linear MS estimates are identical.

The Orthogonality Principle

From (7-66) it follows that

$$E\{[\mathbf{y} - (A\mathbf{x} + B)]\mathbf{x}\} = 0 \tag{7-68}$$

This result can be derived directly from (7-65). Indeed, the MS error e is a function of A and B and it is minimum if $\partial e/\partial A = 0$ and $\partial e/\partial B = 0$. The first equation yields

$$\frac{\partial e}{\partial A} = E\{2[\mathbf{y} - (A\mathbf{x} + B)](-\mathbf{x})\} = 0$$

and (7-68) results. The interchange between expected value and differentiation is equivalent to the interchange of integration and differentiation.

Equation (7-68) states that the optimum linear MS estimate $A\mathbf{x} + B$ of $\mathbf{y}$ is such that the estimation error $\mathbf{y} - (A\mathbf{x} + B)$ is orthogonal to the data $\mathbf{x}$. This is known as the *orthogonality principle.* It is fundamental in MS estimation and will be used extensively. In the following, we reestablish it for the homogeneous case.

Homogeneous linear MS estimation We wish to find a constant a such that, if $\mathbf{y}$ is estimated by $a\mathbf{x}$ the resulting MS error

$$e = E\{(\mathbf{y} - a\mathbf{x})^2\} \tag{7-69}$$

is minimum. We maintain that a must be such that

$$E\{(\mathbf{y} - a\mathbf{x})\mathbf{x}\} = 0 \tag{7-70}$$

PROOF Clearly, e is minimum if $e'(a) = 0$; this yields (7-70). We shall give a second proof: We assume that a satisfies (7-70) and we shall show that e is minimum. With $\bar{a}$ an arbitrary constant,

$$\begin{aligned} E\{(\mathbf{y} - \bar{a}\mathbf{x})^2\} &= E\{[(\mathbf{y} - a\mathbf{x}) + (a - \bar{a})\mathbf{x}]^2\} \\ &= E\{(\mathbf{y} - a\mathbf{x})^2\} + (a - \bar{a})^2 E\{\mathbf{x}^2\} + 2(a - \bar{a})E\{(\mathbf{y} - a\mathbf{x})\mathbf{x}\} \end{aligned}$$

In the above, the last term is zero by assumption and the second term is positive. From this it follows that

$$E\{(\mathbf{y} - \bar{a}\mathbf{x})^2\} \geq E\{(\mathbf{y} - a\mathbf{x})^2\}$$

for any $\bar{a}$, hence, e is minimum.

The linear MS estimate of $\mathbf{y}$ in terms of $\mathbf{x}$ will be denoted by $\hat{E}\{\mathbf{y} \mid \mathbf{x}\}$. Solving (7-70), we conclude that

$$\hat{E}\{\mathbf{y} \mid \mathbf{x}\} = a\mathbf{x} \qquad a = \frac{E\{\mathbf{xy}\}}{E\{\mathbf{x}^2\}} \tag{7-71}$$

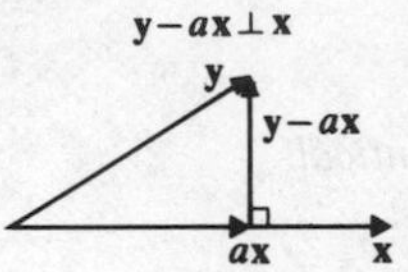

Figure 7-5

MS error Since

$$e = E\{(\mathbf{y} - a\mathbf{x})\mathbf{y}\} - E\{(\mathbf{y} - a\mathbf{x})a\mathbf{x}\} = E\{\mathbf{y}^2\} - E\{(a\mathbf{x})^2\} - 2aE\{(\mathbf{y} - a\mathbf{x})\mathbf{x}\}$$

we conclude with (7-70) that

$$e = E\{(\mathbf{y} - a\mathbf{x})\mathbf{y}\} = E\{\mathbf{y}^2\} - E\{(a\mathbf{x})^2\} \tag{7-72}$$

We note finally that (7-70) is consistent with the orthogonality principle: The error $\mathbf{y} - a\mathbf{x}$ is orthogonal to the data $\mathbf{x}$.

Geometric interpretation of the orthogonality principle In the vector representation of RVs (see Fig. 7-5), the difference $\mathbf{y} - a\mathbf{x}$ is the vector from the point $\mathbf{y}$ to the point $a\mathbf{x}$ on the $\mathbf{x}$ line and the length of that vector equals $\sqrt{e}$. Clearly, this length is minimum if $\mathbf{y} - a\mathbf{x}$ is perpendicular to $\mathbf{x}$ in agreement with (7-70). The right side of (7-72) follows from the Pythagorian theorem and the middle term states that the square of the length of $\mathbf{y} - a\mathbf{x}$ equals the inner product of $\mathbf{y}$ with the error $\mathbf{y} - a\mathbf{x}$.

PROBLEMS

7-1 The RVs $\mathbf{x}$, $\mathbf{y}$ are $N(0; \sigma)$ and independent. Show that, if $\mathbf{z} = |\mathbf{x} - \mathbf{y}|$, then $E\{\mathbf{z}\} = 2\sigma/\sqrt{\pi}$, $E\{\mathbf{z}^2\} = 2\sigma^2$.

7-2 Show that if $\mathbf{x}$ and $\mathbf{y}$ are two independent RVs with $f_x(x) = e^{-x}U(x)$, $f_y(y) = e^{-y}U(y)$, and $\mathbf{z} = (\mathbf{x} - \mathbf{y})U(\mathbf{x} - \mathbf{y})$, then $E\{\mathbf{z}\} = 1/2$.

7-3 Show that for any $\mathbf{x}$, $\mathbf{y}$ real or complex

(*a*) $|E\{\mathbf{xy}\}|^2 < E\{|\mathbf{x}|^2\}E\{|\mathbf{y}|^2\}$;

(*b*) (triangle inequality) $\sqrt{E\{|\mathbf{x} + \mathbf{y}|^2\}} \le \sqrt{E\{|\mathbf{x}|^2\}} + \sqrt{E\{|\mathbf{y}|^2\}}$.

7-4 Show that, if $r_{xy} = 1$, then $\mathbf{y} = a\mathbf{x} + b$.

7-5 Show that, if $E\{\mathbf{x}^2\} = E\{\mathbf{y}^2\} = E\{\mathbf{xy}\}$, then $\mathbf{x} = \mathbf{y}$.

7-6 Show that, if the RV $\mathbf{x}$ is of discrete type taking the values x_n with $P\{\mathbf{x} = x_n\} = p_n$ and $\mathbf{z} = g(\mathbf{x}, \mathbf{y})$, then

$$E\{\mathbf{z}\} = \sum_n E\{g(x_n, \mathbf{y})\}p_n \qquad f_z(z) = \sum_n f_z(z \mid x_n)p_n$$

7-7 The RV $\mathbf{n}$ is Poisson with parameter λ and the RV $\mathbf{x}$ is independent of $\mathbf{n}$. Show that, if $\mathbf{z} = \mathbf{nx}$ and

$$f_x(x) = \frac{\alpha}{\pi(\alpha^2 + x^2)} \qquad \text{then} \qquad \Phi_z(\omega) = \exp\{\lambda e^{-\alpha|\omega|} - \lambda\}$$

7-8 Show that, if $\mathbf{x}$ and $\mathbf{y}$ are $N(0, 0; \sigma, \sigma; r)$, with covariance $C = r\sigma^2$, then

(a) $$\frac{\partial^n f(x, y)}{\partial C^n} = \frac{\partial^{2n} f(x, y)}{\partial x^n \, \partial y^n}$$

(b) $$E\{f_x(\mathbf{x})f_y(\mathbf{y})\} = \frac{1}{2\pi\sigma^2\sqrt{1 - r^2}}$$

7-9 Show that, for any $\mathbf{x}$, $\mathbf{y}$, and $\varepsilon > 0$,

$$P\{|\mathbf{x} - \mathbf{y}| > \varepsilon\} \le \frac{1}{\varepsilon^2} E\{|\mathbf{x} - \mathbf{y}|^2\}$$

7-10 Show that the RVs $\mathbf{x}$ and $\mathbf{y}$ are independent iff for any a and b

$$E\{U(a - \mathbf{x})U(b - \mathbf{y})\} = E\{U(a - \mathbf{x})\}E\{U(a - \mathbf{y})\}$$

7-11 Show that

$$E\{\mathbf{y} \,|\, \mathbf{x} \le 0\} = \frac{1}{F_x(0)} \int_{-\infty}^{0} E\{\mathbf{y} \,|\, x\} f_x(x)\, dx$$

7-12 Show that, if the RVs $\mathbf{x}$ and $\mathbf{y}$ are independent and $\mathbf{z} = \mathbf{x} + \mathbf{y}$, then $f_z(z \,|\, x) = f_y(z - x)$.

7-13 The RVs $\mathbf{x}$, $\mathbf{y}$ are $N(3, 4; 1, 2; 0.5)$. Find $f(y \,|\, x)$ and $f(x \,|\, y)$.

7-14 Show that, for any $\mathbf{x}$ and $\mathbf{y}$, the RVs $\mathbf{z} = F_x(\mathbf{x})$ and $\mathbf{w} = F_y(\mathbf{y} \,|\, \mathbf{x})$ are independent and each is uniform in the interval (0, 1).

7-15 The RVs $\mathbf{x}$ and $\mathbf{y}$ are $N(0, 0; 3, 5; 0.8)$. Find $g(x)$ such that $E\{[\mathbf{y} - g(\mathbf{x})]^2$ is minimum.

7-16 In the approximation of $\mathbf{y}$ by $\varphi(\mathbf{x})$, the "mean cost" $E\{g[\mathbf{y} - \varphi(\mathbf{x})]\}$ results, where $g(x)$ is a given function. Show that, if $g(x)$ is an even concave function as in Fig. P5-20, then the "mean cost" is minimum if $\varphi(x) = E\{\mathbf{y} \,|\, x\}$.

7-17 Show that if $\varphi(x) = E\{\mathbf{y} \,|\, x\}$ is the nonlinear MS estimate of $\mathbf{y}$ in terms of $\mathbf{x}$, then

$$E\{[\mathbf{y} - \varphi(\mathbf{x})]^2\} = E\{\mathbf{y}^2\} - E\{\varphi^2(\mathbf{x})\}$$

7-18 If $\eta_x = \eta_y = 0$, $\sigma_x = \sigma_y = 4$ and $\hat{\mathbf{y}} = 0.2\mathbf{x}$ (linear MS estimate), find $E\{(\mathbf{y} - \hat{\mathbf{y}})^2\}$.

7-19 Show that if the constants A, B, and a are such that $E\{[\mathbf{y} - (A\mathbf{x} + B)]^2\}$ and $E\{[(\mathbf{y} - \eta_x) - a(\mathbf{x} - \eta_x)]^2\}$ are minimum, then $a = A$.

7-20 Given $\eta_x = 4$, $\eta_y = 0$, $\sigma_x = 1$, $\sigma_y = 2$, $r_{xy} = 0.5$, find the parameters A, B, and a that minimize $E\{[\mathbf{y} - (A\mathbf{x} + B)]^2\}$ and $E\{(\mathbf{y} - a\mathbf{x})^2\}$.

7-21 The RVs $\mathbf{x}$, $\mathbf{y}$ are independent, integer-valued with $P\{\mathbf{x} = k\} = p_k$, $P\{\mathbf{y} = k\} = q_k$. Show that

(a) if $\mathbf{z} = \mathbf{x} + \mathbf{y}$, then (discrete-time convolution)

$$P\{\mathbf{z} = n\} = \sum_{k=-\infty}^{\infty} p_{n-k} q_k$$

(b) if the RVs $\mathbf{x}$, $\mathbf{y}$ are Poisson distributed with parameters a and b respectively and $\mathbf{w} = \mathbf{x} - \mathbf{y}$, then

$$P\{\mathbf{w} = n\} = e^{-a-b} \sum_{k=-\infty}^{\infty} \frac{a^{n+k}b^k}{(n+k)!\,k!} = \frac{e^{-a-b}}{2\pi} \int_{-\pi}^{\pi} e^{2(a+b)\cos\omega - jn\omega}\, d\omega$$

CHAPTER

EIGHT

SEQUENCES OF RANDOM VARIABLES

8-1 GENERAL CONCEPTS

A *random vector* is a vector

$$\mathbf{X}: [\mathbf{x}_1, \ldots, \mathbf{x}_n] \tag{8-1}$$

whose components $\mathbf{x}_i$ are RVs.

The probability that $\mathbf{X}$ is in a region D of the n-dimensional space equals the probability masses in D

$$P\{\mathbf{X} \in D\} = \int_D f(X)\, dX \qquad X: [x_1, \ldots, x_n] \tag{8-2}$$

In the above

$$f(X) = f(x_1, \ldots, x_n) = \frac{\partial^n F(x_1, \ldots, x_n)}{\partial x_1, \ldots, \partial x_n} \tag{8-3}$$

is the *joint density* of the RVs $\mathbf{x}_i$ and

$$F(X) = F(x_1, \ldots, x_n) = P\{\mathbf{x} \le x_1, \ldots, \mathbf{x} \le x_n\} \tag{8-4}$$

is their *joint distribution.*

If we substitute in $F(x_1, \ldots, x_n)$ certain variables by ∞, we obtain the joint distribution of the remaining variables. If we integrate $f(x_1, \ldots, x_n)$ with respect to certain variables, we obtain the joint density of the remaining variables. For example

$$F(x_1, x_3) = F(x_1, \infty, x_3, \infty)$$

$$f(x_1, x_3) = \int_{-\infty}^{\infty} \int_{-\infty}^{\infty} f(x_1, x_2, x_3, x_4)\, dx_2\, dx_4 \tag{8-5}$$

Note In the above, we identify various functions in terms of their independent variables. Thus, $f(x_1, x_3)$ is the joint density of the RVs $\mathbf{x}_1$ and $\mathbf{x}_3$ and it is in general *different* from the joint density $f(x_2, x_4)$ of the RVs $\mathbf{x}_2$ and $\mathbf{x}_4$. Similarly, the density $f_i(x_i)$ of the RV $\mathbf{x}_i$ will often be denoted by $f(x_i)$.

Transformations Given k functions

$$g_1(X), \ldots, g_k(X) \qquad X : [x_1, \ldots, x_n]$$

we form the RVs

$$\mathbf{y}_1 = g_1(\mathbf{X}), \ldots, \mathbf{y}_k = g_k(\mathbf{X}) \tag{8-6}$$

The statistics of these RVs can be determined in terms of the statistics of $\mathbf{X}$ as in Sec. 6-3. If $k < n$, then we could determine first the joint density of the n RVs $\mathbf{y}_1, \ldots, \mathbf{y}_k, \mathbf{x}_{k+1}, \ldots, \mathbf{x}_n$ and then use (8-5) to eliminate the **x**'s. If $k > n$, then the RVs $\mathbf{y}_{n+1}, \ldots, \mathbf{y}_k$ can be expressed in terms of $\mathbf{y}_1, \ldots, \mathbf{y}_n$. In this case, the masses in the k space are singular and can be determined in terms of the joint density of $\mathbf{y}_1, \ldots, \mathbf{y}_n$. It suffices, therefore, to assume that $k = n$.

To find the density $f_y(y_1, \ldots, y_n)$ of the random vector $\mathbf{Y} = [\mathbf{y}_1, \ldots, \mathbf{y}_n]$ for a specific set of number $y_1, \ldots, y_n$, we solve the system

$$g_1(X) = y_1, \ldots, g_n(X) = y_n \tag{8-7}$$

If this system has no solutions, then $f_y(y_1, \ldots, y_n) = 0$. If it has a single solution $X : [x_1, \ldots, x_n]$, then

$$f_y(y_1, \ldots, y_n) = \frac{f_x(x_1, \ldots, x_n)}{|J(x_1, \ldots, x_n)|} \tag{8-8}$$

where

$$J(x_1, \ldots, x_n) = \begin{vmatrix} \dfrac{\partial g_1}{\partial x_1} \cdots \dfrac{\partial g_1}{\partial x_n} \\ \cdots\cdots\cdots \\ \dfrac{\partial g_n}{\partial x_1} \cdots \dfrac{\partial g_n}{\partial x_n} \end{vmatrix} \tag{8-9}$$

is the jacobian of the transformation (8-7). If it has several solutions, then we add the corresponding terms as in (6-63).

Independence

The RVs $\mathbf{x}_1, \ldots, \mathbf{x}_n$ are called (mutually) independent if the events $\{\mathbf{x} \le x_1\}, \ldots, \{\mathbf{x} \le x_n\}$ are independent. From this it follows that

$$F(x_1, \ldots, x_n) = F(x_1) \cdots F(x_n)$$
$$f(x_1, \ldots, x_n) = f(x_1) \cdots f(x_n) \qquad (8\text{-}10)$$

Example 8-1 Given n independent RVs with respective densities $f_i(x_i)$, we form the RVs

$$\mathbf{y}_k = \mathbf{x}_1 + \cdots + \mathbf{x}_k \qquad k = 1, \ldots, n$$

We shall determine the joint density of $\mathbf{y}_k$. The system

$$x_1 = y_1, x_1 + x_2 = y_2, \ldots, x_1 + \cdots + x_n = y_n$$

has a unique solution

$$x_1 = y_1, \ldots, x_k = y_k - y_{k-1} \qquad 1 \le k \le n$$

and its jacobian equals 1. Hence [see (8-8) and (8-10)]

$$f_y(y_1, \ldots, y_n) = f_1(y_1) f_2(y_2 - y_1) \cdots f_n(y_n - y_{n-1}) \qquad (8\text{-}11)$$

From (8-10) it follows that any subset of the set $\mathbf{x}_i$ is a set of independent RVs. Suppose, for example, that

$$f(x_1, x_2, x_3) = f(x_1) f(x_2) f(x_3)$$

Integrating with respect to x_3, we obtain $f(x_1, x_2) = f(x_1) f(x_2)$. This shows that the RVs $\mathbf{x}_1$ and $\mathbf{x}_2$ are independent.

We must note, however, that if the RVs $\mathbf{x}_i$ are independent in pairs, they are not necessarily independent. For example, it is possible that

$$f(x_1, x_2) = f(x_1) f(x_2) \qquad f(x_1, x_3) = f(x_1) f(x_3) \qquad f(x_2, x_3) = f(x_2) f(x_3)$$

but $f(x_1, x_2, x_3) \ne f(x_1) f(x_2) f(x_3)$ (see Prob. 8-2).

Reasoning as in (6-29), we can show that if the RVs $\mathbf{x}_i$ are independent, then the RVs

$$\mathbf{y}_1 = g_1(\mathbf{x}_1), \ldots, \mathbf{y}_n = g_n(\mathbf{x}_n)$$

are also independent.

Independent experiments and repeated trials Suppose that

$$\mathscr{S}^n = \mathscr{S}_1 \times \cdots \times \mathscr{S}_n$$

is a combined experiment and the RVs $\mathbf{x}_i$ depend only on the outcomes ζ_i of $\mathscr{S}_i$

$$\mathbf{x}_i(\zeta_1 \cdots \zeta_i \cdots \zeta_n) = \mathbf{x}_i(\zeta_i) \qquad i = 1, \ldots, n$$

If the experiments $\mathscr{S}_i$ are independent, then the RVs $\mathbf{x}_i$ are independent [see also (6-30)]. The following special case is of particular interest.

Suppose that $\mathbf{x}$ is an RV defined on an experiment $\mathscr{S}$ and the experiment is performed n times generating the experiment $\mathscr{S}^n = \mathscr{S} \times \cdots \times \mathscr{S}$. In this experiment, we define the RVs $\mathbf{x}_i$ such that

$$\mathbf{x}_i(\zeta_1 \cdots \zeta_i \cdots \zeta_n) = \mathbf{x}(\zeta_i) \qquad i = 1, \ldots, n \tag{8-12}$$

From this it follows that the distribution $F_i(x_i)$ of $\mathbf{x}_i$ equals the distribution $F_x(x)$ of the RV $\mathbf{x}$. Thus, if an experiment is performed n times, the RVs $\mathbf{x}_i$ defined as in (8-12) are independent and they have the same distribution $F_x(x)$. These RVs are called i.i.d. (independent, identically distributed).

Example 8-2 (order statistics) The *order statistics* of the RVs $\mathbf{x}_i$ are n RV $\mathbf{y}_k$ defined as follows: For a specific outcome ζ, the RVs $\mathbf{x}_i$ take the values $\mathbf{x}_i(\zeta)$. Ordering these numbers, we obtain the sequence

$$\mathbf{x}_{r_1}(\zeta) \le \cdots \le \mathbf{x}_{r_k}(\zeta) \le \cdots \le \mathbf{x}_{r_n}(\zeta)$$

and we define the RV $\mathbf{y}_k$ such that

$$\mathbf{y}_1(\zeta) = \mathbf{x}_{r_1}(\zeta) \le \cdots \le \mathbf{y}_k(\zeta) = \mathbf{x}_{r_k}(\zeta) \le \cdots \le \mathbf{y}_n(\zeta) = \mathbf{x}_{r_n}(\zeta) \tag{8-13}$$

We note that for a specific i, the values $\mathbf{x}_i(\zeta)$ of $\mathbf{x}_i$ occupy different locations in the above ordering as ζ changes.

We maintain that the density $f_k(y)$ of the kth statistic $\mathbf{y}_k$ is given by

$$f_k(y) = \frac{n!}{(k-1)!\,(n-k)!} F_x^{k-1}(y)[1 - F_x(y)]^{n-k} f_x(y) \tag{8-14}$$

where $F_x(x)$ is the distribution of the i.i.d. RVs $\mathbf{x}_i$ and $f_x(x)$ is their density.

PROOF As we know

$$f_k(y)\,dy = P\{y < \mathbf{y}_k \le y + dy\}$$

The event $\mathscr{B} = \{y < \mathbf{y}_k \le y + dy\}$ occurs iff exactly $k - 1$ of the RVs $\mathbf{x}_i$ are less than y and one is in the interval $(y, y + dy)$ (Fig. 8-1). In the original experiment $\mathscr{S}$, the events

$$\mathscr{A}_1 = \{\mathbf{x} \le y\} \qquad \mathscr{A}_2 = \{y < \mathbf{x} \le y + dy\} \qquad \mathscr{A}_3 = \{\mathbf{x} > y + dy\}$$

form a partition and

$$P(\mathscr{A}_1) = F_x(y) \qquad P(\mathscr{A}_2) = f_x(y)\,dy \qquad P(\mathscr{A}_3) = 1 - F_x(y)$$

In the experiment $\mathscr{S}^n$, the event $\mathscr{B}$ occurs iff $\mathscr{A}_1$ occurs $k - 1$ times, $\mathscr{A}_2$ occurs once, and $\mathscr{A}_3$ occurs $n - k$ times. With $k_1 = k - 1$, $k_2 = 1$, $k_3 = n - k$, it follows from (3-38) that

$$P(\mathscr{B}) = \frac{n!}{(k-1)!\,1!\,(n-k)!} P^{k-1}(\mathscr{A}_1)P(\mathscr{A}_2)P^{n-k}(\mathscr{A}_3)$$

and (8-14) results.

We note that

$$f_1(y) = n[1 - F_x(y)]^{n-1} f_x(y) \qquad f_n(y) = nF_x^{n-1}(y) f_x(y)$$

These are the densities of the minimum $\mathbf{y}_1$ and the maximum $\mathbf{y}_n$ of the RVs $\mathbf{x}_i$.

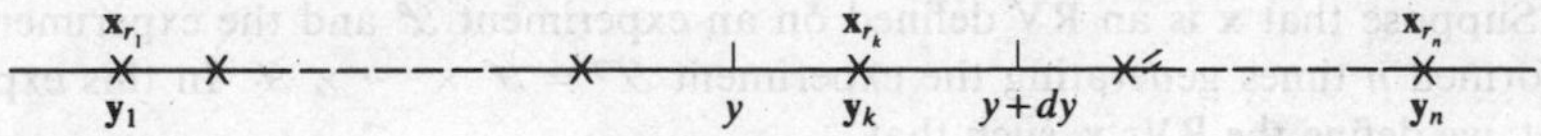

Figure 8-1

SPECIAL CASE If the RVs $\mathbf{x}_i$ are exponential with parameter α:

$$f_x(x) = \alpha e^{-\alpha x}U(x) \qquad F_x(x) = (1 - e^{-\alpha x})U(x)$$

then

$$f_1(y) = n\alpha e^{-\alpha n y}U(y)$$

i.e., their minimum $\mathbf{y}_1$ is also exponential with parameter $n\alpha$.

Example 8-3 A system consists of m components and the time of a failure of the ith component is an RV $\mathbf{x}_i$ with distribution $F_i(x)$. Thus

$$1 - F_i(t) = P\{\mathbf{x}_i > t\}$$

is the probability that the ith component is good at time t. We denote by $\mathbf{n}(t)$ the number of components that are good at time t. Clearly

$$\mathbf{n}(t) = \mathbf{n}_1 + \cdots + \mathbf{n}_m$$

where

$$\mathbf{n}_i = \begin{cases} 1 & \mathbf{x}_i > t \\ 0 & \mathbf{x}_i < t \end{cases} \qquad E\{\mathbf{n}_i\} = 1 - F_i(t)$$

Hence, the mean $E\{\mathbf{n}(t)\} = \eta(t)$ of $\mathbf{n}(t)$ is given by

$$\eta(t) = 1 - F_1(t) + \cdots + 1 - F_m(t) \tag{8-15}$$

We shall assume that the RVs $\mathbf{x}_i$ have the same distribution $F(t)$. In this case,

$$\eta(t) = m[1 - F(t)] \tag{8-16}$$

Failure rate The difference $\eta(t) - \eta(t + dt)$ is the expected number of failures in the interval $(t, t + dt)$. The derivative $-\eta'(t) = mf(t)$ of $-\eta(t)$ is the rate of failure. The ratio

$$\beta(t) = -\frac{\eta'(t)}{\eta(t)} = \frac{f(t)}{1 - F(t)} \tag{8-17}$$

is called *relative expected failure rate*. As we see from (4-50), $\beta(t)$ can also be interpreted as the conditional failure rate of each component in the system.

Assuming that the system is put into operation at $t = 0$, we have $\mathbf{n}(0) = m$, hence, $\eta(0) = E\{\mathbf{n}(0)\} = m$. Solving (8-17) for $\eta(t)$, we obtain

$$\eta(t) = m \exp\left\{-\int_0^t \beta(\tau)\, d\tau\right\}$$

Group independence We say that the group G_x of the RVs $\mathbf{x}_1, \ldots, \mathbf{x}_n$ is independent of the group G_y of the RVs $\mathbf{y}_1, \ldots, \mathbf{y}_k$ if

$$f(x_1, \ldots, x_n, y_1, \ldots, y_k) = f(x_1, \ldots, x_n) f(y_1, \ldots, y_k) \tag{8-18}$$

By suitable integration as in (8-5) we conclude from (8-18) that any subgroup of G_x is independent of any subgroup of G_y. In particular, the RVs $\mathbf{x}_i$ and $\mathbf{y}_j$ are independent for any i and j.

Suppose that $\mathscr{S}$ is a combined experiment $\mathscr{S}_1 \times \mathscr{S}_2$, the RVs $\mathbf{x}_i$ depend only on the outcomes of $\mathscr{S}_1$, and the RVs $\mathbf{y}_j$ depend only on the outcomes of $\mathscr{S}_2$. If the experiments $\mathscr{S}_1$ and $\mathscr{S}_2$ are independent, then the groups G_x and G_y are independent.

We note finally that if the RVs $\mathbf{z}_m$ depend only on the RVs $\mathbf{x}_i$ of G_x and the RVs $\mathbf{w}_r$ depend only on the RVs $\mathbf{y}_j$ of G_y, then the groups G_z and G_w are independent.

Complex random variables The statistics of the RVs

$$\mathbf{z}_1 = \mathbf{x}_1 + j\mathbf{y}_1, \ldots, \mathbf{z}_n = \mathbf{x}_n + j\mathbf{y}_n$$

are determined in terms of the joint density $f(x_1, y_1, \ldots, x_n, y_n)$ of the $2n$ RVs $\mathbf{x}_i$ and $\mathbf{y}_i$. We say that the complex RVs $\mathbf{z}_i$ are independent if

$$f(x_1, y_1, \ldots, x_n, y_n) = f(x_1, y_1) \cdots f(x_n, y_n) \tag{8-19}$$

Mean and Covariance

Extending (7-2) to n RVs, we conclude that the mean of $g(\mathbf{x}_1, \ldots, \mathbf{x}_n)$ equals

$$\int_{-\infty}^{\infty} \cdots \int_{-\infty}^{\infty} g(x_1, \ldots, x_n) f(x_1, \ldots, x_n)\, dx_1 \cdots dx_n \tag{8-20}$$

If the RVs $\mathbf{z}_i = \mathbf{x}_i + j\mathbf{y}_i$ are complex, then the mean of $g(\mathbf{z}_1, \ldots, \mathbf{z}_n)$ equals

$$\int_{-\infty}^{\infty} \cdots \int_{-\infty}^{\infty} g(z_1, \ldots, z_n) f(x_1, y_1, \ldots, x_n, y_n)\, dx_1 \cdots dy_n$$

From the above it follows that (linearity)

$$E\{a_1 g_1(\mathbf{X}) + \cdots + a_n g_n(\mathbf{X})\} = a_1 E\{g_1(\mathbf{X})\} + \cdots + a_n E\{g_n(\mathbf{X})\}$$

for any random vector $\mathbf{X}$ real or complex.

Correlation and covariance matrices The covariance C_{ij} of two real RVs $\mathbf{x}_i$ and $\mathbf{x}_j$ is defined as in (7-6). For complex RVs

$$C_{ij} = E\{(\mathbf{x}_i - \eta_i)(\mathbf{x}_j^* - \eta_j^*)\} = E\{\mathbf{x}_i \mathbf{x}_j^*\} - E\{\mathbf{x}_i\}E\{\mathbf{x}_j^*\}$$

by definition. The variance of $\mathbf{x}_i$ is given by

$$\sigma_i^2 = C_{ii} = E\{|\mathbf{x}_i - \eta_i|^2\} = E\{|\mathbf{x}_i|^2\} - |E\{\mathbf{x}_i\}|^2$$

The RVs $\mathbf{x}_i$ are called (mutually) *uncorrelated* if $C_{ij} = 0$ for $i \neq j$. In this case, if

$$\mathbf{x} = \mathbf{x}_1 + \cdots + \mathbf{x}_n \qquad \text{then} \qquad \sigma_x^2 = \sigma_1^2 + \cdots + \sigma_n^2 \tag{8-21}$$

Example 8-4 The RVs

$$\bar{\mathbf{x}} = \frac{1}{n} \sum_{i=1}^{n} \mathbf{x}_i \qquad \bar{\mathbf{v}} = \frac{1}{n-1} \sum_{i=1}^{n} (\mathbf{x}_i - \bar{\mathbf{x}})^2$$

are by definition the *sample mean* and the *sample variance* respectively of $\mathbf{x}_i$. We shall show that, if the RVs $\mathbf{x}_i$ are uncorrelated with the same mean $E\{\mathbf{x}_i\} = \eta$ and variance $\sigma_i^2 = \sigma^2$, then

$$E\{\bar{\mathbf{x}}\} = \eta \qquad \sigma_{\bar{x}}^2 = \sigma^2/n \tag{8-22}$$

and

$$E\{\bar{\mathbf{v}}\} = \sigma^2 \tag{8-23}$$

Proof The first equation in (8-22) follows from the linearity of expected values and the second from (8-21):

$$E\{\bar{\mathbf{x}}\} = \frac{1}{n} \sum_{i=1}^{n} E\{\mathbf{x}_i\} = \eta \qquad \sigma_{\bar{x}}^2 = \frac{1}{n^2} \sum_{i=1}^{n} \sigma_i^2 = \frac{\sigma^2}{n}$$

To prove (8-23), we observe that

$$E\{(\mathbf{x}_i - \eta)(\bar{\mathbf{x}} - \eta)\} = \frac{1}{n} E\{(\mathbf{x}_i - \eta)[(\mathbf{x}_1 - \eta) + \cdots + (\mathbf{x}_n - \eta)]\}$$

$$= \frac{1}{n} E\{(\mathbf{x}_i - \eta)(\mathbf{x}_i - \eta)\} = \frac{\sigma^2}{n}$$

because the RVs $\mathbf{x}_i$ and $\mathbf{x}_j$ are uncorrelated by assumption. Hence

$$E\{(\mathbf{x}_i - \bar{\mathbf{x}})^2\} = E\{[(\mathbf{x}_i - \eta) - (\bar{\mathbf{x}} - \eta)]^2\} = \sigma^2 + \frac{\sigma^2}{n} - 2\frac{\sigma^2}{n} = \frac{n-1}{n}\sigma^2$$

This yields

$$E\{\bar{\mathbf{v}}\} = \frac{1}{n-1} \sum_{i=1}^{n} E\{(\mathbf{x}_i - \bar{\mathbf{x}})^2\} = \frac{n}{n-1} \frac{n-1}{n} \sigma^2$$

and (8-23) results.

We note that if the RVs $\mathbf{x}_i$ are independent with $E\{\mathbf{x}_i^4\} = m_4$, then (see Prob. 8-4)

$$\sigma_{\bar{v}}^2 = \frac{1}{n}\left(m_4 - \frac{n-3}{n-1}\sigma^4\right)$$

If the RVs $\mathbf{x}_1, \ldots, \mathbf{x}_n$ are independent, they are also uncorrelated. This follows as in (7-14) for real RVs. For complex RVs the proof is similar: If the RVs $\mathbf{z}_1 = \mathbf{x}_1 + j\mathbf{y}_1$ and $\mathbf{z}_2 = \mathbf{x}_2 + j\mathbf{y}_2$ are independent, then $f(x_1, y_1, x_2, y_2) =$

$f(x_1, y_1)f(x_2, y_2)$, hence

$$\int_{-\infty}^{\infty}\cdots\int_{-\infty}^{\infty} z_1 z_2^* f(x_1, y_1, x_2, y_2)\, dx_1\, dy_1\, dx_2\, dy_2$$

$$= \int_{-\infty}^{\infty}\int_{-\infty}^{\infty} z_1 f(x_1, y_1)\, dx_1\, dy_1 \int_{-\infty}^{\infty}\int_{-\infty}^{\infty} z_2^* f(x_2, y_2)\, dx_2\, dy_2$$

This yields $E\{\mathbf{z}_1 \mathbf{z}_2^*\} = E\{\mathbf{z}_1\}E\{\mathbf{z}_2^*\}$ therefore, $\mathbf{z}_1$ and $\mathbf{z}_2$ are uncorrelated.

We note, finally, that if the RVs $\mathbf{x}_i$ are independent, then

$$E\{g_1(\mathbf{x}_1)\cdots g_n(\mathbf{x}_n)\} = E\{g_1(\mathbf{x}_1)\}\cdots E\{g_n(\mathbf{x}_n)\} \tag{8-24}$$

Similarly, if the groups $\mathbf{x}_1, \ldots, \mathbf{x}_n$ and $\mathbf{y}_1, \ldots, \mathbf{y}_k$ are independent, then

$$E\{g(\mathbf{x}_1, \ldots, \mathbf{x}_n)h(\mathbf{y}_1, \ldots, \mathbf{y}_k)\} = E\{g(\mathbf{x}_1, \ldots, \mathbf{x}_n)\}E\{h(\mathbf{y}_1, \ldots, \mathbf{y}_k)\}$$

The correlation matrix We introduce the matrices

$$R_n = \begin{bmatrix} R_{11} & \cdots & R_{1n} \\ \cdots & \cdots & \cdots \\ R_{n1} & \cdots & R_{nn} \end{bmatrix} \qquad C_n = \begin{bmatrix} C_{11} & \cdots & C_{1n} \\ \cdots & \cdots & \cdots \\ C_{n1} & \cdots & C_{nn} \end{bmatrix}$$

where

$$R_{ij} = E\{\mathbf{x}_i \mathbf{x}_j^*\} = R_{ji}^* \qquad C_{ij} = R_{ij} - \eta_i \eta_j^* = C_{ji}^*$$

The first is the *correlation matrix* of the random vector $\mathbf{X}$: $[\mathbf{x}_1, \ldots, \mathbf{x}_n]$ and the second its *covariance matrix*. Clearly

$$R_n = E\{\mathbf{X}^t \mathbf{X}^*\}$$

where $\mathbf{X}^t$ is the transpose of $\mathbf{X}$ (column vector). We shall discuss the properties of the matrix R_n and its determinant Δ_n. The properties of C_n are similar because C_n is the correlation matrix of the "centered" RVs $\mathbf{x}_i - \eta_i$.

Theorem The matrix R_n is *nonnegative definite*. This means that

$$Q = \sum_{i,j} a_i a_j^* R_{ij} = A^t R_n A^* \geq 0 \tag{8-25}$$

for any vector A: $[a_1, \ldots, a_n]$

PROOF It follows readily from the linearity of expected values

$$E\{|a_1 \mathbf{x}_1 + \cdots + a_n \mathbf{x}_n|^2\} = \sum_{i,j} a_i a_j^* E\{\mathbf{x}_i \mathbf{x}_j^*\} \tag{8-26}$$

If (8-25) is strictly positive, i.e., if $Q > 0$ for any A, then R_n is called *positive definite*†. The difference between $Q \geq 0$ and $Q > 0$ is related to the notion of linear dependence.

† We shall use the abbreviation p.d. to indicate that R_n satisfies (8-25). The distinction between $Q \geq 0$ and $Q > 0$ will be understood from the context.

Definition The RVs $\mathbf{x}_i$ are called *linearly independent* if

$$E\{|a_1\mathbf{x}_1 + \cdots + a_n\mathbf{x}_n|^2\} > 0 \tag{8-27}$$

for any A. In this case [see (8-26)], their correlation matrix R_n is positive definite.

The RVs $\mathbf{x}_i$ are called *linearly dependent* if

$$a_1\mathbf{x}_1 + \cdots + a_n\mathbf{x}_n = 0 \tag{8-28}$$

for some A. In this case, the corresponding Q equals zero and the matrix R_n is singular [see also (8-29)].

From the definition it follows that, if the RVs $\mathbf{x}_i$ are linearly independent, then any subset is also linearly independent.

The correlation determinant The determinant Δ_n is real because $R_{ij} = R_{ji}^*$. We shall show that it is also nonnegative

$$\Delta_n \geq 0 \tag{8-29}$$

with equality iff the RVs $\mathbf{x}_i$ are linearly dependent. The familiar inequality $\Delta_2 = R_{11}R_{22} - R_{12}^2 \geq 0$ is a special case [see (7-12)].

Suppose, first, that the RVs $\mathbf{x}_i$ are linearly independent. We maintain that, in this case, the determinant Δ_n and all its principal minors are positive

$$\Delta_k > 0 \qquad k \leq n \tag{8-30}$$

PROOF The above is true for $n = 1$ because $\Delta_1 = R_{11} > 0$. Since any subset of the set $\{\mathbf{x}_i\}$ is linearly independent, we can assume that (8-30) is true for $k \leq n - 1$ and we shall show that $\Delta_n > 0$. For this purpose, we form the system

$$\begin{gathered} R_{11}a_1 + \cdots + R_{1n}a_n = 1 \\ R_{21}a_1 + \cdots + R_{2n}a_n = 0 \\ \cdots\cdots\cdots\cdots\cdots\cdots\cdots \\ R_{n1}a_1 + \cdots + R_{nn}a_n = 0 \end{gathered} \tag{8-31}$$

Solving for a_1, we obtain $a_1 = \Delta_{n-1}/\Delta_n$ where Δ_{n-1} is the correlation determinant of the RVs $\mathbf{x}_2, \ldots, \mathbf{x}_n$. Thus, a_1 is a real number. Multiplying the jth equation by a_j^* and adding, we obtain

$$Q = \sum_{i,j} a_i a_j^* R_{ij} = a_1 = \frac{\Delta_{n-1}}{\Delta_n} \tag{8-32}$$

In the above, $Q > 0$ because the RVs $\mathbf{x}_i$ are linearly independent and the left side of (8-27) equals Q. Furthermore, $\Delta_{n-1} > 0$ by the induction hypothesis, hence $\Delta_n > 0$.

We shall now show that, if the RVs $\mathbf{x}_i$ are linearly dependent, then

$$\Delta_n = 0 \tag{8-33}$$

PROOF In this case, there exists a vector $A \neq 0$ such that $a_1 \mathbf{x}_1 + \cdots + a_n \mathbf{x}_n = 0$. Multiplying by $\mathbf{x}_i^*$ and taking expected values, we obtain

$$a_1 R_{i1} + \cdots + a_n R_{in} = 0 \qquad i = 1, \ldots, n$$

This is a homogeneous system satisfied by the nonzero vector A, hence $\Delta_n = 0$.

We note, finally, that [see (15-161)]

$$\Delta_n \leq R_{11} R_{22} \cdots R_{nn} \tag{8-34}$$

with equality iff the RVs $\mathbf{x}_i$ are (mutually) *orthogonal*, i.e., if the matrix R_n is diagonal.

8-2 CONDITIONAL DENSITIES

Conditional densities can be defined as in Sec. 7-2. We shall discuss various extensions of the equation $f(y \mid x) = f(x, y)/f(y)$. Reasoning as in (7-41), we conclude that the conditional density of the RVs $\mathbf{x}_n, \ldots, \mathbf{x}_{k+1}$ assuming $\mathbf{x}_k, \ldots, \mathbf{x}_1$ is given by

$$f(x_n, \ldots, x_{k+1} \mid x_k, \ldots, x_1) = \frac{f(x_1, \ldots, x_k, \ldots, x_n)}{f(x_1, \ldots, x_k)} \tag{8-35}$$

The corresponding distribution function is obtained by integration

$$F(x_n, \ldots, x_{k+1} \mid x_k, \ldots, x_1) = \int_{-\infty}^{x_n} \cdots \int_{-\infty}^{x_{k+1}} f(\alpha_n, \ldots, \alpha_{k+1} \mid x_k, \ldots, x_1)\, d\alpha_{k+1} \cdots d\alpha_n \tag{8-36}$$

For example

$$f(x_1 \mid x_2, x_3) = \frac{f(x_1, x_2, x_3)}{f(x_2, x_3)} = \frac{dF(x_1 \mid x_2, x_3)}{dx_1}$$

Chain rule From (8-35) it follows that

$$f(x_1, \ldots, x_n) = f(x_n \mid x_{n-1}, \ldots, x_1) \cdots f(x_2 \mid x_1) f(x_1) \tag{8-37}$$

Example 8-5 We have shown that [see (5-19)] if $\mathbf{x}$ is an RV with distribution $F(x)$, then the RV $\mathbf{y} = F(\mathbf{x})$ is uniform in the interval (0, 1). The following is a generalization.

Given n arbitrary RVs $\mathbf{x}_i$, we form the RVs

$$\mathbf{y}_1 = F(\mathbf{x}_1) \qquad \mathbf{y}_2 = F(\mathbf{x}_2 \mid \mathbf{x}_1), \ldots, \mathbf{y}_n = F(\mathbf{x}_n \mid \mathbf{x}_{n-1}, \ldots, \mathbf{x}_1) \tag{8-38}$$

We shall show that these RVs are independent and each is uniform in the interval (0, 1).

PROOF The RVs $\mathbf{y}_i$ are functions of the RVs $\mathbf{x}_i$ obtained with the transformation (8-38). For $0 \le y_i \le 1$, the system

$$y_1 = F(x_1) \quad y_2 = F(x_2 | x_1), \ldots, y_n = F(x_n | x_{n-1}, \ldots, x_1)$$

has a unique solution $x_1, \ldots, x_n$ and its jacobian equals

$$J = \begin{vmatrix} \dfrac{\partial y_1}{\partial x_1} & 0 & 0 & 0 \\ \dfrac{\partial y_2}{\partial x_1} & \dfrac{\partial y_2}{\partial x_2} & 0 \cdots & 0 \\ \cdots & \cdots & \cdots & \cdots \\ \dfrac{\partial y_n}{\partial x_1} & \cdots & \cdots & \dfrac{\partial y_n}{\partial x_n} \end{vmatrix}$$

The above determinant is triangular, hence, it equals the product of its diagonal elements

$$\frac{\partial y_k}{\partial x_k} = f(x_k | x_{k-1}, \ldots, x_1)$$

Inserting into (8-8) and using (8-37), we obtain

$$f(y_1, \ldots, y_n) = \frac{f(x_1, \ldots, x_n)}{f(x_1)f(x_2 | x_1) \cdots f(x_n | x_{n-1}, \ldots, x_1)} = 1$$

in the n-dimensional cube $0 \le y_i \le 1$, and zero otherwise.

From (8-5) and (8-35) it follows that

$$f(x_1 | x_3) = \int_{-\infty}^{\infty} f(x_1, x_2 | x_3)\, dx_2$$

$$f(x_1 | x_4) = \int_{-\infty}^{\infty}\int_{-\infty}^{\infty} f(x_1 | x_2, x_3, x_4) f(x_2, x_3 | x_4)\, dx_2\, dx_3$$

Generalizing, we obtain the following rule for removing variables on the left or on the right of the conditional line: To remove any number of variables on the left of the conditional line we integrate with respect to them. To remove any number of variables to the right of the line we multiply by their conditional density with respect to the remaining variables on the right, and we integrate the product.

The following special case is used extensively (Chapman–Kolmogoroff):

$$f(x_1 | x_3) = \int_{-\infty}^{\infty} f(x_1 | x_2, x_3) f(x_2 | x_3)\, dx_2 \tag{8-39}$$

Discrete type The above rule holds also for discrete type RVs provided that all densities are replaced by probabilities and all integrals by sums. We mention as an example the discrete form of (8-39): If the RVs $\mathbf{x}_1, \mathbf{x}_2, \mathbf{x}_3$ take the values a_i, b_k, c_r respectively, then

$$P\{\mathbf{x}_1 = a_i | \mathbf{x}_3 = c_r\} = \sum_k P\{\mathbf{x}_1 = a_i | b_k, c_r\} P\{\mathbf{x}_2 = b_k | c_r\} \tag{8-40}$$

Conditional Expected Values

The conditional mean of the RVs $g(\mathbf{x}_1, \ldots, \mathbf{x}_n)$ assuming $\mathscr{M}$ is given by the integral in (8-20) provided that the density $f(x_1, \ldots, x_n)$ is replaced by the conditional density $f(x_1, \ldots, x_n \mid \mathscr{M})$. We note, in particular, that [see also (7-50)]

$$E\{\mathbf{x}_1 \mid x_2, \ldots, x_n\} = \int_{-\infty}^{\infty} x_1 f(x_1 \mid x_2, \ldots, x_n)\, dx_1 \tag{8-41}$$

The above is a function of $x_2, \ldots, x_n$; it defines, therefore, the RV $E\{\mathbf{x}_1 \mid \mathbf{x}_2, \ldots, \mathbf{x}_n\}$. Multiplying (8-41) by $f(x_2, \ldots, x_n)$ and integrating, we conclude that

$$E\{E\{\mathbf{x}_1 \mid \mathbf{x}_2, \ldots, \mathbf{x}_n\}\} = E\{\mathbf{x}_1\} \tag{8-42}$$

Reasoning similarly, we obtain

$$\begin{aligned} E\{\mathbf{x}_1 \mid x_2, x_3\} &= E\{E\{\mathbf{x}_1 \mid x_2, x_3, \mathbf{x}_4\}\} \\ &= \int_{-\infty}^{\infty} E\{\mathbf{x}_1 \mid x_2, x_3, x_4\} f(x_4 \mid x_2, x_3)\, dx_4 \end{aligned} \tag{8-43}$$

This leads to the following generalization: To remove any number of variables on the right of the conditional expected value line, we multiply by their conditional density with respect to the remaining variables on the right and we integrate the product.

For example

$$E\{\mathbf{x}_1 \mid x_3\} = \int_{-\infty}^{\infty} E\{\mathbf{x}_1 \mid x_2, x_3\} f(x_2 \mid x_3)\, dx_2 \tag{8-44}$$

and for the discrete case [see (8-40)]

$$E\{\mathbf{x}_1 \mid c_r\} = \sum_k E\{\mathbf{x}_1 \mid b_k, c_r\} P\{\mathbf{x}_2 = b_k \mid c_r\} \tag{8-45}$$

Example 8-6 Given a discrete type RV $\mathbf{n}$ taking the values 1, 2, ... and a sequence of RVs $\mathbf{x}_k$ independent of $\mathbf{n}$, we form the sum

$$\mathbf{s} = \sum_{k=1}^{\mathbf{n}} \mathbf{x}_k \tag{8-46}$$

This sum is an RV specified as follows: For a specific ζ, $\mathbf{n}(\zeta)$ is an integer and $\mathbf{s}(\zeta)$ equals the sum of the numbers $\mathbf{x}_k(\zeta)$ for k from 1 to $\mathbf{n}(\zeta)$. We maintain that if the RVs $\mathbf{x}_k$ have the same mean, then

$$E\{\mathbf{s}\} = \eta E\{\mathbf{n}\} \qquad \text{where} \qquad E\{\mathbf{x}_k\} = \eta \tag{8-47}$$

Clearly, $E\{\mathbf{x}_k \mid \mathbf{n} = n\} = E\{\mathbf{x}_k\}$ because $\mathbf{x}_k$ is independent of $\mathbf{n}$. Hence

$$E\{\mathbf{s} \mid \mathbf{n} = n\} = E\left\{\sum_{k=1}^{n} \mathbf{x}_k \,\middle|\, \mathbf{n} = n\right\} = \sum_{k=1}^{n} E\{\mathbf{x}_k\} = \eta n$$

From this and (7-58) it follows that

$$E\{\mathbf{s}\} = E\{E\{\mathbf{s} \mid \mathbf{n}\}\} = E\{\eta \mathbf{n}\}$$

and (8-47) results.

We show next that if the RVs $\mathbf{x}_k$ are uncorrelated with the same variance σ^2, then

$$E\{\mathbf{s}^2\} = \eta^2 E\{\mathbf{n}^2\} + \sigma^2 E\{\mathbf{n}\} \tag{8-48}$$

Reasoning as above, we have

$$E\{\mathbf{s}^2 \mid \mathbf{n} = n\} = \sum_{i=1}^{n} \sum_{k=1}^{n} E\{\mathbf{x}_i \mathbf{x}_k\} \tag{8-49}$$

where

$$E\{\mathbf{x}_i \mathbf{x}_k\} = \begin{cases} \sigma^2 + \eta^2 & i = k \\ \eta^2 & i \neq k \end{cases}$$

The double sum in (8-49) contains n terms with $i = k$ and $n^2 - n$ terms with $i \neq k$, hence it equals

$$(\sigma^2 + \eta^2)n + \eta^2(n^2 - n) = \eta^2 n^2 + \sigma^2 n$$

This yields (8-48) because

$$E\{\mathbf{s}^2\} = E\{E\{\mathbf{s}^2 \mid \mathbf{n}\}\} = E\{\eta^2 \mathbf{n}^2 + \sigma^2 \mathbf{n}\}$$

Special case The number $\mathbf{n}$ of particles emitted from a substance in t seconds is a Poisson RV with parameter λt. The energy $\mathbf{x}_k$ of the kth particle has a Maxwell distribution with mean $3kT/2$ and variance $3k^2T^2/2$ (see Prob. 8-5). The sum $\mathbf{s}$ in (8-46) is the total emitted energy in t seconds. As we know $E\{\mathbf{n}\} = \lambda t$, $E\{\mathbf{n}^2\} = \lambda^2 t^2 + \lambda t$ [see (5-37)]. Inserting into (8-47) and (8-48), we obtain

$$E\{\mathbf{s}\} = \frac{3kT\lambda t}{2} \qquad \sigma_s^2 = \frac{15k^2T^2\lambda t}{4}$$

8-3 CHARACTERISTIC FUNCTIONS AND NORMALITY

The characteristic function of a random vector is by definition the function

$$\Phi(\Omega) = E\{e^{j\Omega X^t}\} = E\{e^{j(\omega_1 \mathbf{x}_1 + \cdots + \omega_n \mathbf{x}_n)}\} = \mathbf{\Phi}(j\Omega) \tag{8-50}$$

where

$$\mathbf{X}: [\mathbf{x}_1, \ldots, \mathbf{x}_n] \qquad \Omega: [\omega_1, \ldots, \omega_n]$$

As an application, we shall show that if the RVs $\mathbf{x}_i$ are independent with respective densities $f_i(x_i)$, then the density $f_z(z)$ of their sum $\mathbf{z} = \mathbf{x}_1 + \cdots + \mathbf{x}_n$ equals the convolution of their densities

$$f_z(z) = f_1(z) * \cdots * f_n(z) \tag{8-51}$$

Proof Since the RVs $\mathbf{x}_i$ are independent and $e^{j\omega_i \mathbf{x}_i}$ depends only on $\mathbf{x}_i$, we conclude from (8-24) that

$$E\{e^{j(\omega_1 \mathbf{x}_1 + \cdots + \omega_n \mathbf{x}_n)}\} = E\{e^{j\omega_1 \mathbf{x}_1}\} \cdots E\{e^{j\omega_n \mathbf{x}_n}\}$$

Hence

$$\Phi_z(\omega) = E\{e^{j\omega(\mathbf{x}_1 + \cdots + \mathbf{x}_n)}\} = \Phi_1(\omega) \cdots \Phi_n(\omega) \tag{8-52}$$

where $\Phi_i(\omega)$ is the characteristic function of $\mathbf{x}_i$. Applying the convolution theorem for Fourier transforms, we obtain (8-51).

Example 8-7 (*a*) (Bernoulli trials) Using (8-52), we shall rederive the fundamental equation (3-13). We define the RVs $\mathbf{x}_i$ as follows: $\mathbf{x}_i = 1$ if heads shows at the ith trial and $\mathbf{x}_i = 0$ otherwise. Thus

$$P\{\mathbf{x}_i = 1\} = P\{h\} = p \qquad P\{\mathbf{x}_i = 0\} = P\{t\} = q$$
$$\Phi_i(\omega) = pe^{j\omega} + q \tag{8-53}$$

The RV $\mathbf{z} = \mathbf{x}_1 + \cdots + \mathbf{x}_n$ takes the values $0, 1, \ldots, n$ and $\{\mathbf{z} = k\}$ is the event $\{k$ heads in n tossings$\}$. Furthermore

$$\Phi_z(\omega) = E\{e^{j\omega\mathbf{z}}\} = \sum_{k=0}^{n} P\{\mathbf{z} = k\}e^{jk\omega} \tag{8-54}$$

The RVs $\mathbf{x}_i$ are independent because $\mathbf{x}_i$ depends only on the outcomes of the ith trial and the trials are independent. Hence [see (8-52) and (8-53)]

$$\Phi_z(\omega) = (pe^{j\omega} + q)^n = \sum_{k=0}^{n} \binom{n}{k} p^k e^{jk\omega} q^{n-k}$$

Comparing with (8-54), we conclude that

$$P\{\mathbf{z} = k\} = P\{k \text{ heads}\} = \binom{n}{k} p^k q^{n-k} \tag{8-55}$$

(*b*) (Poisson theorem) We shall show that if $p \ll 1$, then

$$P\{\mathbf{z} = k\} \simeq \frac{e^{-np}(np)^k}{k!}$$

as in (3-41). In fact, we shall establish a more general result. Suppose that the RVs $\mathbf{x}_i$ are independent and each takes the values 1 and 0 with respective probabilities p_i and $q_i = 1 - p_i$. If $p_i \ll 1$, then

$$e^{p_i(e^{j\omega} - 1)} \simeq 1 + p_i(e^{j\omega} - 1) = p_i e^{j\omega} + q_i = \Phi_i(\omega)$$

With $\mathbf{z} = \mathbf{x}_1 + \cdots + \mathbf{x}_n$, it follows from (8-52) that

$$\Phi_z(\omega) \simeq e^{p_1(e^{j\omega} - 1)} \cdots e^{p_n(e^{j\omega} - 1)} = e^{a(e^{j\omega} - 1)}$$

where $a = p_1 + \cdots + p_n$. This leads to the conclusion that [see (5-79)] the RV $\mathbf{z}$ is approximately Poisson distributed with parameter a. It can be shown that the result is exact in the limit if

$$p_i \to 0 \qquad \text{and} \qquad p_1 + \cdots + p_n \to a \qquad \text{as} \qquad n \to \infty$$

Normal Vectors

Joint normality of n RVs $\mathbf{x}_i$ can be defined as in (6-15): Their joint density is an exponential whose exponent is a negative quadratic. We give next an equivalent definition that expresses the normality of n RVs in terms of the normality of a single RV.

Definition The RVs $\mathbf{x}_i$ are jointly normal iff the sum

$$a_1 \mathbf{x}_1 + \cdots + a_n \mathbf{x}_n = A\mathbf{X}^t \tag{8-56}$$

is a normal RV for any A.

We shall show that this definition leads to the following conclusions: If the RVs $\mathbf{x}_i$ have zero mean and covariance matrix C, then their joint characteristic function equals

$$\Phi(\Omega) = \exp\{-\tfrac{1}{2}\,\Omega C \Omega^t\} \tag{8-57}$$

Furthermore, their joint density equals

$$f(X) = \frac{1}{\sqrt{(2\pi)^n \Delta}} \exp\{-\tfrac{1}{2}\, X C^{-1} X^t\} \tag{8-58}$$

where Δ is the determinant of C.

Proof From the definition of joint normality it follows that the RV

$$\mathbf{z} = \omega_1 \mathbf{x}_1 + \cdots + \omega_n \mathbf{x}_n = \Omega \mathbf{X}^t \tag{8-59}$$

is normal. Since $E\{\mathbf{x}_i\} = 0$ by assumption, the above yields [see (8-26)]

$$E\{\mathbf{z}\} = 0 \qquad E\{\mathbf{z}^2\} = \sum_{i,j} \omega_i \omega_j C_{ij} = \sigma_z^2$$

Setting $\eta = 0$ and $\omega = 1$ in (5-65), we obtain

$$E\{e^{j\mathbf{z}}\} = \exp\left[-\frac{\sigma_z^2}{2}\right]$$

This yields

$$E\{e^{j\Omega\mathbf{X}^t}\} = \exp\left\{-\frac{1}{2}\sum_{i,j} \omega_i \omega_j C_{ij}\right\} \tag{8-60}$$

as in (8-57). The proof of (8-58) follows from (8-57) and the Fourier inversion theorem.

We note, finally, that if the RVs $\mathbf{x}_i$ are jointly normal and uncorrelated, they are independent. Indeed, in this case, their covariance matrix is diagonal and its diagonal elements equal σ_i^2. Hence, C^{-1} is also diagonal with diagonal elements $1/\sigma_i^2$. Inserting into (8-58), we obtain

$$f(x_1, \ldots, x_n) = \frac{1}{\sigma_1 \cdots \sigma_n \sqrt{(2\pi)^n}} \exp\left\{-\frac{1}{2\sigma_1^2} - \cdots - \frac{1}{2\sigma_n^2}\right\} \tag{8-61}$$

Example 8-8 Using characteristic functions, we shall show that if the RVs $\mathbf{x}_i$ are jointly normal with zero mean, and $E\{\mathbf{x}_i \mathbf{x}_j\} = C_{ij}$, then

$$E\{\mathbf{x}_1 \mathbf{x}_2 \mathbf{x}_3 \mathbf{x}_4\} = C_{12}C_{34} + C_{13}C_{24} + C_{14}C_{23} \tag{8-62}$$

PROOF We expand the exponentials on the left and right side of (8-60) and we show explicitly only the terms containing the factor $\omega_1\omega_2\omega_3\omega_4$:

$$E\{e^{j(\omega_1 \mathbf{x}_1 + \cdots + \omega_4 \mathbf{x}_4)}\} = \cdots + \frac{1}{4!} E\{(\omega_1 \mathbf{x}_1 + \cdots + \omega_4 \mathbf{x}_4)^4\} + \cdots$$

$$= \cdots + \frac{24}{4!} E\{\mathbf{x}_1 \mathbf{x}_2 \mathbf{x}_3 \mathbf{x}_4\}\omega_1\omega_2\omega_3\omega_4$$

$$\exp\left\{-\frac{1}{2}\sum_{i,j}\omega_i\omega_j C_{ij}\right\} = +\frac{1}{2}\left(\frac{1}{2}\sum_{i,j}\omega_i\omega_i C_{ij}\right)^2 + \cdots$$

$$= \cdots + \frac{8}{8}(C_{12}C_{34} + C_{13}C_{24} + C_{14}C_{23})\omega_1\omega_2\omega_3\omega_4$$

Equating coefficients, we obtain (8-62).

Chi and chi-square statistics Given n normal i.i.d. random variables with density

$$f(x_1, \ldots, x_n) = \frac{1}{(\sigma\sqrt{2\pi})^n} e^{-(x_1^2 + \cdots + x_n^2)/2\sigma^2} \tag{8-63}$$

we form the RVs

$$\chi = \sqrt{\mathbf{x}_1^2 + \cdots + \mathbf{x}_n^2} \qquad \mathbf{y} = \chi^2$$

The first is called *chi* statistic and the second *chi-square* statistic with n degrees of freedom. We shall show that

$$f_\chi(\chi) = 2a\chi^{n-1}e^{-\chi^2/2\sigma^2}U(x) \tag{8-64}$$

$$f_y(y) = \frac{1}{2\sqrt{y}} f_\chi(\sqrt{y}) = ay^{-1+n/2}e^{-y/2\sigma^2}U(y) \tag{8-65}$$

where

$$a = \frac{1}{(\sigma\sqrt{2})^n\Gamma(n/2)} \tag{8-66}$$

PROOF The region R_χ of the x_i space such that

$$\chi < \sqrt{x_1^2 + \cdots + x_n^2} < \chi + d\chi$$

is a hypershell with thickness $d\chi$ and inner radius χ. Its volume V_χ equals $d\chi$ times the area $k\chi^{n-1}$ of the inner surface, where k is some constant. Hence, the probability masses in R_χ equal $f(x)V_\chi$ because $f(x)$ is constant in R_χ. This yields

$$f_\chi(\chi)\,d\chi = \frac{k\chi^{n-1}}{(\sigma\sqrt{2\pi})^n} e^{-\chi^2/2\sigma^2} \qquad \chi > 0$$

as in (8-64). Equation (8-65) follows from (8-64) and (5-8). Since $f_y(y)$ is a *gamma density*, a must be given by (8-66). This follows from (4-40) with $c = 1/2\sigma^2$ and $b = -1 + n/2$.

Example 8-9 The three components of the velocity

$$\mathbf{v} = \sqrt{\mathbf{v}_x^2 + \mathbf{v}_y^2 + \mathbf{v}_z^2}$$

of a particle in a gas are normal i.i.d. with zero mean and variance kT/m (see Prob. 8-5). Hence, $\mathbf{v}$ is a chi statistic with three degrees of freedom. Since $\Gamma(3/2) = \sqrt{\pi}/2$ we conclude, setting $n = 3$ in (8-64), that $\mathbf{v}$ has a *Maxwell density*

$$f_v(v) = \sqrt{\frac{2m^3}{\pi k^3 T^3}}\, e^{-mv^2/2kT} \qquad v > 0$$

8-4 RANDOM SEQUENCES AND STOCHASTIC CONVERGENCE

A fundamental problem in the theory of probability is the determination of the asymptotic properties of random sequences. In this section, we introduce the subject, concentrating on the clarification of the underlying concepts. We start with a simple problem.

Suppose that we wish to measure the length a of an object. Due to measurement inaccuracies, the instrument reading is a sum

$$\mathbf{x} = a + \mathbf{v}$$

where $\mathbf{v}$ is the error term. If there are no systematic errors, then $\mathbf{v}$ is an RV with zero mean. In this case, if the standard deviation σ of $\mathbf{v}$ is small compared to a, then the observed value $\mathbf{x}(\zeta)$ of $\mathbf{x}$ at a single measurement is a satisfactory estimate of the unknown length a. In the context of probability, this conclusion can be phrased as follows: The mean of the RV $\mathbf{x}$ equals a and its variance equals σ^2. Applying Tchebycheff's inequality, we conclude that

$$P\{|\mathbf{x} - a| < \varepsilon\} > 1 - \frac{\sigma^2}{\varepsilon^2} \tag{8-67}$$

If, therefore, $\sigma \ll \varepsilon$, then the probability that $|\mathbf{x} - a|$ is less than that ε is close to 1. From this it follows that, "almost certainly" the observed $\mathbf{x}(\zeta)$ is between $a - \varepsilon$ and $a + \varepsilon$, or equivalently, that the unknown a is between $\mathbf{x}(\zeta) - \varepsilon$ and $\mathbf{x}(\zeta) + \varepsilon$. In other words, the reading $\mathbf{x}(\zeta)$ of a single measurement is "almost certainly" a satisfactory estimate of the length a as long as $\sigma \ll a$.

If σ is not small compared to a, then a single measurement does not provide an adequate estimate of a. To improve the accuracy, we perform the measurement a large number of times and we average the resulting readings. The underlying probabilistic model is now a product space

$$\mathscr{S}^n = \mathscr{S} \times \cdots \times \mathscr{S}$$

formed by repeating n times the experiment $\mathscr{S}$ of a single measurement. If the measurements are independent, then the ith reading is a sum

$$\mathbf{x}_i = a + \mathbf{v}_i$$

where the noise components $\mathbf{v}_i$ are independent RVs with zero mean and variance σ^2. This leads to the conclusion that the *sample mean*

$$\bar{\mathbf{x}} = \frac{\mathbf{x}_1 + \cdots + \mathbf{x}_n}{n} \tag{8-68}$$

of the measurements is an RV with mean a and variance σ^2/n. If, therefore, n is so large that $\sigma^2 \ll na^2$, then the value $\bar{\mathbf{x}}(\zeta)$ of the sample mean $\bar{\mathbf{x}}$ in a single performance of the experiment $\mathscr{S}^n$ (consisting of n independent measurements) is a satisfactory estimate of the unknown a.

To find a bound of the error in the estimate of a by $\bar{\mathbf{x}}$, we apply (8-67). To be concrete, we assume that n is so large that $\sigma^2/na^2 = 10^{-4}$, and we ask for the probability that $\mathbf{x}$ is between $0.9a$ and $1.1a$. The answer is given by (8-67) with $\varepsilon = 0.1a$

$$P\{0.9a < \bar{\mathbf{x}} < 1.1a\} \geq 1 - \frac{100\sigma^2}{n} = 0.99$$

Thus, if the experiment is performed $n = 10^4\sigma^2/a^2$ times then "almost certainly" in 99 percent of the cases, the estimate $\bar{\mathbf{x}}$ of a will be between $0.9a$ and $1.1a$.

Convergence Concepts

We introduce next various convergence modes involving sequences of random variables.

Definition A *random sequence* or a *discrete-time random process* is a sequence of RVs

$$\mathbf{x}_1, \ldots, \mathbf{x}_n, \ldots \tag{8-69}$$

For a specific ζ, $\mathbf{x}_n(\zeta)$ is a sequence of numbers that might or might not converge. This suggests that the notion of convergence of a random sequence might be given several interpretations:

Convergence everywhere (e) As we recall, a sequence of numbers x_n tends to a limit x if, given $\varepsilon > 0$, we can find a number n_0 such that

$$|x_n - x| < \varepsilon \quad \text{for every} \quad n > n_0 \tag{8-70}$$

We say that a random sequence $\mathbf{x}_n$ converges everywhere if the sequences of numbers $\mathbf{x}_n(\zeta)$ converges as above for every n. Their limit is a number that depends, in general, on ζ. In other words, the limit of the random sequence $\mathbf{x}_n$ is an RV $\mathbf{x}$

$$\mathbf{x}_n \to \mathbf{x} \quad \text{as} \quad n \to \infty$$

Convergence almost everywhere (a.e) If the set of outcomes ζ such that

$$\lim \mathbf{x}_n(\zeta) = \mathbf{x}(\zeta) \quad \text{as} \quad n \to \infty \tag{8-71}$$

exists, and its probability equals 1, then we say that the sequence $\mathbf{x}_n$ converges almost everywhere (or with probability 1). This is written in the form

$$P\{\mathbf{x}_n \to \mathbf{x}\} = 1 \quad \text{as} \quad n \to \infty \tag{8-72}$$

In the above, $\{\mathbf{x}_n \to \mathbf{x}\}$ is an event consisting of all outcomes ζ such that $\mathbf{x}_n(\zeta) \to \mathbf{x}(\zeta)$.

Convergence in the MS sense (*MS*) The sequence $\mathbf{x}_n$ tends to the RV $\mathbf{x}$ in the MS sense if

$$E\{|\mathbf{x}_n - \mathbf{x}|^2\} \to 0 \quad \text{as} \quad n \to \infty \tag{8-73}$$

This is called *limit in the mean* and it is written in the form

$$\text{l.i.m. } \mathbf{x}_n = \mathbf{x}_n \quad n \to \infty$$

Convergence in probability (*p*) The probability $P\{|\mathbf{x} - \mathbf{x}_n| > \varepsilon\}$ of the event $\{|\mathbf{x} - x_n| > \varepsilon\}$ is a sequence of numbers depending on ε. If this sequence tends to zero

$$P\{|\mathbf{x} - \mathbf{x}_n| > \varepsilon\} \to 0 \quad n \to \infty \tag{8-74}$$

for any $\varepsilon > 0$, then we say that the sequence $\mathbf{x}_n$ tends to the RV $\mathbf{x}$ in probability (or in measure). This is called also stochastic convergence.

Convergence in distribution (*d*) We denote by $F_n(x)$ and $F(x)$ respectively the distribution of the RVs $\mathbf{x}_n$ and $\mathbf{x}$. If

$$F_n(x) \to F(x) \quad n \to \infty \tag{8-75}$$

for every point x of continuity of $F(x)$, then we say that the sequence $\mathbf{x}_n$ tends to the RV $\mathbf{x}$ in distribution. We note that, in this case, the sequence $\mathbf{x}_n(\zeta)$ need not converge for any ζ.

Cauchy criterion As we noted, a deterministic sequence x_n converges if it satisfies (8-70). This definition involves the limit x of x_n. The following theorem, known as Cauchy criterion, establishes conditions for the convergence of x_n that avoid the use of x: If

$$|x_{n+m} - x_n| \to 0 \quad \text{as} \quad n \to \infty \tag{8-76}$$

for any $m > 0$, then the sequence x_n converges.

The above theorem holds also for random sequence. In this case, the limit must be interpreted accordingly. For example, if

$$E\{|\mathbf{x}_{n+m} - \mathbf{x}_n|^2\} \to 0 \quad \text{as} \quad n \to \infty$$

for every $m > 0$, then the random sequence $\mathbf{x}_n$ converges in the MS sense.

Comparison of convergence modes In Fig. 8-2, we show the relationship between various convergence modes. Each point in the rectangle represents a random sequence. The letter on each curve indicates that all sequences in the interior of the curve converge in the stated mode. The shaded region consists of all sequences that do not converge in any sense. The letter d on the outer curve shows that if a sequence converges at all, then it converges also in distribution.

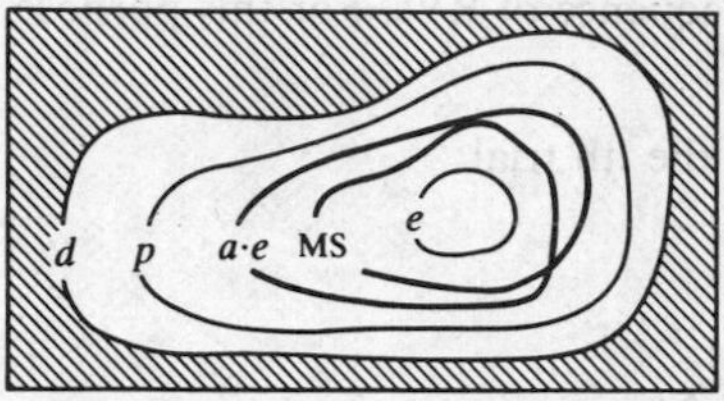

Figure 8-2

The letter e in the innermost region shows that if a sequence converges everywhere, then it converges in any other sense. We comment next on the less obvious comparisons:

If a sequence converges in the MS sense, then it converges also in probability. Indeed, Tchebycheff's inequality yields

$$P\{|\mathbf{x}_n - \mathbf{x}| > \varepsilon\} \leq \frac{E\{|\mathbf{x}_n - \mathbf{x}|^2\}}{\varepsilon^2}$$

If $\mathbf{x}_n \to \mathbf{x}$ in the MS sense, then for a fixed $\varepsilon > 0$ the right side tends to zero, hence the left side also tends to zero as $n \to \infty$ and (8-74) follows. The converse, however, is not necessarily true. If $\mathbf{x}_n$ is not bounded, then $P\{|\mathbf{x}_n - \mathbf{x}| > \varepsilon\}$ might tend to zero but not $E\{|\mathbf{x}_n - \mathbf{x}|^2\}$. If, however, $\mathbf{x}_n$ vanishes outside some interval $(-c, c)$ for every $n > n_0$, then p convergence and MS convergence are equivalent.

It is self-evident that a.e. convergence implies p convergence. We shall show by a heuristic argument that the converse is not true. In Fig. 8-3, we plot the difference $|\mathbf{x}_n - \mathbf{x}|$ as a function of n where, for simplicity, sequences are drawn as curves. Each curve represents, thus, a particular sequence $|\mathbf{x}_n(\zeta) - \mathbf{x}(\zeta)|$. Convergence in probability means that for a *specific* $n > n_0$, only a small percentage of these curves will have ordinates that exceed ε (Fig. 8-3*a*). It is, of course, possible that not even one of these curves will remain less than ε for *every* $n > n_0$. Convergence a.e., on the other hand, demands that most curves will be below ε for every $n > n_0$ (Fig. 8-3*b*).

The law of large numbers (Bernoulli) In Sec. 3-3 we showed that if the probability of an event $\mathscr{A}$ in a given experiment equals p and the number of successes of $\mathscr{A}$ in n trials equals k, then

$$P\left\{\left|\frac{k}{n} - p\right| < \varepsilon\right\} \to 1 \quad \text{as} \quad n \to \infty \tag{8-77}$$

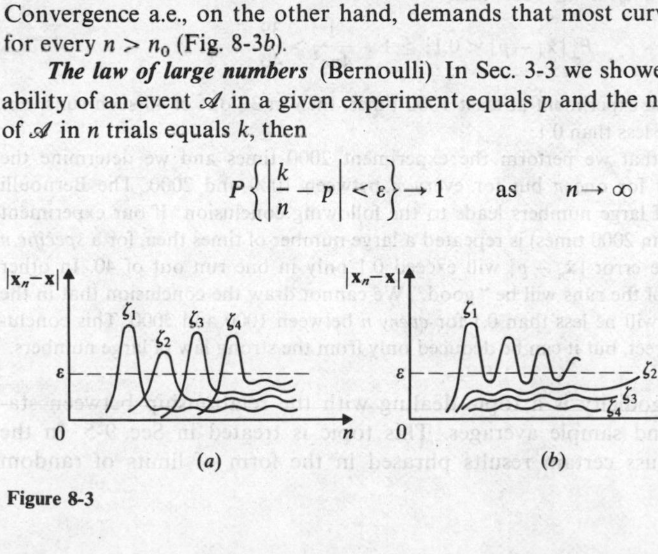

Figure 8-3

We shall reestablish this result as a limit of a sequence of RVs. For this purpose, we introduce the RVs

$$\mathbf{x}_i = \begin{cases} 1 & \text{if } \mathscr{A} \text{ occurs at the } i\text{th trial} \\ 0 & \text{otherwise} \end{cases}$$

We shall show that the sample mean

$$\bar{\mathbf{x}}_n = \frac{\mathbf{x}_1 + \cdots + \mathbf{x}_n}{n}$$

of these RVs tends to p in probability as $n \to \infty$.

PROOF As we know

$$E\{\mathbf{x}_i\} = E\{\bar{\mathbf{x}}_n\} = p \qquad \sigma_{x_i}^2 = pq \qquad \sigma_{\bar{x}_n}^2 = \frac{pq}{n}$$

Furthermore, $pq = p(1 - q) \le 1/4$. Hence [see (5-57)]

$$P\{|\bar{\mathbf{x}}_n - p| < \varepsilon\} \ge 1 - \frac{pq}{n\varepsilon^2} \le 1 - \frac{1}{4n\varepsilon^2} \xrightarrow[n\to\infty]{} 1$$

This reestablishes (8-77) because $\bar{\mathbf{x}}_n(\zeta) = k/n$ if $\mathscr{A}$ occurs k times.

The strong law of large numbers (Borel) It can be shown that $\bar{\mathbf{x}}_n$ tends to p not only in probability, but also with probability 1 (a.e.). This result, due to Borel, is known as the strong law of large numbers. The proof will not be given. We give below only a heuristic explanation of the difference between (8-77) and the strong law of large numbers in terms of relative frequencies.

Frequency interpretation We wish to estimate p within an error $\varepsilon = 0.1$ using as its estimate the sample mean $\bar{\mathbf{x}}_n$. If $n \ge 1000$, then

$$P\{|\bar{\mathbf{x}}_n - p| < 0.1\} \ge 1 - \frac{1}{4n\varepsilon^2} \ge \frac{39}{40}$$

Thus, if we repeat the experiment at least 1000 times, then in 39 out of 40 such runs, our error $|\bar{\mathbf{x}}_n - p|$ will be less than 0.1.

Suppose, now, that we perform the experiment 2000 times and we determine the sample mean $\bar{\mathbf{x}}_n$ not for one n but for every n between 1000 and 2000. The Bernoulli version of the law of large numbers leads to the following conclusion: If our experiment (the tossing of the coin 2000 times) is repeated a large number of times then, for a *specific n* larger than 1000, the error $|\bar{\mathbf{x}}_n - p|$ will exceed 0.1 only in one run out of 40. In other words, 97.5 percent of the runs will be "good." We cannot draw the conclusion that in the good runs the error will be less than 0.1 for *every n* between 1000 and 2000. This conclusion, however, is correct, but it can be deduced only from the strong law of large numbers.

Ergodicity Ergodicity is a topic dealing with the relationship between statistical averages and sample averages. This topic is treated in Sec. 9-5. In the following, we discuss certain results phrased in the form of limits of random sequences.

Markoff's theorem We are given a sequence $\mathbf{x}_i$ of RVs and we form their sample mean

$$\bar{\mathbf{x}}_n = \frac{\mathbf{x}_1 + \cdots + \mathbf{x}_n}{n}$$

Clearly, $\bar{\mathbf{x}}_n$ is an RV whose values $\mathbf{x}_n(\zeta)$ depend on the experimental outcome ζ. We maintain that, if the RVs $\mathbf{x}_i$ are such that the mean $\bar{\eta}_n$ of $\bar{\mathbf{x}}_n$ tends to a limit η and its variance $\bar{\sigma}_n$ tends to zero as $n \to \infty$

$$E\{\bar{\mathbf{x}}_n\} = \bar{\eta}_n \xrightarrow[n\to\infty]{} \eta \qquad \bar{\sigma}_n^2 = E\{(\bar{\mathbf{x}}_n - \bar{\eta}_n)^2\} \xrightarrow[n\to\infty]{} 0 \tag{8-78}$$

then the RV $\bar{\mathbf{x}}_n$ tends to η in the MS sense

$$E\{(\bar{\mathbf{x}}_n - \eta)^2\} \xrightarrow[n\to\infty]{} 0 \tag{8-79}$$

PROOF The proof is based on the simple inequality

$$|\bar{\mathbf{x}}_n - \eta|^2 \le |\bar{\mathbf{x}}_n - \bar{\eta}_n|^2 + |\bar{\eta}_n - \eta|^2$$

Indeed, taking expected values of both sides, we obtain

$$E\{(\bar{\mathbf{x}}_n - \eta)^2\} \le E\{(\mathbf{x}_n - \bar{\eta}_n)^2\} + (\bar{\eta}_n - \eta)^2$$

and (8-79) follows from (8-78).

Corollary (Tchebycheff's condition) If the RVs x_i are uncorrelated and

$$\frac{\sigma_1^2 + \cdots + \sigma_n^2}{n^2} \xrightarrow[n\to\infty]{} 0 \tag{8-80}$$

then $\bar{x}_n \to \eta$ as $n \to \infty$.

PROOF It follows from the theorem because, for uncorrelated RVs, the left side of (8-80) equals $\bar{\sigma}_n^2$.

We note that Tchebycheff's condition (8-80) is satisfied if $\sigma_i < K < \infty$ for every i. This is the case if the RVs $\mathbf{x}_i$ are i.i.d. with finite variance.

Kinchin We mention without proof that if the RVs $\mathbf{x}_i$ are i.i.d., then their sample mean $\bar{\mathbf{x}}_n$ tends to η even if nothing is known about their variance. In this case, however, $\bar{\mathbf{x}}_n$ tends to η in probability only. The following is an application:

Example 8-10 We wish to determine the distribution $F(x)$ of an RV $\mathbf{x}$ defined in a certain experiment. For this purpose we repeat the experiment n times and form the RVs $\mathbf{x}_i$ as in (8-12). As we know, these RVs are i.i.d. and their common distribution equals $F(x)$. We next form the RVs

$$\mathbf{y}_i(x) = \begin{cases} 1 & \text{if} \quad \mathbf{x}_i \le x \\ 0 & \text{if} \quad \mathbf{x}_i > x \end{cases}$$

where x is a fixed number. The RVs $\mathbf{y}_i(x)$ so formed are also i.i.d. and their mean equals

$$E\{\mathbf{y}_i(x)\} = 1 \cdot P\{\mathbf{y}_i = 1\} = P\{\mathbf{x}_i \le x\} = F(x)$$

Applying Kinchin's theorem to $\mathbf{y}_i(x)$, we conclude that

$$\frac{\mathbf{y}_1(x) + \cdots + \mathbf{y}_n(x)}{n} \xrightarrow[n\to\infty]{} F(x) \tag{8-81}$$

in probability. Thus, to determine $F(x)$, we repeat the original experiment n times and count the number of times the RV $\mathbf{x}$ is less than x. If this number equals k and n is sufficiently large, then $F(x) \simeq k/n$. The above is thus a restatement of the relative frequency interpretation (4-2) of $F(x)$ in the form of a limit theorem.

8-5 THE CENTRAL LIMIT THEOREM

The central limit theorem states that if the RVs $\mathbf{x}_i$ are independent, then under general conditions, the density $f(x)$ of their sum

$$\mathbf{x} = \mathbf{x}_1 + \cdots + \mathbf{x}_n$$

properly normalized, tends to a normal curve as $n \to \infty$. In other words, if n is sufficiently large, then (Fig. 8-4*a*)

$$f(x) \simeq \frac{1}{\sigma\sqrt{2\pi}} e^{-(x-\eta)^2/2\sigma^2}. \tag{8-82}$$

If the given RVs are of lattice type and $\mathbf{x}$ takes the values ka with $p_k = P\{\mathbf{x} = ka\}$ then the probabilities p_k equal the samples of a normal curve (Fig. 8-4*b*)

$$P\{\mathbf{x} = ka\} \simeq \frac{1}{\sigma\sqrt{2\pi}} e^{-(ka-\eta)^2/2\sigma^2} \tag{8-83}$$

Note Since $f(x)$ is the convolution of the densities $f_i(x)$ of the RVs $\mathbf{x}_i$

$$f(x) = f_1(x) * \cdots * f_n(x)$$

the central limit theorem can be viewed as a property of convolutions of positive functions.

The theorem does not hold if only a small number m of the given densities are dominant, i.e., if the densities of the other RVs are relatively narrow in the

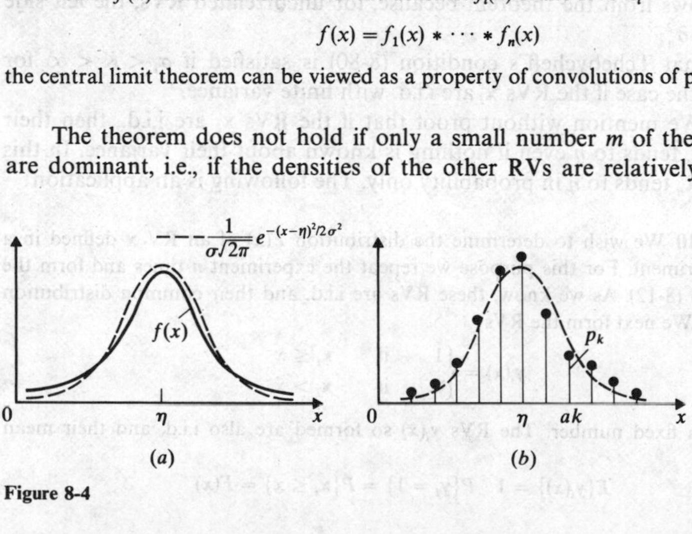

Figure 8-4

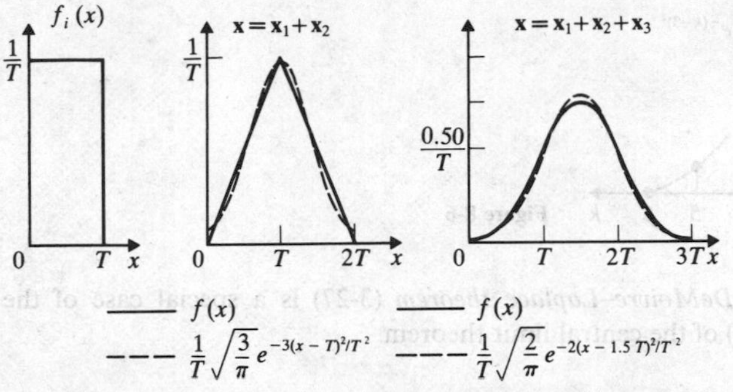

Figure 8-5

sense that, in the evaluation of the total convolution, they can be approximated by impulses. In this case, $f(x)$ is effectively the convolution of only the m dominant densities and it need not be close to a normal curve. As the next example shows, however, if the densities $f_i(x)$ are smooth, then $f(x)$ is nearly normal even when n is small.

Example 8-11 The RVs $\mathbf{x}_i$ are i.i.d. with uniform densities $f_i(x)$ as in Fig. 8-5*a*. We shall determine the density $f(x)$ of their sum and the corresponding normal curve $N(\eta, \sigma^2)$ for $n = 2$ and $n = 3$. Clearly

$$\eta_i = T \qquad \sigma_i^2 = \frac{T^2}{12}$$

Hence, if $n = 2$, then $\eta = 2T$, $\sigma^2 = T^2/6$. In this case, $f(x)$ is a triangle obtained by convolving a pulse with itself (Fig. 8-5*b*).

If $n = 3$, then $\eta = 3T/2$, $\sigma^2 = T^2/4$ and $f(x)$ consists of three parabolic pieces obtained by convolving a triangle with a pulse (Fig. 8-5*c*).

We give next an illustration of the central limit theorem in the context of Bernoulli trials. For this purpose, we consider the zero-one RVs $\mathbf{x}_i$ introduced in Example 8-7, and their sum

$$\mathbf{x} = \mathbf{x}_1 + \cdots + \mathbf{x}_n$$

These RVs are i.i.d. with $E\{\mathbf{x}_i\} = p$, $\sigma_i^2 = pq$. Hence

$$E\{\mathbf{x}\} = np \qquad \sigma_x^2 = npq$$

Furthermore [see (8-55)]

$$P\{\mathbf{x} = k\} = \binom{n}{k} p^k q^{n-k}$$

Inserting into (8-83), we obtain

$$\binom{n}{k} p^k q^{n-k} \simeq \frac{1}{\sqrt{2\pi npq}}\, e^{-(k-np)^2/2npq} \tag{8-84}$$

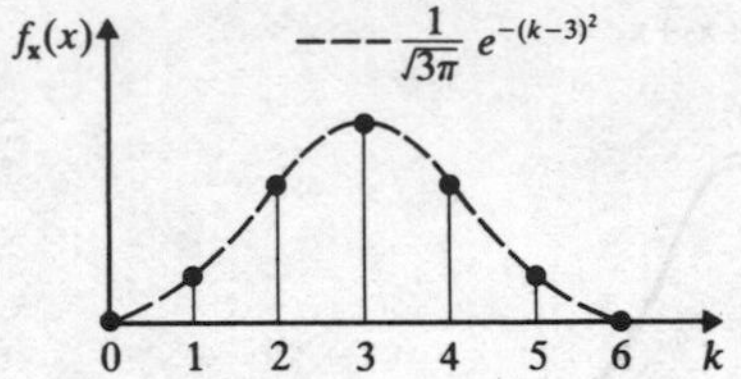

Figure 8-6

This shows that the *DeMoivre–Laplace theorem* (3-27) is a special case of the lattice-type form (8-83) of the central limit theorem.

Example 8-12 A fair coin is tossed six times and $\mathbf{x}_i$ is the zero-one RV associated with the event {heads at the ith tossing}. The probability of k heads in six tossings equals

$$P\{\mathbf{x} = k\} = \binom{6}{k} \frac{1}{2^6} = p_k \qquad \mathbf{x} = \mathbf{x}_1 + \cdots + \mathbf{x}_6$$

In the following table we show the above probabilities and the samples of the normal curve $N(\eta, \sigma^2)$ (Fig. 8-6) where

$$\eta = np = 3 \qquad \sigma^2 = npq = 1.5$$

$k =$	0	1	2	3	4	5	6
p_k	0.016	0.094	0.234	0.312	0.234	0.094	0.016
$N(\eta, \sigma^2)$	0.016	0.086	0.233	0.326	0.233	0.086	0.016

Error correction In the approximation of $f(x)$ by the normal curve $N(\eta, \sigma^2)$, the error

$$\varepsilon(x) = f(x) - \frac{1}{\sigma\sqrt{2\pi}} e^{-x^2/2\sigma^2}$$

results where we assumed, shifting the origin that $\eta = 0$. We shall express this error in terms of the moments

$$m_n = E\{\mathbf{x}^n\}$$

of $\mathbf{x}$ and the *Hermite polynomials*

$$H_k(x) = (-1)e^{x^2/2} \frac{d^k}{dx^k} e^{-x^2/2}$$

$$= x^k - \binom{k}{2} x^{k-2} + 1 \cdot 3 \binom{k}{4} x^{k-4} + \cdots \tag{8-85}$$

These polynomials form a complete orthogonal set on the real line:

$$\int_{-\infty}^{\infty} e^{-x^2/2} H_n(x) H_m(x)\, dx = \begin{cases} n!\sqrt{2\pi} & n = m \\ 0 & n \neq m \end{cases}$$

Hence, $\varepsilon(x)$ can be written as a series

$$\varepsilon(x) = \frac{1}{\sigma\sqrt{2\pi}} e^{-x^2/2\sigma^2} \sum_{k=3}^{\infty} C_k H_k\left(\frac{x}{\sigma}\right) \tag{8-86}$$

The series starts with $k = 3$ because the moments of $\varepsilon(x)$ of order up to two are zero. The coefficients C_n can be expressed in terms of the moments m_n of $\mathbf{x}$. Equating moments of order $n = 3$ and $n = 4$, we obtain [see (5-44)]

$$3!\,\sigma^3 C_3 = m_3 \qquad 4!\,\sigma^4 C_4 = m_4 - 3\sigma^4$$

First-order correction From (8-85) it follows that

$$H_3(x) = x^3 - 3x \qquad H_4(x) = x^4 - 6x^2 + 3$$

Retaining the first nonzero term of the sum in (8-56), we obtain

$$f(x) \simeq \frac{1}{\sigma\sqrt{2\pi}} e^{-x^2/2\sigma^2}\left[1 + \frac{m_3}{6\sigma^3}\left(\frac{x^3}{\sigma^3} - \frac{3x}{\sigma}\right)\right] \tag{8-87}$$

If $f(x)$ is even, then $m_3 = 0$ and (8-86) yields

$$f(x) \simeq \frac{1}{\sigma\sqrt{2\pi}} e^{-x^2/2\sigma^2}\left[1 + \frac{1}{24}\left(\frac{m_4}{\sigma^4} - 3\right)\left(\frac{x^4}{\sigma^4} - \frac{6x^2}{\sigma^2} + 3\right)\right] \tag{8-88}$$

Example 8-13 If the RVs $\mathbf{x}_i$ are i.i.d. with density $f_i(x)$ as in Fig. 8-7*a*, then $f(x)$ consists of three parabolic pieces (see also Example 8-11) and $N(0, 1/4)$ is its normal approximation. Since $f(x)$ is even and $m_4 - 13/80$ (see Prob. 8-4), (8-88) yields

$$f(x) \simeq \sqrt{\frac{2}{\pi}}\, e^{-2x^2}\left(1 - \frac{4x^4}{15} + \frac{2x^2}{5} - \frac{1}{20}\right) \equiv \bar{f}(x)$$

In Fig. 8-7*b*, we show the error $\varepsilon(x)$ of the normal approximation and the first-order correction error $f(x) - \bar{f}(x)$.

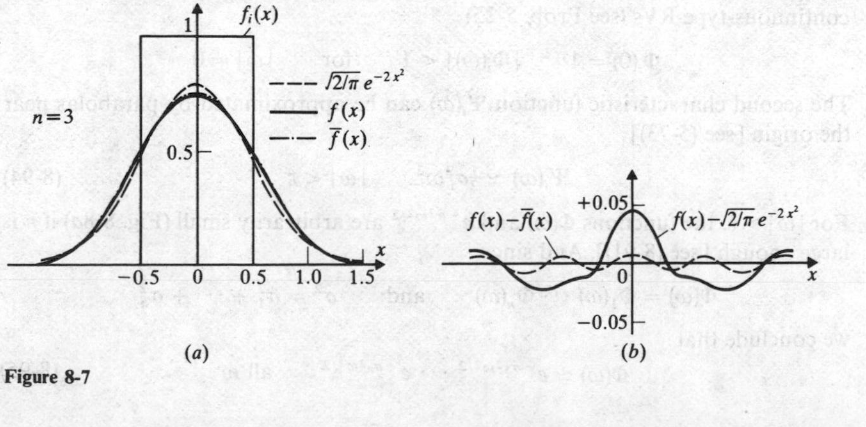

Figure 8-7

On the proof of the central limit theorem The approximation (8-82) becomes an equality in the limit as $n \to \infty$. However, to obtain meaningful results, we must introduce a proper normalization because $\sigma \to \infty$ as $n \to \infty$. Clearly, if

$$\bar{\mathbf{x}} = \frac{\mathbf{x}_1 + \cdots + \mathbf{x}_n}{\sigma} \qquad \text{then} \qquad \sigma_{\bar{x}}^2 = 1 \tag{8-89}$$

Denoting by $\bar{f}(x)$ the density of $\bar{\mathbf{x}}$, we can thus phrase the theorem as a limit

$$\bar{f}(x) \xrightarrow[n \to \infty]{} \frac{1}{\sqrt{2\pi}} e^{-x^2/2} \tag{8-90}$$

As we noted, the theorem is not true always. The following is a set of sufficient conditions:

(*a*)
$$\sigma_1^2 + \cdots + \sigma_n^2 \xrightarrow[n \to \infty]{} \infty \tag{8-91}$$

(*b*) There exists a number $\alpha > 2$ and a finite constant K such that

$$\int_{-\infty}^{\infty} x^\alpha f_i(x)\, dx < K < \infty \qquad \text{for all } i \tag{8-92}$$

These conditions are not the most general. However, they cover a wide range of applications. For example, (8-91) is satisfied if there exists a constant $\varepsilon > 0$ such that $\sigma_i > \varepsilon$ for all i. Condition (8-92) is satisfied if all densities $f_i(x)$ are zero outside a finite interval $(-c, c)$ no matter how large.

Most proofs of the theorem involve characteristic functions. In the following, we sketch the underlying reasoning. At the end of the section, we give a proof based on a more recent result known as the Berry–Esseén theorem.

Characteristic functions The approximation (8-82) can be phrased in terms of the characteristic functions $\Phi_i(\omega)$ and $\Phi(\omega)$ of the RVs $\mathbf{x}_i$ and of their sum $\mathbf{x} = \mathbf{x}_1 + \cdots + \mathbf{x}_n$

$$\Phi(\omega) = \Phi_1(\omega) \cdots \Phi_n(\omega) \simeq e^{-\sigma^2\omega^2/2} \tag{8-93}$$

The justification of the last approximation is based on the following property of continuous-type RVs (see Prob. 5-25)

$$\Phi_i(0) = 1 \qquad |\Phi_i(\omega)| < 1 \qquad \text{for} \qquad |\omega| \neq 0$$

The second characteristic function $\Psi_i(\omega)$ can be approximated by parabolas near the origin [see (5-73)]

$$\Psi_i(\omega) \simeq \tfrac{1}{2}\sigma_i^2\omega^2 \qquad |\omega| < \varepsilon \tag{8-94}$$

For $|\omega| > \varepsilon$, the functions $\Phi(\omega)$ and $e^{-\sigma^2\omega^2/2}$ are arbitrarily small (Fig. 8-8*a*) if n is large enough [see (8-91)]. And since

$$\Phi(\omega) = \Phi_1(\omega) \cdots \Phi_n(\omega) \qquad \text{and} \qquad \sigma^2 = \sigma_1^2 + \cdots + \sigma_n^2$$

we conclude that

$$\Phi(\omega) \simeq e^{-\sigma_1^2\omega^2/2} \cdots e^{-\sigma_n^2\omega^2/2} \qquad \text{all } \omega \tag{8-95}$$

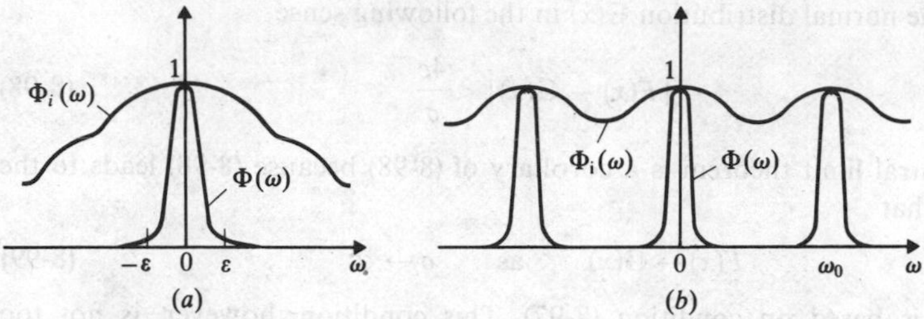

Figure 8-8

Using the above, we can show that the approximation (8-93) properly normalized becomes an equality in the limit as $n \to \infty$. In the proof, condition (8-91) is used to justify the elimination of higher-order terms in (8-94).

The proof of the theorem is simplified if the RVs $\mathbf{x}_i$ are identically distributed. In this case, the characteristic function $\bar{\Phi}(\omega)$ of the normalized sum $\bar{\mathbf{x}}$ [see (8-89)] equals

$$\bar{\Phi}(\omega) = \Phi_i^n\left(\frac{\omega}{\sigma\sqrt{n}}\right)$$

Writing (8-94) in the form

$$\Psi_i\left(\frac{\omega}{\sigma_i\sqrt{n}}\right) = -\frac{\omega^2}{2n} + O(n^{-3/2})$$

we conclude that

$$\bar{\Psi}(\omega) = n\Psi_i(\omega) = -\frac{\omega^2}{2} + O(n^{-1/2}) \to -\frac{\omega^2}{2} \qquad \text{as} \qquad n \to \infty$$

Hence, $\bar{\Phi}(\omega) \to e^{-\omega^2/2}$ and (8-90) results.

Lattice type The preceding reasoning can be applied also to discrete-type RVs. However, in this case the functions $\Phi_i(\omega)$ are periodic (Fig. 8-8*b*) and their product takes significant values only in a small region near the points $\omega = 2\pi n/a$. Using the approximation (8-94) in each of these regions, we obtain

$$\Phi(\omega) \simeq \sum_n e^{-\sigma^2(\omega - \omega_0)^2/2} \qquad \omega_0 = \frac{2\pi}{a} \tag{8-96}$$

As we can see from (10A-1), the inverse of the above yields (8-83).

The Berry-Esseén theorem† This theorem states that if

$$E\{\mathbf{x}_i^3\} \le c\sigma_i^2 \qquad \text{all } i \tag{8-97}$$

where c is some constant, then the distribution $\bar{F}(x)$ of the normalized sum

$$\bar{\mathbf{x}} = \frac{\mathbf{x}_1 + \cdots + \mathbf{x}_n}{\sigma} \qquad \sigma_1^2 + \cdots + \sigma_n^2 = \sigma^2$$

† A. Papoulis, "Narrow-Band Systems and Gaussianity," *IEEE* Transactions on Information Theory, January 1972.

is close to the normal distribution $\mathbb{G}(x)$ in the following sense

$$|\bar{F}(x) - \mathbb{G}(x)| < \frac{4c}{\sigma} \tag{8-98}$$

The central limit theorem is a corollary of (8-98) because (8-98) leads to the conclusion that

$$\bar{F}(x) \to \mathbb{G}(x) \qquad \text{as} \qquad \sigma \to \infty \tag{8-99}$$

This proof is based on condition (8-97). This condition, however, is not too restrictive. It holds, for example, if the RVs $\mathbf{x}_i$ are i.i.d. and their third moment is finite.

We note, finally, that, whereas (8-99) establishes merely the convergence in distribution of $\bar{\mathbf{x}}$ to a normal RV, (8-98) gives also a *bound* of the deviation of $\bar{F}(x)$ from normality.

PROBLEMS

8-1 Show that if $F(x, y, z)$ is a joint distribution, then for any $x_1 \le x_2, y_1 \le y_2, z_1 \le z_2$

$$F(x_2, y_2, z_2) + F(x_1, y_1, z_2) + F(x_1, y_2, z_1) + F(x_2, y_1, z_1)$$
$$- F(x_1, y_2, z_2) - F(x_2, y_1, z_2) - F(x_2, y_2, z_1) - F(x_1, y_1, z_1) \ge 0$$

8-2 The events $\mathscr{A}, \mathscr{B}, \mathscr{C}$ are such that

$$P(\mathscr{A}) = P(\mathscr{B}) = P(\mathscr{C}) = 0.5 \qquad P(\mathscr{A}\mathscr{B}) = P(\mathscr{A}\mathscr{C}) = P(\mathscr{B}\mathscr{C}) = P(\mathscr{A}\mathscr{B}\mathscr{C}) = 0.25$$

Show that the zero-one RVs associated with these events are not independent; they are, however, independent in pairs.

8-3 Show that if the RVs $\mathbf{x}, \mathbf{y}, \mathbf{z}$ are jointly normal and independent in pairs, then they are independent.

8-4 The RVs $\mathbf{x}_i$ are i.i.d. and uniform in the interval $(-0.5, 0.5)$. Show that

$$E\{(\mathbf{x}_1 + \mathbf{x}_2 + \mathbf{x}_3)^4\} = \tfrac{13}{80}$$

8-5 (*a*) Reasoning as in (6-34), show that if the RVs $\mathbf{x}, \mathbf{y}, \mathbf{z}$ are independent and their joint density has spherical symmetry

$$f(x, y, z) = f(\sqrt{\mathbf{x}^2 + \mathbf{y}^2 + \mathbf{z}^2})$$

then they are normal.

(*b*) The components $\mathbf{v}_x, \mathbf{v}_y, \mathbf{v}_z$ of the velocity $\mathbf{v} = \sqrt{\mathbf{v}_x^2 + \mathbf{v}_y^2 + \mathbf{v}_z^2}$ of a particle are independent RVs with zero mean and variance kT/m. Furthermore, their joint density has spherical symmetry. Show that $\mathbf{v}$ has a Maxwell density and

$$E\{\mathbf{v}\} = 2\sqrt{\frac{2kT}{\pi m}} \qquad E\{\mathbf{v}^2\} = \frac{3kT}{m} \qquad E\{\mathbf{v}^4\} = \frac{15k^2T^2}{m^2}$$

8-6 Show that if the RVs $\mathbf{x}, \mathbf{y}, \mathbf{z}$ are such that $r_{xy} = r_{yz} = 1$, then $r_{xz} = 1$.

8-7 Show that

$$E\{\mathbf{x}_1\mathbf{x}_2 \mid \mathbf{x}_3\} = E\{E\{\mathbf{x}_1\mathbf{x}_2 \mid \mathbf{x}_2, \mathbf{x}_3\} \mid \mathbf{x}_3\} = E\{\mathbf{x}_2 E\{\mathbf{x}_1 \mid \mathbf{x}_2, \mathbf{x}_3\} \mid \mathbf{x}_3\}$$

8-8 Show that $\hat{E}\{\mathbf{y} \mid \mathbf{x}_1\} = \hat{E}\{\hat{E}\{\mathbf{y} \mid \mathbf{x}_1, \mathbf{x}_2\} \mid \mathbf{x}_1\}$ where $\hat{E}\{\mathbf{y} \mid \mathbf{x}_1, \mathbf{x}_2\} = a_1\mathbf{x}_1 + a_2\mathbf{x}_2$ is the linear MS estimate of $\mathbf{y}$ in terms of $\mathbf{x}_1$ and $\mathbf{x}_2$.

8-9 Show that if

$$\mathbf{x}_i \geq 0, \qquad E\{\mathbf{x}_i^2\} = M \qquad \text{and} \qquad \mathbf{s} = \sum_{i=1}^{\mathbf{n}} \mathbf{x}_i$$

then

$$E\{\mathbf{s}^2\} \leq ME\{\mathbf{n}^2\}$$

8-10 We denote, by $\mathbf{x}_m$, an RV equal to the number of tossing of a coin until heads shows for the mth time. Show that if $P\{h\} = p$, then $E\{\mathbf{x}_m\} = m/p$.

Hint: $E\{\mathbf{x}_m - \mathbf{x}_{m-1}\} = E\{\mathbf{x}_1\} = p + pq + \cdots + npq^{n-1} + \cdots = 1/p$.

8-11 The number of daily accidents is a Poisson RV $\mathbf{n}$ with parameter a. The probability that a single accident is fatal equals p. Show that the number $\mathbf{m}$ of fatal accidents in one day is a Poisson RV with parameter ap.

Hint:

$$E\{e^{j\omega\mathbf{m}} \mid \mathbf{n} = n\} = \sum_{k=0}^{n} e^{j\omega k}\binom{n}{k}p^k q^{n-k} = (pe^{j\omega} + q)^n$$

8-12 The RVs $\mathbf{x}_k$ are independent with densities $f_k(x)$ and the RV $\mathbf{n}$ is independent of $\mathbf{x}_k$ with $P\{\mathbf{n} = k\} = p_k$. Show that if

$$\mathbf{s} = \sum_{k=1}^{\mathbf{n}} \mathbf{x}_k \qquad \text{then} \qquad f_s(s) = \sum_{k=1}^{\infty} p_k[f_1(s) * \cdots * f_k(s)]$$

8-13 The RVs $\mathbf{x}_i$ are i.i.d. with moment function $\mathbf{\Phi}_x(s) = E\{e^{s\mathbf{x}_i}\}$. The RV $\mathbf{n}$ takes the values 0, 1, ... and its moment function equals $\mathbf{\Gamma}_n(z) = E\{\mathbf{z}^n\}$. Show that if

$$\mathbf{y} = \sum_{i=0}^{\mathbf{n}} \mathbf{x}_i \qquad \text{then} \qquad \mathbf{\Phi}_y(s) = E\{e^{s\mathbf{y}}\} = \mathbf{\Gamma}_n[\mathbf{\Phi}_x(s)]$$

Hint: $E\{e^{s\mathbf{y}} \mid \mathbf{n} = k\} = E\{e^{(\mathbf{x}_0 + \cdots + \mathbf{x}_k)}\} = \Phi_x^k(s)$.

Special case If $\mathbf{n}$ is Poisson with parameter a, then $\mathbf{\Phi}_y(s) = e^{a\mathbf{\Phi}_x(s) - a}$.

8-14 The RVs $\mathbf{x}_i$ are i.i.d. and uniform in the interval (0, 1). Show that if $\mathbf{y} = \max \mathbf{x}_i$, then $F(y) = y^n$ for $0 \leq y \leq 1$.

8-15 We place at random n points in the interval (0, 1) and we denote by $\mathbf{x}$ and $\mathbf{y}$ the distance from the origin to the first and last point respectively. Find $F(x)$, $F(y)$, and $F(x, y)$.

8-16 Show that if the RVs $\mathbf{x}_i$ are independent with variance σ^2 and sample variance $\bar{\mathbf{v}}$ (see Example 8-4), then

$$\sigma_{\bar{v}}^2 = \frac{1}{n}E\{\mathbf{x}^4\} - \frac{n-3}{n-1}\sigma^2$$

8-17 The RVs $\mathbf{x}_i$ are $N(0; \sigma)$ and independent. Using Prob. 7-1, show that if

$$\mathbf{z} = \frac{\sqrt{\pi}}{2n}\sum_{i=1}^{n} |\mathbf{x}_{2i} - \mathbf{x}_{2i-1}| \qquad \text{then} \qquad E\{\mathbf{z}\} = \sigma \qquad \sigma_z^2 = \frac{\pi - 2}{n}\sigma^2$$

8-18 Show that if R is the correlation matrix of the random vector $\mathbf{X} : [\mathbf{x}_1, \ldots, \mathbf{x}_n]$ and R^{-1} is its inverse, then

$$E\{\mathbf{X}R^{-1}\mathbf{X}^t\} = n$$

8-19 Show that if the RVs $\mathbf{x}_i$ are of continuous type and independent, then, for sufficiently large n, the density of $\sin(\mathbf{x}_1 + \cdots + \mathbf{x}_n)$ is nearly equal to the density of $\sin \mathbf{x}$ where $\mathbf{x}$ is an RV uniform in the interval $(-\pi, \pi)$.

8-20 Show that if $a_n \to a$ and $E\{|\mathbf{x}_n - a_n|^2\} \to 0$, then $\mathbf{x}_n \to a$ in the MS sense as $n \to \infty$.

8-21 Using the Cauchy criterion, show that a sequence $\mathbf{x}_n$ tends to a limit in the MS sense iff the limit of $E\{\mathbf{x}_n \mathbf{x}_m\}$ as $n, m \to \infty$ exists.

8-22 An infinite sum is by definition a limit:

$$\sum_{k=1}^{\infty} \mathbf{x}_k = \lim_{n\to\infty} \mathbf{y}_n \qquad \mathbf{y}_n = \sum_{k=1}^{n} \mathbf{x}_k$$

Show that if the RVs $\mathbf{x}_k$ are independent with zero mean and variance σ_k^2, then the sum exists in the MS sense iff

$$\sum_{k=1}^{\infty} \sigma_k^2 < \infty$$

Hint:

$$E\{(\mathbf{y}_{n+m} - \mathbf{y}_n)^2\} = \sum_{k=n+1}^{n+m} \sigma_k^2$$

8-23 The RVs $\mathbf{x}_i$ are i.i.d. with density $ce^{-cx}U(x)$. Show that, if $\mathbf{x} = \mathbf{x}_1 + \cdots + \mathbf{x}_n$, then $f_x(x)$ is an Erlang density.

8-24 Using the central limit theorem, show that for large n

$$\frac{c^n}{(n-1)!} x^{n-1} e^{-cx} \simeq \frac{c}{\sqrt{2\pi n}} e^{-(cx-n)^2/2n} \qquad x > 0$$

8-25 The resistors $\mathbf{r}_1, \mathbf{r}_2, \mathbf{r}_3, \mathbf{r}_4$ are independent RVs and each is uniform in the interval (450; 550). Using the central limit theorem, find $P\{1900 \le \mathbf{r}_1 + \mathbf{r}_2 + \mathbf{r}_3 + \mathbf{r}_4 \le 2100\}$.

8-26 Show that the central limit theorem does not hold if the RVs $\mathbf{x}_i$ have a Cauchy density.

PART TWO

STOCHASTIC PROCESSES

CHAPTER
NINE
GENERAL CONCEPTS

9.1 INTRODUCTION

As we recall, an RV $\mathbf{x}$ is a rule for assigning to every outcome ζ of an experiment $\mathscr{S}$ a *number* $\mathbf{x}(\zeta)$. A stochastic process $\mathbf{x}(t)$ is a rule for assigning to every ζ a *function* $\mathbf{x}(t, \zeta)$. Thus, a stochastic process is a family of time-functions depending on the parameter ζ or, equivalently, a function of t and ζ. The domain of ζ is the set of all experimental outcomes and the domain of t is a set $\mathscr{I}$ of real numbers.

If $\mathscr{I}$ is the real axis, then $\mathbf{x}(t)$ is a *continuous-time* process. If $\mathscr{I}$ is the set of integers then $\mathbf{x}(t)$ is a *discrete-time* process. A discrete-time process is, thus, a sequence of random variables. Such a sequence will be denoted by $\mathbf{x}_n$ as in Sec. 8-4 or, to avoid double indices, by $\mathbf{x}[n]$.

We shall say that $\mathbf{x}(t)$ is a *discrete-state* process if its values are countable. Otherwise, it is a *continuous-state* process.

Most results in this investigation will be phrased in terms of continuous-time processes. Topics dealing with discrete-time processes will be introduced either as illustrations of the general theory, or when their discrete-time version is not self-evident.

We shall use the notation $\mathbf{x}(t)$ to represent a stochastic process omitting, as in the case of random variables, its dependence on ζ. Thus $\mathbf{x}(t)$ has the following interpretations:

1. It is a family (or an *ensemble*) of functions $\mathbf{x}(t, \zeta)$. In this interpretation, t and ζ are variables.

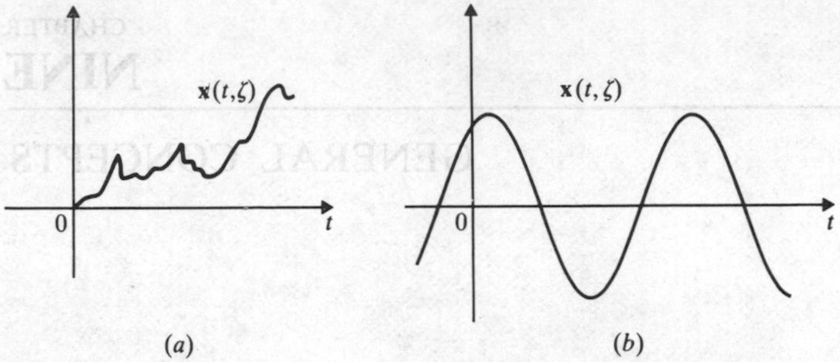

Figure 9-1

2. It is a single time function (or a *sample* of the given process). In this case, t is a variable and ζ is fixed.
3. If t is fixed and ζ is variable, then $\mathbf{x}(t)$ is a random variable equal to the *state* of the given process at time t.
4. If t and ζ are fixed, then $\mathbf{x}(t)$ is a *number*.

It will be understood from the context which of these interpretations holds in a particular case.

Regular and predictable processes A physical example of a stochastic process is the motion of microscopic particles in collision with the molecules in a fluid (*brownian motion*). The resulting process $\mathbf{x}(t)$ consists of the motions of all particles (ensemble). A single realization $\mathbf{x}(t, \zeta_i)$ of this process (Fig. 9-1*a*) is the motion of a specific particle (sample).

Another example is the voltage

$$\mathbf{x}(t) = \mathbf{r} \cos (\omega t + \boldsymbol{\varphi})$$

of an ac generator with random amplitude $\mathbf{r}$ and phase $\boldsymbol{\varphi}$. In this case, the process $\mathbf{x}(t)$ consists of a family of pure sine waves and a single sample is the function (Fig. 9-1*b*)

$$\mathbf{x}(t, \zeta_i) = \mathbf{r}(\zeta_i) \cos [\omega t + \boldsymbol{\varphi}(\zeta_i)]$$

According to our definition, both examples are stochastic processes. There is, however, a fundamental difference between them. The first example (regular) consists of a family of functions that cannot be described in terms of a finite number of parameters. Furthermore, the future of a sample $\mathbf{x}(t, \zeta_i)$ of $\mathbf{x}(t)$ cannot be determined in terms of its past. Finally, under certain conditions, the statistics of a regular process $\mathbf{x}(t)$ can be determined in terms of a single sample (see Sec. 9-6). The second example (predictable) consists of a family of pure sine waves and it is completely specified in terms of the RVs $\mathbf{r}$ and $\boldsymbol{\varphi}$. Furthermore, if $\mathbf{x}(t, \zeta_i)$ is known for $t \le t_o$, then it is determined for $t > t_o$. Finally, a single sample $\mathbf{x}(t, \zeta_i)$ of $\mathbf{x}(t)$ does not specify the properties of the entire process because it depends only on the particular values $\mathbf{r}(\zeta_i)$ and $\boldsymbol{\varphi}(\zeta_i)$ of $\mathbf{r}$ and $\boldsymbol{\varphi}$. A formal definition of regular and predictable processes is given in Sec. 13-4.

Equality We shall say that two stochastic processes $\mathbf{x}(t)$ and $\mathbf{y}(t)$ are equal (everywhere) if their respective samples $\mathbf{x}(t, \zeta_i)$ and $\mathbf{y}(t, \zeta_i)$ are identical for every ζ_i. Similarly, the equality $\mathbf{z}(t) = \mathbf{x}(t) + \mathbf{y}(t)$ means that $\mathbf{z}(t, \zeta_i) = \mathbf{x}(t, \zeta_i) + \mathbf{y}(t, \zeta_i)$ for every ζ_i. Derivatives, integrals, or any other operations involving stochastic processes are defined similarly in terms of the corresponding operations for each sample.

As in the case of limits, the above definition can be relaxed. We give below the meaning of MS equality and in App. 9A we define MS derivatives and integrals.

Two processes $\mathbf{x}(t)$ and $\mathbf{y}(t)$ are equal in the MS sense iff

$$E\{|\mathbf{x}(t) - \mathbf{y}(t)|^2\} = 0 \tag{9-1}$$

for every t. Equality in the MS sense leads to the following conclusions: We denote by $\mathscr{A}_t$ the set of outcomes ζ_i such that $\mathbf{x}(t, \zeta_i) = \mathbf{y}(t, \zeta_i)$ for a *specific* t, and by $\mathscr{A}_\infty$ the set of outcomes ζ_i such that $\mathbf{x}(t, \zeta_i) = \mathbf{y}(t, \zeta_i)$ for *every* t. From (9-1) it follows that $\mathbf{x}(t, \zeta_i) - \mathbf{y}(t, \zeta_i) = 0$ with probability 1, hence $P(\mathscr{A}_t) = P(\mathscr{S}) = 1$. It does not follow, however, that $P(\mathscr{A}_\infty) = 1$. In fact, since $\mathscr{A}_\infty$ is the intersection of all sets $\mathscr{A}_t$ as t ranges over the entire axis, $P(\mathscr{A}_\infty)$ might even equal 0.

Statistics of Stochastic Processes

A stochastic process is a noncountable infinity of random variables, one for each t. For a specific t, $\mathbf{x}(t)$ is an RV with distribution

$$F(x, t) = P\{\mathbf{x}(t) \le x\} \tag{9-2}$$

This function depends on t, and it equals the probability of the event $\{\mathbf{x}(t) \le x\}$ consisting of all outcomes ζ_i such that, at the specific time t, the samples $\mathbf{x}(t, \zeta_i)$ of the given process do not exceed the number x.

The function $F(x, t)$ will be called the *first-order distribution* of the process $\mathbf{x}(t)$. Its derivative with respect to x

$$f(x, t) = \frac{\partial F(x, t)}{\partial x} \tag{9-3}$$

is the *first-order density* of $\mathbf{x}(t)$.

Frequency interpretation If the experiment is performed n times, then n functions $\mathbf{x}(t, \zeta_i)$ are observed, one for each trial (Fig. 9-2). Denoting by $n_t(x)$ the number of trials such that at time t the ordinates of the observed functions do not exceed x (solid lines), we conclude as in (4-2) that

$$F(x, t) \simeq \frac{n_t(x)}{n} \tag{9-4}$$

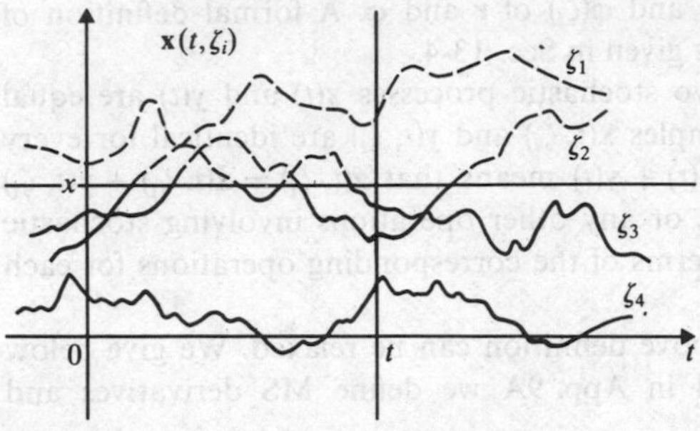

Figure 9-2

The *second-order distribution* of the process $\mathbf{x}(t)$ is the joint distribution

$$F(x_1, x_2; t_1, t_2) = P\{\mathbf{x}(t_1) \le x_1, \mathbf{x}(t_2) \le x_2\} \tag{9-5}$$

of the RVs $\mathbf{x}(t_1)$ and $\mathbf{x}(t_2)$. The corresponding density equals

$$f(x_1, x_2; t_1, t_2) = \frac{\partial^2 F(x_1, x_2; t_1, t_2)}{\partial x_1 \, \partial x_2} \tag{9-6}$$

We note that (consistency conditions)

$$F(x_1; t_1) = F(x_1, \infty; t_1, t_2) \qquad f(x_1, t_1) = \int_{-\infty}^{\infty} f(x_1, x_2; t_1, t_2)\, dx_2$$

as in (6-9) and (6-10).

The *nth-order distribution* of $\mathbf{x}(t)$ is the joint distribution $F(x_1, \ldots, x_n; t_1, \ldots, t_n)$ of the RVs $\mathbf{x}(t_1), \ldots, \mathbf{x}(t_n)$.

Second-order properties For the determination of the statistical properties of a stochastic process, knowledge of the function $F(x_1, \ldots, x_n; t_1, \ldots, t_n)$ is required for every x_i, t_i, and n. However, for many applications, only certain averages are used, in particular, the expected value of $\mathbf{x}(t)$ and of $\mathbf{x}^2(t)$. These quantities can be expressed in terms of the second-order properties of $\mathbf{x}(t)$ defined as follows:

Mean The mean $\eta(t)$ of $\mathbf{x}(t)$ is the expected value of the RV $\mathbf{x}(t)$

$$\eta(t) = E\{\mathbf{x}(t)\} = \int_{-\infty}^{\infty} x f(x, t)\, dx \tag{9-7}$$

Autocorrelation The autocorrelation $R(t_1, t_2)$ of $\mathbf{x}(t)$ is the expected value of the product $\mathbf{x}(t_1)\mathbf{x}(t_2)$

$$R(t_1, t_2) = E\{\mathbf{x}(t_1)\mathbf{x}(t_2)\} = \int_{-\infty}^{\infty}\int_{-\infty}^{\infty} x_1 x_2 f(x_1, x_2; t_1, t_2)\, dx_1\, dx_2 \tag{9-8}$$

The value of $R(t_1, t_2)$ on the diagonal $t_1 = t_2 = t$ is the *average power* of $\mathbf{x}(t)$

$$E\{\mathbf{x}^2(t)\} = R(t, t)$$

The *autocovariance* $C(t_1, t_2)$ of $\mathbf{x}(t)$ is the covariance of the RVs $\mathbf{x}(t_1)$ and $\mathbf{x}(t_2)$

$$C(t_1, t_2) = R(t_1, t_2) - \eta(t_1)\eta(t_2) \tag{9-9}$$

and its value $C(t, t)$ on the diagonal $t_1 = t_2 = t$ equals the variance of $\mathbf{x}(t)$.

Note The following is an explanation of the reason for introducing the function $R(t_1, t_2)$ even in problems dealing only with average power: Suppose that $\mathbf{x}(t)$ is the input to a linear system and $\mathbf{y}(t)$ is the resulting output. In Sec. 9-4 we show that the mean of $\mathbf{y}(t)$ can be expressed in terms of the mean of $\mathbf{x}(t)$. However, the average power of $\mathbf{y}(t)$ cannot be found if only $E\{\mathbf{x}^2(t)\}$ is given. For the determination of $E\{\mathbf{y}^2(t)\}$, knowledge of the function $R(t_1, t_2)$ is required, not just on the diagonal $t_1 = t_2$, but for every t_1 and t_2. The following identity is a simple illustration

$$E\{[\mathbf{x}(t_1) + \mathbf{x}(t_2)]^2\} = R(t_1, t_1) + 2R(t_1, t_2) + R(t_2, t_2)$$

This follows from (9-8) if we expand the square and use the linearity of expected values.

Example 9-1 An extreme example of a stochastic process is a deterministic signal $\mathbf{x}(t) = f(t)$. In this case

$$\eta(t) = E\{f(t)\} = f(t) \qquad R(t_1, t_2) = E\{f(t_1)f(t_2)\} = f(t_1)f(t_2)$$

Example 9-2 Suppose that $\mathbf{x}(t)$ is a process with

$$\eta(t) = 3 \qquad R(t_1, t_2) = 9 + 4e^{-0.2|t_1 - t_2|}$$

We shall determine the mean, the variance, and the covariance of the RVs $\mathbf{z} = \mathbf{x}(5)$ and $\mathbf{w} = \mathbf{x}(8)$.

Clearly, $E\{\mathbf{z}\} = \eta(5) = 3$ and $E\{\mathbf{w}\} = \eta(8) = 3$. Furthermore

$$E\{\mathbf{z}^2\} = R(5, 5) = 13 \qquad E\{\mathbf{w}^2\} = R(8, 8) = 13$$

$$E\{\mathbf{zw}\} = R(5, 8) = 9 + 4e^{-0.6} = 11.195$$

Thus, $\mathbf{z}$ and $\mathbf{w}$ have the same variance $\sigma^2 = 4$ and their covariance equals $C(5, 8) = 4e^{-0.6} = 2.195$.

Example 9-3 The integral

$$\mathbf{s} = \int_a^b \mathbf{x}(t)\, dt$$

of a stochastic process $\mathbf{x}(t)$ is an RV $\mathbf{s}$, and its value $\mathbf{s}(\zeta_i)$ for a specific outcome ζ_i is the area under the curve $\mathbf{x}(t, \zeta_i)$ in the interval (a, b) (see also App. 9A). Interpreting the above as a Riemann integral, we conclude from the linearity of expected values that

$$\eta_s = E\{\mathbf{s}\} = \int_a^b E\{\mathbf{x}(t)\}\, dt = \int_a^b \eta(t)\, dt \tag{9-10}$$

Similarly, since

$$\mathbf{s}^2 = \int_a^b \int_a^b \mathbf{x}(t_1)\mathbf{x}(t_2)\, dt_1\, dt_2$$

we conclude, using again the linearity of expected values, that

$$E\{\mathbf{s}^2\} = \int_a^b \int_a^b E\{\mathbf{x}(t_1)\mathbf{x}(t_2)\}\, dt_1\, dt_2 = \int_a^b \int_a^b R(t_1, t_2)\, dt_1\, dt_2 \qquad (9\text{-}11)$$

Example 9-4 We shall determine the autocorrelation $R(t_1, t_2)$ of the process

$$\mathbf{x}(t) = \mathbf{r} \cos (\omega t + \boldsymbol{\varphi})$$

where we assume that the RVs $\mathbf{r}$ and $\boldsymbol{\varphi}$ are independent and $\boldsymbol{\varphi}$ is uniform in the interval $(-\pi, \pi)$.

Using simple trigonometric identities, we find

$$E\{\mathbf{x}(t_1)\mathbf{x}(t_2)\} = \tfrac{1}{2}E\{\mathbf{r}^2\}E\{\cos \omega(t_1 - t_2) - \cos (\omega t_1 + \omega t_2 + 2\boldsymbol{\varphi})\}$$

and since

$$E\{\cos (\omega t_1 + \omega t_2 + 2\boldsymbol{\varphi})\} = \frac{1}{2\pi}\int_{-\pi}^{\pi} \cos (\omega t_1 + \omega t_2 + 2\varphi)\, d\varphi = 0$$

we conclude that

$$R(t_1, t_2) = \tfrac{1}{2}E\{\mathbf{r}^2\} \cos \omega(t_1 - t_2) \qquad (9\text{-}12)$$

Special Processes

We illustrate the preceding concepts with several basic examples.

Poisson process In Sec. 3-4 we introduced the concept of Poisson points and we showed that these points are specified by the following properties:

P_1: The number $\mathbf{n}(t_1, t_2)$ of the points $\mathbf{t}_i$ in an interval (t_1, t_2) of length $t = t_2 - t_1$ is a Poisson RV with parameter λt

$$P\{\mathbf{n}(t_1, t_2) = k\} = \frac{e^{-\lambda t}(\lambda t)^k}{k!} \qquad (9\text{-}13)$$

P_2: If the intervals (t_1, t_2) and (t_3, t_4) are nonoverlapping, then the RVs $\mathbf{n}(t_1, t_2)$ and $\mathbf{n}(t_3, t_4)$ are independent.

Using the points $\mathbf{t}_i$, we form the stochastic process

$$\mathbf{x}(t) = \mathbf{n}(0, t)$$

shown in Fig. 9-3*a*. This is a discrete-state process consisting of a family of increasing staircase functions with discontinuities at the points $\mathbf{t}_i$.

For a specific t, $\mathbf{x}(t)$ is a Poisson RV with parameter λt, hence

$$E\{\mathbf{x}(t)\} = \eta(t) = \lambda t$$

We shall show that the autocorrelation of $\mathbf{x}(t)$ equals

$$R(t_1, t_2) = \begin{cases} \lambda t_2 + \lambda^2 t_1 t_2 & t_1 \ge t_2 \\ \lambda t_1 + \lambda^2 t_1 t_2 & t_1 \le t_2 \end{cases} \qquad (9\text{-}14)$$

PROOF The above is true for $t_1 = t_2$ because [see (5-36)]

$$E\{\mathbf{x}^2(t)\} = \lambda t + \lambda^2 t^2 \tag{9-15}$$

Since $R(t_1, t_2) = R(t_2, t_1)$, it suffices to prove (9-14) for $t_1 < t_2$. The RVs $\mathbf{x}(t_1)$ and $\mathbf{x}(t_2) - \mathbf{x}(t_1)$ are independent because the intervals $(0, t_1)$ and (t_1, t_2) are non-overlapping. Furthermore, they are Poisson distributed with parameters λt_1 and $\lambda(t_2 - t_1)$ respectively. Hence

$$E\{\mathbf{x}(t_1)[\mathbf{x}(t_2) - \mathbf{x}(t_1)]\} = E\{\mathbf{x}(t_1)\}E\{\mathbf{x}(t_2) - \mathbf{x}(t_1)\} = \lambda t_1 \lambda(t_2 - t_1)$$

Using the identity

$$\mathbf{x}(t_1)\mathbf{x}(t_2) = \mathbf{x}(t_1)[\mathbf{x}(t_1) + \mathbf{x}(t_2) - \mathbf{x}(t_1)]$$

we conclude from the above and (9-15) that

$$R(t_1, t_2) = \lambda t_1 + \lambda^2 t_1^2 + \lambda t_1 \lambda(t_2 - t_1)$$

and (9-14) results.

Nonuniform case If the points $\mathbf{t}_i$ have a nonuniform density $\lambda(t)$ as in (3-54), then the preceding results still hold provided that the product $\lambda(t_2 - t_1)$ is replaced by the integral of $\lambda(t)$ from t_1 to t_2.

Thus

$$E\{\mathbf{x}(t)\} = \int_0^t \lambda(\alpha)\, d\alpha \tag{9-16}$$

and

$$R(t_1, t_2) = \int_0^{t_1} \lambda(t)\, dt \left[1 + \int_0^{t_2} \lambda(t)\, dt\right] \qquad t_1 \le t_2 \tag{9-17}$$

Telegraph signal Using the Poisson points $\mathbf{t}_i$, we form a process $\mathbf{x}(t)$ such that $\mathbf{x}(t) = 1$ if the number of points in the interval $(0, t)$ is even, and $\mathbf{x}(t) = -1$ if this number is odd (Fig. 9-3*b*).

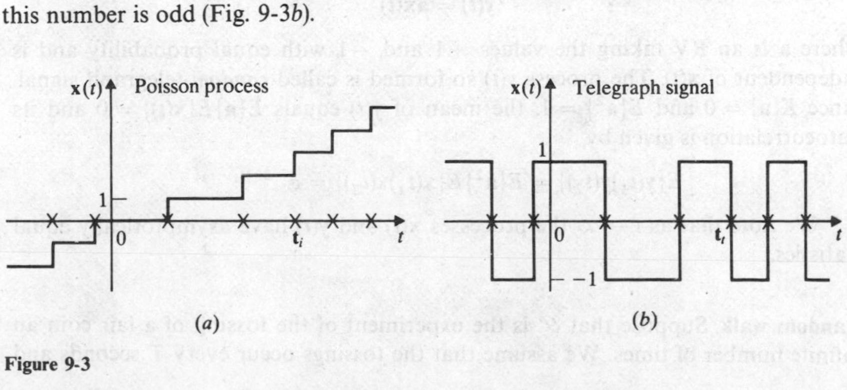

Figure 9-3

Denoting by $p(k)$ the probability that the number of points in the interval $(0, t)$ equals k, we conclude that [see (9-13)]

$$P\{\mathbf{x}(t) = 1\} = p(0) + p(2) + \cdots$$

$$= e^{-\lambda t}\left[1 + \frac{(\lambda t)^2}{2!} + \cdots\right] = e^{-\lambda t}\cosh \lambda t$$

$$P\{\mathbf{x}(t) = -1\} = p(1) + p(3) + \cdots$$

$$= e^{-\lambda t}\left[\lambda t + \frac{(\lambda t)^3}{3!} + \cdots\right] = e^{-\lambda t}\sinh \lambda t$$

Hence

$$E\{\mathbf{x}(t)\} = e^{-\lambda t}(\cosh \lambda t - \sinh \lambda t) = e^{-2\lambda t} \tag{9-18}$$

To determine $R(t_1, t_2)$, we note that, if $\mathbf{x}(t_1) = 1$, then $\mathbf{x}(t_2) = 1$ if the number of points in the interval (t_1, t_2) is even. Hence

$$P\{\mathbf{x}(t_2) = 1 \mid \mathbf{x}(t_1) = 1\} = e^{-\lambda t}\cosh \lambda t \qquad t = |t_2 - t_1|$$

Multiplying by $P\{\mathbf{x}(t_1) = 1\}$, we obtain

$$P\{\mathbf{x}(t_1) = 1, \quad \mathbf{x}(t_2) = 1\} = e^{-\lambda t}\cosh \lambda t e^{-\lambda t_2}\cosh \lambda t_2$$

Similarly

$$P\{\mathbf{x}(t_1) = -1, \mathbf{x}(t_2) = -1\} = e^{-\lambda t}\cosh \lambda t e^{-\lambda t_2}\sinh \lambda t_2$$

$$P\{\mathbf{x}(t_1) = 1, \mathbf{x}(t_2) = -1\} = e^{-\lambda t}\sinh \lambda t e^{-\lambda t_2}\sinh \lambda t_2$$

$$P\{\mathbf{x}(t_1) = -1, \mathbf{x}(t_2) = 1\} = e^{-\lambda t}\sinh \lambda t e^{-\lambda t_2}\cosh \lambda t_2$$

Since the product $\mathbf{x}(t_1)\mathbf{x}(t_2)$ equals 1 of -1, we conclude omitting details that

$$R(t_1, t_2) = e^{-2\lambda|t_1 - t_2|} \tag{9-19}$$

The above process is called *semirandom* telegraph signal because its value $\mathbf{x}(0) = 1$ at $t = 0$ is not random. To remove this certainty, we form the product

$$\mathbf{y}(t) = \mathbf{a}\mathbf{x}(t)$$

where $\mathbf{a}$ is an RV taking the values $+1$ and -1 with equal probability and is independent of $\mathbf{x}(t)$. The process $\mathbf{y}(t)$ so formed is called *random* telegraph signal. Since $E\{\mathbf{a}\} = 0$ and $E\{\mathbf{a}^2\} = 1$, the mean of $\mathbf{y}(t)$ equals $E\{\mathbf{a}\}E\{\mathbf{x}(t)\} = 0$ and its autocorrelation is given by

$$E\{\mathbf{y}(t_1)\mathbf{y}(t_2)\} = E\{\mathbf{a}^2\}E\{\mathbf{x}(t_1)\mathbf{x}(t_2)\} = e^{-2\lambda|t_1 - t_2|}$$

We note that as $t \to \infty$ the processes $\mathbf{x}(t)$ and $\mathbf{y}(t)$ have asymptotically equal statistics.

Random walk Suppose that $\mathscr{S}$ is the experiment of the tossing of a fair coin an infinite number of times. We assume that the tossings occur every T seconds and after each tossing we take instantly a step of length s, to the right if heads shows, to the left if tails shows. The process starts at $t = 0$ and our location at time t is a staircase function with discontinuities at the points $t = n$ (Fig. 9-4*a*). We have thus created a discrete-state stochastic process $\mathbf{x}(t)$ whose samples $\mathbf{x}(t, \zeta)$ depend on the particular sequence of heads and tails. This process is called *random walk*.

Suppose that at the first n tossings we observe k heads and $n - k$ tails. In this case, our walk consists of k steps to the right and $n - k$ steps to the left. Hence, our position at time $t = nT$ is

$$\mathbf{x}(nT) = ks - (n-k)s = ms \qquad m = 2k - n$$

Thus, $\mathbf{x}(nT)$ is an RV taking the values ms, where m equals n, or $n - 2, \ldots,$ or $-n$. Furthermore

$$P\{\mathbf{x}(nT) = ms\} = \binom{n}{k}\frac{1}{2^k} \qquad k = \frac{m+n}{2} \tag{9-20}$$

This is the probability of k heads in n tossings.

We note that $\mathbf{x}(nT)$ can be written as a sum

$$\mathbf{x}(nT) = \mathbf{x}_1 + \cdots + \mathbf{x}_n$$

where $\mathbf{x}_i$ equals the size of the ith step. Thus, the RVs $\mathbf{x}_i$ are independent taking the values $\pm s$ and $E\{\mathbf{x}_i\} = 0$, $E\{\mathbf{x}_i^2\} = s^2$. From this it follows that

$$E\{\mathbf{x}(nT)\} = 0 \qquad E\{\mathbf{x}^2(nT)\} = ns^2 \tag{9-21}$$

Large t As we know, if n is large and k is in the $\sqrt{npq}$ vicinity of np, then [see (3-27)]

$$\binom{n}{k}p^k q^{n-k} \simeq \frac{1}{\sqrt{2\pi npq}} e^{-(k-np)^2/2npq}$$

From this and (9-20) it follows with $p = q = 0.5$ and $m = 2k - n$ that

$$P\{\mathbf{x}(nT) = ms\} \simeq \frac{1}{\sqrt{n\pi/2}} e^{-m^2/2n}$$

for m of the order of $\sqrt{n}$. Hence

$$P\{\mathbf{x}(t) \le ms\} \simeq \mathbb{G}(m/\sqrt{n}) \qquad nT - T < t \le nT \tag{9-22}$$

where $\mathbb{G}(x)$ is the $N(0, 1)$ distribution [see (3-34)].

We note that if $n_1 < n_2 \le n_3 < n_4$ then the increments $\mathbf{x}(n_4 T) - \mathbf{x}(n_3 T)$ and $\mathbf{x}(n_2 T) - \mathbf{x}(n_1 T)$ of $\mathbf{x}(t)$ are independent.

The Wiener process We shall now examine the limiting form of the random walk as $n \to \infty$ or, equivalently, as $T \to 0$. As we have shown

$$E\{\mathbf{x}^2(t)\} = ns^2 = \frac{ts^2}{T} \qquad t = nT$$

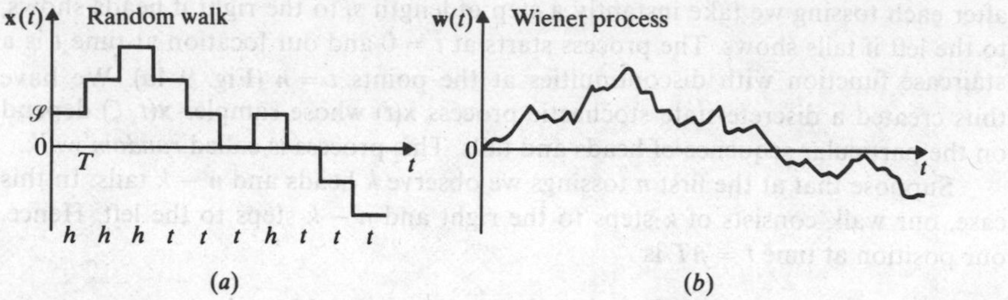

Figure 9-4

Hence, to obtain meaningful results, we shall assume that s tends to zero as $\sqrt{T}$

$$s^2 = \alpha T \tag{9-23}$$

The limit of $\mathbf{x}(t)$ as $T \to 0$ is then a continuous-state process (Fig. 9-4b)

$$\mathbf{w}(t) = \lim \mathbf{x}(t) \qquad T \to 0$$

known as the *Wiener process.*

We shall show that the first-order density $f(w, t)$ of $\mathbf{w}(t)$ is normal with zero mean and variance αt

$$f(\alpha, t) = \frac{1}{\sqrt{2\pi\alpha t}} e^{-w^2/2\alpha t} \tag{9-24}$$

PROOF If $w = ms$ and $t = mT$, then

$$\frac{m}{\sqrt{n}} = \frac{w/s}{\sqrt{t/T}} = \frac{w}{\sqrt{\alpha t}}$$

Inserting into (9-22), we conclude that

$$P\{\mathbf{w}(t) \le w\} = \mathbb{G}\left(\frac{w}{\sqrt{\alpha t}}\right)$$

and (9-24) results.

We show next that the autocorrelation of $\mathbf{w}(t)$ equals

$$R(t_1, t_2) = \alpha \min(t_1, t_2) \tag{9-25}$$

Indeed, if $t_1 < t_2$, then the difference $\mathbf{w}(t_2) - \mathbf{w}(t_1)$ is independent of $\mathbf{w}(t_1)$. Hence

$$E\{[\mathbf{w}(t_2) - \mathbf{w}(t_1)]\mathbf{w}(t_1)\} = E\{[\mathbf{w}(t_2) - \mathbf{w}(t_1)]\}E\{\mathbf{w}(t_1)\} = 0$$

This yields

$$E\{\mathbf{w}(t_1)\mathbf{w}(t_2)\} = E\{\mathbf{w}^2(t_1)\} = \frac{t_1 s^2}{T} = \alpha t_1$$

as in (9-25). The proof is similar if $t_1 > t_2$.

We note finally that if $t_1 < t_2 \le t_3 < t_4$ then the increments $\mathbf{w}(t_4) - \mathbf{w}(t_3)$ and $\mathbf{w}(t_2) - \mathbf{w}(t_1)$ of $\mathbf{w}(t)$ are independent.

9-2 DEFINITIONS

The statistical properties of a real stochastic process $\mathbf{x}(t)$ are completely determined† in terms of its nth-order distribution

$$F(x_1, \ldots, x_n; t_1, \ldots, t_n) = P\{\mathbf{x}(t_1) \le x_1, \ldots, \mathbf{x}(t_n) \le x_n\} \tag{9-26}$$

The joint statistics of two real processes $\mathbf{x}(t)$ and $\mathbf{y}(t)$ are determined in terms of the joint distribution of the RVs

$$\mathbf{x}(t_1), \ldots, \mathbf{x}(t_n), \mathbf{y}(t'_1), \ldots, \mathbf{y}(t'_m)$$

The *complex process* $\mathbf{z}(t) = \mathbf{x}(t) + j\mathbf{y}(t)$ is specified in terms of the joint statistics of the real processes $\mathbf{x}(t)$ and $\mathbf{y}(t)$.

A *vector process* (n-dimensional process) is a family of n stochastic processes.

Correlation and covariance The autocorrelation of a process $\mathbf{x}(t)$, real or complex, is by definition the mean of the product $\mathbf{x}(t_1)\mathbf{x}^*(t_2)$. This function, will be denoted by $R(t_1, t_2)$ or $R_x(t_1, t_2)$ or $R_{xx}(t_1, t_2)$. Thus

$$R_{xx}(t_1, t_2) = E\{\mathbf{x}(t_1)\mathbf{x}^*(t_2)\} \tag{9-27}$$

where the conjugate term is associated with the second variable in $R_{xx}(t_1, t_2)$. From this it follows that

$$R(t_2, t_1) = E\{\mathbf{x}(t_2)\mathbf{x}^*(t_1)\} = R^*(t_1, t_2) \tag{9-28}$$

We note, further, that

$$R(t, t) = E\{|\mathbf{x}(t)|^2\} \ge 0 \tag{9-29}$$

The last two equations are special cases of the following: The autocorrelation $R(t_1, t_2)$ of a stochastic process $\mathbf{x}(t)$ is a *positive definite* (p.d.) function, i.e., for any a_i and a_j

$$\sum_{i,j} a_i a_j^* R(t_i, t_j) \ge 0 \tag{9-30}$$

This is a consequence of the identity

$$0 \le E\left\{\left|\sum_i a_i \mathbf{x}(t_i)\right|^2\right\} = \sum_{i,j} a_i a_j^* E\{\mathbf{x}(t_i)\mathbf{x}^*(t_j)\}$$

We show later that the converse is also true: Given a p.d. function $R(t_1, t_2)$, we can find a process $\mathbf{x}(t)$ with autocorrelation $R(t_1, t_2)$.

Example 9-5 (*a*) If $\mathbf{x}(t) = \mathbf{a}e^{j\omega t}$ then

$$R(t_1, t_2) = E\{\mathbf{a}e^{j\omega t_1}\mathbf{a}^* e^{-j\omega t_2}\} = E\{|\mathbf{a}|^2\}e^{j\omega(t_1 - t_2)}$$

† There are processes (nonseparable) for which this is not true. However, such processes are mainly of mathematical interest.

(*b*) Suppose that the RVs $\mathbf{a}_i$ are uncorrelated with zero mean and variance σ_i^2. If

$$\mathbf{x}(t) = \sum_i \mathbf{a}_i e^{j\omega_i t}$$

then (9-27) yields

$$R(t_1, t_2) = \sum_i \sigma_i^2 e^{j\omega_i(t_1 - t_2)}$$

The *autocovariance* $C(t_1, t_2)$ of a process $\mathbf{x}(t)$ is the covariance of the RVs $\mathbf{x}(t_1)$ and $\mathbf{x}(t_2)$

$$C(t_1, t_2) = R(t_1, t_2) - |\eta(t)|^2 \tag{9-31}$$

In the above, $\eta(t) = E\{\mathbf{x}(t)\}$ is the *mean* of $\mathbf{x}(t)$.

The ratio

$$r(t_1, t_2) = \frac{C(t_1, t_2)}{\sqrt{C(t_1, t_1)C(t_2, t_2)}} \tag{9-32}$$

is the *correlation coefficient*† of the process $\mathbf{x}(t)$.

Note The autocovariance $C(t_1, t_2)$ of a process $\mathbf{x}(t)$ is the autocorrelation of the *centered process*

$$\tilde{\mathbf{x}}(t) = \mathbf{x}(t) - \eta(t)$$

Hence, it is p.d.

The correlation coefficient $r(t_1, t_2)$ of $\mathbf{x}(t)$ is the autocovariance of the *normalized process* $\mathbf{x}(t)/\sqrt{C(t, t)}$, hence it is also p.d. Furthermore [see (7-9)]

$$|r(t_1, t_2)| \le 1 \qquad r(t, t) = 1 \tag{9-33}$$

Example 9-6 If

$$\mathbf{s} = \int_a^b \mathbf{x}(t)\, dt \qquad \text{then} \qquad \mathbf{s} - \eta_s = \int_a^b \tilde{\mathbf{x}}(t)\, dt$$

where $\tilde{\mathbf{x}}(t) = \mathbf{x}(t) - \eta_x(t)$. Using (9-11), we conclude from the above note that

$$\sigma_s^2 = E\{|\mathbf{s} - \eta_s|^2\} = \int_a^b \int_a^b C_x(t_1, t_2)\, dt_1\, dt_2 \tag{9-34}$$

The *cross-correlation* of two processes $\mathbf{x}(t)$ and $\mathbf{y}(t)$ is the function

$$R_{xy}(t_1, t_2) = E\{\mathbf{x}(t_1)\mathbf{y}^*(t_2)\} = R_{yx}^*(t_2, t_1) \tag{9-35}$$

Similarly

$$C_{xy}(t_1, t_2) = R_{xy}(t_1, t_2) - \eta_x(t_1)\eta_y^*(t_2) \tag{9-36}$$

is their *cross-covariance*.

† In optics, $C(t_1, t_2)$ is called *coherence function* and $r(t_1, t_2)$ *complex degree of coherence*. (see P.4).

Two processes $\mathbf{x}(t)$ and $\mathbf{y}(t)$ are called (mutually) *orthogonal* if

$$R_{xy}(t_1, t_2) = 0 \qquad \text{for every } t_1 \text{ and } t_2 \tag{9-37}$$

They are called *uncorrelated* if

$$C_{xy}(t_1, t_2) = 0 \qquad \text{for every } t_1 \text{ and } t_2 \tag{9-38}$$

a-dependent processes In general, the values $\mathbf{x}(t_1)$ and $\mathbf{x}(t_2)$ of a stochastic process $\mathbf{x}(t)$ are statistically dependent for any t_1 and t_2. However, in most cases this dependence decreases as $|t_1 - t_2| \to \infty$. This leads to the following concept: A stochastic process $\mathbf{x}(t)$ is called *a-dependent* if all its values $\mathbf{x}(t)$ for $t < t_o$ and for $t > t_o + a$ are mutually *independent*. From this it follows that

$$C(t_1, t_2) = 0 \qquad \text{for} \qquad |t_1 - t_2| > a \tag{9-39}$$

A process $\mathbf{x}(t)$ is called *correlation a-dependent* if its autocorrelation satisfies (9-39). Clearly, if $\mathbf{x}(t)$ is correlation a-dependent, then any linear combination of its values for $t < t_o$ is uncorrelated with any linear combination of its values for $t > t_o + a$.

White noise We shall say that a process $\mathbf{v}(t)$ is white noise if its values $\mathbf{v}(t_i)$ and $\mathbf{v}(t_j)$ are uncorrelated for every t_i and $t_j \neq t_i$

$$C(t_1, t_2) = 0 \qquad t_i \neq t_j$$

As we explain later, the autocovariance of a nontrivial white noise process must be of the form

$$C(t_1, t_2) = q(t_1)\delta(t_1 - t_2) \qquad q(t) \geq 0 \tag{9-40}$$

If the RVs $\mathbf{v}(t_i)$ and $\mathbf{v}(t_j)$ are not only uncorrelated but also independent, then $\mathbf{v}(t)$ will be called *strictly* white noise.

Unless otherwise stated, it will be assumed that the mean of a white noise process is identically zero.

Example 9-7 Suppose that $\mathbf{v}(t)$ is white noise and

$$\mathbf{x}(t) = \int_0^t \mathbf{v}(\alpha)\, d\alpha \tag{9-41}$$

Inserting (9-40) into (9-11), we obtain

$$E\{\mathbf{x}^2(t)\} = \int_0^t \int_0^t q(t_1)\delta(t_1 - t_2)\, dt_1\, dt_2 = \int_0^t q(t_1)\, dt_1 \tag{9-42}$$

because

$$\int_0^t \delta(t_1 - t_2)\, dt_2 = 1 \qquad \text{for} \qquad 0 < t_1 < t$$

Uncorrelated and independent increments If the increments $\mathbf{x}(t_2) - \mathbf{x}(t_1)$ and $\mathbf{x}(t_4) - \mathbf{x}(t_3)$ of a process $\mathbf{x}(t)$ are uncorrelated (independent) for any $t_1 < t_2 \leq t_3 < t_4$, then we say that $\mathbf{x}(t)$ is a process with uncorrelated (independent) increments. The Poisson process and the Wiener process have independent increments. The integral (9-41) of white noise is a process $\mathbf{x}(t)$ with uncorrelated increments.

Independent processes If two processes $\mathbf{x}(t)$ and $\mathbf{y}(t)$ are such that the RVs $\mathbf{x}(t_1), \ldots, \mathbf{x}(t_n)$ and $\mathbf{y}(t'_1), \ldots, \mathbf{y}(t'_n)$ are mutually independent, then these processes are called independent.

Normal processes A process $\mathbf{x}(t)$ is called normal, if the RVs $\mathbf{x}(t_1), \ldots, \mathbf{x}(t_n)$ are jointly normal for any n and $t_1, \ldots, t_n$.

The statistics of a normal process are completely determined in terms of its mean $\eta(t)$ and autocovariance $C(t_1, t_2)$. Indeed, since

$$E\{\mathbf{x}(t)\} = \eta(t) \qquad \sigma_x^2(t) = C(t, t)$$

we conclude that the first-order density $f(x, t)$ of $\mathbf{x}(t)$ is the normal density $N[\eta(t); \sqrt{C(t, t)}]$.

Similarly, since the function $r(t_1, t_2)$ in (9-32) is the correlation coefficient of the RVs $\mathbf{x}(t_1)$ and $\mathbf{x}(t_2)$, the second-order density $f(x_1, x_2; t_1, t_2)$ of $\mathbf{x}(t)$ is the jointly normal density

$$N[\eta(t_1), \eta(t_2); \sqrt{C(t_1, t_1)}, \sqrt{C(t_2, t_2)}; r(t_1, t_2)]$$

The nth order characteristic function of the process $\mathbf{x}(t)$ is given by [see (8-60)]

$$\exp\left\{j \sum_i \eta(t_i)\omega_i - \frac{1}{2} \sum_{i,k} C(t_i, t_k)\omega_i \omega_k\right\} \tag{9-43}$$

Its inverse $f(x_1, \ldots, x_n; t_1, \ldots, t_n)$ is the nth-order density of $\mathbf{x}(t)$.

Existence theorem Given an arbitrary function $\eta(t)$ and a p.d. function $C(t_1, t_2)$, we can construct a normal process with mean $\eta(t)$ and autocovariance $C(t_1, t_2)$.

This follows if we use in (9-43) the given functions $\eta(t)$ and $C(t_1, t_2)$. The inverse of the resulting characteristic function is a density because the function $C(t_1, t_2)$ is p.d. by assumption.

Example 9-8 Suppose that $\mathbf{x}(t)$ is a normal process with

$$\eta(t) = 3 \qquad C(t_1, t_2) = 4e^{-0.2|t_1 - t_2|}$$

(*a*) Find the probability that $\mathbf{x}(5) \le 2$.

Clearly, $\mathbf{x}(5)$ is a normal RV with mean $\eta(5) = 3$ and variance $C(5, 5) = 4$. Hence

$$P\{\mathbf{x}(5) \le 2\} = \mathbb{G}(-1/2) = 0.309$$

(*b*) Find the probability that $|\mathbf{x}(8) - \mathbf{x}(5)| \le 1$.

The difference $\mathbf{s} = \mathbf{x}(8) - \mathbf{x}(5)$ is a normal RV with mean $\eta(8) - \eta(5) = 0$ and variance

$$C(8, 8) + C(5, 5) - 2C(8, 5) = 8(1 - e^{-0.6}) = 3.608$$

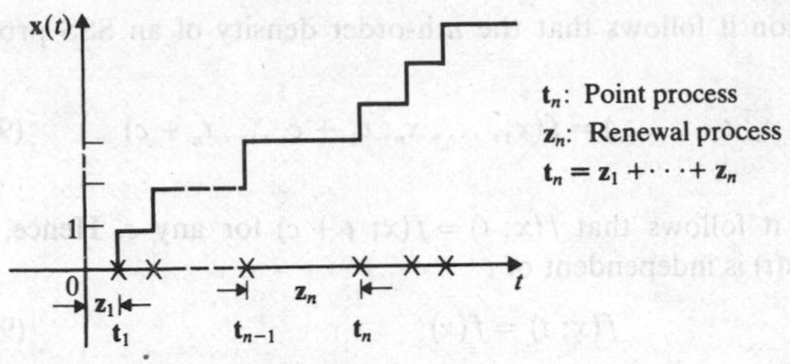

Figure 9-5

Hence

$$P\{|\mathbf{x}(8) - \mathbf{x}(5)| \le 1\} = 2\mathbb{G}(1/1.9) - 1 = 0.4$$

Point and renewal processes A *point process* is a set of random points $\mathbf{t}_i$ on the time axis.

To every point process we can associate a stochastic process $\mathbf{x}(t)$ equal to the number of points $\mathbf{t}_i$ in the interval $(0, t)$. An example is the Poisson process.

To every point process $\mathbf{t}_i$ we can associate a sequence of RVs $\mathbf{z}_n$ such that

$$\mathbf{z}_1 = \mathbf{t}_1 \qquad \mathbf{z}_2 = \mathbf{t}_2 - \mathbf{t}_1 \;\cdots\; \mathbf{z}_n = \mathbf{t}_n - \mathbf{t}_{n-1}$$

where $\mathbf{t}_1$ is the first random point to the right of the origin. This sequence is called a *renewal process*. An example is the life history of light bulbs that are replaced as soon as they fail. In this case, $\mathbf{z}_i$ is the total time the ith bulb is in operation and $\mathbf{t}_i$ is the time of its failure.

We have thus established a correspondence between the following three concepts (Fig. 9-5): (*a*) a point process $\mathbf{t}_i$, (*b*) a discrete-state stochastic process $\mathbf{x}(t)$ increasing in unit steps at the points $\mathbf{t}_i$, (*c*) a renewal process consisting of the RVs $\mathbf{z}_i$ and such that

$$\mathbf{t}_n = \mathbf{z}_1 + \cdots + \mathbf{z}_n$$

This correspondence is developed further in Sec. 12-1.

9-3 STATIONARY PROCESSES

A stochastic process $\mathbf{x}(t)$ is called *strict-sense stationary* (abbreviated SSS) if its statistical properties are invariant to a shift of the origin. This means that the processes $\mathbf{x}(t)$ and $\mathbf{x}(t+c)$ have the same statistics for any c.

Two processes $\mathbf{x}(t)$ and $\mathbf{y}(t)$ are called *jointly stationary* if the joint statistics of $\mathbf{x}(t)$ and $\mathbf{y}(t)$ are the same as the joint statistics of $\mathbf{x}(t + c)$ and $\mathbf{y}(t + c)$ for any c.

A complex process $\mathbf{z}(t) = \mathbf{x}(t) + j\mathbf{y}(t)$ is stationary if the processes $\mathbf{x}(t)$ and $\mathbf{y}(t)$ are jointly stationary.

From the definition it follows that the nth-order density of an SSS process must be such that

$$f(x_1, \ldots, x_n; t_1, \ldots, t_n) = f(x_1, \ldots, x_n; t_1 + c, \ldots, t_n + c) \tag{9-44}$$

for any c.

From the above it follows that $f(x; t) = f(x; t + c)$ for any c. Hence, the first-order density of $\mathbf{x}(t)$ is independent of t

$$f(x; t) = f(x) \tag{9-45}$$

Similarly, $f(x_1, x_2; t_1 + c, t_2 + c)$ is independent of c for any c. This leads to the conclusion that

$$f(x_1, x_2; t_1, t_2) = f(x_1, x_2; \tau) \qquad \tau = t_1 - t_2 \tag{9-46}$$

Thus, the joint density of the RVs $\mathbf{x}(t + \tau)$ and $\mathbf{x}(t)$ is independent of t and it equals $f(x_1, x_2; \tau)$.

Wide sense A stochastic process $\mathbf{x}(t)$ is called *wide-sense stationary* (abbreviated WSS) if its mean is constant

$$E\{\mathbf{x}(t)\} = \eta \tag{9-47}$$

and its autocorrelation depends only on $\tau = t_1 - t_2$

$$E\{\mathbf{x}(t + \tau)\mathbf{x}^*(t)\} = R(\tau) \tag{9-48}$$

Since τ is the distance from t to $t + \tau$, the function $R(\tau)$ can be written in the symmetrical form

$$R(\tau) = E\left\{\mathbf{x}\left(t + \frac{\tau}{2}\right)\mathbf{x}^*\left(t - \frac{\tau}{2}\right)\right\} \tag{9-49}$$

We note in particular that

$$E\{|\mathbf{x}(t)|^2\} = R(0)$$

Thus, the average power of a stationary process is independent of t and it equals $R(0)$.

Example 9-9 Suppose that $\mathbf{x}(t)$ is a WSS stationary process with autocorrelation

$$R(\tau) = Ae^{-\alpha|\tau|}$$

We shall determine the second moment of the RV $\mathbf{x}(8) - \mathbf{x}(5)$. Clearly

$$E\{[\mathbf{x}(8) - \mathbf{x}(5)]^2\} = E\{\mathbf{x}^2(8)\} + E\{\mathbf{x}^2(5)\} - 2E\{\mathbf{x}(8)\mathbf{x}(5)\}$$
$$= R(0) + R(0) - 2R(3) = 2A - 2Ae^{-3\alpha}$$

Note As the above example suggests, the autocorrelation of a stationary process $\mathbf{x}(t)$ can be defined as average power. Assuming for simplicity that $\mathbf{x}(t)$ is real, we conclude from (9-48) that

$$E\{[\mathbf{x}(t + \tau) - \mathbf{x}(t)]^2\} = 2[R(0) - R(\tau)] \tag{9-50}$$

From (9-48) it follows that the autocovariance of a WSS process depends only on $\tau = t_1 - t_2$

$$C(\tau) = R(\tau) - |\eta|^2 \tag{9-51}$$

and its correlation coefficient [see (9-32)] equals

$$r(\tau) = C(\tau)/C(0) \tag{9-52}$$

Thus, $C(\tau)$ is the covariance, and $r(\tau)$ the correlation coefficient of the RVs $\mathbf{x}(t + \tau)$ and $\mathbf{x}(t)$.

Two processes $\mathbf{x}(t)$ and $\mathbf{y}(t)$ are called jointly WSS if each is WSS and their cross-correlation depends only on $\tau = t_1 - t_2$

$$R_{xy}(\tau) = E\{\mathbf{x}(t + \tau)\mathbf{y}^*(t)\} \qquad C_{xy}(\tau) = R_{xy}(\tau) - \eta_x \eta_y^* \tag{9-53}$$

If $\mathbf{x}(t)$ is WSS white noise then [see (9-40)]

$$C(\tau) = q\delta(\tau) \tag{9-54}$$

If $\mathbf{x}(t)$ is an a-dependent process, then $C(\tau) = 0$ for $|\tau| > a$. In this case, the constant a is called the *correlation time* of $\mathbf{x}(t)$. This term is used also for arbitrary processes and it is defined as the ratio

$$\tau_c = \frac{1}{C(0)} \int_0^\infty C(\tau)\, d\tau \tag{9-55}$$

In general $C(\tau) \neq 0$ for every τ. However, for most regular processes

$$C(\tau) \xrightarrow[|\tau| \to \infty]{} 0 \qquad R(\tau) \xrightarrow[|\tau| \to \infty]{} |\eta|^2$$

Example 9-10 If $\mathbf{x}(t)$ is WSS and

$$\mathbf{s} = \int_{-T}^{T} \mathbf{x}(t)\, dt$$

then [see (9-34)]

$$\sigma_s^2 = \int_{-T}^{T} \int_{-T}^{T} C(t_1 - t_2)\, dt_1\, dt_2 = \int_{-2T}^{2T} (2T - |\tau|)C(\tau)\, d\tau \tag{9-56}$$

The last equality follows with $\tau = t_1 = t_2$ (see Fig. 9-6a); the details, however, are omitted [see also (10-44)].

SPECIAL CASES (a) If $C(\tau) = q\delta(\tau)$, then

$$\sigma_s^2 = q \int_{-2T}^{2T} (2T - |\tau|)\delta(\tau)\, d\tau = 2Tq$$

(b) If the process $\mathbf{x}(t)$ is a-dependent and $a \ll T$, then (see Fig. 9-6b), (9-56) yields

$$\sigma_s^2 = \int_{-2T}^{2T} (2T - |\tau|)C(\tau)\, d\tau \simeq 2T \int_{-a}^{a} C(\tau)\, d\tau$$

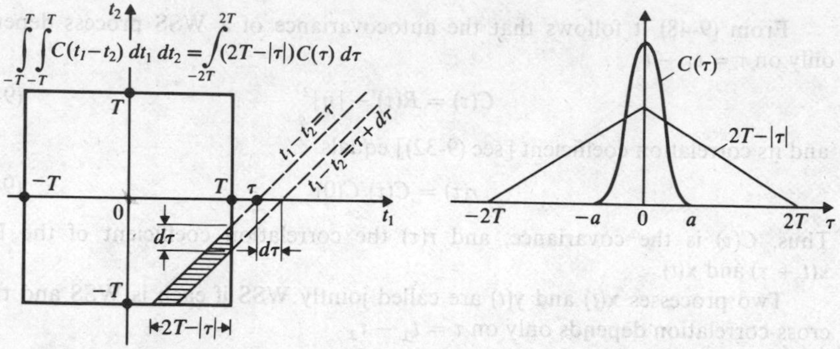

Figure 9-6

This shows that, in the evaluation of the variance of **s**, an a-dependent process with $a \ll T$ can be replaced by white noise with

$$q = \int_{-a}^{a} C(\tau)\,d\tau$$

If a process is SSS, then it is also WSS. This follows readily from (9-45) and (9-46). The converse, however, is not in general true. As we show next, normal processes are an important exception.

Indeed, suppose that $\mathbf{x}(t)$ is a normal WSS process with mean η and autocovariance $C(\tau)$. As we see from (9-43), its nth-order characteristic function equals

$$\exp\left\{ j\eta \sum_i \omega_i - \frac{1}{2}\sum_{i,k} C(t_i - t_k)\omega_i\omega_k \right\} \tag{9-57}$$

This function is invariant to a shift of the origin. And since it determines completely the statistics of $\mathbf{x}(t)$, we conclude that $\mathbf{x}(t)$ is SSS.

Example 9-11 We shall establish necessary and sufficient conditions for the stationarity of the process

$$\mathbf{x}(t) = \mathbf{a}\cos\omega t + \mathbf{b}\sin\omega t \tag{9-58}$$

The mean of this process equals

$$E\{\mathbf{x}(t)\} = E\{\mathbf{a}\}\cos\omega t + E\{\mathbf{b}\}\sin\omega t$$

This function must be independent of t. Hence, the condition

$$E\{\mathbf{a}\} = E\{\mathbf{b}\} = 0 \tag{9-59}$$

is necessary for both forms of stationarity. We shall assume that it holds.

Wide sense The process $\mathbf{x}(t)$ is WSS iff the RVs $\mathbf{a}$ and $\mathbf{b}$ are uncorrelated with equal variance

$$E\{\mathbf{ab}\} = 0 \qquad E\{\mathbf{a}^2\} = E\{\mathbf{b}^2\} = \sigma^2 \tag{9-60}$$

If this holds, then

$$R(\tau) = \sigma^2 \cos \omega\tau \tag{9-61}$$

PROOF If $\mathbf{x}(t)$ is WSS, then

$$E\{\mathbf{x}^2(0)\} = E\{\mathbf{x}^2(\pi/2\omega)\} = R(0)$$

But $\mathbf{x}(0) = \mathbf{a}$ and $\mathbf{x}(\pi/2\omega) = \mathbf{b}$, hence, $E\{\mathbf{a}^2\} = E\{\mathbf{b}^2\}$. Using the above, we obtain

$$\begin{aligned} E\{\mathbf{x}(t+\tau)\mathbf{x}(t)\} &= E\{[\mathbf{a}\cos\omega(t+\tau) + \mathbf{b}\sin\omega(t+\tau)][\mathbf{a}\cos\omega t + \mathbf{b}\sin\omega t]\} \\ &= \sigma^2\cos\omega\tau + E\{\mathbf{ab}\}\sin\omega(2t+\tau) \end{aligned} \tag{9-62}$$

This is independent of t only if $E\{\mathbf{ab}\} = 0$ and (9-60) results.

Conversely, if (9-60) holds, then, as we see from (9-62), the autocorrelation of $\mathbf{x}(t)$ equals $\sigma^2 \cos \omega\tau$, hence, $\mathbf{x}(t)$ is WSS.

Strict sense The process $\mathbf{x}(t)$ is SSS iff the joint density $f(a, b)$ of the RVs $\mathbf{a}$ and $\mathbf{b}$ has circular symmetry

$$f(a, b) = f(\sqrt{a^2 + b^2}) \tag{9-63}$$

PROOF If $\mathbf{x}(t)$ is SSS, then the RV

$$\mathbf{x}(0) = a \qquad \mathbf{x}(\pi/2\omega) = \mathbf{b}$$

and

$$\mathbf{x}(t) = \mathbf{a}\cos\omega t + \mathbf{b}\sin\omega t \qquad \mathbf{x}(t + \pi/2\omega) = \mathbf{b}\cos\omega t - \mathbf{a}\sin\omega t$$

have the same joint density for every t. Hence [see (6-70)], $f(a, b)$ must have circular symmetry.

We shall now show that, if $f(a, b)$ has circular symmetry, then $\mathbf{x}(t)$ is SSS. With τ a given number and

$$\mathbf{a}_1 = \mathbf{a}\cos\omega\tau + \mathbf{b}\sin\omega\tau \qquad \mathbf{b}_1 = \mathbf{b}\cos\omega\tau - \mathbf{a}\sin\omega\tau$$

we form the process

$$\mathbf{x}_1(t) = \mathbf{a}_1\cos\omega t + \mathbf{b}_1\sin\omega t = \mathbf{x}(t+\tau)$$

Clearly, the statistics of $\mathbf{x}(t)$ and $\mathbf{x}_1(t)$ are determined in terms of the joint densities $f(a, b)$ and $f(a_1, b_1)$ of the RVs $\mathbf{a}$, $\mathbf{b}$ and $\mathbf{a}_1$, $\mathbf{b}_1$ respectively. But [see (6-67)] the RVs $\mathbf{a}$, $\mathbf{b}$ and $\mathbf{a}_1$, $\mathbf{b}_1$ have the same joint density. Hence, the processes $\mathbf{x}(t)$ and $\mathbf{x}(t+\tau)$ have the same statistics for every τ.

Corollary If the process $\mathbf{x}(t)$ is SSS and the RVs $\mathbf{a}$ and $\mathbf{b}$ are independent, then they are normal.

PROOF It follows from (9-63) and (6-34).

Example 9-12 (*a*) Given an RV $\boldsymbol{\omega}$ with density $f(\omega)$ and an RV $\boldsymbol{\varphi}$ uniform in the interval $(-\pi, \pi)$ and independent of $\boldsymbol{\omega}$, we form the process

$$\mathbf{x}(t) = a \cos(\boldsymbol{\omega} t + \boldsymbol{\varphi}) \tag{9-64}$$

We shall show that $\mathbf{x}(t)$ is WSS with zero mean and autocorrelation

$$R(\tau) = \frac{a^2}{2} E\{\cos \boldsymbol{\omega}\tau\} = \frac{a^2}{2} \operatorname{Re} \Phi_\omega(\tau) \tag{9-65}$$

where

$$\Phi_\omega(\tau) = E\{e^{j\boldsymbol{\omega}\tau}\} = E\{\cos \boldsymbol{\omega}\tau\} + jE\{\sin \boldsymbol{\omega}\tau\} \tag{9-66}$$

is the characteristic function of $\boldsymbol{\omega}$.

PROOF Clearly [see (7-59)]

$$E\{\cos(\boldsymbol{\omega} t + \boldsymbol{\varphi})\} = E\{E\{\cos(\boldsymbol{\omega} t + \boldsymbol{\varphi}) \mid \boldsymbol{\omega}\}\}$$

From the independence of $\boldsymbol{\omega}$ and $\boldsymbol{\varphi}$ it follows that

$$E\{\cos(\boldsymbol{\omega} t + \boldsymbol{\varphi}) \mid \omega\} = \cos \omega t \, E\{\cos \boldsymbol{\varphi}\} - \sin \omega t \, E\{\sin \boldsymbol{\varphi}\}$$

Hence, $E\{\mathbf{x}(t)\} = 0$ because

$$E\{\cos \boldsymbol{\varphi}\} = \frac{1}{2\pi} \int_{-\pi}^{\pi} \cos \varphi \, d\varphi = 0 \qquad E\{\sin \boldsymbol{\varphi}\} = \frac{1}{2\pi} \int_{-\pi}^{\pi} \sin \varphi \, d\varphi = 0$$

Reasoning similarly, we obtain $E\{\cos(2\boldsymbol{\omega} t + \boldsymbol{\omega}\tau + 2\boldsymbol{\varphi})\} = 0$. And since

$$2 \cos[\boldsymbol{\omega}(t + \tau) + \boldsymbol{\varphi}] \cos(\boldsymbol{\omega} t + \boldsymbol{\varphi}) = \cos \boldsymbol{\omega}\tau + \cos(2\boldsymbol{\omega} t + \boldsymbol{\omega}\tau + 2\boldsymbol{\varphi})$$

we conclude that

$$R(\tau) = a^2 E\{\cos[\boldsymbol{\omega}(t + \tau) + \boldsymbol{\varphi}] \cos(\boldsymbol{\omega} t + \boldsymbol{\varphi})\} = \frac{a^2}{2} E\{\cos \boldsymbol{\omega}\tau\}$$

(*b*) With $\boldsymbol{\omega}$ and $\boldsymbol{\varphi}$ as above, the process

$$\mathbf{z}(t) = e^{j(\boldsymbol{\omega} t + \boldsymbol{\varphi})}$$

is WSS with zero mean and autocorrelation

$$E\{\mathbf{z}(t + \tau)\mathbf{z}^*(t)\} = E\{e^{j\boldsymbol{\omega}\tau}\} = \Phi_\omega(\tau) \tag{9-67}$$

Nearly WSS processes Consider a WSS process $\tilde{\mathbf{x}}(t)$ with zero mean and autocorrelation $\tilde{R}(\tau)$.

(*a*) With $\eta(t)$ an arbitrary function we form the process $\mathbf{x}(t) = \tilde{\mathbf{x}}(t) + \eta(t)$. This process is not WSS because its mean $\eta(t)$ is not constant. However, its autocovariance depends only on $\tau = t_1 - t_2$

$$E\{[\mathbf{x}(t_1) - \eta(t_1)][\mathbf{x}(t_2) - \eta(t_2)]\} = E\{\tilde{\mathbf{x}}_1(t_1)\tilde{\mathbf{x}}(t_2)\} = \tilde{R}(t_1 - t_2)$$

Conversely, if $\mathbf{x}(t)$ is a process such that

$$E\{\mathbf{x}(t)\} = \eta(t) \qquad C_{xx}(t_1, t_2) = C_{xx}(t_1 - t_2)$$

then the centered process $\tilde{\mathbf{x}}(t) = \mathbf{x}(t) - \eta(t)$ is WSS with zero mean and autocorrelation $C_{xx}(\tau)$.

(*b*) With $f(t)$ an arbitrary function, we form the process $\mathbf{y}(t) = f(t)\tilde{\mathbf{x}}(t)$. Clearly, $E\{\mathbf{y}(t)\} = 0$ and

$$R_{yy}(t_1, t_2) = f(t_1)f(t_2)E\{\tilde{\mathbf{x}}(t_1)\tilde{\mathbf{x}}(t_2)\} = f(t_1)f(t_2)\tilde{R}(t_1 - t_2)$$

Thus, $\mathbf{y}(t)$ is not WSS. However, the ratio $\mathbf{y}(t)/f(t)$ is.

Conversely, if $\mathbf{y}(t)$ is a process such that

$$E\{\mathbf{y}(t)\} = 0 \qquad R_{yy}(t_1, t_2) = f(t_1)f(t_2)r(t_1 - t_2)$$

where $r(t_1 - t_2)$ is a p.d. function, then the normalized process $\bar{\mathbf{y}}(t) = \mathbf{y}(t)/f(t)$ is WSS with zero mean and correlation $r(\tau)$.

Special case If $\mathbf{y}(t)$ is white noise with autocorrelation

$$R_{yy}(t_1, t_2) = q(t_1)\delta(t_1 - t_2) \qquad q(t) > 0$$

then the process $\bar{\mathbf{y}}(t) = \mathbf{y}(t)/\sqrt{q(t)}$ is WSS white noise with autocorrelation $\delta(\tau)$.

Other forms of stationarity A process $\mathbf{x}(t)$ is *asymptotically stationary* if the statistics of the RVs $\mathbf{x}(t_1 + c), \ldots, \mathbf{x}(t_n + c)$ do not depend on c if c is large. More precisely, the function

$$f(x_1, \ldots, x_n; t_1 + c, \ldots, t_n + c)$$

tends to a limit (that does not depend on c) as $c \to \infty$. The semirandom telegraph signal is an example.

A process $\mathbf{x}(t)$ is *Nth-order stationary* if (9-44) holds not for every n, but only for $n \le N$.

A process $\mathbf{x}(t)$ is *stationary in an interval* if (9-44) holds for every t_i and $t_i + c$ in this interval.

We say that $\mathbf{x}(t)$ is a process with *stationary increments* if its increments $\mathbf{y}(t) = \mathbf{x}(t + h) - \mathbf{x}(t)$ form a stationary process for every h. The Poisson process and the Wiener process are examples.

Discrete-time processes The discrete-time version of the preceding results is self-evident. We give below several illustrations.

If $\mathbf{x}[n]$ is a discrete-time process, then its mean autocorrelation and auto-covariance are, by definition,

$$\eta[n] = E\{\mathbf{x}[n]\} \qquad R[n_1, n_2] = E\{\mathbf{x}[n_1]\mathbf{x}^*[n_2]\}$$

and

$$C[n_1, n_2] = R[n_1, n_2] - \eta[n_1]\eta^*[n_2]$$

respectively.

A process $\mathbf{x}[n]$ is white noise if the RVs $\mathbf{x}[n_1]$ and $\mathbf{x}[n_2]$ are uncorrelated for every $n_1 \ne n_2$. If $\mathbf{x}[n]$ is white noise, then

$$C[n_1, n_2] = q[n_1]\delta[n_1 - n_2] \qquad \text{where} \qquad \delta[n] = \begin{cases} 1 & n \ne 0 \\ 0 & n = 0 \end{cases}$$

is the delta sequence and $q[n] = E\{\mathbf{x}^2[n]\}$.

If $\mathbf{x}[n]$ is a WSS process, then $E\{\mathbf{x}[n]\} = \eta$ and

$$E\{\mathbf{x}[n+m]\mathbf{x}^*[n]\} = R[m]$$

Sampling Suppose that $\mathbf{x}[n]$ equals the time-samples $\mathbf{x}(nT)$ of a continuous-time process $\mathbf{x}(t)$

$$\mathbf{x}[n] = \mathbf{x}(nT)$$

In this case the statistics of $\mathbf{x}[n]$ can be determined in terms of the statistics of $\mathbf{x}(t)$. We note, in particular, that

$$\eta[n] = \eta(nT) \qquad R[n_1, n_2] = R(n_1 T, n_2 T) \tag{9-68}$$

where $\eta(t)$ is the mean and $R(t_1, t_2)$ the autocorrelation of $\mathbf{x}(t)$.

If $\mathbf{x}(t)$ is SSS, then $\mathbf{x}[n]$ is SSS. If $\mathbf{x}(t)$ is WSS with mean η and autocorrelation $R(\tau)$, then $\mathbf{x}[n]$ is WSS with mean η and autocorrelation

$$R[m] = R(mT) \tag{9-69}$$

The converse of the above is not true. In fact, as we show next, $\mathbf{x}(t)$ might not be stationary even if, not only the sequence $\mathbf{x}(nT)$, but also all sequences $\mathbf{x}(nT + c)$ are stationary for any c.

Cyclostationary Processes

A process $\mathbf{x}(t)$ is called *cyclostationary*† (or *periodically stationary*) if its statistics are invariant to a shift of the time origin by integral multiples of a constant T (period). From this it follows that the distribution function of $\mathbf{x}(t)$ satisfies the equation

$$F(x_1, \ldots, x_n; t_1 + mT, \ldots, t_n + mT) = F(x_1, \ldots, x_n; t_1, \ldots, t_n) \tag{9-70}$$

for any integer m.

A cyclostationary process is not stationary because (9-44) holds not for every c but only for $c = mT$. However, the discrete-time process $\mathbf{x}(nT + c)$ is stationary for any c. This suggests a close relationship between stationary and cyclostationary processes. In fact, if the origin is subjected to a random shift $\boldsymbol{\theta}(t)$ that varies sufficiently slowly, then the resulting process $\mathbf{x}[t - \boldsymbol{\theta}(t)]$ is practically stationary. The following theorem establishes a precise model for this equivalence.

Theorem If $\mathbf{x}(t)$ is a strict-sense (SS) cyclostationary process and $\boldsymbol{\theta}$ is an RV uniform in the interval $(0, T)$ and independent of $\mathbf{x}(t)$, then the process

$$\bar{\mathbf{x}}(t) = \mathbf{x}(t - \boldsymbol{\theta})$$

is SSS and its nth-order distribution equals

$$\bar{F}(x_1, \ldots, x_n; t_1, \ldots, t_n) = \frac{1}{T}\int_0^T F(x_1, \ldots, x_n; t_1 - \alpha, \ldots, t_n - \alpha)\, d\alpha \tag{9-71}$$

† N. A. Gardner and L. E. Franks: Characterization of Cyclostationary Random Signal Processes, *IEEE Transactions in Information Theory*, vol. IT-21, 1975.

PROOF To prove the theorem, it suffices to show that the probability of the event

$$\mathscr{A} = \{\bar{\mathbf{x}}(t_1 + c) \le x_1, \ldots, \bar{\mathbf{x}}(t_n + c) \le x_n\}$$

is independent of c and it equals the integral in (9-71). As we know [see (4-62)]

$$P(\mathscr{A}) = \frac{1}{T}\int_0^T P(\mathscr{A} \mid \boldsymbol{\theta} = \theta)\, d\theta \tag{9-72}$$

Furthermore

$$\begin{aligned} P(\mathscr{A} \mid \boldsymbol{\theta} = \theta) &= P\{\mathbf{x}(t_1 + c - \theta) \le x_1, \ldots, \mathbf{x}(t_n + c - \theta) \le x_n\} \\ &= F(x_1, \ldots, x_n; t_1 + c - \theta, \ldots, t_n + c - \theta) \end{aligned}$$

Inserting into (9-72), we obtain

$$P(\mathscr{A}) = \frac{1}{T}\int_0^T F(x_1, \ldots, x_n; t_1 + c - \theta, \ldots, t_n + c - \theta)\, d\theta$$

and (9-71) results [see (9-70)].

Wide sense A process $\mathbf{x}(t)$ is called wide-sense (WS) cyclostationary if

$$\eta(t + mT) = \eta(t) \tag{9-73}$$

and

$$R(t + mT + \tau, t + mT) = R(t + \tau, t) \tag{9-74}$$

Thus, the autocorrelation $R(t_1, t_2)$ of a WS cyclostationary process is periodic on the diagonal of the $t_1 - t_2$ plane

$$R(t_1 + mT, t_2 + mT) = R(t_1, t_2) \tag{9-75}$$

Note If $\mathbf{x}(t)$ is SS cyclostationary, then

$$f(x; t + mT) = f(x) \qquad f(x_1, x_2; t_1 + mT, t_2 + mT) = f(x_1, x_2; t_1, t_2)$$

Hence, it is also WS cyclostationary. The converse, however, is not true.

Theorem If the process $\mathbf{x}(t)$ is WS cyclostationary, then the shifted process $\bar{\mathbf{x}}(t) = \mathbf{x}(t - \boldsymbol{\theta})$ is WSS with mean

$$\bar{\eta} = \frac{1}{T}\int_0^T \eta(t)\, dt \tag{9-76}$$

and autocorrelation

$$\bar{R}(\tau) = \frac{1}{T}\int_0^T R(t + \tau, t)\, dt \tag{9-77}$$

PROOF From (7-59) and the independence of $\boldsymbol{\theta}$ from $\mathbf{x}(t)$ it follows that

$$E\{\mathbf{x}(t - \boldsymbol{\theta})\} = E\{E\{\mathbf{x}(t - \boldsymbol{\theta}) \mid \boldsymbol{\theta}\}\} = \frac{1}{T}\int_0^T \eta(t - \theta)\, d\theta$$

and (9-76) results because $\eta(t + T) = \eta(T)$. Similarly,

$$E\{\mathbf{x}(t + \tau - \boldsymbol{\theta})\mathbf{x}(t - \boldsymbol{\theta})\} = E\{R(t + \tau - \boldsymbol{\theta}, t - \boldsymbol{\theta})\}$$

$$= \frac{1}{T} \int_0^T R(t + \tau - \theta, t - \theta)\, d\theta$$

and (9-77) follows.

Note Since the function $R(t + \tau, t)$ is periodic in t, the integral in (9-77) can be written as a limit

$$\bar{R}(\tau) = \lim_{c \to \infty} \frac{1}{2c} \int_{-c}^{c} R(t + \tau, t)\, dt \tag{9-78}$$

This limit is defined also for arbitrary processes and it is called the *time-average* autocorrelation of the process $\mathbf{x}(t)$.

Example 9-13 Suppose that $f(t)$ is a periodic function with period T and

$$\mathbf{x}(t) = f(t)$$

In this case, $\mathbf{x}(t)$ is an SS cyclostationary process (deterministic) with mean $f(t)$ and autocorrelation $f(t_1)f(t_2)$. From the preceding discussion it follows that the shifted process

$$\bar{\mathbf{x}}(t) = f(t - \boldsymbol{\theta})$$

is SSS with mean

$$E\{\bar{\mathbf{x}}(t)\} = \frac{1}{T} \int_0^T f(t - \theta)\, d\theta$$

and autocorrelation

$$\bar{R}(\tau) = \frac{1}{T} \int_0^T f(t + \tau) f(t)\, dt$$

Example 9-14 (binary transmission) Suppose that $\mathbf{x}(t)$ is a process taking the values ± 1 with equal probability in each interval

$$T_n\colon (n - 1)T \le t < nT$$

of length T (Fig. 9-7*a*) and that its values in two different intervals are independent.

From the definition it follows that $\mathbf{x}(t)$ is a cyclostationary process with zero mean. Furthermore, $E\{\mathbf{x}(t_1)\mathbf{x}(t_2)\} = 1$ only if the variables t_1 and t_2 are in the same interval T_n; otherwise, $E\{\mathbf{x}(t_1)\mathbf{x}(t_2)\} = 0$. Hence

$$R(t + \tau, t) = \begin{cases} 1 & (n - 1)T \le t,\ t + \tau < nT \\ 0 & \text{otherwise} \end{cases} \tag{9-79}$$

We shall show that the autocorrelation of the shifted process $\mathbf{x}(t - \boldsymbol{\theta})$ is a triangle as in Fig. 9-7*b*

$$\bar{R}(\tau) = 1 - \frac{|\tau|}{T} \qquad |\tau| \le T \tag{9-80}$$

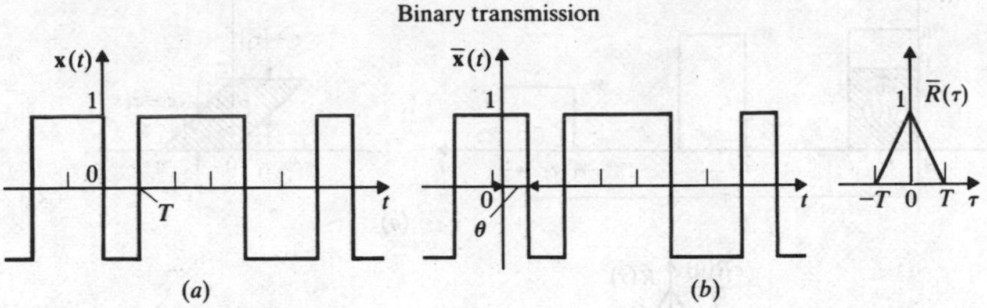

Figure 9-7

PROOF Since the limits of integration in (9-77) are 0 and T, it suffices to assume that $0 \le t \le T$. If $\tau > 0$, then $R(t+\tau, t) = 1$ only if $\tau < T$ and $t < T - \tau$. Hence

$$\bar{R}(\tau) = \frac{1}{T}\int_0^{T-\tau} d\alpha = 1 - \frac{\tau}{T} \qquad 0 \le \tau \le T$$

If $\tau < 0$, then $R(t+\tau, t) = 1$ only if $\tau > -T$ and $t > T + \tau = T - |\tau|$. Hence

$$\bar{R}(\tau) = \frac{1}{T}\int_{|\tau|}^{T-|\tau|} d\alpha = 1 - \frac{|\tau|}{T} \qquad -T \le \tau \le 0$$

and (9-80) results.

Pulse amplitude modulation (PAM) The last two examples are special cases of the PAM process

$$\mathbf{x}(t) = \sum_{n=-\infty}^{\infty} \mathbf{w}_n h(t - nT) \tag{9-81}$$

consisting of a stationary sequence $\mathbf{w}_n$ of RVs with autocorrelation

$$R[m] = E\{\mathbf{w}_{n+m}\mathbf{w}_n\}$$

and an arbitrary function $h(t)$. The general properties of such processes are developed later. In the following, we discuss a special case where we assume that $h(t)$ is a rectangular pulse of width $c < T/2$ as in Fig. 9-8a.

From the stationarity of the sequence $\mathbf{w}_n$ it follows that the PAM process $\mathbf{x}(t)$ is cyclostationary and its shifted version $\bar{\mathbf{x}}(t)$ is stationary. We shall show that the autocorrelation $\bar{R}(\tau)$ of $\bar{\mathbf{x}}(t)$ is a train of triangles of base $2c$ and height $cR[m]/T$ as in Fig. 9-8b

$$\bar{R}(\tau) = \frac{R[m]}{T}(c - |\tau_1|) \qquad |\tau_1| < c \tag{9-82}$$

In the above, $\tau_1 = \tau - mT$ is the distance from τ to the nearest lattice point mT.

PROOF It suffices to assume that $0 \le t < T$. Clearly $|\tau_1| < T/2$, hence the product $\mathbf{x}(t+\tau)\mathbf{x}(t)$ equals $\mathbf{w}_0\mathbf{w}_m$ if the point (t, τ_1) is in the shaded area P of

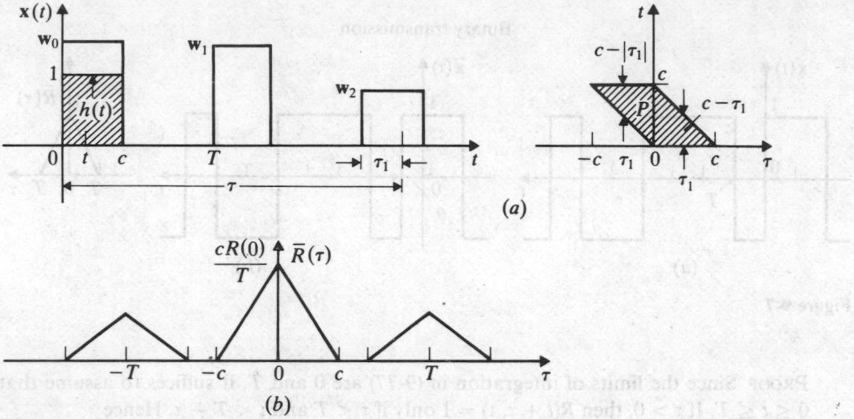

Figure 9-8

Fig. 9-8c and it equals zero otherwise. Hence

$$R(t+\tau, t) = \begin{cases} E\{\mathbf{w}_0\mathbf{w}_m\} = R[m] & (t, \tau_1) \in P \\ 0 & \text{otherwise} \end{cases}$$

Inserting into (9-77), we obtain (9-82).

Mean square periodicity A process $\mathbf{x}(t)$ is called MS periodic if

$$E\{|\mathbf{x}(t+T) - \mathbf{x}(t)|^2\} = 0 \tag{9-83}$$

for every t. From this it follows that, for a specific t,

$$\mathbf{x}(t+T) = \mathbf{x}(t) \tag{9-84}$$

with probability 1. It does not, however, follow that the set of outcomes ζ such that $\mathbf{x}(t+T, \zeta) = \mathbf{x}(t, \zeta)$ for all t has probability 1.

As we see from (9-84) the mean of an MS periodic process is periodic. We shall examine the properties of $R(t_1, t_2)$.

Theorem A process $\mathbf{x}(t)$ is MS periodic iff its autocorrelation is *doubly periodic*, i.e., if

$$R(t_1 + mT, t_2 + nT) = R(t_1, t_2) \tag{9-85}$$

for every integer m and n.

Proof As we know [see (7-12)]

$$E^2\{\mathbf{zw}\} \le E\{\mathbf{z}^2\}E\{\mathbf{w}^2\}$$

With $\mathbf{z} = \mathbf{x}(t_1)$ and $\mathbf{w} = \mathbf{x}(t_2 + T) - \mathbf{x}(t_2)$ the above yields

$$E^2\{\mathbf{x}(t_1)[\mathbf{x}(t_2+T) - \mathbf{x}(t_2)]\} \le E\{\mathbf{x}^2(t_1)\}E\{[\mathbf{x}(t_2+T) - \mathbf{x}(t_2)]^2\}$$

If $\mathbf{x}(t)$ is MS periodic, then the last term above is zero. Equating the left side to zero, we obtain

$$R(t_1, t_2 + T) - R(t_1, t_2) = 0$$

Repeated application of this yields (9-85).

Conversely, if (9-85) is true, then

$$R(t + T, t + T) = R(t + T, t) = R(t, t)$$

Hence

$$E\{[\mathbf{x}(t + T) - \mathbf{x}(t)]^2\} = R(t + T, t + T) + R(t, t) - 2R(t + T, t) = 0$$

therefore $\mathbf{x}(t)$ is MS periodic.

Notes 1. An MS periodic process is WS cyclostationary; however, in general, it is not SS cyclostationary.

2. The autocorrelation $R(t_1, t_2)$ of a cyclostationary process is not, in general, doubly periodic because it does not satisfy (9-85) for any m and n but only for $m = n$.

3. A cyclostationary process is not, therefore, MS periodic. In fact, its values $\mathbf{x}(t + T)$ and $\mathbf{x}(t)$ might not be equal in any sense, even though they have the same statistics.

9-4 SYSTEMS WITH STOCHASTIC INPUTS

Given a stochastic process $\mathbf{x}(t)$, we assign according to some rule to each of its samples $\mathbf{x}(t, \zeta_i)$ a function $\mathbf{y}(t, \zeta_i)$. We have thus created another process

$$\mathbf{y}(t) = T[\mathbf{x}(t)]$$

whose samples are the functions $\mathbf{y}(t, \zeta_i)$. The process $\mathbf{y}(t)$ so formed can be considered as the output of a *system* (transformation) with input the process $\mathbf{x}(t)$. The system is completely specified in terms of the operator T, that is, the rule of correspondence between the samples of the input $\mathbf{x}(t)$ and the output $\mathbf{y}(t)$.

The system is *deterministic* if it operates only on the variable t treating ζ as a parameter. This means that if two samples $\mathbf{x}(t, \zeta_1)$ and $\mathbf{x}(t, \zeta_2)$ of the input are identical in t, then the corresponding samples $\mathbf{y}(t, \zeta_1)$ and $\mathbf{y}(t, \zeta_2)$ of the output are also identical in t.

The system is called *stochastic* if T operates on both variables t and ζ. This means that there exist two outcomes ζ_1 and ζ_2 such that $\mathbf{x}(t, \zeta_1) = \mathbf{x}(t, \zeta_2)$ identically in t by $\mathbf{y}(t, \zeta_1) \neq \mathbf{y}(t, \zeta_2)$.

These classifications are based on the terminal properties of the system. If the system is specified in terms of physical elements or by an equation, then it is deterministic (stochastic) if the elements or the coefficients of the defining equation are deterministic (stochastic).

Throughout this book we shall consider only deterministic systems.

In principle, the statistics of the output of a system can be expressed in terms of the statistics of the input. However, in general this is a complicated problem. We consider next two important special cases.

Memoryless Systems

A system is called memoryless if its output is given by

$$\mathbf{y}(t) = g[\mathbf{x}(t)]$$

where $g(x)$ is a function of x. Thus, at a given time $t = t_1$, the output $\mathbf{y}(t_1)$ depends only on $\mathbf{x}(t_1)$ and not on any other past or future values of $\mathbf{x}(t)$.

From the above it follows that the first-order density $f_y(y; t)$ of $\mathbf{y}(t)$ can be expressed in terms of the corresponding density $f_x(x; t)$ of $\mathbf{x}(t)$ as in Sec. 5-2. We note, in particular, that

$$E\{\mathbf{y}(t)\} = \int_{-\infty}^{\infty} g(x) f_x(x; t)\, dx$$

Similarly, since $\mathbf{y}(t_1) = g[\mathbf{x}(t_1)]$ and $\mathbf{y}(t_2) = g[\mathbf{x}(t_2)]$, the second-order density $f_y(y_1, y_2; t_1, t_2)$ of $\mathbf{y}(t)$ can be determined in terms of the corresponding density $f_x(x_1, x_2; t_1, t_2)$ of $\mathbf{x}(t)$ as in Sec. 6-3. In particular

$$E\{\mathbf{y}(t_1)\mathbf{y}(t_2)\} = \int_{-\infty}^{\infty}\int_{-\infty}^{\infty} x_1 x_2\, g(x_1) g(x_2) f_x(x_1, x_2; t_1, t_2)\, dx_1\, dx_2$$

The nth order density $f_y(y_1, \ldots, y_n; t_1, \ldots, t_n)$ of $\mathbf{y}(t)$ can be determined from the corresponding density of $\mathbf{x}(t)$ as in (8-8) where the underlying transformation is the system

$$\mathbf{y}(t_1) = g[\mathbf{x}(t_1)], \ldots, \mathbf{y}(t_n) = g[\mathbf{x}(t_n)] \tag{9-86}$$

Stationarity Suppose that the input to a memoryless system is an SSS process $\mathbf{x}(t)$. We shall show that the resulting output $\mathbf{y}(t)$ is also SSS.

PROOF To determine the nth order density of $\mathbf{y}(t)$, we solve the system

$$g(x_1) = y_1, \ldots, g(x_n) = y_n \tag{9-87}$$

If this system has a unique solution, then [see (8-8)]

$$f_y(y_1, \ldots, y_n; t_1, \ldots, t_n) = \frac{f_x(x_1, \ldots, x_n; t_1, \ldots, t_n)}{|g'(x_1) \cdots g'(x_n)|} \tag{9-88}$$

From the stationarity of $\mathbf{x}(t)$ it follows that the numerator in (9-88) is invariant to a shift of the time origin. And since the denominator does not depend on t, we conclude that the left side does not change if t_i is replaced by $t_i + c$. Hence, $\mathbf{y}(t)$ is SSS. We can similarly show that this is true even if (9-87) has more than one solution.

Notes 1. If $\mathbf{x}(t)$ is stationary of order N, then $\mathbf{y}(t)$ is stationary of order N.

2. If $\mathbf{x}(t)$ is stationary in an interval, then $\mathbf{y}(t)$ is stationary in the same interval.

3. If $\mathbf{x}(t)$ is WSS stationary, then $\mathbf{y}(t)$ might not be stationary in any sense.

Square-law detector A square-law detector is a memoryless system whose output equals

$$\mathbf{y}(t) = \mathbf{x}^2(t)$$

We shall determine its first- and second-order densities.

If $y > 0$, then the system $y = x^2$ has the two solutions $\pm\sqrt{y}$. Furthermore, $y'(x) = \pm 2\sqrt{y}$, hence

$$f_y(y; t) = \frac{1}{2\sqrt{y}}\left[f_x(\sqrt{y}; t) + f_x(-\sqrt{y}; t)\right]$$

If $y_1 > 0$ and $y_2 > 0$, then the system

$$y_1 = x_1^2 \qquad y_2 = x_2^2$$

has the four solutions $(\pm\sqrt{y_1}, \pm\sqrt{y_2})$. Furthermore, its jacobian equals $\pm 4\sqrt{y_1 y_2}$, hence

$$f_y(y_1, y_2; t_1, t_2) = \frac{1}{4\sqrt{y_1 y_2}} \sum f_x(\pm\sqrt{y_1}, \pm\sqrt{y_2}; t_1, t_2)$$

where the summation has four terms.

We note that, if $\mathbf{x}(t)$ is SSS, then $f_x(x; t) = f_x(x)$ is independent of t and $f_x(x_1, x_2; t_1, t_2) = f_x(x_1, x_2; \tau)$ depends only on $\tau = t_1 - t_2$. Hence, $f_y(y)$ is independent of t and $f_y(y_1, y_2; \tau)$ depends only on $\tau = t_1 - t_2$.

Example 9-15 Suppose that $\mathbf{x}(t)$ is a normal stationary process with zero mean and autocorrelation $R_x(\tau)$. In this case, $f_x(x)$ is normal with variance $R_x(0)$.

If $\mathbf{y}(t) = \mathbf{x}^2(t)$ (Fig. 9-9), then $E\{\mathbf{y}(t)\} = R_x(0)$ and [see (5-9)]

$$f_y(y) = \frac{1}{\sqrt{2\pi R_x(0) y}} e^{-y/2R_x(0)} U(y)$$

We shall show that

$$R_y(\tau) = R_x^2(0) + 2R_x^2(\tau) \tag{9-89}$$

PROOF The RVs $\mathbf{x}(t+\tau)$ and $\mathbf{x}(t)$ are jointly normal with zero mean. Hence [see (7-37)]

$$E\{\mathbf{x}^2(t+\tau)\mathbf{x}^2(t)\} = E\{\mathbf{x}^2(t+\tau)\}E\{\mathbf{x}^2(t)\} + 2E^2\{\mathbf{x}(t+\tau)\mathbf{x}(t)\}$$

and (9-89) results.

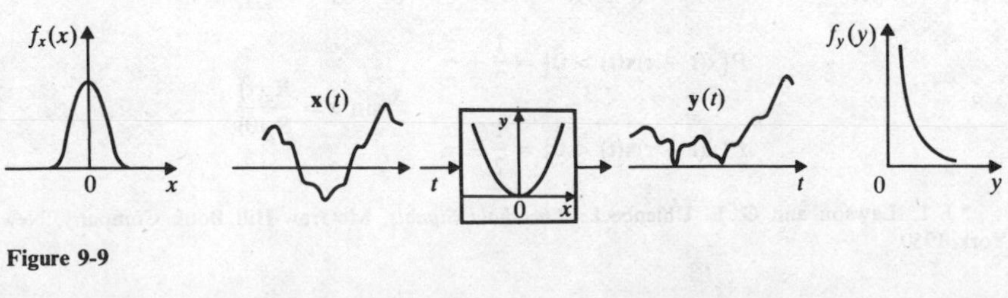

Figure 9-9

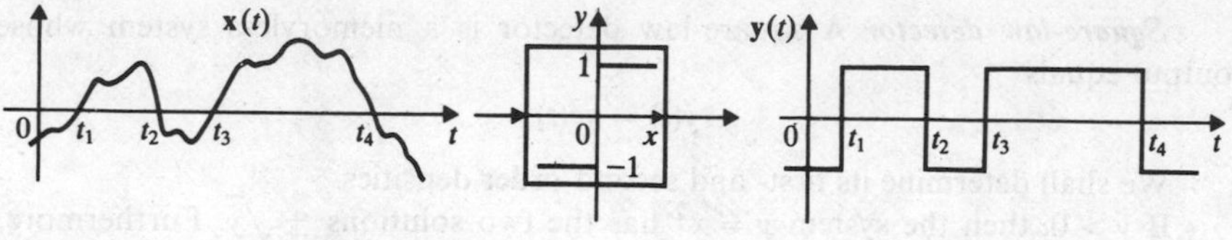

Figure 9-10

We note in particular that

$$E\{\mathbf{y}^2(t)\} = R_y(0) = 3R_x^2(0) \qquad \sigma_y^2 = 2R_x^2(0)$$

Hard limiter Consider a memoryless system with

$$g(x) = \begin{cases} 1 & x > 0 \\ -1 & x < 0 \end{cases}$$

(Fig. 9-10). Its output $\mathbf{y}(t)$ takes the values ± 1 and

$$P\{\mathbf{y}(t) = 1\} = P\{\mathbf{x}(t) > 0\} = 1 - F_x(0)$$

$$P\{\mathbf{y}(t) = -1\} = P\{\mathbf{x}(t) < 0\} = F_x(0)$$

Hence

$$E\{\mathbf{y}(t)\} = 1 \cdot P\{\mathbf{y}(t) = 1\} - 1P\{\mathbf{y}(t) = -1\} = 1 - 2F_x(0)$$

The product $\mathbf{y}(t+\tau)\mathbf{y}(t)$ equals 1 if $\mathbf{x}(t+\tau)\mathbf{x}(t) > 0$ and it equals -1 otherwise. Hence

$$R_y(\tau) = P\{\mathbf{x}(t+\tau)\mathbf{x}(t) > 0\} - P\{\mathbf{x}(t+\tau)\mathbf{x}(t) < 0\} \tag{9-90}$$

Thus, in the probability plane of the RVs $\mathbf{x}(t+\tau)$ and $\mathbf{x}(t)$, $R_y(\tau)$ equals the masses in the first and third quadrants minus the masses in the second and fourth quadrants.

Example 9-16 We shall show that if $\mathbf{x}(t)$ is a normal stationary process, then the autocorrelation of the output of a hard limiter equals

$$R_y(\tau) = \frac{2}{\pi} \arcsin \frac{R_x(\tau)}{R_x(0)} \tag{9-91}$$

This result is known as the *arcsine law*.†

PROOF The RVs $\mathbf{x}(t+\tau)$ and $\mathbf{x}(t)$ are jointly normal with zero mean, variance $R_x(0)$, and correlation coefficient $R_x(\tau)/R_x(0)$. Hence [see (6-47)].

$$P\{\mathbf{x}(t+\tau)\mathbf{x}(t) > 0\} = \frac{1}{2} + \frac{\alpha}{\pi}$$

$$P\{\mathbf{x}(t+\tau)\mathbf{x}(t) < 0\} = \frac{1}{2} - \frac{\alpha}{\pi} \qquad \sin\alpha = \frac{R_x(\tau)}{R_x(0)}$$

† J. L. Lawson and G. E. Uhlenbeck: *Threshold Signals*, McGraw-Hill Book Company, New York, 1950.

Inserting into (9-90), we obtain

$$R_y(\tau) = \frac{1}{2} + \frac{\alpha}{\pi} - \left(\frac{1}{2} - \frac{\alpha}{\pi}\right) = \frac{2\alpha}{\pi}$$

and (9-91) follows.

Linear Systems

The notation

$$\mathbf{y}(t) = L[\mathbf{x}(t)] \qquad (9\text{-}92)$$

will indicate that $\mathbf{y}(t)$ is the output of a *linear* system with input $\mathbf{x}(t)$. This means that

$$L[\mathbf{a}_1 \mathbf{x}_1(t) + \mathbf{a}_2 \mathbf{x}_2(t)] = \mathbf{a}_1 L[\mathbf{x}_1(t)] + \mathbf{a}_2 L[\mathbf{x}_2(t)] \qquad (9\text{-}93)$$

for any $\mathbf{a}_1$, $\mathbf{a}_2$, $\mathbf{x}_1(t)$, $\mathbf{x}_2(t)$

The above is the familiar definition of linearity and it holds also if the coefficients $\mathbf{a}_1$ and $\mathbf{a}_2$ are random variables because, as we have assumed, the system is deterministic, i.e., it operates only on the variable t.

Note If a system is specified by its internal structure or by a differential equation, then (9-93) holds only if $\mathbf{y}(t)$ is the *zero-state* response. The response due to the initial conditions (zero-input response) will not be considered.

A system is called *time-invariant* if its response to $\mathbf{x}(t+c)$ equals $\mathbf{y}(t+c)$. We shall assume throughout that all linear systems under consideration are time-invariant.

It is well known that the output of a linear system is a convolution

$$\mathbf{y}(t) = \mathbf{x}(t) * h(t) = \int_{-\infty}^{\infty} \mathbf{x}(t-\alpha)h(\alpha)\, d\alpha \qquad (9\text{-}94)$$

where

$$h(t) = L[\delta(t)]$$

is its impulse response. In the following, most systems will be specified by (9-94). However, we start our investigation using the operational notation (9-92) to stress the fact that various results based on the next theorem hold also for arbitrary linear operators involving one or more variables.

Fundamental theorem For any linear system

$$E\{L[\mathbf{x}(t)]\} = L[E\{\mathbf{x}(t)\}] \qquad (9\text{-}95)$$

In other words, the mean $\eta_y(t)$ of the output $\mathbf{y}(t)$ equals the response of the system to the mean $\eta_x(t)$ of the input (Fig. 9-11*a*)

$$\eta_y(t) = L[\eta_x(t)] \qquad (9\text{-}96)$$

The above is a simple extension of the linearity of expected values to arbitrary linear operators. In the context of (9-94) it can be deduced if we write the integral as a limit of a sum. This yields

$$E\{\mathbf{y}(t)\} = \int_{-\infty}^{\infty} E\{\mathbf{x}(t-\alpha)\}h(\alpha)\,d\alpha = \eta_x(t) * h(t) \tag{9-97}$$

Frequency interpretation At the ith trial the input to our system is a function $\mathbf{x}(t, \zeta_i)$ yielding as output the function $\mathbf{y}(t, \zeta_i) = L[\mathbf{x}(t, \zeta_i)]$. For large n

$$E\{\mathbf{y}(t)\} \simeq \frac{\mathbf{y}(t, \zeta_1) + \cdots + \mathbf{y}(t, \zeta_n)}{n} = \frac{L[\mathbf{x}(t, \zeta_1)] + \cdots + L[\mathbf{x}(t, \zeta_n)]}{n}$$

From the linearity of the system it follows that the last term above equals

$$L\left[\frac{\mathbf{x}(t, \zeta_1) + \cdots + \mathbf{x}(t, \zeta_n)}{n}\right]$$

This agrees with (9-95) because the fraction is nearly equal to $E\{\mathbf{x}(t)\}$.

Notes 1. From (9-96) it follows that if

$$\tilde{\mathbf{x}}(t) = \mathbf{x}(t) - \eta_x(t) \qquad \tilde{\mathbf{y}}(t) = \mathbf{y}(t) - \eta_y(t)$$

then

$$L[\tilde{\mathbf{x}}(t)] = L[\mathbf{x}(t)] - L[\eta_x(t)] = \tilde{\mathbf{y}}(t) \tag{9-98}$$

Thus, the response of a linear system to the centered input $\tilde{\mathbf{x}}(t)$ equals the centered output $\tilde{\mathbf{y}}(t)$.

2. Suppose that

$$\mathbf{x}(t) = f(t) + \mathbf{v}(t) \qquad E\{\mathbf{v}(t)\} = 0$$

In this case, $E\{\mathbf{x}(t)\} = f(t)$, hence

$$\eta_y(t) = f(t) * h(t)$$

Thus, if $\mathbf{x}(t)$ is the sum of a deterministic signal $f(t)$ and a random component $\mathbf{v}(t)$, then for the determination of the mean of the output we can ignore $\mathbf{v}(t)$ provided that the system is linear and $E\{\mathbf{v}(t)\} = 0$.

Theorem (9-95) can be used to express the joint moments of any order of the output $\mathbf{y}(t)$ of a linear system in terms of the corresponding moments of the input. The following special cases are of fundamental importance in the study of linear systems with stochastic inputs.

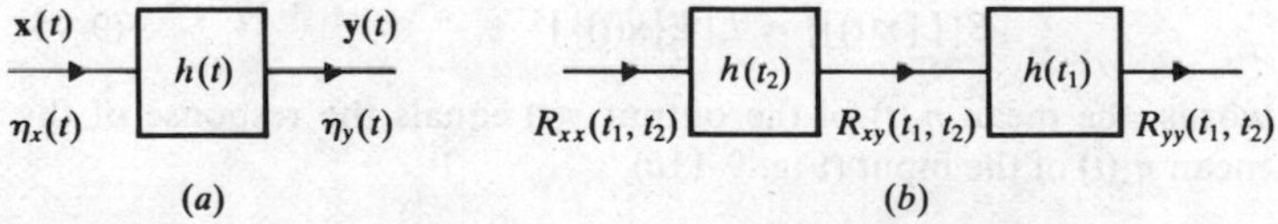

Figure 9-11

Output autocorrelation We wish to express the autocorrelation $R_{yy}(t_1, t_2)$ of the output $\mathbf{y}(t)$ of a linear system in terms of the autocorrelation $R_{xx}(t_1, t_2)$ of the input $\mathbf{x}(t)$. As we shall presently see, it is easier to find first the cross-correlation $R_{xy}(t_1, t_2)$ between $\mathbf{x}(t)$ and $\mathbf{y}(t)$.

Theorem

(*a*)
$$R_{xy}(t_1, t_2) = L_2[R_{xx}(t_1, t_2)] \tag{9-99}$$

In the above notation, L_2 means that the system operates on the variable t_2, treating t_1 as a parameter. In the context of (9-94) this means that

$$R_{xy}(t_1, t_2) = \int_{-\infty}^{\infty} R_{xx}(t_1, t_2 - \alpha)h(\alpha)\, d\alpha \tag{9-100}$$

(*b*)
$$R_{yy}(t_1, t_2) = L_1[R_{xy}(t_1, t_2)] \tag{9-101}$$

In this case, the system operates on t_1

$$R_{yy}(t_1, t_2) = \int_{-\infty}^{\infty} R_{xy}(t_1 - \alpha, t_2)h(\alpha)\, d\alpha \tag{9-102}$$

PROOF Multiplying (9-92) by $\mathbf{x}(t_1)$ and using (9-93), we obtain

$$\mathbf{x}(t_1)\mathbf{y}(t) = L_t[\mathbf{x}(t_1)\mathbf{x}(t)]$$

where L_t means that the system operates on t. Hence [see (9-95)]

$$E\{\mathbf{x}(t_1)\mathbf{y}(t)\} = L_t[E\{\mathbf{x}(t_1)\mathbf{x}(t)\}]$$

and (9-99) follows with $t = t_2$. The proof of (9-101) is similar: We multiply (9-92) by $\mathbf{y}(t_2)$ and use (9-95). This yields

$$E\{\mathbf{y}(t)\mathbf{y}(t_2)\} = L_t[E\{\mathbf{x}(t)\mathbf{y}(t_2)\}]$$

and (9-101) follows with $t = t_1$.

The preceding theorem is illustrated in Fig. 9-11*b*: If $R_{xx}(t_1, t_2)$ is the input to the given system and the system operates on t_2, the output equals $R_{xy}(t_1, t_2)$. If $R_{xy}(t_1, t_2)$ is the input and the system operates on t_1, the output equals $R_{yy}(t_1, t_2)$.

Inserting (9-100) into (9-102), we obtain

$$R_{yy}(t_1, t_2) = \int_{-\infty}^{\infty}\int_{-\infty}^{\infty} R_{xx}(t_1 - \alpha, t_2 - \beta)h(\alpha)h(\beta)\, d\alpha\, d\beta \tag{9-103}$$

This expresses $R_{yy}(t_1, t_2)$ directly in terms of $R_{xx}(t_1, t_2)$. However, conceptually and operationally, it is preferable to find first $R_{xy}(t_1, t_2)$.

Corollary The autocovariance $C_{yy}(t_1, t_2)$ of $\mathbf{y}(t)$ is the autocorrelation of the process $\tilde{\mathbf{y}}(t) = \mathbf{y}(t) - \eta_y(t)$ and, as we see from (9-98), $\tilde{\mathbf{y}}(t)$ equals $L[\tilde{\mathbf{x}}(t)]$. Applying (9-100) and (9-102) to the centered processes $\tilde{\mathbf{x}}(t)$ and $\tilde{\mathbf{y}}(t)$, we obtain

$$\begin{aligned} C_{xy}(t_1, t_2) &= C_{xx}(t_1, t_2) * h(t_2) \\ C_{yy}(t_1, t_2) &= C_{xy}(t_1, t_2) * h(t_1) \end{aligned} \tag{9-104}$$

where the convolutions are in t_2 and t_1 respectively.

Stationarity (*a*) If the input $\mathbf{x}(t)$ to a time-invariant system (the system need not be linear) is SSS, then the output $\mathbf{y}(t)$ is also SSS.

PROOF From the assumption of time-invariance it follows that if

$$\mathbf{y}(t) = L[\mathbf{x}(t)] \qquad \text{then} \qquad \mathbf{y}(t + c) = L[\mathbf{x}(t + c)]$$

If, therefore, the processes $\mathbf{x}(t)$ and $\mathbf{x}(t + c)$ have the same statistics, then the processes $\mathbf{y}(t)$ and $\mathbf{y}(t + c)$ have the same statistics.

(*b*) If the input $\mathbf{x}(t)$ to a linear system is WSS then the output $\mathbf{y}(t)$ is also WSS.

PROOF As we see from (9-97), if η_x is constant then η_y is also constant. Furthermore, if $R_{xx}(t_1, t_2)$ depends only on $t_1 - t_2$, then [see (9-103)] $R_{yy}(t_1, t_2)$ depends only on $t_1 - t_2$. In fact, (9-100) shows that the processes $\mathbf{x}(t)$ and $\mathbf{y}(t)$ are jointly WSS.

(*c*) If $\mathbf{x}(t)$ is stationary of order N, $\mathbf{y}(t)$ might not be stationary in any sense.

(*d*) In the above, we assumed that the process $\mathbf{x}(t)$ is applied to the system at $t = -\infty$. If $\mathbf{x}(t)$ is stationary but it is applied to the system at $t = 0$, that is, if the input equals $\mathbf{x}(t)U(t)$, then the resulting response is not stationary. However, it is asymptotically stationary as $t \to \infty$ provided that the system is stable.

(*e*) The preceding conclusions hold also for cyclostationary processes.

Differentiators A differentiator is a linear system whose output is the derivative of the input

$$L[\mathbf{x}(t)] = \mathbf{x}'(t)$$

We can, therefore, use the preceding results to find the mean and the autocorrelation of $\mathbf{x}'(t)$.

From (9-96) it follows that

$$\eta_{x'}(t) = L[\eta_x(t)] = \eta'_x(t) \tag{9-105}$$

Similarly [see (9-99)]

$$R_{xx'}(t_1, t_2) = L_2[R_{xx}(t_1, t_2)] = \frac{\partial R_{xx}(t_1, t_2)}{\partial t_2} \tag{9-106}$$

because, in this case, L_2 means differentiation with respect to t_2. Finally

$$R_{x'x'}(t_1, t_2) = L_1[R_{xx'}(t_1, t_2)] = \frac{\partial R_{xx'}(t_1, t_2)}{\partial t_1} \tag{9-107}$$

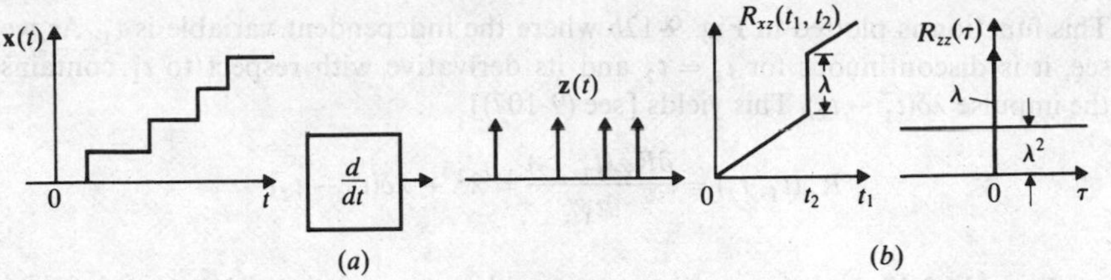

Figure 9-12

Combining, we obtain

$$R_{x'x'}(t_1, t_2) = \frac{\partial^2 R_{xx}(t_1, t_2)}{\partial t_1 \, \partial t_2} \tag{9-108}$$

Stationary processes If $\mathbf{x}(t)$ is WSS, then $\eta_x(t)$ is constant, hence

$$E\{\mathbf{x}'(t)\} = 0$$

Furthermore, since $R_{xx}(t_1, t_2) = R_{xx}(\tau)$, $\tau = t_1 - t_2$

$$\frac{\partial R_{xx}(t_1 - t_2)}{\partial t_2} = -\frac{dR_{xx}(\tau)}{d\tau} \qquad \frac{\partial^2 R_{xx}(t_1 - t_2)}{\partial t_1 \, \partial t_2} = -\frac{d^2 R_{xx}(\tau)}{d\tau^2}$$

Hence

$$R_{xx'}(\tau) = -R'_{xx}(\tau) \qquad R_{x'x'}(\tau) = -R''_{xx}(\tau) \tag{9-109}$$

Poisson impulses If the input $\mathbf{x}(t)$ to a differentiator is a Poisson process, the resulting output $\mathbf{z}(t)$ is a train of impulses (Fig. 9-12)

$$\mathbf{z}(t) = \sum_i \delta(t - \mathbf{t}_i) \tag{9-110}$$

We maintain that $\mathbf{z}(t)$ is a stationary process with mean

$$\eta_z = \lambda \tag{9-111}$$

and autocorrelation

$$R_{zz}(\tau) = \lambda^2 + \lambda\delta(\tau) \tag{9-112}$$

PROOF The first equation follows from (9-105) because $\eta_x(t) = \lambda t$. To prove the second, we observe that [see (9-14)]

$$R_{xx}(t_1, t_2) = \lambda^2 t_1 t_2 + \lambda \min(t_1, t_2) \tag{9-113}$$

And since $\mathbf{z}(t) = \mathbf{x}'(t)$, (9-106) yields

$$R_{xz}(t_1, t_2) = \frac{\partial R_{xx}(t_1, t_2)}{\partial t_2} = \lambda^2 t_1 + \lambda U(t_1 - t_2)$$

This function is plotted in Fig. 9-12*b* where the independent variable is t_1. As we see, it is discontinuous for $t_1 = t_2$ and its derivative with respect to t_1 contains the impulse $\lambda\delta(t_1 - t_2)$. This yields [see (9-107)]

$$R_{zz}(t_1, t_2) = \frac{\partial R_{xz}(t_1, t_2)}{\partial t_1} = \lambda^2 + \lambda\delta(t_1 - t_2)$$

Example 9-17 A stationary process $\mathbf{v}(t)$ with autocorrelation $R_{vv}(\tau) = q\delta(\tau)$ (white noise) is applied at $t = 0$ to a linear system with

$$h(t) = e^{-ct}U(t)$$

We shall show that the autocorrelation of the resulting output $\mathbf{y}(t)$ equals

$$R_{yy}(t_1, t_2) = \frac{q}{2c}(1 - e^{-2ct_1})e^{-c|t_2 - t_1|} \tag{9-114}$$

for $t_1 < t_2$.

PROOF We can use the preceding results if we assume that the input to the system is the process

$$\mathbf{x}(t) = \mathbf{v}(t)U(t)$$

With this assumption, all correlations are zero if $t_1 < 0$ or $t_2 < 0$. For $t_1 > 0$ and $t_2 > 0$

$$R_{xx}(t_1, t_2) = E\{\mathbf{v}(t_1)\mathbf{v}(t_2)\} = q\delta(t_1 - t_2)$$

As we see from (9-99), $R_{xy}(t_1, t_2)$ equals the response of the system to $q\delta(t_1 - t_2)$ considered as function of t_2. Since $\delta(t_1 - t_2) = \delta(t_2 - t_1)$ and $L[\delta(t_2 - t_1)] = h(t_2 - t_1)$ (time-invariance), we conclude that

$$R_{xy}(t_1, t_2) = qh(t_2 - t_1) = qe^{-c(t_2 - t_1)}U(t_2 - t_1)$$

In Fig. 9-13, we show $R_{xy}(t_1, t_2)$ as a function of t_1 and t_2. Inserting into (9-102), we obtain

$$R_{yy}(t_1, t_2) = q\int_0^{t_1} e^{c(t_1 - \alpha - t_2)}e^{-c\alpha}\, d\alpha \qquad t_1 < t_2$$

and (9-114) results.

We note that

$$E\{\mathbf{y}^2(t)\} = R_{yy}(t) = \frac{q}{2c}(1 - e^{-2ct}) = q\int_0^t h^2(\alpha)\, d\alpha \tag{9-115}$$

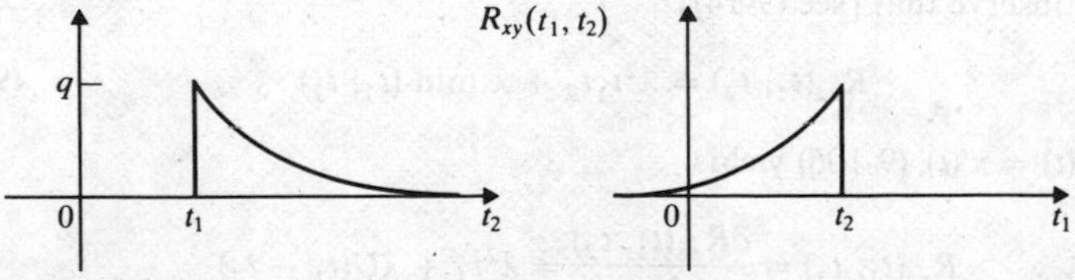

Figure 9-13

Complex processes The preceding results can be readily extended to complex processes and to systems with complex-valued $h(t)$. Reasoning as in the real case, we obtain

$$R_{xy}(t_1, t_2) = R_{xx}(t_1, t_2) * h^*(t_2)$$
$$R_{yy}(t_1, t_2) = R_{xy}(t_1, t_2) * h(t_1) \tag{9-116}$$

Response to white noise We shall determine the average intensity $E\{|\mathbf{y}(t)|^2\}$ of the output of a system driven by white noise. This is a special case of (9-116), however, because of its importance it is stated as a theorem.

Theorem If the input to a linear system is white noise with autocorrelation

$$R_{xx}(t_1, t_2) = q(t_1)\delta(t_1 - t_2)$$

then

$$E\{|\mathbf{y}(t)|^2\} = q(t) * |h(t)|^2 = \int_{-\infty}^{\infty} q(t - \alpha)|h(\alpha)|^2 \, d\alpha \tag{9-117}$$

PROOF From (9-116) it follows that

$$R_{xy}(t_1, t_2) = q(t_1)\delta(t_2 - t_1) * h^*(t_2) = q(t_1)h^*(t_2 - t_1)$$

$$R_{yy}(t_1, t_2) = \int_{-\infty}^{\infty} q(t_1 - \alpha)h^*[t_2 - (t_1 - \alpha)]h(\alpha) \, d\alpha$$

and with $t_1 = t_2 = t$, (9-117) results.

Special cases (*a*) If $\mathbf{x}(t)$ is stationary white noise, then $q(t) = q$ and (9-117) yields

$$E\{\mathbf{y}^2(t)\} = qE \qquad \text{where} \qquad E = \int_{-\infty}^{\infty} |h(t)|^2 \, dt$$

is the energy of $h(t)$.

(*b*) If $h(t)$ is of short duration relative to the variations of $q(t)$, then

$$E\{\mathbf{y}^2(t)\} \simeq q(t) \int_{-\infty}^{\infty} |h(\alpha)|^2 \, d\alpha = Eq(t)$$

This relationship justifies the term *average intensity* used to describe the function $q(t)$.

(*c*) If $R_{vv}(\tau) = q\delta(\tau)$ and $\mathbf{v}(t)$ is applied to the system at $t = 0$, then $q(t) = qU(t)$ and (9-117) yields

$$E\{\mathbf{y}^2(t)\} = q \int_{-\infty}^{t} |h(\alpha)|^2 \, d\alpha$$

Example 9-18 The integral

$$\mathbf{y} = \int_0^t \mathbf{v}(\alpha) \, d\alpha$$

can be considered as the output of a linear system with input $\mathbf{x}(t) = \mathbf{v}(t)U(t)$ and impulse response $h(t) = U(t)$. If, therefore, $\mathbf{v}(t)$ is white noise with average intensity $q(t)$, then $\mathbf{x}(t)$ is white noise with average intensity $q(t)U(t)$ and (9-117) yields

$$E\{\mathbf{y}^2(t)\} = q(t)U(t) * U(t) = \int_0^t q(\alpha)\, d\alpha$$

Low-pass response Suppose that the input to a low-pass system is a real stationary process $\mathbf{x}(t)$ with autocorrelation $R(\tau)$. We shall show that if

$$R(\tau) \simeq 0 \qquad \text{for} \qquad |\tau| > a \tag{9-118}$$

and $h(t)$ is nearly constant in any interval of length a, then

$$E\{\mathbf{y}^2(t)\} \simeq qE \qquad \text{where} \qquad q = \int_{-a}^{a} R(\tau)\, d\tau \tag{9-119}$$

and E is the energy of $h(t)$.

PROOF With $R(t_1, t_2) = R(t_1 - t_2)$, (9-103) yields

$$E\{\mathbf{y}^2(t)\} = \int_{-\infty}^{\infty}\int_{-\infty}^{\infty} R(\beta - \alpha)h(\alpha)h(\beta)\, d\alpha\, d\beta$$

From (9-118) it follows that the integrand is different from zero only on the strip $|\alpha - \beta| < a$. On this strip, $h(\alpha) \simeq h(\beta)$ by assumption. Hence

$$E\{\mathbf{y}^2(t)\} \simeq \int_{-\infty}^{\infty} h^2(\alpha) \int_{-\infty}^{\infty} R(\beta - \alpha)\, d\beta\, d\alpha$$

and (9-119) results because the inner integral equals q.

General moments The moments of any order of the output $\mathbf{y}(t)$ of a linear system can be expressed in terms of the moments of the input $\mathbf{x}(t)$. We shall illustrate with a special case.

We wish to find the moment

$$E\{\mathbf{y}(t_1)\mathbf{y}(t_2)\mathbf{y}(t_3)\}$$

For this purpose, we find a succession of mixed moments involving the input and the output, increasing at each step the number of output factors by one. This yields

$$E\{\mathbf{x}(t_1)\mathbf{x}(t_2)\mathbf{y}(t_3)\} = L_3[E\{\mathbf{x}(t_1)\mathbf{x}(t_2)\mathbf{x}(t_3)\}]$$

$$E\{\mathbf{x}(t_1)\mathbf{y}(t_2)\mathbf{y}(t_3)\} = L_2[E\{\mathbf{x}(t_1)\mathbf{x}(t_2)\mathbf{y}(t_3)\}]$$

$$E\{\mathbf{y}(t_1)\mathbf{y}(t_2)\mathbf{y}(t_3)\} = L_1[E\{\mathbf{x}(t_1)\mathbf{y}(t_2)\mathbf{y}(t_3)\}]$$

We note that for the evaluation of $E\{\mathbf{y}(t_1)\mathbf{y}(t_2)\mathbf{y}(t_3)\}$ for *specific* times t_1, t_2, t_3 the function $E\{\mathbf{x}(t_1)\mathbf{x}(t_2)\mathbf{x}(t_3)\}$ must be known for *every* t_1, t_2, t_3.

Multiple terminals Linear systems with more than one input or output are specified in terms of their impulse response matrix. Once this matrix is specified,

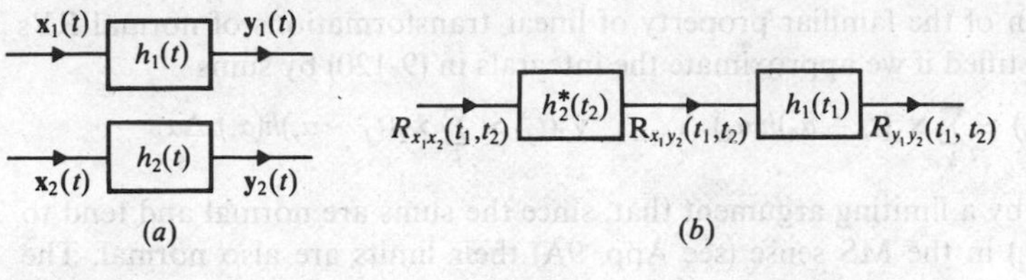

Figure 9-14

the study of their responses can be reduced to the study of several systems with one input and one output. We give next an illustration.

In Fig. 9-14*a* we show two systems with inputs $\mathbf{x}_1(t)$, $\mathbf{x}_2(t)$ and outputs

$$\mathbf{y}_1(t) = \int_{-\infty}^{\infty} \mathbf{x}_1(t-\alpha)h_1(\alpha)\,d\alpha \qquad \mathbf{y}_2(t) = \int_{-\infty}^{\infty} \mathbf{x}_2(t-\alpha)h_2(\alpha)\,d\alpha \tag{9-120}$$

From these equations it follows readily that

$$\mathbf{x}_1(t_1)\mathbf{y}_2^*(t_2) = \int_{-\infty}^{\infty} \mathbf{x}_1(t_1)\mathbf{x}_2^*(t_2-\alpha)h_2^*(\alpha)\,d\alpha$$
$$\mathbf{y}_1(t_1)\mathbf{y}_2^*(t_2) = \int_{-\infty}^{\infty} \mathbf{x}_1(t_1-\alpha)\mathbf{y}_2^*(t_2)h_1(\alpha)\,d\alpha \tag{9-121}$$

Taking expected values, we obtain

$$R_{x_1y_2}(t_1, t_2) = R_{x_1x_2}(t_1, t_2) * h_2^*(t_2) \tag{9-122}$$

$$R_{y_1y_2}(t_1, t_2) = R_{x_1y_2}(t_1, t_2) * h_1(t_1) \tag{9-123}$$

Thus, to find $R_{x_1y_2}(t_1, t_2)$, we use $R_{x_1x_2}(t_1, t_2)$ as the input to the conjugate of the second system operating on the variable t_2. To find $R_{y_1y_2}(t_1, t_2)$, we use $R_{x_1y_2}(t_1, t_2)$ as the input to the first system operating on the variable t_1 (Fig. 9-14*b*).

Example 9-19 Suppose that $\mathbf{x}_1(t) = \mathbf{x}_2(t) = \mathbf{x}(t)$ and

$$\mathbf{y}_1(t) = \mathbf{x}^{(m)}(t) \qquad \mathbf{y}_2(t) = \mathbf{x}^{(n)}(t)$$

In this case, convolving with $h(t)$ means differentiation. Combining (9-122) and (9-123), we conclude, therefore, that

$$R_{x^{(m)}x^{(n)}}(t_1, t_2) = \frac{\partial^{m+n}R_{xx}(t_1, t_2)}{\partial t_1^m\, \partial t_2^n} \tag{9-124}$$

Normal inputs If the input $\mathbf{x}(t)$ to a linear system is a normal process, the resulting output $\mathbf{y}(t)$ is also normal. If the inputs $\mathbf{x}_1(t)$, $\mathbf{x}_2(t)$ to two systems are jointly normal, then the resulting outputs $\mathbf{y}_1(t)$, $\mathbf{y}_2(t)$ are also jointly normal. This is an extension of the familiar property of linear transformations of normal RVs and can be justified if we approximate the integrals in (9-120) by sums

$$\mathbf{y}_1(t_i) \simeq \sum_k \mathbf{x}_1(t_i - \alpha_k)h(\alpha_k)\,\Delta\alpha \qquad \mathbf{y}_2(t_j) \simeq \sum_r \mathbf{x}_2(t_j - \alpha_r)h(\alpha_r)\,\Delta\alpha$$

We can show by a limiting argument that, since the sums are normal and tend to $\mathbf{y}_1(t_i)$ and $\mathbf{y}_2(t_j)$ in the MS sense (see App. 9A) their limits are also normal. The details, however, are omitted.

Differential Equations

A deterministic differential equation with random excitation is an equation of the form

$$a_n\,\mathbf{y}^{(m)}(t) + \cdots + a_0\,\mathbf{y}(t) = \mathbf{x}(t) \tag{9-125}$$

where the coefficients a_k are given numbers and the driver $\mathbf{x}(t)$ is a stochastic process. We shall consider its solution $\mathbf{y}(t)$ under the assumption that the initial conditions are zero. With this assumption, $\mathbf{y}(t)$ is unique (zero-state response) and it satisfies the linearity condition (9-93). We can, therefore, interpret $\mathbf{y}(t)$ as the output of a linear system specified by (9-125).

In general, the determination of the complete statistics of $\mathbf{y}(t)$ is complicated. In the following, we evaluate only its second-order moments using the preceding results. The above system is an operator L specified as follows: Its output $\mathbf{y}(t)$ is a process with zero initial conditions satisfying (9-125).

Mean As we know [see (9-96)] the mean $\eta_y(t)$ of $\mathbf{y}(t)$ is the output of L with input $\eta_x(t)$. Hence, it satisfies the equation

$$a_n\,\eta_y^{(n)}(t) + \cdots + a_0\,\eta_y(t) = \eta_x(t) \tag{9-126}$$

and the initial conditions

$$\eta_y(0) = \cdots = \eta_y^{(n-1)}(0) = 0 \tag{9-127}$$

This result can be established directly: Clearly

$$E\{\mathbf{y}^{(k)}(t)\} = \eta_y^{(k)}(t) \tag{9-128}$$

Taking expected values of both sides of (9-125), and using the above we obtain (9-126). Equation (9-127) follows from (9-128) because $\mathbf{y}^{(k)}(0) = 0$ by assumption.

Correlation To determine $R_{xy}(t_1, t_2)$ we use (9-99)

$$R_{xy}(t_1, t_2) = L_2[R_{xx}(t_1, t_2)]$$

In this case, L_2 means that $R_{xy}(t_1, t_2)$ satisfies the differential equation

$$a_n\,\frac{\partial^n R_{xy}(t_1, t_2)}{\partial t_2^n} + \cdots + a_0\,R_{xy}(t_1, t_2) = R_{xx}(t_1, t_2) \tag{9-129}$$

with the initial conditions

$$R_{xy}(t_1, 0) = \cdots = \frac{\partial^{n-1} R_{xy}(t_1, 0)}{\partial t_2^{n-1}} = 0 \tag{9-130}$$

Similarly, since [see (9-101)]

$$R_{yy}(t_1, t_2) = L_1[R_{xy}(t_1, t_2)]$$

we conclude as above that

$$a_n \frac{\partial^n R_{yy}(t_1, t_2)}{\partial t_1^n} + \cdots + a_0 R_{yy}(t_1, t_2) = R_{xy}(t_1, t_2) \tag{9-131}$$

$$R_{yy}(0, t_2) = \cdots = \frac{\partial^{n-1} R_{yy}(0, t_2)}{\partial t_1^{n-1}} = 0 \tag{9-132}$$

The preceding results can be established directly: From (9-125) it follows that

$$\mathbf{x}(t_1)[a_n \mathbf{y}^{(n)}(t_2) + \cdots + a_0 \mathbf{y}(t_2)] = \mathbf{x}(t_1)\mathbf{x}(t_2)$$

This yields (9-129) because [see (9-124)]

$$E\{\mathbf{x}(t_1)\mathbf{y}^{(k)}(t_2)\} = \partial^k R_{xy}(t_1, t_2)/\partial t_2^k$$

Similarly, (9-131) is a consequence of the identity

$$[a_n \mathbf{y}^{(n)}(t_1) + \cdots + a_0 \mathbf{y}(t_1)]\mathbf{y}(t_2) = \mathbf{x}(t_1)\mathbf{y}(t_2)$$

because

$$E\{\mathbf{y}^{(k)}(t_1)\mathbf{y}(t_2)\} = \partial^k R_{yy}(t_1, t_2)/\partial t_1^k$$

Finally, the expected values of

$$\mathbf{x}(t_1)\mathbf{y}^{(k)}(0) = 0 \qquad \mathbf{y}^{(k)}(0)\mathbf{y}(t_2) = 0$$

yield (9-130) and (9-132).

9-5 ERGODICITY

A central problem in the theory of stochastic processes is the estimation of their various statistics. Suppose, for example, that we wish to determine the mean $\eta(t)$ of a process $\mathbf{x}(t)$. For this purpose, we observe a large number of samples $\mathbf{x}(t, \zeta_i)$ and we use their *ensemble average* as the estimate of $\eta(t)$

$$\eta(t) \simeq \frac{1}{n} \sum_i \mathbf{x}(t, \zeta_i) \tag{9-133}$$

Suppose, however, that we have access only to a single sample $\mathbf{x}(t, \zeta)$ of $\mathbf{x}(t)$. Can we then use its *time-average*

$$\bar{\mathbf{x}} = \lim_{T \to \infty} \frac{1}{2T} \int_{-T}^{T} \mathbf{x}(t, \zeta)\, dt \tag{9-134}$$

as the estimate of $\eta(t)$? This is, of course, impossible if $\eta(t)$ depends on t. However, if $\eta(t) = \eta =$ constant, then under rather general conditions, $\bar{\mathbf{x}}$ equals η. The topic of ergodicity deals with the underlying ideas.

Note The term ergodicity has several interpretations related to the following questions:

(*a*) Under what conditions does the limit in (9-134) exist?

(*b*) If it exists, its value $\bar{\mathbf{x}}(\zeta)$ depends in general on ζ. Under what conditions does it equal the constant η?

(*c*) Suppose that $x(t)$ is a signal representing a physical quantity. Under what conditions can it be considered as a sample of an ergodic process?

The first question is mainly of mathematical interest and it was answered over fifty years ago (Birkhoff†): If the process $\mathbf{x}(t)$ is stationary and $E\{|\mathbf{x}(t)|\} < \infty$, then $\bar{\mathbf{x}}$ exists for almost every ζ.

The third question involves predictions based on past observations and can be answered only inductively (physical interpretation).

In this section, we consider only the second question.

Definition A stochastic process $\mathbf{x}(t)$ is called ergodic if its ensemble averages equal appropriate time averages.

By this we mean that, with probability 1, any statistic of $\mathbf{x}(t)$ can be determined from a single sample $\mathbf{x}(t, \zeta)$. This is a general requirement. In most applications, we are concerned with specific statistics. We start with the estimation of the mean of $\mathbf{x}(t)$. All other cases follow as corollaries because any statistic can be expressed as expected value.

Mean-Ergodic Processes

Given a stochastic process $\mathbf{x}(t)$ with constant mean

$$E\{\mathbf{x}(t)\} = \eta$$

we form the time-average

$$\boldsymbol{\eta}_T = \frac{1}{2T} \int_{-T}^{T} \mathbf{x}(t)\, dt \tag{9-135}$$

We shall say that the process $\mathbf{x}(t)$ is *mean-ergodic* if, with probability 1,

$$\boldsymbol{\eta}_T \to \eta \quad \text{as} \quad T \to \infty \tag{9-136}$$

Clearly, $\boldsymbol{\eta}_T$ is an RV with mean

$$E\{\boldsymbol{\eta}_T\} = \frac{1}{2T} \int_{-T}^{T} E\{\mathbf{x}(t)\}\, dt = \eta \tag{9-137}$$

Denoting its variance by σ_T^2, we conclude that (9-136) is true iff

$$\sigma_T^2 \to 0 \quad \text{as} \quad T \to \infty \tag{9-138}$$

Thus, to establish the ergodicity of $\mathbf{x}(t)$, it suffices to test the validity of (9-138).

† G. D. Birkhoff: "Proof of the Ergodic Theorem," *Proceedings of the National Academy of Science*, U.S.A., vol. 17, 1931.

Theorem A process $\mathbf{x}(t)$ is mean-ergodic iff its autocovariance $C(t_1, t_2)$ is such that

$$\frac{1}{4T^2}\int_{-T}^{T}\int_{-T}^{T} C(t_1, t_2)\, dt_1\, dt_2 \xrightarrow[T\to\infty]{} 0 \qquad (9\text{-}139)$$

PROOF As we know [see (9-34)]

$$\sigma_T^2 = \frac{1}{4T^2}\int_{-T}^{T}\int_{-T}^{T} C(t_1, t_2)\, dt_1\, dt_2 \qquad (9\text{-}140)$$

Hence, (9-138) follows from (9-139).

Example 9-20 Suppose that $\mathbf{x}(t) = \eta + \mathbf{v}(t)$ where $\mathbf{v}(t)$ is white noise with $C(t_1, t_2) = q(t_1)\delta(t_1 - t_2)$. In this case,

$$\sigma_T^2 = \frac{1}{4T^2}\int_{-T}^{T}\int_{-T}^{T} q(t_1)\delta(t_1 - t_2)\, dt_1\, dt_2 = \frac{1}{4T^2}\int_{-T}^{T} q(t)\, dt$$

If, therefore, $q(t)$ is bounded, then $\boldsymbol{\eta}_T \to \eta$.

Corollary A WSS process is mean-ergodic if its autocovariance

$$C(\tau) = R(\tau) - \eta^2$$

is such that

$$\frac{1}{2T}\int_{-2T}^{2T} C(\tau)\left(1 - \frac{|\tau|}{2T}\right) d\tau \xrightarrow[T\to\infty]{} 0 \qquad (9\text{-}141)$$

PROOF It follows from (9-139) because $C(t_1, t_2) = C(t_1 - t_2)$ and (see Fig. 9-6)

$$\int_{-T}^{T}\int_{-T}^{T} C(t_1 - t_2)\, dt_1\, dt_2 = \int_{-2T}^{2T} C(\tau)(2T - |\tau|)\, d\tau$$

Sufficient conditions 1. If $\mathbf{x}(t)$ is WSS and

$$\int_{-\infty}^{\infty} |C(\tau)|\, d\tau < \infty \qquad (9\text{-}142)$$

then $\mathbf{x}(t)$ is mean-ergodic.

PROOF Clearly

$$\frac{1}{2T}\int_{-T}^{T} C(\tau)\left(1 - \frac{|\tau|}{2T}\right) d\tau < \frac{1}{2T}\int_{-T}^{T} |C(\tau)|\, d\tau$$

Hence, if (9-142) holds, then the left side tends to zero as $T \to \infty$.

2. If $C(0) < \infty$ and

$$C(\tau) \to 0 \qquad \text{as} \qquad |\tau| \to \infty \qquad (9\text{-}143)$$

then $\mathbf{x}(t)$ is mean-ergodic. Thus, the process $\mathbf{x}(t)$ is mean ergodic if the RVs $\mathbf{x}(t+\tau)$ and $\mathbf{x}(t)$ are *uncorrelated* for large τ. This is the case for most regular processes.

PROOF If (9-143) holds, then given ε, we can find a constant a such that $|C(\tau)| < \varepsilon$ for every $|\tau| > a$. Hence

$$\int_{-2T}^{2T} C(\tau)\left(1 - \frac{|\tau|}{2T}\right) d\tau < \int_{-a}^{a} |C(\tau)|\, d\tau + \int_{a<|\tau|<2T} |C(\tau)|\, d\tau$$

From (7-11) it follows that $|C(\tau)| \le C(0)$ [see also (10-7)]. Hence, the second integral above is less than $2aC(0)$. The last integral is less than $4\varepsilon T$ because $|C(\tau)| < \varepsilon$ for $a < |\tau| < 2T$. Therefore

$$\frac{1}{2T}\int_{-T}^{T} C(\tau)\left(1 - \frac{|\tau|}{2T}\right) d\tau < \frac{a}{T}\, C(0) + 2\varepsilon \xrightarrow[T\to\infty]{} 2\varepsilon$$

and since this holds for every ε, (9-141) results.

Clearly, (9-143) is true if $\mathbf{x}(t)$ is a-dependent. In this case, $C(\tau) = 0$ for $|\tau| > a$, hence, for $T \gg a$ the variance of $\boldsymbol{\eta}_T$ equals

$$\sigma_T^2 \simeq \frac{1}{2T}\int_{-a}^{a} C(\tau)\, d\tau \tag{9-144}$$

Note Suppose that we use as estimate of η the time-average $\boldsymbol{\eta}_T$. The difference $\boldsymbol{\eta}_T - \eta$ is the resulting error and it is such that [see (5-57)]

$$P\{|\boldsymbol{\eta}_T - \eta| < \varepsilon\} \ge 1 - \frac{\sigma_T^2}{\varepsilon^2}$$

This gives us a guide for the required value of T. However, [see (9-145)], a closer bound is possible if we know the density of $\mathbf{x}(t)$.

Example 9-21 If $C(\tau) = Ae^{-\alpha|\tau|}$, then $C(\tau) \to 0$ as $\tau \to \infty$, hence $\mathbf{x}(t)$ is mean-ergodic. In this case

$$\sigma_T^2 = \frac{A}{T}\int_0^T e^{-\alpha\tau}\left(1 - \frac{\tau}{2T}\right) d\tau = \frac{A}{\alpha T}\left(1 - \frac{1 - e^{-2\alpha T}}{2\alpha T}\right)$$

and if $\alpha T \gg 1$, then $\sigma_T^2 \simeq A/\alpha T$.

If $\mathbf{x}(t)$ is normal, then $\boldsymbol{\eta}_T$ is a normal RV and

$$P\{|\boldsymbol{\eta}_T - \eta| < \varepsilon\} \simeq 2\mathbb{G}\left(\varepsilon\sqrt{\frac{\alpha T}{A}}\right) - 1 \tag{9-145}$$

Example 9-22 Suppose, finally, that $\mathbf{x}(t) = \mathbf{a}$ where $\mathbf{a}$ is an RV. In this case, $\boldsymbol{\eta}_T = \mathbf{a}$ for any T, hence the process $\mathbf{x}(t)$ is not ergodic.

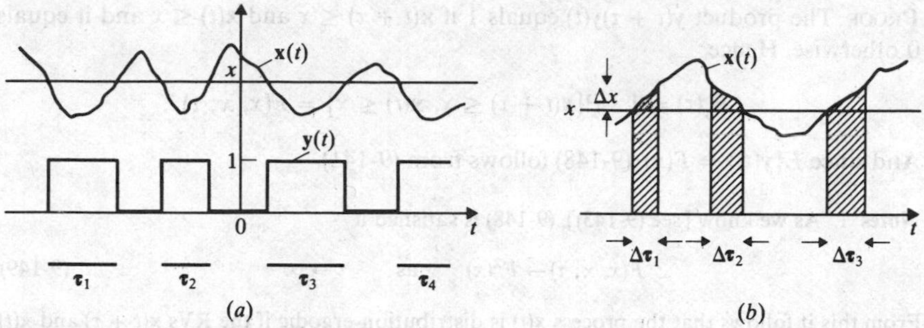

Figure 9-15

Distribution-Ergodic Processes

We wish to determine the distribution function

$$F(x) = P\{\mathbf{x}(t) \le x\}$$

of a stationary process $\mathbf{x}(t)$ in terms of a single sample. For this purpose, we form the process

$$\mathbf{y}(t) = \begin{cases} 1 & \mathbf{x}(t) \le x \\ 0 & \mathbf{x}(t) > x \end{cases}$$

This process takes the values 1 and 0 with probabilities $P\{\mathbf{x}(t) \le x\}$ and $P\{\mathbf{x}(t) > x\}$ respectively. Hence, its mean equals $F(x)$

$$E\{\mathbf{y}(t)\} = P\{\mathbf{x}(t) \le x\} = F(x) \tag{9-146}$$

To determine $F(x)$ it suffices, therefore, to apply the preceding results substituting $\mathbf{y}(t)$ for $\mathbf{x}(t)$.

We shall say that a process $\mathbf{x}(t)$ is distribution-ergodic if the corresponding process $\mathbf{y}(t)$ is mean-ergodic, i.e., if the time-average

$$\mathbf{y}_T = \frac{1}{2T}\int_{-T}^{T} \mathbf{y}(t)\,dt = \frac{\boldsymbol{\tau}_1 + \boldsymbol{\tau}_2 + \cdots + \boldsymbol{\tau}_n}{2T} \tag{9-147}$$

tends to the distribution function $F(x)$ of $\mathbf{x}(t)$ as $T \to \infty$. In the above, $\boldsymbol{\tau}_i$ are the time intervals during which $\mathbf{x}(t)$ is less than x (Fig. 9-15*a*).

Theorem An SSS process $\mathbf{x}(t)$ is distribution-ergodic iff

$$\lim_{T\to\infty} \frac{1}{2T}\int_{-2T}^{2T} \left(1 - \frac{|\tau|}{2T}\right)[F(x, x; \tau) - F^2(x)]\,d\tau = 0 \tag{9-148}$$

where

$$F(x_1, x_2; \tau) = P\{\mathbf{x}(t+\tau) \le x_1, \mathbf{x}(t) \le x_2\}$$

PROOF The product $\mathbf{y}(t+\tau)\mathbf{y}(t)$ equals 1 if $\mathbf{x}(t+\tau) \le x$ and $\mathbf{x}(t) \le x$ and it equals 0 otherwise. Hence

$$R_y(\tau) = 1 \cdot P\{\mathbf{x}(t+\tau) \le x, \mathbf{x}(t) \le x\} = F(x, x; \tau)$$

And since $E\{\mathbf{y}(t)\} = F(x)$, (9-148) follows from (9-141).

Notes 1. As we know [see (9-143)], (9-148) is satisfied if

$$F(x, x; \tau) \to F^2(x) \quad \text{as} \quad \tau \to \infty \tag{9-149}$$

From this it follows that the process $\mathbf{x}(t)$ is distribution-ergodic if the RVs $\mathbf{x}(t+\tau)$ and $\mathbf{x}(t)$ are *independent* for large τ.

2. If $\mathbf{x}(t)$ is distribution-ergodic, then for large T

$$F(x) \simeq \mathbf{y}_T = \frac{1}{2T}(\boldsymbol{\tau}_1 + \cdots + \boldsymbol{\tau}_n) \tag{9-150}$$

Thus, $F(x)$ equals the percentage of time that a single sample of $\mathbf{x}(t)$ is less than the number x.

Density Denoting by $\Delta\boldsymbol{\tau}_i$ the lengths of the time intervals during which $\mathbf{x}(t)$ is between x and $x + dx$ (Fig. 9-15*b*), we conclude from (9-150) that

$$f(x)\,\Delta x \simeq F(x+\Delta x) - F(x) \simeq \frac{1}{2T}\sum_k \Delta\boldsymbol{\tau}_k \tag{9-151}$$

Thus, $f(x)\,\Delta x$ equals the percentage of time that a single sample of a distribution-ergodic process is between x and $x + \Delta x$.

Correlation-Ergodic Processes

We wish to determine the autocorrelation

$$R(\lambda) = E\{\mathbf{x}(t+\lambda)\mathbf{x}(t)\}$$

of a stationary process $\mathbf{x}(t)$ in terms of a single sample. For this purpose, we form the process

$$\mathbf{z}_\lambda(t) = \mathbf{x}(t+\lambda)\mathbf{x}(t)$$

Clearly, $E\{\mathbf{z}_\lambda(\tau)\} = R(\lambda)$, hence, to determine $R(\lambda)$ it suffices to apply the earlier results substituting $\mathbf{z}_\lambda(t)$ for $\mathbf{x}(t)$.

We shall say that a process $\mathbf{x}(t)$ is correlation-ergodic if the corresponding process $\mathbf{z}_\lambda(t)$ is mean-ergodic, i.e., if

$$\mathbf{R}_T = \frac{1}{2T}\int_{-T}^{T} \mathbf{x}(t+\lambda)\mathbf{x}(t)\,dt \xrightarrow[T\to\infty]{} R(\lambda) \tag{9-152}$$

With

$$R_{zz}(\tau) = E\{\mathbf{x}(t+\lambda+\tau)\mathbf{x}(t+\tau)\mathbf{x}(t+\lambda)\mathbf{x}(t)\} \tag{9-153}$$

the autocorrelation of $\mathbf{z}_\lambda(t)$, we conclude that $\mathbf{x}(t)$ is correlation-ergodic iff the autocovariance

$$C_{zz}(\tau) = R_{zz}(\tau) - R^2(\lambda)$$

of $\mathbf{z}_\lambda(t)$ satisfies (9-141).

Notes 1. The integral in (9-152) cannot be evaluated if the available sample of $\mathbf{x}(t)$ is known only in the interval $(-T, T)$ because the term $\mathbf{x}(t+\tau)$ takes values outside this interval [see also (14-40)].

2. If $\mathbf{x}(t)$ is correlation-ergodic, then

$$\frac{1}{2T}\int_{-T}^{T} \mathbf{x}^2(t)\,dt \xrightarrow[T\to\infty]{} E\{\mathbf{x}^2(t)\} \tag{9-154}$$

$$\frac{1}{2T}\int_{-T}^{T} [\mathbf{x}(t+\tau)+\mathbf{x}(t)]^2\,dt \xrightarrow[T\to\infty]{} 2[R(0)+R(\tau)] \tag{9-155}$$

3. To find $C_{zz}(\tau)$, we need, in general, the fourth-order moments of $\mathbf{x}(t)$. However, if $\mathbf{x}(t)$ is *normal* with zero mean, then [see (8-62)]

$$C_{zz}(\tau) = R(\lambda+\tau)R(\lambda-\tau) + R^2(\tau) \tag{9-156}$$

This shows also that, if $R(\tau)\to 0$ then $C_{zz}(\tau)\to 0$ as $\tau\to\infty$, hence the process $\mathbf{x}(t)$ is then correlation-ergodic.

Analog Techniques

We have shown that, under general conditions, the statistics of a stochastic process can be determined from a single sample as time-averages. In the following, we examine the problem of evaluating these averages as responses of appropriate analog systems.

Mean We wish to estimate the mean η_x of a stationary process $\mathbf{x}(t)$. For this purpose, we use $\mathbf{x}(t)$ as the input to a low-pass (LP) filter (Fig. 9-16) with system function

$$W(\omega) = \int_{-\infty}^{\infty} w(t)e^{-j\omega t}\,dt$$

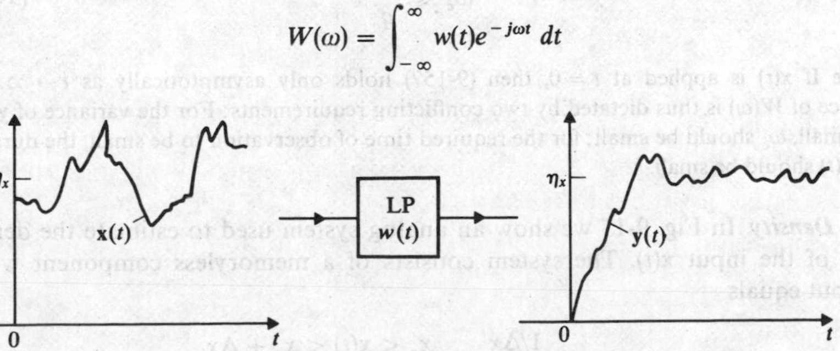

Figure 9-16

The resulting output is the process

$$\mathbf{y}(t) = \int_{-\infty}^{\infty} \mathbf{x}(t-\alpha)w(\alpha)\,d\alpha$$

and our objective is to find $W(\omega)$ such that

$$\mathbf{y}(t) \simeq \eta_x \tag{9-157}$$

This is the case only if

$$\eta_y \simeq \eta_x \qquad \text{and} \qquad \sigma_y \ll \eta_x \tag{9-158}$$

Since

$$\eta_y = \eta_x \int_{-\infty}^{\infty} w(t)\,dt = \eta_x W(0)$$

the first requirement yields $W(0) = 1$.

To meet the second requirement, we must evaluate σ_y. We shall do so under the assumption that the bandwidth ω_c of the LP filter is "small" in the sense that its impulse response $w(t)$ is nearly constant in any interval of length equal to the correlation length of $\mathbf{x}(t)$. With this assumption

$$\sigma_y^2 \simeq qE \qquad \text{where} \qquad q = \int_{-\infty}^{\infty} C_{xx}(\alpha)\,d\alpha \tag{9-159}$$

and

$$E = \int_{-\infty}^{\infty} w^2(t)\,dt = \frac{1}{2\pi}\int_{-\omega_c}^{\omega_c} |W(\omega)|^2\,d\omega \tag{9-160}$$

This follows from (9-119) if we replace $\mathbf{y}(t)$ by its centered process $\mathbf{y}(t) - \eta_y$. Thus, if

$$qE \ll \eta_x^2 \qquad \text{then} \qquad \mathbf{y}(t) \simeq \eta_x \tag{9-161}$$

The above can be expressed in terms of ω_c. Assuming that $|W(\omega)| \le W(0) = 1$, we conclude from (9-160) that $\pi E < \omega_c$. Hence, (9-161) is satisfied if

$$\omega_c \ll \frac{\pi \eta_x^2}{q} \tag{9-162}$$

Note If $\mathbf{x}(t)$ is applied at $t = 0$, then (9-157) holds only asymptotically as $t \to \infty$. The choice of $W(\omega)$ is thus dictated by two conflicting requirements: For the variance of $\mathbf{y}(t)$ to be small, ω_c should be small; for the required time of observation to be small, the duration of $w(t)$ should be small.

Density In Fig. 9-17 we show an analog system used to estimate the density $f(x)$ of the input $\mathbf{x}(t)$. The system consists of a memoryless component whose output equals

$$\mathbf{z}(t) = \begin{array}{ll} 1/\Delta x & x_o < \mathbf{x}(t) < x_o + \Delta x \\ 0 & \text{otherwise} \end{array}$$

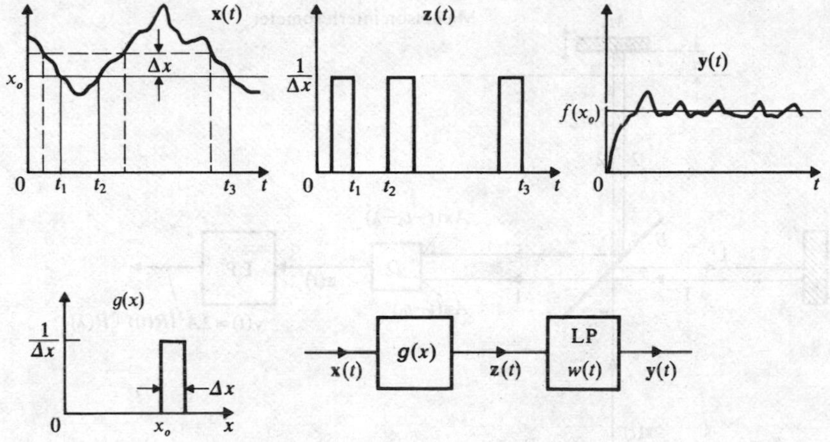

Figure 9-17

in cascade with an LP filter. In this case

$$E\{\mathbf{z}(t)\} = f(x_o) \qquad \mathbf{y}(t) \simeq f(x_o)$$

The variance of the output $\mathbf{y}(t)$ of the filter is given by

$$\sigma_y^2 \simeq E \int_{-\infty}^{\infty} [f(x_o, x_o; \tau) - f^2(x_o)]\, d\tau$$

This follows from (9-159) because the autocovariance of the process $\mathbf{z}(t)$ equals $f(x, x; \tau) - f^2(x)$.

Correlometer In Fig. 9-18 we show two devices that can be used to estimate the correlation $R(\lambda)$ of the input $\mathbf{x}(t)$. The first device consists of a delay element, a multiplier, and an LP filter. The second consists of a delay element, an adder, and a square-law detector followed by an LP filter. The analysis is the same as the

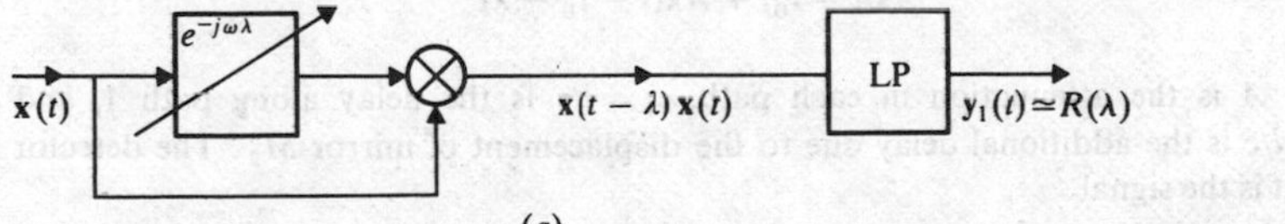

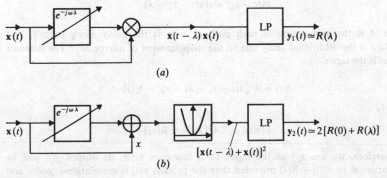

Figure 9-18

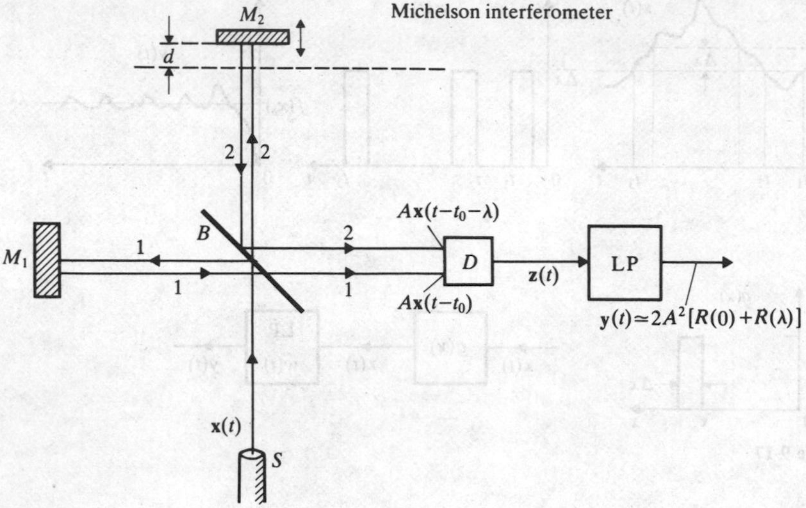

Fugure 9-19

analysis of the mean-estimate of Fig. 9-16 provided that $\mathbf{x}(t)$ is replaced by $\mathbf{x}(t-\lambda)\mathbf{x}(t)$ for the first device and by $[\mathbf{x}(t-\lambda)+\mathbf{x}(t)]^2$ for the second.

Example 9-23 An optical realization of the device of Fig. 9-18*b* is the *Michelson interferometer* shown in Fig. 9-19. It consists of a light source *S*, a beam-splitting surface *B*, and two mirrors. Mirror M_1 is in a fixed position and mirror M_2 is movable. The light from the source *S* is a random signal $\mathbf{x}(t)$ traveling with velocity *c* and it reaches a square-law detector *D* along paths 1 and 2 as shown. The lengths of these paths equal l and $l + 2d$ respectively where *d* is the displacement of mirror M_2 from its equilibrium position.

The signal reaching the detector is thus the sum

$$A\mathbf{x}(t-t_0) + A\mathbf{x}(t-t_0-\lambda)$$

where *A* is the attenuation in each path, $t_o = l/c$ is the delay along path 1, and $\lambda = 2d/c$ is the additional delay due to the displacement of mirror M_2. The detector output is the signal

$$\mathbf{z}(t) = A^2[\mathbf{x}(t-t_o-\lambda)+\mathbf{x}(t-t_o)]^2$$

Clearly

$$E\{\mathbf{z}(t)\} = 2A^2[R(0)+R(\lambda)]$$

If, therefore, we use $\mathbf{z}(t)$ as the input to a low-pass filter, its output $\mathbf{y}(t)$ will be proportional to $R(0)+R(\lambda)$ provided that the process $\mathbf{x}(t)$ is correlation-ergodic and the band of the filter is sufficiently narrow.

APPENDIX 9A CONTINUITY, DIFFERENTIATION, INTEGRATION

In the earlier discussion, we used routinely various limiting operations involving stochastic process, with the tacit assumption that these operations hold for every sample involved. This assumption is, in many cases, unnecessarily restrictive. To give some idea of the notion of limits in a more general sense, we discuss next conditions for the existence of MS limits and we show that these conditions can be phrased in terms of second-order moments (see also Sec. 8-4).

Stochastic continuity A process $\mathbf{x}(t)$ is called MS continuous if

$$E\{[\mathbf{x}(t+\varepsilon)-\mathbf{x}(t)]^2\} \xrightarrow[\varepsilon\to 0]{} 0 \tag{9A-1}$$

Theorem We maintain that $\mathbf{x}(t)$ is MS continuous if its autocorrelation is continuous.

PROOF Clearly

$$E\{[\mathbf{x}(t+\varepsilon)-\mathbf{x}(t)]^2\} = R(t+\varepsilon, t+\varepsilon) - 2R(t+\varepsilon, t) + R(t, t)$$

If, therefore, $R(t_1, t_2)$ is continuous, then the right side tends to zero as $\varepsilon \to 0$ and (9A-1) results.

Note Suppose that (9A-1) holds for *every* t in an interval I. From this it follows that [see (9-1)] almost all samples of $\mathbf{x}(t)$ will be continuous at a *particular* point of I. It does not follow, however, that these samples will be continuous for *every* point in I. We mention as illustrations the Poisson process, and the Wiener process. As we see from (9-14) and (9-25), both processes are MS continuous. However, the samples of the Poisson process are discontinuous at the points $\mathbf{t}_i$ whereas almost all samples of the Wiener process are continuous.

Corollary If $\mathbf{x}(t)$ is MS continuous, then its mean is continuous

$$\eta(t+\varepsilon) \to \eta(t) \qquad \varepsilon \to 0 \tag{9A-2}$$

PROOF As we know

$$E\{[\mathbf{x}(t+\varepsilon)-\mathbf{x}(t)]^2\} \ge E^2\{\mathbf{x}(t+\varepsilon)-\mathbf{x}(t)]\}$$

Hence (9A-2) follows from (9A-1).

The above shows that

$$\lim_{\varepsilon\to 0} E\{\mathbf{x}(t+\varepsilon)\} = E\left\{\lim_{\varepsilon\to 0} \mathbf{x}(t+\varepsilon)\right\} \tag{9A-3}$$

Stochastic differentiation A process $\mathbf{x}(t)$ is MS differentiable if

$$\frac{\mathbf{x}(t+\varepsilon)-\mathbf{x}(t)}{\varepsilon} \xrightarrow[\varepsilon\to 0]{} \mathbf{x}'(t) \tag{9A-4}$$

in the MS sense, i.e., if

$$E\left\{\left[\frac{\mathbf{x}(t+\varepsilon)-\mathbf{x}(t)}{\varepsilon} - \mathbf{x}'(t)\right]^2\right\} \xrightarrow[\varepsilon\to 0]{} 0 \tag{9A-5}$$

Theorem The process $\mathbf{x}(t)$ is MS differentiable if $\partial^2 R(t_1, t_2)/\partial t_1\, \partial t_2$ exists.

PROOF It suffices to show that (Cauchy criterion)

$$E\left\{\left[\frac{\mathbf{x}(t+\varepsilon_1)-\mathbf{x}(t)}{\varepsilon_1} - \frac{\mathbf{x}(t+\varepsilon_2)-\mathbf{x}(t)}{\varepsilon_2}\right]^2\right\} \xrightarrow[\varepsilon_1,\, \varepsilon_2\to 0]{} 0 \tag{9A-6}$$

We use this criterion because, unlike (9A-5), it does not involve the unknown $\mathbf{x}'(t)$. Clearly

$$E\{[\mathbf{x}(t+\varepsilon_1)-\mathbf{x}(t)][\mathbf{x}(t+\varepsilon_2)-\mathbf{x}(t)]\}$$
$$= R(t+\varepsilon_1, t+\varepsilon_2) - R(t+\varepsilon_1, t) - R(t, t+\varepsilon_2) + R(t, t)$$

The right side divided by $\varepsilon_1\varepsilon_2$ tends to $\partial^2 R(t, t)/\partial t\, \partial t$ which, by assumption, exists. Expanding the square in (9A-6) we conclude that its left side tends to

$$\frac{\partial^2 R(t, t)}{\partial t\, \partial t} - 2\frac{\partial^2 R(t, t)}{\partial t\, \partial t} + \frac{\partial^2 R(t, t)}{\partial t\, \partial t} = 0$$

Corollary The above yields

$$E\{\mathbf{x}'(t)\} = E\left\{\lim_{\varepsilon\to 0} \frac{\mathbf{x}(t+\varepsilon)-\mathbf{x}(t)}{\varepsilon}\right\} = \lim_{\varepsilon\to 0} E\left\{\frac{\mathbf{x}(t+\varepsilon)-\mathbf{x}(t)}{\varepsilon}\right\}$$

Note The autocorrelation of a Poisson process $\mathbf{x}(t)$ is discontinuous at the points $\mathbf{t}_i$, hence $\mathbf{x}'(t)$ does not exist at these points. However, as in the case of deterministic signals, it is convenient to introduce random impulses and to interpret $\mathbf{x}'(t)$ as in (9-110).

Stochastic integrals A process $\mathbf{x}(t)$ is MS integrable if the limit

$$\int_a^b \mathbf{x}(t)\, dt = \lim_{\Delta t_i \to 0} \sum_i \mathbf{x}(t_i)\, \Delta t_i \tag{9A-7}$$

exists in the MS sense.

Theorem The process $\mathbf{x}(t)$ is MS integrable if

$$\int_a^b \int_a^b |R(t_1, t_2)|\, dt_1\, dt_2 < \infty \tag{9A-8}$$

PROOF Using again the Cauchy criterion, we must show that

$$E\left\{\left|\sum_i \mathbf{x}(t_i)\,\Delta t_i - \sum_k \mathbf{x}(t_k)\,\Delta t_k\right|^2\right\} \xrightarrow[\Delta t_i,\,\Delta t_k \to 0]{} 0$$

This follows if we expand the square and use the identity

$$E\left\{\sum_i \mathbf{x}(t_i)\,\Delta t_i \sum_k \mathbf{x}(t_k)\,\Delta t_k\right\} = \sum_{i,k} R(t_i, t_k)\,\Delta t_i\,\Delta t_k$$

because the right side tends to the integral of $R(t_1, t_2)$ as Δt_i and Δt_k tend to zero.

Corollary From the above it follows that

$$E\left\{\left|\int_a^b \mathbf{x}(t)\,dt\right|^2\right\} = \int_a^b \int_a^b R(t_1, t_2)\,dt_1\,dt_2 \qquad \text{(9A-9)}$$

as in (9-11).

APPENDIX 9B
SHIFT OPERATORS AND STATIONARY PROCESSES

An SSS process can be generated by a succession of shifts $T\mathbf{x}$ of a single RV $\mathbf{x}$ where T is a one-to-one measure preserving transformation (mapping) of the probability space $\mathscr{S}$ into itself. This difficult topic is of fundamental importance in mathematics. In the following, we give a brief explanation of the underlying concept, limiting the discussion to the discrete-time case.

A *transformation* T of $\mathscr{S}$ into itself is a rule for assigning to each element ζ_i of $\mathscr{S}$ another element of $\mathscr{S}$

$$\tilde{\zeta}_i = T\zeta_i \qquad \text{(9B-1)}$$

called the *image* of ζ_i. The images $\tilde{\zeta}_i$ of all elements ζ_i of a subset $\mathscr{A}$ of $\mathscr{S}$ form another subset

$$\tilde{\mathscr{A}} = T\mathscr{A}$$

of $\mathscr{S}$ called the image of $\mathscr{A}$.

We shall assume that the transformation T has the following properties.

P_1: It is one-to-one. This means that

$$\text{if} \quad \zeta_i \neq \zeta_j \quad \text{then} \quad \tilde{\zeta}_i \neq \tilde{\zeta}_j$$

P_2: It is measure preserving. This means that if $\mathscr{A}$ is an event, then its image $\tilde{\mathscr{A}}$ is also an event and

$$P(\tilde{\mathscr{A}}) = P(\mathscr{A}) \qquad \text{(9B-2)}$$

Suppose that $\mathbf{x}$ is an RV and that T is a transformation as above. The expression $T\mathbf{x}$ will mean another RV

$$\mathbf{y} = T\mathbf{x} \qquad \text{such that} \qquad \mathbf{y}(\tilde{\zeta}_i) = \mathbf{x}(\zeta_i) \tag{9B-3}$$

where ζ_i is the unique inverse of $\tilde{\zeta}_i$. This specifies $\mathbf{y}$ for every element of $\mathscr{S}$ because (see P_1) the set of elements $\tilde{\zeta}_i$ equals $\mathscr{S}$.

The expression $\mathbf{z} = T^{-1}\mathbf{x}$ will mean that $\mathbf{x} = T\mathbf{z}$. Thus

$$\mathbf{z} = T^{-1}\mathbf{x} \qquad \text{iff} \qquad \mathbf{z}(\zeta_i) = \mathbf{x}(\tilde{\zeta}_i)$$

We can define similarly $T^2\mathbf{x} = T(T\mathbf{x}) = T\mathbf{y}$ and

$$T^n\mathbf{x} = T(T^{n-1}\mathbf{x}) = T^{-1}(T^{n+1}\mathbf{x})$$

for any n positive or negative.

From (9B-3) it follows that if, for some ζ_i, $\mathbf{x}(\zeta_i) \le w$, then $\mathbf{y}(\tilde{\zeta}_i) = \mathbf{x}(\zeta_i) \le w$. Hence the event $\{\mathbf{y} \le w\}$ is the image of the event $\{\mathbf{x} \le w\}$. This yields [see (9B-2)]

$$P\{\mathbf{x} \le w\} = P\{\mathbf{y} \le w\} \qquad \mathbf{y} = T\mathbf{x} \tag{9B-4}$$

for any w. We thus conclude that the RVs $\mathbf{x}$ and $T\mathbf{x}$ have the same distribution $F_x(x)$.

Given an RV $\mathbf{x}$ and a transformation T as above, we form the random process

$$\mathbf{x}_0 = \mathbf{x} \qquad \mathbf{x}_n = T^n\mathbf{x} \qquad n = -\infty, \ldots, \infty \tag{9B-5}$$

It follows from (9B-4) that the random variables $\mathbf{x}_n$ so formed have the same distribution. We can similarly show that their joint distributions of any order are invariant to a shift of the origin. Hence the process $\mathbf{x}_n$ so formed is SSS.

It can be shown that the converse is also true: Given an SSS process $\mathbf{x}_n$, we can find an RV $\mathbf{x}$ and a one-to-one measuring preserving transformation of the space $\mathscr{S}$ into itself such that for all essential purposes, $\mathbf{x}_n = T^n\mathbf{x}$. The proof of this difficult result will not be given.

PROBLEMS

9-1 In the fair-coin experiment, we define the process $\mathbf{x}(t)$ as follows: $\mathbf{x}(t) = \sin \pi t$ if heads shows, $\mathbf{x}(t) = 2t$ if tails shows. (*a*) Find $E\{\mathbf{x}(t)\}$. (*b*) Find $F(x, t)$ for $t = 0.25$, $t = 0.5$, and $t = 1$.

9-2 The process $\mathbf{x}(t) = e^{\mathbf{a}t}$ is a family of exponentials depending on the RV $\mathbf{a}$. Express the mean $\eta(t)$, the autocorrelation $R(t_1, t_2)$, and the first-order density $f(x, t)$ of $\mathbf{x}(t)$ in terms of the density $f_a(a)$ of $\mathbf{a}$.

9-3 Find the first-order characteristic function (*a*) of a Poisson process, and (*b*) of a Wiener process.

Answer: (*a*) $e^{\lambda t(e^{j\omega} - 1)}$; (*b*) $e^{\alpha t\omega^2/2}$

9-4 *Two-dimensional random walk.* The coordinates $\mathbf{x}(t)$ and $\mathbf{y}(t)$ of a moving object are two independent random-walk processes with the same s and T as in (9-23). Show that if $\mathbf{z}(t) = \sqrt{\mathbf{x}^2(t) + \mathbf{y}^2(t)}$ is the distance of the object from the origin and $t \gg T$, then for z of the order of $\sqrt{\alpha t}$

$$f_z(z, t) \simeq \frac{z}{\alpha t} e^{-z^2/2\alpha t} U(z) \qquad \alpha = \frac{s^2}{T}$$

9-5 The RV $\mathbf{c}$ is uniform in the interval $(0, T)$. Find $R_x(t_1, t_2)$ if (*a*) $\mathbf{x}(t) = U(t - \mathbf{c})$, (*b*) $\mathbf{x}(t) = \delta(t - \mathbf{c})$.

9-6 The RVs $\mathbf{a}$ and $\mathbf{b}$ are independent $N(0; \sigma)$ and p is the probability that the process $\mathbf{x}(t) = \mathbf{a} - \mathbf{b}t$ crosses the t axis in the interval $(0, T)$. Show that $\pi p = \arctan T$.

Hint: $p = P\{0 \le \mathbf{a}/\mathbf{b} \le T\}$.

9-7 Show that if

$$R_v(t_1, t_2) = q(t_1)\delta(t_1 - t_2) \qquad \text{and} \qquad \mathbf{w}(t) = \int_0^t \mathbf{v}(\tau)\, d\tau$$

then

$$E\{\mathbf{w}^2(t)\} = \int_0^t q(\tau)\, d\tau$$

9-8 The process $\mathbf{x}(t)$ is WSS with autocorrelation $R(\tau)$. (*a*) Show that

$$P\{|\mathbf{x}(t + \tau) - \mathbf{x}(t)| \ge a\} \le 2[R(0) - R(\tau)]/a^2$$

(*b*) Express $P\{|\mathbf{x}(t + \tau) - \mathbf{x}(t)| \ge a\}$ in terms of the second-order density $f(x_1, x_2; \tau)$ of $\mathbf{x}(t)$.

9-9 Show that if $\boldsymbol{\varphi}$ is an RV with $\Phi(\lambda) = E\{e^{j\lambda\boldsymbol{\varphi}}\}$ and $\Phi(1) = \Phi(2) = 0$, then the process $\mathbf{x}(t) = \cos(\omega t + \boldsymbol{\varphi})$ is WSS. Find $E\{\mathbf{x}(t)\}$ and $R_x(\tau)$ if $\boldsymbol{\varphi}$ is uniform in the interval $(-\pi, \pi)$.

9-10 Given a process $\mathbf{x}(t)$ with orthogonal increments and such that $\mathbf{x}(0) = 0$, show that (*a*) $R(t_1, t_2) = R(t_1, t_1)$ for $t_1 \le t_2$, and (*b*) if $E\{[\mathbf{x}(t_1) - \mathbf{x}(t_2)]^2\} = q|t_1 - t_2|$ then the process $\mathbf{y}(t) = [\mathbf{x}(t + \varepsilon) - \mathbf{x}(t)]/\varepsilon$ is WSS and its autocorrelation is a triangle with area q and base 2ε.

9-11 Show that if $R_{xx}(t_1, t_2) = q(t_1)\delta(t_1 - t_2)$ and $\mathbf{y}(t) = \mathbf{x}(t) * h(t)$ then

$$E\{\mathbf{x}(t)\mathbf{y}(t)\} = h(0)q(t).$$

9-12 The process $\mathbf{x}(t)$ is normal with $\eta_x = 0$ and $R_x(\tau) = 4e^{-3|\tau|}$. Find a memoryless system $g(x)$ such that the first-order density $f_y(y)$ of the resulting output $\mathbf{y}(t) = g[\mathbf{x}(t)]$ is uniform in the interval $(6, 9)$.

Answer: $g(x) = 3\mathbb{G}(x/2) + 6$.

9-13 Show that if $\mathbf{x}(t)$ is an SSS process and $\boldsymbol{\varepsilon}$ is an RV independent of $\mathbf{x}(t)$ then the process $\mathbf{y}(t) = \mathbf{x}(t - \boldsymbol{\varepsilon})$ is SSS.

9-14 Show that if $\mathbf{x}(t)$ is a stationary process with derivative $\mathbf{x}'(t)$ then for a given t the RVs $\mathbf{x}(t)$ and $\mathbf{x}'(t)$ are orthogonal and uncorrelated.

9-15 Given a normal process $\mathbf{x}(t)$ with $\eta_x = 0$ and $R_x(\tau) = 4e^{-2|\tau|}$, we form the RVs $\mathbf{z} = \mathbf{x}(t + 1)$, $\mathbf{w} = \mathbf{x}(t - 1)$, (*a*) find $E\{\mathbf{zw}\}$ and $E\{(\mathbf{z} + \mathbf{w})^2\}$, (*b*) find

$$f_z(z) \qquad P\{\mathbf{z} < 1\} \qquad f_{zw}(z, w)$$

9-16 Show that if $\mathbf{x}(t)$ is normal with autocorrelation $R(\tau)$, then

$$P\{\mathbf{x}'(t) \le a\} = \mathbb{G}\left[\frac{-a}{R''(0)}\right]$$

9-17 Show that (*a*) if

$$\mathbf{y}(t) = a\mathbf{x}(ct) \qquad \text{then} \qquad R_y(\tau) = a^2 R_x(c\tau)$$

(*b*) if

$$\mathbf{z}(t) = \lim_{\varepsilon \to 0} \sqrt{\varepsilon}\,\mathbf{x}(\varepsilon t) \qquad \text{then} \qquad R_z(\tau) = q\delta(\tau) \qquad q = \int_{-\infty}^{\infty} R_x(\tau)\,d\tau$$

9-18 Show that if $\mathbf{x}(t)$ is white noise, $h(t) = 0$ outside the interval $(0, T)$ and $\mathbf{y}(t) = \mathbf{x}(t) * h(t)$ then $R_{yy}(t_1, t_2) = 0$ for $|t_1 - t_2| > T$.

9-19 Show that if

$$R_{xx}(t_1, t_2) = q(t_1)\delta(t_1 - t_2) \qquad E\{\mathbf{y}^2(t)\} = I(t)$$

and

$$(a)\ \mathbf{y}(t) = \int_0^t h(t, \alpha)\mathbf{x}(\alpha)\,d\alpha \qquad \text{then} \qquad I(t) = \int_0^t h^2(t, \alpha)q(\alpha)\,d\alpha$$

$$(b)\ \mathbf{y}'(t) + c(t)\mathbf{y}(t) = \mathbf{x}(t) \qquad \text{then} \qquad I'(t) + 2c(t)I(t) = q(t)$$

9-20 Find $E\{\mathbf{y}^2(t)\}$ (*a*) if $R_{xx}(t) = 5\delta(\tau)$ and

$$\mathbf{y}'(t) + 2\mathbf{y}(t) = \mathbf{x}(t) \qquad \text{all } t \tag{i}$$

(*b*) if (i) holds for $t > 0$ only and $\mathbf{y}(t) = 0$ for $t \le 0$.
Hint: Use (9-117).

9-21 The input to a linear system with $h(t) = Ae^{-\alpha t}U(t)$ is a process $\mathbf{x}(t)$ with $R_x(\tau) = N\delta(\tau)$ applied at $t = 0$ and disconnected at $t = T$. Find and sketch $E\{\mathbf{y}^2(t)\}$.
Hint: Use (9-117) with $q(t) = N$ for $0 < t < T$ and zero otherwise.

9-22 Show that if

$$s = \int_0^{10} \mathbf{x}(t)\,dt \qquad \text{then} \qquad E\{\mathbf{s}^2\} = \int_{-10}^{10} (10 - |\tau|)R_x(\tau)\,d\tau$$

Find the mean and variance of $\mathbf{s}$ if $E\{\mathbf{x}(t)\} = 8$, $R_x(\tau) = 64 + 10e^{-2|\tau|}$.

9-23 We are given the data $\mathbf{x}(t) = f(t) + \mathbf{n}(t)$ where $R_n(\tau) = N\delta(\tau)$ and $E\{\mathbf{n}(t) = 0\}$. We wish to estimate the integral

$$g(t) = \int_0^t f(\alpha)\,d\alpha$$

knowing that $g(T) = 0$. Show that if we use as the estimate of $g(t)$ the process $\mathbf{w}(t) = \mathbf{z}(t) - \mathbf{z}(T)t/T$ where

$$\mathbf{z}(t) = \int_0^t \mathbf{x}(\alpha)\,d\alpha \qquad \text{then} \qquad E\{\mathbf{w}(t)\} = g(t) \qquad \sigma_w^2 = Nt\left(1 - \frac{t}{T}\right)$$

9-24 Show that if $\mathbf{x}(t)$ is a normal process with zero mean and $\mathbf{y}(t) = \operatorname{sgn} \mathbf{x}(t)$, then

$$R_y(\tau) = \frac{2}{\pi} \sum_{n=1}^{\infty} \frac{1}{n} \left[J_0(n\pi) - (-1)^n \right] \sin n\pi \frac{R_x(\tau)}{R_x(0)}$$

where $J_0(x)$ is the Bessel function.

Hint: Expand the arcsine in (9-91) into a Fourier series.

9-25 Show that if

$$\mathbf{y}[n] = \mathbf{x}[n] * h[n] = \sum_{k=-\infty}^{\infty} \mathbf{x}[n-k]h[k]$$

is the output of a linear system with input $\mathbf{x}[n]$ and delta response $h[n]$, then

(*a*) $R_{xy}[m_1, m_2] = R_{xx}[m_1, m_2] * h^*[m_2]$

$R_{yy}[m_1, m_2] = R_{xy}[m_1, m_2] * h[m_1]$

(*b*) If $R_{xx}[m_1, m_2] = q[m_1]\delta[m_1 - m_2]$, then

$$E\{|\mathbf{y}[n]|^2\} = q[n] * |h[n]|^2 = \sum_{k=-\infty}^{\infty} q[n-k]|h[k]|^2$$

9-26 The process $\mathbf{x}(t)$ is WSS with $R_{xx}(\tau) = 5\delta(\tau)$ and

$$\mathbf{y}'(t) + 2\mathbf{y}(t) = \mathbf{x}(t) \qquad \text{(i)}$$

Find $E\{\mathbf{y}^2(t)\}$, $R_{xy}(t_1, t_2)$, $R_{yy}(t_1, t_2)$ (*a*) if (i) holds for all t, (*b*) if $\mathbf{y}(0) = 0$ and (i) holds for $t \geq 0$.

9-27 (*a*) Find $E\{\mathbf{y}^2(t)\}$ if $\mathbf{y}(0) = 0$ and

$$\mathbf{y}''(t) + 7\mathbf{y}'(t) + 10\mathbf{y}(t) = \mathbf{x}(t) \qquad R_x(\tau) = 5\delta(\tau)$$

(*b*) Find $E\{\mathbf{y}^2[n]\}$ if $\mathbf{y}[-1] = 0$ and

$$8\mathbf{y}[n] - 6\mathbf{y}[n-1] + \mathbf{y}[n-2] = \mathbf{x}[n] \qquad R_x[m] = 5\delta[m]$$

9-28 The process $\mathbf{x}[n]$ is WSS with $R_{xx}[m] = 5\delta[m]$ and

$$\mathbf{y}[n] - 0.5\mathbf{y}[n-1] = \mathbf{x}[n] \qquad \text{(i)}$$

Find $E\{\mathbf{y}^2[n]\}$, $R_{xy}[m_1, m_2]$, $R_{yy}[m_1, m_2]$ (*a*) if (i) holds for all n, (*b*) if $\mathbf{y}[-1] = 0$ and (i) holds for $n \geq 0$.

9-29 Show that (*a*) if $R_x[m_1, m_2] = q[m_1]\delta[m_1 - m_2]$ and

$$\mathbf{s} = \sum_{n=0}^{N} a_n \mathbf{x}[n] \qquad \text{then} \qquad E\{\mathbf{s}^2\} = \sum_{n=0}^{N} a_n^2 q[n]$$

(*b*) If $R_{xx}(t_1, t_2) = q(t_1)\delta(t_1 - t_2)$ and

$$\mathbf{s} = \int_0^T a(t)\mathbf{x}(t)\, dt \qquad \text{then} \qquad E\{\mathbf{s}^2\} = \int_0^T a^2(t)q(t)\, dt$$

9-30 Find the mean and variance of the RV

$$\mathbf{n}_T = \frac{1}{2T} \int_{-T}^{T} \mathbf{x}(t)\, dt \qquad \text{where} \qquad \mathbf{x}(t) = 10 + 2\delta(\tau)$$

for $T = 5$ and for $T = 100$.

9-31 Show that if a process is distribution-ergodic, then it is also mean-ergodic.
Hint: Use (9-149).

9-32 Show that if $\mathbf{x}(t)$ is normal with $\eta_x = 0$ and $R_x(\tau) = 0$ for $|\tau| > a$, then it is correlation-ergodic.

9-33 Show that the process $\mathbf{a}e^{j(\omega t + \boldsymbol{\varphi})}$ is not correlation ergodic.

9-34 Show that

$$R_{xy}(\lambda) = \lim_{T\to\infty} \frac{1}{2T}\int_{-T}^{T} \mathbf{x}(t+\lambda)\mathbf{y}(t)\,dt$$

iff

$$\lim_{T\to\infty} \frac{1}{2T}\int_{-2T}^{2T} \left(1 - \frac{|\tau|}{2T}\right) E\{\mathbf{x}(t+\lambda+\tau)\mathbf{y}(t+\tau)\mathbf{x}(t+\lambda)\mathbf{y}(t)\}\,d\tau = R_{xy}^2(\lambda)$$

9-35 The process $\mathbf{x}(t)$ is cyclostationary with period T, mean $\eta(t)$, and correlation $R(t_1, t_2)$. Show that if $R(t+\tau, t) \to 0$ as $|\tau| \to \infty$, then

$$\lim_{c\to\infty} \frac{1}{2c}\int_{-c}^{c} \mathbf{x}(t)\,dt = \frac{1}{T}\int_0^T \eta(t)\,dt$$

Hint: The process $\bar{\mathbf{x}}(t) = \mathbf{x}(t - \boldsymbol{\theta})$ is mean-ergodic.

9-36 Show that if

$$C(t+\tau, t) \xrightarrow[\tau\to\infty]{} 0$$

uniformly in t, then $\mathbf{x}(t)$ is mean-ergodic.

9-37 Show that if $\mathbf{x}[n]$ is a WSS process with autovariance $C[m]$ and

$$\boldsymbol{\eta}_N = \frac{1}{2N+1}\sum_{n=-N}^{N} \mathbf{x}[n] \qquad \text{then} \qquad \sigma_{\eta_N}^2 = \frac{1}{2N+1}\sum_{m=-2N}^{2N} C[m]\left(1 - \frac{|m|}{2N+1}\right)$$

CHAPTER

TEN

SPECTRAL ANALYSIS

10-1 CORRELATIONS AND SPECTRA

We shall study the properties of the autocorrelation

$$R(\tau) = E\{\mathbf{x}(t + \tau)\mathbf{x}^*(t)\}$$

of a WSS process $\mathbf{x}(t)$, real or complex, and of the cross-correlation

$$R_{xy}(\tau) = E\{\mathbf{x}(t + \tau)\mathbf{y}^*(t)\}$$

of two jointly WSS processes $\mathbf{x}(t)$ and $\mathbf{y}(t)$. The properties of the covariances

$$C(\tau) = R(\tau) - |\eta|^2 \qquad C_{xy}(\tau) = R_{xy}(\tau) - \eta_x \eta_y^*$$

will follow because these functions are the correlations of the centered processes $\mathbf{x}(t) - \eta_x$ and $\mathbf{y}(t) - \eta_y$.

From the definition it follows readily that

$$R(-\tau) = R^*(\tau) \qquad R_{xy}(-\tau) = R_{yx}^*(\tau) \tag{10-1}$$

And if $\mathbf{x}(t)$ is real, then $R(\tau)$ is real and $R(-\tau) = R(\tau)$.

If the processes $\mathbf{x}(t)$ and $\mathbf{y}(t)$ are jointly WSS, then the process

$$\mathbf{z}(t) = a\mathbf{x}(t) + b\mathbf{y}(t)$$

is WSS and its autocorrelation equals

$$R_{zz}(\tau) = |a|^2 R_{xx}(\tau) + ab^* R_{xy}(\tau) + a^*bR_{yx}(\tau) + |b|^2 R_{yy}(\tau) \tag{10-2}$$

In particular, if

$$\mathbf{z}(t) = a\mathbf{x}(t) \qquad \text{then} \qquad R_{zz}(\tau) = |a|^2 R_{xx}(\tau)$$

If the process $\mathbf{x}(t)$ and $\mathbf{y}(t)$ are real and

$$\mathbf{z}(t) = \mathbf{x}(t) + j\mathbf{y}(t)$$

then (10-2) yields

$$R_{zz}(\tau) = [R_{xx}(\tau) + R_{yy}(\tau)] - j[R_{xy}(\tau) - R_{yx}(\tau)] \tag{10-3}$$

Note In general, the real part $\mathbf{x}(t) = \text{Re } \mathbf{z}(t)$ of a WSS process $\mathbf{z}(t)$ is not WSS. However, if the processes $\mathbf{z}(t)$ and $\mathbf{z}^*(t)$ are jointly WSS, then $\mathbf{x}(t)$ is WSS because $2\mathbf{x}(t) = \mathbf{z}(t) + \mathbf{z}^*(t)$.

Properties As we know [see (9-30)], the function $R(\tau)$ is p.d.

$$\sum_{i,\,k} a_i a_k^* R(\tau_i - \tau_k) \geq 0 \tag{10-4}$$

In the next section we show that this can be expressed in terms of the Fourier transform $S(\omega)$ of $R(\tau)$: For any ω, $S(\omega) \geq 0$.

The following properties of $R(\tau)$ can be derived from (10-4); we shall establish them, however, directly using the inequality [see (7-12) and Prob. 7-3]

$$|E\{\mathbf{zw}\}|^2 \leq E\{|\mathbf{z}|^2\}E\{|\mathbf{w}|^2\} \tag{10-5}$$

With $\mathbf{z} = \mathbf{x}(t + \tau)$, $\mathbf{w} = \mathbf{y}^*(t)$, this yields

$$|E\{\mathbf{x}(t + \tau)\mathbf{y}^*(t)\}|^2 \leq E\{|\mathbf{x}(t + \tau)|^2\}E\{|\mathbf{y}(t)|^2\}$$

Hence

$$|R_{xy}(\tau)|^2 \leq R_{xx}(0)R_{yy}(0) \tag{10-6}$$

From the above it follows with $\mathbf{x}(t) = \mathbf{y}(t)$ that

$$|R(\tau)| \leq R(0) \tag{10-7}$$

For real processes, this can be derived also from the identity

$$0 \leq E\{[\mathbf{x}(t + \tau) \pm \mathbf{x}(t)]^2\} = 2[R(0) \pm R(\tau)]$$

Theorem If $\mathbf{x}(t)$ is a real process and $R(\tau_1) = R(0)$ for some $\tau = \tau_1 \neq 0$, then $R(\tau)$ is periodic with period τ_1

$$R(\tau + m\tau_1) = R(\tau) \qquad \text{all } \tau \tag{10-8}$$

Proof From (10-5) it follows that

$$E^2\{[\mathbf{x}(t + \tau + \tau_1) - \mathbf{x}(t + \tau)]\mathbf{x}(t)\} \leq E\{[\mathbf{x}(t + \tau + \tau_1) - \mathbf{x}(t + \tau)]^2\}E\{\mathbf{x}^2(t)\}$$

Hence

$$[R(\tau + \tau_1) - R(\tau)]^2 \leq 2[R(0) - R(\tau_1)]R(0) \tag{10-9}$$

for every τ and τ_1. If, therefore, $R(\tau_1) = R(0)$, then $R(\tau + \tau_1) = R(\tau)$ for every τ. Repeating this reasoning we conclude that $R(\tau + m\tau_1) = R(\tau)$ for every integer m.

Corollaries 1. If $|R(\tau_1)| = R(0)$, then $R(\tau)$ is of the form $R(\tau) = e^{j\omega_0\tau}w(\tau)$ where $w(\tau + m\tau_1) = w(\tau)$. Furthermore, $\mathbf{x}(t) = e^{j\omega_0 t}\mathbf{y}(t)$ where $\mathbf{y}(t)$ is MS periodic.

PROOF With

$$R(\tau_1) = R(0)e^{j\varphi} \qquad \omega_0 = \varphi/\tau_1 \qquad \mathbf{y}(t) = \mathbf{x}(t)e^{-j\omega_0 t}$$

we have

$$R_{yy}(\tau) = E\{\mathbf{x}(t+\tau)e^{-j\omega_0(t+\tau)}\mathbf{x}^*(t)e^{j\omega_0 t}\} = R(\tau)e^{-j\omega_0\tau}$$

Hence

$$R_{yy}(\tau_1) = e^{-j\omega_0\tau_1}R(\tau_1) = R(0)$$

From this and (10-8) it follows that the function $R_{yy}(\tau) = w(\tau)$ is periodic.

2. If $R(\tau_1) = R(\tau_2) = R(0)$ and the numbers τ_1 and τ_2 are noncommensurate, then $R(\tau)$ = constant.

PROOF The function $R(\tau)$ is periodic with periods τ_1 and τ_2. If, therefore, the ratio τ_1/τ_2 is irrational, then $R(\tau)$ must be constant.

Continuity From (10-9) it follows that if $R(0) < \infty$ and $R(\tau)$ is continuous for $\tau = 0$, then it is continuous for every τ.

PROOF From the continuity of $R(\tau)$ it follows that $R(\tau_1) \to R(0)$ as $\tau_1 \to 0$, hence [see (10-9)] $R(\tau + \tau_1) \to R(\tau)$ as $\tau_1 \to 0$.

Power Spectrum

The *power spectrum* (or *spectral density*) $S(\omega)$ of a process $\mathbf{x}(t)$ is the Fourier transform of its autocorrelation

$$S(\omega) = \int_{-\infty}^{\infty} R(\tau)e^{-j\omega\tau}\,d\tau \tag{10-10}$$

The Fourier inversion formula yields

$$R(\tau) = \frac{1}{2\pi}\int_{-\infty}^{\infty} S(\omega)e^{j\omega\tau}\,d\omega \tag{10-11}$$

Since $R(-\tau) = R^*(\tau)$, we conclude that the power spectrum of a process $\mathbf{x}(t)$, real or complex, is a *real* function of ω.

If $\mathbf{x}(t)$ is real, then $R(\tau)$ is real and even, hence $S(\omega)$ is also real and even. In this case

$$S(\omega) = \int_{-\infty}^{\infty} R(\tau)\cos\omega\tau\,d\tau \qquad R(\tau) = \frac{1}{2\pi}\int_{-\infty}^{\infty} S(\omega)\cos\omega\tau\,d\omega \tag{10-12}$$

The *cross-power spectrum* $S_{xy}(\omega)$ of two processes $\mathbf{x}(t)$ and $\mathbf{y}(t)$ is the Fourier transform of their cross-correlation

$$S_{xy}(\omega) = \frac{1}{2\pi}\int_{-\infty}^{\infty} R_{xy}(\tau)e^{-j\omega\tau}\,d\tau \tag{10-13}$$

The inversion formula yields

$$R_{xy}(\tau) = \frac{1}{2\pi}\int_{-\infty}^{\infty} S_{xy}(\omega)e^{j\omega\tau}\,d\omega \tag{10-14}$$

The function $S_{xy}(\omega)$ is, in general, complex even when $\mathbf{x}(t)$ is real. Furthermore

$$S_{xy}(\omega) = S^*_{yx}(\omega) \tag{10-15}$$

because [see (10-1)] $R_{xy}(\tau) = R^*_{yx}(-\tau)$.

Example 10-1 If $\mathbf{x}(t)$ is the binary transmission process (see Example 9-14), then $R(\tau)$ is a triangle [see (9-80)]

$$R(\tau) = \begin{cases} 1 - \dfrac{|\tau|}{T} & |\tau| < T \\ 0 & |\tau| > T \end{cases}$$

as in Fig. 10-1 and

$$S(\omega) = \frac{4\sin^2(\omega T/2)}{T\omega^2}$$

Example 10-2 If $\mathbf{z}(t)$ is a sum of Poisson impulses

$$\mathbf{z}(t) = \sum_i \delta(t - \mathbf{t}_i)$$

then [see (9-112)]

$$R(\tau) = \lambda^2 + \lambda\delta(\tau)$$

Using the Fourier transform pairs

$$\delta(t) \leftrightarrow 1 \qquad 1 \leftrightarrow 2\pi\delta(\omega)$$

we obtain (Fig. 10-2)

$$S(\omega) = 2\pi\lambda^2\delta(\omega) + \lambda$$

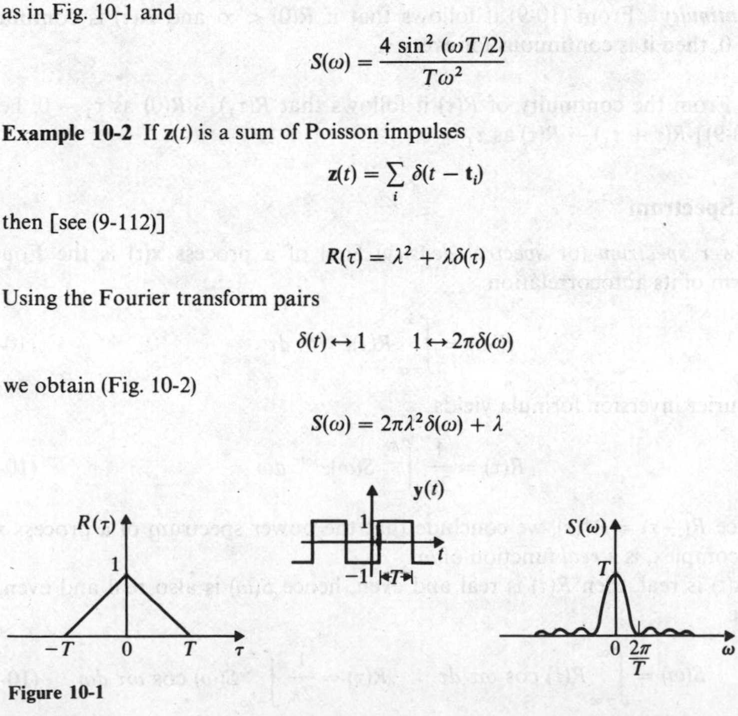

Figure 10-1

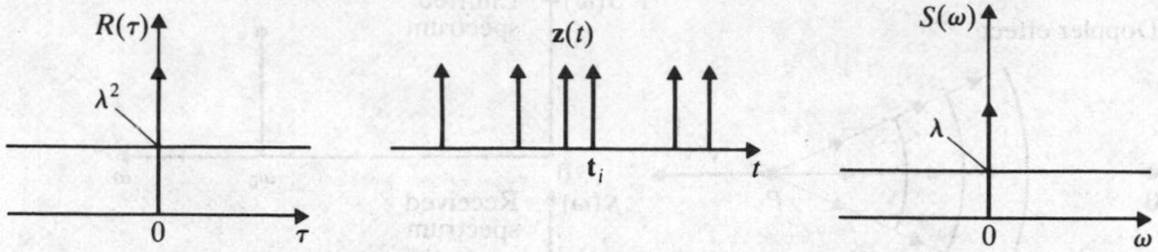

Figure 10-2

Example 10-3 (line spectra) (*a*) Suppose that the RVs $\mathbf{a}_i$ are uncorrelated with zero mean and variance σ_i^2. With

$$\mathbf{x}(t) = \sum_i \mathbf{a}_i e^{j\omega_i t}$$

it follows that (see Example 9-5)

$$R(\tau) = \sum_i \sigma_i^2 e^{j\omega_i \tau} \leftrightarrow S(\omega) = 2\pi \sum_i \sigma_i^2 \delta(\omega - \omega_i)$$

because

$$e^{j\omega_i \tau} \leftrightarrow 2\pi\delta(\omega - \omega_i)$$

(*b*) Suppose now that the RVs $\mathbf{a}_i$ and $\mathbf{b}_i$ are uncorrelated with zero mean and

$$E\{\mathbf{a}_i^2\} = E\{\mathbf{b}_i^2\} = \sigma_i^2$$

With

$$\mathbf{x}(t) = \sum_i (\mathbf{a}_i \cos \omega_i t + \mathbf{b}_i \sin \omega_i t)$$

we conclude as in (9-61) that

$$R(\tau) = \sum_i \sigma_i^2 \cos \omega_i \tau \leftrightarrow S(\omega) = \pi \sum \sigma_i^2 [\delta(\omega - \omega_i) + \delta(\omega + \omega_i)]$$

because

$$2 \cos \omega_i \tau \leftrightarrow \delta(\omega - \omega_i) + \delta(\omega + \omega_i)$$

Example 10-4 Consider the process

$$\mathbf{x}(t) = ae^{j(\boldsymbol{\omega} t - \varphi)} \tag{10-16}$$

where $\boldsymbol{\omega}$ is an RV with density $f_\omega(\omega)$. In this case

$$E\{\mathbf{x}(t + \tau)\mathbf{x}^*(t)\} = a^2 E\{e^{j\boldsymbol{\omega}\tau}\}$$

Hence

$$R(\tau) = a^2 \int_{-\infty}^{\infty} f_\omega(\omega) e^{j\omega\tau} \, d\omega \tag{10-17}$$

Comparing with (10-11) we conclude that

$$S(\omega) = 2\pi a^2 f_\omega(\omega) \tag{10-18}$$

The following is an interesting application.

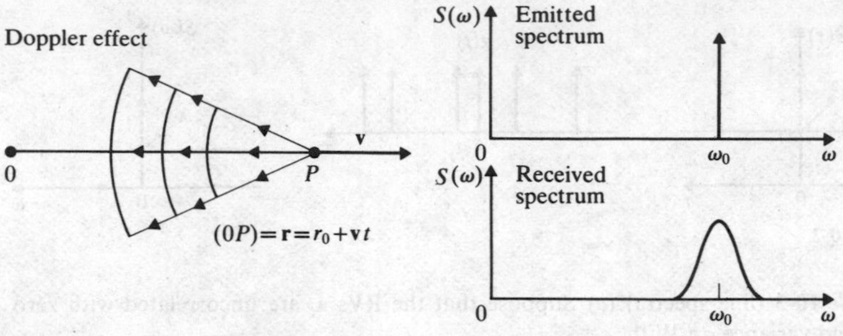

Figure 10-3

Doppler effect A harmonic oscillator located at a point P of the x axis (Fig. 10-3) moves in the x direction with velocity $\mathbf{v}$. The emitted signal equals $e^{j\omega_0 t}$ and the signal received by an observer located at the origin O equals

$$\mathbf{s}(t) = ae^{j\omega_0(t - \mathbf{r}/c)}$$

where $\mathbf{r} = r_0 + \mathbf{v}t$ is the distance from O to P and c is the velocity of propagation. Thus

$$\mathbf{s}(t) = ae^{j(\boldsymbol{\omega} t - \varphi)} \qquad \boldsymbol{\omega} = \omega_0\left(1 - \frac{\mathbf{v}}{c}\right) \qquad \varphi = \frac{r_0\,\omega_0}{c}$$

as in (10-16). Denoting by $f_v(v)$ the density of $\mathbf{v}$, we conclude from (5-6) that

$$f_\omega(\omega) = \frac{c}{\omega_0} f_v\left[\left(1 - \frac{\omega}{\omega_0}\right)c\right]$$

Hence, the spectrum of the received signal equals

$$S(\omega) = \frac{2\pi a^2 c}{\omega_0} f_v\left[\left(1 - \frac{\omega}{\omega_0}\right)c\right] \tag{10-19}$$

This holds also if the motion forms an angle with the x axis provided that $\mathbf{v}$ is replaced by the x component of the velocity.

We note that if $\mathbf{v} = 0$, then $S(\omega) = 2\pi a^2 \delta(\omega - \omega_0)$. Thus, the motion causes broadening of the spectrum of $\mathbf{s}(t)$.

Maxwell spectrum If the emitter is a particle in a gas of temperature T, then (see Prob. 8-5) the $\mathbf{x}$ component of its velocity is a normal RV with zero mean and variance kT/m. Inserting into (10-19), we conclude that the spectrum of the received signal $\mathbf{s}(t)$ is the normal curve

$$S(\omega) = \frac{2\pi a^2 c}{\omega_0\sqrt{2\pi kT/m}} \exp\left\{-\frac{mc^2}{2kT}\left(1 - \frac{\omega}{\omega_0}\right)^2\right\}$$

From this and the pair

$$e^{-\alpha\tau^2} e^{j\omega_0\tau} \leftrightarrow \sqrt{\frac{\pi}{\alpha}}\, e^{-(\omega - \omega_0)^2/4\alpha}$$

it follows that the autocorrelation of $\mathbf{s}(t)$ equals

$$R(\tau) = a^2 \exp\left\{-\frac{kT\omega_0^2\tau^2}{2mc^2}\right\} e^{j\omega_0\tau}$$

Thus, $R(\tau)$ is a modulated sine wave with envelope a normal curve.

Example 10-5 We have shown in Example 9-12 that if $\boldsymbol{\omega}$ is an RV with density $f(\omega)$ and $\boldsymbol{\varphi}$ is an RV uniform in the interval $(-\pi, \pi)$ and independent of $\boldsymbol{\omega}$, then the autocorrelation of the process

$$\mathbf{x}(t) = a \cos(\boldsymbol{\omega} t + \boldsymbol{\varphi}) \tag{10-20}$$

equals

$$R(\tau) = \frac{a^2}{2} E\{\cos \boldsymbol{\omega}\tau\} = \frac{a^2}{2}\int_{-\infty}^{\infty} f(\omega) \cos \omega\tau \, d\omega \tag{10-21}$$

If $f(-\omega) = f(\omega)$ then the above leads to the conclusion that $S(\omega) = \pi a^2 f(\omega)$ [see (10-12)]. If $f(\omega)$ is not even, then, to determine $S(\omega)$, we write (10-21) in the form

$$R(\tau) = \frac{a^2}{4}\int_{-\infty}^{\infty} [f(\omega) + f(-\omega)] e^{j\omega\tau} \, d\omega$$

Hence [see (10-11)]

$$S(\omega) = \frac{\pi a^2}{2}[f(\omega) + f(-\omega)] \tag{10-22}$$

We shall find it convenient to introduce also the Laplace transform† of $R(\tau)$

$$\mathbf{S}(s) = \int_{-\infty}^{\infty} R(\tau)e^{-s\tau} \, d\tau = \mathbf{S}^+(s) + \mathbf{S}^-(s) \tag{10-23}$$

In the above

$$\mathbf{S}^+(s) = \int_0^{\infty} R(\tau)e^{-s\tau} \, d\tau \qquad \mathbf{S}^-(s) = \int_{-\infty}^{0} R(\tau)e^{-s\tau} \, d\tau$$

are the transforms of the causal and anticausal components

$$R^+(\tau) = R(\tau)U(\tau) \qquad R^-(\tau) = R(\tau)U(-\tau)$$

of $R(\tau)$. Thus

$$S(\omega) = \mathbf{S}(j\omega) = \mathbf{S}^+(j\omega) + \mathbf{S}^-(j\omega)$$

If $R(\tau)$ is real, then $\mathbf{S}^-(s) = \mathbf{S}^+(-s)$ and

$$S(\omega) = 2 \operatorname{Re} \mathbf{S}^+(j\omega) \tag{10-24}$$

† Since $\mathbf{S}(j\omega)$ exists by assumption, the region of existence of $\mathbf{S}(s)$ is a vertical strip containing the $j\omega$ axis.

The above can be used to express bilateral transforms in terms of the more familiar unilateral transforms. Consider, for example, the pair

$$e^{-\alpha\tau}U(\tau) \leftrightarrow \frac{1}{s+\alpha}$$

In this case

$$\text{Re}\,\mathbf{S}^{+}(j\omega) = \text{Re}\,\frac{1}{j\omega+\alpha} = \frac{\alpha}{\alpha^2+\omega^2}$$

Hence

$$e^{-\alpha|\tau|} \leftrightarrow \frac{2\alpha}{\alpha^2+\omega^2} \tag{10-25}$$

Thus,

$$\mathbf{S}^{+}(s) = \frac{1}{\alpha+s} \qquad \mathbf{S}^{-}(s) = \frac{1}{\alpha-s} \qquad \mathbf{S}(s) = \frac{2\alpha}{\alpha^2-s^2}$$

Similarly, from the pair

$$e^{-\alpha\tau}\cos\beta\tau U(\tau) \leftrightarrow \frac{s+\alpha}{(s+\alpha)^2+\beta^2}$$

and (10-24) it follows that

$$e^{-\alpha|\tau|}\cos\beta\tau \leftrightarrow \frac{2\alpha(\alpha^2+\beta^2+\omega^2)}{(\alpha^2+\beta^2-\omega^2)^2+4\alpha^2\omega^2} \tag{10-26}$$

Example 10-6 Suppose that $\mathbf{x}(t)$ is a random telegraph signal (Fig. 10-4). In this case, (9-19) yields

$$R(\tau) = e^{-2\lambda|\tau|} \qquad S(\omega) = \frac{4\lambda}{4\lambda^2+\omega^2}$$

Note From (10-11) it follows with $\tau = 0$ that

$$\frac{1}{2\pi}\int_{-\infty}^{\infty} S(\omega)\,d\omega = R(0) = E\{|\mathbf{x}(t)|^2\} \tag{10-27}$$

Thus, the area of $S(\omega)$ is nonnegative and it equals the *average power* of $\mathbf{x}(t)$. This alone does not, of course, justify the term "power spectrum" used to characterize $S(\omega)$. In the course of our discussion we give other justifications.

For deterministic signals, power spectrum is defined as the limit of the integral

$$\mathbf{S}_T(\omega) = \frac{1}{2T}\left|\int_{-T}^{T} \mathbf{x}(t)e^{-j\omega t}\,dt\right|^2 \tag{10-28}$$

If $\mathbf{x}(t)$ is a stochastic process, then [see (14-49)]

$$E\{\mathbf{S}_T(\omega)\} \xrightarrow[T\to\infty]{} S(\omega)$$

We cannot, however, use the limit of $\mathbf{S}_T(\omega)$ as the definition of the power spectrum of $\mathbf{x}(t)$, because, as we show in Sec. 14-2, the variance of $\mathbf{S}_T(\omega)$ does not tend to zero. Thus, $\mathbf{S}_T(\omega)$

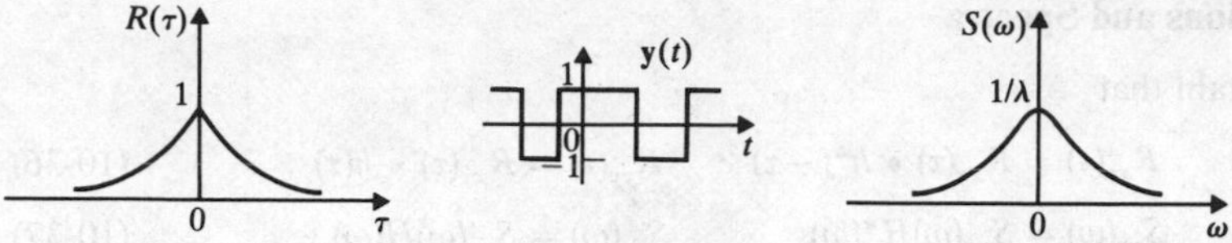

Figure 10-4

is not a satisfactory estimate of $S(\omega)$ no matter how large T is. However, it can be used to estimate a smoothed version of $S(\omega)$ because (see Example 14-2)

$$\int_{\omega_1}^{\omega_2} \mathbf{S}_T(\omega)\, d\omega \xrightarrow[T\to\infty]{} \int_{\omega_1}^{\omega_2} S(\omega)\, d\omega \tag{10-29}$$

10-2 LINEAR SYSTEMS

We are given a linear system with impulse response $h(t)$ and system function

$$H(\omega) = \int_{-\infty}^{\infty} h(t)e^{-j\omega t}\, dt = \mathbf{H}(j\omega)$$

The function $|H(\omega)|^2$ is the *energy spectrum* of the system. Its inverse

$$\rho(t) \leftrightarrow |H(\omega)|^2 = H(\omega)H^*(\omega) \tag{10-30}$$

is called the *deterministic autocorrelation* of $h(t)$. From the Fourier pairs

$$h(t) \leftrightarrow H(\omega) \qquad h^*(-t) \leftrightarrow H^*(\omega) \tag{10-31}$$

and the convolution theorem it follows that

$$\rho(t) = h(t) * h^*(-t) = \int_{-\infty}^{\infty} h(t+\tau)h^*(\tau)\, d\tau \tag{10-32}$$

We now apply to our system a WSS process $\mathbf{x}(t)$. The resulting output is given by

$$\mathbf{y}(t) = \int_{-\infty}^{\infty} \mathbf{x}(t-\alpha)h(\alpha)\, d\alpha = \int_{-\infty}^{\infty} \mathbf{x}(\alpha)h(t-\alpha)\, d\alpha \tag{10-33}$$

Note If the system is *causal*, then $h(t) = 0$ for $t < 0$ and (10-33) yields

$$\mathbf{y}(t) = \int_{0}^{\infty} \mathbf{x}(t-\alpha)h(\alpha)\, d\alpha = \int_{-\infty}^{t} \mathbf{x}(\alpha)h(t-\alpha)\, d\alpha \tag{10-34}$$

We have shown in Sec. 9-4 that $\mathbf{y}(t)$ is WSS with mean

$$\eta_y = \eta_x \int_{-\infty}^{\infty} h(\alpha)\, d\alpha = \eta_x H(0) \tag{10-35}$$

We determine next its autocorrelation $R_{yy}(\tau)$ and power spectrum $S_{yy}(\omega)$.

Correlations and Spectra

We maintain that

$$R_{xy}(\tau) = R_{xx}(\tau) * h^*(-\tau) \qquad R_{yy}(\tau) = R_{xy}(\tau) * h(\tau) \tag{10-36}$$

$$S_{xy}(\omega) = S_{xx}(\omega)H^*(\omega) \qquad S_{yy}(\omega) = S_{xy}(\omega)H(\omega) \tag{10-37}$$

PROOF The two equations in (10-36) follow from (9-100) and (9-102). However, because of their importance they will be rederived. Multiplying the conjugate of (10-33) by $\mathbf{x}(t+\tau)$ and taking expected values, we obtain

$$E\{\mathbf{x}(t+\tau)\mathbf{y}^*(t)\} = \int_{-\infty}^{\infty} E\{\mathbf{x}(t+\tau)\mathbf{x}^*(t-\alpha)\}h^*(\alpha)\,d\alpha$$

But

$$E\{\mathbf{x}(t+\tau)\mathbf{x}^*(t-\alpha)\} = R_{xx}(\tau+\alpha)$$

Hence

$$R_{xy}(\tau) = \int_{-\infty}^{\infty} R_{xx}(\tau+\alpha)h^*(\alpha)\,d\alpha = \int_{-\infty}^{\infty} R_{xx}(\tau-\beta)h^*(-\beta)\,d\beta$$

and the first equation in (10-36) results. Similarly, the product of (10-33) multiplied by $y^*(t-\tau)$ yields

$$E\{\mathbf{y}(t)\mathbf{y}^*(t-\tau)\} = \int_{-\infty}^{\infty} E\{\mathbf{x}(t-\alpha)y^*(t-\tau)\}h(\alpha)\,d\alpha$$

Therefore

$$R_{yy}(\tau) = \int_{-\infty}^{\infty} R_{xy}(\tau-\alpha)h(\alpha)\,d\alpha$$

Transforming (10-36) and using (10-31) and the convolution theorem, we obtain (10-37).

The above can be given the following system interpretation (Fig. 10-5): We connect two systems with impulse responses $h^*(-\tau)$ and $h(\tau)$ in cascade. If the input to the first system equals $R_{xx}(\tau)$, the resulting outputs will equal $R_{xy}(\tau)$ and $R_{yy}(\tau)$ respectively.

Fundamental theorem Combining the two equations in (10-36) and (10-37) we obtain

$$R_{yy}(\tau) = R_{xx}(\tau) * \rho(\tau) \tag{10-38}$$

$$S_{yy}(\omega) = S_{xx}(\omega)\,|\,H(\omega)\,|^2 \tag{10-39}$$

If $h(t)$ is real, then $\mathbf{H}^*(j\omega) = \mathbf{H}(-j\omega)$ and (10-39) yields

$$\mathbf{S}_{yy}(s) = \mathbf{S}_{xx}(s)\mathbf{H}(s)\mathbf{H}(-s) \tag{10-40}$$

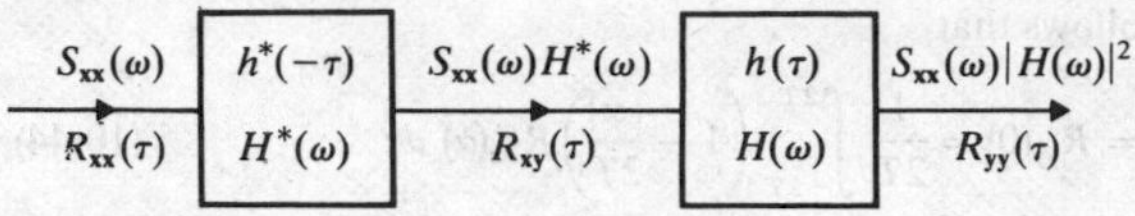

Figure 10-5

Corollary From the above it follows, as in (10-27), that

$$E\{|\mathbf{y}(t)|^2\} = \frac{1}{2\pi}\int_{-\infty}^{\infty} S_{xx}(\omega)|H(\omega)|^2\, d\omega$$

$$= R_{yy}(0) = \int_{-\infty}^{\infty} R_{xx}^*(\alpha)\rho(\alpha)\, d\alpha \tag{10-41}$$

Example 10-7 The *moving average* $\mathbf{y}(t)$ of a process $\mathbf{x}(t)$ is its average in the interval $(t - T, t + T)$

$$\mathbf{y}(t) = \frac{1}{2T}\int_{t-T}^{t+T} \mathbf{x}(\alpha)\, d\alpha$$

Clearly, $\mathbf{y}(t)$ is the output of a system driven by $\mathbf{x}(t)$, with impulse response $h(t)$ a pulse of height $1/2T$ (Fig. 10-6). In this case,

$$H(\omega) = \frac{1}{2T}\int_{-T}^{T} e^{-j\omega t}\, dt = \frac{\sin T\omega}{T\omega}$$

and (10-39) yields

$$S_{yy}(\omega) = S_{xx}(\omega)\frac{\sin^2 T\omega}{T^2\omega^2} \tag{10-42}$$

The autocorrelation $\rho(t)$ of $h(t)$ is a triangle of height $1/2T$ as in Fig. 10-6. Hence [see (10-38)]

$$R_{yy}(\tau) = \frac{1}{2T}\int_{-2T}^{2T}\left(1 - \frac{|\alpha|}{2T}\right)R_{xx}(\tau - \alpha)\, d\alpha \tag{10-43}$$

This result can be used to determine the variance of the RV

$$\boldsymbol{\eta}_T = \frac{1}{2T}\int_{-T}^{T} \mathbf{x}(t)\, dt$$

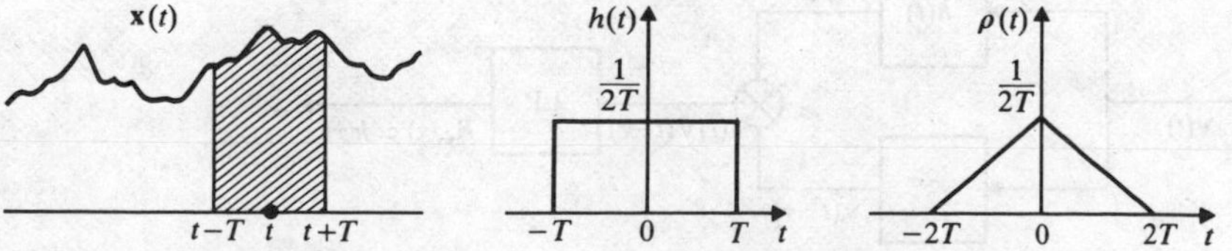

Figure 10-6

Indeed, since $\boldsymbol{\eta}_T = \mathbf{y}(0)$ it follows that

$$E\{\boldsymbol{\eta}_T^2\} = R_{yy}(0) = \frac{1}{2T}\int_{-2T}^{2T}\left(1 - \frac{|\alpha|}{2T}\right)R_{xx}(\alpha)\,d\alpha \tag{10-44}$$

The above integral equals the variance of $\boldsymbol{\eta}_T$ if we change $R_{xx}(\alpha)$ to $C_{xx}(\alpha)$.

Example 10-8 If the input to a system is a white-noise process $\mathbf{v}(t)$ with autocorrelation $R_{vv}(\tau) = \delta(\tau)$, then (10-36) yields [see (10-1)]

$$R_{yv}(\tau) = R_{vy}(-\tau) = \delta(\tau) * h^*(\tau) = h^*(\tau) \tag{10-45}$$

The following is a useful application.

System identification We wish to determine the impulse response $h(t)$ of a real system. For this purpose, we use as input a white-noise process $\mathbf{v}(t)$. As we see from (10-45), the cross-correlation between $\mathbf{v}(t)$ and the output $\mathbf{y}(t)$ equals $h(\tau)$. It suffices, therefore, to find $R_{yv}(\tau)$. Reasoning as in (9-152) we can show that, if T is large, then

$$R_{yv}(\tau) \simeq \frac{1}{T}\int_0^T \mathbf{y}(t)\mathbf{v}(t-\tau)\,dt \tag{10-46}$$

This can be realized in analog form by a minor modification of the correlometer of Fig. 9-18*a*. The resulting system (cross-correlometer) is shown in Fig. 10-7.

Filtering Equation (10-41) is the basis for the extension of the deterministic notion of filtering to systems with stochastic inputs:

$$\begin{array}{llll} \text{If} & S_{xx}(\omega)H(\omega) \equiv 0 & \text{then} & \mathbf{y}(t) = 0 \\ \text{If} & S_{xx}(\omega)H(\omega) \equiv 1 & \text{then} & \mathbf{y}(t) = \mathbf{x}(t) \end{array} \tag{10-47}$$

The first follows readily from (10-41) because if $S_{xx}(\omega)H(\omega) \equiv 0$, then $E\{\mathbf{y}^2(t)\} = 0$. To prove the second, we use (10-37): If $S_{xx}(\omega)H(\omega) \equiv 1$, then $S_{xx}(\omega) = S_{xy}(\omega) = S_{yy}(\omega)$, hence, $R_{xx}(\tau) = R_{xy}(\tau) = R_{yy}(\tau)$. This yields

$$E\{[\mathbf{x}(t) - \mathbf{y}(t)]^2\} = R_{xx}(0) + R_{yy}(0) - 2R_{xy}(0) = 0$$

hence, $\mathbf{x}(t) = \mathbf{y}(t)$ in the MS sense.

Note Theorem (10-39) is the stochastic version of the identity

$$|Y(\omega)|^2 = |X(\omega)|^2 |H(\omega)|^2$$

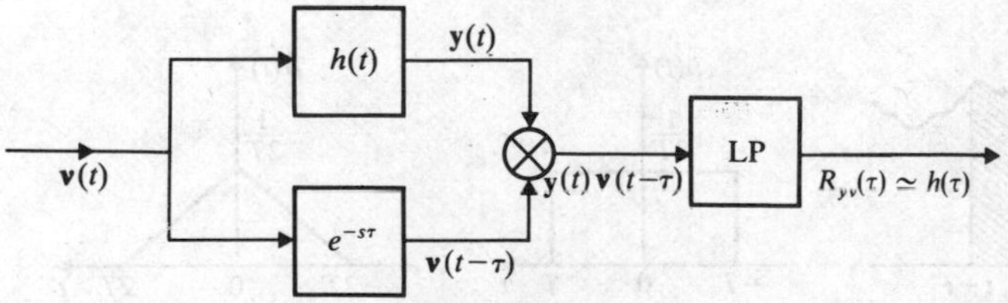

Figure 10-7

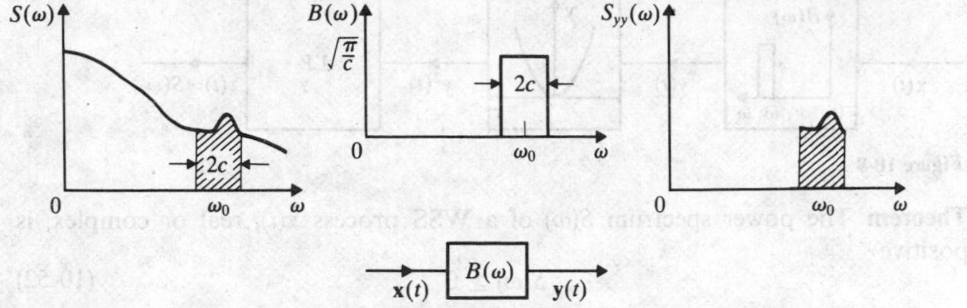

Figure 10-8

relating the energy spectrum $|Y(\omega)|^2$ of the output $y(t)$ of $H(\omega)$ to the energy spectrum $|X(\omega)|^2$ of the deterministic input $x(t)$.

The average power formula (10-41) is the stochastic version of the energy theorem

$$\int_{-\infty}^{\infty} |y(t)|^2 \, dt = \frac{1}{2\pi} \int_{-\infty}^{\infty} |X(\omega)|^2 \, |H(\omega)|^2 \, d\omega \tag{10-48}$$

Localization of power The energy spectrum of a deterministic signal $x(t)$ is a function of ω and its integral equals the energy of $x(t)$. This energy is localized on the ω axis in the sense that, if $x(t)$ is the input to a narrowband system, the energy of the output is proportional to the energy spectrum of $x(t)$. We have shown in (10-27) that the power spectrum $S(\omega)$ of a stochastic process $\mathbf{x}(t)$, defined as the transform of $R(\tau)$, is a function of ω whose integral equals the average power of $\mathbf{x}(t)$. We show next that this power is localized on the ω axis.

For this purpose, we form the ideal bandpass filter

$$B(\omega) = \begin{cases} \sqrt{\pi/c} & \omega_0 - c < \omega < \omega_0 + c \\ 0 & \text{otherwise} \end{cases} \tag{10-49}$$

shown in Fig. 10-8. This filter is centered at ω_0 and its energy equals 1.

Suppose that the input to the above filter is a process $\mathbf{x}(t)$ with power spectrum $S(\omega)$. From (10-41) it follows that the average power of the resulting output $\mathbf{y}(t)$ equals

$$E\{|\mathbf{y}(t)|^2\} = \frac{1}{2c} \int_{\omega_0 - c}^{\omega_0 + c} S(\omega) \, d\omega \tag{10-50}$$

Thus, $E\{|\mathbf{y}(t)|^2\}$ equals the average of $S(\omega)$ in the band of the filter and if c is sufficiently small and $S(\omega)$ is continuous at ω_0 (no lines), then

$$E\{|\mathbf{y}(t)|^2\} \simeq S(\omega_0) \tag{10-51}$$

Note The deterministic version of (10-50) is the identity [see (10-47)]

$$\int_{-\infty}^{\infty} |y(t)|^2 \, dt = \frac{1}{2c} \int_{\omega_0 - c}^{\omega_0 + c} |X(\omega)|^2 \, d\omega \simeq |X(\omega_0)|^2$$

relating the energy of the output of the filter $B(\omega)$ to the energy spectrum $|X(\omega)|^2$ of the input.

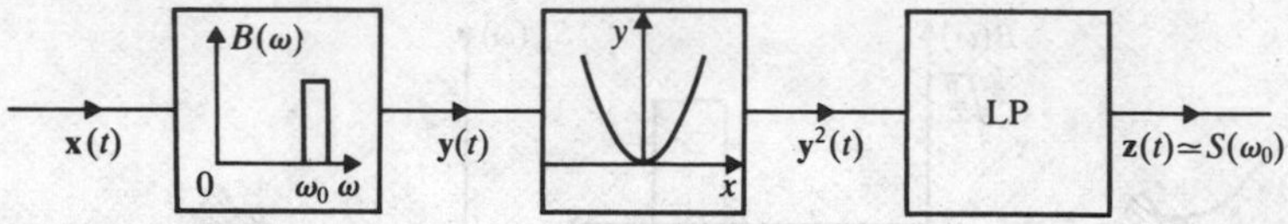

Figure 10-9

Theorem The power spectrum $S(\omega)$ of a WSS process $\mathbf{x}(t)$, real or complex, is positive

$$S(\omega) \geq 0 \tag{10-52}$$

PROOF Since $E\{|\mathbf{y}(t)|^2\} \geq 0$, we conclude from (10-50) that the average of $S(\omega)$ in the interval $(\omega_0 - c, \omega_0 + c)$ is positive. And since this is true for any ω_0 and c, (10-52) results.

Spectrum as local power From (10-50) it follows that

$$S(\omega_0) = \lim_{c \to 0} E\{|\mathbf{y}(t)|^2\} \tag{10-53}$$

This relationship can be used *to define* $S(\omega)$. We have, thus, two equivalent definitions of the power spectrum: (*a*) as the Fourier transform of $R(\tau)$; (*b*) in terms of the average power of the output of a narrowband system.

Spectroscopy Equation (10-51) leads to the following method for determining the power spectrum $S(\omega)$ of a process $\mathbf{x}(t)$: In Fig. 10-9, we show an analog device consisting of a narrowband filter $B(\omega)$ with input $\mathbf{x}(t)$, in series with a square-law detector and an LP filter $W(\omega)$. The input to the LP filter is the detector output $\mathbf{y}^2(t)$ and its output is a process $\mathbf{z}(t)$ with mean

$$E\{\mathbf{z}(t)\} \simeq E\{\mathbf{y}^2(t)\} \simeq S(\omega_0)$$

and small variance. If the band of $W(\omega)$ is sufficiently narrow, then $\mathbf{z}(t) \simeq S(\omega_0)$ (see Sec. 9-5). Varying ω_0, we can thus determine $S(\omega)$ for every ω.

Example 10-9 An optical realization of the above device is the *Fabry–Perot interferometer* shown in Fig. 10-10*a*. The bandpass filter consists of two highly reflective plates P_1 and P_2 distance d apart and the input is a light beam $\mathbf{x}(t)$ with power spectrum $S(\omega)$. The frequency response of the filter is proportional to

$$B(\omega) = \frac{1}{1 - r^2 e^{-j2\omega d/c}} \qquad r \simeq 1$$

where r is the reflection coefficient of each plate and c is the velocity of light in the medium M between the plates. The function $B(\omega)$ is shown in Fig. 10-10*b*. It consists of a sequence of bands centered at

$$\omega_n = \frac{\pi n d}{c}$$

whose bandwidth tends to zero as $r \to 1$. If only the mth band of $B(\omega)$ overlaps with $S(\omega)$ and $r \simeq 1$, then the output $\mathbf{z}(t)$ of the LP filter is proportional to $S(\omega_m)$. To vary ω_m, we can either vary the distance d between the plates or the dielectric constant of the medium M.

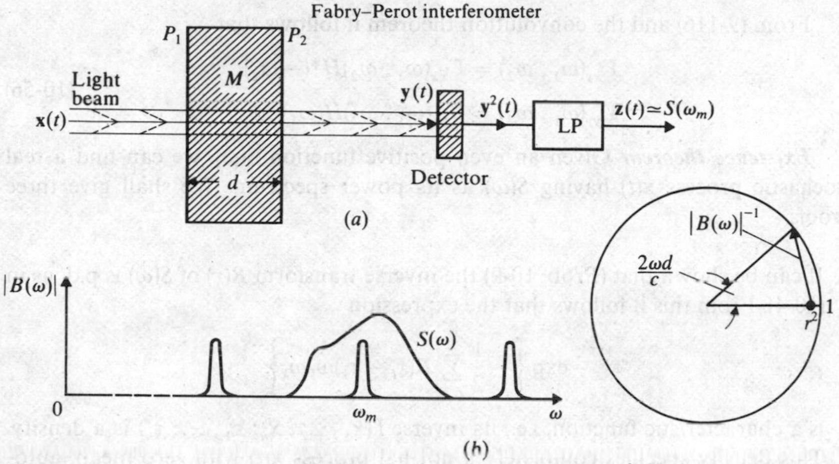

Figure 10-10

Note One might be tempted to use (10-53) as the definition of spectrum even if $\mathbf{x}(t)$ is not stationary. However, this would be only of limited value. For stationary processes, the spectrum has the following properties: (*a*) it simplifies the analysis of linear systems reducing convolution to multiplication [see (10-37)]; (*b*) it is a measure of the localization of power on the frequency axis [see (10-53)]; (*c*) it is related to the spectrum of a single sample of $\mathbf{x}(t)$ [see (10-29)]. If $\mathbf{x}(t)$ is not stationary then, in general, the limit in (10-53) does not have these properties. Cyclostationary processes, however, are an exception.

It can be shown that, if the input to a narrowband filter $B(\omega)$ is a *cyclostationary process* with period T, then

$$E\{|\mathbf{y}(t)|^2\} \xrightarrow[c \to 0]{} \bar{S}(\omega_0)$$

as in (10-53), where now

$$\bar{S}(\omega) = \int_{-\infty}^{\infty} \bar{R}(\tau)e^{-j\omega\tau}\,d\tau \qquad \bar{R}(\tau) = \frac{1}{T}\int_0^T R(t+\tau, t)\,dt \tag{10-54}$$

In this case, $\bar{S}(\omega)$ is the power spectrum of the shifted process $\bar{\mathbf{x}}(t) = \mathbf{x}(t - \boldsymbol{\theta})$ [see (9-77)]. In most applications it is preferable to deal directly with $\bar{\mathbf{x}}(t)$ (see Sec. 11-1).

We conclude with the observation that the analysis of linear systems with nonstationary inputs is simplified if we introduce the two-dimensional transform of their autocorrelation (see P.4):

$$\Gamma(\omega_1, \omega_2) = \int_{-\infty}^{\infty}\int_{-\infty}^{\infty} R(t_1, t_2)e^{-j(\omega_1 t_1 + \omega_2 t_2)}\,dt_1\,dt_2$$

$$R(t_1, t_2) = \frac{1}{4\pi^2}\int_{-\infty}^{\infty}\int_{-\infty}^{\infty} \Gamma(\omega_1, \omega_2)e^{j(\omega_1 t_1 + \omega_2 t_2)}\,d\omega_1\,d\omega_2 \tag{10-55}$$

From (9-116) and the convolution theorem it follows that

$$\Gamma_{xy}(\omega_1, \omega_2) = \Gamma_{xx}(\omega_1, \omega_2)H^*(-\omega_2)$$
$$\Gamma_{yy}(\omega_1, \omega_2) = \Gamma_{xy}(\omega_1, \omega_2)H(\omega_1) \tag{10-56}$$

Existence theorem Given an even positive function $S(\omega)$, we can find a real stochastic process $\mathbf{x}(t)$ having $S(\omega)$ as its power spectrum. We shall give three proofs.

1. It can be shown that (Prob. 10-9) the inverse transform $R(\tau)$ of $S(\omega)$ is p.d. as in (10-4). From this it follows that the expression

$$\exp\left\{-\frac{1}{2}\sum_{i,k} R(\tau_i - \tau_k)\omega_i\omega_k\right\}$$

is a characteristic function, i.e., its inverse $f(x_1, \ldots, x_n; \tau_1, \ldots, \tau_n)$ is a density. This density specifies completely a normal process $\mathbf{x}(t)$ with zero mean, autocorrelation $R(\tau)$, and power spectrum $S(\omega)$.
2. We construct an RV $\boldsymbol{\omega}$ with density $f(\omega) = S(\omega)/\pi a^2$ where $a = 2R(0)$, and a process

$$\mathbf{x}(t) = a\cos(\boldsymbol{\omega}t + \boldsymbol{\varphi})$$

as in (10-20). As we have shown in Example 10-5, the power spectrum of $\mathbf{x}(t)$ equals $\pi a^2 f(\omega) = S(\omega)$.
3. Suppose that $\mathbf{v}(t)$ is *white noise* with

$$R_{vv}(\tau) = q\delta(\tau) \qquad S_{vv}(\omega) = q$$

and $H(\omega)$ is such that $|H(\omega)|^2 = S(\omega)/q$. Using as input to $H(\omega)$ the process $\mathbf{v}(t)$, we obtain as output a process $\mathbf{x}(t)$ with

$$S_{xx}(\omega) = S_{vv}(\omega)|H(\omega)|^2 = S(\omega)$$

Thus, to find $\mathbf{x}(t)$, it suffices to construct a white-noise process $\mathbf{v}(t)$. An example of such a process is the derivative of the Wiener process (see also Sec. 10-5).

In the following examples, we introduce the notion of a system as an auxiliary concept used to simplify the spectral analysis of various functionals of $\mathbf{x}(t)$.

Example 10-10 We wish to find the power spectrum of the derivative $\mathbf{x}'(t)$ of $\mathbf{x}(t)$. Clearly, $\mathbf{x}'(t)$ can be considered as the output of a *differentiator* with input $\mathbf{x}(t)$ and system function $H(\omega) = j\omega$. Hence [see (10-37)]

$$S_{xx'}(\omega) = -j\omega S_{xx}(\omega) \qquad S_{x'x'}(\omega) = \omega^2 S_{xx}(\omega) \tag{10-57}$$

From (10-11) it follows by differentiation that

$$R^{(n)}(\tau) \leftrightarrow (j\omega)^n S(\omega)$$

Applying this to (10-57), we obtain

$$R_{xx'}(\tau) = -R'_{xx}(\tau) \qquad R_{x'x'}(\tau) = -R''_{xx}(\tau)$$

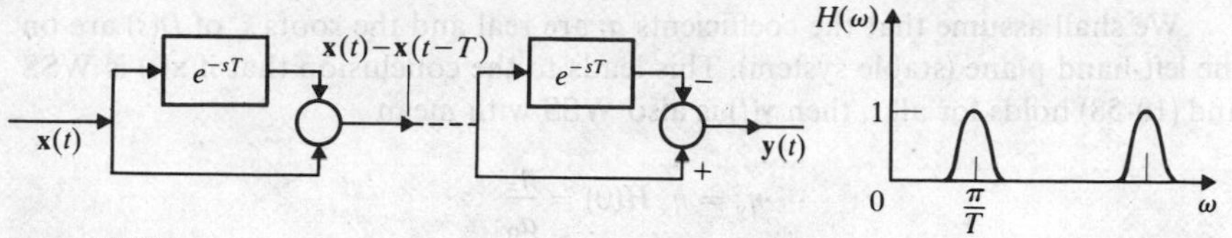

Figure 10-11

Example 10-11 A system with system function $1 - e^{-j\omega T}$ is called a *first-difference* filter because its output equals $\mathbf{x}(t) - \mathbf{x}(t-T)$. Connecting n such systems in cascade (Fig. 10-11) we obtain the *nth-difference* filter

$$H(\omega) = (1 - e^{-j\omega T})^n = 1 - \binom{n}{1} e^{-j\omega T} + \cdots + (-1)^n e^{-jn\omega T}$$

The resulting output

$$\mathbf{y}(t) = \mathbf{x}(t) - \binom{n}{1}\mathbf{x}(t-T) + \cdots + (-1)^n \mathbf{x}(t - nT)$$

is the nth difference of $\mathbf{x}(t)$. From (10-39) it follows that

$$S_{yy}(\omega) = S_{xx}(\omega)|1 - e^{-j\omega T}|^{2n} = S_{xx}(\omega)(2 \sin \omega T/2)^{2n}$$

Example 10-12 We wish to express the second moment of the RV

$$\mathbf{s} = \sum_{n=-N}^{N} \mathbf{x}(nT)$$

in terms of the power spectrum $S(\omega)$ of $\mathbf{x}(t)$. For this purpose we introduce the process

$$\mathbf{y}(t) = \sum_{n=-N}^{N} \mathbf{x}(t + nT)$$

This process is the output of the system

$$H(\omega) = \sum_{n=-N}^{N} e^{jnT\omega} = \frac{\sin (N + 1/2)T\omega}{\sin T\omega/2}$$

with input $\mathbf{y}(t)$. And since $\mathbf{s} = \mathbf{y}(0)$, (10-41) yields

$$E\{\mathbf{s}^2\} = E\{\mathbf{y}^2(0)\} = \frac{1}{2\pi}\int_{-\infty}^{\infty} S(\omega)\frac{\sin^2 (N + 1/2)T\omega}{\sin^2 T\omega/2}\, d\omega$$

Differential Equations and Rational Spectra

The differential equation

$$a_n \mathbf{y}^{(n)}(t) + \cdots + a_0 \mathbf{y}(t) = \mathbf{x}(t) \tag{10-58}$$

defines a system with input $\mathbf{x}(t)$, output $\mathbf{y}(t)$, and system function

$$\mathbf{H}(s) = \frac{1}{D(s)} \qquad D(s) = a_n s^n + \cdots + a_0 \tag{10-59}$$

We shall assume that the coefficients a_i are real and the roots s_i of $D(s)$ are on the left-hand plane (stable system). This leads to the conclusion that if $\mathbf{x}(t)$ is WSS and (10-58) holds for all t, then $\mathbf{y}(t)$ is also WSS with mean

$$\eta_y = \eta_x H(0) = \frac{\eta_x}{a_0}$$

We shall express $R_{yy}(\tau)$ and $S_{yy}(\omega)$ in terms of $R_{xx}(\tau)$ and $S_{xx}(\omega)$. Applying (9-129) and (9-131) to stationary processes, we can show that the functions $R_{xy}(\tau)$ and $R_{yy}(\tau)$ satisfy two differential equations of the form (10-58) (see Prob. 10-17). However, as we show next, $R_{yy}(\tau)$ can be determined directly.

From (10-39) it follows that

$$S_{yy}(\omega) = \frac{S_{xx}(\omega)}{|a_n(j\omega)^n + \cdots + a_0|^2} \tag{10-60}$$

We shall determine the inverse of $S_{yy}(\omega)$ under the assumption that $\mathbf{x}(t)$ is white noise with $S_{xx}(\omega) = 1$. In this case (10-40) yields

$$\mathbf{S}_{yy}(s) = \frac{1}{D(s)D(-s)} \tag{10-61}$$

Clearly, if s_i is a root of $D(s)$, then $-s_i$ is a root of $D(-s_i)$. Assuming that all roots are simple, we conclude that (partial fraction expansion)

$$\mathbf{S}_{yy}(s) = \sum_{i=1}^{n} \frac{c_i}{s - s_i} - \sum_{i=1}^{n} \frac{c_i}{s + s_i} \tag{10-62}$$

Hence [see (10-23)]

$$\mathbf{S}_{yy}^{+}(s) = \sum_{i=1}^{n} \frac{c_i}{s - s_i} \qquad R_{yy}^{+}(\tau) = \sum_{i=1}^{n} c_i e^{s_i \tau} \tag{10-63}$$

This determines $R_{yy}(\tau)$ for all τ because $R_{yy}(-\tau) = R_{yy}(\tau)$.

Note From (10-62) it follows that the denominator of $\mathbf{S}_{yy}^{+}(s)$ equals $D(s)$. This leads to the conclusion that $R_{yy}(\tau)$ satisfies the homogeneous equation

$$a_n R_{yy}^{(n)}(\tau) + \cdots + a_0 R_{yy}(\tau) = 0 \qquad \tau > 0 \tag{10-64}$$

Example 10-13 Suppose that

$$\mathbf{y}''(t) + 5\mathbf{y}'(t) + 6\mathbf{y}(t) = \mathbf{x}(t) \qquad R_{xx}(\tau) = 60\delta(\tau)$$

In this case

$$\mathbf{S}_{yy}(s) = \frac{60}{(s^2 + 5s + 6)(s^2 - 5s + 6)} = \frac{3}{s+2} - \frac{2}{s+3} + \frac{3}{2-s} - \frac{2}{3-s}$$

Hence

$$\mathbf{S}_{yy}^{+}(s) = \frac{3}{s+2} - \frac{2}{s+3} \qquad R_{yy}^{+}(\tau) = (3e^{-2\tau} - 2e^{-3\tau})U(\tau)$$

$$R_{yy}(\tau) = 3e^{-2|\tau|} - 2e^{-3|\tau|}$$

Example 10-14 Suppose, finally, that

$$\mathbf{y}''(t) + 2\alpha\mathbf{y}'(t) + \omega_0^2\mathbf{y}(t) = \mathbf{x}(t) \qquad R_{xx}(\tau) = q\delta(\tau)$$

We now have

$$\mathbf{S}_{yy}(s) = \frac{q}{(s^2 + 2\alpha s + \omega_0^2)(s^2 - 2\alpha s + \omega_0^2)}$$

We shall determine $R_{yy}(\tau)$ under the assumption that $\alpha < \omega_0$. In this case, the roots of $D(s)$ are complex

$$s_{1,2} = -\alpha \pm j\beta \qquad \beta = \sqrt{\omega_0^2 - \alpha^2}$$

Combining the two complex terms of the partial fraction expansion of $\mathbf{S}_{yy}(s)$ into a real quadratic, we obtain

$$\mathbf{S}_{yy}(s) = \frac{q}{4\alpha\omega_0^2}\left(\frac{s + 2\alpha}{s^2 + 2\alpha s + \omega_0^2} + \frac{-s + 2\alpha}{s^2 - 2\alpha s + \omega_0^2}\right)$$

Hence

$$\mathbf{S}^+(s) = \frac{q}{4\alpha\omega_0^2}\frac{s + 2\alpha}{s^2 + 2\alpha s + \omega_0^2}$$

$$R(\tau) = \frac{q}{4\alpha\omega_0^2}e^{-\alpha|\tau|}\left(\cos\beta\tau + \frac{\alpha}{\beta}\sin\beta|\tau|\right)$$

Multiple Terminals

In Fig. 10-12*a* we show two systems with inputs $\mathbf{x}_1(t)$, $\mathbf{x}_2(t)$ and outputs $\mathbf{y}_1(t)$, $\mathbf{y}_2(t)$ as in (9-120). Multiplying the conjugate of the second equation by $\mathbf{x}_1(t + \tau)$ and the first by $\mathbf{y}_2^*(t - \tau)$, we obtain

$$\mathbf{x}_1(t + \tau)\mathbf{y}_2^*(t) = \int_{-\infty}^{\infty} \mathbf{x}_1(t + \tau)\mathbf{x}_2^*(t - \alpha)h_2^*(\alpha)\,d\alpha$$

$$\mathbf{y}_1(t)\mathbf{y}_2^*(t - \tau) = \int_{-\infty}^{\infty} \mathbf{x}_1(t - \alpha)\mathbf{y}_2^*(t - \tau)h_1(\alpha)\,d\alpha$$

Hence

$$R_{x_1y_2}(\tau) = \int_{-\infty}^{\infty} R_{x_1x_2}(\tau + \alpha)h_2^*(\alpha)\,d\alpha = R_{x_1x_2}(\tau) * h_2^*(-\tau)$$

$$R_{y_1y_2}(\tau) = \int_{-\infty}^{\infty} R_{x_1y_2}(\tau - \alpha)h_1(\alpha)\,d\alpha = R_{x_1y_2}(\tau) * h_1(\tau) \tag{10-65}$$

Thus, if we cascade the system $h_2^*(-\tau)$ with the system $h_1(\tau)$ as in Fig. 10-12*b* and use $R_{x_1x_2}(\tau)$ as input, then the output of the first system equals $R_{x_1y_2}(\tau)$ and the output of the second equals $R_{y_1y_2}(\tau)$.

The transforms of (10-65) yield

$$S_{x_1y_2}(\omega) = S_{x_1x_2}(\omega)H_2^*(\omega)$$

$$S_{y_1y_2}(\omega) = S_{x_1y_2}(\omega)H_1(\omega) = S_{x_1x_2}(\omega)H_1(\omega)H_2^*(\omega) \tag{10-66}$$

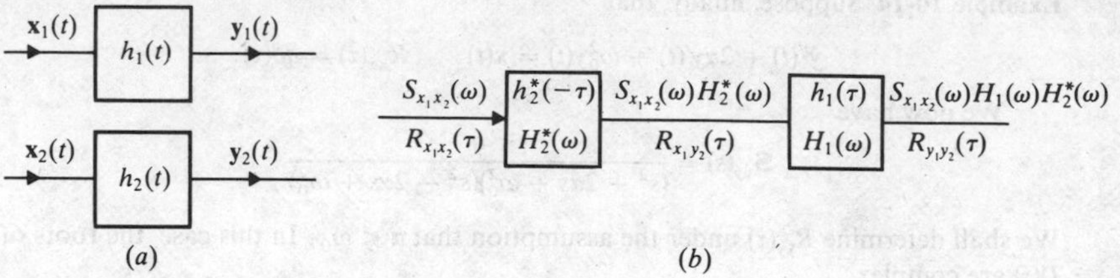

Figure 10-12

Note If the two systems are disjoint, i.e., if $H_1(\omega)H_2(\omega) \equiv 0$, then $R_{y_1y_2}(\tau) = 0$. This shows that the processes $\mathbf{y}_1(t)$ and $\mathbf{y}_2(t)$ are orthogonal. Thus, if $\mathbf{x}(t)$ is a common input to two disjoint systems and its mean is zero, then the resulting responses $\mathbf{y}_1(t)$ and $\mathbf{y}_2(t)$ are uncorrelated. Furthermore, if $\mathbf{x}(t)$ is normal, then the processes $\mathbf{y}_1(t)$ and $\mathbf{y}_2(t)$ are jointly normal; hence, they are independent.

Example 10-15 In the circuit of Fig. 10-13*a*, the generator $\mathbf{e}(t)$ is a random telegraph signal. We shall determine the power spectrum of the voltages $\mathbf{v}_1(t)$ and $\mathbf{v}_2(t)$.

These voltages can be considered as outputs of two linear systems with common input $\mathbf{e}(t)$ and system functions

$$H_1(\omega) = \frac{\alpha + j\omega}{2\alpha + j\omega} \qquad H_2(\omega) = \frac{\alpha}{2\alpha + j\omega} \qquad \alpha = \frac{1}{RC}$$

respectively (Fig. 10-13*b*). Since (see Example 10-6) $S_{ee}(\omega) = 4\lambda/(4\lambda^2 + \omega^2)$, we conclude from (10-39) and (10-66) that

$$S_{v_1v_1}(\omega) = \frac{4\lambda(\alpha^2 + \omega^2)}{(4\lambda^2 + \omega^2)(4\alpha^2 + \omega^2)} \qquad S_{v_2v_2}(\omega) = \frac{4\alpha^2\lambda}{(4\lambda^2 + \omega^2)(4\alpha^2 + \omega^2)}$$

$$S_{v_1v_2}(\omega) = \frac{4\alpha\lambda(\alpha + j\omega)}{(4\lambda^2 + \omega^2)(4\alpha^2 + \omega^2)}$$

Using (10-66), we shall show that

$$\left| \int_a^b S_{xy}(\omega)\, d\omega \right|^2 \le \int_a^b S_{xx}(\omega)\, d\omega \int_a^b S_{yy}(\omega)\, d\omega \qquad \text{(10-67)}$$

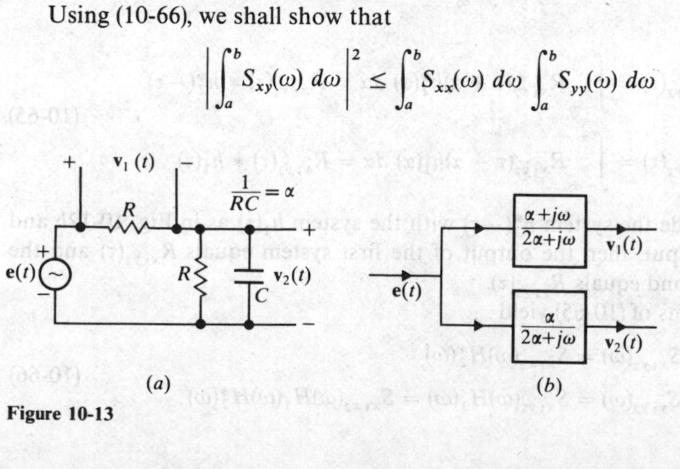

Figure 10-13

PROOF Suppose that $\mathbf{x}(t)$ and $\mathbf{y}(t)$ are the inputs to two ideal filters

$$H_1(\omega) = H_2(\omega) = \begin{cases} 1 & a < \omega < b \\ 0 & \text{otherwise} \end{cases}$$

With $\mathbf{z}(t)$ and $\mathbf{w}(t)$ the resulting outputs, we conclude from (10-66) that

$$S_{zz}(\omega) = \begin{cases} S_{xx}(\omega) & a < \omega < b \\ 0 & \text{otherwise} \end{cases} \qquad R_{zz}(0) = \frac{1}{2\pi}\int_a^b S_{xx}(\omega)\, d\omega$$

Similar relationships hold for $S_{zw}(\omega)$ and $S_{ww}(\omega)$. And since [see (10-6)]

$$|R_{zw}(0)|^2 \le R_{zz}(0)R_{ww}(0)$$

(10-67) results.

From (10-67) it follows that if $S_{xx}(\omega) = 0$ or $S_{yy}(\omega) = 0$ in an interval (a, b), then $S_{xy}(\omega) = 0$ in this interval.

10-3 HILBERT TRANSFORMS, SHOT NOISE, THERMAL NOISE

A system with system function

$$H(\omega) = -j \operatorname{sgn} \omega = \begin{cases} -j & \omega > 0 \\ j & \omega < 0 \end{cases} \tag{10-68}$$

is called *quadrature filter*. The corresponding impulse response equals $1/\pi t$. Thus, $H(\omega)$ is all-pass with $-90°$ phase shift, hence, its response to $\cos \omega t$ equals $\cos(\omega t - 90°) = \sin \omega t$ and its response to $\sin \omega t$ equals $\sin(\omega t - 90°) = -\cos \omega t$.

The response of a quadrature filter to a real process $\mathbf{x}(t)$ is denoted by $\check{\mathbf{x}}(t)$ and it is called the *Hilbert transform* of $\mathbf{x}(t)$. Thus

$$\check{\mathbf{x}}(t) = \mathbf{x}(t) * \frac{1}{\pi t} = \frac{1}{\pi}\int_{-\infty}^{\infty} \frac{\mathbf{x}(\alpha)}{t - \alpha}\, d\alpha \tag{10-69}$$

From (10-37) and (10-15) it follows that (Fig. 10-14)

$$\begin{aligned} S_{x\check{x}}(\omega) &= jS_{xx}(\omega) \operatorname{sgn} \omega = -S_{\check{x}x}(\omega) \\ S_{\check{x}\check{x}}(\omega) &= S_{xx}(\omega) \end{aligned} \tag{10-70}$$

The complex process

$$\mathbf{z}(t) = \mathbf{x}(t) + j\hat{\mathbf{x}}(t)$$

is called the *analytic signal* associated with $\mathbf{x}(t)$. Clearly, $\mathbf{z}(t)$ is the response of the system

$$1 + j(-j \operatorname{sgn} \omega) = 2U(\omega)$$

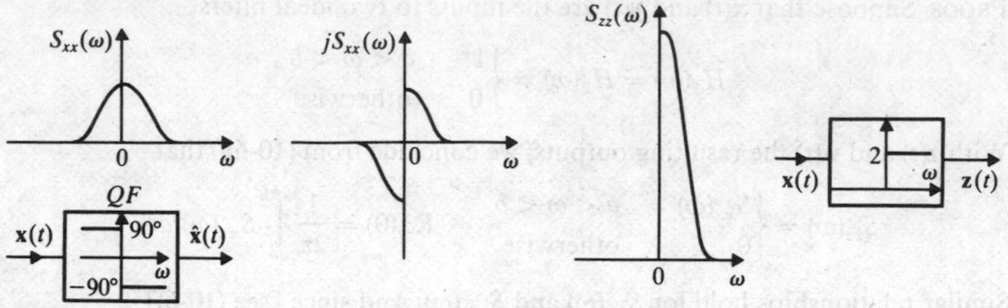

Figure 10-14

with input $\mathbf{x}(t)$. Hence [see (10-39)]

$$S_{zz}(\omega) = 4S_{xx}(\omega)U(\omega) = 2S_{xx}(\omega) + 2jS_{\check{x}x}(\omega) \tag{10-71}$$

$$R_{zz}(\tau) = 2R_{xx}(\tau) + 2jR_{\check{x}x}(\tau) \tag{10-72}$$

Example 10-16 If

$$\mathbf{x}(t) = \mathbf{a} \cos \omega_0 t + \mathbf{b} \sin \omega_0 t$$

as in (9-58), then

$$\check{\mathbf{x}}(t) = -\mathbf{b} \cos \omega_0 t + \mathbf{a} \sin \omega_0 t \qquad \mathbf{z}(t) = (\mathbf{a} - j\mathbf{b})e^{j\omega_0 \tau}$$

Shot Noise

Given a set of Poisson points $\mathbf{t}_i$ with average density λ and a real function $h(t)$, we form the sum

$$\mathbf{s}(t) = \sum_i h(t - \mathbf{t}_i) \tag{10-73}$$

This sum is an SSS process known as *shot noise*. In the following, we discuss its second-order properties. The general statistics are developed in Sec. 12-3.

From the definition it follows that $\mathbf{s}(t)$ can be represented as the output of a linear system (Fig. 10-15) with impulse response $h(t)$ and input the Poisson impulses

$$\mathbf{z}(t) = \sum_i \delta(t - \mathbf{t}_i) \tag{10-74}$$

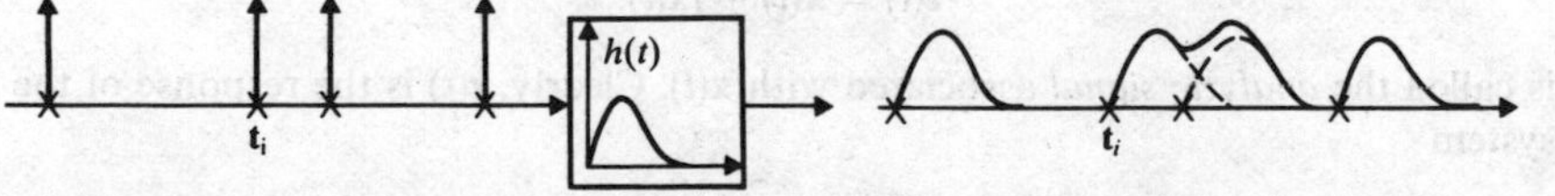

Figure 10-15

This representation agrees with the generation of shot noise in physical problems: The process $\mathbf{s}(t)$ is the output of a dynamic system activated by a sequence of impulses (particle emissions, for example) occurring at the random times $\mathbf{t}_i$.

As we know, $\eta_z = \lambda$, hence

$$E\{\mathbf{s}(t)\} = \lambda \int_{-\infty}^{\infty} h(t)\, dt = \lambda H(0) \tag{10-75}$$

Furthermore, since (see Example 10-2)

$$S_{zz}(\omega) = 2\pi\lambda^2\delta(\omega) + \lambda \tag{10-76}$$

it follows from (10-39) that

$$S_{ss}(\omega) = 2\pi\lambda^2 H^2(0)\delta(\omega) + \lambda |H(\omega)|^2 \tag{10-77}$$

because $|H(\omega)|^2\delta(\omega) = H^2(0)\delta(\omega)$. The inverse of the above yields

$$R_{ss}(\tau) = \lambda^2 H^2(0) + \lambda\rho(\tau) \qquad C_{ss}(\tau) = \lambda\rho(\tau) \tag{10-78}$$

Campbell's theorem The mean η_s and variance σ_s^2 of the shot noise process $\mathbf{s}(t)$ equal

$$\eta_s = \lambda \int_{-\infty}^{\infty} h(t)\, dt \qquad \sigma_s^2 = \lambda\rho(0) = \lambda \int_{-\infty}^{\infty} h^2(t)\, dt \tag{10-79}$$

PROOF It follows from the above because $\sigma_s^2 = C_{ss}(0)$.

Example 10-17 If

$$h(t) = e^{-\alpha t}U(t) \qquad H(\omega) = \frac{1}{\alpha + j\omega}$$

then

$$\eta_s = \frac{\lambda}{\alpha} \qquad \sigma_s^2 = \frac{\lambda}{2\alpha}$$

$$S_{ss}(\omega) = \frac{2\pi\lambda^2}{\alpha^2}\delta(\omega) + \frac{\lambda}{\alpha^2 + \omega^2} \qquad C_{ss}(\tau) = \frac{\lambda}{2\alpha} e^{-\alpha|\tau|}$$

Example 10-18 (electron transit) Suppose that $h(t)$ is a triangle as in Fig. 10-16*a*. Since

$$\int_0^T kt\, dt = \frac{kT^2}{2} \qquad \int_0^T k^2t^2\, dt = \frac{k^2T^3}{3}$$

it follows from (10-79) that

$$\eta_s = \frac{\lambda k T^2}{2} \qquad \sigma_s^2 = \frac{\lambda k^2 T^3}{3}$$

In this case

$$H(\omega) = \int_0^T kte^{-j\omega t}\, dt = e^{-j\omega T/2}\frac{2k \sin \omega T/2}{j\omega^2} - e^{-j\omega T}\frac{kT}{j\omega}$$

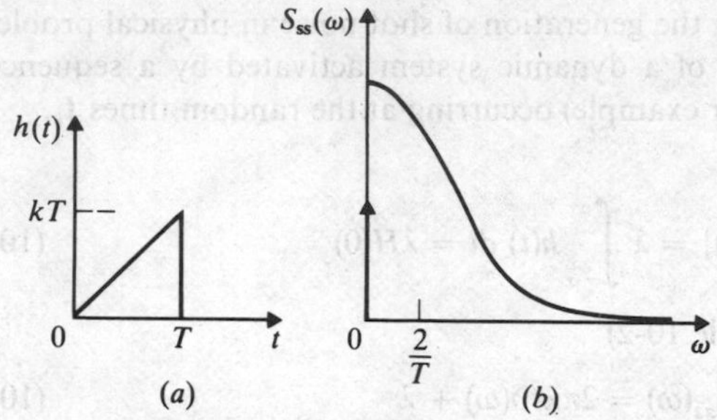

Figure 10-16

Inserting into (10-77), we obtain (Fig. 10-16*b*)

$$S_{ss}(\omega) = 2\pi\eta_s^2\delta(\omega) + \frac{\lambda k^2}{\omega^4}(2 - 2\cos\omega T + \omega^2 T^2 - 2\omega T \sin\omega T)$$

Thermal Noise

Thermal noise is the distribution of voltages and currents in a network due to the thermal electron agitation. In the following, we discuss the statistical properties of thermal noise ignoring the underlying physics. The analysis is based on a model consisting of noiseless reactive elements and noisy resistors.

A noisy resistor is modeled by a noiseless resistor R in series with a voltage source $\mathbf{n}_e(t)$ or in parallel with a current source $\mathbf{n}_i(t) = \mathbf{n}_e(t)/R$ as in Fig. 10-17. It is assumed that $\mathbf{n}_e(t)$ is a normal process with zero mean and flat spectrum

$$S_{n_e}(\omega) = 2kTR \qquad k = 1.37 \times 10^{-23} \text{ Joule-degrees}$$
$$S_{n_i}(\omega) = \frac{S_{n_e}(\omega)}{R^2} = 2kTG \qquad G = \frac{1}{R} \tag{10-80}$$

where T is the absolute temperature of the resistor. Furthermore, the noise sources of all network resistors are independent processes.

Using the above and the properties of linear systems, we shall derive the spectral properties of general network responses starting with an example.

Example 10-19 The circuit of Fig. 10-18 consists of a resistor R and a capacitor C. We shall determine the spectrum of the voltage $\mathbf{v}(t)$ across the capacitor due to thermal noise.

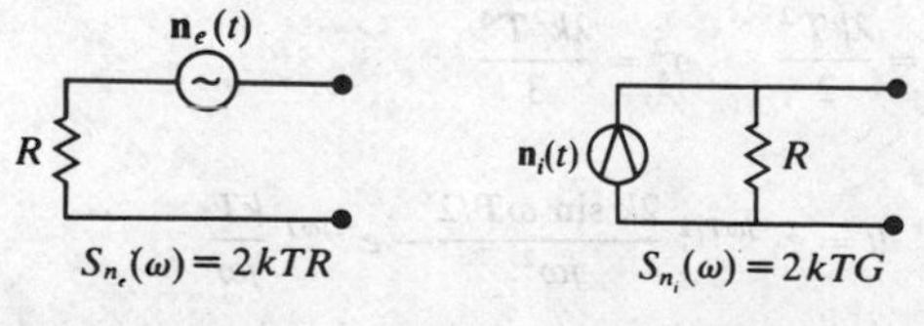

Figure 10-17

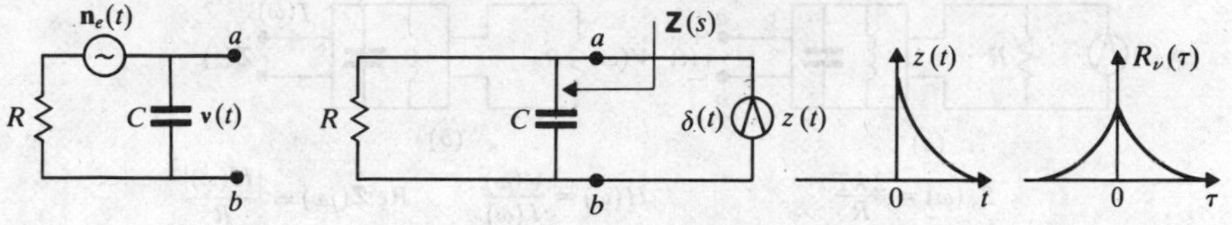

Figure 10-18

The voltage $\mathbf{v}(t)$ can be considered as the output of a system with input the noise voltage $\mathbf{n}_e(t)$ and system function

$$\mathbf{H}(s) = \frac{1}{1 + RCs}$$

Applying (10-39), we obtain [see (10-25)]

$$S_v(\omega) = S_{n_e}(\omega) | H(\omega) |^2 = \frac{2kTR}{1 + \omega^2 R^2 C^2}$$
$$R_v(\tau) = \frac{kT}{C} e^{-|\tau|/RC} \tag{10-81}$$

The following consequences are illustrations of Nyquist's theorem to be discussed presently: We denote by $\mathbf{Z}(s)$ the impedance across the terminals a, b and by $z(t)$ its inverse transform

$$\mathbf{Z}(s) = \frac{R}{1 + RCs} \qquad z(t) = \frac{1}{C} e^{-t/RC} U(t)$$

The function $z(t)$ is the voltage across C due to an impulse current $\delta(t)$ Fig. 10-18b. Comparing with (10-81) we obtain

$$S_v(\omega) = 2kT \text{ Re } \mathbf{Z}(j\omega) \qquad \text{Re } \mathbf{Z}(j\omega) = \frac{R}{1 + \omega^2 R^2 C^2}$$

$$R_v(\tau) = kTz(\tau) \quad \tau > 0 \quad R_v(0) = kTz(0^+)$$

$$E\{\mathbf{v}^2(t)\} = R_v(0) = \frac{kT}{C} \qquad \frac{1}{C} = \lim_{\omega \to \infty} j\omega \mathbf{Z}(j\omega)$$

Given a passive, reciprocal network we denote by $\mathbf{v}(t)$ the voltage across two arbitrary terminals a, b and by $\mathbf{Z}(s)$ the impedance from a to b (Fig. 10-19).

Nyquist theorem The power spectrum of $\mathbf{v}(t)$ equals

$$S_v(\omega) = 2kT \text{ Re } \mathbf{Z}(j\omega) \tag{10-82}$$

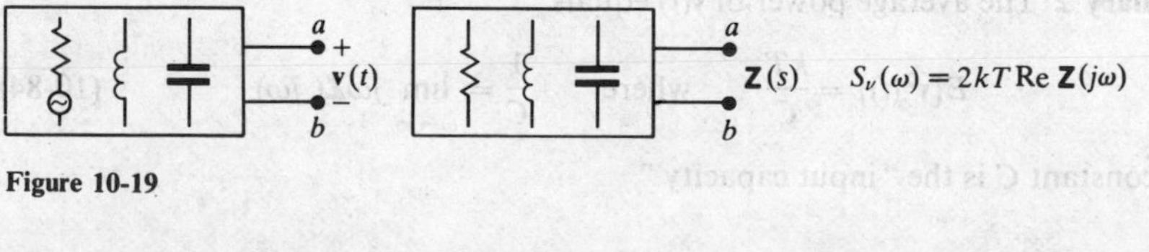

Figure 10-19

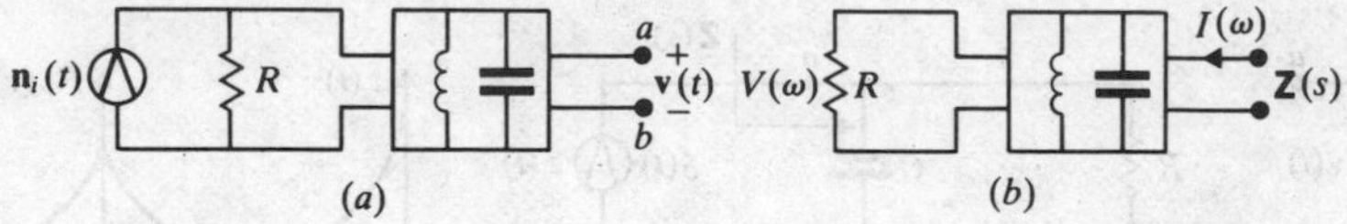

$S_{n_i}(\omega) = \frac{2kT}{R}$ $\qquad H(\omega) = \frac{V(\omega)}{I(\omega)}$ $\qquad \text{Re}\,\mathbf{Z}(j\omega) = \frac{|H(\omega)|^2}{R}$

Figure 10-20

PROOF We shall assume that there is only one resistor in the network. The general case can be established similarly if we use the independence of the noise sources. The resistor is represented by a noiseless resistor in parallel with a current source $\mathbf{n}_i(t)$ and the remaining network contains only reactive elements (Fig. 10-20*a*). Thus, $\mathbf{v}(t)$ is the output of a system with input $\mathbf{n}_i(t)$ and system function $H(\omega)$. From the reciprocity theorem it follows that $H(\omega) = V(\omega)/I(\omega)$ where $I(\omega)$ is the amplitude of a sine wave from a to b (Fig. 10-20*b*) and $V(\omega)$ is the amplitude of the voltage across R. The input power equals $|I(\omega)|^2 \text{ Re } \mathbf{Z}(j\omega)$ and the power delivered to the resistance equals $|V(\omega)|^2/R$. Since the connecting network is lossless by assumption, we conclude that

$$|I(\omega)|^2 \text{ Re } \mathbf{Z}(j\omega) = \frac{|V(\omega)|^2}{R}$$

Hence

$$|H(\omega)|^2 = \frac{|V(\omega)|^2}{|I(\omega)|^2} = R \text{ Re } \mathbf{Z}(j\omega)$$

and (10-82) results because

$$S_v(\omega) = S_{n_i}(\omega)|H(\omega)|^2 \qquad S_{n_i}(\omega) = \frac{2kT}{R}$$

Corollary 1 The autocorrelation of $\mathbf{v}(t)$ equals

$$R_v(\tau) = z(\tau) \qquad \tau > 0 \tag{10-83}$$

where $z(t)$ is the inverse transform of $\mathbf{Z}(s)$.

PROOF Since $\mathbf{Z}(-j\omega) = \mathbf{Z}^*(j\omega)$, it follows from (10-82) that

$$S_v(\omega) = kT[\mathbf{Z}(j\omega) + \mathbf{Z}(-j\omega)]$$

and (10-83) results because the inverse of $\mathbf{Z}(-j\omega)$ equals $z(-t)$ and $z(-t) = 0$ for $t > 0$.

Corollary 2 The average power of $\mathbf{v}(t)$ equals

$$E\{\mathbf{v}^2(t)\} = \frac{kT}{C} \qquad \text{where} \qquad \frac{1}{C} = \lim_{\omega \to \infty} j\omega\mathbf{Z}(j\omega) \tag{10-84}$$

The constant C is the "input capacity."

PROOF As we know (initial value theorem)

$$z(0^+) = \lim s\mathbf{Z}(s) \qquad s \to \infty$$

and (10-84) follows from (10-83) because

$$E\{\mathbf{v}^2(t)\} = R_v(0) = kTz(0^+)$$

Currents From Thévenin's theorem it follows that, terminally, a noisy network is equivalent to a noiseless network with impedance $\mathbf{Z}(s)$ in series with a voltage source $\mathbf{v}(t)$. The power spectrum $S_v(\omega)$ of $\mathbf{v}(t)$ is the right side of (10-82). This leads to the following version of Nyquist's theorem:

The power spectrum of the short-circuit current $\mathbf{i}(t)$ from a to b due to thermal noise equals

$$S_i(\omega) = 2kT \text{ Re } \mathbf{Y}(j\omega) \qquad \mathbf{Y}(s) = \frac{1}{\mathbf{Z}(s)} \tag{10-85}$$

PROOF From Thévenin's theorem it follows that

$$S_i(\omega) = S_v(\omega)|\mathbf{Y}(j\omega)|^2 = \frac{2kT \text{ Re } \mathbf{Z}(j\omega)}{|\mathbf{Z}(j\omega)|^2}$$

and (10-85) results.

The current version of the corollaries is left as exercise.

10-4 DISCRETE-TIME PROCESSES

In this section we present selectively the discrete-time version of earlier results starting with the notion of spectrum.

Suppose that $\mathbf{x}[n]$ is a WSS process with autocorrelation $R[m]$. Its *power spectrum* is by definition the sum

$$S(\omega) = \sum_{m=-\infty}^{\infty} R[m]e^{-jm\omega T} \tag{10-86}$$

Thus, $S(\omega)$ is a periodic function with Fourier series coefficients $R[m]$ and period $2\Omega = 2\pi/T$. The significance of the constant T will be given presently. The function $S(\omega)$ is the *discrete Fourier transform* (DFT) of the sequence $R[m]$.

From (10-86) it follows that

$$R[m] = \frac{1}{2\Omega}\int_{-\Omega}^{\Omega} S(\omega)e^{jm\omega T}\, d\omega \qquad \Omega = \frac{\pi}{T} \tag{10-87}†$$

We note, in particular, that

$$E\{|\mathbf{x}[n]|^2\} = R[0] = \frac{1}{2\Omega}\int_{-\Omega}^{\Omega} S(\omega)\, d\omega \tag{10-88}$$

† The ratio π/T will be denoted also by σ and will be called *Nyquist frequency*. Thus, $\Omega = \sigma = \pi/T$.

We find it convenient to introduce also the z transform† of $R[m]$

$$\mathbf{S}(z) = \sum_{m=-\infty}^{\infty} R[m]z^{-m} \qquad \mathbf{S}(e^{j\omega T}) = S(\omega) \tag{10-89}$$

The term *power spectrum* will be used for $S(\omega)$ and for $\mathbf{S}(z)$.

Example 10-20 If

$$R[m] = a^{|m|} \qquad |a| < 1$$

then

$$\mathbf{S}(z) = \sum_{m=-\infty}^{-1} a^{-m}z^{-m} + \sum_{m=0}^{\infty} a^m z^{-m} = \frac{az}{1-az} + \frac{z}{z-a}$$

Hence

$$\mathbf{S}(z) = \frac{(1-a^2)z}{(z-a)(1-az)} = \frac{a^{-1}-a}{(a^{-1}+a)-(z^{-1}+z)}$$

$$S(\omega) = \frac{a^{-1}-a}{a^{-1}+a-2\cos\omega T}$$

Sampling Suppose that $\mathbf{x}(t)$ is a WSS continuous-time process with autocorrelation $R(\tau)$ and power spectrum $S_c(\omega)$. With T an arbitrary constant, we form the process $\mathbf{x}[n] = \mathbf{x}(nT)$. The autocorrelation $R[m]$ of $\mathbf{x}[n]$ equals the samples of $R(\tau)$

$$R[m] = E\{\mathbf{x}[n+m]\mathbf{x}^*[n]\} = E\{\mathbf{x}(nT+mT)\mathbf{x}^*(nT)\} = R(mT)$$

Using the Poisson sum formula (10A-1), we shall show that the power spectrum $S(\omega)$ of $\mathbf{x}[n]$ equals the sum of $S_c(\omega)$ and all its displacements (Fig. 10-21)

$$S(\omega) = \frac{1}{T}\sum_{n=-\infty}^{\infty} S_c(\omega + 2n\Omega) \qquad \Omega = \frac{\pi}{T} \tag{10-90}$$

Proof Since $R[m] = R(mT)$, we conclude from (10-86) that

$$S(\omega) = \sum_{m=-\infty}^{\infty} R(mT)e^{-jm\omega T}$$

This yields (10-90) if we introduce the following changes in (10A-1): $F(\omega)$ to $2\pi R(\tau)$, $f(x)$ to the Fourier transform $S_c(\omega)$ of $R(\tau)$, and c to 2Ω.

Example 10-21 Consider the process $\mathbf{x}(t) = \mathbf{a}e^{j\omega_0 t}$ where $E\{|\mathbf{a}|^2\} = \sigma^2$. In this case

$$R(\tau) = \sigma^2 e^{j\omega_0\tau} \qquad S_c(\omega) = 2\pi\sigma^2\delta(\omega - \omega_0)$$

† Since $\mathbf{S}(e^{j\omega T})$ exists by assumption, the region of existence of $\mathbf{S}(z)$ is a ring containing the unit circle $|z| = 1$.

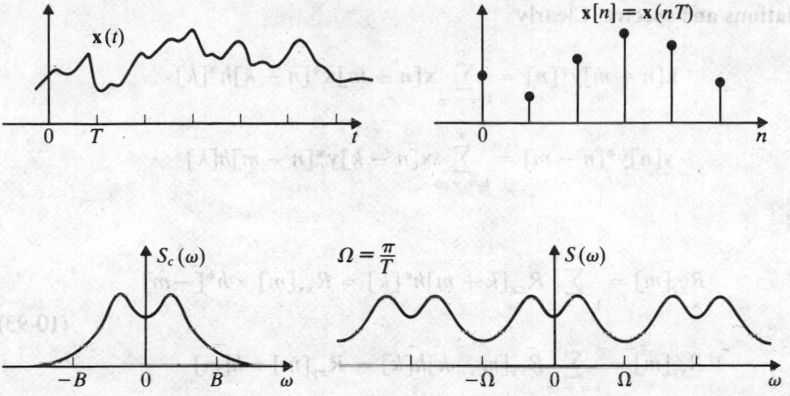

Figure 10-21

With $\mathbf{x}[n] = \mathbf{x}(nT)$ it follows from (10-90) that (Fig. 10-22)

$$S(\omega) = 2\Omega\sigma^2 \sum_{n=-\infty}^{\infty} \delta(\omega - \omega_0 + 2n\Omega) \tag{10-91}$$

Linear Systems

The concepts introduced in Secs. 9-4 and 10-2 hold also for digital systems mutatis mutandis. We determine next the second-order moments of the output $\mathbf{y}[n]$ of a deterministic system driven by a WSS process $\mathbf{x}[n]$. The system is specified in terms of its *delta response* $h[n]$ and system function $\mathbf{H}(z)$

$$h[n] = L\{\delta[n]\} \qquad \mathbf{H}(z) = \sum_{n=-\infty}^{\infty} h[n]z^{-n}$$

Its output equals the *discrete convolution* of $\mathbf{x}[n]$ with $h[n]$

$$\mathbf{y}[n] = \sum_{k=-\infty}^{\infty} \mathbf{x}[n-k]h[k] = \mathbf{x}[n] * h[n] \tag{10-92}$$

From this it follows readily that

$$\eta_y = \eta_x \sum_{k=-\infty}^{\infty} h[k] = \eta_x \mathbf{H}(1)$$

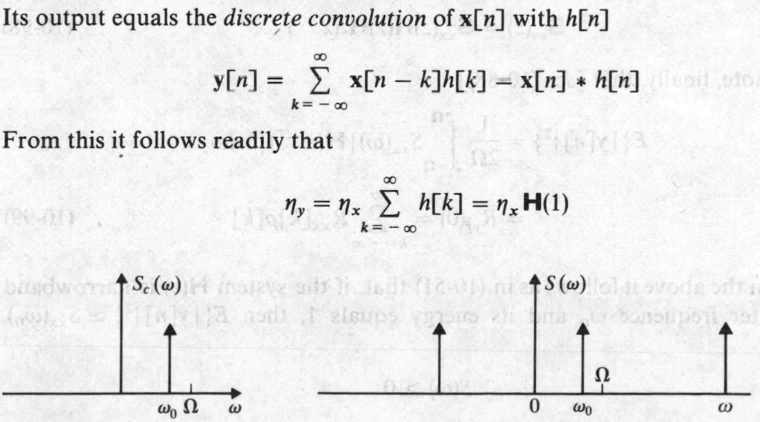

Figure 10-22

Correlations and spectra Clearly

$$\mathbf{x}[n+m]\mathbf{y}^*[n] = \sum_{k=-\infty}^{\infty} \mathbf{x}[n+m]\mathbf{x}^*[n-k]h^*[k]$$

$$\mathbf{y}[n]\mathbf{y}^*[n-m] = \sum_{k=-\infty}^{\infty} \mathbf{x}[n-k]\mathbf{y}^*[n-m]h[k]$$

Hence

$$R_{xy}[m] = \sum_{k=-\infty}^{\infty} R_{xx}[k+m]h^*[k] = R_{xx}[m] * h^*[-m]$$
$$R_{yy}[m] = \sum_{k=-\infty}^{\infty} R_{xy}[m-k]h[k] = R_{xy}[m] * h[m] \tag{10-93}$$

Taking DFT of both sides, we obtain

$$S_{xy}(\omega) = S_{xx}(\omega)\mathbf{H}^*(e^{j\omega T})$$
$$S_{yy}(\omega) = S_{xy}(\omega)\mathbf{H}(e^{j\omega T}) \tag{10-94}$$

With

$$\rho[m] = h[m] * h^*[-m] = \sum_{k=-\infty}^{\infty} h[m+k]h^*[k] \tag{10-95}$$

we conclude that

$$R_{yy}[m] = R_{xx}[m] * \rho[m] \tag{10-96}$$
$$S_{yy}(\omega) = S_{xx}(\omega)|\mathbf{H}(e^{j\omega T})|^2 \tag{10-97}$$

If $h[n]$ is *real*, then $H^*(e^{j\omega T}) = H(e^{-j\omega T})$ and (10-97) yields

$$\mathbf{S}_{vv}(z) = \mathbf{S}_{xx}(z)\mathbf{H}(z)\mathbf{H}(z^{-1}) \tag{10-98}$$

We note, finally, that [see (10-88)]

$$E\{|\mathbf{y}[n]|^2\} = \frac{1}{2\Omega}\int_{-\Omega}^{\Omega} S_{xx}(\omega)|\mathbf{H}(e^{j\omega T})|^2\, d\omega$$
$$= R_{yy}(0) = \sum_{k=-\infty}^{\infty} R_{xx}[k]\rho[k] \tag{10-99}$$

From the above it follows as in (10-51) that, if the system $\mathbf{H}(z)$ is narrowband with center frequence ω_0 and its energy equals 1, then $E\{|\mathbf{y}[n]|^2\} \simeq S_{xx}(\omega_0)$, hence

$$S(\omega) \geq 0$$

for any process real or complex.

Recursion Equations and Rational Spectra

The recursion equation

$$\mathbf{y}[n] + a_1\mathbf{y}[n-1] + \cdots + a_N\mathbf{y}[n-N] = b_0\mathbf{x}[n] + \cdots + b_r\mathbf{x}[n-r] \quad (10\text{-}100)$$

defines a system with input $\mathbf{x}[n]$, output $\mathbf{y}[n]$, and system function

$$\mathbf{H}(z) = \frac{b_0 + \cdots + b_r z^{-r}}{1 + a_1 z^{-1} + \cdots + a_N z^{-N}} = \frac{N(z)}{D(z)} \quad (10\text{-}101)$$

We shall assume that the coefficients of $\mathbf{H}(z)$ are real and that its poles are inside the unit circle (stable system). This leads to the conclusion that if $\mathbf{x}[n]$ is WSS and (10-100) holds for all n, then $\mathbf{y}[n]$ is also WSS with mean $\eta_y = \eta_x \mathbf{H}(1)$.

From (10-97) it follows that the power spectrum of $\mathbf{y}[n]$ equals

$$S_{yy}(\omega) = \frac{|b_0 + \cdots + b_r e^{-jr\omega T}|^2}{|1 + \cdots + a_N e^{-jN\omega T}|^2} \quad (10\text{-}102)$$

White-noise input If $\mathbf{x}[n] = \mathbf{v}[n]$ is white noise with $R_{vv}[m] = \delta[n]$, then $S_{vv}(z) = 1$ and (10-98) yields

$$S_{yy}(z) = \mathbf{H}(z)\mathbf{H}(1/z) = \frac{N(z)N(1/z)}{D(z)D(1/z)} \quad (10\text{-}103)$$

The process $\mathbf{y}[n]$ is then called ARMA (*autoregressive-moving average*). Thus, an ARMA process is a process with rational spectrum.

If $D(z) = 1$, then

$$\mathbf{y}[n] = b_0\mathbf{v}[n] + \cdots + b_r\mathbf{v}[n-r] \quad (10\text{-}104)$$

we then say that $\mathbf{y}[n]$ is an MA (*moving average*) process.

Finally, if $N(z) = 1$, then

$$\mathbf{y}[n] + \cdots + a_N\mathbf{y}[n-N] = \mathbf{v}[n] \quad (10\text{-}105)$$

In this case, $\mathbf{S}(z)$ has only poles and $\mathbf{y}[n]$ is called an AR (*autoregressive*) process.

Line spectra Suppose, finally, that

$$\mathbf{y}[n] + a_1\mathbf{y}[n-1] + \cdots + a_N\mathbf{y}[n-N] = 0 \quad (10\text{-}106)$$

In this case

$$\mathbf{y}[n] = \mathbf{c}_1 z_1^n + \cdots + \mathbf{c}_n z_N^n$$

The roots z_i of $D(z)$ are either inside or on the unit circle. Since $\mathbf{y}[n]$ is stationary, the terms with $|z_i| < 1$ must be zero and the terms with $|z_i| = 1$ must be uncorrelated with zero mean (see Example 10-3). Denoting their variance by σ_i^2, we conclude that

$$R[m] = \sigma_1^2 e^{jm\omega_1 T} + \cdots + \sigma_r^2 e^{jm\omega_r T} \qquad e^{j\omega_i T} = z_i \quad (10\text{-}107)$$

and (10-91) yields

$$S(\omega) = 2\Omega\sigma_1^2\delta(\omega - \omega_1) + \cdots + 2\Omega\sigma_r^2\delta(\omega - \omega_r) \qquad |\omega| < \Omega$$

Thus, if a process $\mathbf{y}[n]$ is AR and $\mathbf{v}[n] = 0$, then its spectrum consists of lines. As we see from (10-106) such a process is predictable (see also Sec. 13-4).

10-5 FACTORIZATION AND INNOVATIONS

Given a stationary process $\mathbf{x}(t)$, we wish to find a minimum-phase† system $\mathbf{\Gamma}(s)$ such that its response to $\mathbf{x}(t)$ is an orthonormal process $\mathbf{i}(t)$

$$\mathbf{i}(t) = \int_0^\infty \gamma(\alpha)\mathbf{x}(t-\alpha)\,d\alpha \qquad R_i(\tau) = \delta(\tau) \tag{10-108}$$

Since $\mathbf{\Gamma}(s)$ is minimum-phase the inverse system

$$\mathbf{L}(s) = \frac{1}{\mathbf{\Gamma}(s)} \tag{10-109}$$

is causal and its response to $\mathbf{i}(t)$ equals $\mathbf{x}(t)$ (see Fig. 10-23):

$$\mathbf{x}(t) = \int_0^\infty l(\alpha)\mathbf{i}(t-\alpha)\,d\alpha \tag{10-110}$$

A process that can be so represented is called *regular*.‡

In the above representation, $\mathbf{\Gamma}(s)$ is called *whitening filter*, the process $\mathbf{i}(t)$ *innovations* of $\mathbf{x}(t)$, and the system $\mathbf{L}(s)$ *innovations filter*. Thus, a regular process is linearly equivalent to its innovations in the sense that each is the output of a linear causal system with input the other. This equivalence is used extensively in Chap. 13.

Note The representation of a regular process as a weighted average of the past values of a white-noise process is an extension of the Gram–Schmidt orthonormalization (see Sec. 13-1).

Since $\mathbf{S}_i(s) = 1$, it follows from (10-40) and Fig. 10-23 that the power spectrum $\mathbf{S}(s)$ of $\mathbf{x}(t)$ can be written as a product

$$\mathbf{S}(s) = \mathbf{L}(s)\mathbf{L}(-s) \qquad \mathbf{S}(j\omega) = |\mathbf{L}(j\omega)|^2 \tag{10-111}$$

where $\mathbf{L}(s)$ is a minimum phase system.

Conversely, if $\mathbf{S}(s)$ can be so factored, then the process $\mathbf{x}(t)$ is regular because then the system $\mathbf{\Gamma}(s) = 1/\mathbf{L}(s)$ is causal and its response $\mathbf{i}(t)$ to $\mathbf{x}(t)$ is white noise

$$\mathbf{S}_i(s) = \mathbf{S}(s)\mathbf{\Gamma}(s)\mathbf{\Gamma}(-s) = 1 \tag{10-112}$$

Thus, a process $\mathbf{x}(t)$ is regular iff its spectrum can be factored as in (10-111). As we show later (Sec. 13-4), not all processes are regular. In the following, we present a general criterion for regularity and we discuss the factorization problem for processes with rational spectra.

† A system $\mathbf{\Gamma}(s)$ is called *minimum-phase* if it is causal, stable, and its inverse $1/\mathbf{\Gamma}(s)$ is also causal and stable. This is equivalent to the condition that the functions $\mathbf{\Gamma}(s)$ and $1/\mathbf{\Gamma}(s)$ are analytic for $\operatorname{Re} s \geq 0$.

‡ In mathematics, the term "regular" has a related but somewhat different meaning.

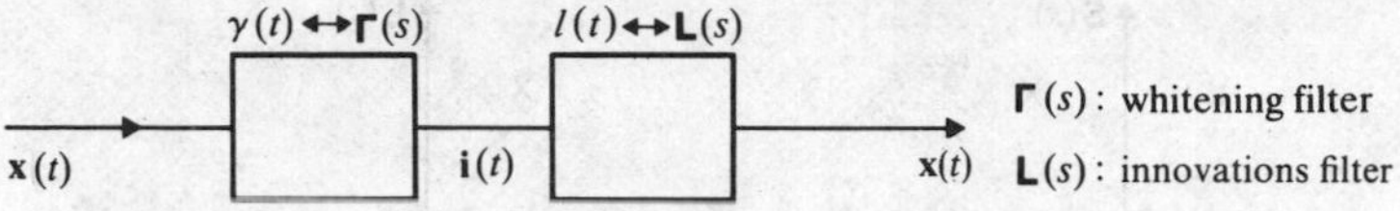

Figure 10-23

The Paley–Wiener condition It can be shown that a process $\mathbf{x}(t)$ with square-integrable spectrum is regular iff its power spectrum $S(\omega)$ satisfies the Paley–Wiener† condition

$$\int_{-\infty}^{\infty} \frac{\ln S(\omega)}{1+\omega^2}\, d\omega < \infty \tag{10-113}$$

This shows that if $S(\omega) = 0$ for every ω in an interval then $\mathbf{x}(t)$ is not regular.

Rational spectra Clearly, all positive rational spectra satisfy (10-113), hence the corresponding processes are regular. In the following, we discuss the factorization of such functions in the form (10-111).

A rational spectrum $S(\omega)$ of a *real* process $\mathbf{x}(t)$ is the ratio of two polynomials in ω^2

$$S(\omega) = \frac{A(\omega^2)}{B(\omega^2)}$$

because $S(-\omega) = S(\omega)$. The corresponding $\mathbf{S}(s)$ equals

$$\mathbf{S}(s) = \frac{A(-s^2)}{B(-s^2)}$$

This shows that, if s_i is a singularity (zero or pole) of $\mathbf{S}(s)$, then $-s_i$ is also a singularity. Furthermore, since the coefficients of the polynomials $A(-s^2)$ and $B(-s^2)$ are real, their roots are either real or complex conjugate.

From the above it follows that the singularities of $\mathbf{S}(s)$ are symmetrical with respect to the $j\omega$ axis (Fig. 10-24*a*) and can be separated into two groups: The "left" group consists of all s_i with Re $s_i < 0$ (Fig. 11-24*b*). The "right" group consists of all other singularities. The minimum phase component $\mathbf{L}(s)$ of $\mathbf{S}(s)$ is a ratio of two polynomials

$$\mathbf{L}(s) = \frac{N(s)}{D(s)}$$

whose zeros and poles are the "left" singularities of $\mathbf{S}(s)$.

† N. Wiener, R. E. A. C. Paley: Fourier Transforms in the Complex Domain, *American Mathematical Society College*, 1934 (see also P.2).

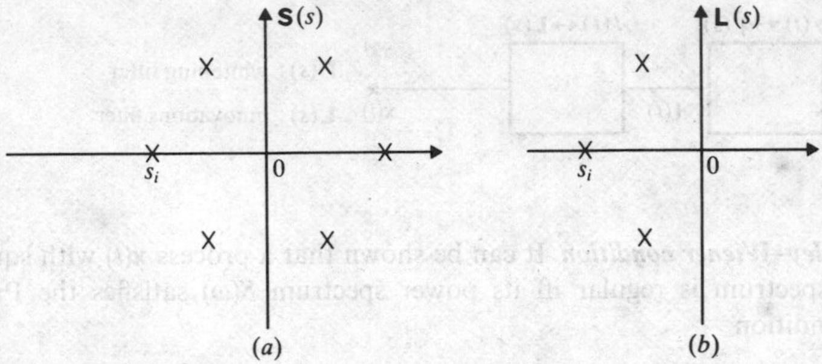

Figure 10-24

Example 10-22 If

$$S(\omega) = \frac{N}{\alpha^2 + \omega^2}$$

then

$$\mathbf{S}(s) = \frac{N}{\alpha^2 - s^2} = \frac{N}{(\alpha + s)(\alpha - s)} \qquad \mathbf{L}(s) = \frac{\sqrt{N}}{\alpha + s}$$

Example 10-23 If

$$S(\omega) = \frac{49 + 25\omega^2}{\omega^4 + 10\omega^2 + 9}$$

then

$$\mathbf{S}(s) = \frac{49 - 25s^2}{(1 - s^2)(9 - s^2)} \qquad \mathbf{L}(s) = \frac{7 + 5s}{(1 + s)(3 + s)}$$

The above approach can be used also for complex singularities. However, in this case it is simpler to factor $\mathbf{S}(s)$ into linear and quadratic terms in s^2 with real coefficients and to determine directly their causal components: Suppose that $\mathbf{S}(s)$ contains the factor

$$s^4 + bs^2 + c \qquad b^2 < 4c$$

We can see by inspection that the corresponding factor in $\mathbf{L}(s)$ equals

$$s^2 + \beta s + \gamma \qquad \gamma = \sqrt{c} \qquad \beta = \sqrt{2\gamma - b}$$

Example 10-24(*a*) Suppose that

$$S(\omega) = \frac{3}{\omega^4 - \omega^2 + 1} \qquad \mathbf{S}(s) = \frac{3}{s^4 + s^2 + 1}$$

In this case, $\mathbf{S}(s)$ has a biquadratic denominator with $b = c = 1$. Hence

$$\gamma = 1 \qquad \beta = 1 \qquad \mathbf{L}(s) = \frac{\sqrt{3}}{s^2 + s + 1}$$

(*b*) If

$$S(\omega) = \frac{1}{\omega^4 + 1} \qquad \mathbf{S}(s) = \frac{1}{s^4 + 1}$$

then $b = 0$, $c = 1$. Hence

$$\gamma = 1 \qquad \beta = \sqrt{2} \qquad \mathbf{L}(s) = \frac{1}{s^2 + \sqrt{2}s + 1}$$

Discrete-Time Processes

A process $\mathbf{x}[n]$ is called *regular* if its spectrum can be written as a product

$$\mathbf{S}(z) = \mathbf{L}(z)\mathbf{L}(z^{-1}) \qquad \mathbf{S}(e^{j\omega T}) = |\mathbf{L}(e^{j\omega T})|^2 \tag{10-114}$$

where $\mathbf{L}(z)$ is a *minimum-phase* system, i.e., the functions $\mathbf{L}(z)$ and

$$\mathbf{\Gamma}(z) = \frac{1}{\mathbf{L}(z)}$$

are analytic for $|z| \geq 1$. From this and (10-98) it follows that if $\mathbf{x}[n]$ is the input to the system $\mathbf{\Gamma}(z)$, then the resulting response $\mathbf{i}[n]$ is white noise

$$\mathbf{i}[n] = \sum_{k=0}^{\infty} \gamma[k]\mathbf{x}[n-k] \qquad R_i[m] = \delta[m] \tag{10-115}$$

Furthermore (Fig. 10-24)

$$\mathbf{x}[n] = \sum_{k=0}^{\infty} l[k]\mathbf{i}[n-k] \tag{10-116}$$

As in the analog case, we shall call the system $\mathbf{\Gamma}(z)$ *whitening filter*, the process $\mathbf{i}[n]$ *innovations* of $\mathbf{x}[n]$, and the system $\mathbf{L}(z)$ *innovations filter*. Thus, a regular process is equivalent to its innovations in the sense that each is the output of a linear causal system with input the other. The following is a general criterion for regularity.

The Paley–Wiener condition It can be shown that a discrete-time process $\mathbf{x}[n]$ is regular if (see P.2)

$$\int_{-\Omega}^{\Omega} |\log S(\omega)|\, d\omega < \infty \tag{10-117}$$

Rational spectra If $\mathbf{S}(z)$ is the power spectrum of a *real* process and it is a rational function of z, then the periodic function $S(\omega) = \mathbf{S}(e^{j\omega T})$ is a rational function of .

$$\cos \omega T = \tfrac{1}{2}(e^{j\omega T} + e^{-j\omega T})$$

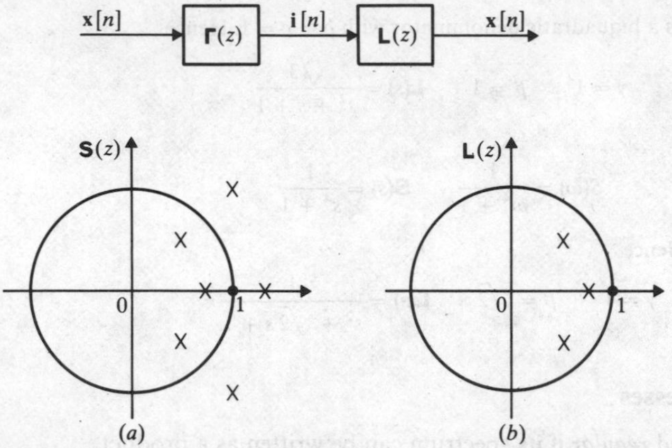

Figure 10-25

because $S(-\omega) = S(\omega)$. From this it follows that $\mathbf{S}(z)$ is a function of $z + 1/z$. This leads to the conclusion that if z_i is a singularity of $\mathbf{S}(z)$, then $1/z_i$ is also a singularity. Furthermore, since the coefficients of $\mathbf{S}(z)$ are real, its singularities are either real or complex conjugate.

The above leads to the conclusion that the singularities of $\mathbf{S}(z)$ are symmetrical with respect to the unit circle (Fig. 10-25*a*) and can be separated into two groups: The "inside" group consisting of all z_i such that $|z_i| < 1$ (Fig. 10-25*b*), and the "outside" group consisting of all other singularities. The minimum-phase component $\mathbf{L}(z)$ of $\mathbf{S}(z)$ is the ratio of two polynomials

$$\mathbf{L}(z) = \frac{N(z)}{D(z)}$$

whose singularities are the "inside" singularities of $\mathbf{S}(z)$.

Example 10-25 If

$$S(\omega) = \frac{5 - 4\cos\omega T}{10 - 6\cos\omega T} \qquad \text{then} \qquad \mathbf{S}(z) = \frac{5 - 2(z + z^{-1})}{10 - 3(z + z^{-1})}$$

The function $\mathbf{S}(z)$ has two zeros, 1/2 and 2, and two poles, 1/3 and 3. Retaining only the "inside" singularities 1/2 and 1/3, we obtain

$$\mathbf{L}(z) = \frac{2z - 1}{3z - 1}$$

The Matched Filter

A basic problem in the applications of stochastic processes is the estimation of a signal $f(t)$ in the presence of an additive interference $\mathbf{v}(t)$ (noise). The available information (data) is the sum

$$\mathbf{x}(t) = f(t) + \mathbf{v}(t)$$

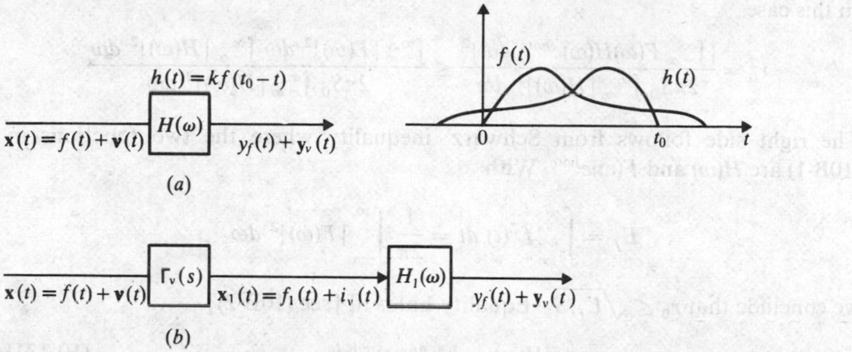

Figure 10-26

and the problem is to establish the presence of $f(t)$ or to estimate its form. The solution of this problem depends on the state of our prior knowledge concerning $f(t)$ and $\mathbf{v}(t)$. We consider the following case: We assume that the signal $f(t)$ is known (within a factor) and that the noise $\mathbf{v}(t)$ is a regular WSS process with known power spectrum $S_v(\omega)$. This problem is typical in radar where we wish to establish the presence and location of a signal $f(t)$ returning from a distant target.

The effectiveness of the detection scheme depends on the ratio r between the value $f(t_0)$ of $f(t)$ at a certain time t_0, and the standard deviation of the noise. This ratio can be small even if the spectrum of $f(t)$ is large compared with $S_v(\omega)$. A common method for increasing r involves a linear processing of the data as in Fig. 10-26. We use as input to a linear system $H(\omega)$ the process $\mathbf{x}(t)$. The resulting output is a sum

$$\mathbf{y}(t) = y_f(t) + \mathbf{y}_v(t)$$

where

$$y_f(t) = \frac{1}{2\pi}\int_{-\infty}^{\infty} F(\omega)H(\omega)e^{j\omega t}\,dt \qquad \text{(10-118)}$$

is due to the signal $f(t)$ and $\mathbf{y}_v(t)$ is due to the noise $\mathbf{v}(t)$. As we know [see (10-41)]

$$E\{\mathbf{y}_v^2(t)\} = \frac{1}{2\pi}\int_{-\infty}^{\infty} S_v(\omega)\,|\,H(\omega)\,|^2\,d\omega \qquad \text{(10-119)}$$

We assume that the Fourier transform $F(\omega)$ of $f(t)$ and the power spectrum $S_v(\omega)$ of $\mathbf{v}(t)$ are known, and our objective is to find the filter $H(\omega)$ so as to maximize the output *signal-to-noise ratio*

$$r_0 = \frac{|\,y_f(t_0)\,|}{\sqrt{E\{\mathbf{y}_v^2(t)\}}} \qquad \text{(10-120)}$$

White noise Suppose, first, that

$$S_v(\omega) = S_0 = \text{constant}$$

In this case

$$r_0^2 = \frac{|\int_{-\infty}^{\infty} F(\omega)H(\omega)e^{j\omega t_0}\, d\omega|^2}{2\pi S_0 \int_{-\infty}^{\infty} |H(\omega)|^2\, d\omega} \le \frac{\int_{-\infty}^{\infty} |F(\omega)|^2\, d\omega \int_{-\infty}^{\infty} |H(\omega)|^2\, d\omega}{2\pi S_0 \int_{-\infty}^{\infty} |H(\omega)|^2\, d\omega}$$

The right side follows from Schwarz' inequality where the two functions in (10B-1) are $H(\omega)$ and $F(\omega)e^{j\omega t_0}$. With

$$E_f = \int_{-\infty}^{\infty} f^2(t)\, dt = \frac{1}{2\pi}\int_{-\infty}^{\infty} |F(\omega)|^2\, d\omega$$

we conclude that $r_0 \le \sqrt{E_f/S_0}$. Equality holds iff [see (10B-2)]

$$H(\omega) = kF^*(\omega)e^{-j\omega t_0} \tag{10-121}$$

The system $H(\omega)$ so obtained is called *matched filter*. Its impulse response equals $h(t) = kf(t - t_0)$ because $F^*(\omega)$ is the transform of $f(-t)$.

Colored noise We shall now show that if the spectrum of the noise $\mathbf{v}(t)$ is an arbitrary function $S_\nu(\omega)$, then the system function of the optimum filter equals

$$H(\omega) = k\frac{F^*(\omega)}{S_\nu(\omega)}e^{-j\omega t_0} \tag{10-122}$$

where, again, k is an arbitrary constant.

This can be shown by a modification of Schwarz' inequality (see Prob. 10-30). We shall prove it, however, using (10-121) and the concept of innovations. For this purpose, we form the whitening filter $\mathbf{\Gamma}_\nu(s)$ of the noise $\mathbf{v}(t)$

$$S_\nu(\omega) = \frac{1}{|\mathbf{\Gamma}_\nu(j\omega)|^2} \tag{10-123}$$

[see (10-111)], and use as input the data $\mathbf{x}(t)$. The resulting output is a sum (Fig. 10-26*b*)

$$\mathbf{x}_1(t) = f_1(t) + \mathbf{i}_\nu(t) \qquad S_{i_\nu}(\omega) = 1$$

where $\mathbf{i}_\nu(t)$ is the innovation of $\mathbf{v}(t)$, and $f_1(t)$ is a deterministic signal with Fourier transform $F_1(\omega) = F(\omega)\mathbf{\Gamma}_\nu(j\omega)$. To maximize the S/N ratio with input $\mathbf{x}(t)$, it suffices to find a filter $H_1(\omega)$ so as to maximize the S/N ratio with input $\mathbf{x}_1(t)$. Since the noise component $\mathbf{i}_\nu(t)$ of $\mathbf{x}_1(t)$ is white, it follows from (10-121) that

$$H_1(\omega) = kF_1^*(\omega)e^{-j\omega t_0} \qquad F_1^*(\omega) = F^*(\omega)\mathbf{\Gamma}_\nu^*(j\omega) \tag{10-124}$$

Cascading with $\mathbf{\Gamma}_\nu(s)$ we obtain

$$H(\omega) = H_1(\omega)\mathbf{\Gamma}_\nu(j\omega) = kF^*(\omega)\mathbf{\Gamma}_\nu^*(j\omega)\mathbf{\Gamma}_\nu(j\omega)$$

and (10-122) results.

10-6 SPECTRAL REPRESENTATION AND FOURIER TRANSFORMS

Spectra are generally associated with trigonometric functions. For deterministic signals, this involves Fourier series and Fourier transforms. For stochastic processes, the notion of spectrum has two interpretations. In the first interpretation the spectrum of a process $\mathbf{x}(t)$ is introduced as a deterministic function $S(\omega)$ equal to the Fourier transform of the correlation $R(\tau)$ of $\mathbf{x}(t)$. In the second, it is defined as the Fourier series or the Fourier transform of the process $\mathbf{x}(t)$ itself. In the earlier sections of this chapter, we dealt with the first interpretation. In this section, we introduce the second. The relationship between the two is discussed in Chap. 14.

Fourier Series

Given a WSS process $\mathbf{x}(t)$, we form the sum

$$\hat{\mathbf{x}}(t) = \sum_{n=-\infty}^{\infty} \mathbf{c}_n e^{jn\omega_0 t} \tag{10-125}$$

where

$$\mathbf{c}_n = \frac{1}{T}\int_{-T/2}^{T/2} \mathbf{x}(t) e^{-jn\omega_0 t}\, dt \qquad \omega_0 = \frac{2\pi}{T} \tag{10-126}$$

We shall show that, if the process $\mathbf{x}(t)$ is MS periodic, then $\mathbf{x}(t) = \hat{\mathbf{x}}(t)$ for every t and the RVs $\mathbf{c}_n$ are uncorrelated. Otherwise, $\mathbf{x}(t) = \hat{\mathbf{x}}(t)$ only for $|t| < T/2$ and the coefficients $\mathbf{c}_n$ are no longer uncorrelated. However, their correlation coefficient tends to zero as $T \to \infty$. In all cases

$$E\{\mathbf{c}_n\} = \frac{1}{T}\int_0^T E\{\mathbf{x}(t)\} e^{-jn\omega_0 t}\, dt = \begin{cases} \eta_x & n = 0 \\ 0 & n \neq 0 \end{cases}$$

MS periodic processes If $\mathbf{x}(t)$ is MS periodic with period T, then its autocorrelation $R(\tau)$ is periodic and can be expanded into a Fourier series

$$R(\tau) = \sum_{n=-\infty}^{\infty} \gamma_n e^{jn\omega_0 \tau} \qquad \gamma_n = \frac{1}{T}\int_0^T R(\tau) e^{-jn\omega_0 \tau}\, d\tau \tag{10-127}$$

We maintain that in this case

$$E\{|\mathbf{x}(t) - \hat{\mathbf{x}}(t)|^2\} = 0 \qquad \text{all } t \tag{10-128}$$

and

$$E\{\mathbf{c}_n \mathbf{c}_m^*\} = \gamma_n \delta[n-m] \tag{10-129}$$

PROOF From (10-126) and the periodicity of $R(\tau)$ it follows that

$$E\{\mathbf{c}_n \mathbf{x}^*(\alpha)\} = \frac{1}{T}\int_0^T R(t-\alpha)e^{-jn\omega_0 t}\,dt = \gamma_n e^{-jn\omega_0\alpha} \tag{10-130}$$

$$E\{\mathbf{c}_n \mathbf{c}_m^*\} = \frac{1}{T}\int_0^T E\{\mathbf{c}_n \mathbf{x}^*(t)\}e^{jm\omega_0 t}\,dt = \begin{cases}\gamma_n & n = m\\ 0 & n \neq 0\end{cases}$$

Hence

$$E\{\mathbf{c}_n \hat{\mathbf{x}}^*(t)\} = \sum_{m=-\infty}^{\infty} E\{\mathbf{c}_n \mathbf{c}_m^*\}e^{-jm\omega_0 t} = \gamma_n e^{-jn\omega_0 t}$$

$$E\{\hat{\mathbf{x}}(t)\mathbf{x}^*(t)\} = \sum_{n=-\infty}^{\infty} E\{\mathbf{c}_n \mathbf{x}^*(t)\}e^{jn\omega_0 t} = \sum_{n=-\infty}^{\infty} \gamma_n$$

$$= R(0) = E\{\hat{\mathbf{x}}^*(t)\mathbf{x}(t)\} = E\{|\mathbf{x}(t)|^2\} = E\{|\hat{\mathbf{x}}(t)|^2\}$$

and (10-128) results.

Nonperiodic processes Suppose now that $\mathbf{x}(t)$ is an arbitrary WSS process. For a given α, the function $R(\tau-\alpha)$ can be expanded into a Fourier series in the interval $|\tau| < T/2$ where T is an arbitrary constant

$$R(\tau-\alpha) = \sum_{n=-\infty}^{\infty} \beta_n(\alpha)e^{jn\omega_0\tau} \qquad |\tau| < \frac{T}{2} \tag{10-131}$$

The coefficients $\beta_n(\alpha)$ are now functions of the parameter α

$$\beta_n(\alpha) = \frac{1}{T}\int_{-T/2}^{T/2} R(\tau-\alpha)e^{-jn\omega_0\tau}\,d\tau \tag{10-132}$$

We maintain that

$$E\{|\mathbf{x}(t) - \hat{\mathbf{x}}(t)|^2\} = 0 \qquad |t| < \frac{T}{2} \tag{10-133}$$

where $\hat{\mathbf{x}}(t)$ is the sum in (10-125). Furthermore

$$E\{\mathbf{c}_n \mathbf{c}_m^*\} = \frac{1}{T}\int_{-T/2}^{T/2} \beta_n(\alpha)e^{jm\omega_0\alpha}\,d\alpha \tag{10-134}$$

PROOF From (10-126) it follows that

$$E\{\mathbf{c}_n \mathbf{x}^*(\alpha)\} = \frac{1}{T}\int_{-T/2}^{T/2} R(t-\alpha)e^{-jn\omega_0 t}\,dt = \beta_n(\alpha)$$

$$E\{\mathbf{c}_n \mathbf{c}_m^*\} = \frac{1}{T}\int_{-T/2}^{T/2} E\{\mathbf{c}_n \mathbf{x}^*(t)\}e^{jm\omega_0 t}\,dt = \frac{1}{T}\int_{-T/2}^{T/2} \beta_n(t)e^{jm\omega_0 t}\,dt \tag{10-135}$$

Hence

$$E\{\hat{\mathbf{x}}(t)\mathbf{x}^*(t)\} = \sum_{n=-\infty}^{\infty} E\{\mathbf{c}_n \mathbf{x}^*(t)\}e^{jn\omega_0 t}$$

$$= \sum_{n=-\infty}^{\infty} \beta_n(t)e^{jn\omega_0 t} = R(0) = E\{\hat{\mathbf{x}}^*(t)\mathbf{x}(t)\}$$

and (10-133) results.

Note If T is large and $\alpha \ll T$, then

$$T\beta_n(\alpha) = \int_{-T/2}^{T/2} R(\tau - \alpha)e^{-jn\omega_0\tau}\,d\tau \simeq \int_{-\infty}^{\infty} R(\tau - \alpha)e^{-jn\omega_0\tau}\,d\tau$$

Since the last integral equals $S(\omega)e^{-jn\omega_0\alpha}$ we conclude from the above and (10-134) that if $R(\tau) \to 0$ as $|\tau| \to \infty$ then

$$TE\{\mathbf{c}_n \mathbf{c}_m^*\} \simeq \frac{S(n\omega_0)}{T}\int_{T/2}^{T/2} e^{j(m-n)\omega_0\alpha}\,d\alpha = \begin{cases} S(n\omega_0) & n = m \\ 0 & n \neq m \end{cases} \tag{10-136}$$

Thus, for large T, the coefficients $\mathbf{c}_n$ of the Fourier series expansion (10-125) of an arbitrary WSS process $\mathbf{x}(t)$ are nearly uncorrelated [see also (10-151)].

The Karhunen–Loève Expansion

The Fourier series is a special case of the expansion of a process $\mathbf{x}(t)$ into a series of the form

$$\hat{\mathbf{x}}(t) = \sum_{n=1}^{\infty} \mathbf{c}_n \varphi_n(t) \qquad 0 < t < T \tag{10-137}$$

where $\varphi_n(t)$ is a set of orthonormal functions in the interval $(0, T)$

$$\int_0^T \varphi_n(t)\varphi_m^*(t)\,dt = \delta[n - m] \tag{10-138}$$

and the coefficients $\mathbf{c}_n$ are RVs given by

$$\mathbf{c}_n = \int_0^T \mathbf{x}(t)\varphi_n^*(t)\,dt \tag{10-139}$$

In the following, we consider the problem of determining a set of orthonormal functions $\varphi_n(t)$ such that: (*a*) the sum in (10-137) equals $\mathbf{x}(t)$; (*b*) the coefficients $\mathbf{c}_n$ are uncorrelated.

To solve this problem we form the *integral equation*

$$\int_0^T R(t_1, t_2)\varphi(t_2)\,dt_2 = \lambda\varphi(t_1) \qquad 0 < t_1 < T \tag{10-140}$$

where $R(t_1, t_2)$ is the autocorrelation of the process $\mathbf{x}(t)$. It is well known from the theory of integral equations that the eigenfunctions $\varphi_n(t)$ of (10-140) are orthonormal as in (10-138) and they satisfy the identity

$$R(t, t) = \sum_{n=1}^{\infty} \lambda_n |\varphi_n(t)|^2 \tag{10-141}$$

where λ_n are the corresponding eigenvalues. This is a consequence of the p.d. character of $R(t_1, t_2)$.

Using the above we shall show that if $\varphi_n(t)$ are the eigenfunctions of (10-140) then

$$E\{|\mathbf{x}(t) - \hat{\mathbf{x}}(t)|^2\} = 0 \qquad 0 < t < T \tag{10-142}$$

and

$$E\{\mathbf{c}_n \mathbf{c}_m^*\} = \lambda_n \delta[n - m] \tag{10-143}$$

PROOF From (10-139) and (10-140) it follows that

$$E\{\mathbf{c}_n \mathbf{x}^*(\alpha)\} = \int_0^T R^*(\alpha, t)\varphi_n^*(t)\, dt = \lambda_n \varphi_n^*(\alpha) \tag{10-144}$$

$$E\{\mathbf{c}_n \mathbf{c}_m^*\} = \lambda_m \int_0^T \varphi_n^*(t)\varphi_m(t)\, dt = \lambda_n \delta[n - m]$$

Hence

$$E\{\mathbf{c}_n \hat{\mathbf{x}}^*(t)\} = \sum_{m=1}^{\infty} E\{\mathbf{c}_n \mathbf{c}_m^*\}\varphi_m^*(t) = \lambda_n \varphi_n^*(t)$$

$$E\{\hat{\mathbf{x}}(t)\mathbf{x}^*(t)\} = \sum_{n=1}^{\infty} \lambda_n \varphi_n(t)\varphi_n^*(t) = R(t, t)$$

$$= E\{\hat{\mathbf{x}}^*(t)\mathbf{x}(t)\} = E\{|\mathbf{x}(t)|^2\} = E\{|\hat{\mathbf{x}}(t)|^2\}$$

and (10-142) results.

It is of interest to note that the converse of the above is also true: If $\varphi_n(t)$ is an orthonormal set of functions and

$$\mathbf{x}(t) = \sum_{n=1}^{\infty} \mathbf{c}_n \varphi_n(t) \qquad E\{\mathbf{c}_n \mathbf{c}_m^*\} = \begin{cases} \sigma_n^2 & n = m \\ 0 & n \neq m \end{cases}$$

then the functions $\varphi_n(t)$ must satisfy (10-140) with $\lambda = \sigma_n^2$.

PROOF From the assumptions it follows that $\mathbf{c}_n$ is given by (10-139). Furthermore

$$E\{\mathbf{x}(t)\mathbf{c}_m^*\} = \sum_{n=1}^{\infty} E\{\mathbf{c}_n \mathbf{c}_m^*\}\varphi_m(t) = \sigma_m^2 \varphi_m(t)$$

$$E\{\mathbf{x}(t)\mathbf{c}_m^*\} = \int_0^T E\{\mathbf{x}(t)\mathbf{x}^*(\alpha)\}\varphi_m(\alpha)\, d\alpha = \int_0^T R(t, \alpha)\varphi_m(\alpha)\, d\alpha$$

This completes the proof.

The sum in (10-137) is called the Karhunen–Loève (K-L) expansion of the process $\mathbf{x}(t)$. In this expansion, $\mathbf{x}(t)$ need not be stationary. If it is stationary, then the origin can be chosen arbitrarily. We shall illustrate with two examples.

Example 10-26 Suppose that the process $\mathbf{x}(t)$ is ideal low-pass with autocorrelation

$$R(\tau) = \frac{\sin a\tau}{\pi\tau}$$

We shall find its K-L expansion. Shifting the origin appropriately, we conclude from (10-140) that the functions $\varphi_n(t)$ must satisfy the integral equation

$$\int_{-T/2}^{T/2} \frac{\sin a(t-\tau)}{\pi(t-\tau)} \varphi_n(\tau)\, d\tau = \lambda_n \varphi_n(t) \qquad (10\text{-}145)$$

The solutions of this equation are known as *prolate-spheroidal* functions† [see also (14-11)].

Example 10-27 We shall determine the K-L expansion (10-137) of the Wiener process $\mathbf{w}(t)$ introduced in Sec. 9-1. In this case [see (9-25)]

$$R(t_1, t_2) = \alpha \min(t_1, t_2) = \begin{cases} \alpha t_2 & t_2 < t_1 \\ \alpha t_1 & t_2 > t_1 \end{cases}$$

Inserting into (10-140), we obtain

$$\alpha \int_0^{t_1} t_2\, \varphi(t_2)\, dt_2 + \alpha t_1 \int_{t_1}^{T} \varphi(t_2)\, dt_2 = \lambda \varphi(t_1) \qquad (10\text{-}146)$$

To solve the above integral equation, we evaluate the appropriate end-point conditions and differentiate twice. This yields

$$\varphi(0) = 0 \qquad \alpha \int_{t_1}^{T} \varphi(t_2)\, dt_2 = \lambda \varphi'(t_1)$$

$$\varphi'(T) = 0 \qquad \lambda \varphi''(t) + \alpha \varphi(t) = 0$$

Solving the last equation, we obtain

$$\varphi_n(t) = \sqrt{\frac{2}{T}} \sin \omega_n t \qquad \omega_n = \sqrt{\frac{\alpha}{\lambda_n}} = \frac{(2n+1)\pi}{2T}$$

Thus, in the interval $(0, T)$, the Wiener process can be written as a sum of sine waves

$$\mathbf{w}(t) = \sqrt{\frac{2}{T}} \sum_{n=1}^{\infty} \mathbf{c}_n \sin \omega_n t \qquad \mathbf{c}_n = \sqrt{\frac{2}{T}} \int_0^T \mathbf{w}(t) \sin \omega_n t\, dt$$

where the coefficients $\mathbf{c}_n$ are uncorrelated with variance $E\{\mathbf{c}_n^2\} = \lambda_n$.

† D. Slepian, H. J. Landau and H. O. Pollack: "Prolate Spheroidal Wave Functions," *Bell System Technical Journal*, vol. 40, 1961.

Fourier Integrals

The Fourier integral of a stochastic process $\mathbf{x}(t)$ is a stochastic process in the variable ω given by

$$\mathbf{X}(\omega) = \int_{-\infty}^{\infty} \mathbf{x}(t)e^{-j\omega t}\,dt \tag{10-147}$$

Applying the Fourier inversion formula to each sample of $\mathbf{x}(t)$, we obtain

$$\mathbf{x}(t) = \frac{1}{2\pi}\int_{-\infty}^{\infty} \mathbf{X}(\omega)e^{j\omega t}\,d\omega \tag{10-148}$$

This shows that an arbitrary process $\mathbf{x}(t)$ can be formally represented by a superposition of exponentials for every t. The properties of $\mathbf{X}(\omega)$ depend on the interpretation of the integral in (10-147). As usual, we shall interpret this integral as an MS limit and we shall relate the second-order moments of $\mathbf{X}(\omega)$ to the two-dimensional Fourier transform

$$\Gamma(u, v) = \int_{-\infty}^{\infty}\int_{-\infty}^{\infty} R(t_1, t_2)e^{-j(ut_1 + vt_2)}\,dt_1\,dt_2 \tag{10-149}$$

of the autocorrelation $R(t_1, t_2)$ of $\mathbf{x}(t)$.

Properties 1. The mean of $\mathbf{X}(\omega)$ is the Fourier transform of the mean of $\mathbf{x}(t)$

$$E\{\mathbf{X}(\omega)\} = \int_{-\infty}^{\infty} E\{\mathbf{x}(t)\}e^{-j\omega t}\,dt \tag{10-150}$$

This follows readily from (10-147).

2. The autocorrelation of $\mathbf{X}(\omega)$ is given by

$$E\{\mathbf{X}(u)\mathbf{X}^*(v)\} = \Gamma(u, -v) \tag{10-151}$$

Indeed, multiplying (10-147) by its conjugate we obtain

$$E\{\mathbf{X}(u)\mathbf{X}^*(v)\} = \int_{-\infty}^{\infty}\int_{-\infty}^{\infty} E\{\mathbf{x}(t_1)\mathbf{x}^*(t_2)\}e^{-j(ut_1 - vt_2)}\,dt_1\,dt_2 \tag{10-152}$$

and (10-151) results.

3. If the process $\mathbf{x}(t)$ is WSS with power spectrum $S(\omega)$, then its transform $\mathbf{X}(\omega)$ is white noise with average intensity $2\pi S(u)$

$$E\{\mathbf{X}(u)\mathbf{X}^*(v)\} = 2\pi S(u)\delta(u - v) \tag{10-153}$$

To prove the above, we observe that if $\mathbf{x}(t)$ is WSS, then $R(t_1, t_2) = R(t_1 - t_2)$. With $t_1 = t_2 + \tau$, (10-149) yields

$$\Gamma(u, v) = \int_{-\infty}^{\infty}\int_{-\infty}^{\infty} R(t_1 - t_2)e^{-j(ut_1 + vt_2)}\,dt_1\,dt_2$$

$$= \int_{-\infty}^{\infty} e^{-j(u+v)t_2}\int_{-\infty}^{\infty} R(\tau)e^{-ju\tau}\,d\tau\,dt_2 = S(u)\int_{-\infty}^{\infty} e^{-j(u+v)t_2}\,dt_2$$

and (10-153) follows from (10-151) because the last integral above equals $2\pi\delta(u+v)$.

4. If $\mathbf{x}(t)$ is white noise with average intensity $q(t)$, then $\mathbf{X}(\omega)$ is WSS with autocorrelation

$$E\{\mathbf{X}(u)\mathbf{X}^*(v)\} = Q(u - v) \tag{10-154}$$

where $Q(\omega)$ is the Fourier transform of $q(t)$

$$Q(\omega) = \int_{-\infty}^{\infty} q(t)e^{-j\omega t}\,dt$$

Indeed, if $\mathbf{x}(t)$ is white noise, then $R(t_1, t_2) = q(t_1)\delta(t_1 - t_2)$. Inserting into (10-149), we obtain

$$\Gamma(u, v) = \int_{-\infty}^{\infty}\int_{-\infty}^{\infty} q(t_1)\delta(t_1 - t_2)e^{-j(ut_1 + vt_2)}\,dt_1\,dt_2$$

$$= \int_{-\infty}^{\infty} q(t_2)e^{-j(u+v)t_2}\,dt_2 = Q(u + v)$$

and (10-154) results. We note in particular that

$$E\{|\mathbf{X}(\omega)|^2\} = Q(0) = \int_{-\infty}^{\infty} q(t)\,dt \tag{10-155}$$

5. The transform $\mathbf{X}(\omega)$ of a process $\mathbf{x}(t)$ is, in general, complex

$$\mathbf{X}(\omega) = \mathbf{A}(\omega) + j\mathbf{B}(\omega) \tag{10-156}$$

We shall express the second-order moments of $\mathbf{A}(\omega)$ and $\mathbf{B}(\omega)$ in terms of the real and imaginary parts of the transform

$$\Gamma(u, v) = \Gamma_r(u, v) + j\Gamma_i(u, v) \tag{10-157}$$

of $R(t_1, t_2)$. Clearly, [see (10-151)]

$$E\{\mathbf{X}(u)\mathbf{X}(v)\} = \Gamma(u, v)$$

Adding and subtracting with (10-151), we obtain

$$2E\{\mathbf{A}(u)\mathbf{A}(v)\} = \Gamma_r(u, -v) + \Gamma_r(u, v) \qquad 2E\{\mathbf{A}(v)\mathbf{B}(u)\} = \Gamma_i(u, v) + \Gamma_i(u, -v)$$

$$2E\{\mathbf{B}(u)\mathbf{B}(v)\} = \Gamma_r(u, -v) - \Gamma_r(u, v) \qquad 2E\{\mathbf{A}(u)\mathbf{B}(v)\} = \Gamma_i(u, v) - \Gamma_i(u, -v) \tag{10-158}$$

6. *Covariance of energy spectrum.* The energy spectrum $|\mathbf{X}(\omega)|^2$ of the process $\mathbf{x}(t)$ is a real process with mean [see (10-151)]

$$E\{|\mathbf{X}(\omega)|^2\} = \Gamma(\omega, -\omega) \tag{10-159}$$

The covariance of $|\mathbf{X}(\omega)|^2$ cannot, in general, be expressed in terms of $\Gamma(u, v)$ because it involves fourth-order moments. The normal case, however, is an exception. We shall show that, if $\mathbf{x}(t)$ is a normal process with zero mean, then

$$\text{cov}\{|\mathbf{X}(u)|^2, |\mathbf{X}(v)|^2\} = |\Gamma(u, -v)|^2 + |\Gamma(u, v)|^2 \tag{10-160}$$

Indeed, the Fourier transform of a normal process with zero mean is normal with zero mean. From this it follows that the processes $\mathbf{A}(\omega)$ and $\mathbf{B}(\omega)$ are jointly normal with zero mean. Hence [see (7-37)]

$$E\{|\mathbf{X}(u)|^2|\mathbf{X}(v)|^2\} - E\{|\mathbf{X}(u)|^2\}E\{|\mathbf{X}(v)|^2\}$$
$$= E\{[\mathbf{A}^2(u) + \mathbf{B}^2(u)][\mathbf{A}^2(v) + \mathbf{B}^2(v)]\} - E\{\mathbf{A}^2(u) + \mathbf{B}^2(u)\}E\{\mathbf{A}^2(v) + \mathbf{B}^2(v)\}$$
$$= 2E^2\{\mathbf{A}(u)\mathbf{A}(v)\} + 2E^2\{\mathbf{B}(u)\mathbf{B}(v)\} + 2E^2\{\mathbf{A}(u)\mathbf{B}(v)\} + 2E^2\{\mathbf{A}(v)\mathbf{B}(u)\}$$

Inserting (10-158) into the above, we obtain (10-160)

Fourier transforms of noisy signals. A common problem in Fourier analysis is the determination of the Fourier transform $F(\omega)$ of a function $f(t)$ in the presence of noise. If we use as the estimate of $F(\omega)$ the Fourier transform of the data

$$\mathbf{x}(t) = f(t) + \mathbf{v}(t)$$

the error due to the noise $\mathbf{v}(t)$ might be large even if $\mathbf{v}(t) \ll f(t)$. To reduce the effect of the noise, we compute the transform of only a segment of the given data

$$\mathbf{X}_a(u) = \int_{-a}^{a} \mathbf{x}(t)e^{-j\omega t}\, dt = F_a(\omega) + \mathbf{X}_v(\omega) \qquad (10\text{-}161)$$

The component $F_a(\omega)$ due to the signal $f(t)$ is a smoothed version of $F(\omega)$ [see (14-2)]

$$F_a(\omega) = \int_{-a}^{a} f(t)e^{-j\omega t}\, dt = \int_{-\infty}^{\infty} F(y)\frac{\sin a(\omega - y)}{(\omega - y)}\, dy$$

approaching $F(\omega)$ as a increases. The component $\mathbf{X}_v(\omega)$ due to the noise is a stochastic process whose variance increases as a increases. In fact, if $\mathbf{v}(t)$ is stationary white noise with autocorrelation $q\delta(\tau)$, then [see (10-155)]

$$E\{|\mathbf{X}_v(\omega)|^2\} = 2qa$$

The optimum choice of a depends on the form of $f(t)$ and the size of q. As we show in Sec. 14-1, the estimate can be improved if we replace in (10-161) the data $\mathbf{x}(t)$ by the product $w(t)\mathbf{x}(t)$ where $w(t)$ is a suitable factor (window).

APPENDIX 10A. THE POISSON SUM FORMULA

If

$$F(\omega) = \int_{-\infty}^{\infty} f(x)e^{-jux}\, dx$$

then for any c

$$\sum_{n=-\infty}^{\infty} f(x + nc) = \frac{1}{c}\sum_{n=-\infty}^{\infty} F(nu_0)e^{jnu_0x} \qquad u_0 = \frac{2\pi}{c} \qquad (10\text{A-}1)$$

Proof Clearly

$$\sum_{n=-\infty}^{\infty} \delta(x + nc) = \frac{1}{c} \sum_{n=-\infty}^{\infty} e^{jnu_0 x} \tag{10A-2}$$

because the left side is periodic and its Fourier series coefficients equal

$$\frac{1}{c} \int_{-c/2}^{c/2} \delta(x) e^{-jnu_0 x} \, dx = \frac{1}{c}$$

Furthermore, $\delta(x + nc) * f(x) = f(x + nc)$ and

$$e^{jnu_0 x} * f(x) = \int_{-\infty}^{\infty} e^{jnu_0 (x-\alpha)} f(\alpha) \, d\alpha = e^{jnu_0 x} F(nu_0)$$

Convolving both sides of (10A-2) with $f(x)$ and using the above, we obtain (10A-1).

APPENDIX 10B. SCHWARZ' INEQUALITY

We shall show that

$$\left| \int_a^b f(x) g(x) \, dx \right|^2 \le \int_a^b |f(x)|^2 \, dx \int_a^b |g(x)|^2 \, dx \tag{10B-1}$$

with equality iff

$$f(x) = kg^*(x) \tag{10B-2}$$

Proof Clearly

$$\left| \int_a^b f(x) g(x) \, dx \right| \le \int_a^b |f(x)| \, |g(x)| \, dx$$

Equality holds only if the product $f(x)g(x)$ is real. This is the case if the angles of $f(x)$ and $g(x)$ are opposite as in (10B-2). It suffices, therefore, to assume that the functions $f(x)$ and $g(x)$ are real. The quadratic

$$I(z) = \int_a^b [f(x) - zg(x)]^2 \, dx = z^2 \int_a^b g^2(x) \, dx - 2z \int_a^b f(x)g(x) \, dx + \int_a^b f^2(x) \, dx$$

is nonnegative for every real z. Hence, its discriminant cannot be positive. This yields (10B-1). If the discriminant of $I(z)$ is zero, then $I(z)$ has a real (double) root. This shows that $I(k) = 0$ for some k, and (10B-2) follows.

PROBLEMS

10-1 Show that $|R_{xy}(\tau)| \le \frac{1}{2}[R_{xx}(0) + R_{yy}(0)]$.

10-2 Show that if the processes $\mathbf{x}(t)$, $\mathbf{y}(t)$ are WSS and $E\{|\mathbf{x}(0) - \mathbf{y}(0)|^2\} = 0$, then $R_{xx}(\tau) \equiv R_{xy}(\tau) \equiv R_{yy}(\tau)$.

Hint: Set $\mathbf{z} = \mathbf{x}(t+\tau)$, $\mathbf{w} = \mathbf{x}^*(t) - \mathbf{y}^*(t)$ in (10-5).

10-3 Find $S(\omega)$ if (*a*) $R(\tau) = e^{-\alpha\tau^2}$, (*b*) $R(\tau) = e^{-\alpha\tau^2} \cos \omega_0 \tau$.

10-4 Show that the power spectrum of an SSS process $\mathbf{x}(t)$ equals

$$S(\omega) = \int_{-\infty}^{\infty}\int_{-\infty}^{\infty} x_1 x_2\, G(x_1, x_2; \omega)\, dx_1\, dx_2$$

where $G(x_1, x_2; \omega)$ is the Fourier transform in the variable τ of the second-order density $f(x_1, x_2; \tau)$ of $\mathbf{x}(t)$.

10-5 Show that if $\mathbf{y}(t) = \mathbf{x}(t+a) - \mathbf{x}(t-a)$, then

$$R_y(\tau) = 2R_x(\tau) - R_x(\tau + 2a) - R_x(\tau - 2a) \qquad S_y(\omega) = 4S_x(\omega)\sin^2 a\omega$$

10-6 Using (10-12), show that

$$R(0) - R(\tau) \ge \frac{1}{4^n}[R(0) - R(2^n\tau)]$$

Hint:

$$1 - \cos\theta = 2\sin^2\frac{\theta}{2} \ge 2\sin^2\frac{\theta}{2}\cos^2\frac{\theta}{2} = \frac{1}{4}(1 - \cos 2\theta)$$

10-7 Show that if $\mathbf{x}(t)$ is a complex WSS process, then

$$E\{|\mathbf{x}(t+\tau) - \mathbf{x}(t)|^2\} = 2\,\mathrm{Re}\,[R(0) - R(\tau)]$$

10-8 The process $\mathbf{x}(t)$ is normal with zero mean and $R_x(\tau) = Ie^{-\alpha|\tau|}\cos\beta\tau$. Show that if $\mathbf{y}(t) = \mathbf{x}^2(t)$, then $R_y(\tau) = I^2 e^{-2\alpha|\tau|}(1 + \cos 2\beta\tau)$. Find $S_y(\omega)$.

10-9 Show that if $R(\tau)$ is the inverse Fourier transform of a function $S(\omega)$ and $S(\omega) \ge 0$, then, for any a_i,

$$\sum_{i,k} a_i a_k^* R(\tau_i - \tau_k) \ge 0$$

Hint:

$$\int_{-\infty}^{\infty} S(\omega)\left|\sum_i a_i e^{j\omega\tau_i}\right|^2 d\omega \ge 0$$

10-10 Find $R(\tau)$ if (*a*) $S(\omega) = 1/(4 + \omega^4)$, (*b*) $S(\omega) = 1/(4 + \omega^2)^2$.

10-11 Show that, for complex systems, (10-39) and (10-97) yield

$$\mathbf{S}_{yy}(s) = \mathbf{S}_{xx}(s)\mathbf{H}(s)\mathbf{H}^*(-s^*) \qquad \mathbf{S}_{yy}(z) = \mathbf{S}_{xx}(z)\mathbf{H}(z)\mathbf{H}^*(1/z^*)$$

10-12 Show that if $\mathbf{x}(t)$ is a WSS process and

$$\mathbf{s} = \frac{1}{n}\sum_{k=1}^{n} \mathbf{x}(kT) \qquad \text{then} \qquad E\{\mathbf{s}^2\} = \frac{1}{2\pi n^2}\int_{-\infty}^{\infty} S_x(\omega)\frac{\sin^2 n\omega T/2}{\sin^2 \omega T/2}\, d\omega$$

10-13 (*Stochastic resonance*). The input to the system

$$\mathbf{H}(s) = \frac{1}{s^2 + 2s + \sqrt{5}}$$

is a WSS process $\mathbf{x}(t)$ with $E\{\mathbf{x}^2(t)\} = 10$. Find $S_x(\omega)$ such that the average power $E\{\mathbf{y}^2(t)\}$ of the resulting output $\mathbf{y}(t)$ is maximum.

Hint: $|\mathbf{H}(j\omega)|$ is maximum for $\omega = \sqrt{3}$.

10-14 Show that if $R_x(\tau) = Ae^{j\omega_0\tau}$, then $R_{xy}(\tau) = Be^{j\omega_0\tau}$ for any $\mathbf{y}(t)$.

Hint: Use (10-67).

10-15 Given a system $H(\omega)$ with input $\mathbf{x}(t)$ and output $\mathbf{y}(t)$, show that

(*a*) if $\mathbf{x}(t)$ is WSS and $R_{xx}(\tau) = e^{j\alpha\tau}$, then

$$R_{yx}(\tau) = e^{j\alpha\tau}H(\alpha) \qquad R_{yy}(\tau) = e^{j\alpha\tau}|H(\alpha)|^2$$

(*b*) if $R_{xx}(t_1, t_2) = e^{j(\alpha t_1 - \beta t_2)}$, then

$$R_{yx}(t_1, t_2) = e^{j(\alpha t_1 - \beta t_2)}H(\alpha) \qquad R_{yy}(t_1, t_2) = e^{j(\alpha t_1 - \beta t_2)}H(\alpha)H^*(\beta)$$

10-16 Show that if $S_{xx}(\omega)S_{yy}(\omega) \equiv 0$, then $S_{xy}(\omega) \equiv 0$.

10-17 The process $\mathbf{x}(t)$ is WSS and

$$\mathbf{y}''(t) + 3\mathbf{y}'(t) + 2\mathbf{y}(t) = \mathbf{x}(t)$$

Show that (*a*)

$$\begin{aligned} R''_{yx}(\tau) + 3R'_{yx}(\tau) + 2R_{yx}(\tau) &= R_{xx}(\tau) \\ R''_{yy}(\tau) + 3R'_{yy}(\tau) + 2R_{yy}(\tau) &= R_{xy}(\tau) \end{aligned} \qquad \text{all } \tau$$

(*b*) If $R_{xx}(\tau) = q\delta(\tau)$, then $R_{yx}(\tau) = 0$ for $\tau < 0$ and for $\tau > 0$

$$R''_{yx}(\tau) + 3R'_{yx}(\tau) + 2R_{yx}(\tau) = 0 \qquad R_{yx}(0) = 0 \qquad R'_{yx}(0^+) = \frac{q}{3}$$

$$R''_{yy}(\tau) + 3R'_{yy}(\tau) + 2R_{yy}(\tau) = 0 \qquad R_{yy}(0) = \frac{q}{12} \qquad R'_{yy}(0) = 0$$

10-18 Show that (*a*) $E\{\mathbf{x}(t)\check{\mathbf{x}}(t)\} = 0$, (*b*) $\check{\check{\mathbf{x}}}(t) = -\mathbf{x}(t)$.

10-19 In the circuit of Fig. P10–19, $\mathbf{n}_e(t)$ is the voltage due to thermal noise. Show that

$$S_v(\omega) = \frac{2kTR}{(1 - \omega^2 LC)^2 + \omega^2 R^2 C^2} \qquad S_i(\omega) = \frac{2kTR}{R^2 + \omega^2 L^2}$$

and verify Nyquist's theorems (10-82) and (10-85).

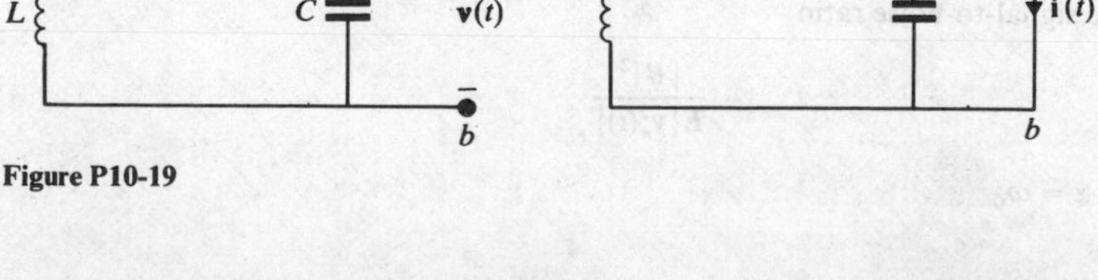

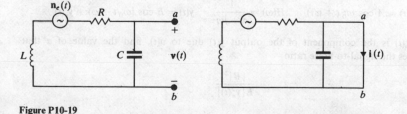

Figure P10-19

10-20 Show that if $\mathbf{s}[n]$ is AR and $\mathbf{v}[n]$ is white-noise orthogonal to $\mathbf{s}[n]$, then the process $\mathbf{x}[n] = \mathbf{s}[n] + \mathbf{v}[n]$ is ARMA. Find $\mathbf{S}_x(z)$ if $R_s[m] = 2^{-|m|}$ and $\mathbf{S}_v(z) = 5$.

10-21 Show that if $\mathbf{x}[n]$ is WSS and $R_x[1] = R_x[0]$, then $R_x[m] = R_x[0]$ for every m.

10-22 Show that if $R[m] = E\{\mathbf{x}[n+m]\mathbf{x}[n]\}$, then

$$3R[0] \geq |4R[1] + 2R[2]|$$

10-23 Given an RV $\boldsymbol{\omega}$ with density $f(\omega)$ such that $f(\omega) = 0$ for $|\omega| > \sigma = \pi/T$, we form the process $\mathbf{x}[n] = Ae^{jn\boldsymbol{\omega}T}$. Show that $S_x(\omega) = 2\sigma A^2 f(\omega)$ for $|\omega| < \sigma$.

10-24 The process $\mathbf{s}[n]$ is WSS and $E\{\mathbf{s}^2[n]\} < \infty$. Show that if

$$\mathbf{s}[n] = \sum_{k=1}^{N} a_k \mathbf{s}[n-k] \qquad \text{then} \qquad \mathbf{s}[n] = \sum_{i=1}^{M} \mathbf{c}_i z_i^n$$

where $z_i = e^{j\omega_i T}$ are the $M \leq N$ roots of the equation $1 = a_1 z^{-1} + \cdots + a_N z^{-N}$ on the unit circle, and $\mathbf{c}_i$ are M arbitrary RVs.

10-25 Find $R_x[m]$ and the whitening filter of $\mathbf{x}[n]$ if

$$S_x(z) = \frac{5 - 2(z + 1/z)}{4 - 3(z + 1/z)}$$

10-26 Find the innovations filter of the process $\mathbf{x}(t)$ if

$$S_x(\omega) = \frac{\omega^4 + 64}{\omega^4 + 10\omega^2 + 9}$$

10-27 Show that if $l_s[n]$ is the delta response of the innovations filter of $\mathbf{s}[n]$, then

$$R_s[0] = \sum_{n=0}^{\infty} l_s^2[n]$$

10-28 (Matched filter). The input to a system $\mathbf{H}(z)$ is the sum $\mathbf{x}[n] = f[n] + \mathbf{v}[n]$ where $f[n]$ is a known sequence with z transform $\mathbf{F}(z)$. We wish to find $\mathbf{H}(z)$ such that the ratio $y_f^2[0]/E\{\mathbf{y}_v^2[n]\}$ of the output $\mathbf{y}[n] = y_f[n] + \mathbf{y}_v[n]$ is maximum. Show that (*a*) if $\mathbf{v}[n]$ is white noise, then $\mathbf{H}(z) = k\mathbf{F}(z^{-1})$, and (*b*) if $\mathbf{H}(z)$ is an FIR filter, i.e., if $\mathbf{H}(z) = a_0 + a_1 z^{-1} + \cdots + a_N z^{-N}$, then its weights a_m are the solutions of the system

$$\sum_{m=0}^{N} R_v[n-m]a_m = kf[-n] \qquad n = 0, \ldots, N$$

10-29 If $R_n(\tau) = N\delta(\tau)$ and

$$\mathbf{x}(t) = A\cos\omega_0 t + \mathbf{n}(t) \qquad H(\omega) = \frac{1}{\alpha + j\omega} \qquad \mathbf{y}(t) = B\cos(\omega_0 t + \varphi) + \mathbf{y}_n(t)$$

where $\mathbf{y}_n(t)$ is the component of the output $\mathbf{y}(t)$ due to $\mathbf{n}(t)$, find the value of α that maximizes the signal-to-noise ratio

$$\frac{|B|^2}{E\{\mathbf{y}_n^2(t)\}}$$

Answer: $\alpha = \omega_0$

10-30(*a*) Show that

$$\left|\int_{-\infty}^{\infty} F(\omega)H(\omega)e^{j\omega t}\,d\omega\right|^2 \le \int_{-\infty}^{\infty} \frac{|F(\omega)|^2}{S_v(\omega)}\,d\omega \int_{-\infty}^{\infty} S_v(\omega)|H(\omega)|^2\,d\omega \qquad \text{(i)}$$

(*b*) Derive the matched filter formula (10-122) using (i).

10-31(*a*) (Cauchy inequality). Show that

$$\left|\sum_i a_i b_i\right|^2 \le \sum_i |a_i|^2 \sum_i |b_i|^2 \qquad \text{(i)}$$

with equality iff $a_i = kb_i^*$.

(*b*) State and solve the discrete-time version of the matched filter problem using (i).

10-32 Show that if $R_x(\tau) = e^{-c|\tau|}$, then the Karhunen–Loève expansion of $\mathbf{x}(t)$ in the interval $(-a, a)$ is the sum

$$\hat{\mathbf{x}}(t) = \sum_{n=1}^{\infty} (\beta_n \mathbf{b}_n \cos \omega_n t + \beta'_n \mathbf{b}'_n \sin \omega'_n t)$$

where

$$\tan a\omega_n = \frac{c}{\omega_n} \qquad \cot a\omega'_n = \frac{-c}{\omega'_n} \qquad \beta_n = (a + c\lambda_n)^{-1/2} \qquad \beta'_n = (a - c\lambda'_n)^{-1/2}$$

$$E\{\mathbf{b}_n^2\} = \lambda_n = \frac{2c}{c^2 + \omega_n^2} \qquad E\{\mathbf{b}_n'^2\} = \lambda'_n = \frac{2c}{c^2 + \omega_n'^2}$$

10-33 Show that if $\mathbf{x}(t)$ is WSS and

$$\mathbf{X}_T(\omega) = \int_{-T/2}^{T/2} \mathbf{x}(t)e^{-j\omega t}\,dt \qquad \text{then} \qquad E\left\{\frac{\partial}{\partial T}|\mathbf{X}_T(\omega)|^2\right\} = \int_{-T}^{T} R_x(\tau)e^{-j\omega\tau}\,d\tau$$

10-34 Find the mean and the variance of the integral

$$\mathbf{X}(\omega) = \int_{-a}^{a} [5\cos 3t + \mathbf{v}(t)]e^{-j\omega t}\,dt$$

if $E\{\mathbf{v}(t)\} = 0$ and $R_v(\tau) = 2\delta(\tau)$.

10-35 Show that if

$$E\{\mathbf{x}_n \mathbf{x}_k\} = \sigma_n^2\,\delta[n-k] \qquad \mathbf{X}(\omega) = \sum_{n=-\infty}^{\infty} \mathbf{x}_n e^{-jn\omega T}$$

and $E\{\mathbf{x}_n\} = 0$, then $E\{\mathbf{X}(\omega)\} = 0$ and

$$E\{\mathbf{X}(u)\mathbf{X}^*(v)\} = \sum_{n=-\infty}^{\infty} \sigma_n^2 e^{-jn(u-v)T}$$

CHAPTER

ELEVEN

APPLICATIONS

11-1 MODULATION†

Given two real jointly WSS processes $\mathbf{a}(t)$ and $\mathbf{b}(t)$ *with zero mean* and a constant ω_0, we form the process

$$\begin{aligned}\mathbf{x}(t) &= \mathbf{a}(t) \cos \omega_0 t - \mathbf{b}(t) \sin \omega_0 t \\ &= \mathbf{r}(t) \cos [\omega_0 t + \boldsymbol{\varphi}(t)]\end{aligned} \tag{11-1}$$

where

$$\mathbf{r}(t) = \sqrt{\mathbf{a}^2(t) + \mathbf{b}^2(t)} \qquad \tan \boldsymbol{\varphi}(t) = \frac{\mathbf{b}(t)}{\mathbf{a}(t)}$$

This process is called modulated with *amplitude modulation* $\mathbf{r}(t)$ and *phase modulation* $\boldsymbol{\varphi}(t)$.

We shall show that $\mathbf{x}(t)$ is WSS iff the processes $\mathbf{a}(t)$ and $\mathbf{b}(t)$ are such that

$$R_{aa}(\tau) = R_{bb}(\tau) \qquad R_{ab}(\tau) = -R_{ba}(\tau) \tag{11-2}$$

PROOF Clearly

$$E\{\mathbf{x}(t)\} = E\{\mathbf{a}(t)\} \cos \omega_0 t - E\{\mathbf{b}(t)\} \sin \omega_0 t = 0$$

† A. Papoulis: "Random Modulation: A Review," *IEEE Transactions on Acoustics, Speech, and Signal Processing*, vol. ASSP-31, 1983.

Furthermore

$$\mathbf{x}(t+\tau)\mathbf{x}(t) = [\mathbf{a}(t+\tau)\cos\omega_0(t+\tau) - \mathbf{b}(t+\tau)\cos\omega_0(t+\tau)]$$
$$[\mathbf{a}(t)\cos\omega_0 t - \mathbf{b}(t)\sin\omega_0 t]$$

Multiplying, taking expected values, and using appropriate trigonometric identities, we obtain

$$\begin{aligned} 2E\{\mathbf{x}(t+\tau)\mathbf{x}(t)\} = {} & [R_{aa}(\tau) + R_{bb}(\tau)]\cos\omega_0\tau + [R_{ab}(\tau) - R_{ba}(\tau)]\sin\omega_0\tau \\ & + [R_{aa}(\tau) - R_{bb}(\tau)]\cos(2t+\tau) \\ & - [R_{ab}(\tau) + R_{ba}(\tau)]\sin\omega_0(2t+\tau) \end{aligned} \tag{11-3}$$

If (11-2) is true, then the above yields

$$R_{xx}(\tau) = R_{aa}(\tau)\cos\omega_0\tau + R_{ab}(\tau)\sin\omega_0\tau \tag{11-4}$$

Conversely, if $\mathbf{x}(t)$ is WSS, then the second and third lines in (11-3) must be independent of t. This is possible only if (11-2) is true.

We introduce the "dual" process

$$\mathbf{y}(t) = \mathbf{b}(t)\cos\omega_0 t + \mathbf{a}(t)\sin\omega_0 t \tag{11-5}$$

This process is also WSS and

$$R_{yy}(\tau) = R_{xx}(\tau) \qquad R_{xy}(\tau) = -R_{yx}(\tau) \tag{11-6}$$

$$R_{xy}(\tau) = R_{ab}(\tau)\cos\omega_0\tau - R_{aa}(\tau)\sin\omega_0\tau \tag{11-7}$$

The above follows from (11-3) if we change one or both factors of the product $\mathbf{x}(t+\tau)\mathbf{x}(t)$ with $y(t+\tau)$ or $\mathbf{y}(t)$.

Complex representation We introduce the processes

$$\begin{aligned} \mathbf{w}(t) &= \mathbf{a}(t) + j\mathbf{b}(t) = \mathbf{r}(t)e^{j\boldsymbol{\varphi}(t)} \\ \mathbf{z}(t) &= \mathbf{x}(t) + j\mathbf{y}(t) = \mathbf{w}(t)e^{j\omega_0 t} \end{aligned} \tag{11-8}$$

Thus

$$\mathbf{x}(t) = \operatorname{Re}\mathbf{z}(t) = \operatorname{Re}[\mathbf{w}(t)e^{j\omega_0 t}]$$

and

$$\mathbf{a}(t) + j\mathbf{b}(t) = \mathbf{w}(t) = \mathbf{z}(t)e^{-j\omega_0 t} \tag{11-9}$$

This yields

$$\begin{aligned} \mathbf{a}(t) &= \mathbf{x}(t)\cos\omega_0 t + \mathbf{y}(t)\sin\omega_0 t \\ \mathbf{b}(t) &= \mathbf{y}(t)\cos\omega_0 t - \mathbf{x}(t)\sin\omega_0 t \end{aligned} \tag{11-10}$$

Correlations and spectra The autocorrelation of the conplex process $\mathbf{w}(t)$ equals

$$R_{ww}(\tau) = E\{[\mathbf{a}(t+\tau) + j\mathbf{b}(t+\tau)][\mathbf{a}(t) - j\mathbf{b}(t)]\}$$

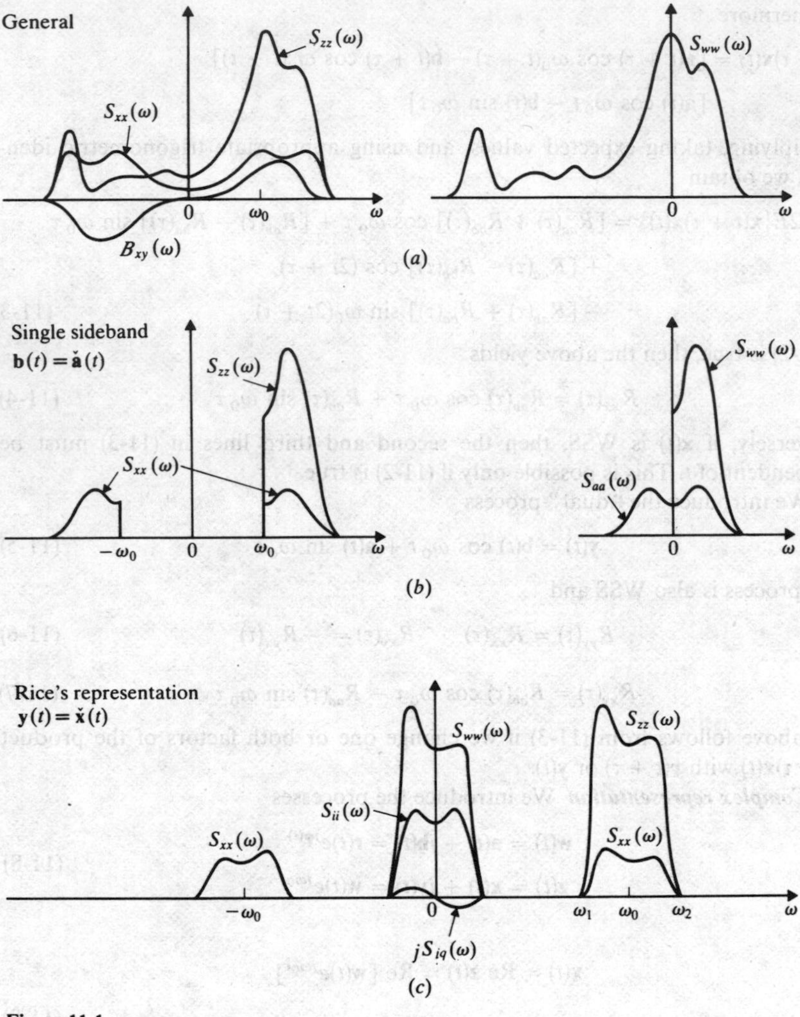

Figure 11-1

Expanding and using (11-2), we obtain

$$R_{ww}(\tau) = 2R_{aa}(\tau) - 2jR_{ab}(\tau) \tag{11-11}$$

Similarly

$$R_{zz}(\tau) = 2R_{xx}(\tau) - 2jR_{xy}(\tau) \tag{11-12}$$

We note, further, that

$$R_{zz}(\tau) = e^{j\omega_0\tau}R_{ww}(\tau) \tag{11-13}$$

From the above it follows that

$$S_{ww}(\omega) = 2S_{aa}(\omega) - 2jS_{ab}(\omega)$$
$$S_{zz}(\omega) = 2S_{xx}(\omega) - 2jS_{xy}(\omega) \tag{11-14}$$

$$S_{zz}(\omega) = S_{ww}(\omega - \omega_0) \tag{11-15}$$

The functions $S_{xx}(\omega)$ and $S_{zz}(\omega)$ are real and positive. Furthermore, [see (11-6) and (10-1)]

$$R_{xy}(-\tau) = -R_{yx}(-\tau) = -R_{xy}(\tau)$$

This leads to the conclusion that the function $-jS_{xy}(\omega) = B_{xy}(\omega)$ is real and (Fig. 11-1*a*)

$$|B_{xy}(\omega)| \le S_{xx}(\omega) \qquad B_{xy}(-\omega) = -B_{xy}(\omega) \tag{11-16}$$

And since $S_{xx}(-\omega) = S_{xx}(\omega)$, we conclude from the second equation in (11-14) that

$$4S_{xx}(\omega) = S_{zz}(\omega) + S_{zz}(-\omega)$$
$$4jS_{xy}(\omega) = S_{zz}(-\omega) - S_{zz}(\omega) \tag{11-17}$$

Single sideband If $\mathbf{b}(t) = \check{\mathbf{a}}(t)$ is the Hilbert transform of $\mathbf{a}(t)$, then [see (10-70)] the constraint (11-2) is satisfied and the first equation in (11-14) yields

$$S_{ww}(\omega) = 4S_{aa}(\omega)U(\omega)$$

(Fig. 11-1*b*) because

$$S_{a\check{a}}(\omega) = jS_{aa}(\omega) \operatorname{sgn} \omega$$

The resulting spectra are shown in Fig. 11-1*b*. We note, in particular, that $S_{xx}(\omega) = 0$ for $|\omega| < \omega_0$.

Rice's representation In (11-1) we assumed that the carrier frequency ω_0 and the processes $\mathbf{a}(t)$ and $\mathbf{b}(t)$ were given. We now consider the converse problem: Given a WSS process $\mathbf{x}(t)$ with zero mean, find a constant ω_0 and two processes $\mathbf{a}(t)$ and $\mathbf{b}(t)$ such that $\mathbf{x}(t)$ can be written in the form (11-1). To do so, it suffices to find the constant ω_0 and the dual process $\mathbf{y}(t)$ [see (11-10)]. This shows that the representation of $\mathbf{x}(t)$ in the form (11-1) is not unique because, not only ω_0 is arbitrary, but also the process $\mathbf{y}(t)$ can be chosen arbitrarily subject only to the constraint (11-6).

The question then arises whether, among all possible representations of $\mathbf{x}(t)$, there is one that is optimum. The answer depends, of course, on the optimality criterion. As we shall presently explain, if $\mathbf{y}(t)$ equals the Hilbert transform $\check{\mathbf{x}}(t)$ of $\mathbf{x}(t)$, then (11-1) is optimum in the sense of minimizing the average rate of variation of the envelope of $\mathbf{x}(t)$.

Hilbert transforms As we know [see (10-70)]

$$R_{\check{x}\check{x}}(\tau) = R_{xx}(\tau) \qquad R_{x\check{x}}(\tau) = -R_{\check{x}x}(\tau) \tag{11-18}$$

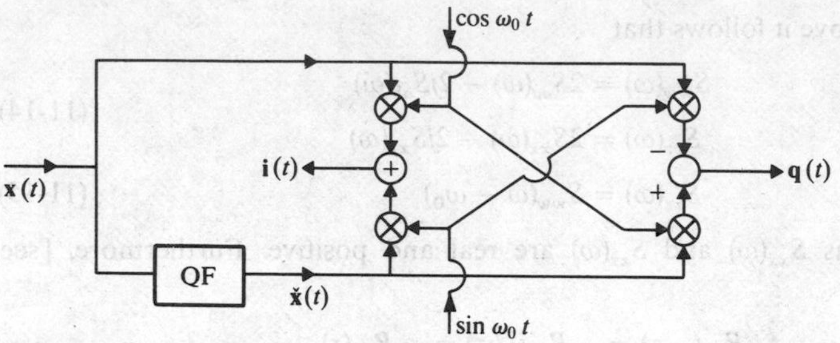

Figure 11-2

We can, therefore, use $\check{\mathbf{x}}(t)$ to form the processes

$$\mathbf{z}(t) = \mathbf{x}(t) + j\check{\mathbf{x}}(t) = \mathbf{w}(t)e^{j\omega_0 t}$$
$$\mathbf{w}(t) = \mathbf{i}(t) + j\mathbf{q}(t) = \mathbf{z}(t)e^{-j\omega_0 t} \tag{11-19}$$

as in (11-8) where now (Fig. 11-1*c*)

$$\mathbf{y}(t) = \check{\mathbf{x}}(t) \qquad \mathbf{a}(t) = \mathbf{i}(t) \qquad \mathbf{b}(t) = \mathbf{q}(t)$$

Inserting into (11-1), we obtain

$$\mathbf{x}(t) = \mathbf{i}(t)\cos\omega_0 t - \mathbf{q}(t)\sin\omega_0 t \tag{11-20}$$

This is known as *Rice's representation.* The process $\mathbf{i}(t)$ is called the *inphase* component and the process $\mathbf{q}(t)$ the *quadrature* component of $\mathbf{x}(t)$. Their realization is shown in Fig. 11-2 [see (11-10)]. These processes depend, not only on $\mathbf{x}(t)$, but also on the choice of the carrier frequency ω_0.

From (10-70) and (11-14) it follows that

$$S_{zz}(\omega) = 4S_{xx}(\omega)U(\omega) \tag{11-21}$$

Bandpass processes A process $\mathbf{x}(t)$ is called bandpass (Fig. 11-1*c*) if its spectrum $S_{xx}(\omega)$ is zero outside an interval (ω_1, ω_2). It is called narrowband or *quasimonochromatic* if its bandwidth $\omega_2 - \omega_1$ is small compared with the center frequency. It is called *monochromatic* if $S_{xx}(\omega)$ is an impulse function. The process $\mathbf{a}\cos\omega_0 t + \mathbf{b}\sin\omega_0 t$ is monochromatic.

The representations (11-1) or (11-20) hold for an arbitrary $\mathbf{x}(t)$. However, they are useful mainly if $\mathbf{x}(t)$ is bandpass. In this case, the complex envelope $\mathbf{w}(t)$ and the processes $\mathbf{i}(t)$ and $\mathbf{q}(t)$ are low-pass because

$$S_{ww}(\omega) = S_{zz}(\omega + \omega_0)$$
$$S_{ii}(\omega) = S_{qq}(\omega) = \tfrac{1}{4}[S_{ww}(\omega) + S_{ww}(-\omega)] \tag{11-22}$$

We shall show that if the process $\mathbf{x}(t)$ is bandpass and $\omega_2 - \omega_1 \le 2\omega_0$, then the inphase component $\mathbf{i}(t)$ and the quadrature component $\mathbf{q}(t)$ can be obtained as

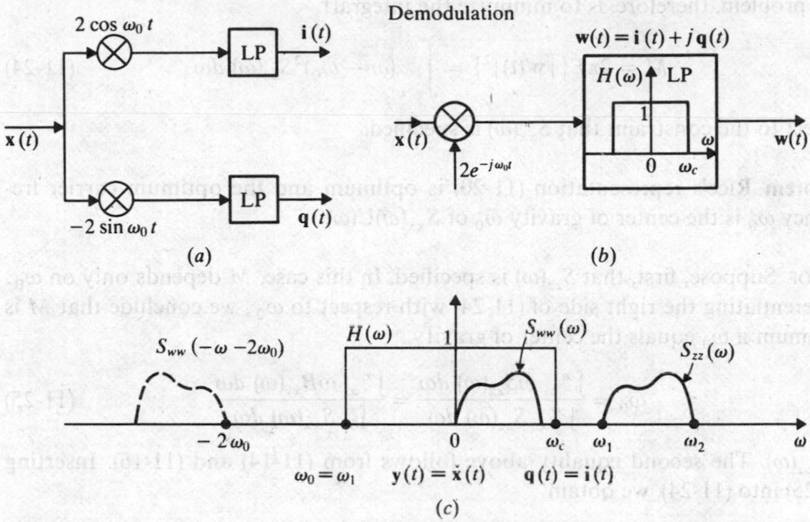

Figure 11-3

responses of the system of Fig. 11-3*a* where the LP filters are ideal with cutoff frequency ω_c such that

$$\omega_2 - \omega_0 < \omega_c < \omega_1 + \omega_0 \tag{11-23}$$

PROOF It suffices to show that (linearity) the response of the system of Fig. 11-3*b* equals $\mathbf{w}(t)$. Clearly

$$2\mathbf{x}(t) = \mathbf{z}(t) + \mathbf{z}^*(t) \qquad \mathbf{w}^*(t) = \mathbf{z}^*(t)e^{j\omega_0 t}$$

Hence

$$2\mathbf{x}(t)e^{-j\omega_0 t} = \mathbf{w}(t) + \mathbf{w}^*(t)e^{-j2\omega_0 t}$$

The spectra of the processes $\mathbf{w}(t)$ and $\mathbf{w}^*(t)e^{-j\omega_0 t}$ equal $S_{ww}(\omega)$ and $S_{ww}(-\omega - 2\omega_0)$ respectively. The first is in the band of the LP filter $H(\omega)$ and the second outside this band. Therefore, the response of the filter equals $\mathbf{w}(t)$ (see (10-47)].

We note, finally, that if $\omega_0 \le \omega_1$, then $S_{ww}(\omega) = 0$ for $\omega < 0$. In this case, $\mathbf{q}(t)$ is the Hilbert transform of $\mathbf{i}(t)$. Since $\omega_2 - \omega_1 \le 2\omega_0$, this is possible only if

$$\omega_2 \le 3\omega_1$$

In Fig. 11-3*c* we show the corresponding spectra for $\omega_0 = \omega_1$.

Optimum envelope We are given an arbitrary process $\mathbf{x}(t)$ and we wish to determine a constant ω_0 and a process $\mathbf{y}(t)$ so that, in the resulting representation (11-1), the complex envelope $\mathbf{w}(t)$ of $\mathbf{x}(t)$ is smooth in the sense of minimizing $E\{|\mathbf{w}'(t)|^2\}$. As we know, the power spectrum of $\mathbf{w}'(t)$ equals

$$\omega^2 S_{ww}(\omega) = \omega^2 S_{zz}(\omega + \omega_0)$$

Our problem, therefore, is to minimize the integral†

$$M = 2\pi E\{|\mathbf{w}'(t)|^2\} = \int_{-\infty}^{\infty} (\omega - \omega_0)^2 S_{zz}(\omega)\, d\omega \tag{11-24}$$

subject to the constraint that $S_{xx}(\omega)$ is specified.

Theorem Rice's representation (11-20) is optimum and the optimum carrier frequency ω_0 is the center of gravity $\bar{\omega}_0$ of $S_{xx}(\omega)U(\omega)$.

PROOF Suppose, first, that $S_{zz}(\omega)$ is specified. In this case, M depends only on ω_0. Differentiating the right side of (11-24) with respect to ω_0, we conclude that M is minimum if ω_0 equals the center of gravity

$$\bar{\omega}_0 = \frac{\int_{-\infty}^{\infty} \omega S_{zz}(\omega)\, d\omega}{\int_{-\infty}^{\infty} S_{zz}(\omega)\, d\omega} = \frac{\int_{-\infty}^{\infty} \omega B_{xy}(\omega)\, d\omega}{\int_0^{\infty} S_{xx}(\omega)\, d\omega} \tag{11-25}$$

of $S_{zz}(\omega)$. The second equality above follows from (11-14) and (11-16). Inserting (11-25) into (11-24), we obtain

$$M = \int_{-\infty}^{\infty} (\omega^2 - \bar{\omega}_0^2) S_{zz}(\omega)\, d\omega = 2\int_{-\infty}^{\infty} (\omega^2 - \bar{\omega}_0^2) S_{xx}(\omega)\, d\omega \tag{11-26}$$

We wish now to choose $S_{zz}(\omega)$ so as to minimize M. Since $S_{xx}(\omega)$ is given, M is minimum if $\bar{\omega}_0$ is maximum. As we see from (11-25), this is the case if $|B_{xy}(\omega)| = S_{xx}(\omega)$ because $|B_{xy}(\omega)| \le S_{xx}(\omega)$. We thus conclude that $-jS_{xy}(\omega) = S_{xx}(\omega) \operatorname{sgn} \omega$ and (11-14) yields

$$S_{zz}(\omega) = 4S_{xx}(\omega)U(\omega)$$

Instantaneous frequency With $\boldsymbol{\varphi}(t)$ as in (11-1), the process

$$\boldsymbol{\omega}_i(t) = \omega_0 + \boldsymbol{\varphi}'(t) \tag{11-27}$$

is called the instantaneous frequency of $\mathbf{x}(t)$. Since

$$\mathbf{z} = \mathbf{r}e^{j(\omega_0 t + \boldsymbol{\varphi})} = \mathbf{x} + j\mathbf{y}$$

we have

$$\mathbf{z}'\mathbf{z}^* = \mathbf{r}\mathbf{r}' + j\mathbf{r}^2\boldsymbol{\omega}_i = (\mathbf{x}' + j\mathbf{y}')(\mathbf{x} - j\mathbf{y}) \tag{11-28}$$

This yields $E\{\mathbf{r}\mathbf{r}'\} = 0$ and

$$E\{\mathbf{r}^2\boldsymbol{\omega}_i\} = \frac{1}{2\pi}\int_{-\infty}^{\infty} \omega S_{zz}(\omega)\, d\omega \tag{11-29}$$

because the cross-power spectrum of $\mathbf{z}'$ and $\mathbf{z}$ equals $j\omega S_{zz}(\omega)$.

† L. Mandel: "Complex Representation of Optical Fields in Coherence Theory," *Journal of the Optical Society of America*, vol. 57, 1967. See also N. M. Blachman, *Noise and its Effect on Communication*, Krieger Publishing Company, Malabar, Florida, 1982.

The instantaneous frequency of a process $\mathbf{x}(t)$ is not a uniquely defined process because the dual process $\mathbf{y}(t)$ is not unique. In Rice's representation $\mathbf{y} = \check{\mathbf{x}}$, hence

$$\boldsymbol{\omega}_i = \frac{\mathbf{x}\check{\mathbf{x}}' - x'\check{\mathbf{x}}}{\mathbf{r}^2} \qquad \mathbf{r}^2 = \mathbf{x}^2 + \check{\mathbf{x}}^2 \tag{11-30}$$

In this case [see (11-21) and (11-25)] the optimum carrier frequency $\bar{\omega}_0$ equals the weighted average of $\boldsymbol{\omega}_i$

$$\bar{\omega}_0 = \frac{E\{\mathbf{r}^2\boldsymbol{\omega}_i\}}{E\{\mathbf{r}^2\}}$$

Frequency Modulation

The process

$$\mathbf{x}(t) = \cos\left[\omega_0 t + \lambda\boldsymbol{\varphi}(t) + \boldsymbol{\varphi}_0\right] \qquad \boldsymbol{\varphi}(t) = \int_0^t \mathbf{c}(\alpha)\,d\alpha \tag{11-31}$$

is FM with instantaneous frequency $\omega_0 + \lambda\mathbf{c}(t)$ and modulation index λ. The corresponding complex processes equal

$$\mathbf{w}(t) = e^{j\lambda\boldsymbol{\varphi}(t)} \qquad \mathbf{z}(t) = \mathbf{w}(t)e^{j(\omega_0 t + \boldsymbol{\varphi}_0)} \tag{11-32}$$

We shall study their spectral properties.

Theorem If the process $\mathbf{c}(t)$ is SSS and the RV $\boldsymbol{\varphi}_0$ is independent of $\mathbf{c}(t)$ and such that

$$E\{e^{j\boldsymbol{\varphi}_0}\} = E\{e^{j2\boldsymbol{\varphi}_0}\} = 0 \tag{11-33}$$

then the process $\mathbf{x}(t)$ is WSS with zero mean. Furthermore

$$\begin{aligned} R_{xx}(\tau) &= \tfrac{1}{2}\,\mathrm{Re}\, R_{zz}(\tau) \\ R_{zz}(\tau) &= R_{ww}(\tau)e^{j\omega_0\tau} \qquad R_{ww}(\tau) = E\{\mathbf{w}(\tau)\} \end{aligned} \tag{11-34}$$

PROOF From (11-33) it follows that $E\{\mathbf{x}(t)\} = 0$ because

$$E\{\mathbf{z}(t)\} = E\{e^{j[\omega_0 t + \boldsymbol{\varphi}(t)]}\}E\{e^{j\boldsymbol{\varphi}_0}\} = 0$$

Furthermore

$$E\{\mathbf{z}(t+\tau)\mathbf{z}(t)\} = E\{e^{j[\omega_0(2t+\tau) + \lambda\boldsymbol{\varphi}(t+\tau) + \lambda\boldsymbol{\varphi}(t)]}\}E\{e^{j2\boldsymbol{\varphi}_0}\} = 0$$

$$E\{\mathbf{z}(t+\tau)\mathbf{z}^*(t)\} = e^{j\omega_0\tau}E\left\{\exp\left[j\lambda\int_t^{t+\tau}\mathbf{c}(\alpha)\,d\alpha\right]\right\} = e^{j\omega_0\tau}E\{\mathbf{w}(\tau)\}$$

The last equality is a consequence of the stationarity of the process $\mathbf{c}(t)$. Since $2\mathbf{x}(t) = \mathbf{z}(t) + \mathbf{z}^*(t)$, we conclude from the above that

$$4E\{\mathbf{x}(t+\tau)\mathbf{x}(t)\} = R_{zz}(\tau) + R_{zz}(-\tau)$$

and (11-34) results because $R_{zz}(-\tau) = R_{zz}^*(\tau)$.

Definitions A process $\mathbf{x}(t)$ is *phase modulated* if the statistics of $\boldsymbol{\varphi}(t)$ are known. In this case, its autocorrelation can be simply found because

$$E\{\mathbf{w}(t)\} = E\{e^{j\lambda\boldsymbol{\varphi}(t)}\} = \Phi_\varphi(\lambda, t) \tag{11-35}$$

where $\Phi_\varphi(\lambda, t)$ is the characteristic function of $\boldsymbol{\varphi}(t)$.

A process $\mathbf{x}(t)$ is *frequency modulated* if the statistics of $\mathbf{c}(t)$ are known. To determine $\Phi_\varphi(\lambda, t)$ we must now find the statistics of the integral of $\mathbf{c}(t)$. However, in general this is not simple. The normal case is an exception because then $\Phi_\varphi(\lambda, t)$ can be expressed in terms of the mean and variance of $\boldsymbol{\varphi}(t)$ and, as we know,

$$\begin{aligned} E\{\boldsymbol{\varphi}(t)\} &= \int_0^t E\{\mathbf{c}(\alpha)\}\, d\alpha = \eta_c\, t \\ E\{\boldsymbol{\varphi}^2(t)\} &= 2\int_0^t R_c(\alpha)(t-\alpha)\, d\alpha \end{aligned} \tag{11-36}$$

For the determination of the power spectrum $S_{xx}(\omega)$ of $\mathbf{x}(t)$ we must find the function $\Phi_\varphi(\lambda, t)$ and its Fourier transform. In general, this is difficult. However, as the next theorem shows, if λ is large, then $S_{xx}(\omega)$ can be expressed directly in terms of the density $f_c(c)$ of $\mathbf{c}(t)$.

Woodward's theorem† If the process $\mathbf{c}(t)$ is continuous and $f_c(c)$ is bounded, then for large λ

$$S_{xx}(\omega) \simeq \frac{\pi}{2\lambda}\left[f_c\!\left(\frac{\omega-\omega_0}{\lambda}\right) + f_c\!\left(\frac{-\omega-\omega_0}{\lambda}\right)\right] \tag{11-37}$$

PROOF If τ_0 is sufficiently small, then $\mathbf{c}(t) \simeq \mathbf{c}(0)$, and

$$\boldsymbol{\varphi}(t) = \int_0^t \mathbf{c}(\alpha)\, d\alpha \simeq \mathbf{c}(0)t \qquad |t| < \tau_0 \tag{11-38}$$

Inserting into (11-35), we obtain

$$E\{\mathbf{w}(\tau)\} \simeq E\{e^{j\lambda\tau\mathbf{c}(0)}\} = \Phi_c(\lambda\tau) \qquad |\tau| < \tau_0 \tag{11-39}$$

where

$$\Phi_c(\mu) = E\{e^{j\mu\mathbf{c}(t)}\}$$

† P. M. Woodward: "The Spectrum of Random Frequency Modulation," *Telecommunications Research*, Great Malvern, Worcs., England, Memo 666, 1952.

is the characteristic function of $\mathbf{c}(t)$. From this and (11-34) it follows that

$$R_{zz}(\tau) \simeq \Phi_c(\lambda\tau)e^{j\omega_0\tau} \qquad |\tau| < \tau_0 \tag{11-40}$$

If λ is sufficiently large, then $\Phi_c(\lambda\tau) \simeq 0$ for $\lambda\tau > \tau_0$ because $\Phi_c(\mu) \to 0$ as $\mu \to \infty$. Hence (11-40) is a satisfactory approximation for every τ in the region where $\Phi_c(\lambda\tau)$ takes significant values. Transforming both sides of (11-40) and using the inversion formula

$$f_c(c) = \frac{1}{2\pi}\int_{-\infty}^{\infty} \Phi_c(\mu)e^{j\mu c}\, d\mu$$

we obtain

$$S_{zz}(\omega) = \int_{-\infty}^{\infty} \Phi_c(\lambda\tau)e^{j\omega_0\tau}e^{-j\omega\tau}\, d\tau = \frac{2\pi}{\lambda} f_c\left(\frac{\omega - \omega_0}{\lambda}\right)$$

and (11-37) follows from (11-17).

Normal processes Suppose now that $\mathbf{c}(t)$ is normal with zero mean. In this case $\boldsymbol{\varphi}(t)$ is also normal with zero mean. Hence [see (11-36)]

$$\Phi_\varphi(\lambda, \tau) = \exp\left\{-\frac{1}{2}\lambda^2\sigma_\varphi^2(\tau)\right\}$$
$$\sigma_\varphi^2(\tau) = 2\int_0^\tau R_c(\alpha)(\tau - \alpha)\, d\alpha \tag{11-41}$$

In general, the Fourier transform of $\Phi_\varphi(\lambda, \tau)$ is found only numerically. However, as we show next, explicit formulas can be obtained if λ is large or small. We introduce the "correlation time" τ_c of $\mathbf{c}(t)$

$$\tau_c = \frac{1}{\rho}\int_0^\infty R_c(\alpha)\, d\alpha \qquad \rho = R_c(0) \tag{11-42}$$

and we select two constants τ_0 and τ_1 such that

$$R_c(\tau) \simeq \begin{cases} 0 & |\tau| > \tau_1 \\ \rho & |\tau| < \tau_0 \end{cases}$$

Inserting into (11-41), we obtain (Fig. 11-4)

$$\sigma_\varphi^2(\tau) \simeq \begin{cases} \rho\tau^2 & |\tau| < \tau_0 \\ 2\rho\tau\tau_c & \tau > \tau_1 \end{cases} \qquad \left.\begin{matrix} e^{-\rho\lambda^2\tau^2/2} \\ e^{-\rho\lambda^2\tau\tau_c} \end{matrix}\right\} \simeq R_{ww}(\tau) \tag{11-43}$$

It is known from the asymptotic properties of Fourier transforms that the behavior of $R_{ww}(\tau)$ for small (large) τ determines the behavior of $S_{ww}(\omega)$ for large (small) ω. Since

$$e^{-\rho\lambda^2\tau^2/2} \leftrightarrow \frac{1}{\lambda}\sqrt{\frac{2\pi}{\rho}}\, e^{-\omega^2/2\rho\lambda^2}$$
$$e^{-\rho\lambda^2\tau_c|\tau|} \leftrightarrow \frac{2\rho\tau_c\lambda^2}{\omega^2 + \rho^2\tau_c^2\lambda^4} \tag{11-44}$$

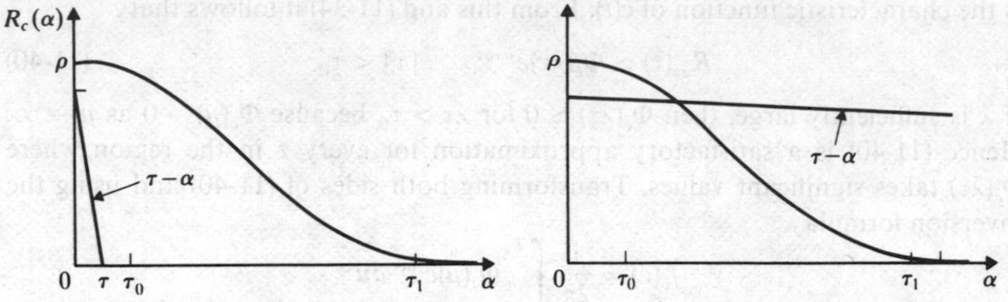

Figure 11-4

we conclude that $S_{ww}(\omega)$ is *normal* near the origin and it is asymptotically *lorenzian* as $\omega \to \infty$. As we show next, these limiting cases give an adequate description of $S_{ww}(\omega)$ for large or small λ.

Wideband FM If λ is such that

$$\rho\lambda^2\tau_0^2 \gg 1$$

then $R_{ww}(\tau) \simeq 0$ for $|\tau| > \tau_0$. This shows that we can use the upper approximation in (11-43) for every significant τ. The resulting spectrum equals

$$S_{ww}(\omega) \simeq \frac{1}{\lambda}\sqrt{\frac{2\pi}{\rho}}\, e^{-\omega^2/2\rho\lambda^2} = \frac{2\pi}{\lambda} f_c\left(\frac{\omega}{\lambda}\right) \tag{11-45}$$

in agreement with Woodward's theorem. The last equality in (11-45) follows because $\mathbf{c}(t)$ is normal with variance $E\{\mathbf{c}^2(t)\} = \rho$.

Narrowband FM If λ is such that

$$\rho\lambda^2\tau_1\tau_c \ll 1$$

then $R_{ww}(\tau) \simeq 1$ for $|\tau| < \tau_1$. This shows that we can use the lower approximation in (11-43) for every significant τ. Hence

$$S_{ww}(\omega) \simeq \frac{2\rho\tau_c\lambda^2}{\omega^2 + \rho^2\tau_c^2\lambda^4} \tag{11-46}$$

Pulse-Amplitude Modulation

We shall determine the spectral properties of the cyclostationary process†

$$\mathbf{c}(t) = \sum_{n=-\infty}^{\infty} \mathbf{c}_{n+1} h(t - nT) \tag{11-47}$$

† N. A. Gardner: "Stationarizable Random Processes," *IEEE Transactions on Information Theory*, vol. IT-24, 1978.

where $\mathbf{c}_n$ is a sequence of RVs and $h(t)$ is a deterministic function. Clearly, $\mathbf{c}(t)$ is not a stationary process, therefore it does not have a spectrum in the sense of (10-10). Its spectrum is defined either as the response of a narrowband system as in (10-51) or, equivalently, as the spectrum of the shifted process

$$\bar{\mathbf{c}}(t) = \mathbf{c}(t - \boldsymbol{\theta})$$

(Fig. 11-5) where $\boldsymbol{\theta}$ is an RV uniform in the interval $(0, T)$ and independent of $\mathbf{c}_n$. We shall use the second approach.

We maintain that if $\mathbf{c}_n$ is a stationary sequence of RVs with autocorrelation

$$R_m = E\{\mathbf{c}_{n+m}\mathbf{c}_n\}$$

then the autocorrelation $\bar{R}(\tau)$ and the power spectrum $\bar{S}(\omega)$ of $\bar{\mathbf{c}}(t)$ are given by

$$\bar{R}(\tau) = \frac{1}{T}\sum_{m=-\infty}^{\infty} R_m \rho(\tau - mT) \qquad \rho(t) = h(t)*h(-t) \tag{11-48}$$

$$\bar{S}(\omega) = \frac{|H(\omega)|^2}{T}\sum_{m=-\infty}^{\infty} R_m e^{-jm\omega T} \qquad H(\omega) = \int_{-\infty}^{\infty} h(t)e^{-j\omega t}\,dt \tag{11-49}$$

PROOF We have shown in Sec. 9-3 that if the function $h(t)$ in (11-47) is a rectangular pulse as in Fig. 9-8*a*, then [see (9-82)] $\bar{R}(\tau)$ is a train of triangles of area R_m/T. As $c \to 0$, the limit of $\mathbf{c}(t)$ is the impulse train

$$\mathbf{v}(t) = \sum_{n=-\infty}^{\infty} \mathbf{c}_{n+1}\delta(t - nT)$$

shown in Fig. 11-5 and the autocorrelation of the shifted process $\bar{\mathbf{v}}(t) = \mathbf{v}(t - \boldsymbol{\theta})$ is the sum

$$\bar{R}_v(\tau) = \frac{1}{T}\sum_{m=-\infty}^{\infty} R_m\,\delta(\tau - mT) \tag{11-50}$$

obtained as the limit of the function $\bar{R}(\tau)$ of Fig. 9-8*b* as $c \to 0$.

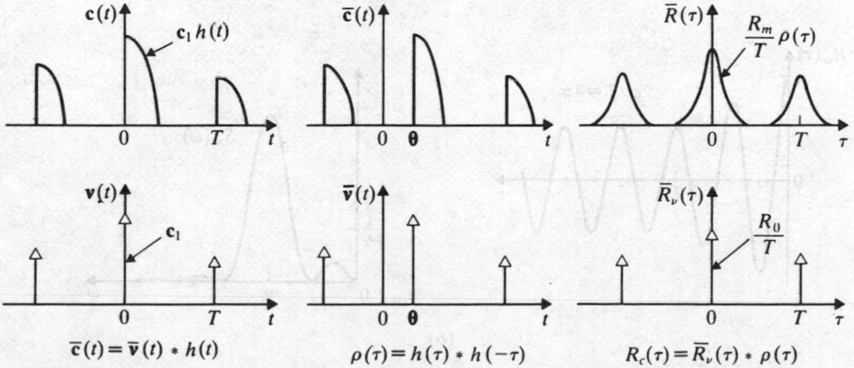

Figure 11-5

To prove (11-48), we note that

$$\mathbf{c}(t) = \mathbf{v}(t)*h(t) \qquad \bar{\mathbf{c}}(t) - \bar{\mathbf{v}}(t)*h(t)$$

Hence [see (10-38)] $\bar{R}(\tau) = \bar{R}_v(\tau)*\rho(\tau)$, and (11-48) results. Transforming both sides of (11-50), we obtain

$$\bar{S}_v(\omega) = \frac{1}{T} \sum_{m=-\infty}^{\infty} R_m e^{-jm\omega T}$$

and (11-49) follows from (10-39).

Frequency-Shift Keying (FSK)

We conclude with a brief discussion of the spectral properties of the FM signal

$$\mathbf{w}(t) = e^{j\boldsymbol{\varphi}(t)} \qquad \boldsymbol{\varphi}(t) = \int_0^t \mathbf{c}(\alpha - \boldsymbol{\theta})\, d\alpha \tag{11-51}$$

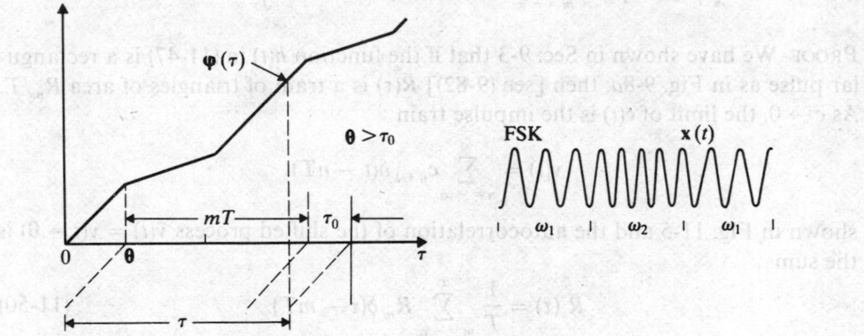

(a)

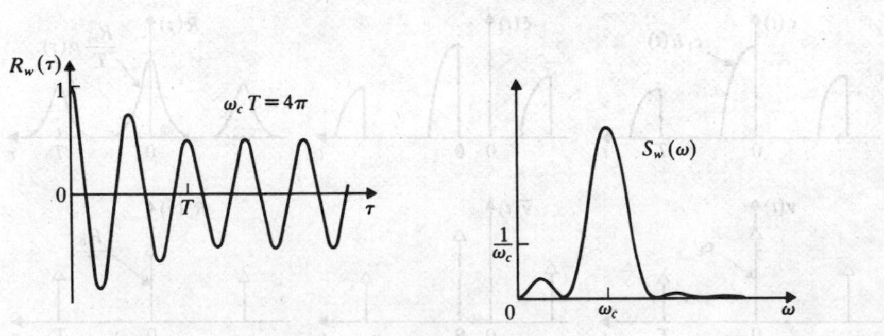

(b)

Figure 11-6

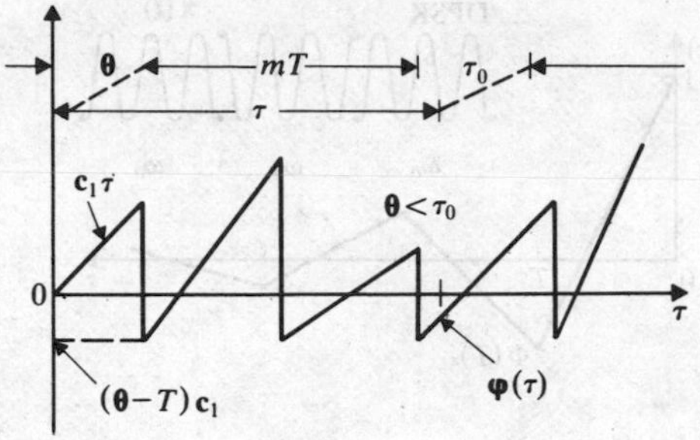

Figure 11-7

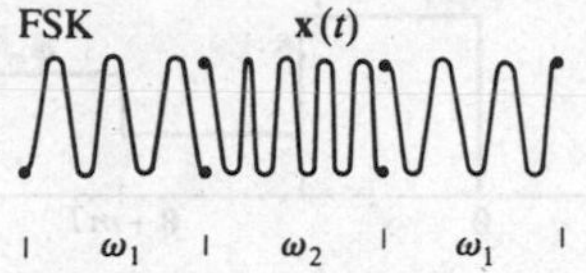

where we assume that $\mathbf{c}(t)$ is a cyclostationary process† as in (11-47) and $\boldsymbol{\theta}$ is an RV uniform in the interval $(0, T)$. From these assumptions it follows that

$$R_w(\tau) = E\{\mathbf{w}(t+\tau)\mathbf{w}^*(t)\} = E\{\mathbf{w}(\tau)\}$$

It suffices, therefore, to find $E\{\mathbf{w}(\tau)\}$.

In most cases of interest, the corresponding real process

$$\mathbf{x}(t) = \cos\,[\omega_0 t + \boldsymbol{\varphi}(t)]$$

consists of pure sine waves in each sampling interval, with frequencies $\omega_n = \omega_0 + \mathbf{c}_n$. This means that the instantaneous frequency $\mathbf{c}(t)$ is a staircase function and the resulting $\boldsymbol{\varphi}(t)$ is a polygon with corners at the sampling points $\boldsymbol{\theta} + mT$ (Fig. 11-6). In the following, we consider only processes with piecewise linear phases, including discontinuities as in Fig. 11-7. The model of Fig. 11-8 is a special case where the corresponding process $\mathbf{x}(t)$ has constant frequency but its phase is discontinuous at the points $\boldsymbol{\theta} + mT$. The size of these discontinuities specifies then the transmitted message (PSK).

We shall base the analysis on the familiar identity [see (7-59)]

$$E\{\mathbf{w}(\tau)\} = E\{E\{\mathbf{w}(\tau)\,|\,\boldsymbol{\theta}\}\} = \frac{1}{T}\int_0^T E\{\mathbf{w}(\tau)\,|\,\theta\}\,d\theta \qquad (11\text{-}52)$$

and, for convenience, we shall express the variable $\tau > 0$ as a sum

$$\tau = mT + \tau_o \qquad 0 \le \tau_o < T$$

where m is an integer.

Continuous-phase FSK We start with a continuous phase as in Fig. 11-6a. In the interval $(0, \boldsymbol{\theta})$, the process $\boldsymbol{\varphi}(t)$ is a straight line with slope $\mathbf{c}_1$. In the interval

† In Sec. 12-4, we develop the spectral properties of FM signals whose instantaneous frequency is a continuous-time Markoff process [see (12-146)].

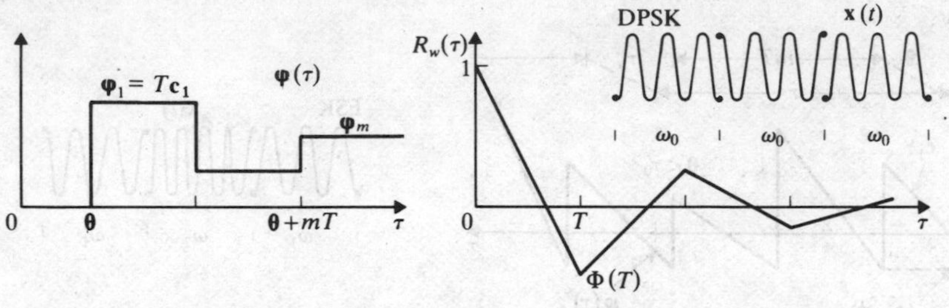

Figure 11-8

$(mT - T + \boldsymbol{\theta},\ mT + \boldsymbol{\theta})$, its slope equals $\mathbf{c}_m$. Omitting details, we conclude from the above that $\boldsymbol{\varphi}(\tau)$ can be written in the following form:

$m = 0$

$$\boldsymbol{\varphi}(\tau) = \begin{cases} \boldsymbol{\theta}\mathbf{c}_1 + (\tau - \boldsymbol{\theta})\mathbf{c}_2 & \boldsymbol{\theta} < \tau \\ \tau\mathbf{c}_1 & \boldsymbol{\theta} > \tau \end{cases}$$

$m > 0$

$$\boldsymbol{\varphi}(\tau) = \boldsymbol{\theta}\mathbf{c}_1 + T\sum_{n=1}^{m} \mathbf{c}_{n+1} + \begin{cases} (\tau_o - \boldsymbol{\theta})\mathbf{c}_{m+2} & \boldsymbol{\theta} < \tau_o \\ (\tau_o - \boldsymbol{\theta})\mathbf{c}_{m+1} & \boldsymbol{\theta} > \tau_o \end{cases}$$

We next make the simplifying assumption† that the RVs $\mathbf{c}_n$ are i.i.d. with characteristic function

$$\Phi(\mu) = E\{e^{j\mu\mathbf{c}_n}\}$$

Setting $\boldsymbol{\theta} = \theta$ in (11-51), we conclude that

$m = 0$

$$E\{\mathbf{w}(\tau)\,|\,\theta\} = \begin{cases} \Phi(\theta)\Phi(\tau - \theta) & \theta < \tau \\ \Phi(\tau) & \theta > \tau \end{cases}$$

$m > 0$

$$E\{\mathbf{w}(\tau)\,|\,\theta\} = \begin{cases} \Phi(\theta)\Phi^m(T)\Phi(\tau_o - \theta) & \theta < \tau_o \\ \Phi(\theta)\Phi^{m-1}(T)\Phi(T + \tau_o - \theta) & \theta > \tau_o \end{cases}$$

To determine $R_w(\tau)$, it suffices to insert the above into (11-52). This yields

$$TR_w(\tau) = \begin{cases} \int_0^\tau \Phi(\theta)\Phi(\tau - \theta)\, d\theta + (T - \tau)\Phi(\tau) & \tau < T \\ \Phi^m(T) \int_0^{\tau_o} \Phi(\theta)\Phi(\tau_o - \theta)\, d\theta & \tau > T \\ \quad + \Phi^{m-1}(T) \int_{\tau_o}^{T} \Phi(\theta)\Phi(T + \tau_o - \theta)\, d\theta & \end{cases} \qquad (11\text{-}53)$$

† For other cases see R. C. Titsworth and L. R. Welch: "Power Spectra of Signals Modulated by Random and Pseudorandom Sequences," *Jet Propulsion Laboratory*, Pasadena, CA, 1961. Also P. Galko and S. Pasupathy, *IEEE Transactions on Information Theory*, vol. IT-27, 1981.

Special case Suppose that the RVs $\mathbf{c}_n$ take the values ω_c and $-\omega_c$ with equal probability. In this case

$$\Phi(\mu) = \tfrac{1}{2}e^{j\omega_c\mu} + \tfrac{1}{2}e^{-j\omega_c\mu} = \cos \omega_c \mu$$

Assuming, further, that T is a multiple of $2\pi/\omega_c$, we obtain $\Phi(T) = \cos \omega_c T = 1$, and (11-53) yields

$$R_w(\tau) = \begin{cases} \left(1 - \dfrac{\tau}{2T}\right) \cos \omega_c \tau + \dfrac{1}{2\omega_c T} \sin \omega_c \tau & \tau < T \\ \tfrac{1}{2} \cos \omega_c \tau & \tau > T \end{cases}$$

This function and its transform $S_w(\omega)$ are shown in Fig. 11-6*b*.

Discontinuous-phase FSK The process $\boldsymbol{\varphi}(t)$ of Fig. 11-7 is discontinuous at the points $mT - \boldsymbol{\theta}$ and its discontinuity equals $T\mathbf{c}_m$ where $\mathbf{c}_m$ is its slope at the preceding sampling interval. The process starts from zero reaching the value $\boldsymbol{\theta}\mathbf{c}_1$ at $t = \boldsymbol{\theta}^-$. Furthermore, $\boldsymbol{\varphi}(\boldsymbol{\theta}^+) = \boldsymbol{\theta}\mathbf{c}_1 - T\mathbf{c}_1 = \boldsymbol{\varphi}(mT + \boldsymbol{\theta}^+)$ for any m. The resulting curve is thus given by

$m = 0$

$$\boldsymbol{\varphi}(\tau) = \begin{cases} (\boldsymbol{\theta} - T)\mathbf{c}_1 + (\tau - \boldsymbol{\theta})\mathbf{c}_2 & \boldsymbol{\theta} < \tau \\ \tau\mathbf{c}_1 & \boldsymbol{\theta} > \tau \end{cases}$$

$m > 0$

$$\boldsymbol{\varphi}(\tau) = \begin{cases} (\boldsymbol{\theta} - T)\mathbf{c}_1 + (\tau_o - \boldsymbol{\theta})\mathbf{c}_{m+2} & \boldsymbol{\theta} < \tau_o \\ (\boldsymbol{\theta} - T)\mathbf{c}_1 + (T + \tau_o - \boldsymbol{\theta})\mathbf{c}_{m+1} & \boldsymbol{\theta} > \tau_o \end{cases}$$

Hence

$m = 0$

$$E\{\mathbf{w}(\tau) \mid \theta\} = \begin{cases} \Phi(\theta - T)\Phi(\tau - \theta) & \theta < \tau \\ \Phi(\tau) & \theta > \tau \end{cases}$$

$m > 0$

$$E\{\mathbf{w}(\tau) \mid \theta\} = \begin{cases} \Phi(\theta - T)\Phi(\tau_o - \theta) & \theta < \tau_o \\ \Phi(\theta - T)\Phi(T + \tau_o - \theta) & \theta > \tau_o \end{cases}$$

and (11-52) yields

$$TR_w(\tau) = \begin{cases} \int_0^\tau \Phi(\theta - T)\Phi(\tau - \theta)\, d\theta + (T - \tau)\Phi(\tau) & \tau < T \\ \int_0^{\tau_o} \Phi(\theta - T)\Phi(\tau_o - \theta)\, d\theta + \int_{\tau_o}^{T} \Phi(\theta - T)\Phi(T + \tau_o - \theta)\, d\theta & \tau > T \end{cases}$$

***Differential phase-shift keying* (DPSK)** We consider, finally, FM signals whose phase $\boldsymbol{\varphi}(t)$ is a staircase with discontinuities at the points $mT + \boldsymbol{\theta}$ equal to $T\mathbf{c}_m$ as in Fig. 11-8. As in the previous cases, we assume that the RVs $\mathbf{c}_n$ are i.i.d.

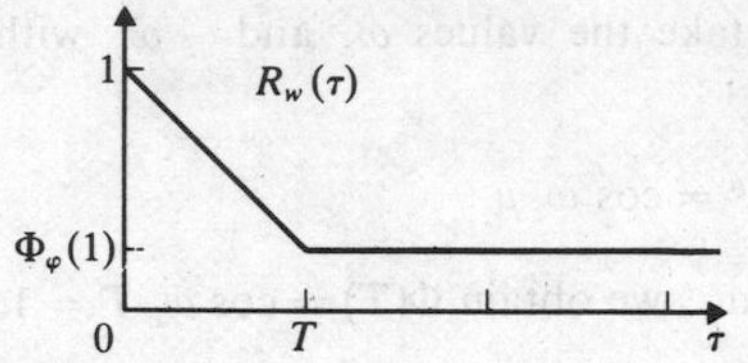

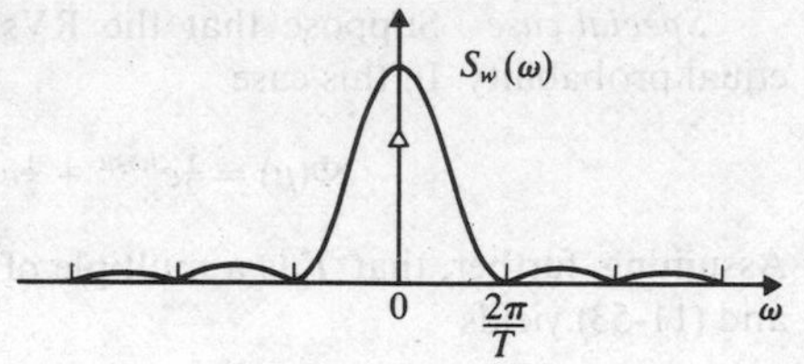

Figure 11-9

Reasoning as before, we obtain the following:

$m = 0$

$$\boldsymbol{\varphi}(\tau) = \begin{cases} T\mathbf{c}_1 & \boldsymbol{\theta} < \tau \\ 0 & \boldsymbol{\theta} > \tau \end{cases} \qquad E\{\mathbf{w}(\tau) \mid \theta\} = \begin{cases} \Phi(T) & \theta < \tau \\ 1 & \theta > \tau \end{cases}$$

$m > 0$

$$\boldsymbol{\varphi}(\tau) = \begin{cases} T(\mathbf{c}_1 + \cdots + \mathbf{c}_{m+1}) & \boldsymbol{\theta} < \tau_o \\ T(\mathbf{c}_1 + \cdots + \mathbf{c}_m) & \boldsymbol{\theta} > \tau_o \end{cases} \qquad E\{\mathbf{w}(\tau) \mid \theta\} = \begin{cases} \Phi^{m+1}(T) & \theta < \tau_o \\ \Phi^m(T) & \theta > \tau_o \end{cases}$$

and (11-52) yields (Fig. 11-8)

$$R_w(\tau) = \Phi^{m+1}(T)\,\frac{\tau_o}{T} + \Phi^m(T)\left(1 - \frac{\tau_o}{T}\right) \tag{11-54}$$

Phase-shift keying (*PSK*) In the preceding, the message was contained in the RVs $\mathbf{c}_n$ which we assumed independent. The values

$$\boldsymbol{\varphi}_n = T(\mathbf{c}_1 + \cdots + \mathbf{c}_n)$$

of the phase $\boldsymbol{\varphi}(t)$ were not independent. We assume now that they are, and we denote by $\Phi_\varphi(\lambda)$ their characteristic function

$$\Phi_\varphi(\lambda) = E\{e^{j\lambda\boldsymbol{\varphi}_n}\}$$

In this case

$$\boldsymbol{\varphi}(\tau) = \begin{cases} \boldsymbol{\varphi}_{m+1} & \boldsymbol{\theta} < \tau \\ 0 & \boldsymbol{\theta} > \tau \end{cases} \qquad E\{\mathbf{w}(\tau) \mid \theta\} = \begin{cases} \Phi_\varphi(1) & \theta < \tau \\ 1 & \theta > \tau \end{cases}$$

Hence

$$R_w(\tau) = \begin{cases} \Phi_\varphi(1)\,\dfrac{\tau}{T} + \left(1 - \dfrac{\tau}{T}\right) & \tau < T \\ \Phi_\varphi(1) & \tau > T \end{cases} \tag{11-55}$$

In Fig. 11-9 we show the function $R_w(\tau)$ and its transform $S_w(\omega)$.

11-2 BANDLIMITED PROCESSES AND SAMPLING THEORY

A process $\mathbf{x}(t)$ is called bandlimited (abbreviated BL) if it has finite power and its spectrum vanishes for $|\omega| > \sigma$

$$R(0) < \infty \qquad S(\omega) = 0 \qquad |\omega| > \sigma \tag{11-56}$$

In this section we establish various identities involving linear functionals of BL processes. To do so, we express the two sides of each identity as responses of linear systems. The underlying reasoning is based on the following:

Theorem Suppose that $\mathbf{w}_1(t)$ and $\mathbf{w}_2(t)$ are the responses of the systems $T_1(\omega)$ and $T_2(\omega)$ to a BL process $\mathbf{x}(t)$ (Fig. 11-10). We shall show that if

$$T_1(\omega) = T_2(\omega) \qquad \text{for} \qquad |\omega| \le \sigma \tag{11-57}$$

then

$$\mathbf{w}_1(t) = \mathbf{w}_2(t) \tag{11-58}$$

PROOF The difference $\mathbf{w}_1(t) - \mathbf{w}_2(t)$ is the response of the system $T_1(\omega) - T_2(\omega)$ to the input $\mathbf{x}(t)$. Since $S(\omega) = 0$ for $|\omega| > \sigma$, we conclude from (10-41) and (11-57) that

$$E\{|\mathbf{w}_1(t) - \mathbf{w}_2(t)|^2\} = \frac{1}{2\pi}\int_{-\sigma}^{\sigma} S(\omega)|T_1(\omega) - T_2(\omega)|^2\,d\omega = 0$$

Hence† $\mathbf{w}_1(t) = \mathbf{w}_2(t)$.

Taylor series If $\mathbf{x}(t)$ is BL, then [see (10-11)]

$$R(\tau) = \frac{1}{2\pi}\int_{-\sigma}^{\sigma} S(\omega)e^{j\omega\tau}\,d\omega \tag{11-59}$$

In the above, the limits of integration are finite and the area $2\pi R(0)$ of $S(\omega)$ is also finite. We can therefore differentiate under the integral sign

$$R^{(n)}(\tau) = \frac{1}{2\pi}\int_{-\sigma}^{\sigma} (j\omega)^n S(\omega)e^{j\omega\tau}\,d\omega \tag{11-60}$$

† All identities in this section are interpreted in the MS sense.

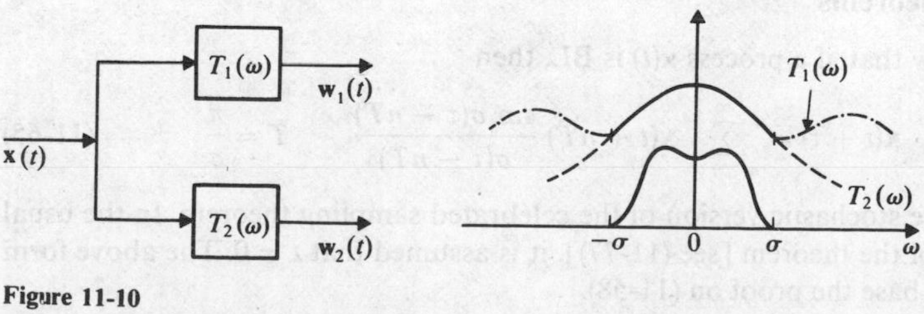

Figure 11-10

This shows that the autocorrelation of a BL process is an entire function, i.e., it has derivatives of any order for every τ. From this it follows that $\mathbf{x}^{(n)}(t)$ exists for any n (see App. 9A).

We maintain that

$$\mathbf{x}(t+\tau) = \sum_{n=0}^{\infty} \mathbf{x}^{(n)}(t)\frac{\tau^n}{n!} \tag{11-61}$$

PROOF We shall prove (11-61) using (11-58). As we know

$$e^{j\omega\tau} = \sum_{n=0}^{\infty} (j\omega)^n \frac{\tau^n}{n!} \qquad \text{all } \omega \tag{11-62}$$

The processes $\mathbf{x}(t+\tau)$ and $\mathbf{x}^{(n)}(t)$ are the responses of the systems $e^{j\omega\tau}$ and $(j\omega)^n$ respectively to the input $\mathbf{x}(t)$. If, therefore, we use as systems $T_1(\omega)$ and $T_2(\omega)$ in (11-57) the two sides of (11-62), the resulting responses will equal the two sides of (11-61). And since (11-62) is true for all ω, (11-61) follows from (11-58).

Bounds Bandlimitedness is often associated with slow variation. The following is an analytical formulation of this association.

If $\mathbf{x}(t)$ is BL, then

$$E\{[\mathbf{x}(t+\tau) - \mathbf{x}(t)]^2\} \le \sigma^2\tau^2 R(0) \tag{11-63}$$

or, equivalently,

$$2[R(0) - R(\tau)] \le \sigma^2\tau^2 R(0) \tag{11-64}$$

PROOF The familiar inequality $|\sin\varphi| \le \varphi$ yields

$$1 - \cos\omega\tau = 2\sin^2\frac{\omega\tau}{2} \le \frac{\omega^2\tau^2}{2}$$

Since $S(\omega) \ge 0$, it follows from the above and (10-12) that

$$R(0) - R(\tau) = \frac{1}{2\pi}\int_{-\sigma}^{\sigma} S(\omega)(1 - \cos\omega\tau)\,d\omega$$

$$\le \frac{1}{2\pi}\int_{-\sigma}^{\sigma} S(\omega)\frac{\omega^2\tau^2}{2}\,d\omega \le \frac{\sigma^2\tau^2}{4\pi}\int_{-\sigma}^{\sigma} S(\omega)\,d\omega = \frac{\sigma^2\tau^2}{2}R(0)$$

as in (11-64).

Sampling Theorems

We shall show that, if a process $\mathbf{x}(t)$ is BL, then

$$\mathbf{x}(t+\tau) = \sum_{n=-\infty}^{\infty} \mathbf{x}(t+nT)\frac{\sin\sigma(\tau - nT)}{\sigma(\tau - nT)} \qquad T = \frac{\pi}{\sigma} \tag{11-65}$$

This is the stochastic version of the celebrated sampling theorem. In the usual formulation of the theorem [see (11-77)], it is assumed that $t = 0$. The above form permits us to base the proof on (11-58).

PROOF We expand the exponential $e^{j\omega\tau}$ into a Fourier series in the interval $(-\sigma, \sigma)$. The coefficients of the expansion equal

$$a_n = \frac{1}{2\sigma}\int_{-\sigma}^{\sigma} e^{j\omega\tau}e^{-jn\omega T}\,d\omega = \frac{\sin \sigma(\tau - nT)}{\sigma(\tau - nT)}$$

Hence

$$e^{j\omega\tau} = \sum_{n=-\infty}^{\infty} e^{jn\omega T}\frac{\sin \sigma(\tau - nT)}{\sigma(\tau - nT)} \qquad |\omega| \le \sigma \tag{11-66}$$

The processes $\mathbf{x}(t + \tau)$ and $\mathbf{x}(t + nT)$ are the responses of the systems $e^{j\omega\tau}$ and $e^{jn\omega T}$ respectively to the input $\mathbf{x}(t)$. If, therefore, we use as systems $T_1(\omega)$ and $T_2(\omega)$ in (11-57) the two sides of (11-66), the resulting responses will equal the two sides of (11-65). And since (11-66) is true for every $|\omega| \le \sigma$, (11-65) follows from (11-58).

Past samples The sampling expansion (11-65) expresses a BL process $\mathbf{x}(t)$ in terms of its samples past and future. We show next that if the sampling rate is tripled, then $\mathbf{x}(t)$ can be expressed in terms of its past samples only.

We maintain that if

$$\bar{T} < \frac{T}{3} = \frac{\pi}{3\sigma}$$

then the process

$$\mathbf{z}_n(t) = n\mathbf{x}(t - \bar{T}) - \binom{n}{2}\mathbf{x}(t - 2\bar{T}) + \cdots + (-1)^n\mathbf{x}(t - n\bar{T}) \tag{11-67}$$

tends to $\mathbf{x}(t)$ as $n \to \infty$.

PROOF The difference $\mathbf{x}(t) - \mathbf{z}_n(t)$ is the process $\mathbf{y}(t)$ in Example 10-11. Hence

$$E\{|\mathbf{x}(t) - \mathbf{z}_n(t)|^2\} = \frac{1}{2\pi}\int_{-\sigma}^{\sigma} S(\omega)\left(2\sin\frac{\omega\bar{T}}{2}\right)^{2n} d\omega \tag{11-68}$$

If $|\omega| \le \sigma$, then

$$\frac{|\omega|\bar{T}}{2} < \frac{|\omega|\pi}{6\sigma} \le \frac{\pi}{6} \qquad \left|2\sin\frac{\omega\bar{T}}{2}\right| < 1$$

hence, the integrand in (11-68) tends to zero as $n \to \infty$. From this it follows that $\mathbf{z}_n(t)$ tends to $\mathbf{x}(t)$.

The above shows again that a BL process is predictable, i.e., it can be expressed in terms of its past.

Generalized sampling expansion† The sampling expansion holds only if $T \le \pi/\sigma$. The following theorem states that if we have access to the samples of the outputs

† A. Papoulis, "Generalized Sampling Expansion," *Transactions on Circuits and Systems*, November 1977.

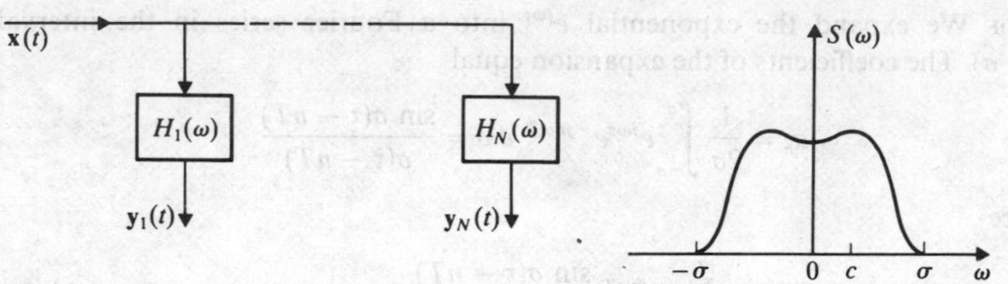

Figure 11-11

$\mathbf{y}_1(t), \ldots, \mathbf{y}_N(t)$ of N linear systems $H_1(\omega), \ldots, H_N(\omega)$ driven by $\mathbf{x}(t)$ (Fig. 11-11) then we can increase the sampling interval from π/σ to $N\pi/\sigma$.

We introduce the constants

$$c = \frac{2\sigma}{N} = \frac{2\pi}{\bar{T}} \qquad \bar{T} = NT \tag{11-69}$$

and the N functions

$$P_1(\omega, t), \ldots, P_N(\omega, t)$$

defined as the solutions of the system

$$\begin{aligned} H_1(\omega)P_1(\omega, \tau) + \cdots + H_N(\omega)P_N(\omega, \tau) &= 1 \\ H_1(\omega + c)P_1(\omega, \tau) + \cdots + H_N(\omega + c)P_N(\omega, \tau) &= e^{jc\tau} \\ \cdots\cdots\cdots\cdots\cdots\cdots\cdots\cdots\cdots\cdots\cdots\cdots \\ H_1(\omega + Nc - c)P_1(\omega, \tau) + \cdots + H_N(\omega + Nc - c)P_N(\omega, \tau) &= e^{j(N-1)c\tau} \end{aligned} \tag{11-70}$$

In the above, ω takes all values in the interval $(-\sigma, -\sigma + c)$ and τ is arbitrary.

We next form the N functions

$$p_k(\tau) = \frac{1}{c}\int_{-\sigma}^{-\sigma+c} P_k(\omega, \tau)e^{j\omega\tau}\, d\omega \qquad 1 \le k \le N \tag{11-71}$$

Theorem

$$\mathbf{x}(t + \tau) = \sum_{n=-\infty}^{\infty} [\mathbf{y}_1(t + n\bar{T})p_1(\tau - n\bar{T}) + \cdots + \mathbf{y}_N(t + n\bar{T})p_N(\tau - n\bar{T})] \tag{11-72}$$

PROOF The process $\mathbf{y}_i(t + n\bar{T})$ is the response of the system $H_i(\omega)e^{jn\bar{T}\omega}$ to the input $\mathbf{x}(t)$. Therefore, if we use as systems $T_1(\omega)$ and $T_2(\omega)$ in Fig. 11-10 the two sides of the identity

$$e^{j\omega\tau} = H_1(\omega)\sum_{n=-\infty}^{\infty} p_1(\tau - n\bar{T})e^{jn\omega\bar{T}} + \cdots + H_N(\omega)\sum_{n=-\infty}^{\infty} p_N(\tau - n\bar{T})e^{jn\omega\bar{T}} \tag{11-73}$$

the resulting responses will equal the two sides of (11-72). To prove (11-72), it suffices therefore to show that (11-73) is true for every $|\omega| \le \sigma$.

The coefficients $H_k(\omega + kc)$ of the system (11-70) are independent of τ and the right side consists of periodic functions of τ with period $\bar{T} = 2\pi/c$ because $e^{jkc\bar{T}} = 1$. Hence, the solutions $P_k(\omega, \tau)$ are periodic

$$P_k(\omega, \tau - n\bar{T}) = P_k(\omega, \tau)$$

From the above and (11-71) it follows that

$$p_k(\tau - n\bar{T}) = \frac{1}{c}\int_{-\sigma}^{-\sigma+c} P_k(\omega, \tau)e^{j\omega(\tau - n\bar{T})}\, d\omega$$

This shows that if we expand the function $P_k(\omega, \tau)e^{j\omega\tau}$ into a Fourier series in the interval $(-\sigma, -\sigma + c)$, the coefficient of the expansion will equal $p_k(\tau - n\bar{T})$. Hence

$$P_k(\omega, \tau)e^{j\omega\tau} = \sum_{n=-\infty}^{\infty} p_k(\tau - n\bar{T})e^{jn\omega\bar{T}} \qquad -\sigma < \omega < -\sigma + c \tag{11-74}$$

Multiplying each of the equations in (11-70) by $e^{j\omega\tau}$ and using (11-74) and the identity

$$e^{jn(\omega + kc)\bar{T}} = e^{jn\omega\bar{T}}$$

we conclude that (11-73) is true for every ω in the interval $(-\sigma, \sigma)$.

Random Sampling

We wish to estimate the Fourier transform $F(\omega)$ of a deterministic signal $f(t)$ in terms of a sum involving the samples of $f(t)$. If we approximate the integral of $f(t)e^{-j\omega t}$ by its Riemann sum, we obtain the estimate

$$F(\omega) \simeq F_*(\omega) \equiv \sum_{n=-\infty}^{\infty} Tf(nT)e^{-jn\omega T} \tag{11-75}$$

From the Poisson sum formula (10A-1) it follows that $F_*(\omega)$ equals the sum of $F(\omega)$ and its displacements

$$F_*(\omega) = \sum_{n=-\infty}^{\infty} F(\omega + 2n\sigma) \qquad \sigma = \frac{\pi}{T}$$

Hence, $F_*(\omega)$ can be used as the estimate of $F(\omega)$ in the interval $(-\sigma, \sigma)$ only if $F(\omega)$ is negligible outside this interval. The difference $F(\omega) - F_*(\omega)$ is called *aliasing error*. In the following, we replace in (11-75) the equidistant samples $f(nT)$ of $f(t)$ by its samples $f(\mathbf{t}_i)$ at a random set of points $\mathbf{t}_i$ and we examine the nature of the resulting error.†

† E. Masry: "Poisson Sampling and Spectral Estimation of Continuous-Time Processes," *IEEE Transactions on Information Theory*, vol. IT-24, 1978. See also F. J. Beutler: "Alias Free Randomly Timed Sampling of Stochastic Processes," *IEEE Transactions on Information Theory*, vol. IT-16, 1970.

Note If $F(\omega) = 0$ for $|\omega| > \sigma$, then $F(\omega)$ does not overlap with its displacements and (11-75) is an equality for $|\omega| < \sigma$. From this it follows that

$$F(\omega) = \sum_{n=-\infty}^{\infty} Tf(nT)e^{-jnT\omega}\pi_\sigma(\omega) \qquad \pi_\sigma(\omega) = \begin{cases} 1 & |\omega| < \sigma \\ 0 & |\omega| > \sigma \end{cases} \tag{11-76}$$

The inverse transform of the above yields

$$f(t) = \sum_{n=-\infty}^{\infty} Tf(nT)\frac{\sin \sigma(t - nT)}{\pi(t - nT)} \tag{11-77}$$

This is the sampling theorem for deterministic signals.

We show next that if $\mathbf{t}_i$ is a Poisson point process with average density λ, then the sum

$$\mathbf{P}(\omega) = \frac{1}{\lambda}\sum_i f(\mathbf{t}_i)e^{-j\omega\mathbf{t}_i} \tag{11-78}$$

is an unbiased estimate of $F(\omega)$ that is, its mean equals $F(\omega)$. Furthermore, if the energy

$$E = \int_{-\infty}^{\infty} f^2(t)\, dt$$

of $f(t)$ is finite, then $\mathbf{P}(\omega) \rightarrow F(\omega)$ as $\lambda \rightarrow \infty$.

To prove the above, we must show that

$$E\{\mathbf{P}(\omega)\} = F(\omega) \qquad \sigma^2_{P(\omega)} = \frac{E}{\lambda} \tag{11-79}$$

Proof Clearly

$$\int_{-\infty}^{\infty} f(t)e^{-j\omega t}\sum_i \delta(t - \mathbf{t}_i)\, dt = \sum_i f(\mathbf{t}_i)e^{-j\omega\mathbf{t}_i} \tag{11-80}$$

Comparing with (11-78), we obtain

$$\mathbf{P}(\omega) = \frac{1}{\lambda}\int_{-\infty}^{\infty} f(t)\mathbf{z}(t)e^{-j\omega t}\, dt \qquad \text{where} \qquad \mathbf{z}(t) = \sum_i \delta(t - \mathbf{t}_i) \tag{11-81}$$

is a Poisson impulse train as in (9-110) with

$$E\{\mathbf{z}(t)\} = \lambda \qquad C_z(t_1, t_2) = \lambda\delta(t_1 - t_2) \tag{11-82}$$

Hence

$$E\{\mathbf{P}(\omega)\} = \frac{1}{\lambda}\int_{-\infty}^{\infty} f(t)E\{\mathbf{z}(t)\}e^{-j\omega t}\, dt = F(\omega)$$

$$\sigma^2_{P(\omega)} = \frac{1}{\lambda^2}\int_{-\infty}^{\infty}\int_{-\infty}^{\infty} f(t_1)f(t_2)\lambda\delta(t_1 - t_2)\, dt_1\, dt_2 = \frac{1}{\lambda}\int_{-\infty}^{\infty} f^2(t_2)\, dt_2$$

and (11-79) results.

From (11-79) it follows that, for a satisfactory estimate of $F(\omega)$, λ must be such that

$$|F(\omega)| \gg \sqrt{\frac{E}{\lambda}} \tag{11-83}$$

Example 11-1 Suppose that $f(t)$ is a sum of sine waves in the interval $(-a, a)$

$$f(t) = \sum_k c_k e^{j\omega_k t} \qquad |t| < a$$

and it equals zero for $|t| > a$. In this case

$$F(\omega) = \sum_k 2c_k \frac{\sin a(\omega - \omega_k)}{\omega - \omega_k} \qquad E \simeq 2a \sum_k |c_k|^2 \tag{11-84}$$

where we neglected cross-products in the evaluation of E. If a is sufficiently large, then

$$F(\omega_k) \simeq 2ac_k$$

This shows that if

$$\sum |c_i|^2 \ll |c_k| \sqrt{2a\lambda} \qquad \text{then} \qquad \mathbf{P}(\omega_k) \simeq F(\omega_k)$$

Thus, with random sampling we can detect line spectra of any frequency even if the average rate λ is small, provided that the observation interval $2a$ is large.

Example 11-2 Suppose now that $F(\omega) = 0$ for $|\omega| > \sigma$. In this case, the average energy spectrum equals $E/2\sigma$. If, therefore, $F(\omega)$ is of the order of $\sqrt{E/2\sigma}$, then for a satisfactory estimate of its value, λ must be such that $E/2\sigma \gg E/\lambda$. This shows that if $F(\omega)$ is BL without sharp peaks, then it can be estimated from random samples only if the average sampling rate λ is larger than the Nyquist rate.

Stochastic signals Suppose now that $\mathbf{x}(t)$ is a stationary process with zero mean. With $\mathbf{z}(t)$ a train of Poisson impulses as in (11-81) and independent of $\mathbf{x}(t)$, we form the process

$$\mathbf{X}_a(\omega) = \frac{1}{\lambda} \int_{-a}^{a} \mathbf{x}(t)\mathbf{z}(t)e^{-j\omega t}\,dt = \sum_{|t_i|<a} \mathbf{x}(t_i)e^{-j\omega t_i} \tag{11-85}$$

We shall show that if $R_x(\tau) \to 0$ as $\tau \to \infty$, then, for sufficiently large a

$$\frac{1}{2a} E\{|\mathbf{X}_a(\omega)|^2\} \simeq S_x(\omega) + \frac{1}{\lambda} R_x(0) \tag{11-86}$$

PROOF Since $R_z(t_1 - t_2) = \lambda^2 + \lambda\delta(t_1 - t_2)$, it follows from (11-85) that

$$E\{|\mathbf{X}_a(\omega)|^2\} = \frac{1}{\lambda^2} \int_{-a}^{a} \int_{-a}^{a} R_x(t_1 - t_2)R_z(t_1 - t_2)e^{-j\omega(t_1 - t_2)}\,dt_1\,dt_2$$

$$= \int_{-a}^{a} e^{j\omega t_2} \int_{-a}^{a} R_x(t_1 - t_2)e^{-j\omega t_1}\,dt_1\,dt_2 + \frac{1}{\lambda} \int_{-a}^{a} R_x(0)\,dt_2 \tag{11-87}$$

This yields (11-86) because, for large a, the inner integral on the right side of (11-87) equals $S_x(\omega)e^{-j\omega t_2}$.

Example 11-3 We wish to estimate the transform $F(\omega)$ of a signal $f(t)$ in terms of the random samples $\mathbf{g}(\mathbf{t}_i)$ of the sum

$$\mathbf{g}(t) = f(t) + \mathbf{v}(t)$$

where $\mathbf{v}(t)$ is zero-mean ideal BL noise

$$S_v(\omega) = \begin{cases} S_0 & |\omega| < \sigma \\ 0 & |\omega| > \sigma \end{cases} \qquad R_v(\tau) = S_0 \frac{\sin \sigma\tau}{\pi\tau}$$

For this purpose we form the sum

$$E\{\mathbf{G}_a(\omega)\} = E\{\mathbf{G}_f(\omega)\} = \int_{-a}^{a} f(t)e^{-j\omega t}\, dt \qquad (11\text{-}88)$$

In the above, $\mathbf{G}_f(\omega)$ and $\mathbf{G}_v(\omega)$ are the contribution to $\mathbf{G}_a(\omega)$ due to $f(t)$ and $\mathbf{v}(t)$ respectively. Since $E\{\mathbf{G}_v(\omega)\} = 0$, (11-79) yields

$$E\{\mathbf{G}_a(\omega)\} = E\{\mathbf{G}_f(\omega)\} = \int_{-a}^{a} f(t)e^{-j\omega t}\, dt$$

The variance of $\mathbf{G}_a(\omega)$ is determined from (11-79) and (11-86) where now $R_v(0) = \sigma S_0/\pi$. With E_a the energy of $f(t)$ in the interval $(-a, a)$, we conclude that

$$\sigma_G^2 = \frac{E_a}{\lambda} + \frac{2a\sigma S_0}{\pi\lambda} + \begin{cases} S_0 & |\omega| < \sigma \\ 0 & |\omega| > \sigma \end{cases}$$

We note that the presence of noise increases the variance of the estimate $\mathbf{G}_a(\omega)$ of $F(\omega)$ even for values of ω outside the band $(-\sigma, \sigma)$ of the noise.

11-3 NORMAL PROCESSES AND BROWNIAN MOTION

As we know, normal processes are completely specified in terms of their mean and autocorrelation. This simplifies the analysis of systems with normal inputs particularly when the systems are linear. We can then apply the results of Chap. 10 to determine not only output correlations, but also complete statistics. In the next four pages we consider various special cases.

We start with an application involving conditional densities. This topic was considered briefly in Sec. 7-3 and is developed further in Sec. 13-1. We shall use the following result: If $\mathbf{x}(t)$ is a normal process with zero mean, then [see (7-42)] the conditional density of $\mathbf{x}(t_1)$ assuming $\mathbf{x}(t_0) = x_0$ is normal with mean

$$E\{\mathbf{x}(t_1) \mid \mathbf{x}(t_0) = x_0\} = ax_0 \qquad a = \frac{R(t_1, t_0)}{R(t_0, t_0)} \qquad (11\text{-}89)$$

and variance

$$\sigma^2_{x(t_1) \mid x_0} = P = R(t_1, t_1) - aR(t_1, t_0)$$

Thus

$$f_{x(t_1)}[x_1 \mid \mathbf{x}(t_0) = x_0] = \frac{1}{\sqrt{2\pi P}}\, e^{-(x_1 - ax_0)^2/2P} \qquad (11\text{-}90)$$

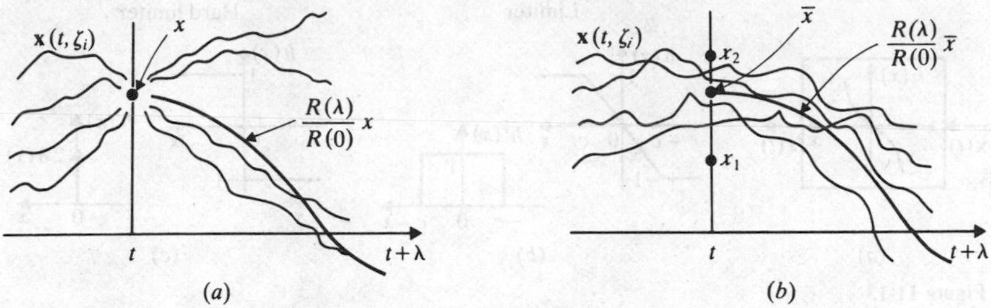

Figure 11-12

Correlometers If the process $\mathbf{x}(t)$ is stationary, then $R(t_1, t_2) = R(t_1 - t_2)$ and (11-89) yields

$$E\{\mathbf{x}(t+\lambda) \mid \mathbf{x}(t) = x\} = ax \qquad a = \frac{R(\lambda)}{R(0)} \tag{11-91}$$

This shows that, if we consider only the samples $\mathbf{x}(t, \zeta_i)$ of $\mathbf{x}(t)$ that at time t pass through the point $\mathbf{x}(t) = x$ as in Fig. 11-12*a*, then their mean at time $t + \lambda$ will be proportional to $R(\lambda)$. This result can be generalized:

We maintain that if D is an arbitrary set of numbers, then

$$E\{\mathbf{x}(t+\lambda) \mid \mathbf{x}(t) \in D\} = a\bar{x} \qquad a = \frac{R(\lambda)}{R(0)} \tag{11-92}$$

where

$$\bar{x} = E\{\mathbf{x}(t) \mid \mathbf{x}(t) \in D\}$$

Indeed, reasoning as in (7-59), we obtain

$$\begin{aligned} E\{\mathbf{x}(t+\lambda) \mid \mathbf{x}(t) \in D\} &= E\{E\{\mathbf{x}(t+\lambda) \mid \mathbf{x}(t)\} \mid \mathbf{x}(t) \in D\} \\ &= E\{a\mathbf{x}(t) \mid \mathbf{x}(t) \in D\} \end{aligned}$$

and (11-92) results. If D is the interval (x_1, x_2), then $\bar{x}$ is the conditional mean of $\mathbf{x}(t)$ assuming that $x_1 < \mathbf{x}(t) < x_2$. Thus, if we consider only the samples $\mathbf{x}(t, \zeta_i)$ of $\mathbf{x}(t)$ that at time t are between x_1 and x_2 as in Fig. 11-12*b*, then their mean at time $t + \lambda$ is proportional to $R(\lambda)$.

Price's theorem† The following theorem simplifies the evaluation of the input–output moments of nonlinear systems with normal inputs.

Given two jointly normal RVs $\mathbf{x}$ and $\mathbf{y}$ with zero mean and an arbitrary function $g(x, y)$, we form the RV $g(\mathbf{x}, \mathbf{y})$. The mean

$$E\{g(\mathbf{x}, \mathbf{y})\} = \int_{-\infty}^{\infty}\int_{-\infty}^{\infty} g(x, y) f(x, y)\, dx\, dy \tag{11-93}$$

† R. Price, "A Useful Theorem for Nonlinear Devices Having Gaussian Inputs," IRE, PGIT, vol. IT-4, 1958.

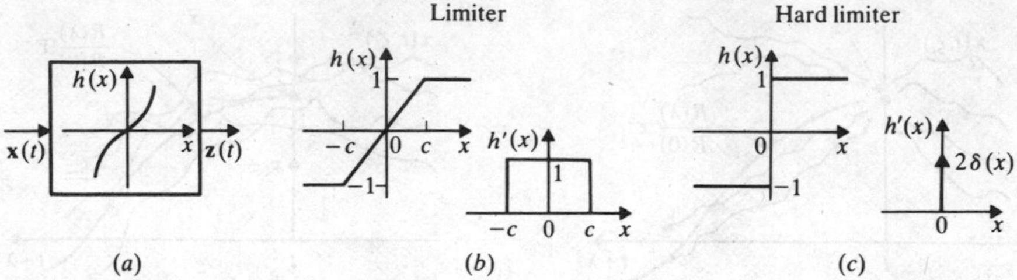

Figure 11-13

of this RV depends on the covariance $C = E\{\mathbf{xy}\}$ of $\mathbf{x}$ and $\mathbf{y}$. We shall show that

$$\frac{\partial^n E\{g(\mathbf{x}, \mathbf{y})\}}{\partial C^n} = E\left\{\frac{\partial^{2n} g(\mathbf{x}, \mathbf{y})}{\partial \mathbf{x}^n \partial \mathbf{y}^n}\right\} \tag{11-94}$$

PROOF The above is a consequence of the following property of normal densities (see Prob. 7-8)

$$\frac{\partial^n f(x, y)}{\partial C^n} = \frac{\partial^{2n} f(x, y)}{\partial x^n \partial y^n} \tag{11-95}$$

Differentiating (11-93) and using (11-95), we obtain

$$\frac{\partial^n E\{g(\mathbf{x}, \mathbf{y})\}}{\partial C^n} = \int_{-\infty}^{\infty}\int_{-\infty}^{\infty} g(x, y)\, \frac{\partial^{2n} f(x, y)}{\partial x^n \partial y^n}\, dx\, dy$$

The right side of (11-94) equals

$$E\left\{\frac{\partial^{2n} g(\mathbf{x}, \mathbf{y})}{\partial \mathbf{x}^n \partial \mathbf{y}^n}\right\} = \int_{-\infty}^{\infty}\int_{-\infty}^{\infty} \frac{\partial^{2n} g(x, y)}{\partial x^n \partial y^n}\, f(x, y)\, dx\, dy$$

Integrating by parts and assuming that the boundary conditions at infinity are zero,† we conclude, omitting details, that the last two integrals are equal. This completes the proof.

Bussgang's theorem If $\mathbf{z}(t)$ is the output of a memoryless system with input a normal process $\mathbf{x}(t)$ (Fig. 11-13*a*):

$$\mathbf{z}(t) = h[\mathbf{x}(t)]$$

then

$$R_{xz}(\tau) = K R_{xx}(\tau) \qquad K = E\{h'[\mathbf{x}(t)]\} \tag{11-96}$$

PROOF If $\mathbf{x} = \mathbf{x}(t)$ and $\mathbf{y} = \mathbf{x}(t + \tau)$, then

$$C = E\{\mathbf{xy}\} = E\{\mathbf{x}(t + \tau)\mathbf{x}(t)\} = R_{xx}(\tau)$$

† This is the case if $|g(x, y)| < Ae^{(x^2 + y^2)}$. See A. Papoulis, Comment on "An Extension of Price's Theorem," *IEEE Transactions Information Theory*, IT-11, 1965.

With $g(x, y) = h(x)y$ the above yields

$$\frac{\partial^2 g(x, y)}{\partial x\, \partial y} = h'(x) \qquad E\{g(\mathbf{x}, \mathbf{y})\} = E\{h[\mathbf{x}(t)]\mathbf{x}(t+\tau)\} = R_{xz}(\tau)$$

Inserting into (11-94) and using the fact that $R_{xz}(\tau) = 0$ if $C = 0$, we obtain

$$\frac{\partial R_{xz}(\tau)}{\partial C} = E\{h'[\mathbf{x}(t)]\} \qquad \text{hence} \qquad R_{xz}(\tau) = E\{h'[\mathbf{x}(t)]\}C$$

and (11-96) results.

Corollary With $\mathbf{w}(t) = \mathbf{z}(t) - K\mathbf{x}(t)$ it follows from (11-96) that

$$h[\mathbf{x}(t)] = K\mathbf{x}(t) + \mathbf{w}(t) \qquad R_{xw}(\tau) = 0$$

Special cases† (*a*) If the system is a *limiter* (Fig. 11-13*b*), then $h'(x)$ is a pulse and

$$K = \int_{-1}^{1} f(x)\, dx = 2\mathbb{G}\left[\frac{1}{R_{xx}(0)}\right] - 1$$

In the above, $f(x)$ is the first-order density of the normal process $\mathbf{x}(t)$.

(*b*) If the system is a *hard limiter* (Fig. 11-13*c*), then $h'(x) = 2\delta(x)$ and

$$K = 2\int_{-\infty}^{\infty} \delta(x) f(x)\, dx = 2f(0) = \sqrt{\frac{2}{\pi R_{xx}(0)}}$$

Brownian Motion

The term "brownian motion" is used to describe the movement of a particle in a liquid, subjected to collision and other forces. Macroscopically, the position $\mathbf{x}(t)$ of the particle can be modeled as a stochastic process satisfying a second-order differential equation:

$$m\mathbf{x}''(t) + f\mathbf{x}'(t) + \kappa\mathbf{x}(t) = \mathbf{F}(t) \tag{11-97}$$

where $\mathbf{F}(t)$ is the collision force, m is the mass of the particle, f is the coefficient of friction, and $\kappa\mathbf{x}(t)$ is an external force which we assume proportional to $\mathbf{x}(t)$. On a macroscopic scale, the process $\mathbf{F}(t)$ can be viewed as normal white noise with zero mean and power spectrum

$$S_F(\omega) = 2kTf \tag{11-98}$$

where T is the temperature of the medium and k is the Boltzmann constant as in (10-80). We note the similarity between (11-97) and the equation

$$L\mathbf{Q}''(t) + R\mathbf{Q}'(t) + \frac{1}{C}\mathbf{Q}(t) = \mathbf{n}_e(t)$$

† H. E. Rowe, "Memoryless Nonlinearities with Gaussian Inputs," BSTJ, vol. 67, no. 7, September 1982.

relating the charge $\mathbf{Q}(t) = C\mathbf{V}(t)$ of the capacity of a series RLC circuit due to the thermal noise source $\mathbf{n}_e(t)$. In fact, if we change the parameters m, f, and κ to L, R, and $1/C$, respectively, then

$$S_F(\omega) = S_{n_e}(\omega) \qquad S_x(\omega) = S_Q(\omega)$$

We can therefore derive the properties of $\mathbf{x}(t)$ from the properties of $\mathbf{Q}(t)$. For example, since [see (10-84)]

$$E\{\mathbf{Q}^2(t)\} = C^2 E\{\mathbf{V}^2(t)\} = kTC$$

we conclude that

$$E\{\mathbf{x}^2(t)\} = R_x(0) = \frac{kT}{\kappa} \tag{11-99}$$

Bound motion If $\kappa \neq 0$, then the particle reaches steady state and $\mathbf{x}(t)$ is a stationary process with power spectrum [see (10-39)]

$$S_x(\omega) = \frac{2kTf}{(\kappa - m\omega^2)^2 + f^2\omega^2}$$

The autocorrelation of $\mathbf{x}(t)$ is the inverse Fourier transform of $S_x(\omega)$, and its form depends on the roots of the polynomial

$$ms^2 + fs + \kappa = 0$$

(see Example 10-14). If these roots are complex

$$s_{1,2} = -\alpha \pm j\beta \qquad \alpha = \frac{f}{2m} \qquad \alpha^2 + \beta^2 = \frac{\kappa}{m}$$

then the motion is oscillatory and

$$R_x(\tau) = \frac{kT}{\kappa} e^{-\alpha|\tau|}\left(\cos \beta\tau + \frac{\alpha}{\beta} \sin \beta|\tau|\right) \tag{11-100}$$

For a specific t, $\mathbf{x}(t)$ is a normal RV with zero mean and variance kT/κ. Hence

$$f_x(x) = \sqrt{\frac{\kappa}{2\pi kT}}\, e^{-\kappa x^2/2kT} \tag{11-101}$$

The conditional density of $\mathbf{x}(t_1)$ assuming $\mathbf{x}(t_0) = x_0$ is also normal as in (11-90) where now

$$a = \frac{R_x(t_1 - t_0)}{R_x(0)} \qquad P = R_x(0)(1 - a^2)$$

Free motion If $\kappa = 0$, then $\mathbf{x}(t)$ is an unbounded nonstationary process approaching the Wiener process as $t \to \infty$ [see (11-106)]. However, the velocity of motion

$$\mathbf{v}(t) = \mathbf{x}'(t)$$

is a stationary process satisfying the equation

$$m\mathbf{v}'(t) + f\mathbf{v}(t) = \mathbf{F}(t) \tag{11-102}$$

From this it follows that

$$S_v(\omega) = \frac{2kTf}{m^2\omega^2 + f^2} \qquad R_v(\tau) = \frac{kT}{m} e^{-f|\tau|/m} \tag{11-103}$$

Furthermore, $\mathbf{v}(t)$ is a normal process with zero mean, variance kT/m, and density

$$f_v(v) = \sqrt{\frac{m}{2\pi kT}}\, e^{-mv^2/2kT} \tag{11-104}$$

The conditional density of $\mathbf{v}(t)$ is also normal with mean

$$E\{\mathbf{v}(t) \mid \mathbf{v}(0) = v_0\} = \frac{R_v(t)}{R_v(0)}\, v_0 = e^{-ft/m} v_0$$

and variance

$$\sigma^2_{v(t)|v_0} = R_v(0) - \frac{R_v^2(t)}{R_v(0)} = \frac{kT}{m}(1 - e^{-2ft/m})$$

Note In physics, (11-102) is called *Langevin's equation*, its stationary solution *Ornstein–Uhlenbeck* process, and the corresponding spectrum *lorenzian*.

The position $\mathbf{x}(t)$ of the particle can be written as the integral of its velocity

$$\mathbf{x}(t) = \int_0^t \mathbf{v}(\alpha)\, d\alpha$$

where we assumed that $\mathbf{x}(0) = 0$. This shows that $\mathbf{x}(t)$ is a normal process with variance

$$E\{\mathbf{x}^2(t)\} = \int_0^t \int_0^t R_v(\alpha - \beta)\, d\alpha\, d\beta = \frac{kT}{m} \int_0^t \int_0^t e^{-f|\alpha-\beta|/m}\, d\alpha\, d\beta$$

$$= \frac{2kT}{f}\left(t - \frac{m}{f} + \frac{m}{f} e^{-ft/m}\right) \tag{11-105}$$

If t is so large that $ft \gg m$, then (Einstein)

$$E\{\mathbf{x}^2(t)\} \simeq 2D^2 t \qquad D^2 = \frac{kT}{f}$$

where D is the *diffusion constant*. This result can be derived directly from (11-97) if we neglect the acceleration term $m\mathbf{x}''(t)$. Indeed, in this case

$$f\mathbf{x}'(t) \simeq \mathbf{F}(t) \qquad \mathbf{x}(t) \simeq \frac{1}{f}\int_0^t \mathbf{F}(\alpha)\, d\alpha \tag{11-106}$$

where $\mathbf{F}(t)$ is white noise with power spectrum $2kTf$. Reasoning as in Example 9-18 we conclude, therefore, that

$$E\{\mathbf{x}^2(t)\} = \frac{2kTft}{f^2} = 2D^2t$$

The Wiener process The Wiener process $\mathbf{w}(t)$ was defined in Sec. 9-1 as the limit of random walk. We define it here as the integral of normal white noise with zero mean

$$\mathbf{w}(t) = \int_0^t \mathbf{v}(\beta)\,d\beta \qquad R_v(\tau) = \alpha\delta(\tau) \tag{11-107}$$

From the above it follows that $\mathbf{w}(t)$ is a normal process with zero mean and variance αt. Furthermore, $\mathbf{w}(t)$ is a process with independent increments. This leads to the conclusion that

$$R_w(t_1, t_2) = \alpha \min(t_1, t_2) \tag{11-108}$$

Indeed, $\mathbf{w}(t_1)\mathbf{w}(t_2) = \mathbf{w}(t_1)[\mathbf{w}(t_2) - \mathbf{w}(t_1)] + \mathbf{w}_1^2(t_1)$. If, therefore, $t_1 < t_2$, then

$$E\{\mathbf{w}(t_1)\mathbf{w}(t_2)\} = E\{\mathbf{w}^2(t_1)\} = \alpha t_1$$

Thus

$$f_{w(t)}(w, t) = \frac{1}{\sqrt{2\pi\alpha t}}\, e^{-w^2/2\alpha t} \tag{11-109}$$

and

$$f_{w(t)}(w \mid \mathbf{w}(t_0) = w_0) = \frac{1}{\sqrt{2\pi\alpha(t - t_0)}}\, e^{-(w-w_0)^2/2\alpha(t-t_0)} \tag{11-110}$$

The last equality follows from (11-90) because

$$a = \frac{R(t, t_0)}{R(t_0, t_0)} = \frac{\alpha t_0}{\alpha t_0} = 1 \qquad P = R(t, t) - aR(t, t_0) = \alpha t - \alpha t_0$$

The diffusion equation Comparing (11-106) with (11-107), we conclude that, for large t, the position $\mathbf{x}(t)$ of a particle in free motion is a Wiener process with $\alpha = 2D^2$. Hence, its conditional density

$$\pi(x, x_0; t, t_0) = f_{x(t)}(x \mid \mathbf{x}(t_0) \simeq x_0)$$

satisfies (11-110) where now $\alpha = 2D^2$. Differentiating (11-110) appropriately, we obtain

$$\frac{\partial \pi}{\partial t} = D^2 \frac{\partial^2 \pi}{\partial x^2} \qquad \frac{\partial \pi}{\partial t_0} = -D^2 \frac{\partial^2 \pi}{\partial x_0^2} \tag{11-111}$$

These relationships are called diffusion equations. In Sec. 12-4, they are rederived in the context of Markoff processes [see (12-164)].

11-4 THE LEVEL-CROSSING PROBLEM

Given a random process $\mathbf{x}(t)$ and a constant a, we denote by $\boldsymbol{\tau}_i$ the time instances when $\mathbf{x}(t)$ crosses the line L_a shown in Fig. 11-14. This line is parallel to the time axis and

$$\mathbf{x}(\boldsymbol{\tau}_i) = a$$

The level-crossing problem is the determination of the statistical properties of the point process $\boldsymbol{\tau}_i$ so formed. A special case is the zero-crossing problem ($a = 0$) when L_a coincides with the t axis.

The level-crossing problem is, in general, complicated. We discuss next only certain aspects that lead to simple results.

Expected number of crossings† We assume that the process $\mathbf{x}(t)$ is stationary and we denote by $\mathbf{n}_a(T)$ the number of points $\boldsymbol{\tau}_i$ in an interval of length T. The following basic theorem expresses the mean of $\mathbf{n}_a(T)$ in terms of the first order density $f_x(x)$ of $\mathbf{x}(t)$ and the conditional mean of its derivative.

Theorem If $\mathbf{x}'(t)$ exists, then

$$E\{\mathbf{n}_a(T)\} = Tf_x(a)E\{\,|\mathbf{x}'(t)|\,\mathbf{x}(t) = a\} \tag{11-112}$$

PROOF We shall prove the theorem using the following property of the impulse function (see P.3): If t_i are all the real zeros of a function $\varphi(t)$, then (Fig. 11-15)

$$\delta[\varphi(t)] = \sum_i \frac{\delta(t - t_i)}{|\varphi'(t_i)|} = \frac{1}{|\varphi'(t)|}\sum_i \delta(t - t_i) \tag{11-113}$$

The zeros of the function $\varphi(t) = \mathbf{x}(t) - a$ are the L_a crossings of $\mathbf{x}(t)$. Thus

$$\varphi(\boldsymbol{\tau}_i) = \mathbf{x}(\boldsymbol{\tau}_i) - a = 0 \qquad \varphi'(t) = \mathbf{x}'(t)$$

† A. Blanc-Lapierre: *Modèles Statistiques pour l'étude de phenomenes de Fluctuations*, Masson et cie, Paris, 1963.

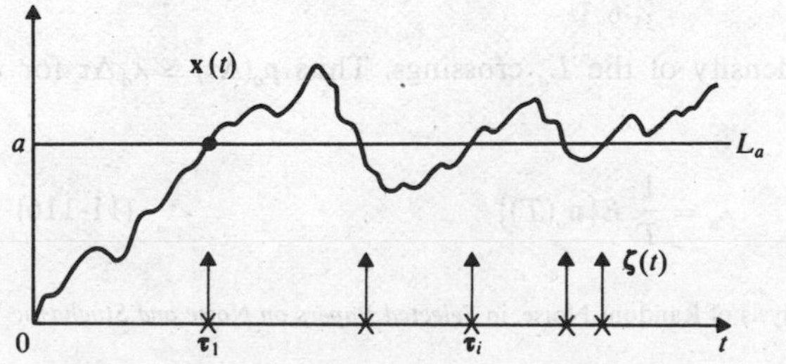

Figure 11-14

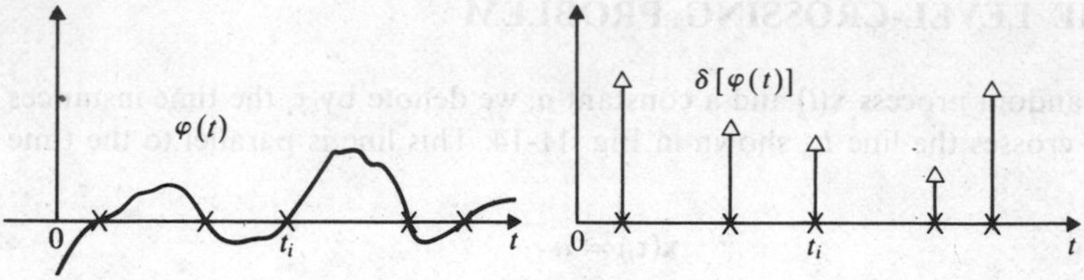

Figure 11-15

Inserting into (11-113), we obtain

$$\sum_i \delta(t - \boldsymbol{\tau}_i) = |\mathbf{x}'(t)|\,\delta[\mathbf{x}(t) - a] \tag{11-114}$$

The sum

$$\boldsymbol{\zeta}(t) = \sum_i \delta(t - \boldsymbol{\tau}_i) \tag{11-115}$$

is a stationary process consisting of a sequence of impulses at the points $\boldsymbol{\tau}_i$. The area of each impulse equals 1 and the number of impulses in the interval $(t, t + T)$ equals $\mathbf{n}_a(T)$. Hence

$$\mathbf{n}_a(T) = \int_t^{t+T} \boldsymbol{\zeta}(\alpha)\, d\alpha \qquad E\{\mathbf{n}_a(T)\} = TE\{\boldsymbol{\zeta}(t)\}$$

To prove (11-112) it suffices, therefore, to find the mean of $\boldsymbol{\zeta}(t)$. As we see from (11-114), the RV $\boldsymbol{\zeta}(t)$ is a function of the RVs $\mathbf{x}(t)$ and $\mathbf{x}'(t)$. Denoting by $f(x, x')$ the joint density of $\mathbf{x}(t)$ and $\mathbf{x}'(t)$, we conclude from (11-114) and (7-2) that

$$E\{\boldsymbol{\zeta}(t)\} = \int_{-\infty}^{\infty} \int_{-\infty}^{\infty} |x'|\,\delta(x - a) f(x, x')\, dx\, dx' = \int_{-\infty}^{\infty} |x'|\, f(a, x')\, dx'$$

This yields (11-112) because $f(a, x') = f_x(a) f(x' \mid a)$.

Level-crossing density We denote by $p_a(\tau)$ the probability that in an interval I_τ of length τ there is one and only one crossing. If $\tau \to 0$ then $p_a(\tau) \to 0$. Furthermore, with the exception of unusual cases, the probability that there are more than one crossing in a small interval τ is small compared to $p_a(\tau)$. Assuming that this is the case for the process $\mathbf{x}(t)$, we introduce the limit

$$\lambda_a = \lim_{\tau \to 0} \frac{1}{\tau}\, p_a(\tau)$$

If this limit exists, it is the density of the L_a crossings. Thus $p_a(\Delta\tau) \simeq \lambda_a \Delta\tau$ for small $\Delta\tau$.

We maintain that†

$$\lambda_a = \frac{1}{T}\, E\{\mathbf{n}_a(T)\} \tag{11-116}$$

† S. O. Rice: Mathematical Analysis of Random Noise, in *Selected Papers on Noise and Stochastic Processes*, Dover, N.Y., 1954.

PROOF If $\Delta\tau$ is small, then

$$P\{\mathbf{n}_a(\Delta\tau) = 1\} = p_a(\Delta\tau) \qquad P\{\mathbf{n}_a(\Delta\tau) = 0\} = 1 - p_a(\Delta\tau)$$

Hence

$$E\{\mathbf{n}_a(\Delta\tau)\} = 1 \cdot P\{\mathbf{n}_a(\Delta\tau) = 1\} = p_a(\Delta\tau) \simeq \lambda_a \Delta\tau$$

From this it follows that (11-116) is true for T small. And since

$$\mathbf{n}_a(T_1 + T_2) = \mathbf{n}_a(T_1) + \mathbf{n}_a(T_2)$$

we conclude that it is true for any T.

We have thus shown that if $\mathbf{x}(t)$ is differentiable, then the level-crossing density λ_a exists and it equals

$$\lambda_a = f_x(a)E\{|\mathbf{x}'(t)|\,|\mathbf{x}(t) = a\} \tag{11-117}$$

Density of maxima The maxima and minima (extrema) of a process $\mathbf{x}(t)$ are the zero crossings of the process $\mathbf{y}(t) = \mathbf{x}'(t)$. From this it follows that the density λ_m of the extrema of $\mathbf{x}(t)$ is obtained from (11-117) if we set $a = 0$ and replace $\mathbf{x}(t)$ by $\mathbf{x}'(t)$. This yields

$$\lambda_m = f_{x'}(0)E\{|\mathbf{x}''(t)|\,|\mathbf{x}'(t) = 0\} \tag{11-118}$$

The density of maxima equals $\lambda_m/2$.

Normal Processes

We shall apply the preceding results to normal processes under the assumption that their mean η_x is zero. This assumption is not essential because the a level crossings of the process $\mathbf{x}(t)$ are identical to the $(a - \eta_x)$-level crossings of the centered process $\mathbf{x}(t) - \eta_x$.

Theorem If $\mathbf{x}(t)$ is a differentiable process with autocorrelation $R(\tau)$, then

$$\lambda_a = f_x(a)E\{|\mathbf{x}'(t)|\} = \frac{1}{\pi}\sqrt{\frac{-R''(0)}{R(0)}}\,e^{-a^2/2R(0)} \tag{11-119}$$

PROOF As we know

$$R_{xx'}(\tau) = -R'(\tau) \qquad R_{x'x'}(\tau) = -R''(\tau)$$

From the existence of $\mathbf{x}'(t)$ it follows that $R''(\tau)$ exists. Therefore, $R'(\tau)$ also exists and $R'(0) = 0$ because $R(\tau)$ is even. Hence

$$E\{\mathbf{x}(t)\mathbf{x}'(t)\} = -R'(0) = 0$$

This shows that the RVs $\mathbf{x}(t)$ and $\mathbf{x}'(t)$ are orthogonal. And since they are normal with zero mean, they are independent. The first equality in (11-119) follows, therefore, from (11-117).

To prove the second equality, we observe that the variance of $\mathbf{x}(t)$ equals $R(0)$ and the variance of $\mathbf{x}'(t)$ equals $-R''(0)$. This yields [see (5-45)]

$$f_x(a) = \frac{1}{\sqrt{2\pi R(0)}} e^{-a^2/2R(0)} \qquad E\{|\mathbf{x}'(t)|\} = \sqrt{\frac{-2R''(0)}{\pi}}$$

and (11-119) results.

Zero-crossing density Denoting by λ_0 the density of the zeros of $\mathbf{x}(t)$, we conclude from (11-119) with $a = 0$ that

$$\lambda_0^2 = \frac{-R''(0)}{\pi^2 R(0)} = \frac{\int_{-\infty}^{\infty} \omega^2 S(\omega)\, d\omega}{\pi^2 \int_{-\infty}^{\infty} S(\omega)\, d\omega} \qquad (11\text{-}120)$$

Example 11-4 If $S(\omega) = 0$ for $|\omega| > \sigma$, then (11-120) yields $\lambda_0 \le \sigma/\pi$ because

$$\int_{-\sigma}^{\sigma} \omega^2 S(\omega)\, d\omega \le \sigma^2 \int_{-\sigma}^{\sigma} S(\omega)\, d\omega$$

If also $S(\omega) = S_0$ for $|\omega| < \sigma$ (ideal low-pass), then

$$\lambda_0 = \frac{\sigma}{\pi\sqrt{3}}$$

Example 11-5(*a*) If $\mathbf{a}$ and $\mathbf{b}$ are two independent RVs with variance σ^2 and

$$\mathbf{x}(t) = \mathbf{a} \cos \omega_0 t + \mathbf{b} \sin \omega_0 t$$

as in Example 9-11, then

$$R_x(\tau) = \sigma^2 \cos \omega_0 \tau \qquad R_x(0) = \sigma^2 \qquad R_x''(0) = -\omega_0^2 \sigma^2$$

and (11-120) yields

$$\lambda_0 = \frac{\omega_0}{\pi}$$

(*b*) If the RVs $\mathbf{a}$ and $\mathbf{b}$ are independent of the process $\mathbf{v}(t)$ and

$$\mathbf{x}(t) = \mathbf{a} \cos \omega_0 t + \mathbf{b} \sin \omega_0 t + \mathbf{v}(t)$$

then

$$R_x(\tau) = \sigma^2 \cos \omega_0 \tau + R_v(\tau)$$

and (11-120) yields

$$\lambda_0 = \frac{1}{\pi} \sqrt{\frac{\omega_0^2 \sigma^2 - R_v''(0)}{\sigma^2 + R_v(0)}}$$

Nondifferentiable processes We have shown that if $\mathbf{x}'(t)$ exists, then the probability $p_0(\tau)$ that there is a zero crossing in a small interval τ equals $\lambda_0 \tau$. If $\mathbf{x}'(t)$ does not exist, then $p_0(\tau)$ is no longer proportional to τ. In the following, we examine the asymptotic form of $p_0(\tau)$ as $\tau \to 0$ under the assumption that $R(\tau)$ has a corner at the origin as in the Ornstein–Uhlenbeck process [see (11-103)].

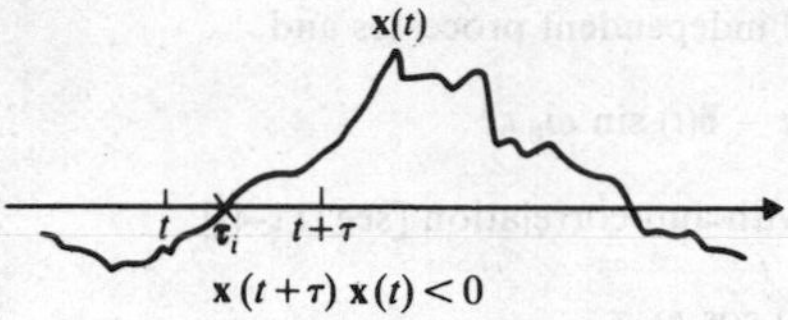

Figure 11-16

Suppose that $R'(\tau)$ is discontinuous at $\tau = 0$ but $R'(0^+)$ exists. In this case

$$R(\tau) = R(0) + R'(0^+)\tau + O(\tau^2) \qquad \tau > 0 \tag{11-121}$$

We shall show that $p_0(\tau)$ is proportional to $\sqrt{\tau}$

$$p_0(\tau) \simeq \frac{1}{\pi}\sqrt{-\frac{2R'(0^+)\tau}{R(0)}} \tag{11-122}$$

PROOF If τ is small, then we can neglect more than one zero crossing in the interval $(t, t + \tau)$. With this assumption, we have one crossing iff $\mathbf{x}(t + \tau)\mathbf{x}(t) < 0$ (Fig. 11-16). Hence

$$p_0(\tau) = E\{\mathbf{x}(t + \tau)\mathbf{x}(t) < 0\}$$

The RVs $\mathbf{x}(t + \tau)$ and $\mathbf{x}(t)$ are jointly normal with correlation coefficient $r(\tau) = R(\tau)/R(0)$. Applying (6-46), we conclude that

$$p_0(\tau) \simeq \frac{\beta}{\pi} \qquad \cos\beta = r(\tau)$$

This yields

$$r(\tau) = \cos\beta = \cos\pi p_0(\tau) \simeq 1 - \frac{\pi^2 p_0^2(\tau)}{2}$$

for $p_0(\tau) \ll 1$. Thus, for small τ

$$\pi p_0(\tau) \simeq \sqrt{2[1 - r(\tau)]} \qquad r(\tau) = \frac{R(\tau)}{R(0)} \tag{11-123}$$

Inserting (11-121) into the above, we obtain (11-122).

Note Using (11-123), we can reestablish (11-120). Indeed, in this case, $R'(0) = 0$ and $R''(0)$ exists. Hence

$$R(\tau) = R(0) + \tfrac{1}{2}R''(0)\tau^2 + O(\tau^3)$$

Inserting into (11-123) we obtain

$$p_0(\tau) \simeq \frac{\tau}{\pi}\sqrt{\frac{-R''(0)}{R(0)}} \tag{11-124}$$

Example 11-6 If $\mathbf{a}(t)$ and $\mathbf{b}(t)$ are two normal independent processes and

$$\mathbf{x}(t) = \mathbf{a}(t) \cos \omega_0 t - \mathbf{b}(t) \sin \omega_0 t$$

as in (11-1), then $\mathbf{x}(t)$ is normal, stationary, with autocorrelation [see (11-4)]

$$R_x(\tau) = R_a(\tau) \cos \omega_0 \tau$$

(*a*) We shall show that, *if* $\mathbf{x}(t)$ differentiable, then its zero crossing density equals

$$\lambda_0 = \sqrt{\bar{\lambda}_0^2 + \frac{\omega_0^2}{\pi^2}} \quad \text{where} \quad \bar{\lambda}_0 = \frac{1}{\pi}\sqrt{\frac{-R_a''(0)}{R(0)}} \tag{11-125}$$

is the zero-crossing density of $\mathbf{a}(t)$. Indeed, in this case, $R_a'(0) = 0$, hence

$$R_x(0) = R_a(0) \qquad R_x'(0) = 0 \qquad R_x''(0) = R_a''(0) - \omega_0^2 R_a(0)$$

and (11-125) follows from (11-120).

(*b*) In the nondifferentiable case, we have

$$R_x(0) = R_a(0) \qquad R_x'(0^+) = R_a'(0^+) \neq 0$$

This shows that the zero-crossing probabilities $p_0(\tau)$ and $\bar{p}_0(\tau)$ of the processes $\mathbf{x}(t)$ and $\mathbf{a}(t)$ are equal [see (11-122)].

Example 11-7 In the preceding discussion we assumed that all processes are stationary. The results can be readily extended to nonstationary processes. We illustrate the nondifferentiable case using as example the Wiener process $\mathbf{w}(t)$.

We shall show that for small τ, the probability $p_0(t; t + \tau)$ that $\mathbf{w}(t)$ crosses the t axis in the interval $(t, t + \tau)$ equals

$$p_0(t, t + \tau) \simeq \frac{1}{\pi}\sqrt{\frac{\tau}{t}} \tag{11-126}$$

Proof Reasoning as in the stationary case we conclude that $p_0(t, t + \tau)$ is given by (11-123) where $r(\tau)$ is the correlation coefficient of the RVs $\mathbf{x}(t + \tau)$ and $\mathbf{x}(t)$. Thus [see (11-93)]

$$r^2(\tau) = \frac{R^2(t + \tau, t)}{R(t + \tau, t + \tau)R(t, t)} = \frac{\alpha^2 t^2}{\alpha(t + \tau)\alpha t} = \frac{t}{t + \tau}$$

Inserting into (11-123) we obtain (11-126) because $\sqrt{t/(t + \tau)} \simeq 1 - \tau/2t$ for small τ.

Density of maxima The extrema of $\mathbf{x}(t)$ are the zeros of $\mathbf{x}'(t)$. Hence, their density λ_m is given by (11-120) if we replace the autocorrelation $R(\tau)$ of $\mathbf{x}(t)$ by the autocorrelation $-R''(\tau)$ of $\mathbf{x}'(t)$. From this it follows that

$$\lambda_m^2 = \frac{-R^{(4)}(0)}{\pi^2 R''(0)} = \frac{\int_{-\infty}^{\infty} \omega^4 S(\omega)\, d\omega}{\pi^2 \int_{-\infty}^{\infty} \omega^2 S(\omega)\, d\omega} \tag{11-127}$$

provided that $\mathbf{x}'(t)$ exists.

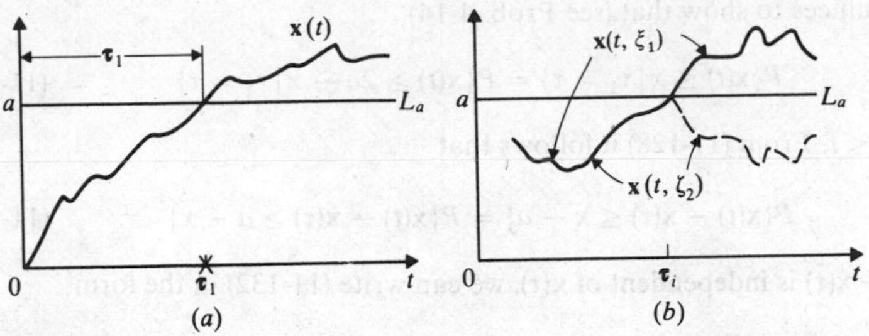

Figure 11-17

Example 11-8(*a*) If $S(\omega) = 0$ for $|\omega| > \sigma$, then the above yields $\lambda_m \le \sigma/\pi$ because

$$\int_{-\sigma}^{\sigma} \omega^4 S(\omega)\, d\omega \le \sigma^2 \int_{-\sigma}^{\sigma} \omega^2 S(\omega)\, d\omega$$

If also $S(\omega) = S_0$ for $|\omega| \le \sigma$, then

$$\lambda_m = \frac{\sigma}{\pi}\sqrt{\frac{3}{5}}$$

(*b*) If $S(\omega) = S_0$ for $\omega_1 < |\omega| < \omega_2$ and zero otherwise (ideal bandpass), then

$$\lambda_m = \frac{1}{\pi}\sqrt{\frac{3(\omega_2^5 - \omega_1^5)}{5(\omega_2^3 - \omega_1^3)}}$$

First passage time We denote by $\boldsymbol{\tau}_1$ the first a-level crossing to the right of the origin (Fig. 11-17*a*). The first passage problem is the determination of the distribution function $F_\tau(\tau, a)$ of the RV $\boldsymbol{\tau}_1$. We shall solve the problem under the assumption that $\mathbf{x}(t)$ is the Wiener process (11-107). We should note, however, that in the solution we make use only of the fact that the increment $\mathbf{x}(t_2) - \mathbf{x}(t_1)$ is *independent* of $\mathbf{x}(t_1)$ and its density is *even*

$$P\{\mathbf{x}(t_2) - \mathbf{x}(t_1) \le w\} = P\{\mathbf{x}(t_2) - \mathbf{x}(t_1) \ge -w\} \tag{11-128}$$

The reflection principle We shall show that the samples $\mathbf{x}(t, \zeta)$ of $\mathbf{x}(t)$ that cross the line L_a continue on symmetrical paths. In other words, if a sample $\mathbf{x}(t, \zeta_1)$ crosses the line L_a at $t = \boldsymbol{\tau}_i$, then there exists another sample $\mathbf{x}(t, \zeta_2)$ that coincides with $\mathbf{x}(t, \zeta_1)$ for $t \le \boldsymbol{\tau}_i$ and for $t > \boldsymbol{\tau}_i$ is the reflection of $\mathbf{x}(t, \zeta_1)$ on the line L_a (Fig. 11-17*b*)

$$\begin{aligned} \mathbf{x}(t, \zeta_1) &= \mathbf{x}(t, \zeta_2) \qquad && t \le \boldsymbol{\tau}_i \\ a - \mathbf{x}(t, \zeta_1) &= \mathbf{x}(t, \zeta_2) - a \qquad && t > \boldsymbol{\tau}_i \end{aligned} \tag{11-129}$$

This result, known as the *reflection principle*, can be stated as follows:

Theorem For any x (less than or greater than a)

$$P\{\mathbf{x}(t) \le x \mid \boldsymbol{\tau}_1 \le t\} = P\{\mathbf{x}(t) \ge 2a - x \mid \boldsymbol{\tau}_1 \le t\} \tag{11-130}$$

PROOF It suffices to show that (see Prob. 4-14)

$$P\{\mathbf{x}(t) \le x \,|\, \boldsymbol{\tau}_1 = \tau\} = P\{\mathbf{x}(t) \ge 2a - x \,|\, \boldsymbol{\tau}_1 = \tau\} \tag{11-131}$$

for every $\tau \le t$. From (11-128) it follows that

$$P\{\mathbf{x}(t) - \mathbf{x}(\tau) \le x - a\} = P\{\mathbf{x}(t) - \mathbf{x}(\tau) \ge a - x\} \tag{11-132}$$

Since $\mathbf{x}(t) - \mathbf{x}(\tau)$ is independent of $\mathbf{x}(\tau)$, we can write (11-132) in the form

$$P\{\mathbf{x}(t) - \mathbf{x}(\tau) \le x - a \,|\, \mathbf{x}(\tau) = a\} = P\{\mathbf{x}(t) - \mathbf{x}(\tau) \ge a - x \,|\, \mathbf{x}(\tau) = a\}$$

This is true for any t and τ, hence it is true for $\tau = \boldsymbol{\tau}_1$. In this case $\{\mathbf{x}(\tau) = a\} = \{\boldsymbol{\tau}_1 = \tau\}$ and (11-131) results because $\mathbf{x}(\boldsymbol{\tau}_1) = a$.

Corollary If $x \le a$, then

$$P\{\mathbf{x}(t) \le x, \boldsymbol{\tau}_1 \le t\} = 1 - F_x(2a - x, t) \tag{11-133}$$

$$P\{\mathbf{x}(t) \le x, \boldsymbol{\tau}_1 > t\} = F_x(x, t) + F_x(2a - x, t) - 1 \tag{11-134}$$

where

$$F_x(x, t) = \mathbb{G}\left(\frac{x}{\sqrt{2\alpha t}}\right)$$

is the first-order distribution of the Wiener process $\mathbf{x}(t)$.

PROOF Multiplying both sides of (11-130) by $P\{\boldsymbol{\tau}_1 \le t\}$ we obtain

$$P\{\mathbf{x}(t) \le x, \boldsymbol{\tau}_1 \le t\} = P\{\mathbf{x}(t) \ge 2a - x, \boldsymbol{\tau}_1 \le t\} \tag{11-135}$$

for every x and t. If $x \le a$, then $\mathbf{x}(t) > 2a - x$ iff $\boldsymbol{\tau}_1 \le t$, hence, the right side of (11-135) equals $P\{\mathbf{x}(t) \ge 2a - x\}$ and (11-133) results. The second equation follows because the sum of the left sides of (11-133) and (11-134) equals $P\{\mathbf{x}(t) \le x\}$.

First passage distribution The distribution $F_\tau(t, a)$ of the RV $\boldsymbol{\tau}_1$ equals

$$P\{\boldsymbol{\tau}_1 \le t\} = P\{\boldsymbol{\tau}_1 \le t, \mathbf{x}(t) \le a\} + P\{\boldsymbol{\tau}_1 \le t, \mathbf{x}(t) > a\} \tag{11-136}$$

Clearly, if $\mathbf{x}(t) > a$, then there must be a crossing prior to time t, hence, $\boldsymbol{\tau}_1 \le t$. From this it follows that

$$P\{\boldsymbol{\tau}_1 \le t, \mathbf{x}(t) > a\} = P\{\mathbf{x}(t) > a\}$$

Setting $x = a$ in (11-133), we obtain

$$P\{\boldsymbol{\tau}_1 \le t, \mathbf{x}(t) \le a\} = 1 - F_x(a, t) = P\{\mathbf{x}(t) > a\}$$

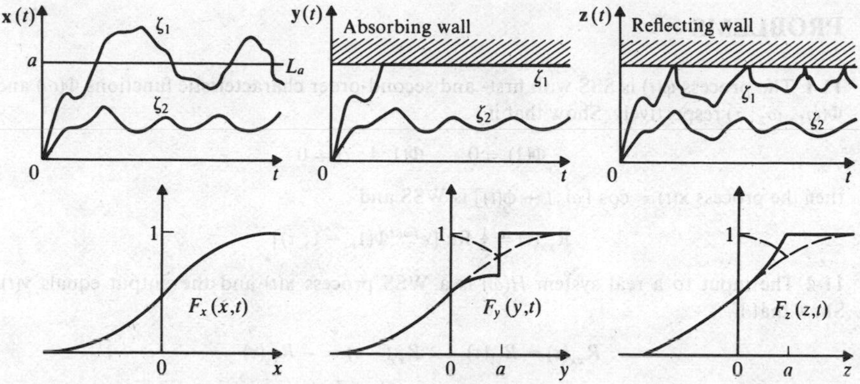

Figure 11-18

Thus, the two terms on the right of (11-136) equal $P\{\mathbf{x}(t) > a\}$. Therefore

$$P\{\boldsymbol{\tau}_1 < t\} = F_\tau(t, a) = 2P\{\mathbf{x}(t) > a\} = 2 - 2F_x(a, t) \qquad (11\text{-}137)$$

Absorbing wall We replace the line L_a by an absorbing barrier. This means that the resulting process $\mathbf{y}(t)$ equals a for every $t > \boldsymbol{\tau}_1$ (Fig. 11-18). We shall show that the distribution function of $\mathbf{y}(t)$ equals

$$F_y(y, t) = F_x(y, t) + F_x(2a - y, t) - 1 \qquad y < a \qquad (11\text{-}138)$$

and, of course, $F_y(y, t) = 1$ for $y \geq a$.

PROOF If $y < a$, then $\mathbf{y}(t, \zeta_i) \leq y$ for some outcome ζ_i iff $\mathbf{x}(t, \zeta_i) \leq y$ *and* $\mathbf{x}(t, \zeta_i)$ does not reach the line L_a prior to time t. Hence

$$\{\mathbf{y}(t) \leq y\} = \{\mathbf{x}(t) \leq y, \boldsymbol{\tau}_1 > t\}$$

and (11-138) follows from (11-134).

Reflecting wall We replace now the line L_a by a reflecting barrier. This means that the resulting process $\mathbf{z}(t)$ equals $\mathbf{x}(t)$ if $\mathbf{x}(t) < a$ and it equals $2a - \mathbf{x}(t)$ if $\mathbf{x}(t) > a$ (Fig. 11-18). We shall show that in this case

$$F_z(z, t) = F_x(z, t) + 1 - F_x(2a - z, t) \qquad z < a \qquad (11\text{-}139)$$

and $F_z(z, t) = 1$ for $z > a$.

PROOF If $z < a$, then $\mathbf{z}(t, \zeta_i) \leq z$ iff either $\mathbf{x}(t, \zeta_i) \leq a$ or $\mathbf{x}(t, \zeta_i) > 2a - z$. Hence

$$\{\mathbf{z}(t) \leq z\} = \{\mathbf{x}(t) \leq z\} + \{\mathbf{x}(t) > 2a - z\}$$

and (11-139) results.

PROBLEMS

11-1 The process $\boldsymbol{\varphi}(t)$ is SSS with first- and second-order characteristic functions $\Phi(\omega)$ and $\Phi(\omega_1, \omega_2; \tau)$ respectively. Show that if

$$\Phi(1) = 0 \qquad \Phi(1, 1; \tau) = 0$$

then the process $\mathbf{x}(t) = \cos[\omega_0 t + \boldsymbol{\phi}(t)]$ is WSS and

$$R_{xx}(\tau) = \tfrac{1}{2}\,\mathrm{Re}\,[e^{j\omega_0\tau}\Phi(1, -1; \tau)]$$

11-2 The input to a real system $H(\omega)$ is a WSS process $\mathbf{x}(t)$ and the output equals $\mathbf{y}(t)$. Show that if

$$R_{xx}(\tau) = R_{yy}(\tau) \qquad R_{xy}(-\tau) = -R_{xy}(\tau)$$

as in (11-6), then $H(\omega) = jB(\omega)$ where $B(\omega)$ is a function taking only the values $+1$ and -1.

Special case: If $\mathbf{y}(t) = \hat{\mathbf{x}}(t)$, then $B(\omega) = -\operatorname{sgn} \omega$.

11-3 Show that if $\check{\mathbf{x}}(t)$ is the Hilbert transform of $\mathbf{x}(t)$ and

$$\mathbf{i}(t) = \mathbf{x}(t)\cos\omega_0 t + \check{\mathbf{x}}(t)\sin\omega_0 t \qquad \mathbf{q}(t) = \check{\mathbf{x}}(t)\cos\omega_0 t - \mathbf{x}(t)\sin\omega_0 t$$

then (Fig. P11-3)

$$S_i(\omega) = S_q(\omega) = \frac{S_w(\omega) + S_w(-\omega)}{2} \qquad S_q(\omega) = \frac{S_w(\omega) - S_w(-\omega)}{2j}$$

where $S_w(\omega) = 4S_x(\omega + \omega_0)U(\omega + \omega_0)$.

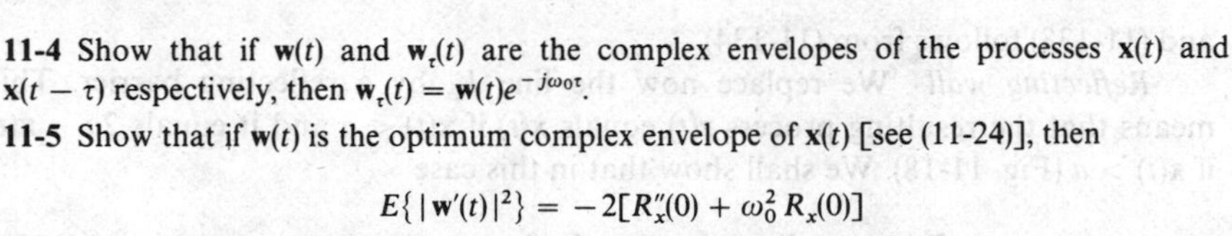

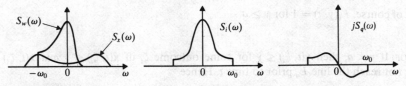

Figure P11-3

11-4 Show that if $\mathbf{w}(t)$ and $\mathbf{w}_\tau(t)$ are the complex envelopes of the processes $\mathbf{x}(t)$ and $\mathbf{x}(t - \tau)$ respectively, then $\mathbf{w}_\tau(t) = \mathbf{w}(t)e^{-j\omega_0\tau}$.

11-5 Show that if $\mathbf{w}(t)$ is the optimum complex envelope of $\mathbf{x}(t)$ [see (11-24)], then

$$E\{|\mathbf{w}'(t)|^2\} = -2[R''_x(0) + \omega_0^2 R_x(0)]$$

11-6 Show that if $\boldsymbol{\theta}$ is an RV uniform in the interval $(0, T)$ and $f(t)$ is a periodic function with period T, then the process $\mathbf{x}(t) = f(t - \boldsymbol{\theta})$ is stationary and

$$S_x(\omega) = \frac{1}{T}\left|\int_0^T f(t)e^{-j\omega t}\,dt\right|^2$$

11-7 In the DPSK process of Fig. 11-8, the RVs $\mathbf{c}_n$ take the values 0 and ω_c/T with $P\{\mathbf{c}_n = 0\} = 0.25$, $P\{\mathbf{c}_n = \omega_c/T\} = 0.75$. Find the spectrum $S_w(\omega)$ of the process $\mathbf{w}(t) = e^{j\boldsymbol{\varphi}(t)}$

11-8 Show that if

$$\varepsilon_N(t) = \mathbf{x}(t) - \sum_{n=-N}^{N} \mathbf{x}(nT)\frac{\sin\sigma(t-nT)}{\sigma(t-nT)} \qquad \sigma = \frac{\pi}{T}$$

then

$$E\{\varepsilon_N^2(t)\} = \frac{1}{2\pi}\int_{-\infty}^{\infty} S(\omega)\left| e^{j\omega t} - \sum_{n=-N}^{N} \frac{\sin\sigma(t-nT)}{\sigma(t-nT)} e^{jn\omega T}\right|^2 d\omega$$

and if $S(\omega) = 0$ for $|\omega| > \sigma$, then $E\{\varepsilon_N^2(t)\} \to 0$ as $N \to \infty$.

11-9 Show that if $\mathbf{x}(t)$ is BL as in (11-56), then† for $|\tau| < \pi/\sigma$

$$\frac{2\tau^2}{\pi^2}|R''(0)| \le R(0) - R(\tau) \le \frac{\tau^2}{2} R''(0)$$

$$E\{[\mathbf{x}(t+\tau) - \mathbf{x}(t)]^2 \ge \frac{4\tau^2}{\pi^2} E\{[\mathbf{x}'(t)]^2\}$$

Hint: If $0 < \varphi < \pi/2$ then $2\varphi/\pi < \sin\varphi < \varphi$.

11-10 A WSS process $\mathbf{x}(t)$ is BL as in (11-56) and its samples $\mathbf{x}(n\pi/\sigma)$ are uncorrelated. Find $S_x(\omega)$ if $E\{\mathbf{x}(t)\} = \eta$ and $E\{\mathbf{x}^2(t)\} = I$.

11-11 Find the power spectrum $S(\omega)$ of a process $\mathbf{x}(t)$ if $S(\omega) = 0$ for $|\omega| > \pi$ and

$$E\{\mathbf{x}(n+m)\mathbf{x}(n)\} = N\delta[n-m]$$

11-12 Show that if $S(\omega) = 0$ for $|\omega| > \sigma$, then

$$R(\tau) \ge R(0)\cos\sigma\tau \qquad \text{for} \quad |\tau| < \pi/2\sigma.$$

11-13 Show that if $\mathbf{x}(t)$ is BL as in (11-56) and $\Delta = 2\pi/\sigma$, then

$$\mathbf{x}(t) = 4\sin^2\frac{\sigma t}{2}\sum_{n=-\infty}^{\infty}\left[\frac{\mathbf{x}(n\Delta)}{(\sigma t - 2n\pi)^2} + \frac{\mathbf{x}'(n\Delta)}{\sigma(\sigma t - 2n\pi)}\right]$$

Hint: Use (11-72) with $N = 2$, $H_1(\omega) = 1$, $H_2(\omega) = j\omega$.

11-14 Find the mean and the variance of $\mathbf{P}(\omega_0)$ if $\mathbf{t}_i$ is a Poisson point process and

$$\mathbf{P}(\omega) = \frac{1}{\lambda}\sum_i \cos\omega_0\mathbf{t}_i\cos\omega\mathbf{t}_i \qquad |\mathbf{t}_i| < a$$

11-15 Show that if $\mathbf{x}(t)$ is a normal process with zero mean and $\mathbf{y}(t) = Ie^{a\mathbf{x}(t)}$, then

$$\eta_y = I\exp\left\{\frac{a^2}{2}R_x(0)\right\} \qquad R_y(\tau) = I^2\exp\{a^2[R_x(0) + R_x(\tau)]\}$$

11-16 The process $\mathbf{x}(t)$ is normal with zero mean. Show that if $\mathbf{y}(t) = \mathbf{x}^2(t)$, then

$$S_y(\omega) = 2\pi R_x^2(0)\delta(\omega) + 2S_x(\omega) * S_x(\omega)$$

Plot $S_y(\omega)$ if $S_x(\omega)$ is (*a*) ideal LP, (*b*) ideal BP.

† A. Papoulis: "An Estimate of the Variation of a Bandlimited Process," *IEEE, PGIT*, 1964.

11-17 Show that if the RVs **x**, **y** are $N(0, 0; \sigma_1, \sigma_2; r)$ then

$$E\{|\mathbf{xy}|\} = \frac{2}{\pi}\int_0^C \arcsin\frac{\mu}{\sigma_1\sigma_2}\,d\mu + \frac{2\sigma_1\sigma_2}{\pi} = \frac{2\sigma_1\sigma_2}{\pi}(\cos\alpha + \alpha\sin\alpha)$$

where $r = \sin\alpha$ and $C = r\sigma_1\sigma_2$.

Hint: Use (11-94) with $g(x, y) = |xy|$.

11-18 A particle in free motion satisfies the equation

$$m\mathbf{x}''(t) + f\mathbf{x}'(t) = \mathbf{F}(t) \qquad S_F(\omega) = 2kTf$$

Show that if $\mathbf{x}(0) = \mathbf{x}'(0) = 0$, then

$$E\{\mathbf{x}^2(t)\} = 2D^2\left(t - \frac{3}{4\alpha} + \frac{1}{\alpha}e^{-2\alpha t} - \frac{1}{4\alpha}e^{-4\alpha t}\right)$$

where $D^2 = kT/f$ and $\alpha = f/2m$.

Hint: Use (9-117) with

$$h(t) = \frac{1}{f}(1 - e^{-2\alpha t})U(t) \qquad q(t) = 2kTf\,U(t)$$

11-19 Given a Wiener process with parameter α, we form the processes

$$\mathbf{x}(t) = \mathbf{w}(t^2) \qquad \mathbf{y}(t) = \mathbf{w}^2(t) \qquad \mathbf{z}(t) = |\mathbf{w}(t)|$$

Show that $\mathbf{x}(t)$ is normal with zero mean. Furthermore, if $t_1 < t_2$, then

$$R_x(t_1, t_2) = \alpha t_1^2 \qquad R_y(t_1, t_2) = \alpha^2 t_1(2t_1 + t_2)$$

$$R_z(t_1, t_2) = \frac{2\alpha}{\pi}\sqrt{t_1 t_2}\,(\cos\theta + \theta\sin\theta) \qquad \sin\theta = \sqrt{\frac{t_1}{t_2}}.$$

11-20 Show that the probability that the Wiener process $\mathbf{w}(t)$ does not cross the line L_a in the interval $(0, t)$ equals $2\mathbb{G}(a/\sqrt{2\alpha t}) - 1$.

11-21 We denote by $P_0(\tau)$ the conditional probability that the number of zeros of a normal process $\mathbf{x}(t)$ in the interval $(t, t + \tau)$, assuming $\mathbf{x}(t) = 0$, is odd. Show that if $E\{\mathbf{x}(t)\} = 0$, then

$$\cos\pi P_0(\tau) = -R'(\tau)\left[-R''(0)R(0) + \frac{R''(0)R^2(\tau)}{R(0)}\right]^{-1/2}$$

11-22 Show that if $f_w(w, t)$ is the first-order density of the Wiener process $\mathbf{w}(t)$ and $f_\tau(\tau, a)$ is the density of the first passage time τ_1 (Fig. 11-17*a*), then†

$$f_\tau(\tau, a) = \frac{a}{\tau}f_w(a, \tau)$$

† A. A. Borovkov: "On the First Passage Time . . .," *Theory of Probability and its Applications*, vol. X, No. 2, 1965.

PART THREE

SELECTED TOPICS

CHAPTER

TWELVE

QUEUEING THEORY, SHOT NOISE, MARKOFF PROCESSES

12-1 POISSON POINTS AND RENEWALS

The term Poisson points is used to describe a point process $\mathbf{t}_i$ specified in terms of the following two properties (see Sec. 3-4):

P_1: The number of points $\mathbf{n}(t_1, t_2)$ in an interval (t_1, t_2) is a Poisson RV with parameter

$$\int_{t_1}^{t_2} \lambda(t)\, dt$$

P_2: If $t_1 < t_2 \le t_3 < t_4$ then the RVs $\mathbf{n}(t_1, t_2)$ and $\mathbf{n}(t_3, t_4)$ are independent.

In this section we consider various properties of these points assuming that the point process is stationary, i.e., that its *mean density* $\lambda(t)$ is constant. The nonstationary case follows readily with a nonlinear transformation of the time axis.

In the stationary case, the statistics of the RV

$$\mathbf{n}_T = \mathbf{n}(t, t + T)$$

depend only on the length T of the interval $(t, t + T)$. Thus, $\mathbf{n}_T$ is a Poisson RV with parameter λT, and its moment function equals [see (5-79)]

$$\mathbf{\Gamma}_{nT}(z) = E\{z^{\mathbf{n}_T}\} = e^{(z-1)\lambda T} \tag{12-1}$$

Ergodicity As we know

$$E\{\mathbf{n}_T\} = \lambda T \qquad \sigma^2_{nT} = \lambda T \tag{12-2}$$

This shows that λ equals the ensemble average of the number of points in a unit interval. We maintain that λ can also be interpreted as time-average. For this purpose, we form the time-average $\boldsymbol{\eta}_T = \mathbf{n}_T/T$. As we see from (12-2), $\boldsymbol{\eta}_T$ is an RV such that

$$E\{\boldsymbol{\eta}_T\} = \lambda \qquad \sigma^2_{\eta T} = \frac{\lambda}{T} \tag{12-3}$$

Hence, $\sigma^2_{\eta T} \to 0$ as $T \to \infty$. From this it follows that

$$\boldsymbol{\eta}_T = \frac{\mathbf{n}_T}{T} \xrightarrow[T\to\infty]{} \lambda \tag{12-4}$$

and $\mathbf{n}_T \simeq \lambda T$ for sufficiently large T.

Poisson Points in Random Intervals

Suppose now that $\mathbf{c}$ is a positive RV and

$$\mathbf{n}_c = \mathbf{n}(t, t + \mathbf{c}) \tag{12-5}$$

is the number of points in an interval of length $\mathbf{c}$. We shall determine its statistics.

If $\mathbf{c} = c$ is a constant, then $\mathbf{n}_c$ is a Poisson RV with parameter λc. Hence

$$E\{\mathbf{n}_c \mid \mathbf{c} = c\} = \lambda c$$

From this and (7-59) it follows that

$$E\{\mathbf{n}_c\} = E\{E\{\mathbf{n}_c \mid \mathbf{c}\}\} = E\{\lambda \mathbf{c}\} = \lambda \eta_c \tag{12-6}$$

Denoting by ρ_c the average number of points in the random interval $\mathbf{c}$, we conclude that $\rho_c = \lambda \eta_c$.

Reasoning similarly, we can find all moments of $\mathbf{n}_c$. In the following, we determine directly the moment function

$$\boldsymbol{\Gamma}_{n_c}(z) = E\{z^{\mathbf{n}_c}\} \tag{12-7}$$

of the discrete-type RV $\mathbf{n}_c$ in terms of the moment function

$$\boldsymbol{\Phi}_c(s) = E\{e^{s\mathbf{c}}\} \tag{12-8}$$

of the continuous-type RV $\mathbf{c}$.

Theorem

$$\boldsymbol{\Gamma}_{n_c}(z) = \boldsymbol{\Phi}_c(\lambda z - \lambda) \tag{12-9}$$

PROOF If $\mathbf{c} = c$, then $\mathbf{n}_c$ is a Poisson RV with parameter λc. Hence [see (12-1)]

$$E\{z^{\mathbf{n}_c} \mid \mathbf{c} = c\} = e^{(z-1)\lambda c}$$

This yields

$$E\{z^{\mathbf{n}_c}\} = E\{E\{z^{\mathbf{n}_c} \mid \mathbf{c}\}\} = E\{e^{(z-1)\lambda\mathbf{c}}\}$$

and (12-9) follows from (12-8) with $s = (z - 1)\lambda$.

Using (12-9) and the moment theorems (5-67) and (5-77), we can express the moments of $\mathbf{n}_c$ in terms of the moments of $\mathbf{c}$. We note in particular that

$$E\{\mathbf{n}_c\} = \Gamma'_{n_c}(1) = \lambda\Phi'_c(0) = \lambda E\{\mathbf{c}\}$$

in agreement with (12-6). Similarly

$$E\{\mathbf{n}_c(\mathbf{n}_c - 1)\} = \Gamma''_{n_c}(1) = \lambda^2\Phi''_c(0) = \lambda^2 E\{\mathbf{c}^2\} \qquad (12\text{-}10)$$

Hence

$$E\{\mathbf{n}_c^2\} = \lambda^2 E\{\mathbf{c}^2\} + \rho_c \qquad \sigma_{n_c}^2 = \lambda^2\sigma_c^2 + \rho_c \qquad (12\text{-}11)$$

In the above, $\rho_c = \lambda\eta_c$ is the average number of points in the random interval $\mathbf{c}$, and σ_c^2 is the variance of $\mathbf{c}$.

Example 12-1 If†

$$f_c(c) = \mu e^{-\mu c} \qquad (12\text{-}12)$$

then $E\{\mathbf{c}\} = 1/\mu$, $E\{\mathbf{c}^2\} = 2/\mu^2$, and

$$\Phi_c(s) = \frac{\mu}{\mu - s}$$

Hence

$$E\{\mathbf{n}_c\} = \frac{\lambda}{\mu} = \rho_c \qquad E\{\mathbf{n}_c^2\} = \frac{2\lambda^2}{\mu^2} + \frac{\lambda}{\mu}$$

and

$$\Gamma_{\mathbf{n}_c}(z) = \frac{\mu}{\mu + \lambda - \lambda z}$$

From this it follows that

$$P\{\mathbf{n}_c = n\} = \frac{\mu}{\mu + \lambda}\left(\frac{\lambda}{\mu + \lambda}\right)^n \qquad n = 0, 1, \ldots$$

Thus, $\mathbf{n}_c$ has a *geometric distribution* with ratio $\lambda/(\mu + \lambda)$.

Example 12-2 If $\mathbf{c} = c$ is a constant, then

$$\Phi_c(s) = e^{cs} \qquad \Gamma_{n_c}(z) = e^{\lambda c(z-1)}$$

Thus, $\mathbf{n}_c$ is a Poisson RV with parameter λc, as it should.

† Since we deal only with positive RVs, we shall assume tacitly that all densities are zero on the negative axis.

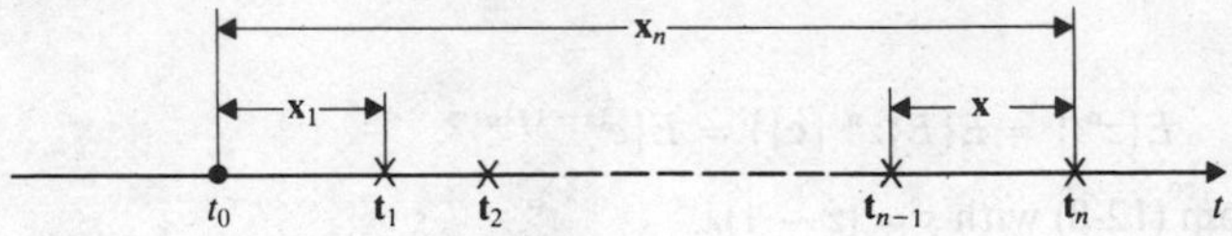

Figure 12-1

Distance between Points

Suppose that $\mathbf{t}_i$ are Poisson points with parameter λ, and t_0 is a fixed point as in Fig. 12-1. We maintain that the distance

$$\mathbf{x}_n = \mathbf{t}_n - t_0$$

from t_0 to the nth random point to the right of t_0 is an RV with density

$$f_n(x) = \frac{\lambda^n}{(n-1)!} x^{n-1} e^{-\lambda x} \tag{12-13}$$

(Erlang). In particular

$$f_1(x) = \lambda e^{-\lambda x} \tag{12-14}$$

Proof With $x > 0$ a given number, the inequality $\{\mathbf{x}_1 \leq x\}$ means that there is at least one random point in the interval $(t_0, t_0 + x)$. Hence

$$F_1(x) = P\{\mathbf{x}_1 \leq x\} = 1 - P\{\mathbf{n}_x = 0\} = 1 - e^{-\lambda x}$$

and (12-14) follows.

Similarly, $\mathbf{x}_n \leq x$ means that there are at least n points in the interval $(t_0, t_0 + x)$. Hence

$$F_n(x) = 1 - \sum_{k=0}^{n-1} P\{\mathbf{n}_x = k\} = 1 - \sum_{k=0}^{n-1} \frac{(\lambda x)^k}{k!} e^{-\lambda x}$$

Differentiating, we obtain (12-13).

We show, next, that the distance

$$\mathbf{x} = \mathbf{x}_n - \mathbf{x}_{n-1} = \mathbf{t}_n - \mathbf{t}_{n-1}$$

between two consecutive points $\mathbf{t}_{n-1}$ and $\mathbf{t}_n$ has an exponential distribution

$$f_x(x) = \lambda e^{-\lambda x} \tag{12-15}$$

Proof From (12-13) and (5-71) it follows that the moment function of $\mathbf{x}_n$ equals

$$\mathbf{\Phi}_n(s) = \frac{\lambda^n}{(\lambda - s)^{n+1}} \tag{12-16}$$

Furthermore, the RVs $\mathbf{x}$ and $\mathbf{x}_{n-1}$ are independent and $\mathbf{x}_n = \mathbf{x} + \mathbf{x}_{n-1}$. Hence, if $\mathbf{\Phi}_x(s)$ is the moment function $\mathbf{x}$, then

$$\mathbf{\Phi}_n(s) = \mathbf{\Phi}_x(s)\mathbf{\Phi}_{n-1}(s)$$

Comparing with (12-16), we obtain $\mathbf{\Phi}_x(s) = \lambda/(\lambda - s)$ and (12-15) results.

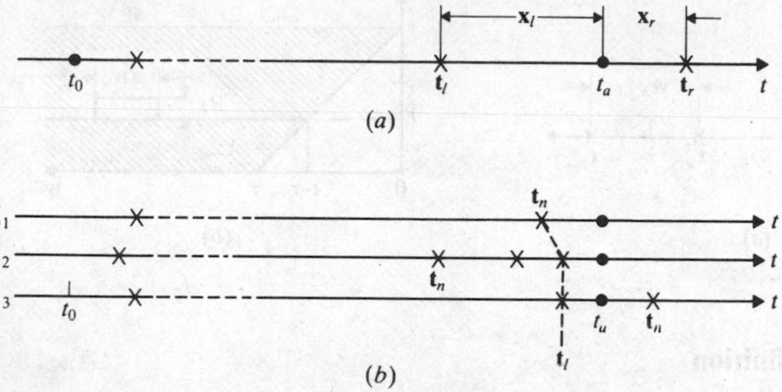

Figure 12-2

An apparent paradox We should stress that the notion of the "distance $\mathbf{x}$ between two consecutive points of a point process" is ambiguous. In Fig. 12-1, we interpreted $\mathbf{x}$ as the distance between $\mathbf{t}_{n-1}$ and $\mathbf{t}_n$ where $\mathbf{t}_n$ was the nth random point to the right of some *fixed* point t_0. This interpretation led to the conclusion that $\mathbf{x}$ is exponential as in (12-15). The same density is obtained if we interpret $\mathbf{x}$ as the distance between consecutive points to the left of t_0. Suppose, however, that $\mathbf{x}$ is interpreted as follows:

Given a fixed point t_a, we denote by $\mathbf{t}_l$ and $\mathbf{t}_r$ the random points nearest to t_a on its left and right respectively (Fig. 12-2*a*). We maintain that the density of the distance

$$\mathbf{x} = \mathbf{t}_r - \mathbf{t}_l$$

between these two points equals

$$f(x) = \lambda^2 x e^{-\lambda x} \tag{12-17}$$

Indeed the RVs

$$\mathbf{x}_l = t_a - \mathbf{t}_l \qquad \text{and} \qquad \mathbf{x}_r = \mathbf{t}_r - t_a$$

are independent with exponential density as in (12-14); furthermore, $\mathbf{x} = \mathbf{x}_r + \mathbf{x}_l$. This yields (12-17) because the convolution of two exponentials is the density in (12-17).

Thus, although $\mathbf{x}$ is again the "distance between two consecutive points," its density is not an exponential. This apparent paradox is a consequence of the ambiguity in the specification of the identity of random points. Suppose, for example, that we identify the points $\mathbf{t}_i$ by their order i, where the count starts from some fixed point t_0, and we observe that in one particular realization of the point process, the point $\mathbf{t}_l$, defined as above, equals $\mathbf{t}_n$. In other realizations of the process, the RVs $\mathbf{t}_l$ might equal some other point in this identification (Fig. 12-2*b*). The same argument shows that the point $\mathbf{t}_r$ does not coincide with the ordered point $\mathbf{t}_{n+1}$ for all realization. Hence, we should not expect that the RV $\mathbf{x} = \mathbf{t}_r - \mathbf{t}_l$ has the same density as the RV $\mathbf{t}_{n+1} - \mathbf{t}_n$.

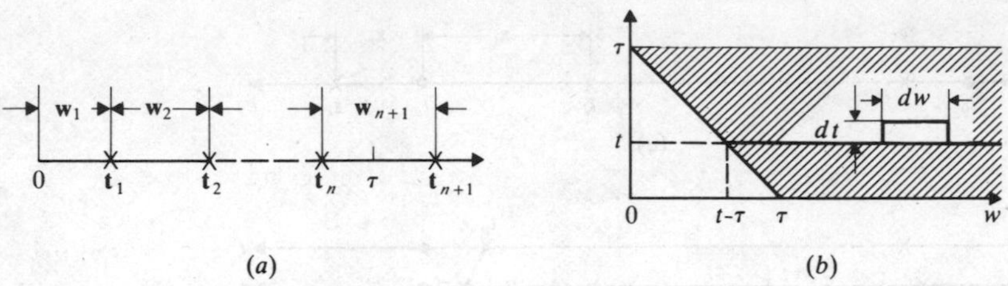

Figure 12-3

Constructive Definition

Given a sequence $\mathbf{w}_n$ of i.i.d. (independent, identically distributed) RVs with density

$$f(w) = \lambda e^{-\lambda w} \qquad (12\text{-}18)$$

we form a set of points $\mathbf{t}_n$ as in Fig. 12-3*a* where $t = 0$ is an arbitrary origin and

$$\mathbf{t}_n = \mathbf{w}_1 + \mathbf{w}_2 + \cdots + \mathbf{w}_n \qquad (12\text{-}19)$$

We maintain that the points so formed are Poisson distributed with parameter λ.

PROOF From the independence of the RVs $\mathbf{w}_n$ it follows that the RVs $\mathbf{t}_n$ and $\mathbf{w}_{n+1}$ are independent, the density $f_n(t)$ of $\mathbf{t}_n$ is given by (12-13)

$$f_n(t) = \frac{\lambda^n}{(n-1)!} t^{n-1} e^{\lambda t} \qquad (12\text{-}20)$$

and the joint density of $\mathbf{t}_n$ and $\mathbf{w}_{n+1}$ equals the product $f_n(t) f(w)$. If $\mathbf{t}_n < \tau$ and $\mathbf{t}_{n+1} = \mathbf{t}_n + \mathbf{w}_{n+1} > \tau$ then there are exactly n points in the interval $(0, \tau)$. As we see from Fig. 12-3*b*, the probability of this event equals

$$\int_0^\tau \int_{\tau - t}^\infty \lambda e^{-\lambda w} \frac{\lambda^n}{(n-1)!} t^{n-1} e^{-\lambda t} \, dw \, dt$$

$$= \int_0^\tau e^{-\lambda(\tau - t)} \frac{\lambda^n}{(n-1)!} t^{n-1} e^{-\lambda t} \, dt = e^{-\lambda \tau} \frac{(\lambda \tau)^n}{n!}$$

Thus, the points $\mathbf{t}_n$ so constructed have property P_1. We can show similarly that they have also property P_2.

Poisson points redefined Poisson points are realistic models for a large class of point processes: photon count, electron emission, telephone calls, data communication, visits to a doctor, arrivals at a park. The reason is that in these and other applications, the properties of the underlying points can be derived from certain general conditions that lead to Poisson distributions. As we show next, these conditions can be stated in a variety of forms that are equivalent to properties P_1 and P_2 of Poisson points, but are easier to check in specific cases.

I. If we place at random N points in an interval of length T where $N \gg 1$, then the resulting point process is nearly Poisson with parameter N/T. This is exact in the limit as N and T tend to infinity [see (3-47)].

II. If the distances $\mathbf{w}_n$ between two consecutive points $\mathbf{t}_{n-1}$ and $\mathbf{t}_n$ of a point process are independent and exponentially distributed, as in (12-18), then this process is Poisson.

The above can be phrased in an equivalent form: If the distance $\mathbf{w}$ from an arbitrary point t_0 to the next point of a point process is an RV whose density does not depend on the choice of t_0, then the process is Poisson. The reason for this equivalence is that this assumption leads to the conclusion that

$$f(w \mid \mathbf{w} \geq t_0) = f(w - t_0) \tag{12-21}$$

and the only function satisfying (12-21) is an exponential (see Example 4-16). In queueing theory, the above is called *Markoff* or *memoryless property*.

III. If the number of points $\mathbf{n}(t, t + dt)$ in an interval $(t, t + dt)$ is such that:

(*a*) $P\{\mathbf{n}(t, t + dt) = 1\}$ is of the order of dt;
(*b*) $P\{\mathbf{n}(t, t + dt) > 1\}$ is of order higher than dt;
(*c*) the above probabilities do not depend on the state of the point process outside the interval $(t, t + dt)$;

then the process is Poisson (see Sec. 12-4).

IV. Suppose, finally, that:

(*a*) $P\{\mathbf{n}(a, b) = k\}$ depends only on k and on the length of the interval (a, b);
(*b*) if the intervals (a_i, b_i) are nonoverlapping, then the RVs $\mathbf{n}(a_i, b_i)$ are independent;
(*c*) $P\{\mathbf{n}(a, b) = \infty\} = 0$.

These conditions lead again to the conclusion that the probability $p_k(\tau)$ of having k points in any interval of length τ equals

$$p_k(\tau) = e^{-\lambda\tau}(\lambda\tau)^k/k! \tag{12-22}$$

The proof is omitted.

Renewal Processes

Consider a stationary point process $\mathbf{t}_i$ such that the RVs

$$\mathbf{c}_i = \mathbf{t}_i - \mathbf{t}_{i-1} \tag{12-23}$$

are i.i.d with distribution $F_c(c)$. With t_0 a fixed point, we form the RV

$$\mathbf{w} = \mathbf{t}_1 - t_0 \tag{12-24}$$

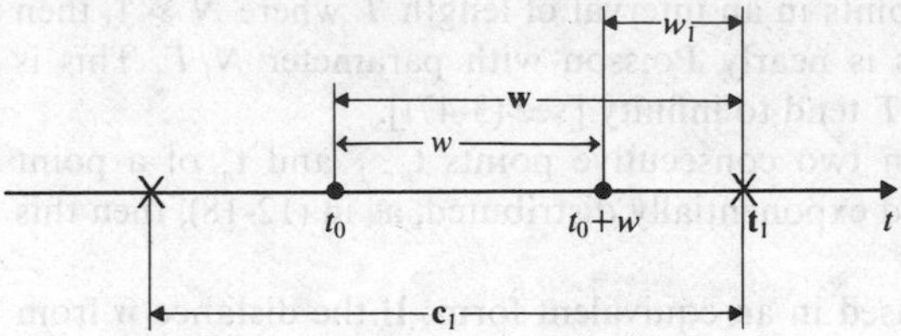

Figure 12-4

where $\mathbf{t}_1$ is the first random point to the right of t_0 (Fig. 12-4). We shall express the density $f_w(w)$ of this RV in terms of $F_c(c)$.

Theorem We maintain that

$$f_w(w) = \frac{1}{\eta_c}[1 - F_c(w)] \tag{12-25}$$

where

$$\eta_c = E\{\mathbf{c}_i\} = \int_0^\infty [1 - F_c(c)]\, dc \tag{12-26}$$

is the mean of $\mathbf{c}_i$ (see Prob. 5-15).

PROOF Given a number w, we define the RV $\mathbf{w}_1$ as the distance from the point $t_w = t_0 + w$ to the next random point $\mathbf{t}_i$ to its right (Fig. 12-3). From the stationarity of the points $\mathbf{t}_i$ it follows that the RVs $\mathbf{w}$ and $\mathbf{w}_1$ have the same density $f_w(w)$. Suppose that the first random point $\mathbf{t}_1$ to the right of t_0 is also to the right of t_w. In this case, there is no random point between t_0 and t_w and

$$\mathbf{w} = \mathbf{t}_1 - t_0 > w \qquad \mathbf{w}_1 = \mathbf{t}_1 - t_w \qquad \mathbf{c}_1 > \mathbf{w}_1 + w$$

From the above it follows that the events

$$\{\mathbf{w} > w\} \qquad \text{and} \qquad \{\mathbf{c}_1 > \mathbf{w}_1 + w\}$$

are equal provided that in the second set we consider only the outcomes such that $\mathbf{c}_1 > \mathbf{w}_1$. Hence

$$P\{\mathbf{w} > w\} = P\{\mathbf{c}_1 > \mathbf{w}_1 + w \mid \mathbf{c}_1 > \mathbf{w}_1\}$$

Reasoning similarly, we conclude that

$$P\{w < \mathbf{w} < w + dw\} = P\{0 < \mathbf{w}_1 < dw, \mathbf{c}_1 > w\} \tag{12-27}$$

The above yields

$$f_w(w)\, dw = f_w(0)\, dw[1 - F_c(w)]$$

because the RVs $\mathbf{w}_1$ and $\mathbf{c}_1$ are independent and

$$P\{\mathbf{c}_1 > w\} = 1 - F_c(w)$$

This completes the proof and it shows that $f_w(0) = 1/\eta_c$.

The theorem is true also if $\mathbf{w}$ is the distance from t_0 to the nearest random point on its left.

Corollary The moment function $\mathbf{\Phi}_w(s)$ of $\mathbf{w}$ equals

$$\mathbf{\Phi}_w(s) = \frac{1}{\eta_c s} [\mathbf{\Phi}_c(s) - 1] \tag{12-28}$$

PROOF Since $F_c'(w) = f_c(w)$ and $F_c(w) = 0$ for $w \leq 0$, we conclude, integrating by parts, that for Re $s < 0$

$$-\int_0^\infty F_c(w)e^{sw}\,dw = \frac{1}{s}\,\mathbf{\Phi}_c(s) \qquad -\int_0^\infty e^{sw}\,dw = \frac{1}{s}$$

and (12-28) follows from (12-25).

Differentiating (12-28) we obtain

$$\mathbf{\Phi}_w'(0) = \frac{\mathbf{\Phi}_c''(0)}{2\eta_c}$$

because $\mathbf{\Phi}_c(0) = 1$. Hence [see (5-67)]

$$\eta_w = \frac{E\{\mathbf{c}^2\}}{2\eta_c} \tag{12-29}$$

Example 12-3 If $f_c(c) = \lambda e^{-\lambda c}$, then

$$F_c(c) = 1 - e^{-\lambda c} \qquad \eta_c = \frac{1}{\lambda}$$

and (12-25) yields $f_w(w) = \lambda e^{-\lambda w} = f_c(w)$. This is the case iff $\mathbf{t}_i$ is a set of Poisson points.

Note With $\mathbf{c}_i$ as in (12-23), we form the point process

$$\boldsymbol{\tau}_i = \mathbf{c}_1 + \mathbf{c}_2 + \cdots + \mathbf{c}_i \tag{12-30}$$

starting from $t = 0$. In general, this process does not have the same statistics as the original process $\mathbf{t}_i$ (it is not even stationary). It is, however, asymptotically equivalent to $\mathbf{t}_i$. The process $\mathbf{t}_i$ is given by

$$\mathbf{t}_i = \mathbf{w} + \mathbf{c}_2 + \cdots + \mathbf{c}_i \tag{12-31}$$

It can, therefore, be constructed in terms of the sequence $\mathbf{c}_i$ and the RV $\mathbf{w}$ specified by (12-25).

12-2 QUEUEING THEORY

The term *queueing* is used to describe a large class of phenomena involving arrivals, waiting, servicing, and departures. In Fig. 12-5 we show a typical model (*queueing system*) whose inputs are certain objects (*units*) identified in terms of

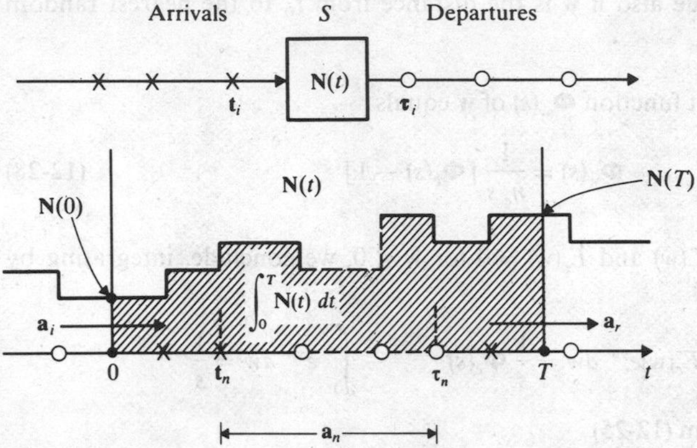

Figure 12-5

their *arrival times* $\mathbf{t}_i$. Each unit stays in the system $\mathbf{a}_i$ seconds (*total system time*) and it departs at the *departure time* $\boldsymbol{\tau}_i$

$$\boldsymbol{\tau}_i = \mathbf{t}_i + \mathbf{a}_i \tag{12-32}$$

The number of units in the system (*state of the system*) at time t will be denoted by $\mathbf{N}(t)$. Thus, $\mathbf{N}(t)$ is a discrete-state process increasing by 1 at $\mathbf{t}_i$ and decreasing by 1 at $\boldsymbol{\tau}_i$.

Our objective in this section is not to develop this involved topic in detail. We plan merely to introduce the main ideas in the context of earlier concepts. We start with a general theorem that does not rely on any special conditions about the interarrival times, the nature of the system, or the properties of $\mathbf{a}_n$. It assumes merely that all processes are SSS with finite second moments.

Little's theorem Suppose that the processes $\mathbf{t}_i$ and $\mathbf{a}_i$ are mean-ergodic

$$\frac{\mathbf{n}_T}{T} \xrightarrow[T\to\infty]{} \lambda \qquad \frac{1}{n}\sum_{k=1}^{n} \mathbf{a}_k \xrightarrow[n\to\infty]{} E\{\mathbf{a}_n\} \tag{12-33}$$

In the above, $\mathbf{n}_T$ is the number of points $\mathbf{t}_i$ in the interval $(0, T)$ and $\lambda = E\{\mathbf{n}_T\}/T$ is the mean density of these points.

We maintain that†

$$E\{\mathbf{N}(t)\} = \lambda E\{\mathbf{a}_i\}. \tag{12-34}$$

In fact, we shall establish the stronger statement that $\mathbf{N}(t)$ is also mean-ergodic

$$\lim_{T\to\infty} \frac{1}{T}\int_0^T \mathbf{N}(t)\,dt = \lambda E\{\mathbf{a}_i\} = E\{\mathbf{N}(t)\} \tag{12-35}$$

Equation (12-34) seems reasonable: The mean $E\{\mathbf{N}(t)\}$ of the number of units in the system equals the mean number λ of arrivals per second multiplied by the

† F. J. Beutler, "Mean Sojourn Times . . . ," *IEEE Transactions Information Theory*, March, 1983.

mean time $E\{\mathbf{a}_i\}$ that each unit remains in the system. It is not, however, always true, although it holds under general conditions.

PROOF We start with the observation that

$$-\sum_{r=1}^{\mathbf{N}(T)} \mathbf{a}_r \le \int_0^T \mathbf{N}(t)\,dt - \sum_{n=1}^{\mathbf{n}_T} \mathbf{a}_n \le \sum_{i=1}^{\mathbf{N}(0)} \mathbf{a}_i \qquad (12\text{-}36)$$

In the above, the terms $\mathbf{a}_n$ of the second sum are due to the $\mathbf{n}_T$ units that arrived in the interval $(0, T)$; the terms $\mathbf{a}_i$ of the last sum are due to the $\mathbf{N}(0)$ units that are in the system at $t = 0$; the terms $\mathbf{a}_r$ of the first sum are due to the $\mathbf{N}(T)$ units that are still in the system at $t = T$. The details of the reasoning that establishes (12-36) are omitted.

As we know (see Prob. 8-9)

$$E\left\{\left(\sum_{k=1}^{\mathbf{N}(t)} \mathbf{a}_k\right)^2\right\} \le E\{\mathbf{N}^2(t)\}E\{\mathbf{a}_k^2\} < \infty \qquad (12\text{-}37)$$

Dividing (12-36) by T we conclude that, if T is sufficiently large, then

$$\frac{1}{T}\int_0^T \mathbf{N}(t)\,dt \simeq \frac{1}{T}\sum_{n=1}^{\mathbf{n}_T} \mathbf{a}_n \qquad (12\text{-}38)$$

because the left and right side of (12-36) tend to zero after the division by T [see (12-37)]. Furthermore, assumption (12-33) yields $\mathbf{n}_T \simeq \lambda T$ and

$$\frac{1}{T}\sum_{n=1}^{\mathbf{n}_T} \mathbf{a}_n \simeq \frac{\lambda}{\mathbf{n}_T}\sum_{n=1}^{\mathbf{n}_T} \mathbf{a}_n \simeq \lambda E\{\mathbf{a}_n\} \qquad (12\text{-}39)$$

Inserting into (12-38), we obtain the first equality in (12-35). The second follows because the mean of the left side equals $E\{\mathbf{N}(t)\}$.

Immediate Service ($M \mid G \mid \infty$)

In general, the system time $\mathbf{a}_n$ is a sum

$$\mathbf{a}_n = \mathbf{b}_n + \mathbf{c}_n$$

where $\mathbf{b}_n$ is the *waiting time* (or *queueing time*) and $\mathbf{c}_n$ is the *service time* of the nth unit. In many applications, $\mathbf{b}_n = 0$ and $\mathbf{a}_n = \mathbf{c}_n$. This is the case if the number of servers is infinite or when no servers are involved (visits to a park, for example). We consider this problem next.

We shall assume that the arrival times $\mathbf{t}_n$ are Poisson points with mean density λ and the service times $\mathbf{c}_n$ are i.i.d. with distribution an arbitrary function $F_c(c)$. In queueing theory this is written in the following form (Kendall)

$$M \mid G \mid \infty$$

The first position in this notation refers to arrivals, the second to service times, and the third to the number of servers in the system. The letter M (for *Markoff* or

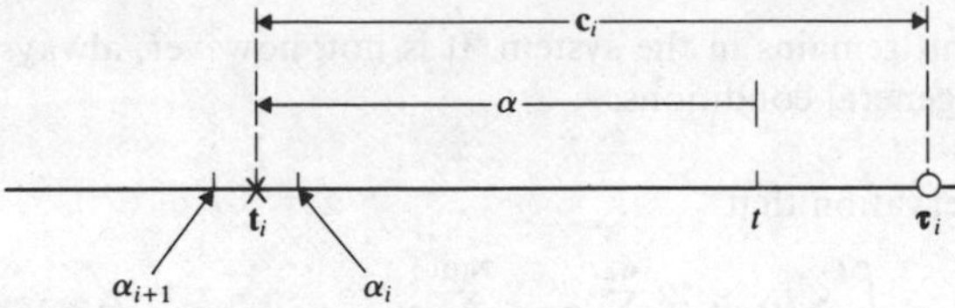

Figure 12-6

memoryless) in the first position means Poisson arrivals; in the second position it means exponentially distributed service times. The letter G (general) indicates that the arrivals or the service times are arbitrary.

Theorem The state $\mathbf{N}(t)$ of a $M \mid G \mid \infty$ system is Poisson distributed

$$P\{\mathbf{N}(t) = k\} = e^{-\rho} \frac{\rho^k}{k!} \tag{12-40}$$

with parameter (*traffic intensity* or *offered load*)

$$\rho = \lambda E\{\mathbf{c}_n\} = \lambda \eta_c \tag{12-41}$$

Proof Using the point t as the origin (Fig. 12-6), we divide the time-axis into consecutive intervals I_i: (α_i, α_{i+1}) of length $\Delta\alpha = \alpha_{i+1} - \alpha_i$. Denoting by $\Delta\mathbf{n}_i$ the number of arrivals in the interval I_i and by $\Delta\mathbf{N}(t, \alpha_i)$ the contribution to the state $\mathbf{N}(t)$ of the system at time t due to these arrivals, we have

$$\mathbf{N}(t) = \lim_{\Delta\alpha \to 0} \sum_i \Delta\mathbf{N}(t, \alpha_i) \tag{12-42}$$

If $\Delta\alpha$ is small, then, within probabilities of order $\Delta\alpha$, the RV $\Delta\mathbf{n}_i$ takes the values 0 or 1 and

$$P\{\Delta\mathbf{n}_i = 1\} = \lambda\, \Delta\alpha \tag{12-43}$$

If $\Delta\mathbf{n}_i = 0$, then $\Delta\mathbf{N}(t, \alpha_i) = 0$; if $\Delta\mathbf{n}_i = 1$, then

$$\Delta\mathbf{N}(t, \alpha_i) = \begin{cases} 1 & \text{if} \quad \mathbf{c}_i > \alpha_i \\ 0 & \text{if} \quad \mathbf{c}_i < \alpha_i \end{cases}$$

where $\mathbf{c}_i$ is the service time of the single unit that enters the system in the interval I_i. Hence

$$P\{\Delta\mathbf{N}(t, \alpha_i) = 1 \mid \Delta\mathbf{n}_i = 1\} = P\{\mathbf{c}_i > \alpha_i\} = 1 - F_c(\alpha_i) \tag{12-44}$$

Multiplying (12-43) and (12-44), we obtain

$$P\{\Delta\mathbf{N}(t, \alpha_i) = 1\} = [1 - F_c(\alpha_i)]\lambda\, \Delta\alpha \tag{12-45}$$

The RV $\Delta\mathbf{N}(t, \alpha_i)$ takes the values 0 or 1 and they are independent because the RVs $\Delta\mathbf{n}_i$ are independent (Poisson points in nonoverlapping intervals.) Hence (see Example 8-7*b*) the sum in (12-42) tends to a Poisson RV with parameter

$$\lambda \int_0^\infty [1 - F_c(\alpha)]\, d\alpha = \lim_{\Delta\alpha \to 0} \sum_{i=1}^{\infty} \lambda\, \Delta\alpha [1 - F_c(\alpha_i)]$$

and (12-40) results because the limit of the sum in (12-42) equals $\mathbf{N}(t)$ and the above integral equals $E\{\mathbf{c}_n\}$ [see (12-26)].

Corollary The traffic intensity ρ equals the average number of units in the system

$$E\{\mathbf{N}(t)\} = \rho = \lambda \eta_c \tag{12-46}$$

PROOF It follows from (12-40) and (5-36).

Example 12-4(*a*) $(M \mid M \mid 1)$ If $\mathbf{c}$ is an exponential with parameter μ, then $\eta_c = 1/\mu$, hence, $\rho = \lambda/\mu$.

(*b*) $(M \mid D \mid 1)$ If $\mathbf{c} = c$ is a constant, then $\eta_c = c$, hence, $\rho = \lambda c$.

Single-Server Queue $(M \mid G \mid 1)$

We conclude the section with a detailed treatment of the $M \mid G \mid 1$ system. In this system, the arrival times $\mathbf{t}_i$ are again Poisson distributed and the service times $\mathbf{c}_i$ are i.i.d. However, there is only one server in the system. The model is described in Fig. 12-7: Suppose that a unit enters the system at $t = \mathbf{t}_i$. Just prior to its arrival, the system contains $\mathbf{N}(\mathbf{t}_i^-)$ units; one of these units is being served and the others are waiting in line. The unit entering at $\mathbf{t}_i$ occupies the last position in the queue (shaded area). Its position in line remains unchanged as other units arrive; the unit advances when a service is completed. It reaches the server (position 0 in line when service starts) at $t = \boldsymbol{\tau}_{i-1}$ and it leaves the system at time $t = \boldsymbol{\tau}_i$

$$\boldsymbol{\tau}_i = \boldsymbol{\tau}_{i-1} + \mathbf{c}_i$$

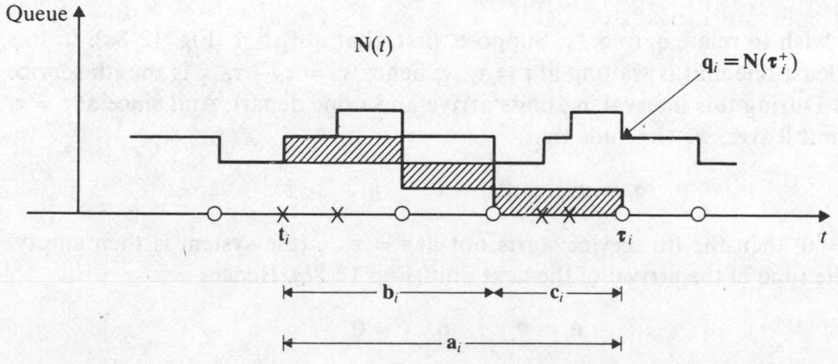

Figure 12-7

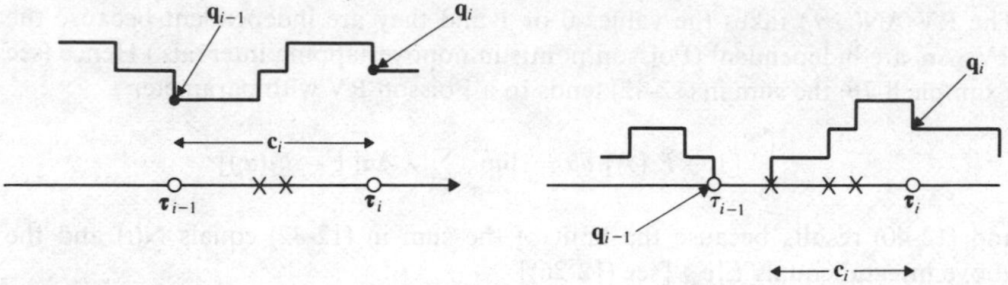

Figure 12-8

where $\mathbf{c}_i$ is the *service time*. Denoting by $\mathbf{b}_i$ the *waiting time* and by $\mathbf{a}_i$ the *system time*, we conclude that

$$\mathbf{b}_i = \boldsymbol{\tau}_{i-1} - \mathbf{t}_i \qquad \mathbf{a}_i = \boldsymbol{\tau}_i - \mathbf{t}_i = \mathbf{b}_i + \mathbf{c}_i$$

This completes the description of the $M\,|\,G\,|\,1$ model.

Our objective is to express the statistics of the various parameters of the system in terms of the distribution $F_c(c)$ of the service time $\mathbf{c}_i$ and the average density λ of the arrival times $\mathbf{t}_i$ under the assumption that $\mathbf{N}(t)$ is *stationary*. As we shall see, the system reaches stationarity only if the mean η_c of $\mathbf{c}_i$ (mean service time) is less than the *average interarrival time* $1/\lambda$.

Note (symmetry) If $\mathbf{N}(t)$ is a stationary $M\,|\,G\,|\,1$ process, then $\mathbf{N}(-t)$ is a $G\,|\,M\,|\,1$ process obtained from $\mathbf{N}(t)$ by interchanging arrivals and departures. This shows that the properties of a $G\,|\,M\,|\,1$ process follow from the properties of the corresponding $M\,|\,G\,|\,1$ process.

The imbedded Markoff chain The ith unit arrives at $t = \mathbf{t}_i$ and departs at $t = \boldsymbol{\tau}_i$ remaining in the system $\mathbf{a}_i = \boldsymbol{\tau}_i - \mathbf{t}_i$ seconds. During the interval $(\mathbf{t}_i, \boldsymbol{\tau}_i)$, $\mathbf{n}_{a_i}$ new units arrive and all other units depart. Hence, $\mathbf{n}_{a_i}$ equals the state $\mathbf{N}(\boldsymbol{\tau}_i^+)$ of the system at $t = \boldsymbol{\tau}_i^+$. This number is denoted by $\mathbf{q}_i$ and is called the Markoff chain imbedded in the process $\mathbf{N}(t)$ (see Sec. 12-4). Thus

$$\mathbf{q}_i = \mathbf{n}_{a_i} = N(\boldsymbol{\tau}_i^+) \tag{12-47}$$

We wish to relate $\mathbf{q}_i$ to $\mathbf{q}_{i-1}$. Suppose, first, that $\mathbf{q}_{i-1} \geq 1$ (Fig. 12-8*a*). In this case, at least one unit is waiting at $t = \boldsymbol{\tau}_{i-1}$, hence, $\mathbf{c}_i = \boldsymbol{\tau}_i - \boldsymbol{\tau}_{i-1}$ is the ith service interval. During this interval, $\mathbf{n}_{c_i}$ units arrive and none depart. And since at $t = \boldsymbol{\tau}_i$ the ith unit leaves, we conclude that

$$\mathbf{q}_i = \mathbf{q}_{i-1} + \mathbf{n}_{c_i} - 1 \qquad \mathbf{q}_{i-1} \geq 1$$

If $\mathbf{q}_{i-1} = 0$, then the ith service starts not at $t = \boldsymbol{\tau}_{i-1}$ (the system is then empty) but at the time of the arrival of the next unit (Fig. 12-8*b*). Hence

$$\mathbf{q}_i = \mathbf{n}_{c_i} \qquad \mathbf{q}_{i-1} = 0$$

where $\mathbf{n}_{c_i}$ equals the arrivals in the *interior* of the service interval.

From the above it follows that $\mathbf{q}_i$ satisfies the recursion equation

$$\mathbf{q}_i = \bar{\mathbf{q}}_{i-1} + \mathbf{n}_{c_i} \tag{12-48}$$

where

$$\bar{\mathbf{q}}_i = \begin{cases} \mathbf{q}_i - 1 & \mathbf{q}_i \geq 1 \\ 0 & \mathbf{q}_i = 0 \end{cases} \tag{12-49}$$

Using (12-48), we shall determine the state probabilities $p_k = P\{\mathbf{q}_i = k\}$ at $\boldsymbol{\tau}_i^+$.

Zero state We maintain that

$$p_0 = 1 - \rho \qquad \rho = \lambda\eta_c \tag{12-50}$$

PROOF Clearly, $\mathbf{n}_{c_i}$ is the number of Poisson points in the random interval $\mathbf{c}_i$, hence [see (12-6)]

$$E\{\mathbf{n}_{c_i}\} = \lambda\eta_c = \rho$$

With

$$\eta_q = \sum_{k=0}^{\infty} kp_k = \sum_{k=1}^{\infty} kp_k$$

the mean of $\mathbf{q}_i$, it follows from (12-49) that

$$E\{\bar{\mathbf{q}}_i\} = \sum_{k=1}^{\infty} (k-1)p_k = \eta_q - \sum_{k=1}^{\infty} p_k = \eta_q - (1 - p_0)$$

And since (stationarity assumption) $E\{\bar{\mathbf{q}}_i\} = E\{\bar{\mathbf{q}}_{i-1}\}$, (12-48) yields

$$\eta_q = \eta_q - (1 - p_0) + \lambda\eta_c$$

and (12-50) results.

Proceeding similarly, we can find the moments of $\mathbf{q}_i$ (see Prob. 12-8). It is simpler, however, to determine directly the moment function

$$\boldsymbol{\Gamma}_q(z) = E\{z^{\mathbf{q}_i}\} = p_0 + \sum_{k=1}^{\infty} p_i z^k \tag{12-51}$$

of the sequence $\mathbf{q}_i$. Using (12-48), we shall show that $\boldsymbol{\Gamma}_q(z)$ can be expressed in terms of the moment function

$$\boldsymbol{\Gamma}_{n_c}(z) = \boldsymbol{\Phi}_c(\lambda z - \lambda)$$

In the above, $\mathbf{n}_{c_i}$ is the number of arrivals in the random interval $\mathbf{c}_i$, $\boldsymbol{\Gamma}_{n_c}(z)$ is the moment function of $\mathbf{n}_{c_i}$, and $\boldsymbol{\Phi}_c(s)$ is the moment function of $\mathbf{c}_i$ [see (12-9)].

Theorem

$$\boldsymbol{\Gamma}_q(z) = \frac{p_0(1 - z)}{1 - z/\boldsymbol{\Phi}_c(\lambda z - \lambda)} \tag{12-52}$$

PROOF The moment function of $\bar{\mathbf{q}}_i$ equals

$$\mathbf{\Gamma}_{\bar{q}}(z) = p_0 + \sum_{k=1}^{\infty} p_k z^{k-1} = p_0 + z^{-1}[\mathbf{\Gamma}_q(z) - p_0] \tag{12-53}$$

From the independence of $\mathbf{q}_i$ and $\mathbf{c}_i$ it follows that $\mathbf{q}_i$ and $\mathbf{n}_{c_i}$ are also independent. Hence [see (12-48)]

$$\mathbf{\Gamma}_q(z) = \mathbf{\Gamma}_{\bar{q}}(z)\mathbf{\Gamma}_{n_c}(z) \tag{12-54}$$

Inserting (12-53) into (12-54), we obtain

$$z\mathbf{\Gamma}_q(z) = [\mathbf{\Gamma}_q(z) + p_0 z - p_0]\mathbf{\Gamma}_{n_c}(z) \tag{12-55}$$

and (12-52) results.

Equation (12-55) can be used to obtain all moments of $\mathbf{q}_i$. We shall use it to determine η_q. As a preparation, however, we reestablish (12-50): Since $\mathbf{\Gamma}_q(1) = 1$, we conclude differentiating (12-55) that

$$1 + \mathbf{\Gamma}_q'(1) = \mathbf{\Gamma}_q'(1) + p_0 + \mathbf{\Gamma}_{n_c}'(1)$$

This yields (12-50) because $\mathbf{\Gamma}_{n_c}'(1) = E\{\mathbf{n}_c\} = \lambda\eta_c$ [see (12-6)].

Pollaczek–Khinchin formula We maintain that the mean of $\mathbf{q}_i$ equals

$$\eta_q = \rho + \frac{\lambda^2 E\{\mathbf{c}_i^2\}}{2(1-\rho)} \tag{12-56}$$

PROOF Differentiating (12-55) twice and setting $z = 1$ we obtain

$$2\mathbf{\Gamma}_q'(1) = 2[\mathbf{\Gamma}_q'(1) + p_0]\mathbf{\Gamma}_{n_c}'(1) + \mathbf{\Gamma}_{n_c}''(1)$$

But [see (12-10)] $\mathbf{\Gamma}_{n_c}''(1) = \lambda^2 E\{\mathbf{c}_i^2\}$, hence

$$2\eta_q = 2(\eta_q + p_0)\lambda\eta_c + \lambda^2 E\{\mathbf{c}_i^2\}$$

and (12-56) results because $\rho = 1 - p_0 = \lambda\eta_c$.

Mean system time and waiting time The process $\mathbf{q}_i$ equals the number of arrivals $\mathbf{n}_{a_i}$ during the service interval $\mathbf{a}_i$. Hence [see (12-6)]

$$E\{\mathbf{q}_i\} = E\{\mathbf{n}_{a_i}\} = \lambda E\{\mathbf{a}_i\}$$

and (12-56) yields

$$E\{\mathbf{a}_i\} = \eta_c + \lambda E\{\mathbf{c}_i^2\}/2p_0 \tag{12-57}$$

This is the *mean system time*. Since $\mathbf{b}_i = \mathbf{a}_i - \mathbf{c}_i$, the *mean waiting time* equals

$$E\{\mathbf{b}_i\} = \frac{\lambda E\{\mathbf{c}_i^2\}}{2p_0} \tag{12-58}$$

Other moments can be found using moment functions. Denoting by $\mathbf{\Phi}_a(s)$ the moment function of $\mathbf{a}_i$, we conclude from (12-9) that $\mathbf{\Gamma}_q(z) = \mathbf{\Phi}_a(\lambda z - \lambda)$. Hence

$$\mathbf{\Phi}_a(s) = \mathbf{\Gamma}_q\left(1 + \frac{s}{\lambda}\right) \tag{12-59}$$

where $\mathbf{\Gamma}_q(z)$ is given by (12-52).

Note As we know (Little's theorem) $E\{\mathbf{N}(t)\} = \lambda E\{\mathbf{a}_i\}$. Comparing with (12-57) and (12-56), we obtain

$$E\{\mathbf{N}(t)\} = E\{\mathbf{N}(\boldsymbol{\tau}_i^+)\} = \lambda\eta_c + \frac{\lambda^2(\eta_c^2 + \sigma_c^2)}{2(1 - \lambda\eta_c)} \tag{12-60}$$

Thus, the mean of $\mathbf{N}(\tau_i^+)$ equals the mean of $\mathbf{N}(t)$ for any t. In the end of the section we shall prove that the processes $\mathbf{N}(t)$ and $\mathbf{N}(\tau_i^+)$ have not only equal means but also identical distributions.

Equation (12-60) shows that, if the mean service time η_c is specified, then the mean $E\{\mathbf{N}(t)\}$ of the number of units in the system is minimum if $\sigma_c = 0$. This is so iff the service time $\mathbf{c}_i$ is constant.

Example 12-5 $(M \mid M \mid 1)$ If $F_c(c) = 1 - e^{-\mu c}$, then

$$\mathbf{\Phi}_c(s) = \frac{\mu}{\mu - s} \qquad E\{\mathbf{c}_i\} = \frac{1}{\mu} \qquad E\{\mathbf{c}_i^2\} = \frac{2}{\mu^2}$$

Thus, $\rho = \lambda/\mu$, $p_0 = 1 - \lambda/\mu$ and (12-52) yields

$$\mathbf{\Gamma}_q(z) = \frac{\mu - \lambda}{\mu - \lambda z} = \frac{p_0}{1 - \rho z}$$

This shows that $\mathbf{q}_i$ has a geometric distribution with ratio the traffic intensity ρ

$$P\{\mathbf{q}_i = k\} = p_0\,\rho^k \qquad E\{\mathbf{q}_i\} = \frac{\rho}{1 - \rho} = \frac{\lambda}{\mu - \lambda}$$

We note, finally, that [see (12-59)]

$$\mathbf{\Phi}_a(s) = \frac{\mu p_0}{\mu p_0 - s} \qquad F_a(a) = 1 - e^{-\mu p_0 a}$$

Thus, the system time $\mathbf{a}_i$ is exponential with parameter μp_0 and the mean system time equals

$$E\{\mathbf{a}_i\} = \frac{1}{\mu p_0} = \frac{1}{\mu - \lambda}$$

Example 12-6 $(M \mid D \mid 1)$ If $\mathbf{c} = c$ is a constant then

$$\mathbf{\Phi}_c(s) = e^{cs} \qquad E\{\mathbf{c}\} = \mathrm{C} \qquad E\{\mathbf{c}^2\} = c^2$$

$\rho = \lambda c$, $p_0 = 1 - \rho$ and (12-56) yields

$$E\{\mathbf{q}_i\} = \frac{\rho(2 - \rho)}{2(1 - \rho)}$$

Finally [see (12-52) and (12-59)]

$$\mathbf{\Gamma}_q(z) = \frac{p_0(1 - z)}{1 - ze^{-\lambda c(z-1)}} \qquad \mathbf{\Phi}_a(s) = \frac{p_0\, s/\lambda}{(1 + s/\lambda)e^{-cs} - 1}$$

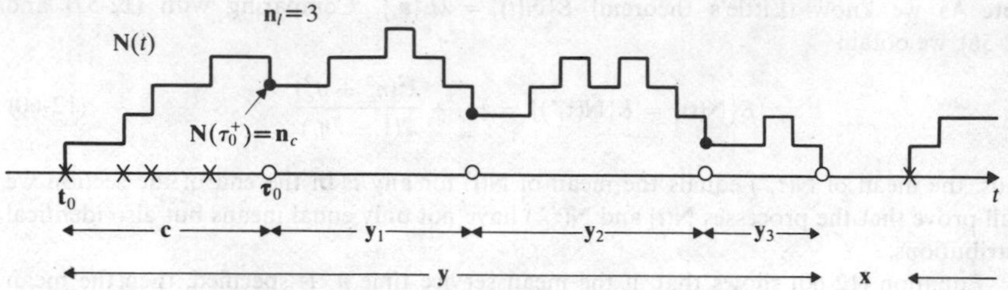

Figure 12-9

Busy period The state of a queueing system is characterized by a succession of *idle periods*, when $\mathbf{N}(t) = 0$, and *busy periods*, when there is at least one unit in the system. We denote by $\mathbf{x}$ the length of an idle period and by $\mathbf{y}$ the length of a busy period. Clearly, $\mathbf{x}$ is exponentially distributed as in (12-14) because there are no arrivals during an idle period. Our objective is to determine the properties of $\mathbf{y}$.

The busy period starts with the arrival of a unit at $t = \mathbf{t}_0$ into an empty system. The unit is served instantly and it departs at $t = \boldsymbol{\tau}_0$. The difference $\mathbf{c} = \boldsymbol{\tau}_0 - \mathbf{t}_0$ is its service time. Denoting by $\mathbf{n}_c$ the number of arrivals in the first service interval $(\mathbf{t}_0, \boldsymbol{\tau}_0)$ we conclude that

$$\mathbf{N}(\mathbf{t}_0^+) = 1 \qquad N(\boldsymbol{\tau}_0^+) = \mathbf{n}_c$$

The busy period $\mathbf{y}$ equals, thus, the interval of time from $t = \mathbf{t}_0$ when $\mathbf{N}(t_0^+) = 1$ until the moment $t = t_0 + \mathbf{y}$ when $\mathbf{N}(t) = \mathbf{N}(t_0^+) - 1 = 0$ *for the first time*. But the variations of $\mathbf{N}(t)$ during that interval do not depend on its initial value $\mathbf{N}(t_0^+)$. Hence, a busy period $\mathbf{y}_i$ can be characterized statistically as follows: It is an RV equal to the time interval from $t = \boldsymbol{\tau}_i$, when a service period begins, to $t = \boldsymbol{\tau}_i + \mathbf{y}_i$ when $\mathbf{N}(t) = \mathbf{N}(\boldsymbol{\tau}_i^+) - 1$ *for the first time.*

From the above it follows that the RVs $\mathbf{y}_1$, $\mathbf{y}_2$, $\mathbf{y}_3$ in Fig. 12-9 are independent and each is a busy period. Furthermore, the total number of these RVs equals $\mathbf{n}_c$ because, at the start of $\mathbf{y}_1$, the state of the system equals $\mathbf{N}(\boldsymbol{\tau}_0^+) = \mathbf{n}_c$ and at the end of the total busy period $\mathbf{y}$, $\mathbf{N}(t)$ equals zero. Hence

$$\mathbf{y} = \mathbf{c} + \mathbf{y}_1 + \cdots + \mathbf{y}_{\mathbf{n}_c} \tag{12-61}$$

Mean busy period Using (12-61), we shall show that

$$\eta_y = \frac{\eta_c}{p_0} \qquad p_0 = 1 - \lambda\eta_c \tag{12-62}$$

PROOF Since $E\{\mathbf{n}_c\} = \lambda\eta_c$, it follows from (8-47) that

$$E\left\{\sum_{i=1}^{\mathbf{n}_c} \mathbf{y}_i\right\} = E\{\mathbf{n}_c\}E\{\mathbf{y}_i\} = \lambda\eta_c\,\eta_y \tag{12-63}$$

Inserting into (12-61), we obtain $\eta_y = \eta_c + \lambda\eta_c\,\eta_y$ and (12-62) results.

Mean number of units served The number of units served in a busy period equals the number of arrivals $\mathbf{n}_y$ in the random interval $\mathbf{y}$. From (12-6) and (12-62) it follows that the mean of this number equals

$$E\{\mathbf{n}_y\} = \lambda\eta_y = \frac{\rho}{1-\rho} \tag{12-64}$$

Moment function We shall show that the moment function $\mathbf{\Phi}_y(s)$ of the busy period $\mathbf{y}$ satisfies the functional equation

$$\mathbf{\Phi}_y(s) = \mathbf{\Phi}_c[s + \lambda\mathbf{\Phi}_y(s) - \lambda] \tag{12-65}$$

PROOF If $\mathbf{c} = c$, then $\mathbf{n}_c$ is a Poisson RV with parameter λc. And since the RVs $\mathbf{y}$ and $\mathbf{y}_i$ have the same statistics, we conclude that (see Prob. 8-13)

$$E\{\exp[s(\mathbf{y}_1 + \cdots + \mathbf{y}_{n_c})]\} = e^{\lambda c \mathbf{\Phi}_y(s) - \lambda c}$$

From this and (12-61) it follows that

$$E\{e^{s\mathbf{y}}\} = E\{E\{e^{s\mathbf{y}} \mid \mathbf{c}\}\} = E\{e^{[s+\lambda\mathbf{\Phi}_y(s)-\lambda]\mathbf{c}}\}$$

and (12-65) results.

We note that (12-65) does not give $\mathbf{\Phi}_y(s)$ explicitly. It can, however, be used to determine the moments of $\mathbf{y}$. For example, its derivative at $s = 0$ yields

$$\mathbf{\Phi}'_y(0) = [1 + \lambda\mathbf{\Phi}'_y(0)]\mathbf{\Phi}'_c(0)$$

in agreement with (12-62).

Ergodicity The process $\mathbf{N}(t)$ and the sequences $\mathbf{N}(\mathbf{t}_i)$ and $\mathbf{N}(\boldsymbol{\tau}_i)$ are ergodic. This means that statistical averages can be expressed as time averages. The proof is a consequence of the fact that the state of the system is a succession of idle and busy periods and its behavior in each period does not depend on what happens elsewhere in time. The details, however, are omitted. We note the following consequences:

1. The state of the system $\mathbf{N}(t_i^-)$ just before a unit arrives is an ergodic sequence of RVs, hence, the probability that $\mathbf{N}(\mathbf{t}_i^-) = k$ equals the number of times this occurs in a large interval T divided by the number of points $\mathbf{t}_i$ in this interval. This number equals also the number of times $\mathbf{N}(\boldsymbol{\tau}_i^+)$ equals k because every time $\mathbf{N}(t)$ crosses the line $\mathbf{N} = k$ increasing, it crosses it also decreasing. Hence

$$P\{\mathbf{N}(\mathbf{t}_i^-) = k\} = P\{\mathbf{N}(\boldsymbol{\tau}_i^+) = k\} \tag{12-66}$$

In other words, the processes $\mathbf{N}(\mathbf{t}_i^-)$ and $\mathbf{N}(\boldsymbol{\tau}_i^+)$ have the same statistics.

2. The expected values $\eta_x = E\{\mathbf{x}_i\}$, $\eta_y = E\{\mathbf{y}_i\}$ of the idle period and busy period intervals $\mathbf{x}_i$ and $\mathbf{y}_i$ are given by

$$\eta_x \simeq \frac{T_x}{T} \qquad \eta_y \simeq \frac{T_y}{T}$$

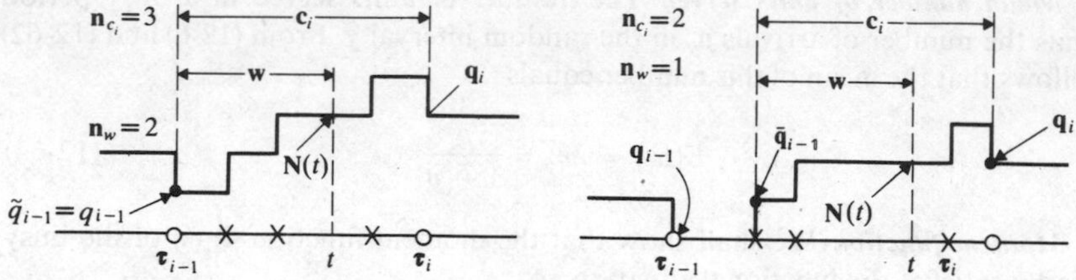

Figure 12-10

where T_x and T_y are the total times the system is idle or busy. Furthermore, $P\{\mathbf{N}(t) = 0\} \simeq T_x/T$. This leads to the conclusion that

$$P\{\mathbf{N}(t) = 0\} = \frac{\eta_x}{\eta_x + \eta_y} = \frac{1/\lambda}{1/\lambda + \eta_c/p_0} = \frac{p_0}{p_0 + \lambda\eta_c}$$

because $\eta_x = 1/\lambda$ and $\eta_y = \eta_c/p_0$ [see (12-62)]. And since $p_0 = 1 - \lambda\eta_c = P\{N(\tau_i^+) = 0\}$, we conclude that

$$P\{\mathbf{N}(t) = 0\} = P\{N(\tau_i^+) = 0\} = p_0 \qquad P\{\mathbf{N}(t) > 0\} = \rho \tag{12-67}$$

State statistics We shall, finally, show that the process $\mathbf{N}(t)$ and the sequence $\mathbf{q}_i = \mathbf{N}(\tau_i^+)$ have the same distribution

$$P\{\mathbf{N}(t) = k\} = P\{\mathbf{q}_i = k\} = p_k \qquad k \geq 0 \tag{12-68}$$

PROOF If we eliminate all idle periods and connect all busy periods together, contracting the t axis, we obtain a renewal process specified in terms of the service times $\mathbf{c}_i = \boldsymbol{\tau}_i - \boldsymbol{\tau}_{\tau_i - 1}$. With t an arbitrary constant, we denote by $\mathbf{w}$ the distance from t to the nearest point $\boldsymbol{\tau}_{i-1}$ on its left and by $\mathbf{n}_w$ the number of arrivals in the interval $(\boldsymbol{\tau}_{i-1}, t)$. From (12-28) and (12-9) it follows that

$$\mathbf{\Phi}_w(s) = \frac{1}{\eta_c s}[\mathbf{\Phi}_c(s) - 1]$$

$$\mathbf{\Gamma}_{n_w}(z) = \mathbf{\Phi}_w(\lambda z - \lambda) = \frac{\mathbf{\Gamma}_{n_c}(z) - 1}{\rho(z-1)} \tag{12-69}$$

This holds for every t in a busy period, i.e., when $\mathbf{N} = \mathbf{N}(t) \neq 0$. Returning to the original time (Fig. 12-10), we observe that, if $\mathbf{N} \neq 0$, then

$$\mathbf{N} = \tilde{\mathbf{q}}_{i-1} + \mathbf{n}_w \qquad \text{where} \qquad \tilde{\mathbf{q}}_i = \begin{cases} \mathbf{q}_i & \mathbf{q}_i > 0 \\ 1 & \mathbf{q}_i = 0 \end{cases} \tag{12-70}$$

Since $\mathbf{n}_w$ is independent of $\tilde{\mathbf{q}}_i$, the above yields

$$E\{z^{\mathbf{N}} \mid \mathbf{N} \neq 0\} = \mathbf{\Gamma}_{\tilde{q}}(z)\mathbf{\Gamma}_{n_w}(z)$$

where

$$\mathbf{\Gamma}_{\tilde{q}}(z) = p_0 z + \sum_{k=1}^{\infty} p_k z^k = p_0 z + \mathbf{\Gamma}_q(z) - p_0 \tag{12-71}$$

As we know, $P\{\mathbf{N} = 0\} = p_0$, $P\{\mathbf{N} > 0\} = \rho$. Hence

$$\Gamma_N(z) = E\{z^{\mathbf{N}}\} = E\{z^{\mathbf{N}} \mid \mathbf{N} = 0\}p_0 + E\{z^{\mathbf{N}} \mid \mathbf{N} \geq 0\}\rho$$

From this and (12-55) it follows, after some manipulations, that $\Gamma_N(z) = \Gamma_q(z)$ and (12-68) results.

12-3 SHOT NOISE

Shot noise is the process

$$\mathbf{s}(t) = \sum_i h(t - \mathbf{t}_i) \tag{12-72}$$

where $h(t)$ is a given function and $\mathbf{t}_i$ is a set of Poisson points. The process $\mathbf{s}(t)$ is the output of a system with impulse response $h(t)$ and input a Poisson impulse train (Fig. 12-11). If $h(t) = U(t)$, then $\mathbf{s}(t)$ is a Poisson process (Fig. 9-3a). If $h(t)$ is a pulse of width c, then $\mathbf{s}(t)$ is the queueing process $M \mid D \mid \infty$ of Example 12-4b. In Sec. 10-3, we determined the second-order properties of the shot noise. In the following, we evaluate its general statistics.

Density function We start with the determination of the density $f_s(x)$ of $\mathbf{s}(t)$

$$f_s(x)\, dx = P\{x < \mathbf{s}(t) \leq x + dx\}$$

under the assumption that $h(t)$ is of finite duration

$$h(t) = 0 \qquad \text{for} \qquad t < 0 \qquad \text{and} \qquad t > T \tag{12-73}$$

Denoting by $\mathbf{n}_T$ the number of Poisson points in the interval $(t - T, t)$, we have [see (4-58)]

$$f_s(x) = \sum_{k=0}^{\infty} f_s(x \mid \mathbf{n}_T = k)P\{\mathbf{n}_T = k\} \tag{12-74}$$

In the above

$$P\{\mathbf{n}_T = k\} = \frac{e^{-\lambda T}(\lambda T)^k}{k!}$$

To find $f_s(x)$, it suffices, therefore, to find the conditional density $f_s(x \mid \mathbf{n}_T = k)$ assuming that there are k points in the interval $(t - T, t)$. The evaluation of this density is based on the following property of Poisson points [see (3-51)]:

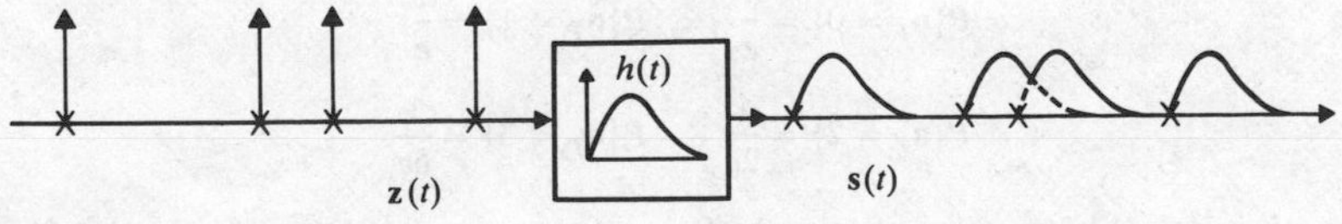

Figure 12-11

If it is known that there are exactly k points in an interval (t_1, t_2), then these points have the same statistics as k arbitrary points placed at random in this interval. In other words, the k points can be assumed to be k independent RVs uniform in the interval (t_1, t_2).

From the above and (12-73) it follows that the conditional density $f_s(x \mid \mathbf{n}_T = 1)$ equals the density $g_1(x)$ of the process

$$\mathbf{x}_1(t) = h(t - \mathbf{t}_1) = h(\boldsymbol{\tau}_1) \qquad \boldsymbol{\tau}_1 = t - \mathbf{t}_1$$

where $\mathbf{t}_1$ is uniform in the interval $(t - T, t)$ or, equivalently, $\boldsymbol{\tau}_1$ is uniform in the interval $(0, T)$. The function $g_1(x)$ is independent of t and can be found with the techniques of Sec. 5-2.

Similarly, $f_s(x \mid \mathbf{n}_T = 2)$ equals the density of the process

$$\mathbf{x}_2(t) = h(t - \mathbf{t}_1) + h(t - \mathbf{t}_2) = h(\boldsymbol{\tau}_1) + h(\boldsymbol{\tau}_2)$$

where $\boldsymbol{\tau}_1$ and $\boldsymbol{\tau}_2$ are two independent RVs uniform in the interval $(0, T)$. Hence, the RVs $h(\boldsymbol{\tau}_1)$ and $h(\boldsymbol{\tau}_2)$ have the same density $g_1(x)$ and they are independent because they are functions of the independent RVs $\boldsymbol{\tau}_1$ and $\boldsymbol{\tau}_2$. From this it follows that [see (6-39)] the density $g_2(x)$ of $\mathbf{x}_2(t)$ is the convolution

$$g_2(x) = g_1(x) * g_1(x)$$

Reasoning similarly, we conclude that

$$f_s(x \mid \mathbf{n}_T = k) = g_k(x) = g_1(x) * \cdots * g_1(x) \qquad \text{(12-75)}$$

We note, finally, that if $\mathbf{n}_T = 0$, then $\mathbf{s}(t) = 0$, therefore

$$f_s(x \mid \mathbf{n}_T = 0) = g_0(x) = \delta(x)$$

Inserting into (12-74), we obtain

$$f_s(x) = e^{-\lambda T} \sum_{k=0}^{\infty} \frac{g_k(x)(\lambda T)^k}{k!} \qquad \text{(12-76)}$$

This formula is useful mainly for "*low density*" shot noise i.e., when λT is of the order of 1.

Example 12-7 Suppose that $h(t)$ is a trapezoid as in Fig. 12-12. In this case

$$P\{\mathbf{x}_1(t) \le x\} = P\{\boldsymbol{\tau} \ge (1.5 - x)T\} = x - 0.5$$

for $0.5 \le x \le 1.5$ and 0 otherwise. Hence, $g_1(x)$ is uniform in the interval (0.5, 1.5), $g_2(x)$ is a triangle in the interval (1, 3), and $g_3(x)$ consists of three parabola pieces in the interval (1.5, 4.5). Assuming that $\lambda = 1/T$, we have

$$P\{\mathbf{n}_T = 0\} = \frac{1}{e} \qquad P\{\mathbf{n}_T = 1\} = \frac{1}{e}$$

$$P\{\mathbf{n}_T = 2\} = \frac{1}{2e} \qquad P\{\mathbf{n}_T = 3\} = \frac{1}{6e}$$

Inserting into (12-76) and neglecting higher-order terms, we obtain the density $f_s(x)$ shown in Fig. 12-12.

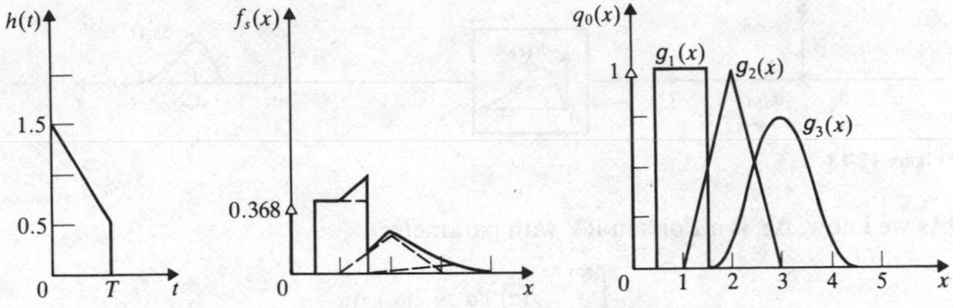

(a) (b) (c)

Figure 12-12

Moment function The moment function of $\mathbf{s}(t)$ equals

$$\mathbf{\Psi}(s) = \int_{-\infty}^{\infty} e^{sx} f_s(x)\, dx = \exp\left\{\lambda \int_0^T [e^{sh(\alpha)} - 1]\, d\alpha\right\} \tag{12-77}$$

PROOF This is a special case of (12-80) but can be established directly: As we have seen, $\mathbf{s}(t)$ can be written as a sum

$$\mathbf{s}(t) = \sum_{i=0}^{\mathbf{n}_T} h(\boldsymbol{\tau}_i) \tag{12-78}$$

where $\mathbf{n}_T$ is a Poisson RV with parameter λT and $\boldsymbol{\tau}_i$ are independent RVs uniform in the interval $(0, T)$. Reasoning as in Prob. 8-13, we obtain (12-77) because

$$E\{e^{sh(\boldsymbol{\tau}_i)}\} = \frac{1}{T}\int_0^T e^{sh(\alpha)}\, d\alpha \tag{12-79}$$

General Properties

Suppose now that the points $\mathbf{t}_i$ are nonuniform with mean density $\lambda(t)$ and that the function $h(t)$ is arbitrary. We shall show that the second moment function of the resulting *nonhomogeneous* shot noise process $\mathbf{s}(t)$ equals

$$\mathbf{\Psi}(s) = \ln \mathbf{\Phi}(s) = \int_{-\infty}^{\infty} \lambda(\alpha)[e^{h(t-\alpha)s} - 1]\, d\alpha \tag{12-80}$$

PROOF We divide the time axis into consecutive intervals I_i: (α_i, α_{i+1}) of length $\Delta\alpha = \alpha_{i+1} - \alpha_i$ as in Fig. 12-13 and we denote by $\Delta\mathbf{n}_i$ the number of points $\mathbf{t}_i$ in I_i. If $\Delta\alpha$ is sufficiently small, then the contribution $\Delta\mathbf{s}_i(t)$ to $\mathbf{s}(t)$ due to these points is given by [see (12-72)]

$$\Delta\mathbf{s}_i(t) \simeq h(t - \alpha_i)\, \Delta\mathbf{n}_i \tag{12-81}$$

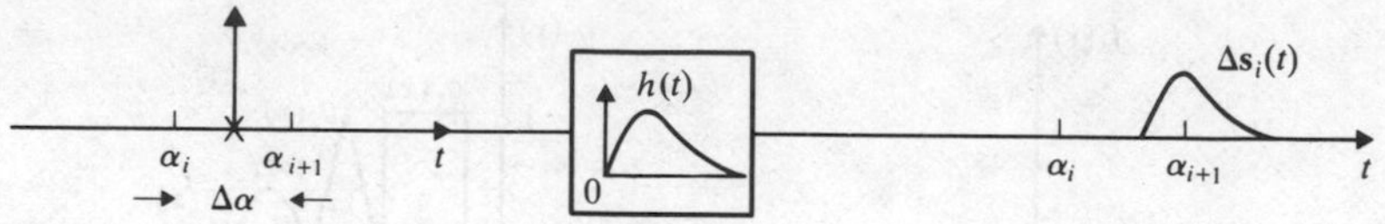

Figure 12-13

As we know, $\Delta\mathbf{n}_i$ is a Poisson RV with parameter

$$\int_{\alpha_i}^{\alpha_i+\Delta\alpha} \lambda(\alpha)\, d\alpha \simeq \lambda(\alpha_i)\, \Delta\alpha$$

Hence, the moment function of $\Delta\mathbf{s}_i(t)$ equals

$$\Delta\mathbf{\Phi}_i(s) \simeq E\{e^{sh(t-\alpha_i)\Delta\mathbf{n}_i}\} = \exp\{\lambda(\alpha_i)\,\Delta\alpha[e^{h(t-\alpha_i)s} - 1]\} \tag{12-82}$$

Furthermore

$$\mathbf{s}(t) = \sum_i \Delta\mathbf{s}_i(t) \simeq \sum_i h(t-\alpha_i)\,\Delta\mathbf{n}_i$$

And since the RVs $\Delta\mathbf{n}_i$ are independent, we conclude that

$$\mathbf{\Psi}(s) = \sum_i \ln \Delta\mathbf{\Phi}_i(s) \simeq \sum_i \lambda(\alpha_i)\,\Delta\alpha[e^{h(t-\alpha_i)s} - 1] \tag{12-83}$$

As $\Delta\alpha \to 0$, the last sum tends to the integral in (12-80).

Generalization of Campbell's theorem The mean η_s and the variance σ_s^2 of the nonhomogeneous shot noise process $\mathbf{s}(t)$ are given by

$$\eta_s = \int_{-\infty}^{\infty} \lambda(\alpha)h(t-\alpha)\, d\alpha \qquad \sigma_s^2 = \int_{-\infty}^{\infty} \lambda(\alpha)h^2(t-\alpha)\, d\alpha \tag{12-84}$$

PROOF Expanding the exponential in (12-80) into a series and integrating termwise, we obtain (12-84) because [see (5-73)]

$$\eta_s = \mathbf{\Psi}'(0) \qquad \sigma_s^2 = \mathbf{\Psi}''(0)$$

Joint characteristic functions The joint second moment function $\mathbf{\Psi}$ of the n RVs

$$\mathbf{s}(t_1), \ldots, \mathbf{s}(t_n)$$

equals

$$\mathbf{\Psi}(s_1, \ldots, s_n) = \int_{-\infty}^{\infty} \lambda(\alpha)(e^{\beta} - 1)\, d\alpha \tag{12-85}$$

where

$$\beta = s_1 h(t_1 - \alpha) + \cdots + s_n h(t_n - \alpha)$$

PROOF We assume for simplicity that $n = 2$. Proceeding as in (12-81) we denote by $\Delta\mathbf{n}_i$ the number of impulses in the interval I_i of Fig. 12-12, and by

$$\Delta\mathbf{s}_i(t_1) \simeq h(t_1 - \alpha_i)\,\Delta\mathbf{n}_i \qquad \Delta\mathbf{s}_i(t_2) \simeq h(t_2 - \alpha_i)\,\Delta\mathbf{n}_i$$

the response of the system at $t = t_1$ and $t = t_2$ respectively due to these impulses. The joint moment function $\mathbf{\Phi}$ of the RVs $\Delta\mathbf{s}_i(t_1)$ and $\Delta\mathbf{s}_i(t_2)$ equals

$$E\{e^{[s_1 h(t_1 - \alpha_i) + s_2 h(t_2 - \alpha_i)]\,\Delta\mathbf{n}_i}\}$$

This is of the same form as the second term in (12-82) if we replace the exponent $sh(t - \alpha_i)$ by the sum $s_1 h(t_1 - \alpha_i) + s_2 h(t_2 - \alpha_i)$. Hence, as in (12-83),

$$\mathbf{\Psi}(s_1, s_2) \simeq \sum_i \lambda(\alpha_i)[e^{s_1 h(t_1 - \alpha_i) + s_2 h(t_2 - \alpha_i)} - 1]\,\Delta\alpha$$

and with $\Delta\alpha \to 0$, (12-85) results.

Covariance We shall use (12-85) to determine the autocovariance $C(t_1, t_2)$ of $\mathbf{s}(t)$. As we know [see (7-36)], $C(t_1, t_2)$ is the coefficient of $s_1 s_2$ in the series expansion of $\mathbf{\Psi}(s_1, s_2)$ about the origin. Expanding the term e^{β} in (12-85), we conclude that

$$C(t_1, t_2) = \int_{-\infty}^{\infty} \lambda(\alpha) h(t_1 - \alpha) h(t_2 - \alpha)\, d\alpha \tag{12-86}$$

If $\lambda(t) = \lambda =$ constant, then, with $t_1 - t_2 = \tau$, the above yields

$$C(\tau) = \lambda \int_{-\infty}^{\infty} h(\tau + \alpha) h(\alpha)\, d\alpha \tag{12-87}$$

in agreement with (10-78).

High density and normality We shall show that if the density λ of stationary shot noise is large compared with the time constants of $h(t)$, then $\mathbf{s}(t)$ is approximately normal. For this purpose, we introduce the normalizations

$$\lambda_0 = \frac{\lambda}{k} \qquad h_0(t) = \sqrt{k}\, h(t) \qquad \beta_0 = \sqrt{k}\, \beta$$

and examine the form of $\mathbf{\Psi}(s_1, s_2)$ as $k \to \infty$. Clearly

$$\lambda(e^{j\beta} - 1) = \lambda_0\left(\sqrt{k}\,\beta_0 + \frac{\beta_0^2}{2!} + \frac{\beta_0^3}{\sqrt{k}} + \cdots\right)$$

Neglecting negative powers of k, we conclude from (12-85) that

$$\mathbf{\Psi}(s_1, \ldots, s_2) \simeq \lambda_0 \int_{-\infty}^{\infty} \left(\sqrt{k}\,\beta_0 + \frac{\beta_0^2}{2}\right) d\alpha \tag{12-88}$$

This shows that $\mathbf{\Psi}$ is a quadratic function of s_i because β_0 is linear in s_i. Hence, $\mathbf{s}(t)$ is nearly normal. This result is exact in the limit as $k \to \infty$ if the linear term is omitted (centering).

Intensity of shot noise The square

$$\mathbf{I}(t) = \mathbf{s}^2(t)$$

of the shot noise $\mathbf{s}(t)$ is a stationary process with mean

$$\eta_I = E\{\mathbf{s}^2(t)\} = \lambda \int_{-\infty}^{\infty} h^2(t)\,dt + \lambda^2 \left[\int_{-\infty}^{\infty} h(t)\,dt\right]^2 \qquad (12\text{-}89)$$

We shall determine its autocorrelation $R_I(\tau)$ under the assumption that

$$H(0) = \int_{-\infty}^{\infty} h(t)\,dt = 0 \qquad (12\text{-}90)$$

With this assumption, the mean and autocorrelation of $\mathbf{s}(t)$ are given by

$$\eta_s = 0 \qquad R_s(\tau) = \lambda \int_{-\infty}^{\infty} h(\tau + \alpha)h(\alpha)\,d\alpha \qquad (12\text{-}91)$$

We maintain that

$$R_I(\tau) = \lambda^2 E^2 + 2R_s^2(\tau) + \lambda \int_{-\infty}^{\infty} h^2(\tau + \alpha)h^2(\alpha)\,d\alpha \qquad (12\text{-}92)$$

where

$$E = \int_{-\infty}^{\infty} h^2(t)\,dt$$

is the energy of $h(t)$.

Proof The second moment function $\mathbf{\Psi}(s_1, s_2)$ of the RVs $\mathbf{s}(t_1)$ and $\mathbf{s}(t_2)$ equals the integral in (12-85) where

$$\beta = s_1 h(t_1 - \alpha) + s_2 h(t_2 - \alpha) \qquad \lambda(t) = \lambda$$

As we know [see (7-34)] the coefficient of $s_1^2 s_2^2$ in the expansion of the moment function $\mathbf{\Phi}(s_1, s_2)$ equals

$$\frac{E\{\mathbf{s}^2(t_1)\mathbf{s}^2(t_2)\}}{4}$$

We introduce the functions

$$\gamma_n = \frac{\lambda}{n!} \int_{-\infty}^{\infty} \beta^n\,d\alpha$$

Since

$$e^\beta - 1 = \beta + \frac{\beta^2}{2} + \cdots$$

and $\gamma_1 = 0$ [see (12-90)], we conclude that

$$\mathbf{\Psi}(s_1, s_2) = \gamma_2 + \gamma_3 + \gamma_4 + \cdots$$

$$\mathbf{\Phi}(s_1, s_2) = 1 + \gamma_2 + \gamma_3 + \gamma_4 + \cdots + \frac{(\gamma_2 + \gamma_3 + \gamma_4 + \cdots)^2}{2} + \cdots$$

In this expansion, only the sum $\gamma_4 + \gamma_2^2/2$ will have terms in $s_1^2 s_2^2$. Furthermore

$$\gamma_4 = \cdots + \lambda \frac{s_1^2 s_2^2}{4!} \int_{-\infty}^{\infty} \binom{4}{2} h^2(t_1 - \alpha) h^2(t_2 - \alpha)\, d\alpha + \cdots$$

$$\gamma_2^2 = \cdots + \lambda^2 \frac{s_1^2 s_2^2}{2} \int_{-\infty}^{\infty} h^2(t_1 - \alpha)\, d\alpha \int_{-\infty}^{\infty} h^2(t_2 - \alpha)\, d\alpha$$

$$+ \lambda^2 s_1^2 s_2^2 \left[\int_{-\infty}^{\infty} h(t_1 - \alpha) h(t_2 - \alpha)\, d\alpha \right]^2$$

and with $t_1 = t_2 + \tau$, (12-92) results.

Power spectrum The power spectrum $S_s(\omega)$ of $\mathbf{s}(t)$ equals [see (12-91)]

$$S_s(\omega) = \lambda |H(\omega)|^2 \tag{12-93}$$

Furthermore

$$\lambda^2 E^2 \leftrightarrow 2\pi\lambda^2 E^2 \delta(\omega) \qquad 2\pi R_s^2(\tau) \leftrightarrow S_s(\omega) * S_s(\omega)$$

The integral in (12-92) equals $h^2(t) * h^2(-t)$. And since

$$2\pi h^2(t) \leftrightarrow H(\omega) * H(\omega)$$

we conclude from (12-92) that the power spectrum of the intensity $\mathbf{I}(t)$ of $\mathbf{s}(t)$ equals

$$2\pi\lambda^2 E^2 \delta(\omega) + \frac{\lambda^2}{\pi} |H(\omega)|^2 * |H(\omega)|^2 + \frac{\lambda}{4\pi^2} |H(\omega) * H(\omega)|^2 \tag{12-94}$$

High density If λ is sufficiently large, then the third term above can be neglected. In this case, $\mathbf{s}(t)$ is nearly normal and the power spectrum of $\mathbf{s}^2(t)$ equals the sum of the first two terms in (12-94) [see also (9-89)].

Low density If λ is small, then the first two terms in (12-94) can be neglected. In this case

$$\mathbf{I}(t) = \mathbf{s}^2(t) \simeq \sum_i h^2(t - \mathbf{t}_i) \tag{12-95}$$

because the probability that the terms $h(t - \mathbf{t}_i)$ and $h(t - \mathbf{t}_k)$ have a significant overlap is negligible. This means that the intensity $\mathbf{I}(t)$ of $\mathbf{s}(t)$ is approximately a shot noise process generated by $h^2(t)$.

12-4 MARKOFF PROCESSES

A Markoff process is a stochastic process whose past has no influence on the future if its present is specified. This means the following:

If $t_{n-1} < t_n$, then

$$P\{\mathbf{x}(t_n) \le x_n | \mathbf{x}(t), t \le t_{n-1}\} = P\{\mathbf{x}(t_n) \le x_n | \mathbf{x}(t_{n-1})\} \tag{12-96}$$

From this it follows that if

$$t_1 < t_2 < \cdots < t_n$$

then

$$P\{\mathbf{x}(t_n) \le x_n \mid \mathbf{x}(t_{n-1}), \ldots, \mathbf{x}(t_1)\} = P\{\mathbf{x}(t_n) \le x_n \mid \mathbf{x}(t_{n-1})\} \tag{12-97}$$

The above definition holds also for discrete-time processes if $\mathbf{x}(t_n)$ is replaced by $\mathbf{x}_n$.

In this section, we develop various properties of Markoff processes, concentrating on three classes: (*a*) discrete-time, discrete-state; (*b*) continuous-time, discrete-state; (*c*) continuous-time, continuous-state. We start with a brief discussion of certain general properties phrasing the results in terms of discrete-time, continuous-state processes.

1. From (12-97) it follows that

$$f(x_n \mid x_{n-1}, \ldots, x_1) = f(x_n \mid x_{n-1}) \tag{12-98}$$

Applying the chain rule (8-37) to the above, we obtain

$$f(x_1, \ldots, x_n) = f(x_n \mid x_{n-1}) f(x_{n-1} \mid x_{n-2}) \cdots f(x_2 \mid x_1) f(x_1) \tag{12-99}$$

Conversely, if (12-99) is true for all n, then the process $\mathbf{x}_n$ is Markoff because, in this case,

$$f(x_n \mid x_{n-1}, \ldots, x_1) = \frac{f(x_1, \ldots, x_{n-1}, x_n)}{f(x_1, \ldots, x_{n-1})} = f(x_n \mid x_{n-1}) \tag{12-100}$$

2. From (12-99) it follows that

$$E\{\mathbf{x}_n \mid \mathbf{x}_{n-1}, \ldots, \mathbf{x}_1\} = E\{\mathbf{x}_n \mid \mathbf{x}_{n-1}\} \tag{12-101}$$

3. A Markoff process is also Markoff if time is reversed

$$f(x_n \mid x_{n+1}, \ldots, x_{n+k}) = f(x_n \mid x_{n+1}) \tag{12-102}$$

PROOF The left side of (12-102) equals

$$\frac{f(x_{n,\,n+1}, \ldots, x_{n+k})}{f(x_{n+1}, \ldots, x_{n+k})} = \frac{f(x_{n+1} \mid x_n)}{f(x_{n+1})} f(x_n)$$

And since

$$f(x_{n+1} \mid x_n) f(x_n) = f(x_n, x_{n+1}) = f(x_n \mid x_{n+1}) f(x_{n+1})$$

(12-102) results.

4. If the present is specified, then the past is independent of the future in the following sense: If $k < m < n$, then

$$f(x_n, x_k \mid x_m) = f(x_n \mid x_m) f(x_k \mid x_m) \tag{12-103}$$

PROOF From (12-99) it follows that

$$f(x_n, x_k \mid x_m) = \frac{f(x_n, x_m, x_k)}{f(x_m)} = \frac{f(x_n \mid x_m) f(x_m \mid x_k)}{f(x_m)} f(x_k)$$

and (12-103) results.

The above relationship can be used to express conditional densities involving the past and the future, in terms of the conditional (*transition*) densities $f(x_k \mid x_{k+1})$. For example

$$f(x_m \mid x_n, x_k) = \frac{f(x_k \mid x_m)}{f(x_k \mid x_n)} f(x_m \mid x_n) \tag{12-104}$$

Example 12-8 If $\mathbf{x}_n$ satisfies the recursion equation

$$\mathbf{x}_{n+1} - a(\mathbf{x}_n, n) = \mathbf{v}_n \tag{12-105}$$

and $\mathbf{v}_n$ is strictly white noise, then $\mathbf{x}_n$ is Markoff.

PROOF The process $\mathbf{x}_{n+1}$ is determined in terms of $\mathbf{x}_n$ and $\mathbf{v}_n$, hence it is independent of $\mathbf{x}_k$ for $k < n$.

SPECIAL CASE (generalized random walk) If $\mathbf{x}_0 = 0$ and $a(\mathbf{x}_n, n) = \mathbf{x}_n$, then

$$\mathbf{x}_n = \mathbf{x}_{n-1} + \mathbf{v}_n = \mathbf{v}_1 + \mathbf{v}_2 + \cdots + \mathbf{v}_n$$

Thus, the sum of independent RVs is a Markoff sequence.

Homogeneous processes From the chain rule (12-99) it follows that the statistics of any order of a Markoff process can be determined in terms of the conditional densities $f(x_n \mid x_{n-1})$ and the first-order density $f(x_n)$. If the process $\mathbf{x}_n$ is *stationary*, then the functions $f(x_n)$ and $f(x_n \mid x_{n-1})$ are invariant to a shift of the origin. In this case, the statistics of $\mathbf{x}_n$ are completely determined in terms of the second-order density

$$f(x_1, x_2) = f(x_2 \mid x_1) f(x_1)$$

A Markoff process $\mathbf{x}_n$ is called *homogeneous* if the conditional density $f(x_n \mid x_{n-1})$ is invariant to a shift of the origin but the first-order density $f(x_n)$ might depend on n. In general, a homogeneous process is not stationary. However, in many cases, it tends to a stationary process as $n \to \infty$.

The Chapman–Kolmogoroff equation The conditional density $f(x_n \mid x_k)$ can be expressed in terms of $f(x_n \mid x_m)$ and $f(x_m \mid x_k)$ for any $n > m > k$

$$f(x_n \mid x_k) = \int_{-\infty}^{\infty} f(x_n \mid x_m) f(x_m \mid x_k)\, dx_m \tag{12-106}$$

This follows from (8-39) because $f(x_n \mid x_m, x_k) = f(x_n \mid x_m)$.

Discrete-Time Markoff Chains

A discrete-time Markoff chain is a Markoff process $\mathbf{x}_n$ having a countable number of states a_i. A Markoff chain is specified in terms of its *state probabilities*

$$p_i[n] = P\{\mathbf{x}_n = a_i\} \qquad i = 1, 2, \ldots \tag{12-107}$$

and the *transition probabilities*

$$\pi_{ij}[n_1, n_2] = P\{\mathbf{x}_{n_2} = a_j \mid \mathbf{x}_{n_1} = a_i\} \tag{12-108}$$

As we know [see (7-48)]

$$\sum_j \pi_{ij}[n_1, n_2] = 1 \qquad \sum_i p_i[k]\pi_{ij}[k, n] = p_j[n] \tag{12-109}$$

Furthermore, if $n_1 < n_2 < n_3$, then

$$\pi_{ij}[n_1, n_3] = \sum_r \pi_{ir}[n_1, n_2]\pi_{rj}[n_2, n_3] \tag{12-110}$$

This is the discrete form of the Chapman–Kolmogoroff equation and it follows readily from (8-40).

Homogeneous chains If the process $\mathbf{x}_n$ is homogeneous, then the transition probabilities in (12-108) depend only on the difference $m = n_2 - n_1$. Thus

$$\pi_{ij}[m] = P\{\mathbf{x}_{n+m} = a_j \mid \mathbf{x}_n = a_i\} \tag{12-111}$$

Setting $n_2 - n_1 = k$, $n_3 - n_2 = n$ in (12-110), we obtain

$$\pi_{ij}[n + k] = \sum_r \pi_{ir}[k]\pi_{rj}[n] \tag{12-112}$$

For a finite-state Markoff chain, the above can be written in a vector form

$$\Pi[n + k] = \Pi[n]\Pi[k] \tag{12-113}$$

where $\Pi[n]$ is a Markoff matrix with elements $\pi_{ij}[n]$. This yields

$$\Pi[n] = \Pi^n \qquad \text{where} \qquad \Pi = \Pi[1] \tag{12-114}$$

is the one-step transition matrix with elements $\pi_{ij} = \pi_{ij}[1]$. The above is the solution of the first-order recursion equation [see (12-113)]

$$\Pi[n + 1] = \Pi[n]\Pi \tag{12-115}$$

The matrix Π is shown schematically in Fig. 12-14. The circles in the diagram represent the states a_i of the process and the number on each segment, from a_i to a_j, the transition probabilities π_{ij}. The number on the loop from a_i to a_i equals π_{ii}. This loop (dashed line) can be omitted because the sum of the row elements of Π (segments leaving a state, including the loop) equals 1 [see (12-109)].

$$\Pi = \begin{bmatrix} \pi_{11} & \pi_{12} & \cdots & \pi_{1N} \\ \pi_{21} & \pi_{22} & \cdots & \pi_{2N} \\ \cdots & \cdots & \cdots & \cdots \\ \pi_{N1} & \pi_{N2} & \cdots & \pi_{NN} \end{bmatrix}$$

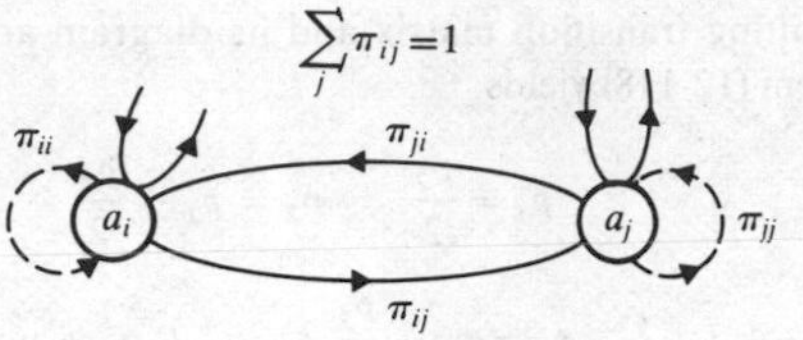

Figure 12-14

Writing (12-109) in vector form and using (12-114), we conclude that

$$P[n] = \cdots = P[n-k]\Pi^k = \cdots = P[0]\Pi^n \tag{12-116}$$

where $P[n]$ is a vector whose elements are the state probabilities $p_i[n]$.

In general, $P[n]$ depends on n. However, if the initial state vector

$$P[1] = P = [p_1, \ldots, p_N] \tag{12-117}$$

is such that $P[2] = P$, then $P[n] = P$ for all n. In this case, the homogeneous process $\mathbf{x}_n$ is also stationary and its state vector P is the solution of the system

$$P\Pi = P \qquad \sum_i p_i = 1 \tag{12-118}$$

The state probability vector P of a stationary Markoff chain is thus an eigenvector of its transition matrix Π and the corresponding eigenvalue equals 1 [see (12-109)].

If the initial state $P[1]$ of a homogeneous chain $\mathbf{x}_n$ does not equal P, then $\mathbf{x}_n$ is not stationary. In this case, certain of its states might never be reached. The details of the underlying theory will not be discussed. We note only that, if Π^n tends to a limit as $n \to \infty$, then $\mathbf{x}_n$ is asymptotically stationary. This is the case if all the elements π_{ij} of Π are strictly positive (Prob. 12-16).

Example 12-9 In the random walk experiment (Sec. 9-1), we place two reflecting walls at $x = 2s$ and $x = -2s$ as in Fig. 12-15. The resulting motion $\mathbf{x}(t)$ between the two walls generates a homogeneous Markoff chain $\mathbf{x}_n = \mathbf{x}(nT)$ taking the values

$$-2s \qquad -s \qquad 0 \qquad s \qquad 2s$$

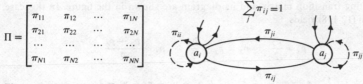

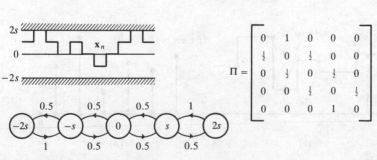

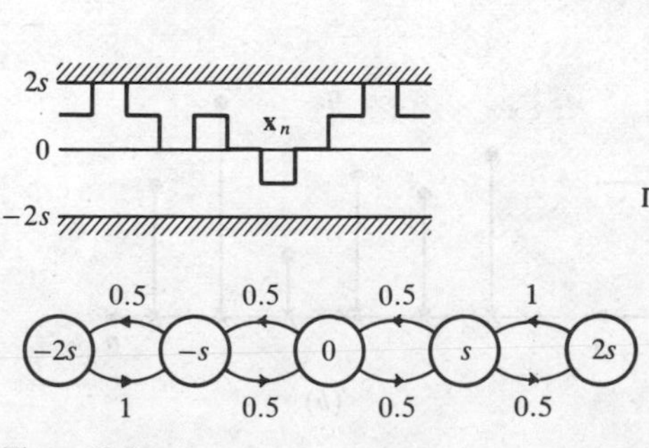

$$\Pi = \begin{bmatrix} 0 & 1 & 0 & 0 & 0 \\ \frac{1}{2} & 0 & \frac{1}{2} & 0 & 0 \\ 0 & \frac{1}{2} & 0 & \frac{1}{2} & 0 \\ 0 & 0 & \frac{1}{2} & 0 & \frac{1}{2} \\ 0 & 0 & 0 & 1 & 0 \end{bmatrix}$$

Figure 12-15

The resulting transition matrix and its diagram are shown in the figure. In this case, the system (12-118) yields

$$p_1 = \frac{p_2}{2} \qquad p_2 = p_1 + \frac{p_3}{2} \qquad p_3 = \frac{p_2 + p_4}{2}$$

$$p_4 = p_5 + \frac{p_3}{2} \qquad p_1 + p_2 + p_3 + p_4 + p_5 = 1$$

Solving, we obtain

$$p_1 = p_3 = \frac{1}{8} \qquad p_2 = p_3 = p_4 = \frac{1}{4}$$

These are the initial state probabilities that generate a stationary process.

Continuous-Time Markoff Chains

A continuous-time Markoff chain is a Markoff process $\mathbf{x}(t)$ consisting of a family of staircase functions (discrete states) with discontinuities at the random points $\mathbf{t}_n$ (Fig. 12-16*a*). The values

$$\mathbf{q}_n = \mathbf{x}(t_n^+) \tag{12-119}$$

of $\mathbf{x}(t)$ at these points (Fig. 12-16*b*) form a discrete-state Markoff sequence called the *Markoff chain imbedded* in the process $\mathbf{x}(t)$.

A discrete-state stochastic process is called *semi-Markoff* if it is not Markoff but the imbedded sequence $\mathbf{q}_n$ is a Markoff chain. An example is the queueing process $\mathbf{N}(t)$ of Sec. 12-2.

A Markoff chain $\mathbf{x}(t)$ is specified in terms of the underlying point process $\mathbf{t}_n$ and the imbedded Markoff chain $\mathbf{q}_n$.

We denote by

$$p_i(t) = P\{\mathbf{x}(t) = a_i\} \tag{12-120}$$

the *state probabilities* of $\mathbf{x}(t)$ and by

$$\pi_{ij}(t_1, t_2) = P\{\mathbf{x}(t_2) = a_j \mid \mathbf{x}(t_1) = a_i\} \tag{12-121}$$

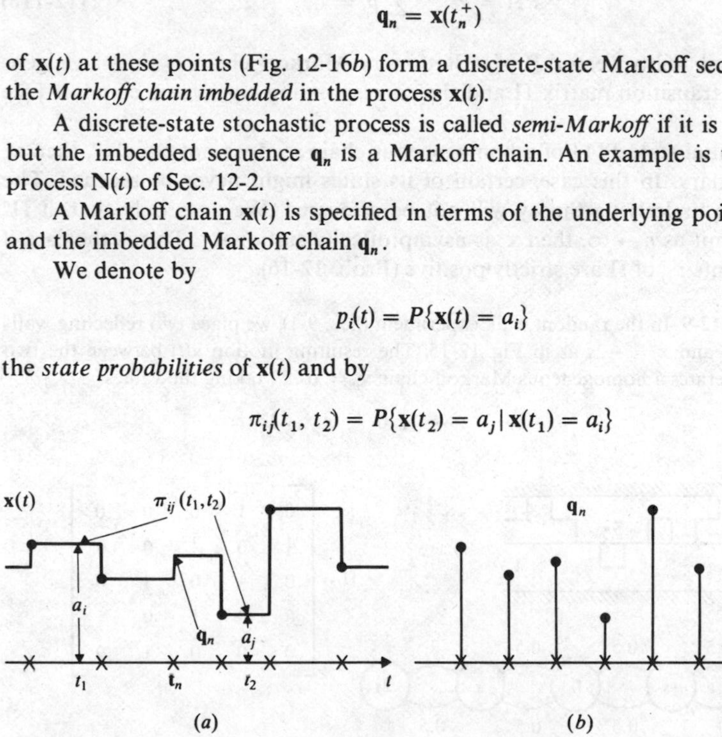

Figure 12-16

its *transition probabilities.* These functions are such that

$$\sum_j \pi_{ij}(t_1, t_2) = 1 \qquad \sum_i p_i(t_1)\pi_{ij}(t_1, t_2) = p_j(t_2) \tag{12-122}$$

and they satisfy the Chapman–Kolmogoroff equation

$$\pi_{ij}(t_1, t_3) = \sum_r \pi_{ir}(t_1, t_2)\pi_{rj}(t_2, t_3) \qquad t_1 < t_2 < t_3 \tag{12-123}$$

In specific problems the functions $\pi_{ij}(t_1, t_2)$ are not given directly. As we shall presently see, however, they can be determined in terms of the transition probability rates to be presently defined. For simplicity, we shall consider only *homogeneous processes.*

A Markoff process $\mathbf{x}(t)$ is homogeneous if its transition probabilities depend on the difference $\tau = t_2 - t_1$

$$\pi_{ij}(\tau) = P\{\mathbf{x}(t + \tau) = a_j \mid \mathbf{x}(t) = a_i\} \qquad \tau \geq 0 \tag{12-124}$$

From the above and (12-123) it follows with $\alpha = t_3 - t_2$ that

$$\pi_{ij}(\tau + \alpha) = \sum_r \pi_{ir}(\tau)\pi_{rj}(\alpha) \tag{12-125}$$

This is the Chapman–Kolmogoroff equation for continuous-time Markoff chains and it can be written in a vector form

$$\Pi(\tau + \alpha) = \Pi(\tau)\Pi(\alpha) \qquad \tau, \alpha \geq 0 \tag{12-126}$$

where $\Pi(\tau)$ is a matrix with elements $\pi_{ij}(\tau)$.

Probability rates In the discrete-time case, we showed that the matrix $\Pi[n]$ satisfies the recursion equation (12-115) and can be determined in terms of the one-step transition matrix Π. We show next that the transition matrix $\Pi(\tau)$ of a continuous-time chain $\mathbf{x}(t)$ satisfies a differential equation and can be determined in terms of the matrix

$$\Pi'(0^+) = \Lambda \equiv \begin{bmatrix} \lambda_{11}, \ldots, \lambda_{1n} \\ \cdots\cdots\cdots\cdots \\ \lambda_{n1}, \ldots, \lambda_{nn} \end{bmatrix} \tag{12-127}$$

whose elements $\lambda_{ij} = \pi'_{ij}(0^+)$ are the derivatives from the right of the elements $\pi_{ij}(\tau)$ of $\Pi(\tau)$. These derivatives will be called the *transition probability rates* of $\mathbf{x}(t)$. Clearly

$$\sum_j \lambda_{ij} = 0 \qquad \text{because} \qquad \sum_j \pi_{ij}(\tau) = 1$$

and since

$$\pi_{ij}(0) = \delta[i - j] = \begin{cases} 1 & i = j \\ 0 & i \neq j \end{cases} \tag{12-128}$$

we conclude with $\mu_i = -\lambda_{ii}$ that

$$\mu_i = \sum_j{}' \lambda_{ij} \geq 0 \qquad \lambda_{ij} > 0 \qquad i \neq j \tag{12-129}$$

The prime indicates summation for every $j \neq i$.

In the above, we have assumed that $\pi_{ij}(\tau)$ is differentiable at $\tau = 0^+$. This is so only if the probability that there is one discontinuity point in the interval $(t, t + \Delta t)$ is of the order of Δt

$$P\{\mathbf{x}(t + \Delta t) = a_j \mid \mathbf{x}(t) = a_i\} = \begin{cases} 1 - \mu_i \,\Delta t & i = j \\ \lambda_{ij} \,\Delta t & i \neq j \end{cases} \tag{12-130}$$

The Kolmogoroff equations Differentiating (12-126) with respect to α and setting $\alpha = 0$, we obtain

$$\Pi'(\tau) = \Pi(\tau)\Lambda \qquad \Pi(0) = 1 \tag{12-131}$$

This is a system of linear differential equations with constant coefficients and its initial condition $\Pi(0)$ is the identity matrix [see (12-128)]. Solving, we obtain

$$\Pi(\tau) = e^{\Lambda t} \tag{12-132}$$

We have thus expressed $\Pi(\tau)$ in terms of the transition rate matrix Λ.

The state probabilities $p_i(t)$ satisfies a similar system: Denoting by $P(t)$ a vector with elements $p_i(t)$, we conclude from (12-122) that

$$P(t + \tau) = P(t)\Pi(\tau) \tag{12-133}$$

Differentiating with respect to τ and setting $\tau = 0$, we obtain

$$P'(t) = P(t)\Lambda \tag{12-134}$$

This is a system of N equations of the form

$$p_i'(t) = -\mu_i p_i(t) + \sum_j{}' \lambda_{ji} p_j(t) \tag{12-135}$$

Its formal solution is a vector exponential

$$P(t) = P(0)e^{\Lambda t} \tag{12-136}$$

We have, thus, expressed $P(t)$ in terms of Λ and the initial state probabilities $p_i(0)$.

If $\mathbf{x}(t)$ is stationary, then $p_i(t) = p_i$ = constant. Hence [see (12-135)]

$$\mu_i p_i = \sum_j{}' \lambda_{ji} p_j \qquad \sum_i p_i = 1 \tag{12-137}$$

This is a system expressing the state probabilities of a stationary process in terms of the transition rates λ_{ij}.

Example 12-10 (generalized telegraph signal) Suppose that $\mathbf{x}(t)$ takes two values (Fig. 12-17):

$$a_1 = A \qquad a_2 = -A$$

and

$$\begin{aligned} P\{\mathbf{x}(t + \Delta t) = A \mid \mathbf{x}(t) = A\} &= 1 - \mu_1 \,\Delta t = \pi_{11}(\Delta t) \\ P\{\mathbf{x}(t + \Delta t) = -A \mid \mathbf{x}(t) = -A\} &= 1 - \mu_2 \,\Delta t = \pi_{22}(\Delta t) \end{aligned} \tag{12-138}$$

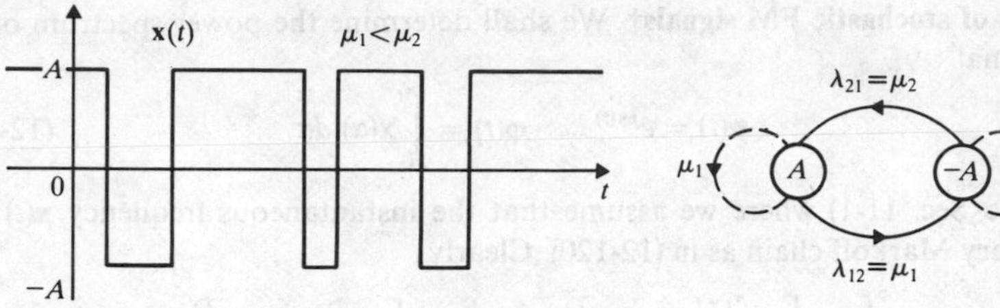

Figure 12-17

In this case, $\lambda_{12} = \mu_1$, $\lambda_{21} = \mu_2$. Inserting into (12-135), we obtain

$$p_1'(t) + \mu_1 p_1(t) = \mu_2 p_2(t)$$

And since $p_2(t) = 1 - p_1(t)$, we conclude that

$$p_1(t) = \frac{\mu_2}{\mu_1 + \mu_2}\left[1 - e^{-(\mu_1+\mu_2)t}\right] + p_1(0)e^{-(\mu_1+\mu_2)t}$$

We note that

$$p_1(t) \xrightarrow[t\to\infty]{} \frac{\mu_2}{\mu_1+\mu_2} = p_1 \qquad p_2(t) \xrightarrow[t\to\infty]{} \frac{\mu_1}{\mu_1+\mu_2} = p_2$$

The transition probabilities are determined from (12-131)

$$\pi_{11}'(\tau) + \mu_1 \pi_{11}(\tau) = \mu_2 \pi_{12}(\tau) \qquad \pi_{11}(0) = 1$$

$$\pi_{22}'(\tau) + \mu_2 \pi_{22}(\tau) = \mu_1 \pi_{21}(\tau) \qquad \pi_{22}(0) = 1$$

where

$$\pi_{12}(\tau) = 1 - \pi_{11}(\tau) \qquad \pi_{21}(\tau) = 1 - \pi_{22}(\tau)$$

The above yields

$$\pi_{11}(\tau) = p_1 + p_2 e^{-(\mu_1+\mu_2)\tau}$$
$$\pi_{21}(\tau) = p_2 + p_1 e^{-(\mu_1+\mu_2)\tau} \tag{12-139}$$

Mean and autocorrelation The process $\mathbf{x}(t)$ is asymptotically stationary with

$$P\{\mathbf{x}(t) = a_i\} = p_i \qquad a_i = A \qquad a_2 = -A$$

and

$$P\{\mathbf{x}(t+\tau) = a_j,\, \mathbf{x}(t) = a_i\} = p_i \pi_{ij}(\tau) \qquad i, j = 1, 2$$

From this it follows that

$$E\{\mathbf{x}(t)\} = \eta = p_1 A - p_2 A$$

$$R(\tau) = \eta^2 + 4A^2 p_1 p_2 e^{-(\mu_1+\mu_2)|\tau|}$$

If $\mu_1 = \mu_2 = \lambda$, then $\eta = 0$ and $R(\tau) = e^{-2\lambda|\tau|}$ as in (9-19). In this case only, the discontinuity points $\mathbf{t}_i$ of $\mathbf{x}(t)$ are Poisson.

Spectra of stochastic FM signals† We shall determine the power spectrum of the FM signal

$$\mathbf{w}(t) = e^{j\boldsymbol{\varphi}(t)} \qquad \boldsymbol{\varphi}(t) = \int_0^t \mathbf{x}(\alpha)\, d\alpha \tag{12-140}$$

(see also Sec. 11-1) where we assume that the instantaneous frequency $\mathbf{x}(t)$ is a stationary Markoff chain as in (12-120). Clearly

$$R(\tau) = E\left\{\exp\left[j\int_t^{t+\tau} \mathbf{x}(\alpha)\, d\alpha\right]\right\} = E\left\{\exp\left[j\int_0^{\tau} \mathbf{x}(\alpha)\, d\alpha\right]\right\} = E\{\mathbf{w}(\tau)\}$$

We introduce the conditional correlations

$$R_{ik}(\tau) = E\{\mathbf{w}(\tau) \mid \mathbf{x}(0) = a_i, \mathbf{x}(\tau) = a_k\}\pi_{ik}(\tau) \tag{12-141}$$

where $\pi_{ik}(\tau)$ are the transition probabilities defined in (12-124). Clearly

$$P\{\mathbf{x}(0) = a_i, \mathbf{x}(\tau) = a_k\} = p_i \pi_{ik}(\tau)$$

hence

$$R(\tau) = \sum_{i,k} p_i R_{ik}(\tau) \tag{12-142}$$

To determine $R(\tau)$ it suffices, therefore, to find $R_{ik}(\tau)$.

Theorem For any $\tau > 0$ and $\nu > 0$

$$R_{ik}(\tau + \nu) = \sum_m R_{im}(\tau) R_{mk}(\nu) \tag{12-143}$$

PROOF In the following, the conditions

$$\mathbf{x}(0) = a_i \qquad \mathbf{x}(\tau) = a_m \qquad \mathbf{x}(\tau + \nu) = a_k$$

will be abbreviated as a_i, a_m, and a_k respectively. Reasoning as in (12-104), we obtain

$$P\{\mathbf{x}(\tau) = a_m \mid a_i, a_k\} = \frac{\pi_{im}(\tau)\pi_{mk}(\tau)}{\pi_{ik}(\tau + \nu)} \tag{12-144}$$

Furthermore [see (8-43)]

$$E\{\mathbf{w}(\tau + \nu) \mid a_i, a_k\} = \sum_m E\{\mathbf{w}(\tau + \nu) \mid a_i, a_m, a_k\} P\{\mathbf{x}(\tau) = a_m \mid a_i, a_k\} \tag{12-145}$$

If $\mathbf{x}(\tau) = a_m$ is specified, then the integrals

$$\int_0^{\tau} \mathbf{x}(\alpha)\, d\alpha \qquad \int_{\tau}^{\tau+\nu} \mathbf{x}(\alpha)\, d\alpha$$

† R. Kubo: "A Stochastic Theory of Line-Shape and Relaxation," Scottish Universities Summer School, D. ter Haar, ed., Plenum Press, New York, 1961. See also A. Papoulis: "Spectra of Stochastic FM Signals" in *Proceedings of Transactions of 9th Prague Conference on Information Theory*, 1982.

are conditionally independent. Hence

$$E\left\{\exp\left[j\int_{\tau}^{\tau+\nu}\mathbf{x}(\alpha)\,d\alpha\bigg|_{a_i,\,a_m,\,a_k}\right]\right\}$$

$$= E\left\{\exp\left[j\int_{0}^{\tau}\mathbf{x}(\alpha)\,d\alpha\bigg|_{a_i,\,a_m}\right]\right\}E\left\{\exp\left[j\int_{\tau}^{\tau+\nu}\mathbf{x}(\alpha)\,d\alpha\bigg|_{a_m,\,a_k}\right]\right\}$$

From the stationarity of $\mathbf{x}(t)$ it follows that the last term equals

$$E\left\{\exp\left[j\int_{0}^{\nu}\mathbf{x}(\alpha)\,d\alpha\bigg|_{\mathbf{x}(0)=a_m,\,\mathbf{x}(\nu)=a_k}\right]\right\}$$

Inserting into (12-145) and using (12-144), we obtain (12-143).

The Kolmogoroff equations Differentiating (12-143) with respect to ν and setting $\nu = 0$, we obtain

$$R'_{ik}(\tau) = \sum_m R_{im}(\tau)R'_{mk}(0^+) \qquad \tau > 0 \tag{12-146}$$

Initial conditions To determine $R_{ik}(\tau)$ it suffices, therefore, to find the values of $R_{ik}(\tau)$ and its derivative at $\tau = 0^+$. We maintain that

$$R_{ik}(0^+) = \begin{cases}1 & i = k\\ 0 & i \neq k\end{cases} \qquad R'_{ik}(0^+) = \begin{cases}ja_i - \mu_i & i = k\\ \lambda_{ik} & i \neq k\end{cases} \tag{12-147}$$

PROOF For small τ,

$$\int_0^{\tau}\mathbf{x}(\alpha)\,d\alpha \simeq \mathbf{x}(0)\tau \qquad \pi_{ik}(\tau) \simeq \begin{cases}1 - \mu_i\tau & i = k\\ \lambda_{ik} & i \neq k\end{cases}$$

Neglecting terms of the order of τ^2, we conclude that

$$R_{ii}(\tau) \simeq E\left\{\exp\left[j\mathbf{x}(0)\tau\bigg|_{a_i}\right]\right\}\pi_{ii}(\tau) \simeq e^{ja_i\tau}(1 - \mu_i\tau) \simeq 1 + ja_i\tau - \mu_i\tau$$

Furthermore, for $i \neq k$

$$R_{ik}(\tau) \simeq e^{ja_i\tau}\lambda_{ik}\tau \simeq \lambda_{ik}\tau$$

and (12-147) follows.

Combining (12-147) and (12-146) we obtain

$$R'_{ik}(\tau) = (ja_k - \mu_k)R_{ik}(\tau) + \sum_m{}' \lambda_{mk}R_{im}(\tau) \tag{12-148}$$

This yields $R_{ik}(\tau)$. The autocorrelation $R(\tau)$ of $\mathbf{w}(t)$ is determined from (12-142). Since the coefficients of (12-148) are constant, we conclude that the power spectrum of an FM signal, whose instantaneous frequency is a finite-state Markoff chain, is rational.

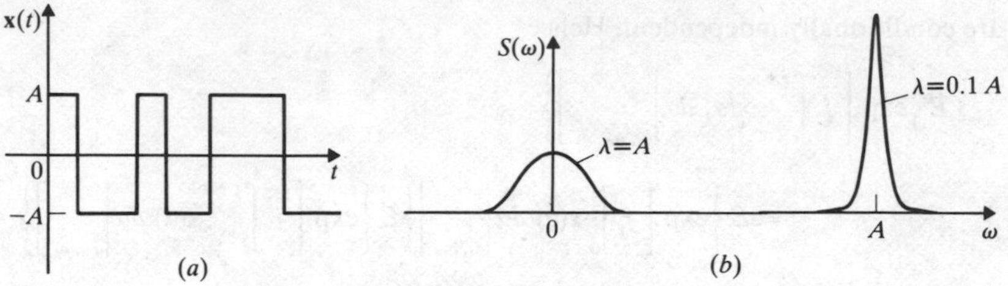

Figure 12-18

Example 12-11 Suppose that $\mathbf{x}(t)$ is a symmetrical telegraph signal (Fig. 12-18*a*). In this case

$$a_1 = -a_2 = A \qquad \mu_1 = \mu_2 = \lambda$$

and (12-148) yields

$$R'_{11}(\tau) = (jA - \lambda)R_{11}(\tau) + \lambda R_{12}(\tau) \qquad R_{11}(0) = 1$$

$$R'_{12}(\tau) = \lambda R_{11}(\tau) - (jA + \lambda)R_{12}(\tau) \qquad R_{12}(0) = 0$$

Denoting by $\mathbf{S}_{ik}^{+}(s)$ the Laplace transform of $R_{ik}^{+}(\tau)$, we conclude from the above that

$$s\mathbf{S}_{11}^{+}(s) - 1 = (jA - \lambda)\mathbf{S}_{11}^{+}(s) + \lambda\mathbf{S}_{12}^{+}(s)$$

$$s\mathbf{S}_{12}^{+}(s) = \lambda\mathbf{S}_{11}^{+}(s) - (jA + \lambda)\mathbf{S}_{12}^{+}(s)$$

Hence

$$\mathbf{S}_{11}^{+}(s) = \frac{s + jA + \lambda}{D(s)} \qquad \mathbf{S}_{12}^{+}(s) = \frac{\lambda}{D(s)}$$

where

$$D(s) = s^2 + 2\lambda s + A^2$$

Reasoning similarly, we find

$$\mathbf{S}_{21}^{+}(s) = \frac{\lambda}{D(s)} \qquad \mathbf{S}_{22}^{+}(s) = \frac{s - jA + \lambda}{D(s)}$$

And since $p_1 = p_2 = 0.5$, (12-142) yields

$$\mathbf{S}^{+}(s) = \frac{s + 2\lambda}{D(s)} \qquad S(\omega) = 2 \text{ Re } \mathbf{S}^{+}(j\omega)$$

In Fig. 12-18*b*, we plot $S(\omega)$ for $\lambda = A$ and $\lambda = 0.1A$. We note that the discontinuity points (zero crossings) of $\mathbf{x}(t)$ are Poisson distributed and their average density equals λ.

Birth processes A birth process is a Markoff chain $\mathbf{x}(t)$ consisting of a family of increasing staircase functions (Fig. 12-19). The process $\mathbf{x}(t)$ (population size) takes the values 1, 2, 3, ... and it increases by 1 at the discontinuity points $\mathbf{t}_i$ (birth

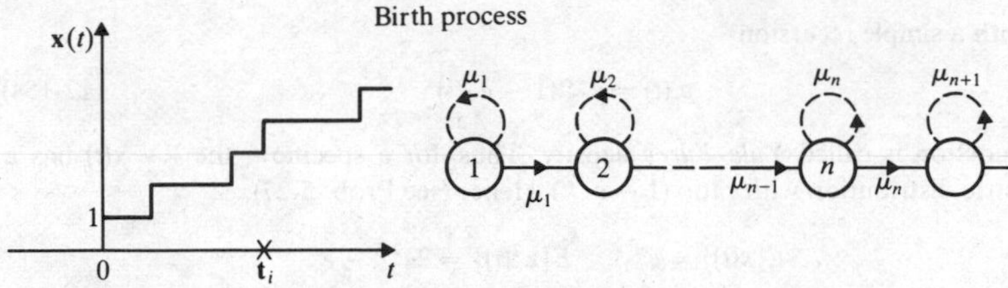

Figure 12-19

times). From the definition it follows that the transition rates λ_{ij} are different from zero only if $i = j$ or $i = j - 1$. Thus

$$-\lambda_{ii} = \mu_i \qquad \lambda_{i(i+1)} = \mu_i \qquad \lambda_{ij} = 0 \qquad \text{otherwise}$$

hence, the process is specified in terms of the parameter μ_i. Clearly [see (12-130)]

$$\begin{aligned} P\{\mathbf{x}(t+\Delta t) = n \mid \mathbf{x}(t) = n\} &= 1 - \mu_n \Delta t \qquad n \ge 1 \\ P\{\mathbf{x}(t+\Delta t) = n \mid \mathbf{x}(t) = n-1\} &= \mu_{n-1} \Delta t \qquad n > 1 \end{aligned} \tag{12-149}$$

Hence

$$p_1(t+\Delta t) = p_1(t)(1 - \mu_1 \Delta t)$$

$$p_n(t+\Delta t) = p_n(t)(1 - \mu_n \Delta t) + p_{n-1}(t)\mu_{n-1} \Delta t \qquad n > 1$$

This yields

$$\begin{aligned} p_1'(t) &= -\mu_1 p_1(t) \\ p_n'(t) &= -\mu_n p_n(t) + \mu_{n-1} p_{n-1}(t) \qquad n > 1 \end{aligned} \tag{12-150}$$

in agreement with (12-135).

Note The difference $\mathbf{x}(t_2) - \mathbf{x}(t_1)$ equals the number of discontinuity points $\mathbf{t}_i$ in the interval (t_1, t_2). This shows that a birth process is completely specified in terms of the point process $\mathbf{t}_i$.

Example 12-12 If the birthrate is proportional to the population size n

$$\mu_n = nc \tag{12-151}$$

then $\mathbf{x}(t)$ is called *simple birth process* (the constant c is the birthrate per person). We shall determine $p_n(t)$ under the (unrealistic!) assumption that (12-151) holds for every $n \ge 1$ and that $\mathbf{x}(0) = 1$. In this case

$$p_1(0) = 1 \qquad p_n(0) = 0 \qquad n > 1 \tag{12-152}$$

Setting $\mu_n = nc$ in (12-150), we obtain

$$\begin{aligned} p_1'(t) + cp_1(t) &= 0 \\ p_n'(t) + ncp_n(t) &= (n-1)cp_{n-1}(t) \qquad n > 1 \end{aligned} \tag{12-153}$$

The above yields

$$p_1(t) = p_1(0)e^{-ct} = e^{-ct}$$

and with a simple recursion

$$p_n(t) = e^{-ct}(1 - e^{-ct})^{n-1} \qquad (12\text{-}154)$$

This function is called *Yule–Furry* density. Thus, for a specific t, the RV $\mathbf{x}(t)$ has a geometric distribution with ratio $(1 - e^{-ct})$. Hence (see Prob. 5-27)

$$E\{\mathbf{x}(t)\} = e^{ct} \qquad E\{\mathbf{x}^2(t)\} = 2e^{2ct} - e^{ct}$$

Example 12-13 We now assume that the rate of increase of $\mathbf{x}(t)$ is independent of its present state

$$\mu_n = \lambda = \text{constant}$$

As we shall presently see, the resulting $\mathbf{x}(t)$ is a *Poisson process.* We assume again that $\mathbf{x}(0) = 1$. Setting $\mu_n = \lambda$ in (12-150), we obtain

$$p_1'(t) + \lambda p_1(t) = 0 \qquad p_1(0) = 1$$

$$p_n'(t) + \lambda p_n(t) = \lambda p_{n-1}(t) \qquad p_n(0) = 0 \qquad n > 1$$

This yields

$$p_{n+1}(t) = \frac{e^{-\lambda t}(\lambda t)^n}{n!}$$

The above is the probability that $\mathbf{x}(t) = n + 1$ and it equals the probability that the number of points $\mathbf{x}(t) - \mathbf{x}(0)$ in the interval $(0, t)$ equals n.

Birth–death processes Suppose now that a Markoff chain takes the values 0, 1, 2, ... and its discontinuities equal +1 or −1. (Fig. 12-20). We then say that $\mathbf{x}(t)$ is a birth–death process. In this case, λ_{ij} is different from zero only if $i = j$ or $j - 1$ or $j + 1$. Hence, $\mathbf{x}(t)$ is specified in terms of the two parameters

$$\alpha_i = \lambda_{i(i+1)} \qquad \beta_i = \lambda_{i(i-1)}$$

Thus, $-\lambda_{ii} = \mu_i = \alpha_i + \beta_i$ and

$$P\{\mathbf{x}(t + \Delta t) = n \mid \mathbf{x}(t) = n - 1\} = \alpha_{n-1}\,\Delta t$$

$$P\{\mathbf{x}(t + \Delta t) = n \mid \mathbf{x}(t) = n\} = (\alpha_n + \beta_n)[1 - (\alpha_n + \beta_n)\,\Delta t]$$

$$P\{\mathbf{x}(t + \Delta t) = n \mid \mathbf{x}(t) = n + 1\} = \beta_{n+1}\,\Delta t$$

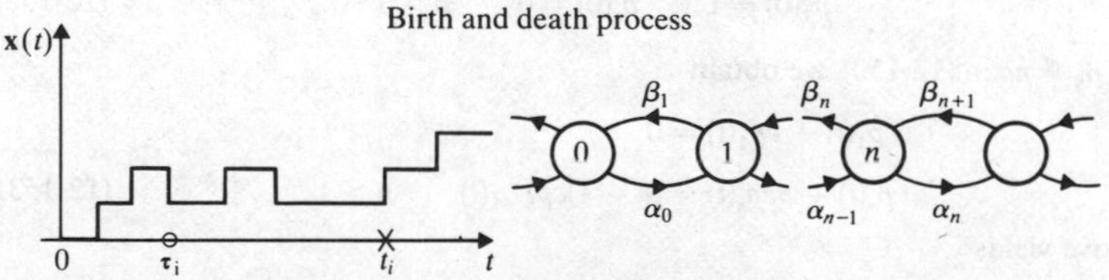

Figure 12-20

From (12-135) or, directly from the above, it follows that

$$\begin{aligned} p_0'(t) + \alpha_0 p_0(t) &= \beta_1 p_1(t) \\ p_n'(t) + (\alpha_n + \beta_n)p_n(t) &= \alpha_{n-1}p_{n-1}(t) + \beta_{n+1}p_{n+1}(t) \qquad n > 0 \end{aligned} \tag{12-155}$$

Example 12-14 ($M \mid M \mid 1$ queue) A queueing process $\mathbf{N}(t)$ is not, in general, Markoff. The $M \mid M \mid 1$ case, however, is an exception. In this case, the arrival times $\mathbf{t}_i$ are Poisson with average density λ and the service time $\mathbf{c}_i$ is an RV with density $\mu e^{-\mu c}$ where $\mu > \lambda$. We maintain that the resulting $\mathbf{N}(t)$ is a birth–death process.

The probability that a unit will arrive in the interval $(t, t + \Delta t)$ equals $\lambda\,\Delta t$ (property P_1 of Poisson points). We shall show that the probability that a unit will depart in the interval $(t, t + \Delta t)$ equals $\mu\,\Delta t$. Indeed, denoting by $\boldsymbol{\tau}_i$ the first departure point to the right of t and by $\mathbf{c}_i$ the corresponding service time, we conclude from (4-54) that

$$P\{t < \mathbf{c}_i \le t + \Delta t\} \simeq f_c(0)\,\Delta t = \mu\,\Delta t$$

no matter when this service started.

We have thus shown that

$$\begin{aligned} P\{\mathbf{N}(t + \Delta t) = n \mid \mathbf{N}(t) = n - 1\} &= \lambda\,\Delta t \\ P\{\mathbf{N}(t + \Delta t) = n - 1 \mid \mathbf{N}(t) = n\} &= \mu\,\Delta t \end{aligned} \tag{12-156}$$

Thus, $\mathbf{N}(t)$ is a birth–death process with $\alpha_n = \lambda$ and $\beta_n = \mu$. We shall determine its state probabilities p_n for the stationary case.

In this case, $p_n'(t) = 0$ and (12-155) yields

$$\lambda p_0 = \mu p_1 \qquad (\lambda + \mu)p_n = \lambda p_{n-1} + \mu p_{n+1}$$

From this it follows readily that

$$p_n = p_0\left(\frac{\lambda}{\mu}\right)^n \qquad p_0 = 1 - \frac{\lambda}{\mu}$$

as in Example 12-5.

Continuous-State Processes

A continuous-state Markoff process $\mathbf{x}(t)$ is specified in terms of its first-order density

$$p(x, t) = P\{\mathbf{x}(t) \le x\}$$

and the conditional (transition) density

$$\pi(x, x_0; t, t_0) = f_{\mathbf{x}(t)}(x \mid \mathbf{x}(t_0) = x_0) \qquad t > t_0$$

These functions are such that, if $t_0 < t < t_1$, then

$$\begin{aligned} &\int_{-\infty}^{\infty} p(x, t)\,dx = 1 \\ &p(x, t) = \int_{-\infty}^{\infty} p(x_0, t_0)\pi(x, x_0; t, t_0)\,dx_0 \end{aligned} \tag{12-157}$$

and

$$\int_{-\infty}^{\infty} \pi(x, x_0; t, t_0)\, dx = 1$$
$$\pi(x, x_0; t, t_0) = \int_{-\infty}^{\infty} \pi(x, x_1; t, t_1)\pi(x_1, x_0; t_1, t_0)\, dx_1 \tag{12-158}$$

as in (12-122) and (12-123).

Furthermore

$$\pi(x, x_0; t, t_0) \xrightarrow[t \to t_0]{} \delta(x - x_0) \tag{12-159}$$

We shall show that the function $\pi(x, x_0; t, t_0)$ can be determined in terms of the slopes η and σ^2 of the *conditional mean* $a(x_0; t, t_0)$ and the *conditional variance* $b(x_0; t, t_0)$ of $\mathbf{x}(t)$ assuming $\mathbf{x}(t_0) = x_0$, defined as follows:

$$a(x_0; t, t_0) = \int_{-\infty}^{\infty} x\pi(x, x_0; t, t_0)\, dx$$
$$b(x_0; t, t_0) = \int_{-\infty}^{\infty} (x - a)^2 \pi(x, x_0; t, t_0)\, dx \tag{12-160}$$

We assume that these functions are differentiable from the right and we denote by $\eta(x_0, t_0)$ and $\sigma^2(x_0, t_0)$ respectively their slopes at $t = t_0$ (Fig. 12-21)

$$\eta(x_0, t_0) = \frac{\partial}{\partial t} a(x_0; t_0, t)\bigg|_{t=t_0^+}$$
$$\sigma^2(x_0; t_0) = \frac{\partial}{\partial t} b(x_0; t_0, t)\bigg|_{t=t_0^+} \tag{12-161}$$

Clearly [see (12-159) and (12-160)]

$$a(x_0; t_0, t_0) = x_0 \qquad b(x_0; t_0, t_0) = 0$$

Hence, for $\Delta t > 0$,

$$a(x_0; t_0 + \Delta t, t_0) \simeq x_0 + \eta(x_0, t_0)\, \Delta t$$
$$b(x_0; t_0 + \Delta t, t_0) \simeq \sigma^2(x_0, t_0)\, \Delta t \tag{12-162}$$

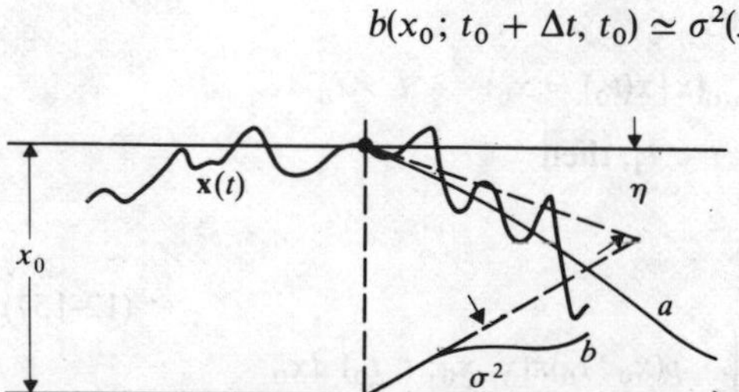

Figure 12-21

If the process $\mathbf{x}(t)$ is homogeneous, then the function $\pi(x, x_0; t, t_0)$ depends on $\tau = t - t_0$. In this case, the slopes $\eta(x_0)$ and $\sigma^2(x_0)$ of a and b are independent of t_0.

From (12-162) and the definition of the functions a and b it follows that

$$E\{d\mathbf{x}(t) \mid \mathbf{x}(t) = x\} = \eta(x, t)\, dt$$
$$E\{[d\mathbf{x}(t) - \eta(\mathbf{x}, t)\, dt]^2 \mid x\} = \sigma^2(x, t)\, dt \tag{12-163}$$

As we show in the next example, these equations can often be used to determine $\eta(x, t)$ and $\sigma^2(x, t)$ directly in terms of the specifications of the process $\mathbf{x}(t)$.

Example 12-15 Consider the nonlinear stochastic differential equation

$$\frac{d\mathbf{x}(t)}{dt} + \beta(\mathbf{x}, t) = \frac{d\mathbf{w}(x, t)}{dt}$$

where $\mathbf{w}(x, t)$ is a process with independent increments and such that

$$E\{d\mathbf{w}(x, t)\} = 0 \qquad E\{[d\mathbf{w}(x, t)]^2\} = \gamma(x, t)\, dt$$

The solution of the above equation is a Markoff process [see also (12-105)]. Clearly

$$E\{d\mathbf{x}(t) \mid x\} = -\beta(x, t)\, dt$$
$$E\{[d\mathbf{x}(t) + \beta(\mathbf{x}, t)\, dt]^2 \mid x\} = E\{[d\mathbf{w}(x, t)]^2 \mid x\} = \gamma(x, t)\, dt$$

Hence, $\eta(x, t) = -\beta(x, t)$, $\sigma^2(x, t) = \gamma(x, t)$.

The diffusion equations We shall show that the conditional density $\pi = \pi(x, x_0; t, t_0)$ satisfies the *diffusion equations*

$$\frac{\partial \pi}{\partial t} + \frac{\partial}{\partial x}[\eta(x, t)\pi] - \frac{1}{2}\frac{\partial^2}{\partial x^2}[\sigma^2(x, t)\pi] = 0$$
$$\frac{\partial \pi}{\partial t_0} + \eta(x_0, t_0)\frac{\partial \pi}{\partial x_0} + \frac{1}{2}\sigma^2(x_0, t_0)\frac{\partial^2 \pi}{\partial x_0^2} = 0 \tag{12-164}$$

The first is called *forward* (or *Fokker–Planck*) and the second *backward*.

PROOF If $f(x)$ is a density with mean η and "small" variance, then [see (5-55)]

$$\int_{-\infty}^{\infty} g(\xi) f(\xi)\, d\xi \simeq g(\eta) + \frac{\sigma^2}{2} g''(\eta) \tag{12-165}$$

From (12-158) it follows with $x_1 = \xi$ and $t_1 = t_0 + \varepsilon$ that

$$\pi(x, x_0; t, t_0) \simeq \int_{-\infty}^{\infty} \pi(x, \xi; t, t_0 + \varepsilon)\pi(\xi, x_0; t_0 + \varepsilon, t_0)\, d\xi$$

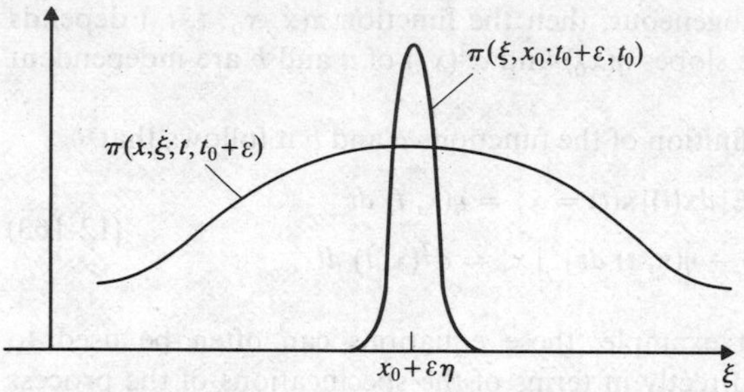

Figure 12-22

In the above, $\pi(\xi, x_0; t_0 + \varepsilon, t_0)$ is a density in the variable ξ and its mean and variance equal [see (12-160) and (12-162)]

$$a(x_0; t_0 + \varepsilon, t_0) \simeq x_0 + \varepsilon\eta(x_0, t_0) \qquad b(x; t_0 + \varepsilon, t_0) \simeq \varepsilon\sigma^2(x_0, t_0)$$

Therefore, with

$$g(\xi) = \pi(x, \xi; t, t_0 + \varepsilon) \qquad f(\xi) = \pi(\xi, x_0; t_0 + \varepsilon, t_0)$$

(12-165) yields (Fig. 12-22)

$$\pi(x, x_0; t, t_0) \simeq \pi(x, x_0 + \varepsilon\eta; t, t_0 + \varepsilon) + \frac{\varepsilon\sigma^2}{2}\frac{\partial^2}{\partial x_0^2}\pi(x, x_0 + \varepsilon; t, t_0 + \varepsilon)$$

within $O(\varepsilon^2)$. Expanding the right side into a power series in ε and retaining only linear terms, we obtain the second equation in (12-164). The proof of the first is similar.

Corollary The first-order density $p = p(x, t)$ of the process $\mathbf{x}(t)$ satisfies the Fokker–Planck equation

$$\frac{\partial p}{\partial t} + \frac{\partial}{\partial x}[\eta(x, t)p] - \frac{1}{2}\frac{\partial^2}{\partial x^2}[\sigma^2(x, t)p] = 0 \qquad (12\text{-}166)$$

PROOF It follows if we express the function $p(x, t)$ in terms of the integral in (12-157) and use (12-164).

Example 12-16 The velocity $\mathbf{v}(t)$ of a particle in brownian motion satisfies Langevin's equation

$$\mathbf{v}'(t) + \beta\mathbf{v}(t) = \mathbf{w}'(t)$$

where $\mathbf{w}(t)$ is a process with orthogonal increments and such that

$$E\{d\mathbf{w}(t)\} = 0 \qquad E\{[d\mathbf{w}(t)]^2\} = \gamma\, dt$$

Hence (see Example 12-15) $\mathbf{v}(t)$ is a Markoff process with $\eta(x, t) = -\beta v$, $\sigma^2(x, t) = \gamma$, and (12-166) yields

$$\frac{\partial p}{\partial t} = \beta\frac{\partial(vp)}{\partial v} + \frac{\gamma}{2}\frac{\partial^2 p}{\partial v^2} \qquad (12\text{-}167)$$

where $p = p(v, t)$ is the density of $\mathbf{v}(t)$.

Solution of the Fokker–Planck equation We shall solve the forward equation in (12-164) under the assumption that the conditional density $\pi(x, x_0; t, t_0)$ does not depend explicitly on the state $\mathbf{x}(t_0) = x_0$ of $\mathbf{x}(t)$ but only on the increment $u = x - x_0$. In this case [see (12-160)]

$$a(x_0; t, t_0) = \int_{-\infty}^{\infty} (u + x_0)\pi(u; t, t_0)\, du = a_1(t, t_0) + x_0$$

Inserting into the second integral in (12-160), we conclude that the conditional variance $b = b(t, t_0)$ does not depend on x_0. From the above it follows that the slopes $\eta(t_0)$ and $\sigma^2(t_0)$ of a and b are independent of x_0. This simplifies the form of the forward equation

$$\frac{\partial \pi}{\partial t} + \eta(t)\frac{\partial \pi}{\partial x} - \frac{1}{2}\sigma^2(t)\frac{\partial^2 \pi}{\partial x^2} = 0 \tag{12-168}$$

The solution $\pi(u; t, t_0)$ of (12-168) is a density in the variable u. In fact, it is the density of the increment $\mathbf{x}(t) - \mathbf{x}(t_0)$ of $\mathbf{x}(t)$ under the assumption that $\mathbf{x}(t_0) = x_0$. We shall show that this density is normal with

$$\text{Mean:} \quad \int_{t_0}^{t} \eta(\tau)\, d\tau \qquad \text{Variance:} \quad \int_{t_0}^{t} \sigma^2(\tau)\, d\tau \tag{12-169}$$

PROOF With $\mathbf{\Phi}(s, t)$ the bilateral Laplace transform in the variable u of the function $\pi(u; t, t_0)$, it follows from (12-168) that

$$\frac{\partial \mathbf{\Phi}}{\partial t} = -\eta(t)s\mathbf{\Phi} + \frac{\sigma^2(t)s^2}{2}\mathbf{\Phi} \tag{12-170}$$

The function $\mathbf{\Phi}(s, t)$, evaluated at $t = t_0$, is the transform of the function $\pi(u; t_0, t_0) = \delta(u)$ [see (12-159)]. Hence, $\mathbf{\Phi}(s, t_0) = 1$ and (12-170) yields

$$\ln \mathbf{\Phi}(s, t) = -s\int_{t_0}^{t} \eta(\tau)\, d\tau + \frac{s^2}{2}\int_{t_0}^{t} \sigma^2(\tau)\, d\tau \tag{12-171}$$

This shows that $\mathbf{\Phi}(s, t)$ is the moment function of a normal density with mean and variance as in (12-169).

PROBLEMS

12-1 Passengers arrive at a terminal boarding the next bus. The times of their arrival are Poisson with density $\lambda = 1$ per minute. The times of departure of each bus are Poisson with density $\mu = 2$ per hour. (*a*) Find the mean number of passengers in each bus. (*b*) Find the mean number of passengers in the first bus that leaves after 9 A.M.

Answer: (*a*) 30; (*b*) 60.

12-2 Passengers arrive at a terminal after 9 A.M. The times of their arrival are Poisson with mean density $\lambda = 1$ per minute. The time interval from 9 A.M. to the departure of the next bus is an RV **c**. Find the mean number of passengers in this bus (*a*) if **c** has an exponential density with mean $\eta_c = 30$ min, (*b*) if **c** is uniform between 0 and 60 min.

Answer: (*a*) 30; (*b*) 30.

12-3 The point process $\mathbf{t}_i$ is stationary and the RVs $\mathbf{c}_i = \mathbf{t}_i - \mathbf{t}_{i-1}$ are uniform in the interval (0, *a*). Show that if $\mathbf{t}_1$ is the first point to the right of a fixed point t_0, then $E\{\mathbf{t}_1 - t_0\} = a/3$.

12-4(*a*) The RVs $\mathbf{c}_i$ are i.i.d. and $E\{e^{j\omega \mathbf{c}_i}\} = \Phi_c(\omega)$. The process $\mathbf{n}(t)$ is Poisson with parameter λ and independent of $\mathbf{c}_i$. Show that, if (Fig. P12-4*a*)

$$\mathbf{x}(t) = \sum_{i=1}^{\mathbf{n}(t)} \mathbf{c}_i \qquad \text{then } E\{e^{j\omega \mathbf{x}(t)}\} = \exp\{\lambda t[\Phi_c(\omega) - 1]\}$$

Special case If the RV $\mathbf{c}_i$ takes the values 1 and 0 (Fig. P12-4*b*) and $P\{\mathbf{c}_i = 1\} = p$, then $\mathbf{x}(t)$ is a Poisson process with parameter λp (see also Prob. 8-11).

Hint: $E\{e^{j\omega \mathbf{x}(t)} \mid \mathbf{n}(t) = n\} = \Phi_c^n(\omega)$.

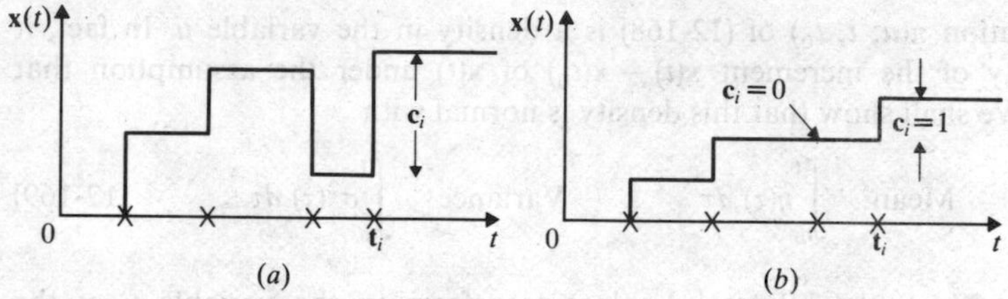

Figure P12-4

(*b*) Using the above, show that, if $\mathbf{t}_i$ is a Poisson point process with mean density λ and $\boldsymbol{\tau}_i$ is a process obtained by eliminating at random a subset of $\mathbf{t}_i$, then $\boldsymbol{\tau}_i$ is a Poisson point process with mean density λp where p is the probability that a point of $\mathbf{t}_i$ is not eliminated.

12-5(*a*) Show that if $\mathbf{t}_n$ is a Poisson point process, then the process $\mathbf{t}_{2n}$ consisting of every other point of $\mathbf{t}_n$ is not Poisson.

(*b*) Show that if $\boldsymbol{\alpha}_n$ and $\boldsymbol{\beta}_n$ are two independent Poisson point processes with densities λ_α and λ_β respectively, then the process $\mathbf{t}_n$ consisting of all the points of $\boldsymbol{\alpha}_n$ and $\boldsymbol{\beta}_n$ is Poisson with density $\lambda_\alpha + \lambda_\beta$.

12-6 Visitors enter a park at Poisson times with mean density $\lambda = 2$ per minute. Each visitor stays in the park **c** minutes where **c** is an RV uniform between 30 and 90 min. Find the mean and the variance of the number $\mathbf{N}(t)$ of visitors in the park.

12-7 In the $M \mid M \mid 1$ queue (Example 12-4), **y** is the busy period and $\boldsymbol{\Phi}_y(s)$ is its moment function. Show that

$$\lambda \boldsymbol{\Phi}_y^2(s) - (\lambda + \mu - s)\boldsymbol{\Phi}_y(s) + \mu = 0 \qquad \boldsymbol{\Phi}_y(s) \xrightarrow[|s \to \infty]{} 0$$

12-8(*a*) With $\bar{q}_i$ as in (12-49), show that

$$E\{\bar{\mathbf{q}}_i^2\} = E\{\mathbf{q}_i^2\} - 2\eta_q + \rho \qquad \text{(i)}$$

(*b*) Prove the Pollaczek–Khinchin formula (12-56), using (i) and the identity

$$E\{\mathbf{q}_i^2\} = E\{\bar{\mathbf{q}}_i^2\} + E\{\mathbf{n}_{ci}^2\} + 2E\{\bar{\mathbf{q}}_i\}E\{\mathbf{n}_{ci}\}$$

12-9 In a single-server queueing system, the arrival times $\mathbf{t}_i$ are Poisson with mean density $\lambda = 9$ per hour. Find the mean of the following: the service time $\mathbf{c}$, the waiting time $\mathbf{b}$, the system time $\mathbf{a}$, the idle period $\mathbf{x}$, the busy period $\mathbf{y}$, and the number $\mathbf{n}_y$ of units served during a busy period. Consider two cases: the density of the service time $\mathbf{c}$ is (*a*) uniform between 4 and 8 min; (*b*) it equals $\mu c e^{-\mu c}$ where $\mu = 1/3$.

12-10 Show that if, in an $M \mid M \mid 1$ queue, the service time density equals $\mu e^{-\mu c}$, then the distance from a fixed point t_0 to the next departure point is an RV $\mathbf{z}$ with density $\lambda e^{-\lambda z}$.

Hint: The probability that t_0 is a point of a busy period equals λ/μ.

12-11 Find the probability $P\{\mathbf{s}(t) \geq 2\}$ of the shot noise process

$$\mathbf{s}(t) = \sum_i h(t - t_i) \qquad h(t) = 4U(t) - 3U(t-1) - U(t-2)$$

where $\mathbf{t}_i$ are Poisson points with $\lambda = 2$.

12-12 The shot noise process $\mathbf{s}(t)$ is a train of triangles

$$\mathbf{s}(t) = \sum_i h(t - \mathbf{t}_i) \qquad h(t) = \begin{cases} 5(2 - |t|) & |t| < 2 \\ 0 & |t| > 2 \end{cases}$$

and the points $\mathbf{t}_i$ are Poisson with $\lambda = 0.01$. (*a*) Find its power spectrum $S_s(\omega)$. (*b*) Find its first-order density. (Note that $2\lambda \ll 1$.)

12-13 The points $\mathbf{t}_i$ are Poisson with density λ and

$$\mathbf{s}(t) = \sum_i h(t - \mathbf{t}_i) \qquad h(t) = e^{-\alpha t}U(t)$$

(*a*) Find the mean, the variance, and the power spectrum of $\mathbf{s}(t)$.
(*b*) Find the power spectrum of the process $\mathbf{y}(t) = \mathbf{s}'(t)$ for $\lambda \gg \alpha$ and for $\lambda \ll \alpha$.

12-14 The RVs $\mathbf{x}_n$ are i.i.d. taking the values $+1$ and -1 with $P\{\mathbf{x}_n = 1\} = 0.6$ and $P\{\mathbf{x}_n = -1\} = 0.4$. Show that the process $\mathbf{y}_n = \mathbf{x}_n + \mathbf{x}_{n-1}$ is a stationary Markoff chain and find its state probabilities p_i and transition probabilities $\pi_{ij}[m]$.

12-15 Show that if $\mathbf{x}(0) = 0$ and $\mathbf{x}(t)$ is a process with independent increments, then it is Markoff.

12-16 Given a two-state Markoff chain $\mathbf{x}_n$ taking the values 1 and 0 with state probability vector $P[n]$ and transition matrix Π. Show that, if

$$\Pi = \begin{bmatrix} \frac{2}{3} & \frac{1}{3} \\ \frac{1}{3} & \frac{2}{3} \end{bmatrix} \qquad \text{then} \qquad \Pi^n \xrightarrow[n\to\infty]{} \begin{bmatrix} \frac{1}{2} & \frac{1}{2} \\ \frac{1}{2} & \frac{1}{2} \end{bmatrix} \qquad P[n] \xrightarrow[n\to\infty]{} \left[\frac{1}{2}, \frac{1}{2}\right]$$

Find $P[2]$ and $P[3]$ if $\mathbf{x}_1 = 0$.

12-17 Show that, if $\mathbf{x}(t)$ is a discrete-state Markoff process taking the values a_i and

$$P\{\mathbf{x}(t) = a_i\} = p_i(t) \qquad P\{\mathbf{x}(t_2) = a_j \mid \mathbf{x}(t_1) = a_i\} = \pi_{ij}(t_1, t_2)$$

then its autocorrelation equals

$$R_x(t_1, t_2) = \sum_{i,j} a_i a_j \pi_{ij}(t_1, t_2) p_i(t_1)$$

12-18 Show that if $\pi_{ij}(t_1, t_2)$ are the transition probabilities of a Markoff chain $\mathbf{x}(t)$ and

$$P\{\mathbf{x}(t + \Delta\tau) = a_i \mid \mathbf{x}(t) = a_i\} \simeq 1 - \mu(t)\,\Delta t$$

$$P\{\mathbf{x}(t + \Delta\tau) = a_j \mid \mathbf{x}(t) = a_i\} = \lambda_{ij}\,\Delta t$$

then

$$\frac{\partial \pi_{ji}(t, t_0)}{\partial t} = -\mu_i(t)\pi_{ji}(t, t_0) + \sum_k{}' \lambda_{ki}(t)\pi_{jk}(t, t_0)$$

$$\frac{\partial \pi_{ji}(t, t_0)}{\partial t_0} = \pi_{ji}(t, t_0)\mu_j(t_0) + \sum_k{}' \pi_{ki}(t, t_0)\lambda_{jk}(t_0)$$

12-19 The telegraph signal $\mathbf{x}(t)$ of Example 12-10 is stationary with $\mu_2 = 3\mu_1 = 6$ and $A = 100$.

(*a*) Find its mean η_x and autocorrelation $R_x(\tau)$.

(*b*) Find the power spectrum $S_w(\omega)$ of the FM signal

$$\mathbf{w}(t) = e^{j\boldsymbol{\varphi}(t)} \qquad \boldsymbol{\varphi}(t) = \int_0^t \mathbf{x}(\alpha)\,d\alpha$$

(*c*) Show that $\mathbf{w}(t)$ satisfies the time-varying stochastic differential equation

$$\mathbf{w}'(t) + j\mathbf{x}(t)\mathbf{w}(t) = 0 \qquad \mathbf{w}(0) = 1$$

Find $E\{\mathbf{w}(t)\}$ and $R_w(t_1, t_2)$.

12-20 Show that the distribution function

$$F(x, x_0; t, t_0) = P\{\mathbf{x}(t) < x \mid \mathbf{x}(t_0) = x_0\} = \int_{-\infty}^{x} \pi(\xi, x_0; t, t_0)\,d\xi$$

of a Markoff process satisfies the backward diffusion equation

$$\frac{\partial F}{\partial t_0} + \eta(x_0, t_0)\frac{\partial F}{\partial x_0} + \frac{1}{2}\sigma^2(x_0, t_0)\frac{\partial^2 F}{\partial x_0^2} = 0$$

CHAPTER

THIRTEEN

MEAN SQUARE ESTIMATION

13-1 THE ORTHOGONALITY PRINCIPLE

We are given n RVs

$$\mathbf{x}_1, \mathbf{x}_2, \ldots, \mathbf{x}_n$$

and we wish to find n constants

$$a_1, a_2, \ldots, a_n$$

such that, if we estimate the RV $\mathbf{s}$ by the sum†

$$\hat{\mathbf{s}} = a_1\mathbf{x}_1 + \cdots + a_n\mathbf{x}_n \equiv \hat{E}\{\mathbf{s} \mid \mathbf{x}_1, \ldots, \mathbf{x}_n\} \tag{13.1}$$

the MS value

$$P = E\{|\mathbf{s} - (a_1\mathbf{x}_1 + \cdots + a_n\mathbf{x}_n)|^2\} \tag{13-2}$$

of the resulting error

$$\boldsymbol{\varepsilon} = \mathbf{s} - \hat{\mathbf{s}} = \mathbf{s} - (a_1\mathbf{x}_1 + \cdots + a_n\mathbf{x}_n) \tag{13-3}$$

is minimum.

This is the homogeneous linear MS estimate of the RV $\mathbf{s}$ (*signal*) in terms of the RVs $\mathbf{x}_i$ (*data*). In this chapter, we deal with various aspects of this problem. The solution is based on the following fundamental result known also as the *orthogonality principle*:

† We should stress that the notation $\hat{E}\{\ \}$ does not indicate expected values. It indicates only optimum linear MS estimates.

Projection theorem The MS error P is minimum if the constants a_i are such that the error ε is orthogonal to the data

$$E\{[\mathbf{s} - (a_1\mathbf{x}_1 + \cdots + a_n\mathbf{x}_n)]\mathbf{x}_i^*\} = 0 \qquad i = 1, \ldots, n \tag{13-4}$$

We shall give two proofs. In the first proof we assume that all variables are real.

PROOF 1 The MS error P is a function of the constants a_i and it is minimum if

$$\frac{\partial P}{\partial a_i} = E\{2[\mathbf{s} - (a_1\mathbf{x}_1 + \cdots + a_n\mathbf{x}_n)](-\mathbf{x}_i)\} = 0$$

This yields (13-4).

PROOF 2 We shall show that, if the constants a_i satisfy (13-4), then the resulting P is minimum. From (13-4) it follows that

$$E\{[\mathbf{s} - (a_1\mathbf{x}_1 + \cdots + a_n\mathbf{x}_n)](c_1\mathbf{x}_1^* + \cdots + c_n\mathbf{x}_n^*)\} = 0 \tag{13-5}$$

for any $c_1, \ldots, c_n$. If $\bar{a}_i$ is an arbitrary set of constants, then

$$\begin{aligned}
&E\{|\mathbf{s} - (\bar{a}_1\mathbf{x}_1 + \cdots + \bar{a}_n\mathbf{x}_n)|^2\} \\
&\quad = E\{|[\mathbf{s} - (a_1\mathbf{x}_1 + \cdots + a_n\mathbf{x}_n)] + [(a_1 - \bar{a}_1)\mathbf{x}_1 + \cdots + (a_n - \bar{a}_n)\mathbf{x}_n]|^2\} \\
&\quad = E\{|\mathbf{s} - (a_1\mathbf{x}_1 + \cdots + a_n\mathbf{x}_n)|^2\} + E\{|(a_1 - \bar{a}_1)\mathbf{x}_1 + \cdots + (a_n - \bar{a}_n)\mathbf{x}_n|^2\}
\end{aligned}$$

because the two brackets in the middle line are orthogonal [see (13-5)]. The last line is minimum if $\bar{a}_i = a_i$.

The Yule–Walker equations Setting $i = 1, \ldots, n$ in (13-4), we obtain

$$\begin{aligned}
R_{11}a_1 + R_{12}a_2 + \cdots + R_{1n}a_n &= R_{01} \\
R_{21}a_1 + R_{22}a_2 + \cdots + R_{2n}a_n &= R_{02} \\
\cdots\cdots\cdots\cdots\cdots\cdots\cdots\cdots \\
R_{n1}a_1 + R_{n2}a_2 + \cdots + R_{nn}a_n &= R_{0n}
\end{aligned} \tag{13-6}$$

where

$$R_{ij} = E\{\mathbf{x}_i\mathbf{x}_j^*\} \qquad R_{0j} = E\{\mathbf{s}\mathbf{x}_j^*\}$$

This is a system of n equations and its solution yields the constants a_i. If the data $\mathbf{x}_i$ are *linearly independent* [see (8-30)], then the determinant Δ of the coefficients R_{ij} is positive. In this case, (13-6) has a unique solution. If the data are linearly dependent, then they can be expressed as a linear combination of a subset of $m < n$ linearly independent components. The estimate $\hat{\mathbf{s}}$ can then be written as a linear sum involving only the m reduced data.

MS error From (13-4) it follows that

$$\mathbf{s} - \hat{\mathbf{s}} \perp c_1\mathbf{x}_1 + \cdots + c_n\mathbf{x}_n$$

for any $c_1, \ldots, c_n$; therefore, $\mathbf{s} - \hat{\mathbf{s}} \perp \hat{\mathbf{s}}$. This can be used to simplify the expression for the minimum MS error P

$$P = E\{|\mathbf{s} - \hat{\mathbf{s}}|^2\} = E\{(\mathbf{s} - \hat{\mathbf{s}})\mathbf{s}^*\} \tag{13-7}$$

Inserting (13-1) into (13-7), we obtain

$$P = R_{00} - (a_1 R_{01} + \cdots + a_n R_{0n}) \qquad R_{00} = E\{|\mathbf{s}|^2\} \tag{13-8}$$

Vector formulation We introduce the vectors

$$\mathbf{X}: [\mathbf{x}_1, \ldots, \mathbf{x}_n] \qquad A: [a_1, \ldots, a_n]$$

With $\mathbf{X}^t$ the transpose of $\mathbf{X}$, the data correlation matrix equals

$$R = E\{\mathbf{X}^t \mathbf{X}^*\}$$

Furthermore

$$\hat{\mathbf{s}} = A\mathbf{X}^t = \mathbf{X}A^t \tag{13-9}$$

and

$$E\{|\hat{\mathbf{s}}|^2\} = E\{A\mathbf{X}^t\overset{*}{\mathbf{X}}\overset{*}{A^t}\} = AR\overset{*}{A^t}$$

Finally, with $R_0: [R_{01}, \ldots, R_{0n}]$ (13-6) yields

$$AR = R_0 \qquad A = R_0 R^{-1}$$

Geometric interpretation In the representation of RVs as vectors in an abstract space, the sum $\hat{\mathbf{s}} = a_1\mathbf{x}_1 + \cdots + a_n\mathbf{x}_n$ is a vector in the subspace S_n of the data $\mathbf{x}_i$ and the error $\boldsymbol{\varepsilon} = \mathbf{s} - \hat{\mathbf{s}}$ is the vector from $\mathbf{s}$ to $\hat{\mathbf{s}}$ as in Fig. 13-1*a*. The projection theorem states that the length of $\boldsymbol{\varepsilon}$ is minimum if $\boldsymbol{\varepsilon}$ is orthogonal to $\mathbf{x}_i$, that is, if it is perpendicular to the data subspace S_n. The estimate $\hat{\mathbf{s}}$ is, thus, the "projection" of $\mathbf{s}$ on S_n.

If $\mathbf{s}$ is a vector in S_n, then $\hat{\mathbf{s}} = \mathbf{s}$ and $P = 0$. In this case, the $n + 1$ RVs $\mathbf{s}, \mathbf{x}_1, \ldots, \mathbf{x}_n$ are linearly dependent and the determinant Δ_{n+1} of their correlation matrix is zero.

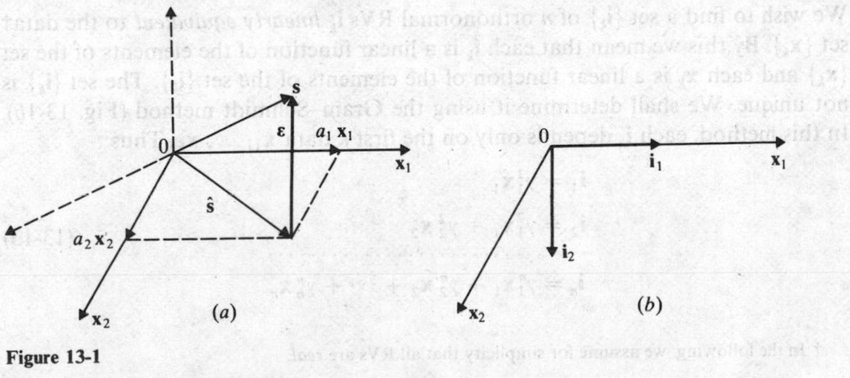

Figure 13-1

If $\mathbf{s}$ is perpendicular to S_n, then $\hat{\mathbf{s}} = 0$ and $P = E\{|\mathbf{s}|^2\}$. This is the case if $\mathbf{s}$ is orthogonal to all the data $\mathbf{x}_i$, that is, if $R_{0j} = 0$ for $j \neq 0$.

We note, finally, that if the data are mutually orthogonal, then their correlation matrix R is diagonal and (13-6) yields

$$a_i = \frac{R_{0i}}{R_{ii}} = \frac{E\{\mathbf{s}\mathbf{x}_i^*\}}{E\{|\mathbf{x}_i|^2\}} \tag{13-10}$$

From the above it follows that the determination of the estimate (projection) $\hat{\mathbf{s}}$ of $\mathbf{s}$ is simplified if $\hat{\mathbf{s}}$ is expressed in terms of an orthogonal data base. As we shall presently see, this can be done with a linear transformation of the data.

Linear nonhomogeneous estimation The estimate (13-1) can be improved if a constant is added to the sum. The problem now is to determine $n + 1$ parameters α_k such that if

$$\hat{\mathbf{s}} = \alpha_0 + \alpha_1 \mathbf{x}_1 + \cdots + \alpha_n \mathbf{x}_n \tag{13-11}$$

then the resulting MS error is minimum. This problem can be reduced to the homogeneous case if we replace the term α_0 by the product $\alpha_0 \mathbf{x}_0$ where $\mathbf{x}_0 = 1$. Applying (13-4) to the enlarged data set

$$\mathbf{x}_0, \mathbf{x}_1, \ldots, \mathbf{x}_n \qquad \text{where} \qquad E\{\mathbf{x}_0 \mathbf{x}_i\} = \begin{cases} E\{\mathbf{x}_i\} = \eta_i & i \neq 0 \\ 1 & i = 0 \end{cases}$$

we obtain

$$\begin{aligned} \alpha_0 + \eta_1 \alpha_1 \;\; + \cdots + \eta_n \alpha_n \;\; &= \eta_s \\ \eta_1^* \alpha_0 + R_{11}\alpha_1 + \cdots + R_{1n}\alpha_n &= R_{01} \\ \cdots\cdots\cdots\cdots\cdots\cdots\cdots\cdots\cdots\cdots\cdots\cdots \\ \eta_n^* \alpha_0 + R_{n1}\alpha_1 + \cdots + R_{nn}\alpha_n &= R_{0n} \end{aligned} \tag{13-12}$$

We note that, if $\eta_s = \eta_i = 0$, then (13-12) reduces to (13-6). This yields $\alpha_0 = 0$ and $\alpha_n = a_n$.

Orthonormal Data Transformation

We wish to find a set $\{\mathbf{i}_k\}$ of n orthonormal RVs $\mathbf{i}_k$ *linearly equivalent* to the data† set $\{\mathbf{x}_k\}$. By this we mean that each $\mathbf{i}_k$ is a linear function of the elements of the set $\{\mathbf{x}_k\}$ and each $\mathbf{x}_k$ is a linear function of the elements of the set $\{\mathbf{i}_k\}$. The set $\{\mathbf{i}_k\}$ is not unique. We shall determine it using the Gram–Schmidt method (Fig. 13-1*b*). In this method, each $\mathbf{i}_k$ depends only on the first k data $\mathbf{x}_1, \ldots, \mathbf{x}_k$. Thus

$$\begin{aligned} \mathbf{i}_1 &= \gamma_1^1 \mathbf{x}_1 \\ \mathbf{i}_2 &= \gamma_1^2 \mathbf{x}_1 + \gamma_2^2 \mathbf{x}_2 \\ &\cdots\cdots\cdots\cdots\cdots\cdots\cdots\cdots\cdots \\ \mathbf{i}_n &= \gamma_1^n \mathbf{x}_1 + \gamma_2^n \mathbf{x}_2 + \cdots + \gamma_n^n \mathbf{x}_n \end{aligned} \tag{13-13}$$

† In the following, we assume for simplicity that all RVs are *real*.

In the notation γ_r^k, k is a superscript identifying the kth equation and r is a subscript taking the values 1 to k.

The coefficient γ_1^1 is obtained from the normalization condition

$$E\{\mathbf{i}_1^2\} = (\gamma_1^1)^2 R_{11} = 1$$

To find the coefficients γ_1^2 and γ_2^2 we observe that $\mathbf{i}_2 \perp \mathbf{x}_1$ because $\mathbf{i}_2 \perp \mathbf{i}_1$ by assumption. From this it follows that

$$E\{\mathbf{i}_2 \mathbf{x}_1\} = 0 = \gamma_1^2 R_{11} + \gamma_2^2 R_{21}$$

The condition $E\{\mathbf{i}_2^2\} = 1$ yields a second equation.

Similarly, since $\mathbf{i}_k \perp \mathbf{i}_r$ for $r < k$, we conclude from (13-13) that $\mathbf{i}_k \perp \mathbf{x}_r$ if $r < k$. Multiplying the kth equation in (13-13) by $\mathbf{x}_r$ and using the above, we obtain

$$E\{\mathbf{i}_k \mathbf{x}_r\} = 0 = \gamma_1^k R_{1r} + \cdots + \gamma_k^k R_{kr} \qquad 1 \le r \le k-1 \tag{13-14}$$

This is a system of $k - 1$ equations for the k unknowns $\gamma_1^k, \ldots, \gamma_k^k$. The condition $E\{\mathbf{i}_k^2\} = 1$ yields one more equation.

The system (13-13) can be written in a vector form

$$\mathbf{I} = \mathbf{X}\Gamma \tag{13-15}$$

where $\mathbf{I}$ is a row vector with elements $\mathbf{i}_k$. Solving for $\mathbf{X}$ we obtain

$$\begin{aligned} \mathbf{x}_1 &= l_1^1 \mathbf{i}_1 \\ \mathbf{x}_2 &= l_1^2 \mathbf{i}_1 + l_2^2 \mathbf{i}_2 \\ &\cdots\cdots\cdots\cdots\cdots\cdots \\ \mathbf{x}_n &= l_1^n \mathbf{i}_1 + l_2^n \mathbf{i}_2 + \cdots + l_n^n \mathbf{i}_n \end{aligned} \qquad X = \mathbf{I}\Gamma^{-1} = \mathbf{I}L \tag{13-16}$$

In the above, the matrix Γ and its inverse are upper triangular

$$\Gamma = \begin{bmatrix} \gamma_1^1 & \gamma_1^2 & \cdots & \gamma_1^n \\ & \gamma_2^2 & \cdots & \gamma_2^n \\ & & \cdots\cdots & \\ 0 & & & \gamma_n^n \end{bmatrix} \qquad L = \begin{bmatrix} l_1^1 & l_1^2 & \cdots & l_1^n \\ & l_2^2 & \cdots & l_2^n \\ & & \cdots\cdots & \\ 0 & & & l_n^n \end{bmatrix}$$

Since $E\{\mathbf{i}_k \mathbf{i}_j\} = \delta[i-j]$ by construction, we conclude that

$$E\{\mathbf{I}^t\mathbf{I}\} = 1 = E\{\Gamma^t \mathbf{X}^t \mathbf{X}\Gamma\} = \Gamma^t E\{\mathbf{X}^t\mathbf{X}\}\Gamma \tag{13-17}$$

where 1 is the identity matrix. Hence

$$\Gamma^t R \Gamma = 1 \qquad R = L^t L \qquad R^{-1} = \Gamma\Gamma^t \tag{13-18}$$

We have, thus, expressed the matrix R and its inverse R^{-1} as products of an upper triangular and a lower triangular matrix (see also Cholesky factorization, Sec. 13-4).

The orthonormal base $\{\mathbf{i}_n\}$ in (13-13) is the finite version of the innovations process $\mathbf{i}[n]$ introduced in Sec. 10-5. The matrices Γ and L correspond to the whitening filter and to the innovations filter respectively and the factorization (13-18) corresponds to the spectral factorization (10-115).

From the linear equivalence of the sets $\{\mathbf{i}_k\}$ and $\{\mathbf{x}_k\}$ it follows that the estimate (13-1) of the RV $\mathbf{s}$ can be expressed in terms of the set $\{\mathbf{i}_k\}$

$$\hat{\mathbf{s}} = b_1 \mathbf{i}_1 + \cdots + b_n \mathbf{i}_n = B\mathbf{I}^t$$

where again the coefficients b_k are such that

$$\mathbf{s} - \hat{\mathbf{s}} \perp \mathbf{i}_k \qquad 1 \le k \le n$$

This yields [see (13-17)]

$$E\{(\mathbf{s} - B\mathbf{I}^t)\mathbf{I}\} = 0 = E\{\mathbf{s}\mathbf{I}\} - B$$

from which it follows that

$$B = E\{\mathbf{s}\mathbf{I}\} = E\{\mathbf{s}\mathbf{X}\Gamma\} = R_0 \Gamma \qquad (13\text{-}19)$$

Returning to the estimate (13-1) of $\mathbf{s}$, we conclude that

$$\hat{\mathbf{s}} = B\mathbf{I}^t = B\Gamma^t\mathbf{X}^t = A\mathbf{X}^t \qquad A = B\Gamma^t \qquad (13\text{-}20)$$

This simplifies the determination of the vector A if the matrix Γ is known.

Nonlinear MS Estimation and Normality

The nonlinear MS estimation problem involves the determination of a function $c(\mathbf{x}_1, \ldots, \mathbf{x}_n)$ of the data $\mathbf{x}_k$ such that if the RV $\mathbf{s}$ is estimated by this function, the resulting MS error

$$e = E\{[\mathbf{s} - c(\mathbf{x}_1, \ldots, \mathbf{x}_n)]^2\}$$

is minimum. We maintain that e is minimum if $c(\mathbf{x}_1, \ldots, \mathbf{x}_n)$ equals the conditional mean of $\mathbf{s}$ assuming the data

$$c(x_1, \ldots, x_n) = E\{\mathbf{s} \mid x_1, \ldots, x_n\} = \int_{-\infty}^{\infty} s f(s \mid x_1, \ldots, x_n)\, ds \qquad (13\text{-}21)$$

Indeed, from (8-42) it follows that

$$e = E\{E\{[\mathbf{s} - c(\mathbf{x}_1, \ldots, \mathbf{x}_n)]^2 \mid \mathbf{x}_1, \ldots, \mathbf{x}_n\}\}$$

Since all quantities are positive, the above is minimum if the conditional MS error

$$E\{[\mathbf{s} - c(x_1, \ldots, x_n)]^2 \mid x_1, \ldots, x_n\}$$

is minimum. This is the case if $c(x_1, \ldots, x_n)$ is given by (13-21) [see also (7-64)].

Normality We shall show that if the RVs under consideration are *normal with zero mean*, then the nonlinear estimate (13-21) equals the linear estimate (13-1)

$$c(\mathbf{x}_1, \ldots, \mathbf{x}_n) = \hat{\mathbf{s}} = a_1 \mathbf{x}_1 + \cdots + a_n \mathbf{x}_n \qquad (13\text{-}22)$$

PROOF The linear estimation error $\varepsilon = \mathbf{s} - \hat{\mathbf{s}}$ is orthogonal to the data and $E\{\varepsilon\} = 0$. Hence, ε is *independent* of the data. From this it follows that

$$E\{\varepsilon \mid \mathbf{x}_1, \ldots, \mathbf{x}_n\} = E\{\varepsilon\} = 0$$

Using the linearity of conditional expected values, we conclude from the above that

$$0 = E\{\mathbf{s} - \hat{\mathbf{s}} \mid \mathbf{x}_1, \ldots, \mathbf{x}_n\} = E\{\mathbf{s} \mid \mathbf{x}_1, \ldots, \mathbf{x}_n\} - E\{\hat{\mathbf{s}} \mid \mathbf{x}_1, \ldots, \mathbf{x}_n\}$$

and (13-22) results because

$$E\{\mathbf{s} \mid \mathbf{x}_1, \ldots, \mathbf{x}_n\} = c(\mathbf{x}_1, \ldots, \mathbf{x}_n) \quad \text{and} \quad E\{\hat{\mathbf{s}} \mid \mathbf{x}_1, \ldots, \mathbf{x}_n\} = \hat{\mathbf{s}}$$

Conditional densities We have shown that the error $\mathbf{s} - \hat{\mathbf{s}}$ is independent of the data $\mathbf{x}_k$. Hence

$$E\{(\mathbf{s} - \hat{\mathbf{s}})^2 \mid \mathbf{x}_1, \ldots, \mathbf{x}_n\} = E\{(\mathbf{s} - \hat{\mathbf{s}})^2\} = P$$

Thus, the conditional mean of $\mathbf{s}$ assuming the data equals its MS estimate $\hat{\mathbf{s}}$ [see (13-22)] and the conditional variance equals the MS error P. And since $f(s \mid x_1, \ldots, x_n)$ is normal, we conclude that

$$f(s \mid x_1, \ldots, x_n) = \frac{1}{\sqrt{2\pi P}} e^{-[s-(a_1x_1 + \cdots + a_nx_n)]^2/2P} \qquad (13\text{-}23)$$

This result simplifies the evaluation of conditional densities involving normal RVs.

Example 13-1 The RVs $\mathbf{x}_1$ and $\mathbf{x}_2$ are jointly normal with zero mean. We shall determine their conditional density $f(x_2 \mid x_1)$. As we know [see (7-71)]

$$E\{\mathbf{x}_2 \mid x_1\} = ax_1 \qquad a = \frac{R_{12}}{R_{11}}$$

$$\sigma^2_{x_2|x_1} = P = E\{(\mathbf{x}_2 - a\mathbf{x}_1)\mathbf{x}_2\} = R_{22} - aR_{12}$$

Hence

$$f(x_2 \mid x_1) = \frac{1}{\sqrt{2\pi P}} e^{-(x_2 - ax_1)^2/2P}$$

Example 13-2 We now wish to find the conditional density $f(x_3 \mid x_1, x_2)$. In this case

$$E\{\mathbf{x}_3 \mid x_1, x_2\} = a_1 x_1 + a_2 x_2$$

where the constants a_1 and a_2 are determined from (13-6)

$$R_{11}a_1 + R_{12}a_2 = R_{13} \qquad R_{12}a_1 + R_{22}a_2 = R_{23}$$

Furthermore [see (13-22) and (13-7)]

$$\sigma^2_{x_3|x_1, x_2} = P = R_{33} - (R_{13}a_1 + R_{23}a_2)$$

And since $f(x_3 \mid x_1, x_2)$ is a one-dimensional normal density, we conclude that

$$f(x_3 \mid x_1, x_2) = \frac{1}{\sqrt{2\pi P}} e^{-(x_3 - a_1x_1 - a_2x_2)^2/2P}$$

Example 13-3 In this example, we shall find the two-dimensional density $f(x_2, x_3 \mid x_1)$. This involves the evaluation of five parameters [see (6-15)]: two conditional means, two conditional variances, and the conditional covariance of the RVs $\mathbf{x}_2$ and $\mathbf{x}_3$ assuming $\mathbf{x}_1$.

The first four parameters are determined as in Example 13-1:

$$E\{\mathbf{x}_2 \mid x_1\} = \frac{R_{12}}{R_{11}} x_1 \qquad E\{\mathbf{x}_3 \mid x_1\} = \frac{R_{13}}{R_{11}} x_1$$

$$\sigma^2_{x_2|x_1} = R_{22} - \frac{R_{12}^2}{R_{11}} \qquad \sigma^2_{x_3|x_1} = R_{33} - \frac{R_{13}^2}{R_{11}}$$

The conditional covariance

$$C_{x_2x_3|x_1} = E\left\{\left(\mathbf{x}_2 - \frac{R_{12}}{R_{11}}\mathbf{x}_1\right)\left(\mathbf{x}_3 - \frac{R_{13}}{R_{11}}\mathbf{x}_1\right)\middle| \mathbf{x}_1 = x_1\right\} \tag{13-24}$$

is found as follows: We know that the errors $\mathbf{x}_2 - R_{12}\mathbf{x}_1/R_{11}$ and $\mathbf{x}_3 - R_{13}\mathbf{x}_1/R_{11}$ are independent of $\mathbf{x}_1$. Hence, the condition $\mathbf{x}_1 = x_1$ in (13-24) can be removed. Expanding the product, we obtain

$$C_{x_2x_3|x_1} = R_{23} - \frac{R_{12}R_{13}}{R_{11}}$$

This completes the specification of $f(x_2, x_3 \mid x_1)$.

13-2 STOCHASTIC PROCESSES†

The projection theorem can be readily extended to stochastic processes. We give next a formal justification omitting mathematical subtleties.

We wish to estimate the present value $\mathbf{s}(t)$ of a stochastic process in terms of the values $\mathbf{x}(\xi)$ of another process $\mathbf{x}(t)$ specified for every ξ in an interval $a \le \xi \le b$. The available data form now a noncountable infinity of RVs and the desirable linear estimate $\hat{\mathbf{s}}(t)$ of $\mathbf{s}(t)$ is an integral

$$\hat{\mathbf{s}}(t) = \hat{E}\{\mathbf{s}(t) \mid \mathbf{x}(\xi), a \le \xi \le b\} = \int_a^b h(\alpha)\mathbf{x}(\alpha)\, d\alpha \tag{13-25}$$

where $h(\alpha)$ is a function to be determined. Approximating the integral by a sum, we obtain

$$\hat{\mathbf{s}}(t) \simeq \sum_{k=1}^{n} h(\alpha_k)\mathbf{x}(\alpha_k)\, \Delta\alpha$$

† N. Wiener, "Extrapolation, Interpolation, and Smoothing of Stationary Time Series," MIT Press, 1950. J. Makhoul: "Linear Prediction: A Tutorial Review," *Proceedings of the IEEE*, vol. 63, 1965. T. Kailath: A View of Three Decades of Linear Filtering Theory, *IEEE Transactions Information Theory*, vol. IT-20, 1974.

Thus, the estimate of $\mathbf{s}(t)$ is a sum as in (13-1) where $a_k = h(\alpha_k)\,\Delta\alpha$. Applying (13-4), we conclude that $h(\alpha_k)$ must be such that

$$E\left\{\left[\mathbf{s}(t) - \sum_{k=1}^{n} h(\alpha_k)\mathbf{x}(\alpha_k)\,\Delta\alpha\right]\mathbf{x}(\xi_j)\right\} \simeq 0 \qquad 1 \le j \le n$$

Hence

$$R_{sx}(t, \xi_j) \simeq \sum_{k=1}^{n} h(\alpha_k) R_{xx}(\alpha_k, \xi_j)\,\Delta\alpha \tag{13-26}$$

With $\Delta\alpha \to 0$ this yields

$$R_{sx}(t, \xi) = \int_a^b h(\alpha) R_{xx}(\alpha, \xi)\,d\alpha \qquad a \le \xi \le b \tag{13-27}$$

The unknown function $h(\alpha)$ is the solution of the *integral equation* (13-27). This equation is the continuous version of the Yule–Walker system (13-6). We note that (13-27) can be obtained directly from (13-25) if we apply (13-4) to the estimation error $\mathbf{s}(t) - \hat{\mathbf{s}}(t)$. This yields

$$E\left\{\left[\mathbf{s}(t) - \int_a^b h(\alpha)\mathbf{x}(\alpha)\,d\alpha\right]\mathbf{x}(\xi)\right\} = 0 \qquad a \le \xi \le b \tag{13-28}$$

and (13-27) results. The MS error $P = E\{(\mathbf{s} - \hat{\mathbf{s}})\mathbf{s}\}$ of the estimation of $\mathbf{s}(t)$ by the integral in (13-25) equals

$$P = E\left\{\left[\mathbf{s}(t) - \int_a^b h(\alpha)\mathbf{x}(\alpha)\,d\alpha\right]\mathbf{s}(t)\right\}$$

$$= R_{ss}(0) - \int_a^b h(\alpha) R_{sx}(t, \alpha)\,d\alpha \tag{13-29}$$

In general, the integral equation (13-27) can be solved only numerically. In fact, if we assign to the variable ξ the values ξ_j and we approximate the integral by a sum, we obtain the system (13-26). In this chapter, we consider various special cases that lead to explicit solutions. Unless stated otherwise, it will be assumed that all processes are WSS and real.

We shall use the following terminology:

If the time t in (13-25) is in the interior of the data interval (a, b), then the estimate $\hat{\mathbf{s}}(t)$ of $\mathbf{s}(t)$ will be called *smoothing*.

If t is outside this interval and $\mathbf{x}(t) = \mathbf{s}(t)$ (no noise), then $\hat{\mathbf{s}}(t)$ is a *predictor* of $\mathbf{s}(t)$. If $t > b$, then $\hat{\mathbf{s}}(t)$ is a "forward predictor"; if $t < a$, it is a "backward predictor."

Finally, if t is outside the data interval and $\mathbf{x}(t) \ne \mathbf{s}(t)$, then the estimate is called *filtering and prediction*.

In this section, we present a number of simple estimation problems involving a finite number of data. More elaborate problems are considered in subsequent sections. For continuous time processes, the orthogonality principle will be stated directly as in (13-28).

1. *Prediction* (*a*) We wish to estimate the future value $\mathbf{s}(t+\lambda)$ of a stationary process $\mathbf{s}(t)$ in terms of its present value

$$\hat{\mathbf{s}}(t+\lambda) = \hat{E}\{\mathbf{s}(t+\lambda) \mid \mathbf{s}(t)\} = a\mathbf{s}(t)$$

From (13-4) and (13-7) it follows with $n = 1$ that

$$E\{[\mathbf{s}(t+\lambda) - a\mathbf{s}(t)]\mathbf{s}(t)\} = 0 \qquad a = \frac{R(\lambda)}{R(0)}$$

$$P = E\{[\mathbf{s}(t+\lambda) - a\mathbf{s}(t)]\mathbf{s}(t+\lambda)\} = R(0) - aR(\lambda)$$

Special case If

$$R(\tau) = Ae^{-\alpha|\tau|} \qquad \text{then} \qquad a = e^{-\alpha\lambda}$$

We note that in this case, the difference $\mathbf{s}(t+\lambda) - a\mathbf{s}(t)$ is orthogonal to $\mathbf{s}(t-\xi)$ for every $\xi \geq 0$

$$\begin{aligned} E\{[\mathbf{s}(t+\lambda) - a\mathbf{s}(t)]\mathbf{s}(t-\xi)\} &= R(\lambda+\xi) - aR(\xi) \\ &= Ae^{-\alpha(\lambda+\xi)} - Ae^{-\alpha\lambda}e^{-\alpha\xi} = 0 \end{aligned}$$

This shows that $a\mathbf{s}(t)$ is the estimate of $\mathbf{s}(t+\lambda)$ in terms of its entire past. Such a process is called *wide-sense Markoff.*

(*b*) We shall now find the estimate of $\mathbf{s}(t+\lambda)$ in terms of $\mathbf{s}(t)$ and $\mathbf{s}'(t)$

$$\hat{\mathbf{s}}(t+\lambda) = a_1\mathbf{s}(t) + a_2\mathbf{s}'(t)$$

The orthogonality condition (13-4) yields

$$\mathbf{s}(t+\lambda) - \hat{\mathbf{s}}(t+\lambda) \perp \mathbf{s}(t), \mathbf{s}'(t)$$

Using the identities

$$R'(0) = 0 \qquad R_{ss'}(\tau) = -R'(\tau) \qquad R_{s's'}(\tau) = -R''(\tau)$$

we obtain

$$a_1 = R(\lambda)/R(0) \qquad a_2 = R'(\lambda)/R''(0)$$

$$P = E\{[\mathbf{s}(t+\lambda) - a_1\mathbf{s}(t) - a_2\mathbf{s}'(t)]\mathbf{s}(t+\lambda)\} = R(0) - a_1R(\lambda) + a_2R'(\lambda)$$

If λ is small, then

$$R(\lambda) \simeq R(0) \qquad R'(\lambda) \simeq R'(0) + R''(0)\lambda \simeq R''(0)\lambda$$

$$a \simeq 1 \qquad b \simeq \lambda \qquad \mathbf{s}(t+\lambda) \simeq \mathbf{s}(t) + \lambda\mathbf{s}'(t)$$

2. *Filtering* We shall estimate the present value of a process $\mathbf{s}(t)$ in terms of the present value of another process $\mathbf{x}(t)$

$$\hat{\mathbf{s}}(t) = \hat{E}\{\mathbf{s}(t) \mid \mathbf{x}(t)\} = a\mathbf{x}(t)$$

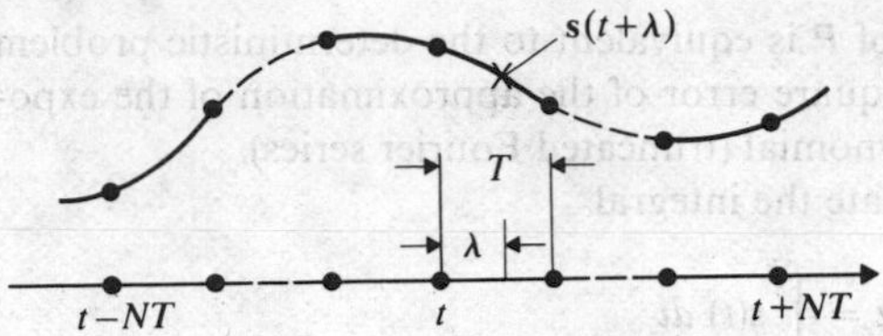

Figure 13-2

From (13-4) and (13-7) it follows that

$$E\{[\mathbf{s}(t) - a\mathbf{x}(t)]\mathbf{x}(t)] = 0 \qquad a = R_{sx}(0)/R_{xx}(0)$$
$$P = E\{[\mathbf{s}(t) - a\mathbf{x}(t)]\mathbf{s}(t)\} = R_{ss}(0) - aR_{xx}(0)$$

3. *Interpolation* We wish to estimate the value $\mathbf{s}(t+\lambda)$ of a process $\mathbf{s}(t)$ in terms of its $2N+1$ samples $\mathbf{s}(t+kT)$ that are nearest to t (Fig. 13-2)

$$\hat{\mathbf{s}}(t+\lambda) = \sum_{k=-N}^{N} a_k \mathbf{s}(t+kT) \qquad 0 < \lambda < T \tag{13-30}$$

The orthogonality principle now yields

$$E\left\{\left[\mathbf{s}(t+\lambda) - \sum_{k=-N}^{N} a_k \mathbf{s}(t+kT)\right]\mathbf{s}(t+nT)\right\} = 0 \qquad |n| \le N$$

from which it follows that

$$\sum_{k=-N}^{N} a_k R(kT - nT) = R(\lambda - nT) \qquad -N \le n \le N \tag{13-31}$$

The MS value P of the estimation error

$$\boldsymbol{\varepsilon}_N(t) = \mathbf{s}(t+\lambda) - \sum_{k=-N}^{N} a_k \mathbf{s}(t+kT) \tag{13-32}$$

equals

$$P = E\{\boldsymbol{\varepsilon}_N(t)\mathbf{s}(t+\lambda)\} = R(0) - \sum_{k=-N}^{N} a_k R(\lambda - kT) \tag{13-33}$$

Interpolation as deterministic approximation The error $\boldsymbol{\varepsilon}_N(t)$ can be considered as the output of the system

$$E_N(\omega) = e^{j\omega\lambda} - \sum_{k=-N}^{N} a_k e^{jkT\omega}$$

(error filter) with input $\mathbf{s}(t)$. Denoting by $S(\omega)$ the power spectrum of $\mathbf{s}(t)$, we conclude from (10-41) that

$$P = E\{\boldsymbol{\varepsilon}_N^2(t)\} = \frac{1}{2\pi}\int_{-\infty}^{\infty} S(\omega)\left|e^{j\omega\lambda} - \sum_{k=-N}^{N} a_k e^{jkT\omega}\right|^2 d\omega \tag{13-34}$$

This shows that the minimization of P is equivalent to the deterministic problem of minimizing the weighted mean square error of the approximation of the exponential $e^{j\omega\lambda}$ by a trigonometric polynomial (truncated Fourier series).

4. *Quadrature* We shall estimate the integral

$$\mathbf{z} = \int_0^b \mathbf{s}(t)\, dt$$

of a process $\mathbf{s}(t)$ in terms of its $N + 1$ samples $\mathbf{s}(nT)$

$$\hat{\mathbf{z}} = a_0\, \mathbf{s}(0) + a_1 \mathbf{s}(T) + \cdots + a_N\, \mathbf{s}(NT) \qquad T = \frac{b}{N}$$

Applying (13-4), we obtain

$$E\left\{\left[\int_0^b \mathbf{s}(t)\, dt - \hat{\mathbf{z}}\right]\mathbf{s}(kT)\right\} = 0 \qquad 0 \le k \le N$$

Hence

$$\int_0^T R(t - kT)\, dt = a_0\, R(kT) + \cdots + a_N R(kT - NT) \qquad 0 \le k \le N$$

This is a system of $N + 1$ equations and its solution yields the coefficients a_k.

13-3 SMOOTHING

We wish to estimate the present value of a process $\mathbf{s}(t)$ in terms of the values $\mathbf{x}(\xi)$ of the sum

$$\mathbf{x}(t) = \mathbf{s}(t) + \mathbf{v}(t)$$

available for every ξ from $-\infty$ to ∞. The desirable estimate

$$\hat{\mathbf{s}}(t) = \hat{E}\{\mathbf{s}(t) \mid \mathbf{x}(\xi),\ -\infty < \xi < \infty\}$$

will be written in the form

$$\hat{\mathbf{s}}(t) = \int_{-\infty}^{\infty} h(\alpha)\mathbf{x}(t - \alpha)\, d\alpha \qquad (13\text{-}35)$$

In this notation, $h(\alpha)$ is independent of t and $\hat{\mathbf{s}}(t)$ can be considered as the output of a linear time-invariant noncausal system with input $\mathbf{x}(t)$ and impulse response $h(t)$. Our problem is to find $h(t)$.

Clearly

$$\mathbf{s}(t) - \hat{\mathbf{s}}(t) \perp \mathbf{x}(\xi) \qquad \text{all } \xi$$

Setting $\xi = t - \tau$, we obtain

$$E\left\{\left[\mathbf{s}(t) - \int_{-\infty}^{\infty} h(\alpha)\mathbf{x}(t - \alpha)\, d\alpha\right]\mathbf{x}(t - \tau)\right\} = 0 \qquad \text{all } \tau$$

This yields

$$R_{sx}(\tau) = \int_{-\infty}^{\infty} h(\alpha)R_{xx}(\tau - \alpha)\, d\alpha \qquad \text{all } \tau \tag{13-36}$$

Thus, to determine $h(t)$, we must solve the above integral equation. This equation can be solved easily because it holds for all τ and the integral is a convolution of $h(\tau)$ with $R_{xx}(\tau)$. Taking transforms of both sides, we obtain $S_{sx}(\omega) = H(\omega)S_{xx}(\omega)$. Hence

$$H(\omega) = \frac{S_{sx}(\omega)}{S_{xx}(\omega)} \tag{13-37}$$

The resulting system is called *noncausal Wiener filter*.

The MS estimation error P equals

$$P = E\left\{\left[\mathbf{s}(t) - \int_{-\infty}^{\infty} h(\alpha)\mathbf{x}(t-\alpha)\, d\alpha\right]\mathbf{s}(t)\right\}$$

$$= R_{ss}(0) - \int_{-\infty}^{\infty} h(\alpha)R_{sx}(\alpha)\, d\alpha = \frac{1}{2\pi}\int_{-\infty}^{\infty} [S_{ss}(\omega) - H^*(\omega)S_{sx}(\omega)]\, d\omega \tag{13-38}$$

If the signal $\mathbf{s}(t)$ and the noise $\mathbf{v}(t)$ are *orthogonal*, then

$$S_{sx}(\omega) = S_{ss}(\omega) \qquad S_{xx}(\omega) = S_{ss}(\omega) + S_{vv}(\omega)$$

Hence (Fig. 13-3)

$$H(\omega) = \frac{S_{ss}(\omega)}{S_{ss}(\omega) + S_{vv}(\omega)} \qquad P = \frac{1}{2\pi}\int_{-\infty}^{\infty} \frac{S_{ss}(\omega)S_{vv}(\omega)}{S_{xx}(\omega) + S_{vv}(\omega)}\, d\omega \tag{13-39}$$

If the spectra $S_{ss}(\omega)$ and $S_{vv}(\omega)$ do not overlap, then $H(\omega) = 1$ in the band of the signal and $H(\omega) = 0$ in the band of the noise. In this case, $P = 0$.

Example 13-4 If

$$S_{ss}(\omega) = \frac{N_0}{\alpha^2 + \omega^2} \qquad S_{vv}(\omega) = N \qquad S_{sv}(\omega) = 0$$

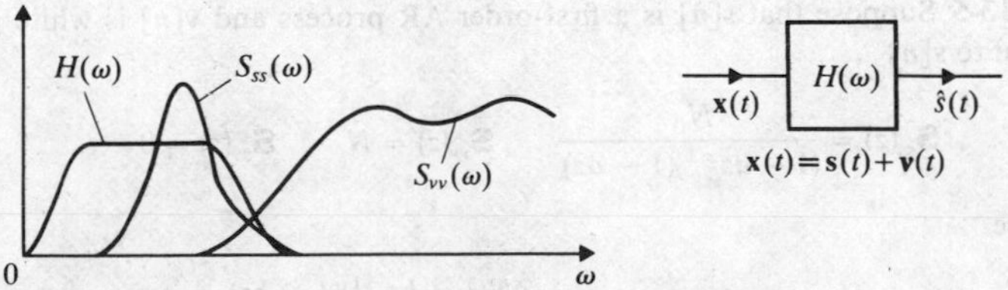

Figure 13-3

then (13-39) yields

$$H(\omega) = \frac{N_0}{N_0 + N(\alpha^2 + \omega^2)} \qquad h(t) = \frac{N_0}{2\beta N} e^{-\beta|t|}$$

$$P = \frac{1}{2\pi}\int_{-\infty}^{\infty} \frac{N_0}{\beta^2 + \omega^2}\, d\omega = \frac{N_0}{2\beta} \qquad \beta^2 = \alpha^2 + \frac{N_0}{N}$$

Discrete-Time Processes

The noncausal estimate $\hat{\mathbf{s}}[n]$ of a discrete-time process in terms of the data

$$\mathbf{x}[n] = \mathbf{s}[n] + \mathbf{v}[n]$$

is the output

$$\hat{\mathbf{s}}[n] = \sum_{k=-\infty}^{\infty} h[k]\mathbf{x}[n-k]$$

of a linear time-invariant noncausal system with input $\mathbf{x}[n]$ and delta response $h[n]$. The orthogonality principle yields

$$E\left\{\left(\mathbf{s}[n] - \sum_{k=-\infty}^{\infty} h[k]\mathbf{x}[n-k]\right)\mathbf{x}[n-m]\right\} = 0 \qquad \text{all } m$$

Hence

$$R_{sx}[m] = \sum_{k=-\infty}^{\infty} h[k]R_{xx}[m-k] = 0 \qquad \text{all } m \tag{13-40}$$

Taking transforms of both sides, we obtain

$$\mathbf{H}(z) = \frac{\mathbf{S}_{sx}(z)}{\mathbf{S}_{xx}(z)} \tag{13-41}$$

The resulting MS error equals

$$P = E\left\{\left[\mathbf{s}[n] - \sum_{k=-\infty}^{\infty} h[k]\mathbf{x}[n-k]\right]\mathbf{s}[n]\right\}$$

$$= R_{ss}(0) - \sum_{k=-\infty}^{\infty} h[k]R_{sx}[k] = \frac{1}{2\sigma}\int_{-\sigma}^{\sigma} [S_{ss}(\omega) - \mathbf{H}(e^{-j\omega T})S_{sx}(\omega)]\, d\omega$$

Example 13-5 Suppose that $\mathbf{s}[n]$ is a first-order AR process and $\mathbf{v}[n]$ is white noise orthogonal to $\mathbf{s}[n]$

$$\mathbf{S}_{ss}(z) = \frac{N_0}{(1 - az^{-1})(1 - az)} \qquad \mathbf{S}_{vv}(z) = N \qquad \mathbf{S}_{sv}(z) = 0$$

In this case

$$\mathbf{S}_{xx}(z) = \mathbf{S}_{ss}(z) + N = \frac{aN(1 - bz^{-1})(1 - bz)}{b(1 - az^{-1})(1 - az)}$$

where

$$0 < b < a < 1 \qquad b + b^{-1} = a + a^{-1} + \frac{N_0}{aN}$$

Hence

$$\mathbf{H}(z) = \frac{bN_0}{aN(1 - bz^{-1})(1 - bz)} \qquad h[n] = cb^{|n|} \qquad c = \frac{bN_0}{aN(1 - b^2)}$$

$$P = \frac{N_0}{1 - a^2}\left[1 - c \sum_{k=-\infty}^{\infty} (ab)^{|k|}\right] = \frac{bN_0}{a(1 - b^2)}$$

13-4 PREDICTION

The r-step predictor of a process $\mathbf{s}[n]$ is the estimate of $\mathbf{s}[n + r]$ in terms of $\mathbf{s}[n]$ and its past

$$\hat{\mathbf{s}}[n + r] = \hat{E}\{\mathbf{s}[n + r] \mid \mathbf{s}[n - k], k \geq 0\} = \sum_{k=0}^{\infty} h[k]\mathbf{s}[n - k]$$

This estimate is the output of a linear time-invariant causal system with input $\mathbf{s}[n]$ and delta response $h[n]$. From (13-4) it follows that

$$E\left\{\left(\mathbf{s}[n + r] - \sum_{k=0}^{\infty} h[k]\mathbf{s}[n - k]\right)\mathbf{s}[n - m]\right\} = 0 \qquad m \geq 0$$

Hence

$$R_{ss}[m + r] = \sum_{k=0}^{\infty} h[k]R_{ss}[m - k] \qquad m \geq 0 \tag{13-42}$$

The above system cannot be solved directly with transform techniques because, unlike (13-40), the two sides of (13-42) are equal not for every m but only for $m \geq 0$. To solve it, we could proceed as follows: The sequence

$$y[m] = R_{ss}[m + r] - \sum_{k=0}^{\infty} h[k]R_{ss}[m - k]$$

equals the difference of the two sides of (13-42), hence $y[m] = 0$ for $m \geq 0$. This leads to the conclusion that its z transforms

$$\mathbf{Y}(z) = z^r\mathbf{S}_{ss}(z) - \mathbf{H}(z)\mathbf{S}_{ss}(z)$$

is analytic for $|z| < 1$. Similarly, the z transform $\mathbf{H}(z)$ of $h[n]$ is analytic for $|z| > 1$ because $h[n] = 0$ for $n < 0$. The above leads to the unique determination of the functions $\mathbf{H}(z)$ and $\mathbf{Y}(z)$ (see Prob. 13-19), however, the proof is based on the properties of analytic functions. In the following, we give a simpler solution based on the concept of *innovations*. This approach is an extension of (13-20) to infinitely many data.

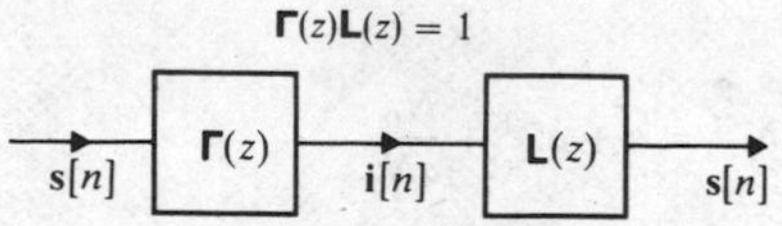

Figure 13-4

As we have shown in Sec. 10-5, a *regular* process $\mathbf{s}[n]$ is *linearly equivalent* with its innovations $\mathbf{i}[n]$ in the sense that each is the response of a causal system with input the other (Fig. 13-4)

$$\mathbf{s}[n] = \sum_{k=0}^{\infty} l[k]\mathbf{i}[n-k] \qquad \mathbf{i}[n] = \sum_{k=0}^{\infty} \gamma(k)\mathbf{s}[n-k] \tag{13-43}$$

The process $\mathbf{i}[n]$ is orthonormal, i.e., it is white noise with autocorrelation $\delta[m]$, and the sequences $l[n]$ and $\gamma[n]$ are the delta responses of the innovations filter $\mathbf{L}(z)$ and the whitening filter $\mathbf{\Gamma}(z) = 1/\mathbf{L}(z)$ respectively. These filters can be determined by factoring the power spectrum $\mathbf{S}(z)$ of $\mathbf{s}[n]$ as in (10-115)

$$\mathbf{S}(z) = \mathbf{L}(z)\mathbf{L}(z^{-1}) \tag{13-44}$$

Using the above equivalence, we shall determine the predictor of a regular process $\mathbf{s}[n]$ starting with the case $r = 1$.

The one-step predictor Shifting the origin one unit, we phrase our problem as follows: We wish to estimate the present value of a process $\mathbf{s}[n]$ in terms of its entire past. The desired estimate is the sum

$$\hat{\mathbf{s}}_1[n] = \hat{E}\{\mathbf{s}[n] \mid \mathbf{s}[n-k],\ k \geq 1\} = \sum_{k=1}^{\infty} a_k \mathbf{s}[n-k] \tag{13-45}$$

and our problem is to determine the coefficients a_k. The system

$$\mathbf{H}_1(z) = \sum_{k=1}^{\infty} a_k z^{-k}$$

we shall call the *predictor* of $\mathbf{s}[n]$ and the system

$$\mathbf{E}_1(z) = 1 - \mathbf{H}_1(z)$$

the *error filter* (Fig. 13-5). The estimation error

$$\boldsymbol{\varepsilon}_1[n] = \mathbf{s}[n] - \hat{\mathbf{s}}_1[n]$$

is the output of the filter $\mathbf{E}_1(z)$ with input $\mathbf{s}[n]$.

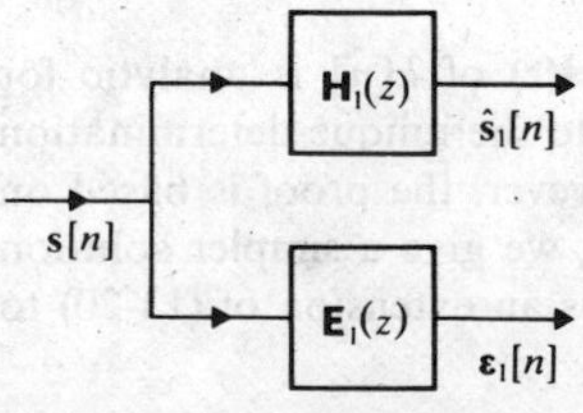

Figure 13-5

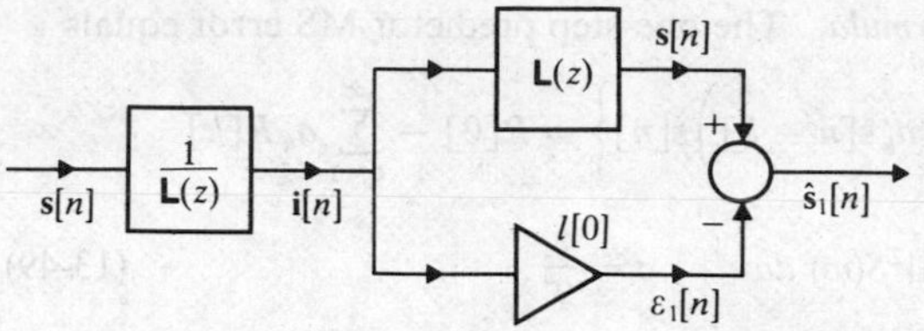

Figure 13-6

Since the processes $\mathbf{s}[n]$ and $\mathbf{i}[n]$ are linearly equivalent, $\hat{\mathbf{s}}_1[n]$ can be expressed in terms of $\mathbf{i}[n]$. We maintain, in fact, that

$$\hat{\mathbf{s}}_1[n] = \sum_{k=1}^{\infty} l[k]\mathbf{i}[n-k] \tag{13-46}$$

PROOF It suffices to show that, if $\hat{\mathbf{s}}_1[n]$ equals the above sum, the resulting error is orthogonal to $\mathbf{i}[n-k]$ for every $k > 1$. As we see from (13-43), the error equals

$$\boldsymbol{\varepsilon}_1[n] - \mathbf{s}[n] - \hat{\mathbf{s}}_1[n] = l[0]\mathbf{i}[n] \tag{13-47}$$

Hence, $\boldsymbol{\varepsilon}_1[n]$ is orthogonal to $\mathbf{i}[n-k]$ for $k > 0$ because $\mathbf{i}[n]$ is white noise. This completes the proof (orthogonality principle).

From the above it follows that the processes $\boldsymbol{\varepsilon}_1[n]$ and $\hat{\mathbf{s}}_1[n]$ are the responses of the systems $l[0]$ and $\mathbf{L}(z) - l[0]$ respectively (Fig. 13-6) with input $\mathbf{i}[n]$. To complete the specification of $\mathbf{H}_1(z)$, we must express $\mathbf{i}[n]$ in terms of $\mathbf{s}[n]$. Since $\mathbf{i}[n]$ is the output of the whitening filter $1/\mathbf{L}(z)$ with input $\mathbf{s}[n]$, we conclude, cascading with $1/\mathbf{L}(z)$, that (Fig. 13-7)

$$\mathbf{E}_1(z) = \frac{l[0]}{\mathbf{L}(z)} \qquad \mathbf{H}_1(z) = 1 - \frac{l[0]}{\mathbf{L}(z)} \tag{13-48}$$

To find $l[0]$, we can use the initial value theorem.

$$l[0] = \lim_{z \to \infty} \mathbf{L}(z)$$

Thus, to determine the predictor $\mathbf{H}_1(z)$ of a regular process, it suffices to find the function $\mathbf{L}(z)$. This involves the factorization of $\mathbf{S}(z)$ as in (13-44). Examples of factorization are given in Sec. 10-5.

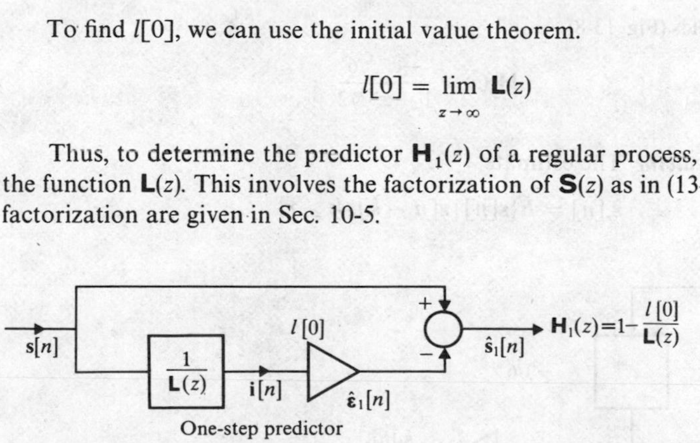

Figure 13-7

Kolmogoroff's MS error formula The one-step predictor MS error equals

$$P_1 = E\left\{\left(\mathbf{s}[n] - \sum_{k=1}^{\infty} a_k \mathbf{s}[n-k]\right)\mathbf{s}[n]\right\} = R[0] - \sum_{k=1}^{\infty} a_k R[k]$$

$$= \frac{1}{2\sigma}\int_{-\sigma}^{\sigma} |\mathbf{E}_1(e^{j\omega T})|^2 S(\omega)\, d\omega \qquad \sigma = \frac{\pi}{T} \tag{13-49}$$

This error can be expressed directly in terms of $S(\omega)$. Indeed, as we know

$$\boldsymbol{\varepsilon}_1[n] = l[0]\mathbf{i}[n] \qquad E\{\mathbf{i}^2[n]\} = 1$$

This yields

$$P_1 = l^2[0]$$

But $\mathbf{L}(z)$ is a minimum-phase system and

$$S(\omega) = |\mathbf{L}(e^{j\omega T})|^2$$

Hence [see (13A-1)]

$$P_1 = \exp\left\{\frac{1}{2\sigma}\int_{-\sigma}^{\sigma} \ln S(\omega)\, d\omega\right\} \tag{13-50}$$

The above result is known as Kolmogoroff's formula.

Example 13-6 If

$$S(\omega) = \frac{5 - 4\cos \omega T}{10 - 6\cos \omega T}$$

as in Example 10-24, then

$$\mathbf{L}(z) = \frac{2z-1}{3z-1} \qquad l[0] = \tfrac{2}{3}$$

and (13-48) yields (Fig. 13-8)

$$\mathbf{H}_1(z) = \frac{-z^{-1}/6}{1 - z^{-1}/2}$$

The r-step predictor The estimate

$$\hat{\mathbf{s}}_r[n] = \hat{E}\{\mathbf{s}[n] \mid \mathbf{s}[n-k],\ k \ge r\}$$

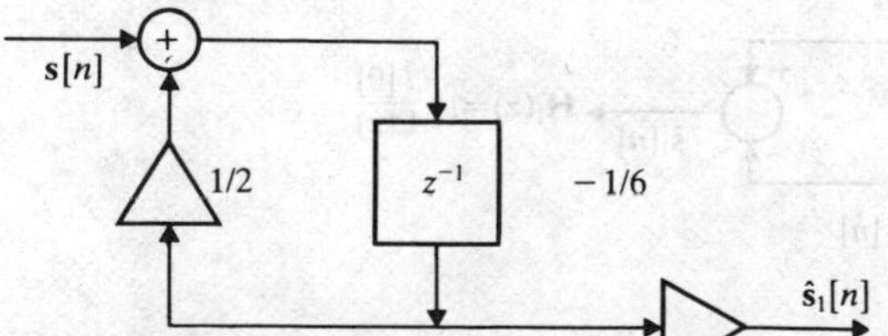

Figure 13-8

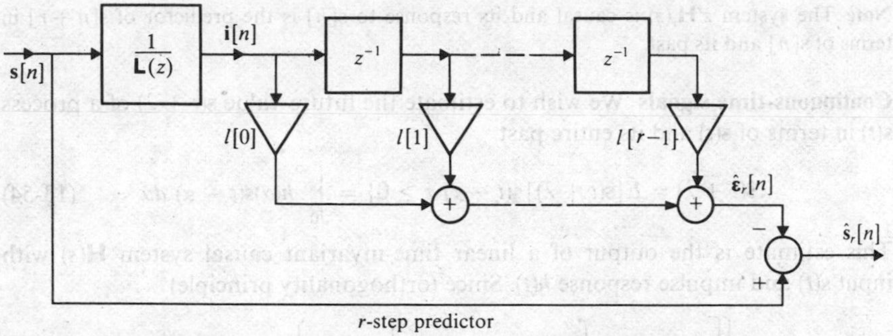

Figure 13-9

of $\mathbf{s}[n]$ in terms of $\mathbf{s}[n-r]$ and its past can be found similarly. Reasoning as in (13-46), we conclude that

$$\hat{\mathbf{s}}_r[n] = \sum_{k=r}^{\infty} l[k]\mathbf{i}[n-k] \tag{13-51}$$

because the resulting error

$$\boldsymbol{\varepsilon}_r[n] = \sum_{k=0}^{r-1} l[k]\mathbf{i}[n-k] \tag{13-52}$$

is orthogonal to $\mathbf{i}[n-k]$ for $k \geq r$.

To complete the specification of the r-step predictor $\mathbf{H}_r(z)$, we must express $\mathbf{i}[n]$ in terms of $\mathbf{s}[n]$. This yields the system (Fig. 13-9)

$$\mathbf{H}_r(z) = 1 - \frac{1}{\mathbf{L}(z)} \sum_{k=0}^{r-1} l[k]z^{-k} \tag{13-53}$$

The MS estimation error equals

$$P_r = E\{\boldsymbol{\varepsilon}_r^2[n]\} = \sum_{k=0}^{r-1} l^2[k]$$

The above follows from (13-52) because $\mathbf{i}[n]$ is white noise (see Prob. 9-29*a*).

Example 13-7 If

$$\mathbf{S}(z) = \frac{N_0}{(1-az^{-1})(1-az)} \qquad R[m] = \frac{N_0}{1-a^2}\, a^{|m|}$$

as in Example 10-20, then

$$\mathbf{L}(z) = \frac{\sqrt{N_0}}{1-az^{-1}} \qquad l[n] = \sqrt{N_0}\, a^n U[n]$$

$$\mathbf{H}_r(z) = 1 - \frac{1-az^{-1}}{\sqrt{N_0}} \sum_{k=0}^{r-1} \sqrt{N_0}\, a^k z^{-k} = a^r z^{-r} \qquad \hat{\mathbf{s}}_r[n] = a^r \mathbf{s}[n-r]$$

$$P_r = N_0 \sum_{k=0}^{r-1} a^{2k} = \frac{N_0(1-a^{2r})}{1-a^2}$$

Note The system $z^r\mathbf{H}_r(z)$ is causal and its response to $\mathbf{s}[n]$ is the predictor of $\mathbf{s}[n+r]$ in terms of $\mathbf{s}[n]$ and its past.

Continuous-time signals We wish to estimate the future value $\mathbf{s}(t+\lambda)$ of a process $\mathbf{s}(t)$ in terms of $\mathbf{s}(t)$ and its entire past

$$\hat{\mathbf{s}}(t+\lambda) = \hat{E}\{\mathbf{s}(t+\lambda) \mid \mathbf{s}(t-\tau)\ \tau \geq 0\} = \int_0^{\infty} h(\alpha)\mathbf{s}(t-\alpha)\, d\alpha \tag{13-54}$$

This estimate is the output of a linear time-invariant causal system $\mathbf{H}(s)$ with input $\mathbf{s}(t)$ and impulse response $h(t)$. Since (orthogonality principle)

$$E\left\{\left[\mathbf{s}(t+\lambda) - \int_0^{\infty} h(\alpha)\mathbf{s}(t-\alpha)\, d\alpha\right]\mathbf{s}(t-\tau)\right\} = 0 \qquad \tau \geq 0$$

we conclude as in (13-27) that

$$R(\tau+\lambda) = \int_0^{\infty} h(\alpha)R(\tau-\alpha)\, d\alpha \qquad \tau \geq 0 \tag{13-55}$$

This is the celebrated *Wiener–Hopf* equation. The resulting MS error P equals

$$P = E\{[\mathbf{s}(t+\lambda) - \hat{\mathbf{s}}(t+\lambda)]\mathbf{s}(t+\lambda)\} = R(0) - \int_0^{\infty} h(\alpha)R(\lambda+\alpha)\, d\alpha$$

Note The Wiener–Hopf equation can be given the following interpretation: The response $y(\tau)$ of the predictor $\mathbf{H}(s)$ to the input $R(\tau)$ equals

$$y(\tau) = \int_0^{\infty} R(\tau-\alpha)h(\alpha)\, d\alpha$$

Hence [see (13-55)]

$$y(\tau) = R(\tau+\lambda) \quad \text{for} \quad \tau \geq 0 \qquad \text{and} \qquad y(-\lambda) = R(0) - P$$

as in Fig. 13-10. In the same figure, we show also the response $q(\tau) = R(\tau+\lambda) - y(\tau)$ of the error filter

$$\mathbf{E}(s) = e^{\lambda s} - \mathbf{H}(s) = e^{\lambda s} - \int_0^{\infty} h(t)e^{-st}\, dt$$

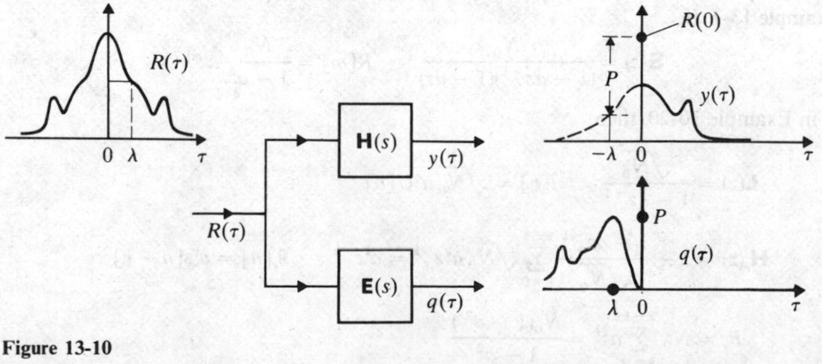

Figure 13-10

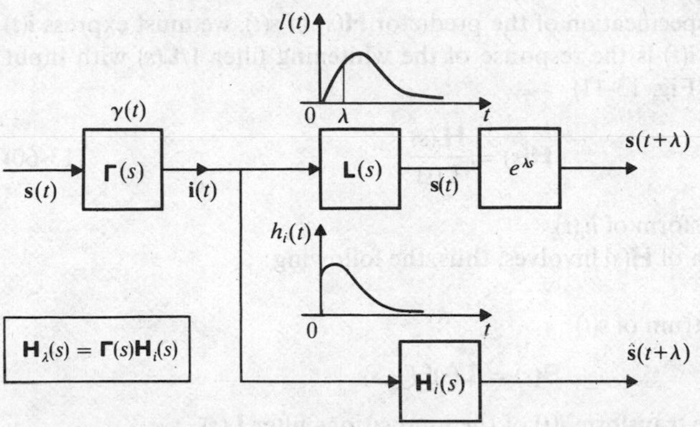

Figure 13-11

to $R(\tau)$. In this case

$$q(\tau) = 0 \quad \text{for} \quad \tau \geq 0 \qquad \text{and} \qquad q(-\lambda) = P$$

The Wiener–Hopf equation cannot be solved directly with transform techniques because, unlike (13-36), it does not hold for every τ. We shall solve it using innovations.†

Changing t to $t + \lambda$ in (10-110), we obtain

$$\mathbf{s}(t + \lambda) = \int_0^\infty l(\alpha)\mathbf{i}(t + \lambda - \alpha)\, d\alpha \tag{13-56}$$

In Fig. 13-11, we show the signal $\mathbf{s}(t + \lambda)$ as the response of the system $\mathbf{L}(s)e^{\lambda s}$ with input $\mathbf{i}(t)$.

As we know, the processes $\mathbf{s}(t)$ and $\mathbf{i}(t)$ are linearly equivalent. From this it follows that $\hat{\mathbf{s}}(t + \lambda)$ can be expressed in terms of $\mathbf{i}(t)$ and its past. We maintain, in fact, that

$$\hat{\mathbf{s}}(t + \lambda) = \int_\lambda^\infty l(\alpha)\mathbf{i}(t + \lambda - \alpha)\, d\alpha = \int_0^\infty l(\beta + \lambda)\mathbf{i}(t - \beta)\, d\beta \tag{13-57}$$

Indeed, the difference

$$\boldsymbol{\varepsilon}(t) = \mathbf{s}(t + \lambda) - \hat{\mathbf{s}}(t + \lambda) = \int_0^\lambda l(\alpha)\mathbf{i}(t + \lambda - \alpha)\, d\alpha \tag{13-58}$$

depends only on the values of $i(t + \tau)$ for $0 \leq \tau \leq \lambda$; hence, it is orthogonal to the past of $\mathbf{i}(t)$. From (13-57) it follows that $\hat{\mathbf{s}}(t + \lambda)$ is the response of a causal filter with input $\mathbf{i}(t)$ and impulse response

$$h_i(t) = l(t + \lambda)U(t) \tag{13-59}$$

† T. Kailath: "An Innovations Approach to Detection and Estimation Theory," *Proceeding of the IEEE*, vol. 58, 1970.

To complete the specification of the predictor $\mathbf{H}(s)$ of $\mathbf{s}(t)$, we must express $\mathbf{i}(t)$ in terms of $\mathbf{s}(t)$. Since $\mathbf{i}(t)$ is the response of the whitening filter $1/\mathbf{L}(s)$ with input $\mathbf{s}(t)$, we conclude that (Fig. 13-11)

$$\mathbf{H}(s) = \frac{\mathbf{H}_i(s)}{\mathbf{L}(s)} \tag{13-60}$$

where $\mathbf{H}_i(s)$ is the transform of $h_i(t)$.

The determination of $\mathbf{H}(s)$ involves, thus, the following:

(*a*) We factor the spectrum of $\mathbf{s}(t)$

$$\mathbf{S}(s) = \mathbf{L}(s)\mathbf{L}(-s)$$

(*b*) We find the inverse transform $l(t)$ of the innovations filter $\mathbf{L}(s)$.
(*c*) We shift $l(t)$ and truncate as in (13-59).
(*d*) We find the transform $\mathbf{H}_i(s)$ of $h_i(t)$ and insert into (13-60).

MS error Since $\mathbf{i}(t)$ is white noise, we conclude from (13-58) and Prob. 9-29*b* that

$$P = E\left\{\left|\int_0^\lambda l(\alpha)\mathbf{i}(t+\lambda-\alpha)\,d\alpha\right|^2\right\} = \int_0^\lambda l^2(\alpha)\,d\alpha$$

Rational spectra If $\mathbf{S}(s)$ is rational, then $\mathbf{L}(s)$ is also rational. Assuming that its poles are simple, we obtain

$$\mathbf{L}(s) = \frac{N(s)}{D(s)} = \sum_i \frac{c_i}{s - s_i}$$

$$l(t) = \sum_i c_i e^{s_i t} U(t) \qquad l(t+\lambda)U(t) = \sum_i c_i e^{s_i \lambda} e^{s_i t} U(t) \tag{13-61}$$

$$\mathbf{H}_i(s) = \sum_i \frac{c_i e^{s_i \lambda}}{s - s_i} = \frac{N_1(s)}{D(s)} \qquad \mathbf{H}(s) = \frac{\mathbf{H}_i(s)}{\mathbf{L}(s)} = \frac{N_1(s)}{N(s)}$$

If $\mathbf{S}(s)$ is *all-pole*, then $N(s) = 1$ and (13-61) yields

$$\mathbf{H}(s) = N_1(s) = b_0 + b_1 s + \cdots + b_n s^n$$

In this case

$$\hat{\mathbf{s}}(t+\lambda) = b_0\,\mathbf{s}(t) + b_1\mathbf{s}'(t) + \cdots + b_n\mathbf{s}^{(n)}(t)$$

Example 13-8 If

$$\mathbf{S}(s) = \frac{N_0}{\alpha^2 - s^2} \qquad \text{then} \qquad \mathbf{L}(s) = \frac{\sqrt{N_0}}{s+\alpha}$$

$$\mathbf{H}_i(s) = \frac{\sqrt{N_0}}{s+\alpha} \qquad \mathbf{H}(s) = e^{-\alpha\lambda} \qquad h(t) = e^{-\alpha\lambda}\delta(t)$$

Thus, $\hat{\mathbf{s}}(t+\lambda) = e^{-\alpha\lambda}\mathbf{s}(t)$.

Example 13-9 If

$$\mathbf{S}(s) = \frac{49 - 25s^2}{(1 - s^2)(9 - s^2)} \qquad \text{and} \qquad \lambda = \ln 2$$

then

$$\mathbf{L}(s) = \frac{7 + 5s}{(1 + s)(3 + s)} = \frac{1}{s + 1} + \frac{4}{s + 3}$$

and (13-61) yields

$$\mathbf{H}_i(s) = \frac{e^{-\lambda}}{s + 1} + \frac{4e^{-3\lambda}}{s + 3} = \frac{s + 2}{(s + 1)(s + 3)}$$

$$\mathbf{H}(s) = \frac{s + 2}{5s + 7} \qquad h(t) = \frac{1}{5}\,\delta(t) + \frac{3}{25}\,e^{-1.4t}U(t)$$

We note that, in this case,

$$\hat{E}\{\mathbf{s}(t + \lambda) \mid \mathbf{s}(\xi),\ \xi \le t\} = 0.2\mathbf{s}(t) + \hat{E}\{\mathbf{s}(t + \lambda) \mid \mathbf{s}(\xi),\ \xi < t\}$$

Note In all MS estimation problems, only second-order moments are used. If, therefore, two processes have the same autocorrelation, then their predictors are identical. This suggests the following derivation of the Wiener–Hopf equation: Suppose that $\boldsymbol{\omega}$ is an RV with density $f(\omega)$ and $\mathbf{z}(t) = e^{j\boldsymbol{\omega}t}$. Clearly

$$R_{zz}(\tau) = E\{e^{j\boldsymbol{\omega}(t+\tau)}e^{-j\boldsymbol{\omega}t}\} = \int_{-\infty}^{\infty} f(\omega)e^{j\omega\tau}\,d\omega$$

From this it follows that the power spectrum of $\mathbf{z}(t)$ equals $2\pi f(\omega)$. If, therefore, $\mathbf{s}(t)$ is a process with power spectrum $S(\omega) = 2\pi f(\omega)$, then its predictor $h(t)$ will equal the predictor of $\mathbf{z}(t)$

$$\hat{\mathbf{z}}(t + \lambda) = \hat{E}\{e^{j\boldsymbol{\omega}(t+\lambda)} \mid e^{j\boldsymbol{\omega}(t-\alpha)},\ \alpha \ge 0\} = \int_0^{\infty} h(\alpha)e^{j\boldsymbol{\omega}(t-\alpha)}\,d\alpha$$

$$= e^{j\boldsymbol{\omega}t}\int_0^{\infty} h(\alpha)e^{-j\boldsymbol{\omega}\alpha}\,d\alpha = e^{j\boldsymbol{\omega}t}H(\boldsymbol{\omega})$$

And since $\mathbf{z}(t + \lambda) - \hat{\mathbf{z}}(t + \lambda) \perp \mathbf{z}(t - \tau)$, for $\tau \ge 0$, we conclude from the above that

$$E\{[e^{j\boldsymbol{\omega}(t+\lambda)} - e^{j\boldsymbol{\omega}t}H(\boldsymbol{\omega})]e^{-j\boldsymbol{\omega}(t-\tau)}\} = 0 \qquad \tau \ge 0$$

Hence

$$\int_{-\infty}^{\infty} f(\omega)[e^{j\omega(\tau+\lambda)} - e^{j\omega\tau}H(\omega)]\,d\omega = 0 \qquad \tau \ge 0$$

This yields (13-55) because the inverse transform of $f(\omega)e^{j\omega(\tau+\lambda)}$ equals $R(\tau + \lambda)$ and the inverse transform of $f(\omega)e^{j\omega\tau}H(\omega)$ equals the integral in (13-55).

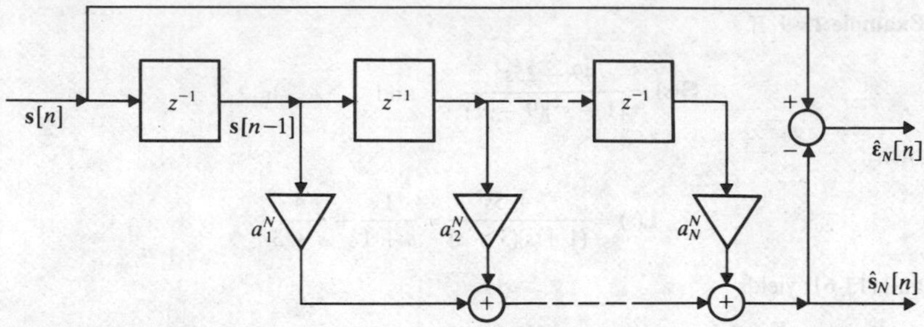

Figure 13-12

Levinson's Algorithm and Wold's Decomposition

We consider now the problem of estimating the present value $\mathbf{s}[n]$ of a real discrete-time stochastic process in terms of its N most recent past values

$$\hat{\mathbf{s}}_N[n] = \hat{E}\{\mathbf{s}[n] \mid \mathbf{s}[n-k], 1 \le k \le N\} = \sum_{k=1}^{N} a_k^N \mathbf{s}[n-k] \tag{13-62}$$

This is the one-step *forward* predictor of $\mathbf{s}[n]$ of order N and

$$\mathbf{H}_N(z) = \sum_{k=1}^{N} a_k^N z^{-k} \tag{13-63}$$

is the predictor filter. The superscript N in a_k^N identifies the order of the predictor. The system $\mathbf{H}_N(z)$ is an FIR (finite impulse response) filter realized in Fig. 13-12 as a tapped delay line of length N. In the figure we show also the resulting *forward error* $\hat{\boldsymbol{\varepsilon}}_N[n]$

$$\hat{\boldsymbol{\varepsilon}}_N[n] = \mathbf{s}[n] - \hat{\mathbf{s}}_N[n] \qquad P_N = E\{\hat{\boldsymbol{\varepsilon}}_N^2[n]\} \tag{13-64}$$

The above problem is a special case of the problem considered in Sec. 13-1. We shall reexamine it, however, in the context of stochastic processes. We do so not only because the problem is an important special case in the theory of prediction but also because, as we shall see, the underlying reasoning clarifies the connection between linear MS estimation and spectral decomposition.

The $N + 1$ unknowns a_k^N and P_N can be determined from (13-4) and (13-7)

$$E\{(\mathbf{s}[n] - \hat{\mathbf{s}}_N[n])\mathbf{s}[n-k]\} = 0 \qquad 1 \le k \le N$$
$$P_N = E\{(\mathbf{s}[n] - \hat{\mathbf{s}}_N[n])\mathbf{s}[n]\}$$

Inserting (13-62) into the above, we obtain the system (Yule–Walker equations)

$$\begin{aligned}
P_N + R[1]a_1^N &+ R[2]a_2^N + \cdots + R[N]a_N^N = R[0] \\
R[0]a_1^N &+ R[1]a_2^N + \cdots + R[N-1]a_N^N = R[1] \\
R[1]a_1^N &+ R[0]a_2^N + \cdots + R[N-2]a_N^N = R[2] \\
&\cdots\cdots\cdots\cdots\cdots\cdots\cdots\cdots \\
R[N-1]a_1^N &+ R[N-2]a_2^N + \cdots + R[0]a_N^N = R[N]
\end{aligned} \tag{13-65}$$

whose solution yields a_k^N and P_N. Thus

$$P_N = \frac{\Delta_{N+1}}{\Delta_N} \tag{13-66}$$

where Δ_N is the determinant of the correlation matrix R_N formed with the coefficients of the last N equation. The matrix R_N is *Teoplitz*, i.e., its elements on each diagonal are identical. As we shall presently show, this property of R_N simplifies the evaluation of its inverse and it leads to a simple recursive method for determining the constants a_k^N.

Note The Yule–Walker equations (13-65) can be given the following interpretation: The response $y[m]$ of the predictor $\mathbf{H}_N(z)$ to the input $R[m]$ equals

$$y[m] = \sum_{k=1}^{N} R[m-k]a_k^N$$

Comparing with (13-65), we conclude that

$$y[m] = R[m] \quad \text{for} \quad 1 \le m \le N \qquad \text{and} \qquad y[0] = R[0] - P_N \tag{13-67}$$

as in Fig. 13-13. In the same figure, we show also the response $\hat{q}_N[m] = R[m] - y[m]$ of the error filter

$$\mathbf{E}_N(z) = 1 - \mathbf{H}_N(z) = 1 - \sum_{k=1}^{N} a_k^N z^{-k}$$

to $R[m]$. In this case

$$\hat{q}_N[m] = 0 \quad \text{for} \quad 1 \le m \le N \qquad \text{and} \qquad \hat{q}_N[0] = P_N \tag{13-68}$$

Conversely, if the response of an FIR filter of length $N+1$ to the input $R[m]$ satisfies (13-68), then its coefficients satisfy the Yule–Walker equations.

We mention also, for later use, that

$$y[N+1] = \sum_{k=1}^{N} R[N+1-k]a_k^N \qquad \hat{q}_N[N+1] = R[N+1] - y[N+1] \tag{13-69}$$

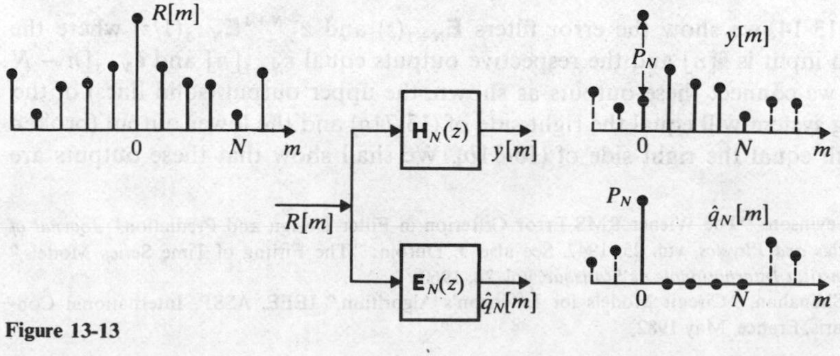

Figure 13-13

Levinson's algorithm† We shall show that the system (13-65) can be solved recursively. For this purpose, we form the *backward predictor* of $\mathbf{s}[n]$, that is, its estimate in terms of its N most recent future values

$$\check{\mathbf{s}}_N[n] = \hat{E}\{\mathbf{s}[n] \mid \mathbf{s}[n+k], 1 \le k \le N\} = \sum_{k=1}^{N} a_k^N \mathbf{s}[n+k] \qquad (13\text{-}70)$$

The coefficients in (13-62) and (13-70) are identical because the backward predictor of $\mathbf{s}[n]$ is the forward predictor of $\mathbf{s}[-n]$ and the processes $\mathbf{s}[n]$ and $\mathbf{s}[-n]$ have the same autocorrelation. We introduce also the *backward error* $\check{\boldsymbol{\varepsilon}}_N[n]$ and its MS value

$$\check{\boldsymbol{\varepsilon}}_N[n] = \mathbf{s}[n] - \check{\mathbf{s}}_N[n] \qquad P_N = E\{\check{\boldsymbol{\varepsilon}}_N^2[n]\}$$

The process $\check{\boldsymbol{\varepsilon}}_N[n]$ is the response of the backward error filter

$$1 - \sum_{k=1}^{N} a_k^N z^k = \mathbf{E}_N(z^{-1})$$

to the input $\mathbf{s}[n]$.

Theorem The forward and backward errors satisfy the equations

$$\hat{\boldsymbol{\varepsilon}}_N[n] = \hat{\boldsymbol{\varepsilon}}_{N-1}[n] - K_N \check{\boldsymbol{\varepsilon}}_{N-1}[n-N] \qquad (13\text{-}71a)$$

$$\check{\boldsymbol{\varepsilon}}_N[n-N] = \check{\boldsymbol{\varepsilon}}_{N-1}[n-N] - K_N \hat{\boldsymbol{\varepsilon}}_{N-1}[n] \qquad (13\text{-}71b)$$

where K_N is a constant to be determined.

Proof The right sides of the above equations are orthogonal to $\mathbf{s}[n-k]$ for $1 \le k \le N-1$. To prove the theorem, it suffices to choose K_N such that

$$\hat{\boldsymbol{\varepsilon}}_{N-1}[n] - K_N \check{\boldsymbol{\varepsilon}}_{N-1}[n-N] \perp \mathbf{s}[n-N]$$

We shall give another proof‡ based on the note leading to (13-68):

The two equations in (13-71) are equivalent to the error filter equations

$$\mathbf{E}_N(z) = \mathbf{E}_{N-1}(z) - K_N z^{-N} \mathbf{E}_{N-1}(1/z) \qquad (13\text{-}72a)$$

$$z^{-N}\mathbf{E}_N(1/z) = z^{-N}\mathbf{E}_{N-1}(1/z) - K_N \mathbf{E}_{N-1}(z) \qquad (13\text{-}72b)$$

In Fig. 13-14, we show the error filters $\mathbf{E}_{N-1}(z)$ and $z^{-N+1}\mathbf{E}_{N-1}(1/z)$ where the common input is $\hat{\mathbf{s}}[n]$ and the respective outputs equal $\hat{\boldsymbol{\varepsilon}}_{N-1}[n]$ and $\check{\boldsymbol{\varepsilon}}_{N-1}[n-N+1]$. If we connect these outputs as shown, the upper output (solid lines) of the resulting system will equal the right side of (13-71*a*) and the lower output (broken lines) will equal the right side of (13-71*b*). We shall show that these outputs are

† N. Levinson: "The Wiener RMS Error Criterion in Filter Design and Prediction," *Journal of Mathematics and Physics*, vol. 25, 1947. See also J. Durbin: "The Fitting of Time Series Models," *Revue L'Institut Internationale de Statisque*, vol. 28, 1960.

‡ W. Shanahan, "Circuit Models for Levinson's Algorithm," IEEE, ASSP, International Conference, Paris, France, May 1982.

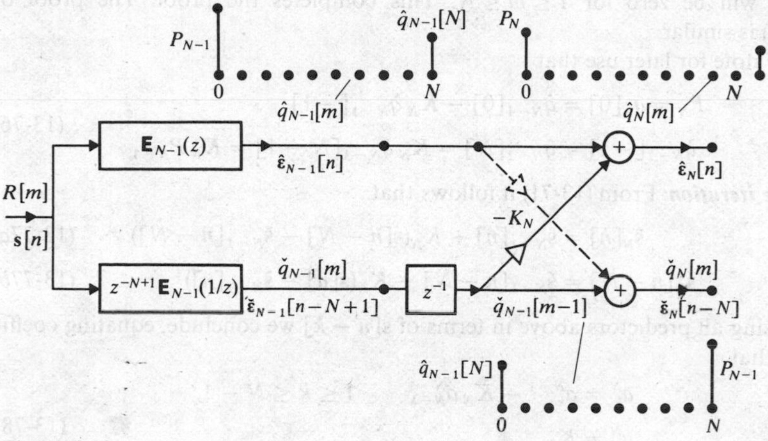

Figure 13-14

$\hat{\varepsilon}_N[n]$ and $\check{\varepsilon}_N[n-N]$ respectively. For this purpose, we introduce the error filters $\mathbf{E}_N(z)$ and $z^{-N}\mathbf{E}_N(1/z)$ and we denote by $\hat{q}_N[m]$ and $\check{q}_N[m]$ their respective outputs when the input is the autocorrelation $R[m]$ of $s[n]$ (the time variable is now m). To prove (13-72*a*), it suffices to show that (see Fig. 13-14)

$$\hat{q}_N[m] = \hat{q}_{N-1}[m] - K_N \check{q}_{N-1}[m-1] \tag{13-73}$$

In fact [see (13-68) and following comment], it is enough to show that the right side of (13-73) is zero for $1 \le m \le N$. Clearly

$$\hat{q}_N[m] = R[m] - \sum_{k=1}^{N} R[m-k]a_k^N$$

$$\check{q}_N[m] = R[m-N] - \sum_{k=1}^{N} R[m-N+k]a_k^N$$

Since $R[m] = R[-m]$, this yields

$$\hat{q}_N[m] = \check{q}_N[N-m]$$

Changing N to $N-1$, we conclude from (13-68), (13-69), and the above, that

$$\hat{q}_{N-1}[m] = \check{q}_{N-1}[N-1-m] = 0 \qquad 1 \le m \le N-1$$

$$\hat{q}_{N-1}[0] = \check{q}_{N-1}[N-1] = P_{N-1} \tag{13-74}$$

$$\hat{q}_{N-1}[N] = \check{q}_{N-1}[-1] = R[N] - \sum_{k=1}^{N-1} R[N-k]a_k^{N-1}$$

From the first equation in (13-74) it follows that $\check{q}_{N-1}[m-1] = 0$ for $1 \le m \le N-1$, hence, the right side of (13-73) is zero for $1 \le m \le N-1$. If, therefore, we choose K_N such that

$$\hat{q}_{N-1}[N] - K_N \check{q}_{N-1}[N-1] = 0 \tag{13-75}$$

then it will be zero for $1 \le m \le N$. This completes the proof. The proof of (13-72*b*) is similar

We note for later use that

$$\begin{aligned} P_N = \hat{q}_N[0] &= \hat{q}_{N-1}[0] - K_N \check{q}_{N-1}[-1] \\ \check{q}_{N-1}[-1] &= \hat{q}_{N-1}[N] = K_N \check{q}_{N-1}[N-1] = K_N P_{N-1} \end{aligned} \tag{13-76}$$

The iteration From (13-71) it follows that

$$\hat{\mathbf{s}}_N[n] = \hat{\mathbf{s}}_{N-1}[n] + K_N(\mathbf{s}[n-N] - \check{\mathbf{s}}_{N-1}[n-N]) \tag{13-77a}$$

$$\check{\mathbf{s}}_N[n-N] = \check{\mathbf{s}}_{N-1}[n-N] + K_N(\mathbf{s}[n] - \hat{\mathbf{s}}_{N-1}[n]) \tag{13-77b}$$

Expressing all predictors above in terms of $\mathbf{s}[n-k]$ we conclude, equating coefficients, that

$$\begin{aligned} a_k^N &= a_k^{N-1} - K_N a_{N-k}^{N-1} \qquad 1 \le k \le N-1 \\ a_N^N &= K_N \end{aligned} \tag{13-78}$$

where K_N is such that [see (13-74) and (13-75)]

$$P_{N-1} K_N = R[N] - \sum_{k=1}^{N-1} R[N-k] a_k^{N-1} \tag{13-79}$$

We have, thus, expressed the constants a_k^N and K_N in terms of the predictor parameters of order $N-1$. To continue the iteration, we must find also P_N. Since $\hat{q}_{N-1}[0] = P_{N-1}$, (13-76) yields

$$P_N = (1 - K_N^2) P_{N-1} \tag{13-80}$$

The iteration starts with

$$P_0 = E\{\mathbf{s}^2[n]\} = R[0]$$

and for $N = 1$ it yields

$$P_0 K_1 = R[1] \qquad P_1 = (1 - K_1^2) P_0$$

Lattice filters† The error filters

$$\begin{aligned} \mathbf{E}_N(z) &= 1 - (a_1^N z^{-1} + \cdots + a_N^N z^{-N}) \\ z^{-N} \mathbf{E}_N(1/z) &= -(a_N^N + a_{N-1}^N z^{-1} + \cdots + a_1^N z^{-N+1}) + z^{-N} \end{aligned} \tag{13-81}$$

are FIR filters and can be realized by tapped-delay lines as in Fig. 13-12. However, they can also be realized by the lattice structure shown in Fig. 13-15*a*. This follows readily if the construction of Fig. 13-14 is repeated starting with $N = 0$ and input

$$\mathbf{s}[n] = \hat{\boldsymbol{\varepsilon}}_0[n] = \check{\boldsymbol{\varepsilon}}_0[n]$$

† J. Makhoul: "Stable and Efficient Lattice Methods for Linear Predictions," *IEEE Transactions on Acoustics, Speech, and Signal Processing*, vol. ASSP-26, 1978.

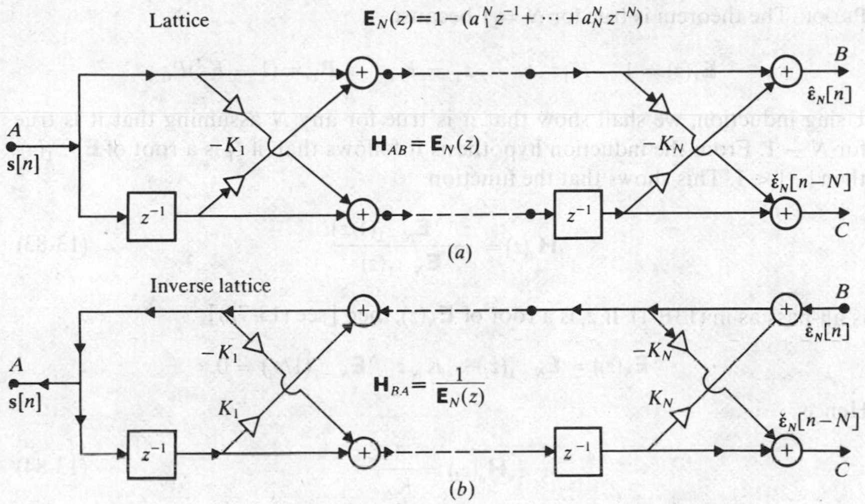

Figure 13-15

It then follows that the upper output (point B) equals $\hat{\varepsilon}_N[n]$ and the lower output (point C) equals $\check{\varepsilon}_N[n-N]$. The system function from A to B equals $\mathbf{E}_N(z)$ and from A to C it equals $z^{-N}\mathbf{E}_N(1/z)$ as in (13-81).

The above realization of the error filters has various advantages that will be discussed later. We mention here only that, with the simple modification shown in Fig. 13-15b, we obtain a system with input at B and two outputs at the points A and C. If the input at B equals $\hat{\varepsilon}_N[n]$, then the output at A equals s$[n]$ and the output at C equals $\check{\varepsilon}_N[n-N]$. This is so because the resulting equations

$$\hat{\varepsilon}_{N-1}[n] = \hat{\varepsilon}_N[n] + K_N \check{\varepsilon}_{N-1}[n-N]$$

$$\check{\varepsilon}_N[n-N] = \check{\varepsilon}_{N-1}[n-N] - K_N \hat{\varepsilon}_{N-1}[n]$$

are equivalent to the two equations in (13-71). The system so formed is called *inverse lattice* and its system function from B to A equals

$$\mathbf{H}_{AB}(z) = \frac{1}{\mathbf{E}_N(z)}$$

As we show next, the function $1/\mathbf{E}_N(z)$ is *stable*.

Properties of the error filter We maintain that the roots z_i of the error filter $\mathbf{E}_N(z)$ are either *all* inside the unit circle or *all* on the unit circle.

Theorem

If $P_N > 0$, then $|z_i| < 1$ for every i (13-82a)

If $P_{N-1} > 0$ and $P_N = 0$, then $|z_i| = 1$ for every i (13-82b)

PROOF The theorem is true for $N = 1$ because

$$\mathbf{E}_1(z) = 1 - K_1 z^{-1} \qquad z_1 = K_1 \qquad P_1 = (1 - K_1^2)P_0$$

Using induction, we shall show that it is true for any N assuming that it is true for $N - 1$. From the induction hypothesis it follows that, if ζ_i is a root of $\mathbf{E}_{N-1}(z)$, then $|\zeta_i| < 1$. This shows that the function

$$\mathbf{H}_a(z) = \frac{z^{-N}\mathbf{E}_{N-1}(1/z)}{\mathbf{E}_{N-1}(z)} \tag{13-83}$$

is all-pass as in (13B-1). If z_i is a root of $\mathbf{E}_N(z)$, then [see (13-72)]

$$\mathbf{E}_N(z_i) = \mathbf{E}_{N-1}(z_i) - K_N z_i^{-N}\mathbf{E}_{N-1}(1/z_i) = 0$$

Hence

$$\mathbf{H}_a(z_i) = \frac{1}{K_N} \tag{13-84}$$

And since $P_N = (1 - K_N^2)P_{N-1}$, we conclude that:

If $P_N > 0$, then $|K_N| < 1$ and $|\mathbf{H}_a(z_i)| > 1$

If $P_N = 0$ and $P_{N-1} > 0$, then $|K_N| = 1$ and $|\mathbf{H}_a(z_i)| = 1$

Hence, (13-82) follows from (13B-2).

We shall use the above theorem to relate the spectral properties of the process $\mathbf{s}[n]$ to the properties of the MS error sequence P_N. As N increases, this sequence is nonincreasing and bounded from below because $P_N \geq 0$. Hence, it has a limit

$$P_0 \geq P_1 \geq \cdots \geq P_N \xrightarrow[N\to\infty]{} P \tag{13-85}$$

We start with two important special cases.

Wide-sense Markoff processes and AR signals Suppose, first, that

$$P_{M-1} > P_M = P_{M+1} > 0 \tag{13-86}$$

Setting $N = M + 1, M + 2, \ldots$ in (13-80), we conclude that

$$P_N = P_M \qquad \text{and} \qquad K_N = 0 \qquad \text{for} \qquad N > M \tag{13-87}$$

Thus, if $P_M = P_{M+1}$, then the Levinson algorithm terminates for $N = M$ and $\hat{\mathbf{s}}_M[n] = \hat{\mathbf{s}}_N[n]$ for every $N > M$. This shows that the predictor $\hat{\mathbf{s}}_M[n]$ of $\mathbf{s}[n]$ of order M is also the predictor of $\mathbf{s}[n]$ in terms of its entire past

$$\hat{E}\{\mathbf{s}[n] \mid \mathbf{s}[n-k], 1 \leq k \leq M\} = \hat{E}\{\mathbf{s}[n] \mid \mathbf{s}[n-k], k \geq 1\} \tag{13-88}$$

A process with this property is called *wide-sense Markoff* of order M.

If (13-88) holds, then the error $\hat{\varepsilon}_M[n]$ is white noise because it is orthogonal to $\mathbf{s}[n-k]$ for every $k \geq 1$ and $\hat{\varepsilon}_M[n-k]$ depends only on $\mathbf{s}[n-k]$ and its past. This shows that the process $\mathbf{s}[n]$ is autoregressive

$$\mathbf{s}[n] - a_1^M \mathbf{s}[n-1] - \cdots - a_M^M \mathbf{s}[n-M] = \hat{\varepsilon}_M[n] \qquad R_{\varepsilon\varepsilon}[m] = P_M \delta[m] \tag{13-89}$$

Furthermore

$$\mathbf{S}(z) = \mathbf{L}(z)\mathbf{L}(1/z) \qquad \mathbf{L}(z) = \frac{\sqrt{P_M}}{\mathbf{E}_M(z)} \tag{13-90}$$

where $\mathbf{E}_M(z) = 1 - a_1^M z^{-1} - \cdots - a_M^M z^{-M}$ is the error filter. As we know [see (13-82*a*)], if $\mathbf{E}_M(z_i) = 0$, then $|z_i| < 1$.

From the above it follows that the function $\mathbf{L}(z)$ is minimum-phase (innovations filter), the process $\mathbf{s}[n]$ is regular, and the error $\hat{\varepsilon}_M[m]$ is proportional to its *innovations*.

We have, thus, the following equivalent characterizations of a WS (wide-sense) Markoff process: (*a*) it satisfies (13-88); (*b*) it is AR; (*c*) $P_N = P_M$ for $N > M$; (*d*) $K_N = 0$ for $N > M$; (*e*) its spectrum is *all-pole* as in (13-90); (*f*) $\Delta_{N+1}/\Delta_N =$ constant for $N > M$ [see (13-66)].

Extrapolation Multiplying (13-89) by $\mathbf{s}[n-m]$, we obtain

$$R[m] - a_1^M R[m-1] - \cdots - a_M^M R[m-M] = 0 \qquad m > M \tag{13-91}$$

because $\hat{\varepsilon}_M[n] \perp \mathbf{s}[n-m]$ for $m \geq 1$. From this and (13-65) it follows that $R[m]$ can be determined recursively for any m if its first $M+1$ values are given. It shows also that

$$R[m] = \sum_{i=1}^{m} c_i z_i^{|m|} \qquad |z_i| < 1 \tag{13-92}$$

where z_i are the roots of the error filter $\mathbf{E}_M(z)$.

Note If a process $s[n]$ is strict-sense Markoff as in (12-101), then

$$E\{\mathbf{s}[n] \mid \mathbf{s}[n-k], k \geq 1\} = E\{\mathbf{s}[n] \mid \mathbf{s}[n-1]\}$$

If it is also normal, then the above conditional expected values equal also the corresponding linear MS estimates of $\mathbf{s}[n]$ [see (13-22)]. This shows that if a process is normal and SS (strict-sense) Markoff, then it is also WS Markoff. However, in general, an SS Markoff process is not WS Markoff. The following is a counter example: The process $\mathbf{s}[n] = \mathbf{z}^n$ is SS Markoff because $\mathbf{s}[n-1]$ determines uniquely the RV $\mathbf{z}$ and, therefore, it determines also $\mathbf{s}[n-k]$ for every k. However, it is not WS Markoff because the estimate of $\mathbf{z}^n$ in terms of $\mathbf{z}^{n-1}, \mathbf{z}^{n-2}, \ldots$ is better than the estimate in terms of $\mathbf{s}[n-1] = \mathbf{z}^{n-1}$ only.

Predictable processes and line spectra We now assume that

$$P_{M-1} > P_M = 0 \tag{13-93}$$

In this case [see (13-80)]

$$|K_M| = 1 \qquad \text{and} \qquad P_{M+1} = 0$$

hence, as in (13-87),

$$P_N = 0 \qquad \text{and} \qquad K_N = 0 \qquad \text{for} \qquad N > M$$

The above shows that $\mathbf{s}[n]$ satisfies (13-89) where now $P_M = 0$. In other words

$$\mathbf{s}[n] = a_1^M \mathbf{s}[n-1] + \cdots + a_n^M \mathbf{s}[n-M] \tag{13-94}$$

in the MS sense. This shows that $\mathbf{s}[n]$ can be expressed in terms of its M past values. A process with this property is called *predictable* (or *deterministic-like*).

Reasoning as in the AR case, we conclude that $R[m]$ satisfies the recursion equation (13-91). However the roots z_i of the characteristic polynomial $\mathbf{E}_M(z)$ are now all on the unit circle [see (13-82*b*)]

$$z_i = e^{j\varphi_i}$$

Hence

$$R[m] = \sum_{i=1}^{M} \alpha_i e^{jmT\omega_i} \qquad \omega_i = \frac{\varphi_i}{T} \tag{13-95}$$

and

$$S(\omega) = \frac{2\pi}{T} \sum_{i=1}^{M} \alpha_i \delta(\omega - \omega_i) \qquad |\omega| < \frac{\pi}{T} \tag{13-96}$$

Thus, the power spectrum of a predictable process consists of lines.

We maintain that the converse is also true: If $S(\omega)$ consists of lines as in (13-96), then $\mathbf{s}[n]$ is predictable.

PROOF We form the polynomial

$$D(z) = \prod_{i=1}^{M} (1 - e^{j\varphi_i} z^{-1}) = 1 - b_1 z^{-1} - \cdots - b_M z^{-M}$$

and the process

$$\mathbf{y}[n] = \mathbf{s}[n] - b_1 \mathbf{s}[n-1] - \cdots - b_M \mathbf{s}[n-M] \tag{13-97}$$

Clearly, $\mathbf{y}[n]$ is the output of a system with input $\mathbf{s}[n]$ and system function $D(z)$. Therefore

$$E\{\mathbf{y}^2[n]\} = \frac{1}{2\sigma} \int_{-\sigma}^{\sigma} |D(e^{j\omega T})|^2 S(\omega)\, d\omega = 0$$

The first equality follows from (10-97); the second follows because $D(e^{j\omega_i T}) = 0$ and $S(\omega)$ consists of lines at $\omega = \omega_i$. This shows that $\mathbf{y}[n] = 0$; hence, the right side of (13-97) is also zero.

We have, thus, the following characterizations of a predictable process: (*a*) $P_M = 0$; (*b*) $|K_M| = 1$; (*c*) all zeros of $\mathbf{E}_M(z)$ are on the unit circle; (*d*) $S(\omega)$ consists of lines.

Note The predictability condition $P_M = 0$ can be expressed in terms of the determinant Δ_N of R_N [see (13-66)]

$$\Delta_{M+1} = 0 \qquad \Delta_N > 0 \qquad \text{for} \qquad N \geq M \tag{13-98}$$

A process $\mathbf{s}[n]$ is predictable even if $P_N > 0$ for every finite N provided that

$$\lim P_N = 0 \qquad N \to \infty \tag{13-99}$$

In this case, $\mathbf{s}[n]$ is predictable in terms of its entire past and $S(\omega)$ consists of infinitely many lines.

General processes Suppose, finally, that

$$\lim P_N = P \neq 0 \qquad N \to \infty \tag{13-100}$$

The limit of $\hat{\mathbf{s}}_N[n]$ as $N \to \infty$ is the one-step predictor of $\mathbf{s}[n]$ in terms of its entire past

$$\hat{\mathbf{s}}_1[n] = \hat{E}\{\mathbf{s}[n] \mid \mathbf{s}[n-k], k \geq 1\} = \sum_{k=1}^{\infty} a_k \mathbf{s}[n-k] \tag{13-101}$$

We should note that $\hat{\mathbf{s}}_1[n]$ is now the one-step predictor of $\mathbf{s}[n]$ of infinite order as in (13-45). The corresponding error filter equals

$$\mathbf{E}_1(z) = 1 - \sum_{k=1}^{\infty} a_k z^{-k} \qquad a_k = \lim_{N \to \infty} a_k^N$$

We maintain that the error

$$\boldsymbol{\varepsilon}_1[n] = \mathbf{s}[n] - \hat{\mathbf{s}}_1[n] = \lim_{N \to \infty} \hat{\boldsymbol{\varepsilon}}_N[n]$$

is white noise. Indeed, $\boldsymbol{\varepsilon}_1[n] \perp \mathbf{s}[n-k]$ for $k \geq 1$ and $\boldsymbol{\varepsilon}_1[n-k]$ depends linearly on $\mathbf{s}[n-k]$ and its past, hence, $\boldsymbol{\varepsilon}_1[n] \perp \boldsymbol{\varepsilon}_1[n-k]$ for $k \neq 0$. From this it follows that if

$$\mathbf{i}[n] = \frac{\boldsymbol{\varepsilon}_1[n]}{\sqrt{P}} \qquad \text{then} \qquad R_{ii}[m] = \delta[m] \tag{13-102}$$

The process $\mathbf{i}[n]$ is determined in terms of $\mathbf{s}[n]$ and its past (Fig. 13-16). However, in general, $\mathbf{s}[n]$ cannot be expressed in terms of $\mathbf{i}[n]$ and its past.

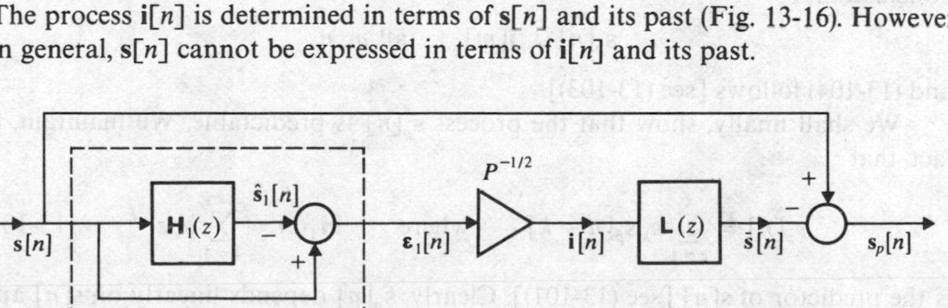

Figure 13-16

Projection Extending the results of Sec. 13-1 to infinitely many RVs, we form the MS estimate of $\mathbf{s}[n]$ in terms of the orthonormal process $\mathbf{i}[n]$ and its past

$$\bar{\mathbf{s}}[n] = E\{\mathbf{s}[n] \mid \mathbf{i}[n-k], k \geq 0\} = \sum_{k=0}^{\infty} l[k]\mathbf{i}[n-k] \qquad (13\text{-}103)$$

The process $\bar{\mathbf{s}}[n]$ so formed is the "projection" of $\mathbf{s}[n]$ on the subspace of $\mathbf{i}[n]$ and its past.

Clearly, $\bar{\mathbf{s}}[n]$ is a regular process and $\mathbf{i}[n]$ is its innovations. Thus, $\bar{\mathbf{s}}[n]$ is the response of the innovations filter

$$\mathbf{L}(z) = \sum_{k=0}^{\infty} l[k]z^{-k}$$

with input $\mathbf{i}[n]$ and, as we see from Fig. 13-15, it is also the response of the system $\mathbf{L}(z)\mathbf{E}_1(z)/\sqrt{P}$ to the input $\mathbf{s}[n]$. If $\mathbf{s}[n]$ is a regular process, then $\bar{\mathbf{s}}[n] = \mathbf{s}[n]$.

Theorem The difference

$$\mathbf{s}_p[n] = \mathbf{s}[n] - \bar{\mathbf{s}}[n]$$

is a predictable process orthogonal to $\bar{\mathbf{s}}[n]$

$$\mathbf{s}_p[n] \perp \bar{\mathbf{s}}[m] \qquad \text{all } m, n \qquad (13\text{-}104)$$

Proof The process $\mathbf{s}_p[n]$ is the error of the estimation of $\mathbf{s}[n]$ in terms of $\mathbf{i}[n]$ and its past. Hence

$$\mathbf{s}_p[n] \perp \mathbf{i}[n-k] \qquad k \geq 0 \qquad (13\text{-}105)$$

Furthermore

$$\mathbf{s}[n-k], \bar{\mathbf{s}}[n-k], \mathbf{s}_p[n-k] \perp \mathbf{i}[n] \qquad k \geq 1 \qquad (13\text{-}106)$$

because $\bar{\mathbf{s}}[n-k]$ depends linearly on $\mathbf{i}[n-k]$ and its past and $\mathbf{s}[n-k]$ is orthogonal to $\boldsymbol{\varepsilon}_1[n] = \sqrt{P}\,\mathbf{i}[n]$ for $k \geq 1$. Combining (13-105) and (13-106), we conclude that

$$\mathbf{s}_p[n] \perp \mathbf{i}[m] \qquad \text{all } m, n$$

and (13-104) follows [see (13-103)].

We shall finally, show that the process $\mathbf{s}_p[n]$ is predictable. We maintain, in fact, that

$$\mathbf{s}_p[n] = \sum_{k=1}^{\infty} a_k \mathbf{s}_p[n-k] \qquad \text{where} \qquad \mathbf{H}_1(z) = \sum_{k=1}^{\infty} a_k z^{-k} \qquad (13\text{-}107)$$

is the predictor of $\mathbf{s}[n]$ [see (13-101)]. Clearly, $\mathbf{s}_p[n]$ depends linearly on $\mathbf{s}[n]$ and its past (see Fig. 13-16). Furthermore, the output of the error filter $\mathbf{E}_1(z) = 1 - \mathbf{H}_1(z)$ with input $\mathbf{s}[n]$ equals $\boldsymbol{\varepsilon}_1[n]$. From this it follows that the output $\mathbf{q}[n]$ of $\mathbf{E}_1(z)$ with input $\mathbf{s}_p[n]$ (Fig. 13-17) depends linearly on $\boldsymbol{\varepsilon}_1[n]$ and its past. But $\mathbf{q}[n]$

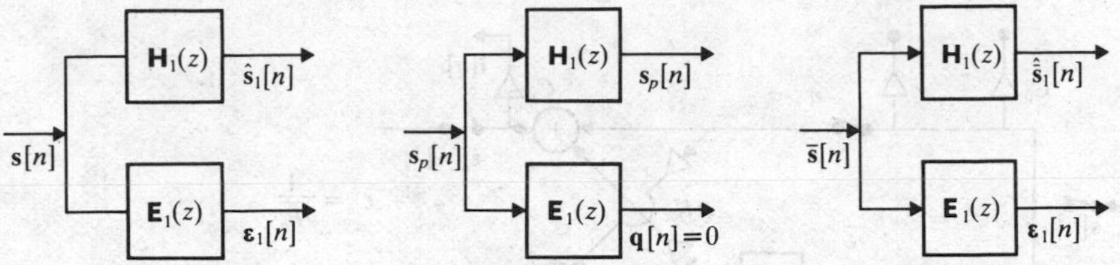

Figure 13-17

depends also linearly on $\mathbf{s}_p[n]$ and its past because the filter $\mathbf{E}_1(z)$ is causal. Furthermore, $\mathbf{s}_p[n] \perp \boldsymbol{\varepsilon}[m] = \sqrt{P}\,\mathbf{i}[m]$ for any m. Hence, $\mathbf{q}[n] = 0$ and (13-107) results.

We note that the response of $\mathbf{E}_1(z)$ to $\bar{\mathbf{s}}[n] = \mathbf{s}[n] - \mathbf{s}_p[n]$ equals $\boldsymbol{\varepsilon}_1[n] - \mathbf{q}[n] = \boldsymbol{\varepsilon}_1[n]$. This shows that $\mathbf{H}_1(z)$ is the predictor of $\bar{\mathbf{s}}[n]$ [see (13-106)].

Wold's decomposition The preceding discussion leads to the conclusion that an arbitrary WSS process $\mathbf{s}[n]$ can be written as a sum

$$\mathbf{s}[n] = \bar{\mathbf{s}}[n] + \mathbf{s}_p[n] \tag{13-108}$$

where $\bar{\mathbf{s}}[n]$ is a regular process and $\mathbf{s}_p[n]$ is a predictable process orthogonal to $\bar{\mathbf{s}}[n]$. Furthermore, the predictor $\mathbf{H}_1(z)$ of $\mathbf{s}[n]$ is also a predictor of $\mathbf{s}[n]$ and $\mathbf{s}_p[n]$. This result is known as Wold's decomposition.

The above shows that the power spectrum $S(\omega)$ of $\mathbf{s}[n]$ consists of two components

$$S(\omega) = \bar{S}(\omega) + S_p(\omega)$$

due to $\bar{\mathbf{s}}[n]$ and $\mathbf{s}_p[n]$ respectively. The first component is given by

$$\bar{S}(\omega) = \left| \sum_{n=0}^{\infty} l[n] e^{-jn\omega T} \right|^2$$

The above sum is convergent because [see (13-103)]

$$\sum_{n=0}^{\infty} l^2[n] = E\{\bar{\mathbf{s}}^2[n]\} \le R[0]$$

The second component consists of lines

$$S_p(\omega) = 2\sigma \sum_i \alpha_i \delta(\omega - \omega_i) \qquad |\omega| < \sigma$$

where $e^{j\omega_i T}$ are the roots of the error filter $\mathbf{E}_1(z)$ on the unit circle.

Causal Data†

We wish to establish the present value of a regular process $\mathbf{s}[n]$ in terms of its finite past, starting from some origin. The data are now available from 0 to $n - 1$

† B. Friedlander: "Lattice Filters for Adaptive Processing, "*Proceedings of the IEEE*, vol. 70, 1982.

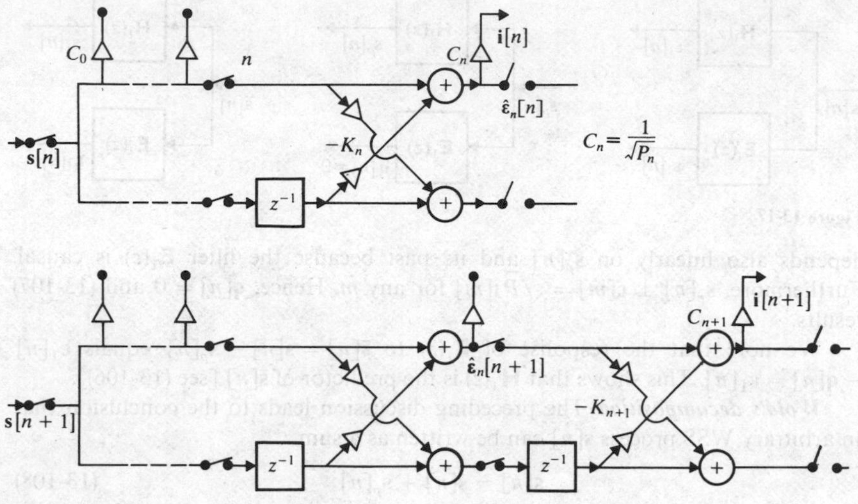

Figure 13-18

and the desired estimate is given by

$$\hat{\mathbf{s}}_n[n] = \hat{E}\{\mathbf{s}[n] \mid \mathbf{s}[n-k],\ 1 \le k \le n\} = \sum_{k=1}^{n} a_k^n \mathbf{s}[n-k] \qquad (13\text{-}109)$$

Unlike the fixed-order estimate $\hat{\mathbf{s}}_N[n]$ considered earlier, the order n of the above estimate is not constant. Furthermore, the values a_k^n of the coefficients of the filter specified by (13-109) depend on n. Thus, the estimator of the process $\mathbf{s}[n]$ in terms of its causal past is a linear *time-varying filter*. If it is realized by a tapped-delay line as in Fig. 13-12, the number of the taps increases and the values of the weights change as n increases.

The coefficients a_k^n of $\hat{\mathbf{s}}_n[n]$ can be determined recursively from Levinson's algorithm where now $N = n$. Introducing the backward estimate $\check{\mathbf{s}}[n]$ of $\mathbf{s}[n]$ in terms of its n most recent future values, we conclude from (13-77) that

$$\begin{aligned} \hat{\mathbf{s}}_n[n] &= \hat{\mathbf{s}}_{n-1}[n] + K_n(\mathbf{s}[0] - \check{\mathbf{s}}_{n-1}[0]) \\ \check{\mathbf{s}}_n[0] &= \check{\mathbf{s}}_{n-1}[0] + K_n(\mathbf{s}[n] - \hat{\mathbf{s}}_{n-1}[n]) \end{aligned} \qquad (13\text{-}110)$$

In Fig. 13-18, we show the *normalized lattice* realization of the error filter $\mathbf{E}_n(z)$ where we use as upper output the process

$$\mathbf{i}[n] = \frac{1}{\sqrt{P_n}} \varepsilon_n[n] \qquad E\{\mathbf{i}^2[n]\} = 1 \qquad (13\text{-}111)$$

The filter is formed by switching "on" successively a new lattice section starting from the left. This filter is again time-varying, however, unlike the tapped-delay line realization, the elements of each section remain unchanged as n increases. We should point out that, whereas $\varepsilon_k[n]$ is the value of the upper response of the kth section at time n, the process $\mathbf{i}[n]$ does not appear at a fixed position. It is the output of the last section that is switched "on" and as n increases, the point where $\mathbf{i}[n]$ is observed changes.

We conclude with the observation that if the process $\mathbf{s}[n]$ is AR of order M [see (13-89)], then the lattice stops increasing for $n > M$, realizing, thus, the time invariant system $\mathbf{E}_M(z)/\sqrt{P_M}$. The corresponding inverse lattice (see Fig. 13-15b) realizes the all-pole system

$$\frac{\sqrt{P_M}}{\mathbf{E}_M(z)}$$

We shall now show that the output $\mathbf{i}[n]$ of the normalized lattice is white noise

$$R_{ii}[m] = \delta[m] \tag{13-112}$$

Indeed, as we know, $\hat{\varepsilon}_n[n] \perp \mathbf{s}[n-k]$ for $1 \le k \le n$. Furthermore, $\hat{\varepsilon}_{n-k}[n-r]$ depends linearly only on $\mathbf{s}[n-r]$ and its past values. Hence

$$\hat{\varepsilon}_n[n] \perp \hat{\varepsilon}_{n-1}[n-1] \tag{13-113}$$

This yields (13-112) because $P_n = E\{\varepsilon_n^2[n]\}$.

Note In a lattice of fixed length, the output $\hat{\varepsilon}_N[n]$ is not white noise and it is not orthogonal to $\hat{\varepsilon}_{N-1}[n]$. However, the random variables $\hat{\varepsilon}_N[n]$ and $\hat{\varepsilon}_{N-1}[n-1]$ are orthogonal.

Kalman innovations† The output $\mathbf{i}[n]$ of the time-varying lattice of Fig. 13-18 is an orthonormal process that depends linearly on $\mathbf{s}[n-k]$. Denoting by γ_k^n the response of the lattice at time n to the input $\mathbf{s}[n] = \delta[n-k]$, we obtain

$$\begin{aligned} \mathbf{i}[0] &= \gamma_0^0\,\mathbf{s}[0] \\ \mathbf{i}[1] &= \gamma_0^1\mathbf{s}[0] + \gamma_1^1\mathbf{s}[1] \\ &\cdots\cdots\cdots\cdots\cdots\cdots \\ \mathbf{i}[n] &= \gamma_0^n\,\mathbf{s}[0] + \cdots + \gamma_k^n\mathbf{s}[k] + \cdots + \gamma_n^n\mathbf{s}[n] \end{aligned} \tag{13-114}$$

or in vector form

$$\mathbf{I}_{n+1} = \mathbf{S}_{n+1}\Gamma_{n+1} \qquad \Gamma_{n+1} = \begin{bmatrix} \gamma_0^0 & \gamma_0^1 & \cdots & \gamma_0^n \\ & \gamma_1^1 & \cdots & \gamma_1^n \\ & & \cdots\cdots & \\ 0 & & & \\ & & & \gamma_n^n \end{bmatrix}$$

where $\mathbf{S}_{n+1}$ and $\mathbf{I}_{n+1}$ are vectors with components

$$\mathbf{s}[0], \ldots, \mathbf{s}[n] \qquad \text{and} \qquad \mathbf{i}[0], \ldots, \mathbf{i}[n]$$

respectively.

† T. Kailath, A. Vieira, and M. Morf: "Inverses of Toeplitz Operators, Innovations, and Orthogonal Polynomials," *SIAM Review*, vol. 20, no. 1, 1978.

From the above it follows that if

$$\mathbf{s}[n] = \delta[n-k] \qquad \text{then} \qquad \mathbf{i}[n] = \gamma_k^n \qquad n \geq k$$

This shows that to determine the delta response of the lattice of Fig. 13-18, we use as input the delta sequence $\delta[n-k]$ and we observe the moving output $\mathbf{i}[n]$ for $n \geq k$.

The elements γ_k^n of the triangular matrix Γ_{n+1} can be expressed in terms of the weights a_k^n of the causal predictor $\hat{\mathbf{s}}_n[n]$. Since

$$\hat{\boldsymbol{\varepsilon}}_n[n] = \mathbf{s}[n] - \hat{\mathbf{s}}_n[n] = \sqrt{P_n}\,\mathbf{i}[n]$$

it follows from (13-109) that

$$\gamma_n^n = \frac{1}{\sqrt{P_n}} \qquad \gamma_{n-k}^n = \frac{-1}{\sqrt{P_n}}\,a_k^n \qquad k \geq 1$$

The inverse of the lattice of Fig. 13-18 is obtained by reversing the flow direction of the upper line and the sign of the downward weights $-K_n$ as in Fig. 13-15*b*. The turn-on switches close again in succession starting from the left, and the input $\mathbf{i}[n]$ is applied at the terminal of the section that is connected last. The output at A is thus given by

$$\begin{aligned} \mathbf{s}[0] &= l_0^0\mathbf{i}[0] \qquad\qquad \mathbf{S}_{n+1} = \mathbf{I}_{n+1}L_{n+1} \\ \mathbf{s}[1] &= l_0^1\mathbf{i}[0] + l_1^1\mathbf{i}[1] \\ &\cdots\cdots\cdots\cdots \\ \mathbf{s}[n] &= l_0^n\mathbf{i}[0] + \cdots + l_n^n\mathbf{i}[n] \end{aligned} \qquad (13\text{-}115)$$

From this it follows that if

$$\mathbf{i}[n] = \delta[n-k] \qquad \text{then} \qquad \mathbf{s}[n] = l_k^n \qquad n \geq k$$

Thus, to determine the delta response l_k^n of the inverse lattice, we use as moving input the delta sequence $\delta[n-k]$ and we observe the left output $\mathbf{s}[n]$ for $n \geq k$.

From the preceding discussion it follows that the random vector $\mathbf{S}_n$ is linearly equivalent to the orthonormal vector $\mathbf{I}_n$. Thus, Eqs. (13-114) and (13-115) correspond to the Gram–Schmidt orthonormalization equations (13-13) and (13-16) of Sec. 13-1. Applying the terminology of Sec. 10-4 to causal signals, we shall call the process $\mathbf{i}[n]$ the *Kalman innovations* of $\mathbf{s}[n]$ and the lattice filter and its inverse *Kalman whitening* and *Kalman innovations* filters respectively. These filters are *time-varying* and their transition matrices equal Γ_n and L_n respectively. Their elements can be expressed in terms of the parameters K_n and P_n of Levinson's algorithm because these parameters specify completely the filters.

Cholesky factorization We maintain that the correlation matrix R_n and its inverse can be written as products

$$R_n = L_n^t L_n \qquad R_n^{-1} = \Gamma_n \Gamma_n^t \qquad (13\text{-}116)$$

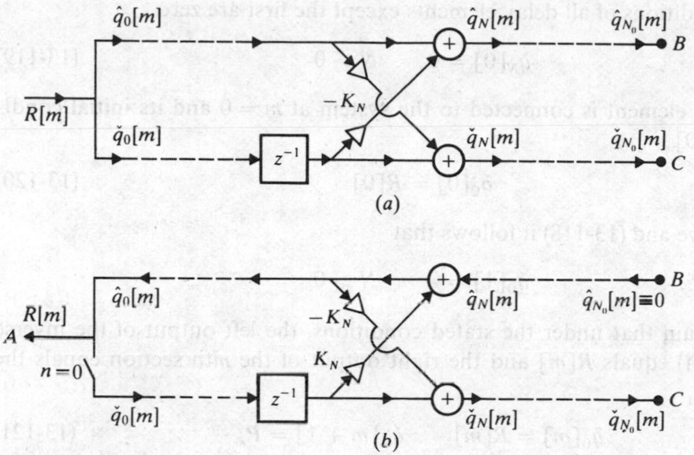

Figure 13-19

where Γ_n and L_n are the triangular matrices introduced earlier. Indeed, from the orthonormality of $\mathbf{I}_n$ and the definition of R_n it follows that

$$E\{\mathbf{I}_n^t \mathbf{I}_n\} = 1_n \qquad E\{\mathbf{S}_n^t \mathbf{S}_n\} = R_n$$

where 1_n is the identity matrix. Since $\mathbf{I}_n = \mathbf{S}_n \Gamma_n$ and $\mathbf{S}_n = \mathbf{I}_n L_n$, the above yields

$$\Gamma_n^t R_n \Gamma_n = 1_n \qquad L_n^t 1_n L_n = R_n$$

and (13-116) results.

Autocorrelation as lattice response We shall determine the autocorrelation $R[m]$ of the process $\mathbf{s}[n]$ in terms of the Levinson parameters K_N and P_N. For this purpose, we form a lattice of order N_0 and we denote by $\hat{q}_N[m]$ and $\check{q}_N[m]$ respectively its upper and lower responses (Fig. 13-19*a*) to the input $R[m]$. As we see from the figure

$$\hat{q}_{N-1}[m] = \hat{q}_N[m] + K_N \check{q}_{N-1}[m-1] \tag{13-117a}$$

$$\check{q}_N[m] = \check{q}_{N-1}[m-1] - K_N \hat{q}_{N-1}[m] \tag{13-117b}$$

$$\hat{q}_0[m] = \check{q}_0[m] = R[m] \tag{13-117c}$$

Using the above, we shall show that $R[m]$ can be determined as the response of the inverse lattice† of Fig. 13-19*b* provided that the following boundary and initial conditions are satisfied: The input to the system (point B) is identically zero

$$\hat{q}_{N_0}[m] = 0 \qquad \text{all } m \tag{13-118}$$

† E. A. Robinson and S. Treitel: "Maximum entropy and the relationship of the partial autocorrelation to the reflection coefficients of a layered system," *IEEE Transactions on Acoustics, Speech, and Signal Process, vol. ASSP-28*, no. 2, 1980.

The initial conditions of all delay elements except the first are zero

$$\check{q}_N[0] = 0 \qquad N > 0 \tag{13-119}$$

The first delay element is connected to the system at $m = 0$ and its initial condition equals $R[0]$

$$\check{q}_0[0] = R[0] \tag{13-120}$$

From the above and (13-118) it follows that

$$\check{q}_0[1] = 0 \qquad N > 0$$

We maintain that under the stated conditions, the left output of the inverse lattice (point A) equals $R[m]$ and the right output of the mth section equals the MS error P_{m+1}

$$\hat{q}_0[m] = R[m] \qquad \check{q}_m[m+1] = P_m \tag{13-121}$$

PROOF The proof is based on the fact that the responses of the lattice of Fig. 13-18a satisfy (13-74)

$$\hat{q}_N[m] = \check{q}_N[m] = 0 \qquad 1 \le m \le N-1 \tag{13-122}$$

$$\check{q}_N[N] = P_N \tag{13-123}$$

From (13-117) it follows that, if we know $\hat{q}_N[m]$ and $\check{q}_{N-1}[m-1]$, then we can find $\hat{q}_{N-1}[m]$ and $\check{q}_N[m]$. By a simple induction, this leads to the conclusion that if $\hat{q}_{N_0}[m]$ is specified for every m (boundary conditions) and $\check{q}_N[1]$ is specified for every N (initial conditions), then all responses of the lattice are determined uniquely. The two systems of Fig. 13-19 satisfy the same equations (13-117) and, as we noted, they have identical initial and boundary conditions. Hence, all their responses are identical. This yields (13-121).

13-5 FILTERING AND PREDICTION

In this section, we consider the problem of estimating the future value $\mathbf{s}(t+\lambda)$ of a stochastic process $\mathbf{s}(t)$ (signal) in terms of the present and past values of a regular process $\mathbf{x}(t)$ (signal plus noise)

$$\hat{\mathbf{s}}(t+\lambda) = \hat{E}\{\mathbf{s}(t+\lambda) \mid \mathbf{x}(t-\tau), \tau \ge 0\} = \int_0^\infty h_x(\alpha)\mathbf{x}(t-\alpha)\,d\alpha \tag{13-124}$$

Thus, $\hat{\mathbf{s}}(t+\lambda)$ is the output of a linear time-invariant causal system $\mathbf{H}_x(s)$ with input $\mathbf{x}(t)$. To determine $\mathbf{H}_x(s)$, we use the orthogonality principle

$$E\left\{\left[\mathbf{s}(t+\lambda) - \int_0^\infty h_x(\alpha)\mathbf{x}(t-\alpha)\,d\alpha\right]\mathbf{x}(t-\tau)\right\} = 0 \qquad \tau \ge 0$$

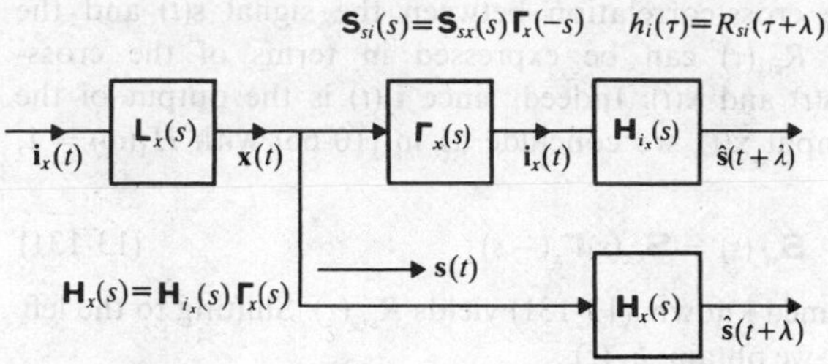

Figure 13-20

This yields the *Wiener–Hopf* equation

$$R_{sx}(\tau + \lambda) = \int_0^\infty h_x(\alpha)R_{xx}(\tau - \alpha)\, d\alpha \qquad \tau \geq 0 \tag{13-125}$$

The solution $h_x(t)$ of (13-125) is the impulse response of the prediction and filtering system known as *Wiener filter*. If $\mathbf{x}(t) = \mathbf{s}(t)$, then $h_x(t)$ is a pure predictor as in (13-54). If $\lambda = 0$, then $h_x(t)$ is a pure filter.

To solve (13-125), we express $\mathbf{x}(t)$ in terms of its innovations $\mathbf{i}_x(t)$ (Fig. 13-20)

$$\mathbf{x}(t) = \int_0^\infty l_x(\alpha)\mathbf{i}_x(t - \alpha)\, d\alpha \qquad R_{ii}(\tau) = \delta(\tau) \tag{13-126}$$

where $l_x(t)$ is the impulse response of the innovations filter $\mathbf{L}_x(s)$ obtained by factoring the spectrum of $\mathbf{x}(t)$ as in (10-111):

$$\mathbf{S}_{xx}(s) = \mathbf{L}_x(s)\mathbf{L}_x(-s) \tag{13-127}$$

As we know, the processes $\mathbf{i}_x(t)$ and $\mathbf{x}(t)$ are linearly equivalent; hence, the estimate $\hat{\mathbf{s}}(t + \lambda)$ can be expressed as the output of a causal filter $\mathbf{H}_{i_x}(s)$ with input $\mathbf{i}_x(t)$

$$\hat{\mathbf{s}}(t + \lambda) = \int_0^\infty h_{i_x}(\alpha)\mathbf{i}_x(t - \alpha)\, d\alpha \tag{13-128}$$

To determine $h_{i_x}(t)$, we use the orthogonality principle

$$E\left\{\left[\mathbf{s}(t + \lambda) - \int_0^\infty h_{i_x}(\alpha)\mathbf{i}_x(t - \alpha)\, d\alpha\right]\mathbf{i}_x(t - \tau)\right\} = 0 \qquad \tau \geq 0$$

Since $\mathbf{i}_x(t)$ is white noise, the above yields

$$R_{si_x}(\tau + \lambda) = \int_0^\infty h_{i_x}(\alpha)\delta(\tau - \alpha)\, d\alpha = h_{i_x}(\tau) \qquad \tau \geq 0 \tag{13-129}$$

This determines $h_{i_x}(\tau)$ for all τ because $h_{i_x}(\tau) = 0$ for $\tau < 0$

$$h_{i_x}(\tau) = R_{si_x}(\tau + \lambda)U(\tau) \tag{13-130}$$

In the above, $R_{si_x}(\tau)$ is the cross-correlation between the signal $\mathbf{s}(t)$ and the process $\mathbf{i}_x(t)$. The function $R_{si_x}(\tau)$ can be expressed in terms of the cross-correlation $R_{sx}(\tau)$ between $\mathbf{s}(t)$ and $\mathbf{x}(t)$. Indeed, since $\mathbf{i}_x(t)$ is the output of the whitening filter $\mathbf{\Gamma}_x(s)$ with input $\mathbf{x}(t)$, we conclude as in (10-66) with $H_1(\omega) = 1$, $H_2^*(\omega) = \mathbf{\Gamma}_x(-j\omega)$, that

$$\mathbf{S}_{si_x}(s) = \mathbf{S}_{sx}(s)\mathbf{\Gamma}_x(-s) \tag{13-131}$$

Thus, since $\mathbf{S}_{sx}(s)$ is assumed known, (13-131) yields $R_{si_x}(\tau)$. Shifting to the left and truncating as in (13-130), we obtain $h_{i_x}(\tau)$.

To complete the specification of $\mathbf{H}_x(s)$, we multiply the transform $\mathbf{H}_{i_x}(s)$ of the function $h_{i_x}(t)$ so obtained with $\mathbf{\Gamma}_x(s)$ (see Fig. 13-20)

$$\mathbf{H}_x(s) = \mathbf{H}_{i_x}(s)\mathbf{\Gamma}_x(s) \tag{13-132}$$

The function $\mathbf{H}_{i_x}(s)$ can be determined directly from (13-131): As we know (shifting theorem) the transform of $R_{si_x}(\tau + \lambda)$ equals

$$\mathbf{S}_\lambda(s) = \mathbf{S}_{si_x}(s)e^{\lambda s} = \mathbf{S}_{sx}(s)\mathbf{\Gamma}_x(-s)e^{\lambda s} \tag{13-133}$$

To find $\mathbf{H}_{i_x}(s)$, it suffices to write $\mathbf{S}_\lambda(s)$ as a sum

$$\mathbf{S}_\lambda(s) = \mathbf{S}_\lambda^+(s) + \mathbf{S}_\lambda^-(s) \tag{13-134}$$

where $\mathbf{S}_\lambda^+(s)$ is analytic in the left-hand s plane and $\mathbf{S}_\lambda^-(s)$ is analytic in the right-hand s plane. Since the inverse transforms of the function $\mathbf{S}_\lambda^+(s)$ and $\mathbf{S}_\lambda^-(s)$ equal $R_{si_x}(\tau + \lambda)U(\tau)$ and $R_{si_x}(\tau + \lambda)U(-\tau)$ respectively, we conclude from (13-130) that (see also Note on next page)

$$\mathbf{H}_{i_x}(s) = \mathbf{S}_\lambda^+(s) \tag{13-135}$$

Thus, to determine $\mathbf{H}_x(s)$, we proceed as follows: (*a*) we factor $\mathbf{S}_{xx}(s)$ as in (13-127)—this yields $\mathbf{\Gamma}_x(s)$; (*b*) we evaluate $\mathbf{S}_{si_x}(s)$ from (13-131) and form the function $\mathbf{S}_\lambda(s)$; (*c*) we decompose $\mathbf{S}_\lambda(s)$ as (13-134) and obtain $\mathbf{H}_{i_x}(s)$ from (13-135); (*d*) we insert $\mathbf{H}_{i_x}(s)$ into (13-132).

If the function $\mathbf{S}_\lambda(s)$ is rational, then the decomposition (13-134) can be accomplished by expanding $\mathbf{S}_{si_x}(s)$ into partial fractions. Assuming that $\mathbf{S}_{si_x}(s)$ is a proper fraction with simple poles, we obtain

$$\mathbf{S}_{si_x}(s) = \sum_i \frac{a_i}{s - s_i} + \sum_k \frac{b_k}{s - z_k} \qquad \begin{matrix} \text{Re } s_i < 0 \\ \text{Re } z_k > 0 \end{matrix} \tag{13-136}$$

The inverse of the second sum is zero for $\tau > 0$. If, therefore, it is shifted to the left, it will remain zero for $\tau > 0$. This shows that only the first sum will contribute to the term $R_{si_x}(\tau + \lambda)U(\tau)$. In other words

$$R_{si_x}(\tau + \lambda)U(\tau) = [a_1 e^{s_1(\tau+\lambda)} + \cdots + a_n e^{s_n(\tau+\lambda)}]U(\tau)$$

The transform of the above yields

$$\mathbf{S}_\lambda^+(s) = \frac{a_1 e^{s_1\lambda}}{s - s_1} + \cdots + \frac{a_n e^{s_n\lambda}}{s - s_n} \tag{13-137}$$

Example 13-10 Suppose that $\mathbf{x}(t) = \mathbf{s}(t) + \mathbf{v}(t)$ and

$$S_{ss}(\omega) = \frac{N_0}{\alpha^2 + \omega^2} \qquad S_{vv}(\omega) = N \qquad S_{sv}(\omega) = 0 \tag{13-138}$$

as in Example 13-4. In this case, $S_{sx}(s) = S_{ss}(s)$ and

$$\mathbf{S}_{xx}(s) = \frac{N_0}{\alpha^2 - s^2} + N = N\frac{\beta^2 - s^2}{\alpha^2 - s^2} \qquad \beta^2 = \alpha^2 + \frac{N_0}{N}$$

Hence

$$\mathbf{L}_x(s) = \sqrt{N}\,\frac{s+\beta}{s+\alpha} \qquad \mathbf{\Gamma}_x(-s) = \frac{1}{\sqrt{N}}\,\frac{\alpha - s}{\beta - s} \tag{13-139}$$

Inserting into (13-131) and expanding into partial fractions, we obtain

$$\mathbf{S}_{si_x}(s) = \frac{N_0}{\alpha^2 - s^2}\,\frac{\alpha - s}{(\beta - s)\sqrt{N}} = \frac{A}{s+\alpha} - \frac{A}{s-\beta} \qquad A = \frac{N_0}{(\alpha+\beta)\sqrt{N}}$$

and with $s_1 = -\alpha$, (13-137) yields

$$\mathbf{S}_\lambda^+(s) = \frac{A}{s+\alpha}\,e^{-\alpha\lambda}$$

Hence

$$\mathbf{H}_x(s) = \mathbf{S}_\lambda^+(s)\mathbf{\Gamma}_x(s) = \frac{\beta - \alpha}{s+\beta}\,e^{-\alpha\lambda} \tag{13-140}$$

Note In the decomposition (13-134) of $\mathbf{S}_\lambda(s)$, the functions $\mathbf{S}_\lambda^+(s)$ and $\mathbf{S}_\lambda^-(s)$ are unique within an additive constant. This causes an ambiguity in the determination of $h_{i_x}(t)$. The ambiguity is removed if we impose the condition that

$$\mathbf{S}_\lambda^-(\infty) = 0$$

In the pure filtering case ($\lambda = 0$), the resulting $h_x(t)$ might contain impulses at the origin. This is acceptable because, by assumption, the estimate $\hat{\mathbf{s}}(t)$ of $\mathbf{s}(t)$ is a functional of the past *and* the present value of the data $\mathbf{x}(t)$.

Filtering white noise In the pure filtering problem, the determination of the estimates $\mathbf{H}_x(s)$ can be simplified if $R_{ss}(0) < \infty$ and $\mathbf{v}(t)$ is white noise orthogonal to the signal as in (13-138). We maintain, in fact, that in this case

$$\mathbf{H}_x(s) = 1 - \sqrt{N}\,\mathbf{\Gamma}_x(s) \tag{13-141}$$

where $\mathbf{\Gamma}_x(s)$ is the whitening filter of $\mathbf{x}(t)$.

Proof From the above assumptions it follows that $\mathbf{S}_{ss}(\infty) = 0$, hence

$$\mathbf{S}_{sx}(s) = \mathbf{S}_{ss}(s) = \mathbf{S}_{xx}(s) - N = \mathbf{L}_x(s)\mathbf{L}_x(-s) - N$$

$$\mathbf{S}_{sx}(\infty) = 0 \qquad \mathbf{S}_{xx}(\infty) = N \qquad \mathbf{L}_x(\pm\infty) = \sqrt{N}$$

Inserting into (13-131), we obtain

$$\mathbf{S}_{si_x}(s) = \mathbf{L}_x(s) - N\mathbf{\Gamma}_x(-s) = \mathbf{L}_x(s) + K - N\mathbf{\Gamma}_x(-s) - K$$

From the preceding note it follows that the constant K must be such that the noncausal component of $\mathbf{S}_{si_x}(s)$ satisfies the infinity condition $-N\mathbf{\Gamma}_x(-\infty) - K = 0$. And since $\mathbf{\Gamma}_x(-\infty) = 1/\mathbf{L}_x(-\infty) = 1/\sqrt{N}$, (13-141) follows from (13-132).

Example 13-11 We shall determine the pure filter of the process in Example 13-10. From (13-139) and (13-141) it follows that

$$\mathbf{H}_x(s) = 1 - \frac{\alpha + s}{\beta + s} = \frac{\beta - \alpha}{s + \beta} \qquad h_x(t) = (\beta - \alpha)e^{-\beta t}U(t)$$

in agreement with (13-140). We note that the resulting MS error equals

$$P = E\left\{\left[\mathbf{s}(t) - \int_0^\infty h_x(\alpha)\mathbf{x}(t-\alpha)\,d\alpha\right]\mathbf{s}(t)\right\} = \frac{N_0}{\alpha + \beta}$$

Discrete-Time Processes

We shall state briefly the discrete-time version of the preceding results. Our problem now is the determination of the future value $\mathbf{s}[n+r]$ of a stochastic process in terms of the present and past values of another process $\mathbf{x}[n]$

$$\hat{\mathbf{s}}_r[n+r] = \sum_{k=0}^{\infty} h_x^r[k]\mathbf{x}[n-k] \tag{13-142}$$

In this case

$$\mathbf{s}[n+r] - \hat{\mathbf{s}}_r[n+r] \perp \mathbf{x}[n-m] \qquad m \ge 0$$

hence

$$R_{sx}[m+r] = \sum_{k=0}^{\infty} h_x^r[k]R_{xx}[m-k] \qquad m \ge 0 \tag{13-143}$$

This is the discrete version of the Wiener–Hopf equation (13-125).

To determine $h_x^r[n]$, we proceed as in the continuous-time case: We express $\hat{\mathbf{s}}_r[n+r]$ in terms of the innovations $\mathbf{i}_x[n]$ of $\mathbf{x}[n]$ (Fig. 13-21)

$$\hat{\mathbf{s}}_r[n+r] = \sum_{k=0}^{\infty} h_{i_x}^r[k]\mathbf{i}_x[n-k] \tag{13-144}$$

From this and (13-4) it follows that

$$R_{si_x}[m+r] = \sum_{k=0}^{\infty} h_{i_x}^r[k]\delta[m-k] = h_{i_x}[m] \qquad m \ge 0$$

because $R_{i_x}[m] = \delta[m]$. Hence

$$h_{i_x}^r[m] = R_{si_x}[m+r]U[m] \qquad \text{all } m \tag{13-145}$$

The function $R_{si_x}[m]$ can be expressed in terms of $R_{sx}[m]$ as in (13-131)

$$\mathbf{S}_{si_x}(z) = \mathbf{S}_{sx}(z)\mathbf{\Gamma}_x(z^{-1}) \tag{13-146}$$

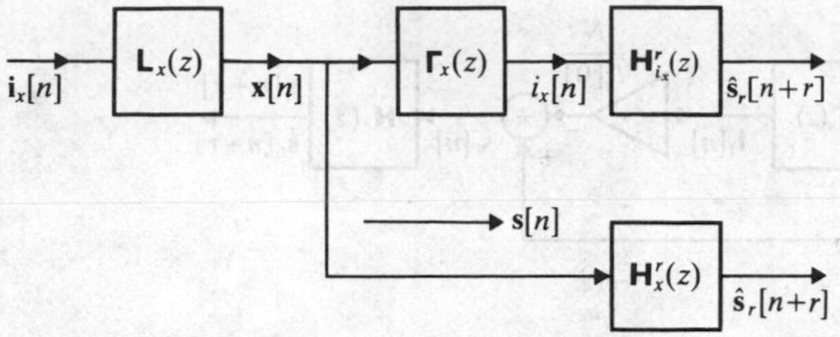

Figure 13-21

Thus, the transform of $R_{si_x}[m + r]$ equals

$$\mathbf{S}_r(z) = z^r\mathbf{S}_{si_x}(z) = z^r\mathbf{S}_{sx}(z)\mathbf{\Gamma}_x(z^{-1}) \tag{13-147}$$

The function $\mathbf{S}_r(z)$ is then written as a sum

$$\mathbf{S}_r(z) = \mathbf{S}_r^+(z) + \mathbf{S}_r^-(z) \tag{13-148}$$

where $\mathbf{S}_r^+(z)$ is analytic for $|z| > 1$ and $\mathbf{S}_r^-(z)$ is analytic for $|z| < 1$. Furthermore the inverse of $\mathbf{S}_r^-(z)$ at the origin is zero. Thus, $\mathbf{S}_r^+(z)$ is the transform of the causal function $R_{si_x}[m + r]U[m]$. And since $\mathbf{i}_x[n]$ is the response of the whitening filter $\mathbf{\Gamma}_x(z)$ with input $\mathbf{x}[n]$, we conclude from (13-145) that

$$\mathbf{H}^r_x(z) = \mathbf{H}^r_{i_x}(z)\mathbf{\Gamma}_x(z) = \mathbf{S}_r^+(z)\mathbf{\Gamma}_x(z) \tag{13-149}$$

Example 13-12 We shall determine the one-step predictor $\hat{\mathbf{s}}_1[n + 1]$ of the process $\mathbf{s}[n]$ where

$$\mathbf{S}_{ss}(z) = \frac{N_0}{(1 - az^{-1})(1 - az)} \qquad \mathbf{S}_{vv}(z) = N \qquad \mathbf{S}_{sv}(z) = 0$$

In this case (see Example 13-5)

$$\mathbf{L}_x(z) = \sqrt{\frac{Na}{b}}\,\frac{1 - bz^{-1}}{1 - az^{-1}}$$

From (13-147) it follows with $r = 1$ that

$$z\mathbf{S}_{si_x}(z) = \frac{zN_0\sqrt{b/Na}}{(1 - az^{-1})(1 - bz)} = \frac{Aaz}{z - a} - \frac{Az/b}{z - 1/b} \qquad A = (a - b)\sqrt{\frac{N}{ab}}$$

Since $0 < a < 1$ and $1/b > 1$, we conclude from the above that $\mathbf{S}_1^+(z) = Aaz/(z - a)$ and (13-149) yields

$$\mathbf{H}^1_x(z) = (a - b)\,\frac{z}{z - b} \qquad h^1_x[n] = (a - b)b^nU[n]$$

We discuss presently a more direct method for determining $\mathbf{H}^r_x(z)$ [see (13-155) below].

White noise We shall examine the nature of the predictor $\mathbf{H}^r_x(z)$ of $\mathbf{s}[n + r]$ under the assumption that the noise is white and orthogonal to the signal

$$R_{vv}(\tau) = N\delta(\tau) \qquad R_{sv}(\tau) = 0 \tag{13-150}$$

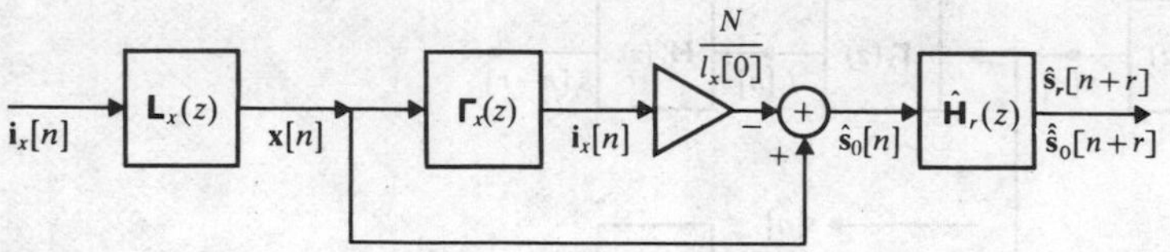

Figure 13-22

Pure filter We maintain that the estimator $\mathbf{H}_x^0(z)$ of $\mathbf{s}[n]$ equals (Fig. 13-22)

$$\mathbf{H}_x^0(z) = 1 - \frac{D}{\mathbf{L}_x(z)} \qquad D = \frac{N}{l_x[0]} \tag{13-151}$$

PROOF From (13-150) it follows that

$$\mathbf{S}_{sx}(z) = \mathbf{S}_{ss}(z) = \mathbf{S}_{xx}(z) - N = \mathbf{L}_x(z)\mathbf{L}_x(z^{-1}) - N$$

Inserting into (13-146), we obtain

$$\mathbf{S}_{si_x}(z) = \mathbf{L}_x(z) - N\boldsymbol{\Gamma}_x(z^{-1}) \tag{13-152}$$

We wish to find the causal part of the above, including the value of its inverse at $n = 0$. Since the inverse z transform of $\boldsymbol{\Gamma}_x(1/z)$ is zero for $n > 0$ and for $n = 0$ it equals $\boldsymbol{\Gamma}_x(0)$, we conclude that

$$\mathbf{H}_{i_x}^0(z) = \mathbf{L}_x(z) - N\boldsymbol{\Gamma}_x(0) \tag{13-153}$$

Multiplying by $\boldsymbol{\Gamma}_x(z)$, we obtain (13-151) because $\boldsymbol{\Gamma}_x(0) = 1/l_x[0]$.

Filtering and prediction We shall now show that the estimate $\hat{\mathbf{s}}_r[n+r]$ of $\mathbf{s}[n+r]$ equals the pure predictor $\hat{\hat{\mathbf{s}}}_0[n+r]$ of the estimate $\hat{\mathbf{s}}_0[n]$ of $\mathbf{s}[n]$ (Fig. 13-22)

$$\hat{\mathbf{s}}_r[n+r] = \hat{\hat{\mathbf{s}}}_0[n+r] = \hat{E}\{\hat{\mathbf{s}}_0[n+r] \mid \hat{\mathbf{s}}_0[n-k], k \geq 0\} \tag{13-154}$$

PROOF From (13-147) and (13-152) it follows that

$$\mathbf{S}_r(z) = z^r[\mathbf{L}_x(z) - N\boldsymbol{\Gamma}_x(z^{-1})]$$

But the inverse of $z^r\boldsymbol{\Gamma}_x(1/z)$ is zero for $n \geq 0$. Hence, $\mathbf{S}_r^+(z)$ is the causal part of $z^r\mathbf{L}_x(z)$. Inserting into (13-149), we obtain

$$\mathbf{H}_x^r(z) = z^r\left(\mathbf{L}_x(z) - \sum_{k=0}^{r-1} l_x[k]z^{-k}\right)\boldsymbol{\Gamma}_x(z)$$

$$= z^r\left(1 - \frac{\sum_{k=0}^{r-1} l_x[k]z^{-k}}{\mathbf{L}_x(z)}\right) \tag{13-155}$$

As we see from Fig. 13-22, the innovations filter of $\hat{\mathbf{s}}_0[n]$ equals $\mathbf{L}_x(z)\mathbf{H}_x^0(z)$. To determine the pure predictor $\hat{\mathbf{H}}_r(z)$ of $\hat{\mathbf{s}}_0[n+r]$ it suffices, therefore to multiply

(13-53) by z^r (we are predicting now the future) and to replace the function $\mathbf{L}(z)$ by $\mathbf{L}_x(z)\mathbf{H}_x^0(z)$. This yields

$$\hat{\mathbf{H}}_r(z) = z^r\left(1 - \frac{\sum_{k=0}^{r-1} l_x[k]z^{-k} - D}{\mathbf{L}_x(z) - D}\right)$$

because the inverse of $\mathbf{L}_x(z) - D$ equals $l_x[n] - D\delta[n]$. Comparing with (13-155), we conclude that

$$\mathbf{H}_x^r(z) = \mathbf{H}_x^0(z)\hat{\mathbf{H}}_r(z)$$

The preceding discussion leads to the following important consequences of the white-noise assumption (13-150):

1. The innovations $\mathbf{i}_x[n]$ of $\mathbf{x}[n]$ are proportional to the difference $\mathbf{x}[n] - \hat{\mathbf{s}}_0[n]$

$$\mathbf{x}[n] - \hat{\mathbf{s}}_0[n] = D\mathbf{i}_x[n] \qquad D = \frac{N}{l_x[0]} \tag{13-156}$$

Indeed, $\mathbf{x}[n] - \hat{\mathbf{s}}_0[n]$ is the output of the filter

$$\mathbf{L}_x(z) - [\mathbf{L}_x(z) - D] = D$$

with input $\mathbf{i}_x[n]$ (Fig. 13-23*a*). Thus, the process $\mathbf{i}_x[n]$ can be realized simply by a feedback system (Fig. 13-23*b*) involving merely the filter $\mathbf{H}_{i,r}^0(z)$.

2. The r-step filtering and prediction estimate $\hat{\mathbf{s}}_0[n+r]$ can be obtained by cascading the pure filter $\mathbf{H}_x^0(z)$ of $\mathbf{s}[n]$ with the pure predictor $\hat{\mathbf{H}}_r(z)$ of $\hat{\mathbf{s}}_0[n+r]$.
3. If the signal $\mathbf{s}[n]$ is an ARMA process, then its estimate $\hat{\mathbf{s}}_0[n]$ is also an ARMA process.

Indeed, if $\mathbf{L}_x(z) = A(z)/B(z)$ is rational, then [see (13-151)], the filter $\mathbf{H}_x^0(z)$ is also rational. Furthermore, the denominator $B(z)$ of $\mathbf{L}_x(z)$ is the same as the denominator of the forward component $\mathbf{L}_x(z) - D$ of the feedback realization of $\mathbf{H}_x^0(z)$ shown in Fig. 13-23*b*.

As we shall presently see, these results are central in the development of Kalman filters.

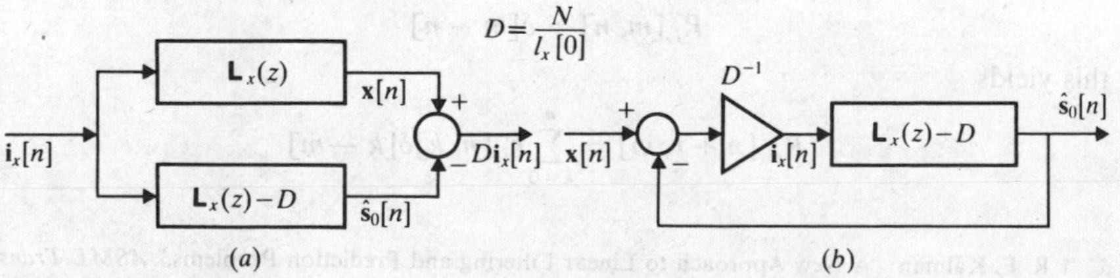

Figure 13-23

13-6 KALMAN FILTERS†

In this section we extend the preceding results to nonstationary processes with causal data and we show that the results can be simplified if the noise is white and the signal an ARMA process. The estimate $\hat{\mathbf{s}}_r[n+r]$ of $\mathbf{s}[n+r]$ in terms of the data

$$\mathbf{x}[n] = \mathbf{s}[n] + \mathbf{v}[n]$$

takes now the form

$$\hat{\mathbf{s}}_r[n+r] = \hat{E}\{\mathbf{s}[n+r] \mid \mathbf{x}[k], 0 \le k \le n\} = \sum_{k=0}^{n} h_x^r[n, k]\mathbf{x}[k] \qquad \text{(13-157)}$$

Thus, $\hat{\mathbf{s}}_r[n+r]$ is the output of a causal, time-varying system with input $\mathbf{x}[n]U[n]$, and our problem is to find its delta response $h_x^r[n, k]$.

As we know

$$\mathbf{s}[n+r] - \hat{\mathbf{s}}_r[n+r] \perp \mathbf{x}[m] \qquad 0 \le m \le n$$

This yields

$$R_{sx}[n+r, m] = \sum_{k=0}^{n} h_x^r[n, k]R_{xx}[k, m] \qquad 0 \le m \le n \qquad \text{(13-158)}$$

Thus, $h_x^r[n, k]$ must be such that its response to $R_{xx}[n, m]$ (the time variable is n) equals $R_{sx}[n+r, m]$ for every $0 \le m \le n$. For a specific n, this yields $n+1$ equations for the $n+1$ unknowns $h_x^r[n, k]$.

To simplify the determination of $h_x^r[n, k]$, we shall express the desired estimate $\hat{\mathbf{s}}_r[n+r]$ in terms of the *Kalman innovations* [see (13-114)]

$$\mathbf{i}_x[n] = \sum_{k=0}^{n} \gamma_x[n, k]\mathbf{x}[k] \qquad \text{(13-159)}$$

of the process $\mathbf{x}[n]U[n]$ where $\gamma_x[n, k]$ is the Kalman whitening filter. The process $\mathbf{i}_x[n]$ is orthonormal and, if the data are linearly independent, then the processes $\mathbf{x}[n]$ and $\mathbf{i}_x[n]$ are linearly equivalent. This leads to the conclusion that $\hat{\mathbf{s}}_r[n+r]$ can be expressed in terms of $\mathbf{i}_x[n]$ and its past (Fig. 13-24)

$$\hat{\mathbf{s}}_r[n+r] = \sum_{k=0}^{n} h_{i_x}^r[n, k]\mathbf{i}_x[k] \qquad \text{(13-160)}$$

To determine $h_{i_x}^r[n, k]$, we apply the orthogonality principle. Since

$$R_{i_x}[m, n] = \delta[m-n]$$

this yields

$$R_{si_x}[n+r, m] = \sum_{k=0}^{n} h_{i_x}^r[n, k]\delta[k-m]$$

† R. E. Kalman: "A New Approach to Linear Filtering and Prediction Problems," *ASME Transactions*, vol. 82D, 1960.

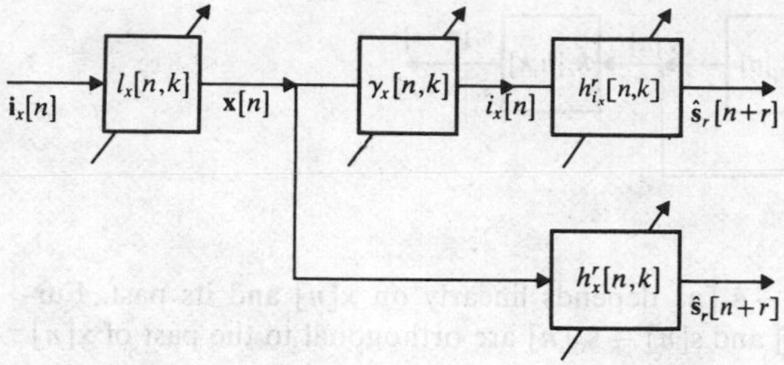

Figure 13-24

Hence

$$h^r_{i_x}[n, m] = R_{si_x}[n + r, m] \qquad 0 \le m \le n \tag{13-161}$$

This function can be expressed in terms of the cross-correlation $R_{sx}[m, n]$. Multiplying (13-159) by $\mathbf{s}[m]$ we obtain

$$R_{si_x}[m, n] = \sum_{k=0}^{n} \gamma_k[n, k] R_{sx}[m, k] \tag{13-162}$$

Thus, for a specific m, $R_{si_x}[m, n]$ is the response of the Kalman whitening filter of $\mathbf{x}[n]$ to the function $R_{sx}[m, n]$ where n is the variable. To complete the specification of $\hat{\mathbf{s}}_r[n + r]$, we must cascade the filter $h^r_{i_x}[n, m]$ with the whitening filter $\gamma_x[n, k]$ as in Fig. 13-24.

ARMA Signals in White Noise

In the numerical implementation of the above, we are faced with two problems: (*a*) the realization of the Kalman innovations process $\mathbf{i}_x[n]$; (*b*) the determination of the sum in (13-160). In general, these problems are complex, involving storage capacity and number of computations proportional to n. However, as we show next, under certain realistic assumptions the problem can be simplified drastically.

Assumption 1 The noise is white and orthogonal to the signal

$$R_{\nu\nu}[m, n] = N_n \delta[m - n] \qquad R_{s\nu}[m, n] = 0 \tag{13-163}$$

This leads to the following conclusions:

Property 1 The difference $\mathbf{x}[n] - \hat{\mathbf{s}}_0[n]$ is proportional to the Kalman innovations $\mathbf{i}_x[n]$ of the data $\mathbf{x}[n]$

$$\mathbf{x}[n] - \hat{\mathbf{s}}_0[n] = D_n \mathbf{i}_x[n] \tag{13-164a}$$

$$D_n^2 = E\{|\mathbf{x}[n] - \hat{\mathbf{s}}_0[n]|^2\} \tag{13-164b}$$

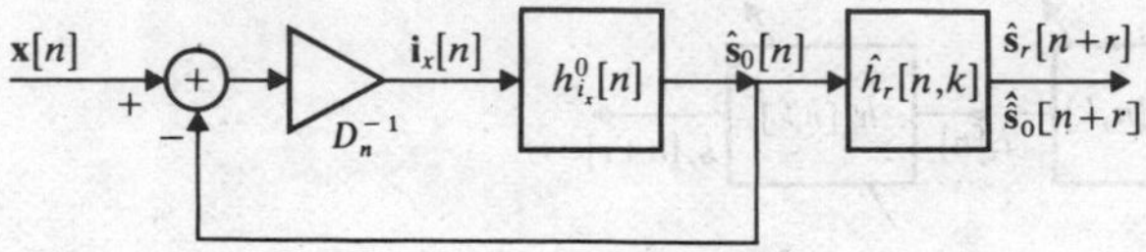

Figure 13-25

PROOF The difference $\mathbf{x}[n] - \hat{\mathbf{s}}_0[n]$ depends linearly on $\mathbf{x}[n]$ and its past. Furthermore, the processes $\mathbf{v}[n]$ and $\mathbf{s}[n] - \hat{\mathbf{s}}_0[n]$ are orthogonal to the past of $\mathbf{x}[n]$. Hence

$$\mathbf{x}[n] - \hat{\mathbf{s}}_0[n] = \mathbf{s}[n] - \hat{\mathbf{s}}_0[n] + \mathbf{v}[n] \perp \mathbf{x}[k] \qquad k < n$$

From this it follows that the process $\mathbf{x}[n] - \hat{\mathbf{s}}_0[n]$ is white noise and

$$\mathbf{x}[n] - \hat{\mathbf{s}}_0[n] \perp \mathbf{i}_x[k] \qquad 0 \le k \le n-1$$

because the processes $\mathbf{x}[k]$ and $\mathbf{i}_x[k]$ are linearly equivalent. And since $\mathbf{x}[n] - \hat{\mathbf{s}}_0[n]$ depends linearly on $\mathbf{i}_x[k]$ for $0 \le k \le n$, (13-164*a*) results. Equation (13-164*b*) is a consequence of the requirement that $E\{\mathbf{i}_x^2[n]\} = 1$.

Property 1 shows that the process $\mathbf{i}_x[n]$ can be realized simply by the feedback system of Fig. 13-25. This eliminates the need for designing the whitening filter $\gamma_x[n, k]$.

Property 2 The estimate $\hat{\mathbf{s}}_r[n+r]$ of $\mathbf{s}[n+r]$ equals the pure predictor $\hat{\hat{\mathbf{s}}}_0[n+r]$ of the estimate $\hat{\mathbf{s}}_0[n]$ of $\mathbf{s}[n]$ (Fig. 13-25)

$$\hat{\mathbf{s}}_r[n+r] = \hat{\hat{\mathbf{s}}}_0[n+r] = \sum_{k=0}^{n} \hat{h}_r[n, k]\hat{\mathbf{s}}_0[k] \tag{13-165}$$

provided that

$$E\{\hat{\mathbf{s}}_0[n]\mathbf{i}_x[n]\} = E\{\mathbf{x}[n]\mathbf{i}_x[n]\} - D_n \ne 0 \tag{13-166}$$

PROOF The process $\hat{\mathbf{s}}_0[n]$ is linearly dependent on $\mathbf{x}[n]$ and its past. Condition (13-166) means that the component of $\hat{\mathbf{s}}_0[n]$ in the $\mathbf{i}_x[n]$ direction is not zero. Hence, the processes $\hat{\mathbf{s}}_0[n]$ and $\mathbf{x}[n]$ are linearly equivalent. And since

$$\hat{\mathbf{s}}_0[n+r] - \hat{\hat{\mathbf{s}}}_0[n+r] \perp \hat{\mathbf{s}}_0[k] \qquad 0 \le k \le n$$

we conclude that

$$\hat{\hat{\mathbf{s}}}_0[n+r] - \hat{\mathbf{s}}_0[n+r] \perp \mathbf{x}[k] \qquad 0 \le k \le n$$

Furthermore

$$\mathbf{s}[n+r] - \hat{\mathbf{s}}_0[n+r] \perp \mathbf{x}[k] \qquad 0 \le k \le n+r$$

because $\hat{\mathbf{s}}_0[n+r]$ is the estimate of $\mathbf{s}[n+r]$ in terms of $\mathbf{x}[k]$ for $0 \le k \le n+r$. Finally

$$\mathbf{s}[n+r] - \hat{\hat{\mathbf{s}}}_0[n+r] = (\mathbf{s}[n+r] - \hat{\mathbf{s}}_0[n+r]) + (\hat{\mathbf{s}}_0[n+r] - \hat{\hat{\mathbf{s}}}_0[n+r])$$

Hence

$$\mathbf{s}[n+r] - \hat{\mathbf{s}}_0[n+r] \perp \mathbf{x}[k] \qquad 0 \le k \le n$$

and (13-165) results.

This property shows that filtering and prediction can be reduced to a cascade of a pure filter and a pure predictor.

Assumption 2 The signal $\mathbf{s}[n]$ is a time-varying ARMA process (Fig. 13-26*a*)

$$\mathbf{s}[n] - a_1^n \mathbf{s}[n-1] - \cdots - a_M^n \mathbf{s}[n-M] = \sum_{k=0}^{M-1} b_k^n \zeta[n-k]$$
$$R_{\zeta\zeta}[m, n] = V_n \delta[m-n] \tag{13-167}$$

Property 3 The estimate $\hat{\mathbf{s}}_0[n]$ of $\mathbf{s}[n]$ is also an ARMA process

$$\hat{\mathbf{s}}_0[n] - a_1^n \hat{\mathbf{s}}_0[n-1] - \cdots - a_M^n \hat{\mathbf{s}}_0[n-M] = \sum_{k=0}^{M-1} c_k^n \mathbf{i}_x[n-k] \tag{13-168}$$

where the coefficients a_k^n are the same as in (13-167) and the coefficients c_k^n are M constants to be determined.

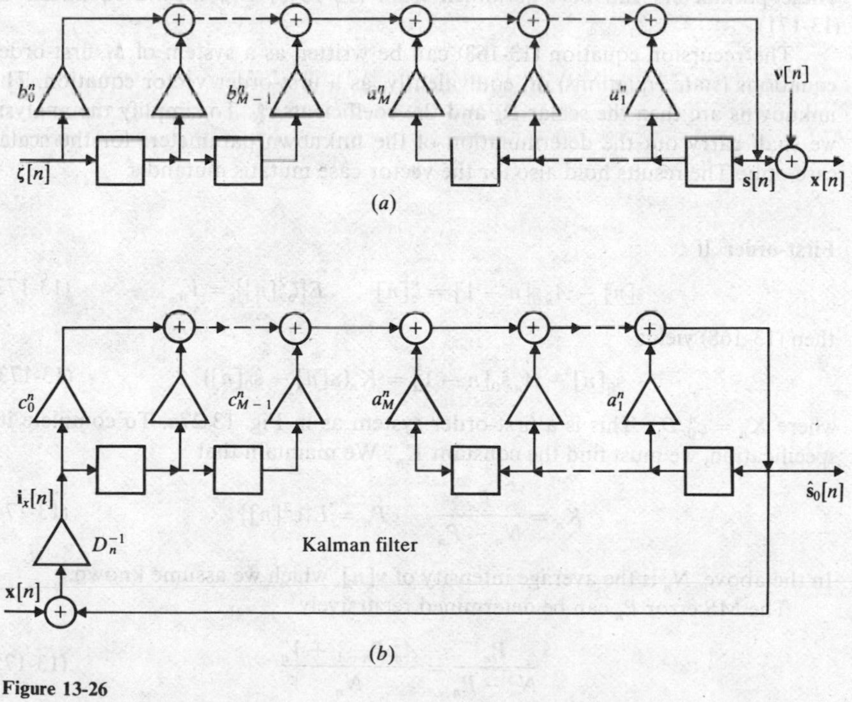

Figure 13-26

PROOF We assume that the above is true for all past estimates $\hat{\mathbf{s}}_0[n-k]$ and we shall prove that if $\hat{\mathbf{s}}_0[n]$ is given by (13-168), then it is the estimate of $\mathbf{s}[n]$. It suffices to show that if the constants c_k^n are suitably chosen, then the resulting error satisfies the orthogonality principle

$$\boldsymbol{\varepsilon}[n] = \mathbf{s}[n] - \hat{\mathbf{s}}_0[n] \perp \mathbf{x}[r] \qquad 0 \le r \le n \tag{13-169}$$

Subtracting (13-168) from (13-167), we obtain

$$\boldsymbol{\varepsilon}[n] = \sum_{k=1}^{M} a_k^n \boldsymbol{\varepsilon}[n-k] + \sum_{k=0}^{M-1} (b_k^n \boldsymbol{\zeta}[n-k] - c_k^n \mathbf{i}_x[n-k]) \tag{13-170}$$

But

$$\boldsymbol{\zeta}[n_1], \mathbf{i}_x[n_1] \perp \mathbf{x}[r] \qquad \text{for} \qquad r < n_1$$

and $\boldsymbol{\varepsilon}[n-k] \perp \mathbf{x}[r]$ for $r \le n-k$ (induction hypothesis). Hence, (13-169) is true for $r \le n - M$. It suffices, therefore, to select the M constants c_k^n such that

$$E\{\boldsymbol{\varepsilon}[n]\mathbf{x}[r]\} = 0 \qquad n - M + 1 \le r \le n \tag{13-171}$$

We have thus expressed $\hat{\mathbf{s}}_0[n]$ in terms of $\mathbf{i}_x[n]$. To complete the specification of the filter, we use (13-164*a*). This yields the feedback system of Fig. 13-26*b* involving $M + 1$ unknown parameters: the constant D_n and the M coefficients c_k^n. These parameters can be determined from (13-164*b*) and the M equations in (13-171).

The recursion equation (13-168) can be written as a system of M first-order equations (state equations) or, equivalently, as a first-order vector equation. The unknowns are then the scalar D_n and the coefficients c_k^n. To simplify the analysis we shall carry out the determination of the unknown parameters for the scalar case only. The results hold also for the vector case mutatis mutandis.

First-order If

$$\mathbf{s}[n] - A_n \mathbf{s}[n-1] = \boldsymbol{\zeta}[n] \qquad E\{\boldsymbol{\zeta}^2[n]\} = V_n \tag{13-172}$$

then (13-168) yields

$$\hat{\mathbf{s}}_0[n] - A_n \hat{\mathbf{s}}_0[n-1] = K_n(\mathbf{x}[n] - \hat{\mathbf{s}}_0[n]) \tag{13-173}$$

where $K_n = c_0^n / D_n$. This is a first-order system as in Fig. 13-27*a*. To complete its specification, we must find the constant K_n. We maintain that

$$K_n = \frac{P_n}{N_n - P_n} \qquad P_n = E\{\boldsymbol{\varepsilon}^2[n]\} \tag{13-174}$$

In the above, N_n is the average intensity of $\mathbf{v}[n]$, which we assume known.

The MS error P_n can be determined recursively

$$\frac{P_n}{N_n - P_n} = \frac{A_n^2 P_{n-1} + V_n}{N_n} \tag{13-175}$$

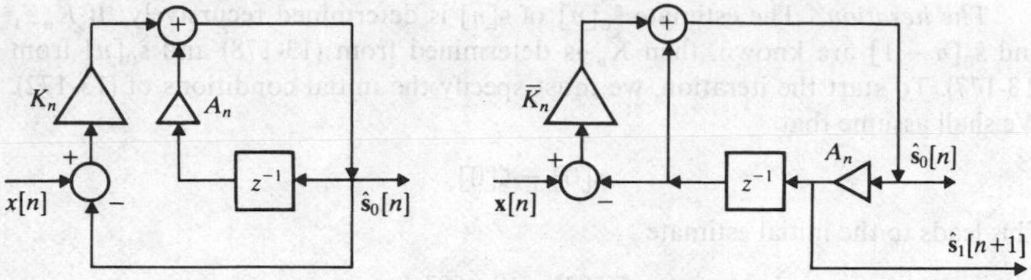

(a) (b)

Figure 13-27

PROOF Multiplying the data $\mathbf{x}[n] = \mathbf{s}[n] + \mathbf{v}[n]$ by the error

$$\varepsilon[n] = \mathbf{s}[n] - \hat{\mathbf{s}}_0[n] = \mathbf{x}[n] - \hat{\mathbf{s}}_0[n] - \mathbf{v}[n]$$

and using the orthogonality condition (13-169), we obtain

$$E\{\varepsilon[n]\mathbf{x}[n]\} = 0 = P_n + E\{\varepsilon[n]\mathbf{v}[n]\}$$

From (13-172) and (13-173) it follows that

$$\varepsilon[n] = A_n \varepsilon[n-1] + \zeta[n] - K_n(\varepsilon[n] + \mathbf{v}[n])$$

$$(1 + K_n)\varepsilon[n] = A_n \varepsilon[n-1] + \zeta[n] - K_n \mathbf{v}[n] \tag{13-176}$$

Hence

$$(1 + K_n)E\{\varepsilon[n]\mathbf{v}[n]\} = -K_n E\{\mathbf{v}^2[n]\}$$

and (13-174) results.

To prove (13-175), we multiply each side of (13-176) by each side of the identity $\mathbf{s}[n] = A_n \mathbf{s}[n-1] + \zeta[n]$. This yields

$$(1 + K_n)P_n = A_n^2 P_{n-1} + V_n$$

Since $1 + K_N = N_n/(N_n - P_n)$, the above yields (13-175).

Note Using (13-172), we can readily show that

$$\hat{\mathbf{s}}_0[n] - A_n \hat{\mathbf{s}}_0[n-1] = \bar{K}_n(\mathbf{x}[n] - A_n \hat{\mathbf{s}}_0[n-1]) \tag{13-177}$$

where

$$\bar{K}_n = \frac{P_n}{N_n} = \frac{A_n^2 P_{n-1} + V_n}{A_n^2 P_{n-1} + V_n + N_n} \tag{13-178}$$

The corresponding system is shown in Fig. 13-27*b*. In the same diagram, we show also the realization of the one-step predictor

$$\hat{\mathbf{s}}_1[n+1] = \hat{\hat{\mathbf{s}}}_0[n+1] = A_n \hat{\mathbf{s}}_0[n]$$

of $\mathbf{s}[n+1]$. This follows readily from (13-165) because the process $\hat{\mathbf{s}}_0[n]$ is AR; hence, its pure predictor equals $A_n \hat{\mathbf{s}}_0[n]$.

The iteration The estimate $\hat{\mathbf{s}}_0[n]$ of $\mathbf{s}[n]$ is determined recursively: If $\bar{K}_{n-1}$ and $\hat{\mathbf{s}}_0[n-1]$ are known, then $\bar{K}_n$ is determined from (13-178) and $\hat{\mathbf{s}}_0[n]$ from (13-177). To start the iteration, we must specify the initial conditions of (13-172). We shall assume that

$$\mathbf{s}[0] = \zeta[0]$$

This leads to the initial estimate

$$\hat{\mathbf{s}}_0[0] = \bar{K}_0 \mathbf{x}[0]$$

from which it follows that

$$\bar{K}_0 = \frac{E\{\mathbf{s}[0]\mathbf{x}[0]\}}{E\{\mathbf{x}^2[0]\}} = \frac{E\{\zeta^2[0]\}}{E\{\zeta^2[0]\} + E\{\mathbf{v}^2[0]\}}$$

Hence

$$\bar{K}_0 = \frac{V_0}{V_0 + N_0} \qquad P_0 = \frac{V_0 N_0}{V_0 + N_0} \tag{13-179}$$

Linearization Equation (13-175) and its equivalent (13-178) are nonlinear. However, each can be replaced by two linear equations. Indeed, if F_n and G_n are two sequences such that

$$\begin{aligned} F_n &= A_n^2 F_{n-1} + V_n G_{n-1} \qquad & F_0 &= V_0 N_0 \\ N_n G_n &= A_n^2 F_{n-1} + (V_n + N_n) G_{n-1} \qquad & G_0 &= V_0 + N_0 \end{aligned} \tag{13-180}$$

then

$$P_n = \frac{F_n}{G_n}$$

Example 13-13 We shall determine the noncausal, the causal, and the Kalman estimate of a process $s[n]$ in terms of the data $\mathbf{x}[k] = \mathbf{s}[k] + \mathbf{v}[k]$, and the corresponding MS error P. We assume that the process $\mathbf{s}[n]$ satisfies the equation

$$\mathbf{s}[n] - 0.8\mathbf{s}[n-1] = \zeta[n]$$

and that

$$R_{\zeta\zeta}[m] = 0.36\delta[m] \qquad R_{\zeta\nu}[m] = 0 \qquad R_{\nu\nu}[m] = \delta[n-m]$$

This is a special case of the process considered in Example 13-5 with

$$a = 0.8 \qquad N = 1 \qquad N_0 = 0.36 \qquad b = 0.5$$

Hence

$$\mathbf{S}_{ss}(z) = \frac{0.36}{(1 - 0.8z^{-1})(1 - 0.8z)} \qquad R_{ss}[m] = 0.8^{|m|}$$

$$\mathbf{S}_{xx}(z) = \mathbf{L}_x(z)\mathbf{L}_x(z^{-1}) \qquad \mathbf{L}_x(z) = \sqrt{1.6}\,\frac{z - 0.5}{z - 0.8}$$

(*a*) *Smoothing*: $\mathbf{x}[k]$ is available for all k. In this case the solution is obtained from Example 13-5 with $b = 0.5$ and $c = 0.375$

$$h[n] = 0.3 \times 0.5^{|n|} \qquad P = 0.3$$

(*b*) *Causal filter*: $\mathbf{x}[k]$ is available for $k \le n$. The unknown filter is determined from (13-151) where now $l_x[0] = \sqrt{1.6}$

$$\mathbf{H}_x^0(z) = 1 - \frac{z - 0.8}{1.6(z - 0.5)} = \frac{0.375z}{z - 0.5} \qquad h[n] = 0.375 \times 0.5^n U[n]$$

This shows that the estimate $\hat{\mathbf{s}}[n]$ of $\mathbf{s}[n]$ satisfies the recursion equation

$$\hat{\mathbf{s}}[n] - 0.5\hat{\mathbf{s}}[n-1] = 0.375\mathbf{x}[n] \qquad n \ge 0$$

The resulting MS error equals

$$P = R_{ss}[0] - \sum_{k=0}^{\infty} R_{ss}[k]h[k] = 0.375$$

(*c*) *Kalman filter*: $\mathbf{x}[k]$ is available for $0 \le k \le n$. Our process is a special case of (13-172) with

$$A_n = 0.8 \qquad V_n = E\{\zeta^2[n]\} = 0.36 \qquad N_n = E\{\mathbf{v}^2[n]\} = 1$$

Inserting into (13-180), we obtain

$$F_n = 0.64F_{n-1} + 0.36G_{n-1} \qquad F_0 = 0.36$$

$$G_n = 0.64F_{n-1} + 1.36G_{n-1} \qquad G_0 = 1.36$$

This is a system of linear recursion equations and can be readily solved with z transforms. Since

$$\bar{K}_n = \frac{P_n}{N} = \frac{F_n}{G_n}$$

and $N = 1$, the solution yields

$$\bar{K}_n = P_n = \frac{0.48z_1^n - 0.12z_2^n}{1.28z_1^n + 0.08z_2^n} \qquad z_1 = 1.6 \quad z_2 = 0.4$$

In particular

$n =$	0	1	2	3	4
$P_n \simeq$	0.3	0.357	0.371	0.374	0.375

Thus, although the number of the available data increases as n increases, the MS error P_n also increases. The reason is that $\mathbf{s}[n]$ is a nonstationary process with initial second moment $V_0 = 0.36$ because $\mathbf{s}[0] = \zeta[0]$, and, as n increases, $E\{\mathbf{s}^2[n]\}$ approaches the value 1.

We note, finally, that

$$\bar{K}_n = P_n \xrightarrow[n \to \infty]{} \frac{0.48}{1.28} = 0.375$$

and (13-177) yields

$$\hat{\mathbf{s}}_0[n] - 0.8\hat{\mathbf{s}}_0[n-1] = 0.375\mathbf{x}[n] - 0.3\hat{\mathbf{s}}_0[n-1]$$

The above shows that, if the process $\mathbf{s}[n]$ is WSS, then its Kalman filter approaches the causal Wiener filter as $n \to \infty$. This is the case for any P_0 because the limit of F_n/G_n as $n \to \infty$ equals 0.375 regardless of the initial conditions.

Example 13-14 We wish to estimate the RV $\mathbf{s}$ in terms of the sum

$$\mathbf{x}[n] = \mathbf{s} + \mathbf{v}[n] \qquad \text{where} \qquad E\{\mathbf{s}\mathbf{v}[n]\} = 0, \qquad R_{vv}[m, n] = N\delta[m - n]$$

The estimate $\hat{\mathbf{s}}_0[n]$ in terms of the data $\mathbf{x}[n]$ can be obtained as the output of a Kalman filter if we consider the RV $\mathbf{s}$ as a stochastic process satisfying trivially (13-172)

$$\mathbf{s}[n] = \mathbf{s}[n-1] + \boldsymbol{\zeta}[n] \qquad \mathbf{s}[-1] = 0$$

$$\boldsymbol{\zeta}[n] = \begin{cases} \mathbf{s} & n = 0 \\ 0 & n > 0 \end{cases} \qquad V_n = \begin{cases} E\{\mathbf{s}^2\} = M & n = 0 \\ 0 & n > 0 \end{cases}$$

In this case, $A_n = 1$, $N_n = N$, and (13-180) yields

$$F_n = F_{n-1} \qquad NG_n = F_{n-1} + NG_{n-1} \qquad F_0 = MN \qquad G_0 = M + N$$

Solving, we obtain

$$F_n = MN \qquad G_n = M + N + Mn$$

Hence

$$\hat{\mathbf{s}}_0[n] = \frac{N + Mn}{M + N + Mn}\,\hat{\mathbf{s}}_0[n-1] + \frac{M}{M + N + Mn}\,\mathbf{x}[n]$$

Continuous-Time Processes

We wish, finally, to determine the estimate

$$\hat{\mathbf{s}}_0(t) = \hat{E}\{\mathbf{s}(t) \mid \mathbf{x}(\tau), 0 \le \tau \le t\} \tag{13-181}$$

of a continuous-time process $\mathbf{s}(t)$ in terms of the data

$$\mathbf{x}(t) = \mathbf{s}(t) + \mathbf{v}(t) \tag{13-182}$$

The solution of this problem parallels the discrete-time solution if recursion equations are replaced by differential equations and sums by integrals. It might be instructive, however, to rederive the principal results using a different approach.

To avoid repetition, we start directly with the white-noise assumption

$$R_{vv}(t, \tau) = N(\tau)\delta(t - \tau) \qquad N(\tau) > 0 \qquad R_{sv}(t, \tau) = 0 \tag{13-183}$$

and we show that the process

$$\mathbf{w}(t) = \mathbf{x}(t) - \hat{\mathbf{s}}_0(t)$$

is white noise with autocorrelation

$$R_{ww}(t, \tau) = N(\tau)\delta(t - \tau) \tag{13-184}$$

PROOF As we know

$$\boldsymbol{\varepsilon}(t) = \mathbf{s}(t) - \hat{\mathbf{s}}_0(t) \perp \mathbf{x}(\tau) \qquad \mathbf{v}(t) \perp \mathbf{x}(\tau)$$

for $\tau < t$. Furthermore, $\mathbf{w}(\tau)$ depends linearly on $\mathbf{x}(\tau)$ and its past. Hence

$$\mathbf{w}(t) = \boldsymbol{\varepsilon}(t) + \mathbf{v}(t) \perp \mathbf{w}(\tau) \qquad \tau < t \tag{13-185}$$

To complete the proof of (13-184), we shall assume that $\mathbf{s}_0(t)$ is continuous from the left

$$\hat{\mathbf{s}}_0(t^-) = \hat{\mathbf{s}}_0(t)$$

This is not true at the origin if $\mathbf{s}(0) \neq 0$. However, for sufficiently large t, the effect of the initial condition can be neglected. From the above it follows that

$$P(t) = E\{\boldsymbol{\varepsilon}^2(t)\} < \infty$$

and since $\boldsymbol{\varepsilon}(t) \perp \mathbf{v}(\tau)$ for $\tau > t$, we conclude that

$$R_{ww}(t, \tau) = R_{vw}(t, \tau) = R_{vv}(t, \tau) = N(\tau)\delta(t - \tau)$$

Using a limit argument, we can show that, as in the discrete-time case, the normalized process $\mathbf{w}(t)/\sqrt{N(t)}$ is the Kalman innovations of $\mathbf{x}(t)$. The details, however, will be omitted. This leads to the conclusion that $\hat{\mathbf{s}}_0(t)$ can be expressed in terms of $\mathbf{w}(t)$ [see also (13-160)]

$$\hat{\mathbf{s}}_0(t) = \int_0^t h_w(t, \alpha)\mathbf{w}(\alpha)\, d\alpha \tag{13-186}$$

Since $\mathbf{s}(t) - \hat{\mathbf{s}}_0(t) \perp \mathbf{w}(\tau)$ for $\tau \le t$, we conclude from the above and (13-184) that

$$R_{sw}(t, \tau) = \int_0^t h_w(t, \alpha)N(\alpha)\delta(\tau - \alpha)\, d\alpha = h_w(t, \tau)N(\tau)$$

and (13-186) yields

$$\hat{\mathbf{s}}_0(t) = \int_0^t \frac{1}{N(\alpha)} R_{sw}(t, \alpha)\mathbf{w}(\alpha)\, d\alpha \tag{13-187}$$

We note that [see (13-185) and (13-183)]

$$R_{sw}(t, t) = E\{\mathbf{s}(t)[\boldsymbol{\varepsilon}(t) + \mathbf{v}(t)]\} = P(t)$$

Wide-sense Markoff processes Using the above, we shall show that, if the signal $\mathbf{s}(t)$ is WS Markoff, i.e., if it satisfies a differential equation driven by white noise, then its estimate $\hat{\mathbf{s}}_0(t)$ satisfies a similar equation. For simplicity, we consider the first-order case

$$\mathbf{s}'(t) + A(t)\mathbf{s}(t) = \boldsymbol{\zeta}(t) \qquad R_{\zeta\zeta}(t, \tau) = V(\tau)\delta(t - \tau) \tag{13-188}$$

The Kalman–Bucy equations † We maintain that

$$\hat{\mathbf{s}}_0'(t) + A(t)\hat{\mathbf{s}}_0(t) = K(t)[\mathbf{x}(t) - \hat{\mathbf{s}}_0(t)] \qquad (13\text{-}189)$$

where

$$K(t) = \frac{P(t)}{N(t)} \qquad (13\text{-}190)$$

Furthermore, the MS error $P(t)$ satisfies the *Riccati equation*

$$P'(t) + 2A(t)P(t) = V(t) - \frac{1}{N(t)}P^2(t) \qquad (13\text{-}191)$$

PROOF Multiplying the differential equation in (13-188) by $\mathbf{w}(\tau)$, we obtain

$$\frac{\partial}{\partial t}R_{sw}(t, \tau) + A(t)R_{sw}(t, \tau) = 0 \qquad \tau < t \qquad (13\text{-}192)$$

We next equate the derivatives of both sides of (13-187)

$$\hat{\mathbf{s}}_0'(t) = \frac{1}{N(t)}R_{sw}(t, t)\mathbf{w}(t) + \int_0^t \frac{1}{N(\alpha)}\frac{\partial}{\partial t}R_{sw}(t, \alpha)\mathbf{w}(\alpha)\,d\alpha$$

Finally, we multiply (13-187) by $A(t)$ and add with the above. This yields (13-189) because, as we see from (13-192), the sum of the two integrals is zero.

To prove (13-191), we use the following version of (9-117): If $\mathbf{z}(t)$ is a process with $E\{\mathbf{z}^2(t)\} = I(t)$ and such that

$$\mathbf{z}'(t) + B(t)\mathbf{z}(t) = \boldsymbol{\xi}(t) \qquad R_{\xi\xi}(t, \tau) = Q(\tau)\delta(t - \tau) \qquad (13\text{-}193)$$

then (see Prob. 9-19*b*)

$$I'(t) + 2B(t)I(t) = Q(t) \qquad (13\text{-}194)$$

Returning to (13-189) we observe, subtracting from (13-188), that the estimation error $\boldsymbol{\varepsilon}(t)$ satisfies the equation

$$\boldsymbol{\varepsilon}'(t) + [A(t) + K(t)]\boldsymbol{\varepsilon}(t) = \boldsymbol{\zeta}(t) - K(t)\mathbf{v}(t)$$

In the above, the right side $\boldsymbol{\xi}(t) = \boldsymbol{\zeta}(t) - K(t)\mathbf{v}(t)$ is white noise as in (13-193) with

$$Q(\tau) = V(\tau) + K^2(\tau)N(\tau)$$

Hence, the function $P(t) = E\{\boldsymbol{\varepsilon}^2(t)\}$ satisfies (13-194) where $B(t) = A(t) + K(t)$. This yields

$$P'(t) + 2[A(t) + K(t)]P(t) = V(t) + K^2(t)N(t)$$

and (13-191) results.

† R. E. Kalman and R. C. Bucy: "New Results in Linear Filtering and Prediction Theory," *ASME Transactions*, vol. 83D, 1961.

Linearization We shall now show that the nonlinear equation (13-191) is equivalent to two linear equations. For this purpose, we introduce the functions $F(t)$ and $G(t)$ such that

$$P(t) = \frac{F(t)}{G(t)} \tag{13-195}$$

Clearly

$$F'(t) = P'(t)G(t) + P(t)G'(t)$$

and (13-191) yields

$$F'(t) + A(t)F(t) - V(t)G(t) = P(t)\left[G'(t) - A(t)G(t) - \frac{F(t)}{N(t)}\right]$$

This is satisfied if

$$\begin{aligned} F'(t) &= -A(t)F(t) + V(t)G(t) \\ G'(t) &= \frac{F(t)}{N(t)} + A(t)G(t) \end{aligned} \tag{13-196}$$

To solve the above system, we must specify $F(0)$ and $G(0)$. Setting arbitrarily $G(0) = 1$, we obtain $F(0) = P(0)$ where

$$P(0) = E\{\mathbf{s}^2(0)\}$$

is the initial value of the MS error $P(t)$. The determination of the Kalman filter depends, thus, on the second moment of $\mathbf{s}(0)$.

Example 13-15 We shall determine the noncausal, the causal, and the Kalman estimate of a process $\mathbf{s}(t)$ in terms of the data $\mathbf{x}(t) = \mathbf{s}(t) + \mathbf{v}(t)$, and the corresponding MS error P. We assume that $\mathbf{s}(t)$ satisfies the equation

$$\mathbf{s}'(t) + 2\mathbf{s}(t) = \boldsymbol{\zeta}(t)$$

and that

$$R_{\zeta\zeta}(\tau) = 12\delta(\tau) \qquad R_{s\nu}(\tau) = 0 \qquad R_{\nu\nu}(\tau) = \delta(\tau)$$

This is a special case of the process considered in Example 13-10 with

$$a = 2 \qquad N = 1 \qquad N_0 = 12 \qquad \beta = 4$$

Hence

$$S_{ss}(\omega) = \frac{12}{4 + \omega^2} \qquad R_{ss}(\tau) = 3e^{-2|\tau|}$$

$$S_{xx}(\omega) = \frac{16 + \omega^2}{4 + \omega^2} \qquad \mathbf{L}_x(s) = \frac{s + 4}{s + 2}$$

(a) *Smoothing:* $\mathbf{x}(\xi)$ is available for all ξ. In this case, (13-39) yields

$$H(\omega) = \frac{12}{16 + \omega^2} \qquad h(t) = \frac{3}{2}e^{-4|\tau|}$$

The MS error is obtained from (13-38)

$$P = 3 - \frac{9}{2}\int_{-\infty}^{\infty} e^{-4|\tau|}e^{-2|\tau|}\,d\tau = 1.5$$

(*b*) *Causal filter:* $\mathbf{x}(\xi)$ is available for $\xi \le t$. The unknown filter is specified in Example 13-11 with

$$\alpha = 2 \qquad \beta = 4 \qquad N = 12$$

Thus

$$\mathbf{H}_x(s) = \frac{2}{s+4} \qquad h_x(t) = 2e^{-4t}U(t) \qquad P = 2$$

This shows that the estimate $\hat{\mathbf{s}}(t)$ of $\mathbf{s}(t)$ satisfies the differential equation

$$\hat{\mathbf{s}}'(t) + 4\hat{\mathbf{s}}(t) = 2\mathbf{x}(t)$$

(*c*) *Kalman filter:* $\mathbf{x}(\xi)$ is available for $0 \le \xi \le t$. Our process is a special case of (13-188) with

$$A(t) = 2 \qquad V(t) = 12 \qquad N(t) = 1$$

Hence [see (13-196)]

$$F'(t) = -2F(t) + 12G(t) \qquad G'(t) = F(t) + 2G(t)$$

To solve this system, we must know $P(0)$.

Case 1 If $\mathbf{s}(0) = 0$, then $P(0) = 0$. In this case, $F(0) = 0$, $G(0) = 1$. Inserting the solution of the above system into (13-190), we obtain

$$K(t) = P(t) = \frac{6e^{4t} - 6e^{-4t}}{3e^{4t} + e^{-4t}} \xrightarrow[t\to\infty]{} 2$$

Case 2 We now assume that $\mathbf{s}(t)$ is the stationary solution of the differential equation specifying $\mathbf{s}(t)$. In this case, $E\{\mathbf{s}^2(0)\} = 3$, hence, $P(0) = F(0) = 3$ and

$$K(t) = P(t) = \frac{18e^{4t} + 6e^{-4t}}{9e^{4t} - e^{-4t}} \xrightarrow[t\to\infty]{} 2$$

Thus, in both cases, the solution $\hat{\mathbf{s}}_0(t)$ of the Kalman–Busy equation (13-189) tends to the solution of the causal Wiener filter

$$\hat{\mathbf{s}}_0'(t) + 2\hat{\mathbf{s}}_0(t) = 2\mathbf{x}(t) - 2\hat{\mathbf{s}}_0(t)$$

as $t \to \infty$.

Example 13-16 We wish to estimate the RV $\mathbf{s}$ in terms of the sum

$$\mathbf{x}(t) = \mathbf{s} + \mathbf{v}(t) \qquad E\{\mathbf{s}\mathbf{v}(t)\} = 0 \qquad R_{vv}(\tau) = N\delta(\tau)$$

This is a special case of (13-188) if

$$A(t) = 0 \qquad \mathbf{s}(t) = \mathbf{s} \qquad \zeta(t) = 0 \qquad N(t) = N$$

In this case, $V(t) = 0$, $P(0) = E\{\mathbf{s}^2\} \equiv M$ and (13-196) yields

$$F'(t) = 0 \qquad G'(t) = \frac{F(t)}{N} \qquad F(0) = M \qquad G(0) = 1$$

Hence

$$F(t) = M \qquad G(t) = 1 + \frac{Mt}{N}$$

Inserting into (13-189), we obtain

$$\hat{\mathbf{s}}_0'(t) + \frac{M}{N + Mt}\,\hat{\mathbf{s}}_0(t) = \frac{M}{N + Mt}\,\mathbf{x}(t)$$

13-7 ADAPTIVE FILTERS

Up to now, we assumed that the second-order moments of all processes were known and we expressed the coefficients of the various estimators in terms of these moments. In this section, we seek estimates that are not based on knowledge of prior statistics but rely only on the data. We examine two approaches involving filters with time-varying parameters. In the first, the variation is expressed in terms of the data and the instantaneous values of the estimation error. In the second, the coefficients are determined in terms of time-average estimates of the unknown moments.

The problem under consideration is phrased as follows: We wish to estimate a process $\mathbf{y}[n]$ in terms of the sum (Fig. 13-28*a*)

$$\hat{\mathbf{y}}[n] = \sum_{k=0}^{N-1} a_k[n]\mathbf{x}[n-k] \tag{13-197}$$

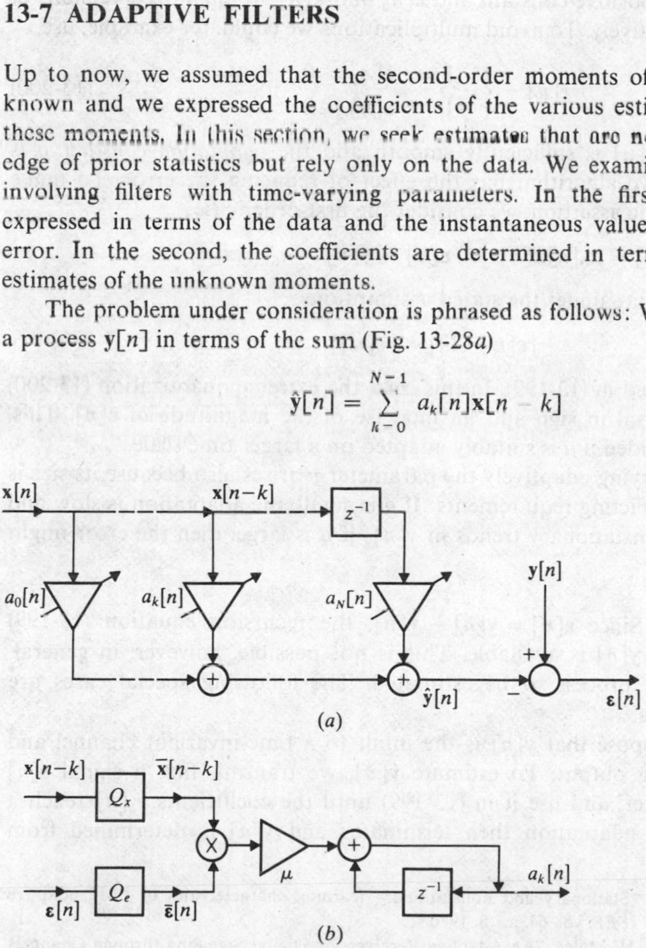

Figure 13-28

In this estimate, the coefficients $a_k[n]$ are time-dependent and they are so chosen as to reduce somehow the instantaneous error

$$\boldsymbol{\varepsilon}[n] = \mathbf{y}[n] - \hat{\mathbf{y}}[n] \tag{13-198}$$

Our objective is to determine $a_k[n]$.

The Widrow Algorithm†

We evaluate $a_k[n]$ recursively according to the following rule (Fig. 13-28*b*)

$$a_k[n] = a_k[n-1] + \mu\bar{\boldsymbol{\varepsilon}}[n]\bar{\mathbf{x}}[n-k] \tag{13-199}$$

In the above, μ is a positive constant and $\bar{\boldsymbol{\varepsilon}}[n]$ and $\bar{\mathbf{x}}[n]$ are quantized versions of $\boldsymbol{\varepsilon}[n]$ and $\mathbf{x}[n]$ respectively. To avoid multiplications we could, for example, use

$$\bar{\varepsilon}[n] = \begin{cases} 1 & \boldsymbol{\varepsilon}[n] > 0 \\ -1 & \boldsymbol{\varepsilon}[n] < 0 \end{cases} \tag{13-200}$$

If the process $\mathbf{y}[n]$ is sufficiently smooth and the *adaptation constant* μ is small, then the above algorithm has the effect of reducing the error. To make plausible this heuristic assertion, we consider the first-order case

$$\hat{\mathbf{y}}[n] = a_0[n]\mathbf{x}[n] \qquad \boldsymbol{\varepsilon}[n] = \mathbf{y}[n] - a_0[n]\mathbf{x}[n]$$

We can readily see that, under the stated assumptions,

$$|\boldsymbol{\varepsilon}[n+1]| < |\boldsymbol{\varepsilon}[n]|$$

even if (13-200) is used in (13-199). In this case, the extreme quantization (13-200) might cause a reversal in sign and an increase of the magnitude of $\boldsymbol{\varepsilon}[n]$. This, however, can be avoided if μ is suitably adapted on a larger time scale.

The need for varying adaptively the parameter μ arises also because its size is dictated by two conflicting requirements: If μ is small, the adaptation is slow and might not follow nonstationary trends in $y[n]$; if μ is large, then the error might increase.

Realizable models‡ Since $\boldsymbol{\varepsilon}[n] = \mathbf{y}[n] - \hat{\mathbf{y}}[n]$, the recursion equation (13-199) can be used only if $\mathbf{y}[n]$ is available. This is not possible, however, in general, because $\mathbf{y}[n]$ is the process to be estimated. The following special cases are important exceptions.

Pilot signal Suppose that $\mathbf{y}[n]$ is the input to a time-invariant channel and $\mathbf{x}[n]$ is the resulting output. To estimate $\mathbf{y}[n]$, we transmit first a signal $s[n]$ known to the receiver, and use it in (13-199) until the coefficients $a_k[n]$ reach a limiting value. The adaptation then terminates and $\mathbf{y}[n]$ is determined from

† B. Widrow, et al.: "Stationary and nonstationary learning characteristics of LMS adaptive filters," *Proceedings of the IEEE*, vol. 64, no. 8, 1976.

‡ J. G. Proakis and J. H. Miller: "An Adaptive Receiver for Digital Signaling through Channels with Intersymbol Interference," *IEEE Transactions on Information Theory*, vol. IT-15, 1969.

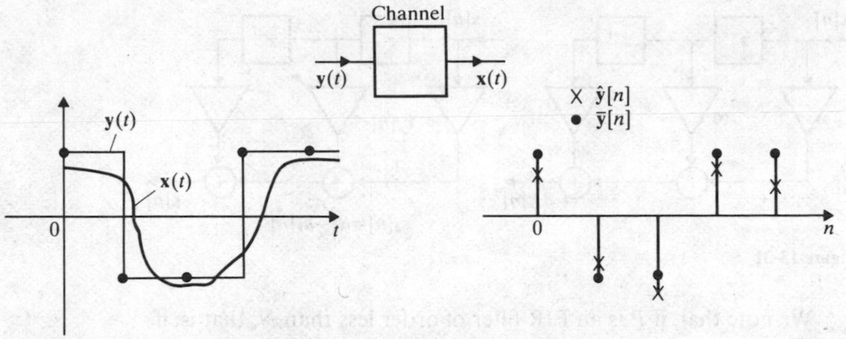

Figure 13-29

(13-197). The method can be used also for time-varying channels provided that the variation is slow relative to the adaptation time. In this case, the transmission of $\mathbf{y}[n]$ is interrupted periodically and the pilot signal is retransmitted.

Binary signals Suppose that the signal $\mathbf{y}[n]$ equals $+1$ or -1 as in Fig. 13-29. In this case, we use as the estimate of $\mathbf{y}[n]$ the quantized signal

$$\bar{\mathbf{y}}[n] = \begin{cases} 1 & \text{if} \quad \hat{\mathbf{y}}[n] > 0 \\ -1 & \text{if} \quad \hat{\mathbf{y}}[n] < 0 \end{cases} \tag{13-201}$$

If at some point the estimate is correct, then $\mathbf{y}[n]$ is available and can be used in (13-199) to update the coefficients $a_k[n]$. The method fails only if the error $\varepsilon[n]$ exceeds 1.

Prediction If the desired estimate is a pure predictor, then $\mathbf{x}[n] = \mathbf{s}[n]$ and $\mathbf{y}[n] = \mathbf{s}[n+1]$. In this case, we can use the observed error

$$\varepsilon[n-1] = \mathbf{s}[n] - \mathbf{s}[n-1]$$

to update the coefficients for the next estimate.

System identification We wish to approximate an unknown plant P by a linear FIR filter. For this purpose, we apply a stationary process $\mathbf{x}[n]$ to the system P, and we obtain as output the process $\mathbf{y}[n]$. We then determine the estimate $\hat{\mathbf{y}}[n]$ of $\mathbf{y}[n]$ as in (13-197). The resulting FIR filter is the desired approximation of the plant P (Fig. 13-30).

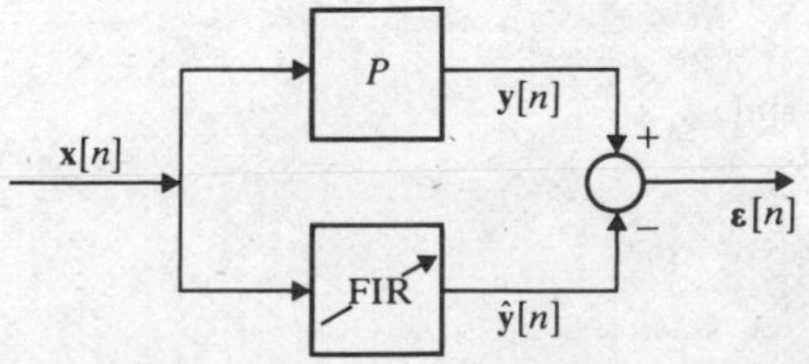

Figure 13-30

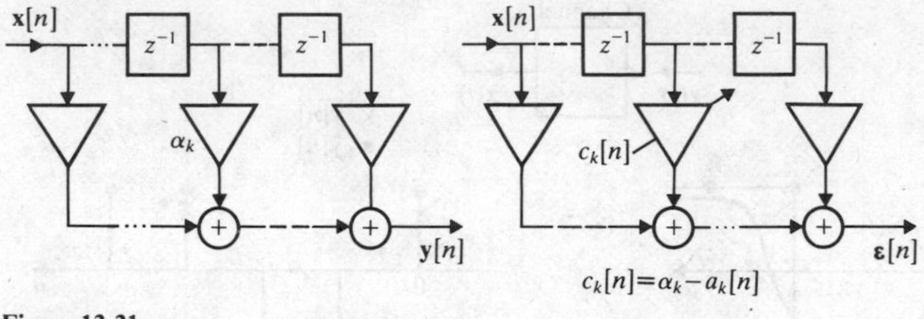

Figure 13-31

We note that, if P is an FIR filter of order less than N, that is, if

$$\mathbf{y}[n] = \sum_{k=0}^{N-1} \alpha_k \, \mathbf{x}[n-k] \tag{13-202}$$

then

$$a_k[n] \to \alpha_k \quad \text{as} \quad n \to \infty$$

In this case, the error can be written in the form (Fig. 13-31)

$$\boldsymbol{\varepsilon}[n] = \sum_{k=0}^{N-1} c_k[n]\mathbf{x}[n-k]$$

where

$$c_k[n] = \alpha_k - a_k[n] \to 0 \quad \text{as} \quad n \to 0 \tag{13-203}$$

The coefficients $c_k[n]$ satisfy now the equation

$$c_k[n] = c_k[n-1] + \mu\bar{\boldsymbol{\varepsilon}}[n]\bar{\mathbf{x}}[n-k]$$

involving only the signal $\mathbf{x}[n]$.

Echo cancelling A related application is the echo-cancelling problem in telephone communications. A simplified form is shown in Fig. 13-32: The signal $\mathbf{x}[n]$

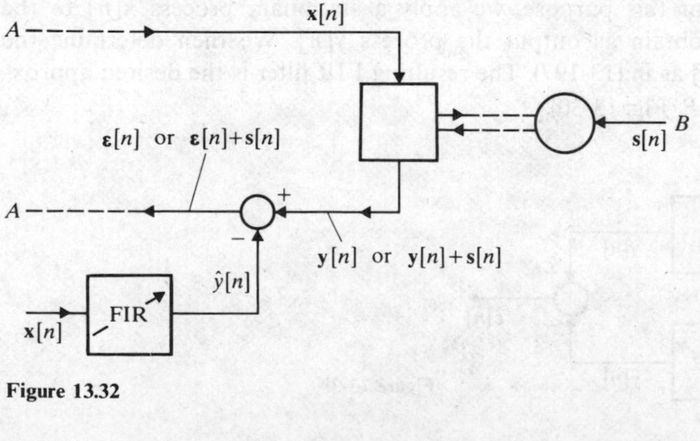

Figure 13.32

is the incoming call to a substation, originating at point A with destination point B. The return signal is the sum

$$\mathbf{r}[n] = \mathbf{y}[n] + \mathbf{s}[n]$$

where $\mathbf{s}[n]$ is the call coming from point B and $\mathbf{y}[n]$ is an attenuated and distorted version of $\mathbf{x}[n]$. The purpose of the filter is to remove the *echo* $\mathbf{y}[n]$ from the return signal $\mathbf{r}[n]$.

The above is similar to the identification problem if we interpret $\mathbf{y}[n]$ as the output of a system (plant) with input $\mathbf{x}[n]$. However, in the echo problem, $\mathbf{y}[n]$ is known only when $\mathbf{s}[n] \doteq 0$. We can, therefore, use (13-199) only if point B is silent. This can be readily sensed with a threshold device because usually the echo $\mathbf{y}[n]$ is small compared to the return signal $\mathbf{s}[n]$.

Note The Widrow filter is simple, it requires no knowledge of prior statistics, and it can be used for stationary and quasi-stationary processes. However, the properties of $\hat{\mathbf{y}}[n]$ are not known although the topic has been investigated extensively. The available results are mostly in the form of asymptotic statements valid only for special cases. The reason is that the problem is inherently difficult involving the solution of a system of nonlinear stochastic recursion equations.

Time-Average Estimates

The estimate (13-197) of $\mathbf{y}[n]$ can be written in vector form

$$\hat{\mathbf{y}}[n] = A[n]\mathbf{X}^t[n] \tag{13-204}$$

In the above, $A[n]$ is a vector with elements $a_k[n]$ and $\mathbf{X}^t[n]$ is the transpose of a random vector process with elements $\mathbf{x}[n-k]$. In Sec. 13-4 we showed that for WSS processes, $\hat{\mathbf{y}}[n]$ is optimum if $A[n]$ is a constant vector given by

$$A = R^{-1}B \tag{13-205}$$

[see (13-165)] where R is the correlation matrix of $\mathbf{X}[n]$, and B is the cross-correlation vector between $\mathbf{y}[n]$ and $\mathbf{X}[n]$

$$R = E\{\mathbf{X}^t[n]\mathbf{X}[n]\} \qquad B = E\{\mathbf{X}^t[n]\mathbf{y}[n]\} \tag{13-206}$$

This result is useful if R and B are known. In the following, we determine $A[n]$ in terms of estimates of R and B based on past observations only. In this development, no knowledge of prior statistics is assumed and the results hold also for nonstationary processes. However, the required numerical operations are not simple.

Time-average MS optimum We start with the problem of estimating a signal $y[n]$ in terms of the present value of $x[n]$

$$\hat{y}[n] = a[n]x[n] \tag{13-207}$$

The signals $x[n]$ and $y[n]$ can be deterministic or random. We shall determine the coefficient $a[n]$ using as optimality criterion the minimization of the weighted time-average MS error

$$e_n = \sum_{k=1}^{n} c\alpha^k(y[n-k] - a[n]x[n-k])^2 \tag{13-208}$$

where $0 < \alpha < 1$ and

$$\sum_{k=1}^{n} c\alpha^k = 1 \qquad c = \frac{1-\alpha}{\alpha(1-\alpha^n)}$$

The weights α^k are introduced to reduce the effect of the distant past on the estimate of $y[n]$.

Clearly, e_n depends on $a[n]$ and it is minimum if

$$\frac{de_n}{da[n]} = 0 = -2c \sum_{k=1}^{n} \alpha^k \Big(y[n-k] - a[n]\mathbf{x}[n-k] \Big) \mathbf{x}[n-k]$$

This yields

$$a[n] = \frac{\bar{R}[n]}{\bar{B}[n]} \tag{13-209}$$

where

$$\bar{R}[n] = c \sum_{k=1}^{n} \alpha^k x^2[n-k]$$
$$\bar{B}[n] = c \sum_{k=1}^{n} \alpha^k x[n-k]y[n-k] \tag{13-210}$$

If $x[n]$ and $y[n]$ are samples of WSS processes then $\bar{R}[n]$ and $\bar{B}[n]$ are time-average estimates of $R_{xx}[0]$ and $R_{xy}[0]$. Furthermore

$$\bar{R}[n] \to R_{xx}[0] \qquad \bar{B}[n] \to R_{xy}[0]$$

as $n \to \infty$.

Vectors The vector $A[n]$ of the estimate (13-204) of $y[n]$ is determined similarly: We form the MS error

$$e_n = \sum_{k=1}^{n} c\alpha^k(y[n-k] - A[n]X^t[n-k])^2 \tag{13-211}$$

and set all its derivatives with respect to the N components $a_i[n]$ of $A[n]$ equal to zero. This yields

$$A[n] = \bar{R}^{-1}[n]\bar{B}[n] \tag{13-212}$$

where

$$\bar{R}[n] = c \sum_{k=1}^{n} \alpha^k X^t[n-k]X[n-k] \tag{13-213}$$

is an N-by-N matrix and

$$\bar{B}[n] = c \sum_{k=1}^{n} \alpha^k X^t[n-k]y[n-k] \tag{13-214}$$

is a vector.

If the processes $\mathbf{x}[n]$ and $\mathbf{y}[n]$ are WSS then

$$\bar{R}[n] \to R \qquad \bar{B}[n] \to B \qquad A[n] \to R^{-1}B$$

as $n \to \infty$. In other words, $A[n]$ approaches the statistical MS optimum (13-205). This is true also for nonstationary processes provided that their statistics are nearly constant in any interval of length n_0 and $\alpha^{n_0} \ll 1$. The choice of α is thus dictated by two conflicting requirements: It must be large for time-averages to approach statistical averages; it must be small for time-averages to follow statistical variations.

Recursion The time-average method of estimation involves the computationally complex problem of evaluating the matrix $\bar{R}[n]$ and its inverse for every n. The computations can, however, be simplified if they are carried out recursively: From (13-213) and (13-214) it follows readily that

$$\begin{aligned} \bar{R}[n+1] &= \alpha \bar{R}[n] + cX^t[n]X[n] \\ \bar{B}[n+1] &= \alpha \bar{B}[n] + cX^t[n]y[n] \end{aligned} \tag{13-215}$$

Inserting into (13-212), we conclude, after some straightforward manipulations, that the vector $A[n]$ satisfies the recursion equation

$$A[n+1] = A[n] + c\bar{R}^{-1}[n+1]X^t[n]\varepsilon[n] \tag{13-216}$$

where

$$\varepsilon[n] = y[n] - A[n]X^t[n] \tag{13-217}$$

The inverse of the matrix $\bar{R}[n]$ can also be determined recursively

$$\alpha \bar{R}^{-1}[n+1] = \bar{R}^{-1}[n] - \frac{c\bar{R}^{-1}[n]X^t[n]X[n]\bar{R}^{-1}[n]}{\alpha + cX[n]\bar{R}^{-1}[n]X^t[n]} \tag{13-218}$$

The above equations simplify the evaluation of $A[n]$ but the required operations are still not trivial. We note that Widrow's algorithm (13-199) is a special case of (13-216) obtained by replacing the matrix $\bar{R}^{-1}[n+1]$ by a constant.

APPENDIX 13A
MINIMUM-PHASE FUNCTIONS

A function

$$\mathbf{H}(z) = \sum_{n=0}^{\infty} h_n z^{-n}$$

is called minimum-phase, if it is analytic and its inverse $1/\mathbf{H}(z)$ is also analytic for $|z| \geq 1$. We shall show that if $\mathbf{H}(z)$ is minimum-phase, then

$$\ln h_0^2 = \frac{1}{2\pi} \int_{-\pi}^{\pi} \ln |\mathbf{H}(e^{j\varphi})|^2 \, d\varphi \qquad \text{(13A-1)}$$

Proof Using the identity $|\mathbf{H}(e^{j\varphi})|^2 = \mathbf{H}(e^{j\varphi})\mathbf{H}(e^{-j\varphi})$, we conclude with $e^{j\varphi} = z$, that

$$\int_{-\pi}^{\pi} \ln |\mathbf{H}(e^{j\varphi})|^2 \, d\varphi = \oint \frac{1}{jz} \ln [\mathbf{H}(z)\mathbf{H}(z^{-1})] \, dz$$

where the path of integration is the unit circle. We note further, changing z to $1/z$, that

$$\oint \frac{1}{z} \ln \mathbf{H}(z) \, dz = \oint \frac{1}{z} \ln \mathbf{H}(z^{-1}) \, dz$$

To prove (13A-1), it suffices, therefore, to show that

$$\ln |h_0| = \frac{1}{2\pi j} \oint \frac{1}{z} \ln \mathbf{H}(z) \, dz$$

This follows readily because $\mathbf{H}(z)$ tends to h_0 as $z \to \infty$ and the function $\ln \mathbf{H}(z)$ is analytic for $|z| \geq 1$ by assumption.

APPENDIX 13B
ALL-PASS FUNCTIONS

The unit circle is the locus of points N such that (see Fig. 13-33*a*)

$$\frac{(NA)}{(NB)} = \frac{|e^{j\varphi} - 1/z_i^*|}{|e^{j\varphi} - z_i|} = \frac{1}{|z_i|} \qquad |z_i| < 1$$

From this it follows that, if

$$\mathbf{F}(z) = \frac{zz_i^* - 1}{z - z_i} \qquad |z_i| < 1$$

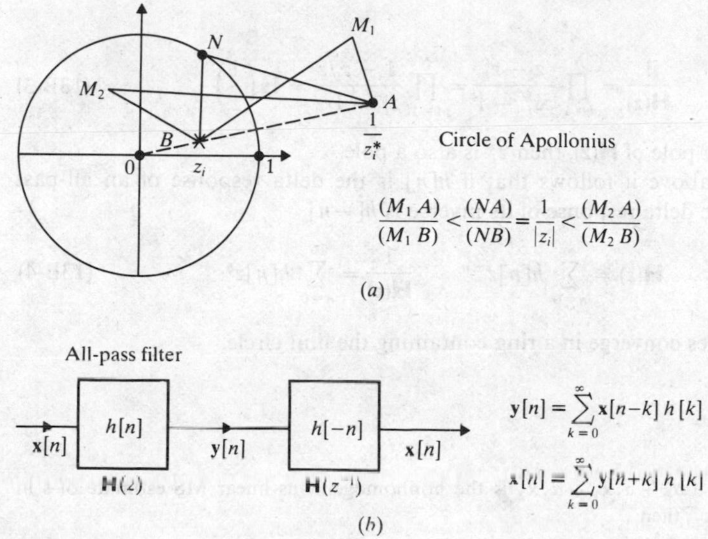

Figure 13-33

then $|\mathbf{F}(e^{j\varphi})| = 1$. Furthermore, $|\mathbf{F}(z)| > 1$ for $|z| < 1$ and $|\mathbf{F}(z)| < 1$ for $|z| > 1$ because $\mathbf{F}(z)$ is continuous and

$$|\mathbf{F}(0)| = \frac{1}{|z_i|} > 1 \qquad |\mathbf{F}(\infty)| = |z_i^*| < 1$$

Multiplying N bilinear fractions of the above form we conclude that, if

$$\mathbf{H}(z) = \prod_{i=1}^{N} \frac{zz_i^* - 1}{z - z_i} \qquad |z_i| < 1 \tag{13B-1}$$

then

$$|\mathbf{H}(z)| \begin{cases} > 1 & |z| < 1 \\ = 1 & |z| = 1 \\ < 1 & |z| > 1 \end{cases} \tag{13B-2}$$

A system with system function $\mathbf{H}(z)$ as in (13B-1) is called *all-pass*. Thus, an all-pass system is stable, causal, and

$$|\mathbf{H}(e^{j\omega T})| = 1$$

Furthermore

$$\frac{1}{\mathbf{H}(z)} = \prod_{i=1}^{N} \frac{z - z_i}{zz_i^* - 1} = \prod_{i=1}^{N} \frac{1 - z_i/z}{z_i^* - 1/z} = \mathbf{H}\left(\frac{1}{z}\right) \tag{13B-3}$$

because if z_i is a pole of $\mathbf{H}(z)$, then z_i^* is also a pole.

From the above it follows that if $h[n]$ is the delta response of an all-pass system, then the delta response of its inverse is $h[-n]$

$$\mathbf{H}(z) = \sum_{n=0}^{\infty} h[n]z^{-n} \qquad \frac{1}{\mathbf{H}(z)} = \sum_{n=0}^{\infty} h[n]z^{n} \tag{13B-4}$$

where both series converge in a ring containing the unit circle.

PROBLEMS

13-1 Show that, if $\alpha_0 + \alpha_1 \mathbf{x}_1 + \alpha_2 \mathbf{x}_2$ is the nonhomogeneous linear MS estimate of $\mathbf{s}$ in terms of $\mathbf{x}_1$ and $\mathbf{x}_2$, then

$$\hat{E}\{\mathbf{s} - \eta_s \mid \mathbf{x}_1 - \eta_1, \mathbf{x}_2 - \eta_2\} = \alpha_1(\mathbf{x} - \eta_1) + \alpha_2(\mathbf{x} - \eta_2)$$

13-2 If $\mathbf{y} = \mathbf{x}^3$, find the nonlinear and linear MS estimate of $\mathbf{y}$ in terms of $\mathbf{x}$ and the resulting MS errors.

13-3 Show that

$$\hat{E}\{\mathbf{y} \mid \mathbf{x}_1\} = \hat{E}\{\hat{E}\{\mathbf{y} \mid \mathbf{x}_1, \mathbf{x}_2\} \mid \mathbf{x}_1\}$$

13-4 If $R_s(\tau) = Ie^{-|\tau|/T}$ and

$$\hat{E}\{\mathbf{s}(t - T/2) \mid \mathbf{s}(t), \mathbf{s}(t - T)\} = a\mathbf{s}(t) + b\mathbf{s}(t - T)$$

find the constants a, b and the MS error.

13-5 Show that if $\hat{\mathbf{z}} = a\mathbf{s}(0) + b\mathbf{s}(T)$ is the MS estimate of

$$\mathbf{z} = \int_0^T \mathbf{s}(t)\, dt \qquad \text{then} \qquad a = b = \frac{\int_0^T R_s(\tau)\, d\tau}{R_s(0) + R_s(T)}$$

13-6 Show that if $\mathbf{x}(t) = \mathbf{s}(t) + \mathbf{v}(t)$, $R_{sv}(\tau) = 0$ and

$$\hat{E}\{\mathbf{s}'(t) \mid \mathbf{x}(t), \mathbf{x}(t - \tau)\} = a\mathbf{x}(t) + b\mathbf{x}(t - \tau)$$

then for small τ, $a = -b \simeq R''_{ss}(0)/\tau R''_{xx}(0)$.

13-7 The position of a particle in underdamped harmonic motion is a normal process with autocorrelation as in (11-100). Show that its conditional density assuming $\mathbf{x}(0) = x_o$ and $\mathbf{x}'(0) = \mathbf{v}(0) = v_o$ equals

$$f_{x(t)}(x \mid x_o, v_o) = \frac{1}{\sqrt{2\pi P}} e^{-(x - ax_o - bv_o)^2/2P}$$

Find the constants a, b, and P.

13-8 Show that, if $S_x(\omega) = 0$ for $|\omega| > \sigma = \pi/T$, then the linear MS estimate of $\mathbf{x}(t)$ in terms of its samples $\mathbf{x}(nT)$ equals

$$\hat{E}\{\mathbf{x}(t) \mid \mathbf{x}(nT),\ n = -\infty, \ldots, \infty\} = \sum_{n=-\infty}^{\infty} \frac{\sin(\sigma t - nT)}{\sigma t - n\pi} \mathbf{x}(nT)$$

and the MS error equals zero.

13-9 Show that if

$$\hat{E}\{\mathbf{s}(t+\lambda) \mid \mathbf{s}(t), \mathbf{s}(t-\tau)\} = \hat{E}\{\mathbf{s}(t+\lambda) \mid \mathbf{s}(t)\}$$

then $R_s(\tau) = Ie^{-\alpha|\tau|}$.

13-10 Given n independent RVs $\mathbf{x}_i$ with mean η and variance σ_i^2, find n constant a_i such that if $\mathbf{z} = a_1\mathbf{x}_1 + \cdots + a_n\mathbf{x}_n$, then $E\{\mathbf{z}\} = \eta$ and σ_z^2 is minimum.

13-11 A random sequence $\mathbf{x}_n$ is called *martingale* if

$$E\{\mathbf{x}_n \mid \mathbf{x}_{n-1}, \ldots, \mathbf{x}_1\} = \mathbf{x}_{n-1}$$

Show that if the RVs $\mathbf{y}_n$ are *independent*, then their sum $\mathbf{x}_n = \mathbf{y}_1 + \cdots + \mathbf{y}_n$ is a martingale.

13-12 A random sequence $\mathbf{x}_n$ is called *wide-sense martingale* if

$$\hat{\mathbf{E}}\{\mathbf{x}_n \mid \mathbf{x}_{n-1}, \ldots, \mathbf{x}_1\} = \mathbf{x}_{n-1}$$

(*a*) Show that a sequence $\mathbf{x}_n$ is WS martingale iff it can be written as a sum $\mathbf{x}_n = \mathbf{y}_1 + \cdots + \mathbf{y}_n$ where the RVs $\mathbf{y}_n$ are *orthogonal*.

(*b*) Show that if the sequence $\mathbf{x}_n$ is WS martingale, then

$$E\{\mathbf{x}_n^2\} \geq E\{\mathbf{x}_{n-1}^2\} \geq \cdots \geq E\{\mathbf{x}_1^2\}$$

Hint: $\mathbf{x}_n = \mathbf{x}_n - \mathbf{x}_{n-1} + \mathbf{x}_{n-1}$ and $\mathbf{x}_n - \mathbf{x}_{n-1} \perp \mathbf{x}_{n-1}$.

13-13 Find the noncausal estimators $H_1(\omega)$ and $H_2(\omega)$ respectively of a process $\mathbf{s}(t)$ and its derivative $\mathbf{s}'(t)$ in terms of the data $\mathbf{x}(t) = \mathbf{s}(t) + \mathbf{v}(t)$ where

$$R_s(\tau) = A\frac{\sin^2 a\tau}{\tau^2} \qquad R_v(\tau) = N\delta(\tau) \qquad R_{sv}(\tau) = 0$$

13-14 We denote by $H_s(\omega)$ and $H_y(\omega)$ respectively the noncausal estimators of the input $\mathbf{s}(t)$ and the output $\mathbf{y}(t)$ of the system $T(\omega)$ in terms of the data $\mathbf{x}(t)$ (Fig. P13-14). Show that $H_y(\omega) = H_x(\omega)T(\omega)$.

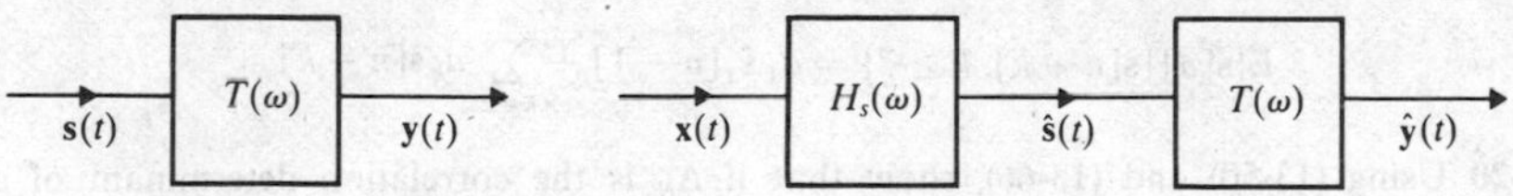

Figure P13-14

13-15 We wish to find the estimate $\hat{\mathbf{s}}(t)$ of the random telegraph signal $\mathbf{s}(t)$ in terms of the sum $\mathbf{x}(t) = \mathbf{s}(t) + \mathbf{v}(t)$ and its past, where

$$R_s(\tau) = e^{-2\lambda|\tau|} \qquad R_v(\tau) = N\delta(\tau) \qquad R_{sv}(\tau) = 0$$

Show that

$$\hat{\mathbf{s}}(t) = (c - 2\lambda)\int_0^\infty \mathbf{x}(t-\alpha)e^{-c\alpha}\,d\alpha \qquad c = 2\lambda\sqrt{1 + \frac{1}{\lambda N}}$$

13-16 Show that if $S(\omega) = 1/(1 + \omega^4)$, then the predictor of $\mathbf{s}(t)$ in terms of its entire past equals $\hat{\mathbf{s}}(t + \lambda) = b_0\,\mathbf{s}(t) + b_1\,\mathbf{s}'(t)$ where

$$b_0 = e^{-\lambda/\sqrt{2}}\cos\frac{\lambda}{\sqrt{2}} + \sin\frac{\lambda}{\sqrt{2}} \qquad b_1 = \sqrt{2}\,e^{-\lambda/\sqrt{2}}\sin\frac{\lambda}{\sqrt{2}}$$

13-17 (*a*) Find a function $h(t)$ satisfying the integral equation (Wiener–Hopf)

$$\int_0^\infty h(\alpha)R(\tau - \alpha)\,d\alpha = R(\tau + \ln 2) \qquad \tau \geq 0 \qquad R(\tau) = \frac{3}{2}e^{-\tau} + \frac{11}{3}e^{-3\tau}$$

(*b*) The function $\mathbf{H}(s)$ is rational with poles in the left-hand plane. The function $\mathbf{Y}(s)$ is rational with poles in the right-hand plane. Find $\mathbf{H}(s)$ and $\mathbf{Y}(s)$ if

$$[\mathbf{H}(s) - 2^s]\frac{49 - 25s^2}{9 - 10s^2 + s^4} = \mathbf{Y}(s)$$

(*c*) Discuss the relationship between (*a*) and (*b*).

13-18 (*a*) Find a sequence h_n satisfying the system

$$\sum_{k=0}^{\infty} h_k R_{m-k} = R_{m+1} \qquad m \geq 0 \qquad R_m = \frac{1}{2^m} + \frac{1}{3^m}$$

(*b*) The function $\mathbf{H}(z)$ is rational with poles in the unit circle. The function $\mathbf{Y}(z)$ is rational with poles outside the unit circle. Find $\mathbf{H}(z)$ and $\mathbf{Y}(z)$ if

$$[\mathbf{H}(z) - z]\frac{70 - 25(z + z^{-1})}{6(z + z^{-1})^2 - 35(z + z^{-1}) + 50} = \mathbf{Y}(z)$$

(*c*) Discuss the relationship between (*a*) and (*b*).

13-19 We have shown that the one-step predictor $\hat{\mathbf{s}}_1[n]$ of an AR process of order m in terms of its entire past equals [see (13-88)]

$$\hat{E}\{\mathbf{s}[n]\,|\,\mathbf{s}[n-k],\, k \geq 1\} = \sum_{k=1}^{m} a_k\,\mathbf{s}[n-k]$$

Show that its two-step predictor $\hat{\mathbf{s}}_2[n]$ is given by

$$\hat{E}\{\mathbf{s}[n]\,|\,\mathbf{s}[n-k],\, k \geq 2\} = a_1\,\hat{\mathbf{s}}_1[n-1] + \sum_{k=2}^{m} a_k\,\mathbf{s}[n-k]$$

13-20 Using (13-50) and (13-66), show that if Δ_N is the correlation determinant of a process $\mathbf{s}[n]$, then

$$\lim_{N\to\infty} \ln\frac{\Delta_{N+1}}{\Delta_N} = \lim_{N\to\infty}\frac{\ln \Delta_N}{N} = \frac{1}{2\sigma}\int_{-\sigma}^{\sigma} \ln S_s(\omega)\,d\omega$$

Hint:

$$\frac{1}{N}\sum_{n=1}^{N}\ln\frac{\Delta_{n+1}}{\Delta_n} = \frac{1}{N}\ln \Delta_{N+1} - \frac{1}{N}\ln \Delta_1 \to \lim_{N\to\infty}\ln\frac{\Delta_{N+1}}{\Delta_N}$$

13-21 Show that if $\hat{\varepsilon}_N[n]$ is the forward prediction error and $\check{\varepsilon}_N[n]$ is the backward prediction error of a process $\mathbf{s}[n]$, then (*a*) $\hat{\varepsilon}_N[n] \perp \hat{\varepsilon}_{N+m}[n+m]$, (*b*) $\check{\varepsilon}_N[n] \perp \check{\varepsilon}_{N+m}[n-m]$, (*c*) $\hat{\varepsilon}_N[n] \perp \check{\varepsilon}_{N+m}[n-N-m]$.

13-22 Find the predictor

$$\hat{\mathbf{s}}_N[n] = E\{\mathbf{s}[n] \mid \mathbf{s}[n-k],\ 1 \le k \le N\}$$

of a process $\mathbf{s}[n]$ and realize the error filter $\mathbf{E}_N(z)$ as an FIR filter (Fig. 13-12) and as a lattice filter (Fig. 13-15) for $N = 1$, 2, and 3 if

$$R_s[m] = \begin{cases} 5(3 - |m|) & |m| < 3 \\ 0 & |m| \ge 3 \end{cases}$$

13-23 The lattice filter of a process $\mathbf{s}[n]$ is shown in Fig. P13-23 for $N = 3$. Find the corresponding FIR filter for $N = 1$, 2, and 3 and the values of $R[m]$ for $|m| \le 3$ if $R[0] = 5$.

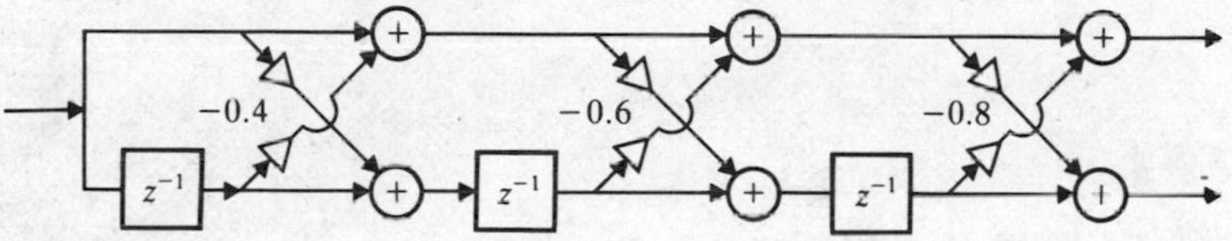

Figure P13-23

13-24 If $\mathbf{x}(t) = \mathbf{s}(t) + \mathbf{v}(t)$

$$R_s(\tau) = 5e^{-0.2|\tau|} \qquad R_v(\tau) = 5\delta(\tau) \qquad R_{sv}(\tau) = 0$$

Find the following MS estimates and the corresponding MS errors: (*a*) the noncausal filter of $\mathbf{s}(t)$; (*b*) the causal filter of $\mathbf{s}(t)$; (*c*) the estimate of $\mathbf{s}(t+2)$ in terms of $\mathbf{s}(t)$ and its past; (*d*) the estimate of $\mathbf{s}(t+2)$ in terms of $\mathbf{x}(t)$ and its past.

13-25 If $\mathbf{x}[n] = \mathbf{s}[n] + \mathbf{v}[n]$

$$R_s[m] = 5 \times 0.8^{|m|} \qquad R_v[m] = 5\delta[m] \qquad R_{sv}[m] = 0$$

Find the following MS estimates and the corresponding MS errors: (*a*) the noncausal filter of $\mathbf{s}[n]$; (*b*) the causal filter of $\mathbf{s}[n]$; (*c*) the estimate of $\mathbf{s}[n+1]$ in terms of $\mathbf{s}[n]$ and its past; (*d*) the estimate of $\mathbf{s}[n+1]$ in terms of $\mathbf{x}[n]$ and its past.

13-26 Find the Kalman estimate

$$\hat{\mathbf{s}}_0[n] = E\{\mathbf{s}[n] \mid \mathbf{s}[k] + \mathbf{v}[k],\ 0 \le k \le n\}$$

of $\mathbf{s}[n]$ and the MS error $P_n = E\{(\mathbf{s}[n] - \hat{\mathbf{s}}_0[n])^2\}$ if

$$R_s[m] = 5 \times 0.8^{|m|} \qquad R_v[m] = 5\delta[m] \qquad R_{sv}[m] = 0$$

13-27 Find the Kalman estimate

$$\hat{\mathbf{s}}_0(t) = E\{\mathbf{s}(t) \mid \mathbf{s}(\tau) + \mathbf{v}(\tau),\ 0 \le \tau \le t\}$$

of $\mathbf{s}(t)$ and the MS error $P(t) = E\{[\mathbf{s}(t) - \hat{\mathbf{s}}_0(t)]^2\}$ if

$$R_s(\tau) = 5e^{-0.2|\tau|} \qquad R_v(\tau) = 5\delta(\tau) \qquad R_{sv}(\tau) = 0$$

CHAPTER

FOURTEEN

SPECTRAL ESTIMATION

14-1 DETERMINISTIC DATA

In this chapter, we examine the problem of estimating the power spectrum

$$S(\omega) = \int_{-\infty}^{\infty} R(\tau)e^{-j\omega\tau}\, d\tau$$

of a random process $\mathbf{x}(t)$ in terms of specified data. The problem is presented in two parts depending on the available data:

I. *Deterministic* We are given the values of $R(\tau)$ for every τ in an interval $(-a, a)$ (Fig. 14-1*a*).

II. *Stochastic* We are given the values of a single sample of $\mathbf{x}(t)$ for every t in an interval $(-T, T)$ (Fig. 14-1*b*).

The discrete-time version of the problem is similar: We wish to estimate the power spectrum of a process $\mathbf{x}[n]$ in terms of specified data. The problem is deterministic if we are given the values of $R[m]$ for $|m| \le N$. It is stochastic if we are given M consecutive values of a single sample of $\mathbf{x}[n]$. To avoid duplication, we discuss either the discrete or the continuous form of each method.

The separation of the problem as above is a convenient idealization. In the applications, the function $R(\tau)$ is not known exactly; it is determined either as an ensemble average of the product $\mathbf{x}(t+\tau)\mathbf{x}(t)$ or as a time average from a single long sample of $\mathbf{x}(t)$. In the deterministic case, the restriction on the size of a is dictated not by the length of the sample, but by other measurement constraints. In the Michelson interferometer, for example, the time of observation is long but

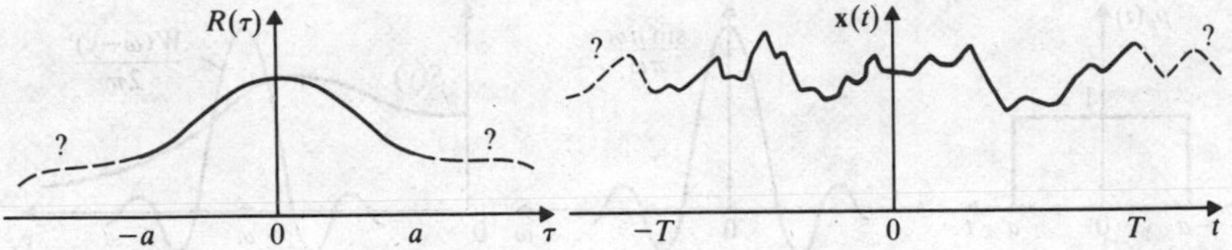

Figure 14-1

$R(\tau)$ can be determined only for $|\tau| < a$ where a is proportional to the maximum displacement of the moving mirror.

In this section, we deal with part I in its idealized form, ignoring the method by which $R(\tau)$ is obtained. Our problem is thus deterministic: we wish to find the Fourier transform of a p.d. function $R(\tau)$ knowing only a segment of $R(\tau)$. We shall present two methods of solution. In the first method (windows), we replace the unknown part of $R(\tau)$ by zero (Fig. 14-2*a*) and we use as the estimate of $S(\omega)$ the Fourier transform of the product $R(\tau)w(\tau)$ involving the known part of $R(\tau)$ and a factor $w(\tau)$ introduced to reduce the error due to the truncation of $R(\tau)$. In the second method (extrapolation), we assume that $S(\omega)$ belongs to a class of functions that can be specified in terms of certain parameters, and we use the given data to determine these parameters. The choice of this class depends on prior knowledge about $S(\omega)$ and on the complexity of the underlying equations. We shall consider three such classes: (*a*) bandlimited spectra; (*b*) all-pole models; (*c*) spectra consisting of lines.

Windows

Suppose that we use as the estimate of $S(\omega)$ the Fourier transform $S_a(\omega)$ of the given segment of $R(\tau)$

$$S_a(\omega) = \int_{-a}^{a} R(\tau)e^{-j\omega\tau}\, d\tau \tag{14-1}$$

Clearly, $S_a(\omega)$ is the transform of the product $R(\tau)p_a(\tau)$ where $p_a(\tau)$ is a rectangular pulse (Fig. 14-3*a*). Since

$$p_a(t) = \begin{cases} 1 & |t| < a \\ 0 & |t| > a \end{cases} \longleftrightarrow \frac{2 \sin a\omega}{\omega} = P(\omega)$$

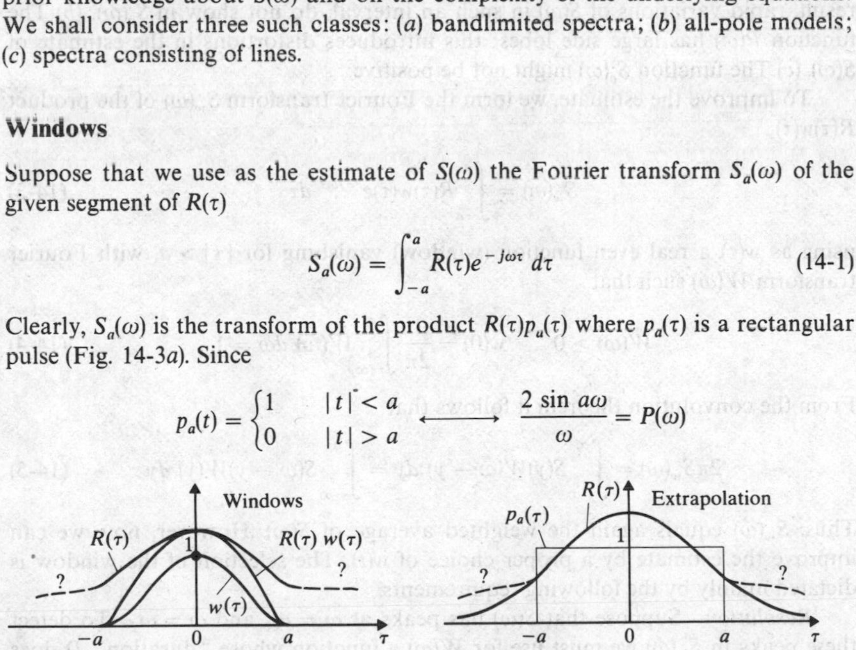

Figure 14-2

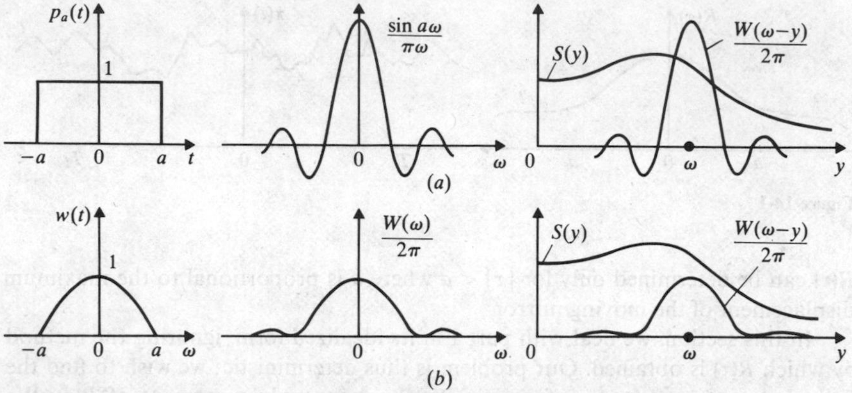

Figure 14-3

it follows from the frequency convolution theorem that

$$S_a(\omega) = \frac{1}{2\pi} S(\omega) * P(\omega) = \int_{-\infty}^{\infty} S(y) \frac{\sin a(\omega - y)}{\pi(\omega - y)} dy \tag{14-2}$$

This estimate is not good for several reasons: (*a*) The function $S_a(\omega)$ is the weighted average of the values of $S(\omega)$ in an interval of the order of $2\pi/a$. As a result, rapid variations of $S(\omega)$ in such an interval, do not show in $S_a(\omega)$. (*b*) The function $P(\omega)$ has large side lobes; this introduces distortions in the estimate of $S(\omega)$. (*c*) The function $S_a(\omega)$ might not be positive.

To improve the estimate, we form the Fourier transform $S_w(\omega)$ of the product $R(\tau)w(\tau)$

$$S_w(\omega) = \int_{-a}^{a} R(\tau)w(\tau)e^{-j\omega\tau} d\tau \tag{14-3}$$

using as $w(\tau)$ a real even function (window) vanishing for $|\tau| > a$, with Fourier transform $W(\omega)$ such that

$$W(\omega) \geq 0 \qquad w(0) = \frac{1}{2\pi} \int_{-\infty}^{\infty} W(\omega)\, d\omega = 1 \tag{14-4}$$

From the convolution theorem it follows that

$$2\pi S_w(\omega) = \int_{-\infty}^{\infty} S(y)W(\omega - y)\, dy = \int_{-\infty}^{\infty} S(\omega - y)W(y)\, dy \tag{14-5}$$

Thus, $S_w(\omega)$ equals again the weighted average of $S(\omega)$. However, now we can improve the estimate by a proper choice of $w(t)$. The selection of the window is dictated mainly by the following requirements.

Resolution Suppose that $S(\omega)$ has peaks at $\omega = \omega_1$ and $\omega = \omega_2$. To detect these peaks in $S_w(\omega)$ we must use for $W(\omega)$ a function whose "duration" D does not exceed $\omega_2 - \omega_1$. A reasonable measure of D is the width of the main lobe of

$W(\omega)$. We should note that the critical factor that determines the width of the main lobe of $W(\omega)$ is not the shape of $w(t)$ but the length $2a$ of the data interval (uncertainty principle).

Leakage If one of the peaks $S(\omega_1)$ of $S(\omega)$ is small compared to a neighboring peak $S(\omega_2)$, it might not show in $S_a(\omega)$ even if $\omega_2 - \omega_1$ exceeds D. The extent of this interference depends on the size of the side lobes of $W(\omega)$ and can be reduced if $W(\omega)$ attenuates rapidly as ω increases.

The appropriate choice of $w(t)$ is based on the following property of Fourier transforms: If $w(t) \leftrightarrow W(\omega)$ and $w(t)$ is an ordinary function (no impulses), then $W(\omega) \to 0$ as $\omega \to \infty$. If $w(t)$ is continuous, then $\omega W(\omega) \to 0$ as $\omega \to \infty$. If $w^{(k)}(t)$ exists and is continuous, then $\omega^{k+1} W(\omega) \to 0$ as $\omega \to \infty$. From the above it follows that, if we wish $W(\omega)$ to approach zero faster than $1/\omega^{k+1}$, we must select $w(t)$ such that $w^{(k)}(t)$ is continuous. And since $w(t) = 0$ for $|t| > a$, this requires that $w(t)$ and its first k derivatives vanish at the endpoints of the interval $(-a, a)$

$$w(\pm a) = w'(\pm a) = \cdots = w^{(k)}(\pm a) = 0 \tag{14-6}$$

In Table 14-1, we show a number of typical window pairs. As we see from the figures, these examples satisfy (14-6) for $k = 0$, 1, and 2, and their main lobe is of the order of π/a.

We called the last pair minimum bias for the following reason†: Suppose that $W(\omega)$ is smooth in the sense that it can be approximated by a parabola in any interval of the order of π/a:

$$S(\omega - y) \simeq S(\omega) - yS'(\omega) + \frac{y^2}{2} S''(\omega)$$

Inserting into (14-5) and using (14-4) and the evenness of $W(\omega)$, we conclude that the estimation error $S(\omega) - S_a(\omega)$ equals

$$S(\omega) - \frac{1}{2\pi} \int_{-\infty}^{\infty} S(y) W(\omega - y)\, dy \simeq \frac{S''(\omega)}{4\pi} \int_{-\infty}^{\infty} y^2 W(y)\, dy$$

From this it follows that to minimize the error, we must select $W(\omega)$ such that its second moment is minimum subject to the constraint (14-4). As we show in the cited reference, this is the case if

$$w(t) = \frac{2}{a} \left[\cos \frac{2\pi t}{a}\, p_{a/2}(t) \right] * \left[\cos \frac{2\pi t}{a}\, p_{a/2}(t) \right] \tag{14-7}$$

Performing the convolution, we obtain the last example in Table 14-1.

A general class of windows In certain applications, it is required that $W(\omega)$ have certain special properties, specified zeros, for example. The appropriate $W(\omega)$ cannot, however, be selected at will. It must belong to the class of functions with time-limited inverse. It would be desirable, therefore, to know whether a

† A. Papoulis: "Minimum Bias Windows for High Resolution Spectral Estimates," *IEEE Transactions on Information Theory*, vol. IT-19, pp. 9–12, 1973.

Table 14-1

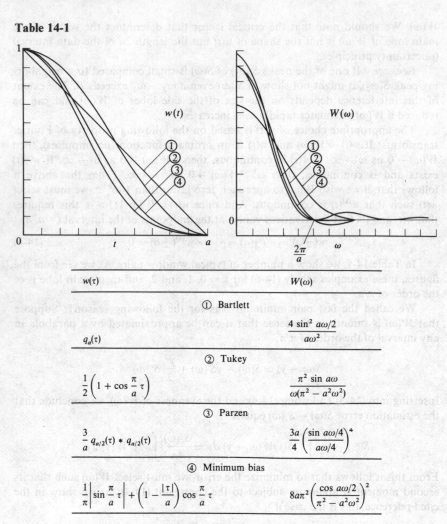

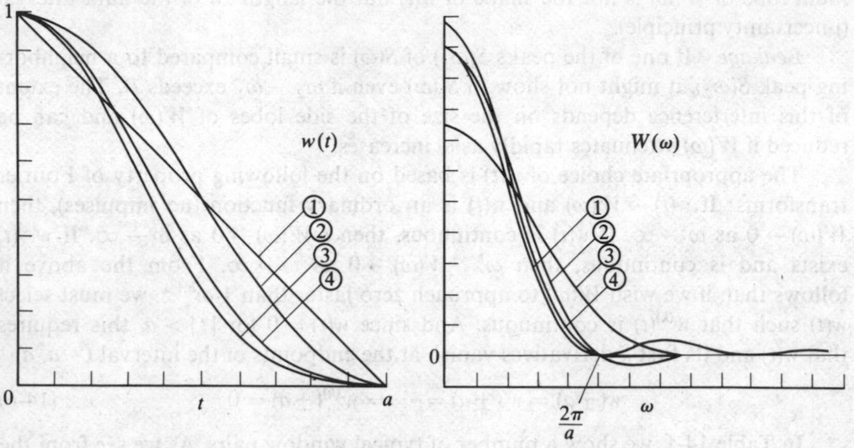

$w(\tau)$	$W(\omega)$
① Bartlett	
$q_a(\tau)$	$\dfrac{4\sin^2 a\omega/2}{a\omega^2}$
② Tukey	
$\dfrac{1}{2}\left(1+\cos\dfrac{\pi}{a}\tau\right)$	$\dfrac{\pi^2 \sin a\omega}{\omega(\pi^2 - a^2\omega^2)}$
③ Parzen	
$\dfrac{3}{a}\, q_{a/2}(\tau) * q_{a/2}(\tau)$	$\dfrac{3a}{4}\left(\dfrac{\sin a\omega/4}{a\omega/4}\right)^4$
④ Minimum bias	
$\dfrac{1}{\pi}\left\lvert\sin\dfrac{\pi}{a}\tau\right\rvert + \left(1-\dfrac{\lvert\tau\rvert}{a}\right)\cos\dfrac{\pi}{a}\tau$	$8a\pi^2\left(\dfrac{\cos a\omega/2}{\pi^2 - a^2\omega^2}\right)^2$

given $W(\omega)$ belongs to this class without actually computing its inverse. The next theorem gives us a criterion.

The Paley–Wiener theorem† If $\mathbf{W}(s)$ is an entire function (i.e., it is analytic for every s) and there exist two constants A and a such that

$$|\mathbf{W}(s)| < Ae^{a|s|} \qquad \text{for all } s \tag{14-8}$$

then its inverse transform $w(t)$ is time-limited:

$$w(t) = 0 \qquad \text{for} \qquad |t| > a$$

† R. P. Boas: *Entire Functions*, Academic Press, New York, 1954.

The proof of this difficult theorem is given in P.2.

Example 14-1 The windows of Table 14-1 are special cases of the following class of time-limited functions: Suppose that $N(\omega)$ is an even or odd polynomial in ω, and $P(\omega)$ is an even or odd polynomial in $e^{j\omega T}$ whose highest harmonic equals $e^{jM\omega T}$. If

$$W(\omega) = \frac{N(\omega)P(\omega)}{(\omega^2 - \omega_1^2) \cdots (\omega^2 - \omega_n^2)} \tag{14-9}$$

and $P(\omega)$ is such that $P(\pm\omega_i) = 0$, then the function $\mathbf{W}(s)$ is entire and it satisfies the Paley–Wiener condition (14-8) with $a = MT$. Hence, its inverse $w(t)$ is zero for $|t| > MT$.

Bandlimited Extrapolation

We shall extrapolate $R(\tau)$ under the assumption that

$$S(\omega) = 0 \qquad \text{for} \qquad |\omega| > \sigma \tag{14-10}$$

In this case, $R(\tau)$ is an entire function [see (11-60)], hence it can be expanded into a Taylor series converging for every τ. This shows that a BL function is in principle uniquely determined from its values in a finite interval no matter how small. However, in a real problem, the derivatives of $R(\tau)$ cannot be computed exactly. Furthermore, its Taylor series expansion must be truncated causing, thus, unacceptable errors for large τ. We give next two methods of extrapolation that in certain cases lead to realistic results.

Prolate spheroidal functions It is well known that (see P.2) the eigenfunctions $\varphi_n(t)$ of the integral equation

$$\int_{-a}^{a} \varphi_n(\tau) \frac{\sin \sigma(t-\tau)}{\pi(t-\tau)} \, d\tau = \lambda_n \varphi_n(t) \tag{14-11}$$

are orthogonal in the intervals $(-\infty, \infty)$ and $(-a, a)$

$$\int_{-\infty}^{\infty} \varphi_n(t)\varphi_k(t) \, dt = \delta[n-k] \qquad \int_{-a}^{a} \varphi_n(t)\varphi_k(t) \, dt = \lambda_n \delta[n-k] \tag{14-12}$$

Furthermore, these functions form a complete set in the class of BL functions.

From the above it follows that $R(\tau)$ can be expanded into a series

$$R(\tau) = \sum_{n=0}^{\infty} c_n \varphi_n(\tau) \qquad \text{all } \tau \tag{14-13}$$

where [see (14-12)]

$$c_n = \frac{1}{\lambda_n} \int_{-a}^{a} R(\tau)\varphi_n(\tau) \, d\tau \tag{14-14}$$

Thus, c_n can be computed if $R(\tau)$ is known for $|\tau| < a$. Inserting into (14-13), we obtain $R(\tau)$ for all τ.

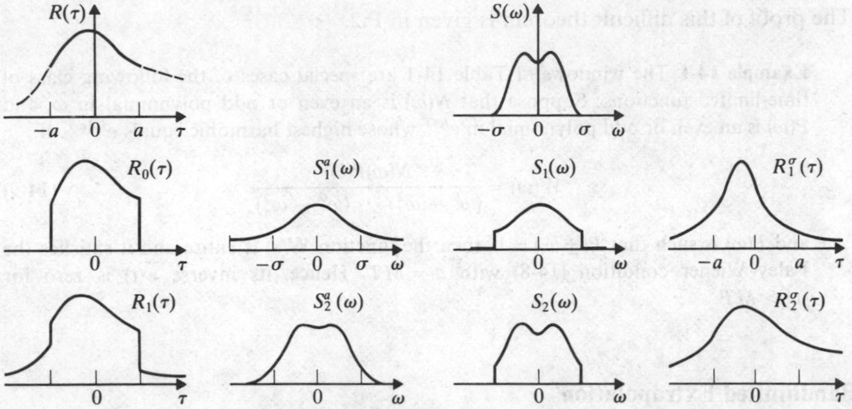

Figure 14-4

An iteration method† In the problem under consideration, the available information about the pair $R(\tau) \leftrightarrow S(\omega)$ is given partially in the time domain: $R(\tau)$ is known for $|\tau| < a$, and partially in the frequency domain: $S(\omega)$ is known for $|\omega| > \sigma$; in fact, $S(\omega) = 0$ for $|\omega| > \sigma$. We shall present an iteration method for determining $S(\omega)$ based on successive utilization of this information at each iteration step.

Step one We form the Fourier transform $S_1^a(\omega)$ of the given segment $R_0(\tau) = R(\tau)p_a(\tau)$ of $R(\tau)$. We know that $S_1^a(\omega)$ is not the true spectrum because it is not *BL*. Replacing its values for $|\omega| > \sigma$ by zero, we form the function

$$S_1(\omega) = S_1^a(\omega)p_\sigma(\omega) \tag{14-15}$$

and its inverse transform $R_1^\sigma(\tau)$. This completes the first iteration (Fig. 14-4).

Step two We know that $R_1^\sigma(\tau)$ is not the true correlation because for $|\tau| < a$ it is not equal to the given segment of $R(\tau)$. Replacing that segment of $R_1^\sigma(\tau)$ with the known part of $R(\tau)$, we obtain the function

$$R_1(\tau) = R_1^\sigma(\tau) + [R(\tau) - R_1^\sigma(\tau)]p_a(\tau) = \begin{cases} R(\tau) & |\tau| < a \\ R_1^\sigma(\tau) & |\tau| > a \end{cases} \tag{14-16}$$

We next form the transform $S_2^a(\omega)$ of $R_1(\tau)$, we truncate it forming the function $S_2(\omega)$, we take the inverse transform $R_2^\sigma(\tau)$, and we correct it as in (14-16) forming the function $R_2(\tau)$. And so we continue.

Thus, at the nth iteration step, we correct in the time domain and truncate in the frequency domain. These operations are shown in Fig. 14-4 and can be described by a single equation

$$R_n(\tau) = \{R_n^\sigma(\tau) + [R(\tau) - R_n^\sigma(\tau)]p_a(\tau)\} * \frac{\sin \sigma\tau}{\pi\tau} \tag{14-17}$$

† A. Papoulis: "A New Method of Image Restoration," Joint Services Technical Advisory Committee, Report no. 39, 1973–74. R. W. Gerchberg, "Super-Resolution through Error Energy Reduction," *Optica Acta*, vol. 21, 1974.

The use of the available information in the τ and ω domain results in a reduction of the energy of the error $S(\omega) - S_n(\omega)$ twice at each iteration step. This follows from Parseval's formula

$$\frac{1}{2\pi}\int_{-\sigma}^{\sigma} [S(\omega) - S_n(\omega)]^2 \, d\omega = \int_{-\infty}^{\infty} [R(\tau) - R_n^{\sigma}(\tau)]^2 \, d\tau$$

$$> \int_{-\infty}^{\infty} [R(\tau) - R_n(\tau)]^2 \, d\tau$$

$$= \frac{1}{2\pi}\int_{-\infty}^{\infty} [S(\omega) - S_n^{a}(\omega)]^2 \, d\omega$$

$$> \frac{1}{2\pi}\int_{-\sigma}^{\sigma} [S(\omega) - S_{n+1}(\omega)]^2 \, d\omega \qquad (14\text{-}18)$$

The above suggest that $S_n(\omega)$ tends to $S(\omega)$ as $n \to \infty$. It can be shown that† this is, indeed, the case.

Maximum Entropy (ME) and Line Spectra‡

The following two methods are based on the results of Sec. 13-4. As we know [see Kolmogoroff's formula (13-50)], the MS error of the estimate

$$\hat{\mathbf{s}}[n] = \hat{E}\{\mathbf{s}[n] \mid \mathbf{s}[n-k],\ k \geq 1\} \qquad (14\text{-}19)$$

of $\mathbf{s}[n]$ in terms of its entire past equals

$$P = \exp\left\{\frac{1}{2\sigma}\int_{-\sigma}^{\sigma} \ln S(\omega)\, d\omega\right\} \qquad (14\text{-}20)$$

where

$$S(\omega) = \sum_{m=-\infty}^{\infty} R[m] e^{-jm\omega T} \qquad (14\text{-}21)$$

is the power spectrum of $\mathbf{s}[n]$. In the method of ME (abbreviated MEM), the objective is to maximize the integral in (14-20) subject to the constraint that the $N + 1$ values

$$R[0], R[1], \ldots, R[N] \qquad (14\text{-}22)$$

of $R[m]$ are given (see Sec. 15-4). It is clear from (14-20) that the integral is maximum if the MS error P is maximum. This shows that the MEM can be interpreted as follows:

† A. Papoulis: "A New Algorithm in Spectral Analysis and Bandlimited Extrapolation," *IEEE Transactions on Circuits and Systems*, vol. CAS-22, 1975.

‡ A. Papoulis: "Maximum Entropy and Spectral Estimation: A Review," *IEEE Transactions on Acoustics, Speech, and Signal Processing*, vol. ASSP-29, 1981.

Consider the class C_N of all WSS processes whose autocorrelations equal the given data (14-22). All these processes have the same predictor $\hat{\mathbf{s}}_N[n]$ of order N and the same MS predictor error P_N because P_N depends only on the values of $R[m]$ for $m \leq N$ [see (13-65)]. However, the infinite-past MS prediction error P is not the same because P depends on $R[m]$ for $m > N$. And since

$$P \leq P_N \tag{14-23}$$

we conclude that P is maximum if the unspecified values of $R[m]$ are such that $P = P_N$. As we know, this is the case if the corresponding process $\mathbf{s}[n]$ is AR as in (13-89)

$$\mathbf{s}[n] - a_1\mathbf{s}[n-1] - \cdots - a_N\mathbf{s}[n-N] = \zeta[n] \tag{14-24}$$

Thus, the MEM estimate $S_{\text{MEM}}(\omega)$ of the unknown spectrum equals the spectrum of an autoregressive process:†

$$S_{\text{MEM}}(\omega) = \frac{P_N}{|1 - a_1 e^{-j\omega T} - \cdots - a_N e^{-jN\omega T}|^2} \tag{14-25}$$

as in (13-90) and it is specified in terms of the $N+1$ parameters P_N and a_k. As we know, these parameters satisfy the Yule–Walker equations (13-65). They can, therefore, be determined either directly from (13-65) or, recursively, from Levinson's algorithm (13-78).

The corresponding autocorrelation is a sum as in (13-92)

$$R_{\text{MEM}}[m] = \sum_{i=1}^{N} c_i z_i^{|m|} \qquad |z_i| < 1 \tag{14-26}$$

where z_i are the roots of the characteristic polynomial

$$\mathbf{E}_N(z) = 1 - a_1 z^{-1} - \cdots - a_N z^{-N} \tag{14-27}$$

The MEM estimate $R_{\text{MEM}}[m]$ of the unknown part $m > N$ of the correlation $R[m]$ can be determined also from the extrapolation formula (13-91)

$$R_{\text{MEM}}[m] = \sum_{k=1}^{N} a_k R_{\text{MEM}}[m-k] \tag{14-28}$$

The MEM has a solution only if $P_N > 0$ or, equivalently, if the correlation determinant Δ_{N+1} is strictly positive. In this case, all the roots z_i of $\mathbf{E}_N(z)$ are inside the unit circle and the spectrum $S_{\text{MEM}}(\omega)$ is positive and finite. It has sharp peaks if $|z_i| \simeq 1$ as in Fig. 14-5a, however, it does not contain lines. If $\Delta_{N+1} = 0$, then the MEM has no meaningful solution. However, the given data $R[m]$ satisfy again (14-26) where now all the roots z_i are on the unit circle.

† We should point out that the approximation of the unknown spectrum by a rational function can be improved if the underlying process is ARMA. However, unlike the AR case, the evaluation of its unknown parameters is not, in general, simple. See, for example, J. A. Cadzow and R. L. Moses: "An Adaptive ARMA Spectral Estimator, Parts 1 and 2," *Proceedings 1st ASSP Workshop on Spectral Estimation*, McMarten University, Canada, 1982.

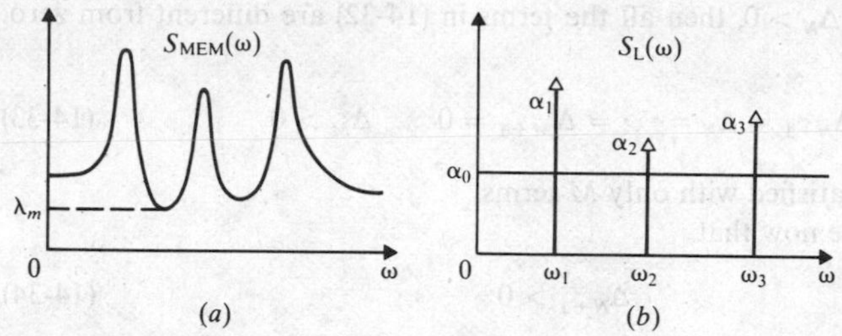

Figure 14-5

We note, finally, that the MEM can be viewed as a method of fitting the data $R[m]$ by a sum of the form (14-26) where the $2N$ parameters c_i and z_i are such that the sum is a positive definite sequence and it satisfies the $N + 1$ constraints (14-22).

Line spectra and Caratheodory's theorem We shall now show that the data can be fitted by the sum

$$R_L[m] = \alpha_0 \, \delta[m] + \sum_{i=1}^{N} \alpha_i \, e^{jm\omega_i T} \tag{14-29}$$

where ω_i are N real constants and α_i are $N + 1$ nonnegative constants. In other words, the N values $R[1], \ldots, R[N]$ of an arbitrary autocorrelation can be fitted by a sum of sines and cosines. The corresponding spectrum equals

$$S_L(\omega) = \alpha_0 + \frac{2\pi}{T} \sum_{i=1}^{N} \alpha_i \, \delta(\omega - \omega_i) \qquad |\omega| < \frac{\pi}{T} \tag{14-30}$$

(Fig. 14-5*b*). Thus, the above can be viewed as a method of estimating the unknown spectrum $S(\omega)$ by a spectrum consisting only of lines as in (13-96).

PROOF The representation of $R[m]$ as a sum of the form (14-29) is due to Caratheodory†. We shall derive it here, however, as a simple consequence of the properties of predictable processes.

Case 1 If

$$\Delta_{N+1} = 0 \tag{14-31}$$

then the process $\mathbf{s}[n]$ is predictable and its spectrum consists of lines as in (13-96). This shows that the sum

$$R_L[m] = \sum_{i=1}^{N} \alpha_i \, e^{jm\omega_i T} \tag{14-32}$$

fits the data and it establishes (14-29) with $\alpha_0 = 0$.

† U. Grenander and G. Szego, *Toeplitz Forms and Their Applications*, Berkley University Press, 1958. See also V. F. Pisarenko, "The Retrieval of Harmonics," *Geophysical Journal of the Royal Astronomical Society*, 1973.

We note that if $\Delta_N > 0$, then all the terms in (14-32) are different from zero. If, however,

$$\Delta_{N+1} = \Delta_N = \cdots = \Delta_{M+1} = 0 \qquad \Delta_M > 0 \tag{14-33}$$

then (14-29) can be satisfied with only M terms.

Case 2 Suppose now that

$$\Delta_{N+1} > 0 \tag{14-34}$$

We shall show that, by a mere reduction of $R[0]$, this case can be reduced to (14-31). For this purpose, we form a new set of data

$$R_\alpha[m] = R[m] - \alpha\delta[m] = \begin{cases} R[m] & m > 0 \\ R[0] - \alpha & m = 0 \end{cases} \tag{14-35}$$

The corresponding determinant equals

$$\Delta_{N+1}(\alpha) = \begin{bmatrix} R[0]-\alpha & R[1] & \cdots & R[N] \\ R[1] & R[0]-\alpha & \cdots & R[N-1] \\ \cdots & \cdots & \cdots & \cdots \\ R[N] & R[N-1] & \cdots & R[0]-\alpha \end{bmatrix} \tag{14-36}$$

This determinant is a continuous function of α and for $\alpha = 0$ it is positive. If, therefore, we increase α starting from zero, we will reach a value α_0 such that

$$\Delta_{N+1}(\alpha_0) = 0 \qquad \Delta_{N+1}(\alpha) > 0 \qquad 0 \le \alpha < \alpha_0 \tag{14-37}$$

The matrix of the corresponding data $R[m] - \alpha_0\, \delta[m]$ is, therefore, nonnegative definite and its determinant is zero. Hence (case 1), $R[m] - \alpha_0\, \delta[m]$ can be written as a sum of the form (14-32). This completes the proof.

We note that α_0 is the minimum eigenvalue of Δ_{N+1}. If it is simple, then all terms of the sum in (14-29) are different from zero. If, however, it has multiplicity $N - M + 1$, then, as in case 1 [see (14-33)], only M terms will be different from zero.

Note We denote by λ_m the minimum of $S_{\text{MEM}}(\omega)$. Since $S_{\text{MEM}}(\omega) - \lambda > 0$ for $0 \le \lambda < \lambda_m$, we conclude that the correlation matrix of the data $R[m] - \lambda\, \delta[m]$ is positive definite. This leads to the conclusion that $\alpha_0 \ge \lambda_m$.

Hidden periodicities If $\mathbf{v}[n]$ is white noise and $\mathbf{c}_i$ are mutually orthogonal RVs, then the spectrum of the process

$$\mathbf{s}[n] = \mathbf{v}[n] + \sum_{i=1}^{N} \mathbf{c}_i\, e^{jn\omega_i T} \tag{14-38}$$

is of the form of $S_L(\omega)$. Thus, the above method can be used to detect hidden periodicities in the presence of white noise.

14-2 STOCHASTIC DATA

We are given a segment $\mathbf{x}_T(t)$ of a sample of a real process $\mathbf{x}(t)$ and we wish to estimate its power spectrum $S(\omega)$. For this purpose, we estimate first the autocorrelation $R(\tau)$ of $\mathbf{x}(t)$. To do so, we write $R(\tau)$ in a symmetrical form

$$R(\tau) = E\left\{\mathbf{x}\left(t + \frac{\tau}{2}\right)\mathbf{x}\left(t - \frac{\tau}{2}\right)\right\} \tag{14-39}$$

and use the product $\mathbf{x}(t + \tau/2)\mathbf{x}(t - \tau/2)$ as the integrand in (9-152). However, since $\mathbf{x}(t)$ is known for $|t| < T$ only, the above product is known for $|t| < T - |\tau|/2$ (Fig. 14-6). We must, therefore, change the limits of integration accordingly. The interval of integration is now $2T - |\tau|$ and the resulting estimate $\mathbf{R}^T(\tau)$ of $R(\tau)$ equals

$$\mathbf{R}^T(\tau) = \frac{1}{2T - |\tau|}\int_{-T+|\tau|/2}^{T-|\tau|/2} \mathbf{x}\left(t + \frac{\tau}{2}\right)\mathbf{x}\left(t - \frac{\tau}{2}\right) dt \tag{14-40}$$

The Fourier transform $\mathbf{S}^T(\omega)$ of $\mathbf{R}^T(\tau)$

$$\mathbf{S}^T(\omega) = \int_{-2T}^{2T} \mathbf{R}^T(\tau)e^{-j\omega\tau}\, d\tau \tag{14-41}$$

is not a reliable estimate of $S(\omega)$. Although $\mathbf{R}^T(\tau)$ is an unbiased estimate of $R(\tau)$

$$E\{\mathbf{R}^T(\tau)\} = R(\tau) \qquad |\tau| < 2T \tag{14-42}$$

$\mathbf{S}^T(\omega)$ is not an unbiased estimate of $S(\omega)$ because its inverse is zero for $|\tau| > 2T$. However, this is not the main problem. In (14-40), the length of integration equals $2T - |\tau|$ and it approaches zero as $|\tau|$ approaches $2T$. Hence [see (9-140)], the variance of $\mathbf{R}^T(\tau)$ is large for $|\tau|$ close to $2T$. And since all values of τ are used in (14-41), the variance of $\mathbf{S}^T(\omega)$ is large for *any* T. To reduce this effect, we multiply $\mathbf{R}^T(\tau)$ by a weight (lag window) that approaches zero as $|\tau|$ approaches the constant $2T$. This reduces the variance but it increases the bias of the estimate. In the following, we develop the underlying theory using as estimate of $R(\tau)$ a modified form of (14-40)

$$\mathbf{R}_T(\tau) = \frac{1}{2T}\int_{-T+|\tau|/2}^{T-|\tau|/2} \mathbf{x}\left(t + \frac{\tau}{2}\right)\mathbf{x}\left(t - \frac{\tau}{2}\right) dt \tag{14-43}$$

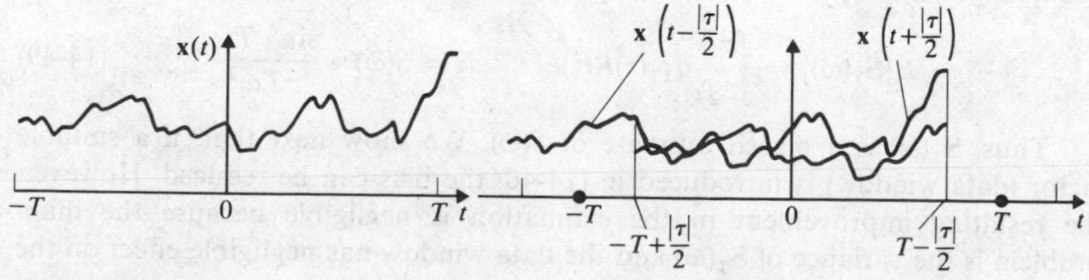

Figure 14-6

We do so because, unlike $\mathbf{S}^T(\omega)$, the Fourier transform of $\mathbf{R}_T(\tau)$

$$\mathbf{S}_T(\omega) = \int_{-2T}^{2T} \mathbf{R}_T(\tau)e^{-j\omega\tau}\, d\tau \tag{14-44}$$

can be expressed directly in terms of $\mathbf{x}_T(t)$. Indeed, the integral in (14-43) equals the convolution of $\mathbf{x}_T(t)$ with $\mathbf{x}_T(-t)$, hence,

$$\mathbf{R}_T(\tau) = \frac{1}{2T}\, \mathbf{x}_T(t) * \mathbf{x}_T(-t) \tag{14-45}$$

From this and the convolution theorem it follows that

$$\mathbf{S}_T(\omega) = \frac{1}{2T}\, |\mathbf{X}_T(\omega)|^2 \qquad \mathbf{X}_T(\omega) = \int_{-T}^{T} \mathbf{x}(t)e^{-j\omega t}\, dt \tag{14-46}$$

The process $\mathbf{S}_T(\omega)$ so formed is called *sample spectrum* or *periodogram.*

We note that, whereas $\mathbf{S}^T(\omega)$ might be negative, all samples of the periodogram $\mathbf{S}_T(\omega)$ are positive for every ω. From this it follows that, unlike $\mathbf{R}^T(\tau)$, all samples of the weighted estimate $\mathbf{R}_T(\tau)$ of $R(\tau)$ are *positive definite.* It is true, of course, that the estimate is biased because

$$E\{\mathbf{R}_T(\tau)\} = \left(1 - \frac{|\tau|}{2T}\right) R(\tau) \qquad |\tau| < 2T \tag{14-47}$$

However, the error due to the factor $1 - |\tau|/2T$ is small in the region $|\tau| \ll T$ of interest. Furthermore, it is partly offset by a reduction in the variance of the estimate.

Note For a specific τ, no matter how large, the estimate $\mathbf{R}_T(\tau)$ tends to $R(\tau)$ as $T \to \infty$. However, its transform $\mathbf{S}_T(\omega)$ does not tend to the transform $S(\omega)$ of $R(\tau)$. The reason is that the convergence of $\mathbf{R}_T(\tau)$ to $R(\tau)$ is not uniform in τ, that is, there is no T, no matter how large, such that $\mathbf{R}_T(\tau) \simeq R(\tau)$ for every τ.

Bias We shall use the notation $p_a(t)$ for a rectangular pulse and $q_a(t)$ for a triangular pulse. Thus

$$q_{2T}(\tau) = \left(1 - \frac{|\tau|}{2T}\right) p_{2T}(\tau) \leftrightarrow \frac{2\sin^2 T\omega}{T\omega^2} \tag{14-48}$$

From (14-47) and the convolution theorem it follows that

$$E\{\mathbf{S}_T(\omega)\} = \int_{-2T}^{2T} q_{2T}(\tau)R(\tau)e^{-j\omega\tau}\, d\tau = S(\omega) * \frac{\sin^2 T\omega}{\pi T\omega^2} \tag{14-49}$$

Thus, $\mathbf{S}_T(\omega)$ is a biased estimate of $S(\omega)$. We show next that, if a suitable factor (data window) is introduced in (14-46) the bias can be reduced. However, the resulting improvement in the estimation is negligible because the main problem is the variance of $\mathbf{S}_T(\omega)$ and the data window has negligible effect on the variance.

Data window We form the weighted sample spectrum

$$\mathbf{S}_c(\omega) = \frac{1}{2T} |\mathbf{X}_c(\omega)|^2 \qquad \mathbf{X}_c(\omega) = \int_{-T}^{T} c(t)\mathbf{x}(t)e^{-j\omega t}\, dt \tag{14-50}$$

where $c(t)$ is a real even function (data window) with Fourier transform $C(\omega)$ and such that $c(t) = 0$ for $|t| > T$. We maintain that

$$E\{\mathbf{S}_c(\omega)\} = \frac{1}{4\pi T} S(\omega) * C^2(\omega) \tag{14-51}$$

PROOF The autocorrelation of the product $c(t)\mathbf{x}_T(t)$ is the function

$$c(t_1)c(t_2)R(t_1 - t_2)$$

and its Fourier transform equals

$$\Gamma_c(u, v) = \int_{-T}^{T}\int_{-T}^{T} c(t_1)c(t_2)R(t_1 - t_2)e^{-j(ut_1 + vt_2)}\, dt_1\, dt_2 \tag{14-52}$$

From the above and the two-dimensional convolution theorem (see P.4) it follows with some effort that

$$\Gamma_c(u, v) = \frac{1}{2\pi}\int_{-\infty}^{\infty} C(u - \alpha)C(v + \alpha)S(\alpha)\, d\alpha \tag{14-53}$$

Since $C(-\omega) = C(\omega)$ the above yields [see (10-159)]

$$E\{|\mathbf{X}_c(\omega)|^2\} = \Gamma_c(\omega, -\omega) = \frac{1}{2\pi}\int_{-\infty}^{\infty} C^2(\omega - \alpha)S(\alpha)\, d\alpha \tag{14-54}$$

and (14-51) results.

Comparing with (14-5), we conclude that $C^2(\omega)$ must have the properties of the window $W(\omega)$ introduced in Sec. 14-1. Conversely, if $W(\omega)$ is the square of a function $C(\omega)$ with time-limited inverse $c(t)$, then $c(t)$ can be used as a data window. Among the examples of Table 14-1, the frequency functions 1, 3, and 4 have this property. The corresponding data windows equal

$$c(t) = p_T(t) \qquad c(t) = q_T(t) \qquad c(t) = \cos\frac{\pi t}{2T}\, p_T(t)$$

respectively.

Variance For the determination of the variance of the sample spectrum, knowledge of the fourth-order moments of the process $\mathbf{x}(t)$ is required. This information is not generally available. However, it can be shown that, for the evaluation of the variance V of $\mathbf{S}_c(\omega)$, the difference between the fourth-order moments of $\mathbf{x}(t)$ and of a normal process with the same spectrum can be neglected if T is sufficiently large (see P.2). We shall assume, therefore, that $\mathbf{x}(t)$ is normal.

From (14-50) and (10-160) it follows that

$$\operatorname{var} \mathbf{S}_c(\omega) = \frac{1}{4T^2} \{\Gamma_c^2(\omega, -\omega) + \Gamma_c^2(\omega, \omega)\} \tag{14-55}$$

Comparing with (14-54), we conclude that

$$\operatorname{var} \mathbf{S}_c(\omega) \geq \frac{1}{4T^2} \Gamma_c^2(\omega, -\omega) = E^2\{\mathbf{S}_c(\omega)\} \tag{14-56}$$

This shows that $\mathbf{S}_c(\omega)$ is not a reliable estimate of $S(\omega)$ no matter how large T is or how $c(t)$ is chosen. It shows also that the improvement resulting from the use of a data window is negligible. Its presence does not reduce V and its effect on the bias is small. If $c(t) = 1$, then $\mathbf{S}_c(\omega) = \mathbf{S}_T(\omega)$ and $C^2(\omega)$ equals the Bartlett window [see (14-49)] which is reasonable. Furthermore, as we show next, the bias of the estimate is due primarily to the spectral window $W(\omega)$ that must be introduced to reduce V. We shall assume, therefore, in the following that $c(t) = 1$.

Lag window and spectral window We shall use as estimate of $S(\omega)$ the *smoothed spectrum*

$$\mathbf{S}_w(\omega) = \frac{1}{2\pi} \int_{-\infty}^{\infty} \mathbf{S}_T(\omega - y) W(y)\, dy \tag{14-57}$$

where $W(\omega)$ is a real, even function such that

$$\frac{1}{2\pi} \int_{-\infty}^{\infty} W(\omega)\, d\omega = 1 \tag{14-58}$$

but otherwise *arbitrary*. This function is called *spectral window* and its inverse $w(\tau)$ *lag window*. The lag window need not be time-limited.

The smoothed spectrum can be expressed in terms of the inverse transform $\mathbf{R}_T(\tau)$ of the sample spectrum $\mathbf{S}_T(\omega)$. Since $\mathbf{S}_w(\omega)$ equals the convolution of $\mathbf{S}_T(\omega)$ with $W(\omega)$ and $\mathbf{R}_T(\tau) = 0$ for $|\tau| > 2T$, we conclude that

$$\mathbf{S}_w(\omega) = \int_{-2T}^{2T} w(\tau) \mathbf{R}_T(\tau) e^{-j\omega\tau}\, d\tau \tag{14-59}$$

Mean and variance of smoothed spectrum From (14-57) and (14-49) it follows that

$$E\{\mathbf{S}_w(\omega)\} = S(\omega) * \frac{\sin^2 T\omega}{\pi T \omega^2} * \frac{W(\omega)}{2\pi} \tag{14-60}$$

To find the variance of $\mathbf{S}_w(\omega)$ we must determine the autocovariance of $\mathbf{S}_T(\omega)$. Assuming that $\mathbf{x}(t)$ is normal, we proceed as follows: Setting $C(\omega) = 2 \sin T\omega/\omega$ in (14-53), we determine $\Gamma(u, v)$. Inserting into (10-160), we obtain the autocovariance of the process $2T\mathbf{S}_T(\omega)$. The variance of $\mathbf{S}_w(\omega)$ can then be determined

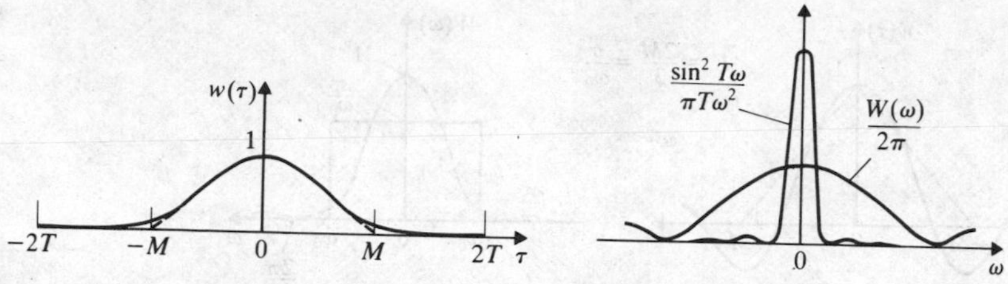

Figure 14-7

from (9-103) mutatis mutandis. We state the final result omitting the rather involved details (see P.2): If T is sufficiently large, then

$$\operatorname{var} \mathbf{S}_w(\omega) \simeq \begin{cases} \dfrac{E_w}{T} S^2(\omega) & \omega = 0 \\ \dfrac{E_w}{2T} S^2(\omega) & |\omega| \gg \dfrac{1}{T} \end{cases} \tag{14-61}$$

where

$$E_w = \int_{-\infty}^{\infty} w^2(\tau)\, d\tau = \frac{1}{2\pi} \int_{-\infty}^{\infty} W^2(\omega)\, d\omega \tag{14-62}$$

is the energy of the window.

Window selection If we make the reasonable assumption that $0 \le w(t) \le 1$, then the duration of $w(t)$ is of the order of E_w and the duration of $W(\omega)$ is of the order of π/E_w. From this and (14-61) it follows that if $E_w \ll T$, then the variance of the estimate is small compared to $S^2(\omega)$. Furthermore, the term $\sin^2 T\omega/\pi T\omega^2$ in (14-60) is of short duration relative to π/E_w (Fig. 14-7). Hence

$$E\{\mathbf{S}_w(\omega)\} \simeq \frac{1}{2\pi} S(\omega) * W(\omega) \tag{14-63}$$

The above leads to the conclusion that for high resolution E_w must be large but for small variance E_w must be small compared to T. Both requirements can be met if T is sufficiently large.

In principle, $w(t)$ need not be time-limited. However, since its energy equals E_w and $w(0) = 1$, we can assume that $w(t) \ll 1$ for $|t| > M$ where M is a constant of the order of E_w. If, therefore, we require that $w(t) = 0$ for $|t| > M$, the resulting restriction in the choice of windows will not be significant but the interval of integration in (14-59) will be reduced from $4T$ to $2M$. With M so determined, the shape of the window is selected as in Sec. 14-1.

Computational routine We determine $\mathbf{S}_T(\omega)$ either from (14-44) or from (14-46). We determine $\mathbf{S}_w(\omega)$ either from (14-57) or from (14-59). However, in general, the evaluation of Fourier transforms is simpler than the evaluation of convolutions. It is preferable, therefore, to evaluate $\mathbf{S}_w(\omega)$ using only multiplications and Fourier transforms (FFT algorithms). This involves the following: We compute $\mathbf{S}_T(\omega)$ from (14-46) (one FFT and one multiplication). We find its inverse

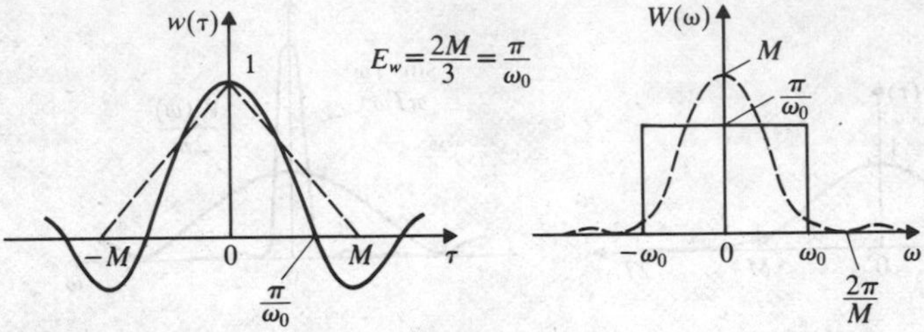

Figure 14-8

transform $\mathbf{R}_T(\tau)$ (one FFT). We form the product $w(\tau)\mathbf{R}_T(\tau)$ and compute its Fourier transform $\mathbf{S}_w(\tau)$ (one multiplication and one FFT).

Example 14-2 The moving average

$$\mathbf{S}_w(\omega) = \frac{1}{2\omega_0}\int_{-\omega_0}^{\omega_0} \mathbf{S}_T(\omega - y)\,dy$$

is a special case of (14-57) obtained with $W(\omega)$ equal to a rectangular pulse. In this case

$$E\{\mathbf{S}_w(\omega)\} = \frac{1}{2\omega_0}\int_{-\omega_0}^{\omega_0} S(\omega - y)\,dy$$

$$E_w = \frac{1}{2\pi}\int_{-\omega_0}^{\omega_0}\frac{\pi^2}{\omega_0^2}\,d\omega = \frac{\pi}{\omega_0} \qquad \operatorname{var}\mathbf{S}_w(\omega) = \frac{\pi}{2\omega_0 T}S^2(\omega)$$

In Fig. 14-8, we show the above window and the Bartlett window (dashed lines) with the same energy.

Burg's Method†

We are given the M values of a sample of a real process $\mathbf{x}[n]$ and we wish to estimate its power spectrum $S(\omega)$. Burg's method is a formal adaptation of the MEM to this problem. The estimate of $S(\omega)$ is again an all-pole model whose coefficients are determined recursively from Levinson's algorithm. However, the algorithm is modified as follows.

1. Equation (13-79)

$$P_{N-1}K_N = R[N] - \sum_{k=1}^{N-1} a_k^{N-1}R[N-k] \tag{14-64}$$

relating K_N to the data $R[m]$ is replaced by equation

$$P_{N-1}K_N = E\{\hat{\boldsymbol{\varepsilon}}_{N-1}[n]\check{\boldsymbol{\varepsilon}}_{N-1}[n-N]\} \tag{14-65}$$

† J. P. Burg: Maximum entropy spectral analysis, presented at the International Meeting of the Society for the Exploration of Geophysics, Orlando, FL, 1967.

involving the errors $\hat{\varepsilon}_{N-1}[n]$ and $\check{\varepsilon}_{N-1}[n-N]$ and their MS value

$$P_{N-1} = E\{\hat{\varepsilon}^2_{N-1}[n]\} = E\{\check{\varepsilon}^2_{N-1}[n-N]\} \tag{14-66}$$

2. All ensemble averages are changed to time averages.

The equivalence between (14-64) and (14-65) can be established as follows: Using (13-71a), we write P_N in the form

$$P_N = E\{(\hat{\varepsilon}_{N-1}[n] - K_N \check{\varepsilon}_{N-1}[n-N])^2\}$$

As we know, P_N depends on the constants a_k^N and these constants are so chosen as to minimize P_N. Since $K_N = a_N^N$, we conclude that

$$\frac{\partial P_N}{\partial K_N} = E\{-2(\hat{\varepsilon}_{N-1}[n] - K_N \check{\varepsilon}_{N-1}[n-N])\check{\varepsilon}_{N-1}[n-N]\} = 0$$

and (14-65) results.

The algorithm We use as estimate of $S(\omega)$ the process

$$\mathbf{S}_N(\omega) = \frac{\mathbf{P}_N}{|1 - \mathbf{a}_1^N e^{-j\omega T} - \cdots - \mathbf{a}_N^N e^{-jN\omega T}|^2} \tag{14-67}$$

and we determine its parameters recursively as in (13-78) and (13-86)

$$\begin{aligned} \mathbf{a}_k^N &= \mathbf{a}_k^{N-1} - \mathbf{K}_N \mathbf{a}_{N-k}^{N-1} \\ \mathbf{a}_N^N &= \mathbf{K}_N \qquad\qquad \mathbf{P}_N = (1 - \mathbf{K}_N^2)\mathbf{P}_{N-1} \end{aligned} \tag{14-68}$$

To find $\mathbf{K}_N$, we express $\mathbf{P}_{N-1}$ as the average of the forward and backward MS errors (they are equal), and we change the ensemble averages to time averages. Cancelling the term $M - N + 1$, we obtain [see (14-65)]

$$\mathbf{K}_N = \frac{\sum_{n=N+1}^{M} \hat{\boldsymbol{\varepsilon}}_{N-1}[n]\check{\boldsymbol{\varepsilon}}_{N-1}[n-N]}{\frac{1}{2}\sum_{n=N+1}^{M} (\hat{\boldsymbol{\varepsilon}}^2_{N-1}[n] + \check{\boldsymbol{\varepsilon}}^2_{N-1}[n-N])} \tag{14-69}$$

where

$$\begin{aligned} \hat{\boldsymbol{\varepsilon}}_N[n] &= \mathbf{x}[n] - \mathbf{a}_1^N \mathbf{x}[n-1] - \cdots - \mathbf{a}_N^N \mathbf{x}[n-N] \\ \check{\boldsymbol{\varepsilon}}_N[n] &= \mathbf{x}[n] - \mathbf{a}_1^N \mathbf{x}[n+1] - \cdots - \mathbf{a}_N^N \mathbf{x}[n+N] \end{aligned} \tag{14-70}$$

The denominator in (14-69) is the estimate of P_{N-1}. It is written as the average of the estimates of the two terms in (14-66) because the resulting $\mathbf{K}_N$ satisfies (14-71). To avoid overflow beyond the data interval $1 \le n \le M$, we used $N + 1$ as the lower summation limit in (14-69).

The algorithm starts with

$$\mathbf{P}_0 = \frac{1}{M}\sum_{n=1}^{M} \mathbf{x}^2[0] \qquad \hat{\boldsymbol{\varepsilon}}_0[n] = \check{\boldsymbol{\varepsilon}}_0[n] = \mathbf{x}[n]$$

and for $N = 1$ it yields

$$\mathbf{K}_1 = \frac{\sum_{n=2}^{M} \mathbf{x}[n]\mathbf{x}[n-1]}{\frac{1}{2}\sum_{n=2}^{M} (\mathbf{x}^2[n] + \mathbf{x}^2[n-1])} \qquad \mathbf{P}_1 = (1 - \mathbf{K}_1^2)\mathbf{P}_0$$

We note that as N increases, the order of $\mathbf{S}_N(\omega)$ increases. However, the estimation does not necessarily improve. The optimum value of N is dictated by two conflicting requirements: For a satisfactory approximation of the unknown spectrum by an all-pole function, N should be large. However, if N is large, then the estimate is not reliable because the number of terms in (14-69) decreases as N increases, causing an increase in the variance of $\mathbf{K}_N$. To determine the optimum N we must find, first, the statistical properties of $\mathbf{S}_N(\omega)$. This, however, is a difficult problem. In most applications, N is chosen empirically.

Stability The coefficients $\mathbf{a}_k^N$ of Burg's algorithm are RVs, hence, the roots $\mathbf{z}_i$ of the polynomial

$$\mathbf{E}_N(z) = 1 - \mathbf{a}_1^N z^{-1} - \cdots - \mathbf{a}_N^N z^{-N}$$

are RVs depending on the given sample $\mathbf{x}[n, \zeta]$. We maintain, however, that $|\mathbf{z}_i| \le 1$ for every ζ.

PROOF It suffices to show that [see (13-84)]

$$|\mathbf{K}_N| \le 1 \tag{14-71}$$

As we know (Prob. 10-31)

$$\left(\sum x_n y_n\right)^2 \le \sum x_n^2 \sum y_n^2 \le \tfrac{1}{4}\left(\sum x_n^2 + \sum y_n^2\right)^2$$

Applying the above to the numerator of (14-69), we obtain (14-71).

PROBLEMS

14-1(*a*) Show that if we use as estimate of the power spectrum $S(\omega)$ of a discrete-time process $\mathbf{x}[n]$ the function

$$S_w(\omega) = \sum_{m=-N}^{N} w_m R[m] e^{-jm\omega T}$$

then

$$S_w(\omega) = \frac{1}{2\sigma}\int_{-\sigma}^{\sigma} S(y) W(\omega - y)\, dy \qquad W(\omega) = \sum_{-N}^{N} w_n e^{-jn\omega T}$$

(*b*) Find $W(\omega)$ if $N = 10$ and $w_n = 1 - |n|/10$.

14-2 Show that the function

$$W(\omega) = 2\omega \sin \pi\omega \sum_{n=0}^{m} \frac{c_n}{\omega^2 - n^2}$$

is a window as in (14-9) and its inverse equals

$$w(t) = \begin{cases} c_0 - c_1 \cos t + \cdots c_m \cos mt & |t| < \pi \\ 0 & |t| > \pi \end{cases}$$

14-3(*a*) Show that if $R[0] = 8$ and $R[1] = 4$, then the MEM estimate of $S(\omega)$ equals

$$S_{\text{MEM}}(\omega) = \frac{6}{|1 - 0.5e^{-j\omega T}|^2}$$

14-4 Find the maximum entropy estimate $S_{\text{MEM}}(\omega)$ and the line-spectral estimate $S_L(\omega)$ of a process $\mathbf{x}[n]$ if

$$R[0] = 13 \qquad R[1] = 5 \qquad R[2] = 2$$

14-5 Show that if $\mathbf{x}(t)$ is a zero-mean normal process with sample spectrum

$$\mathbf{S}_T(\omega) = \frac{1}{2T}\left|\int_{-T}^{T} \mathbf{x}(t)e^{-j\omega t}\, dt\right|^2$$

then

$$E^2\{\mathbf{S}_T(\omega)\} \le \text{var}\ \mathbf{S}_T(\omega) \le 2E^2\{\mathbf{S}_T(\omega)\}$$

The right side is an equality if $\omega = 0$. The left side is an approximate equality if $T \gg 1/\omega$.
Hint: Use (10-160).

14-6 Show that if

$$\mathbf{R}_T(\tau) = \frac{1}{2T}\int_{-T+|\tau|/2}^{T-|\tau|/2} \mathbf{x}\left(t + \frac{\tau}{2}\right)\mathbf{x}\left(t - \frac{\tau}{2}\right) dt$$

is the estimate of the autocorrelation $R(\tau)$ of a zero-mean normal process, then

$$\sigma_{R_T}^2 = \frac{1}{2T}\int_{-2T+|\tau|}^{2T-|\tau|} [R^2(\alpha) + R(\alpha + \tau)R(\alpha - \tau)]\left(1 - \frac{|\tau| + |\alpha|}{2T}\right) d\alpha$$

14-7 Show that the weighted sample spectrum

$$\mathbf{S}_c(\omega) = \frac{1}{2T}\left|\int_{-T}^{T} c(t)\mathbf{x}(t)e^{-j\omega t}\, dt\right|^2$$

of a process $\mathbf{x}(t)$ is the Fourier transform of the function

$$\mathbf{R}_c(\tau) = \frac{1}{2T}\int_{-T+|\tau|/2}^{T-|\tau|/2} c\left(t + \frac{\tau}{2}\right)c\left(t - \frac{\tau}{2}\right)\mathbf{x}\left(t + \frac{\tau}{2}\right)\mathbf{x}\left(t - \frac{\tau}{2}\right) dt$$

CHAPTER

FIFTEEN

ENTROPY

15-1 INTRODUCTION

As we have noted in Chap. 1, the probability $P(\mathscr{A})$ of an event $\mathscr{A}$ can be interpreted as a measure of our uncertainty about the occurrence or nonoccurrence of $\mathscr{A}$ in a single performance of the underlying experiment $\mathscr{S}$. If $P(\mathscr{A}) \simeq 0.999$, then we are almost certain that $\mathscr{A}$ will occur; if $P(\mathscr{A}) = 0.1$, then we are reasonably certain that $\mathscr{A}$ will not occur; our uncertainty is maximum if $P(\mathscr{A}) = 0.5$. In this chapter, we consider the problem of assigning a measure of uncertainty to the occurrence or nonoccurrence not of a single event of $\mathscr{S}$, but of any event $\mathscr{A}_i$ of a partition $\mathfrak{A}$ of $\mathscr{S}$ where, as we recall, a partition is a collection of mutually exclusive events whose union equals $\mathscr{S}$ (Fig. 15-1). The measure of uncertainty about $\mathfrak{A}$ will be denoted by $H(\mathfrak{A})$ and will be called the *entropy of the partition* $\mathfrak{A}$.

Historically, the functional $H(\mathfrak{A})$ was derived from a number of postulates based on our heuristic understanding of uncertainty. The following is a typical set of such postulates:†

1. $H(\mathfrak{A})$ is a continuous function of $p_i = P(\mathscr{A}_i)$.
2. If $p_1 = \cdots = p_N = 1/N$, then $H(\mathfrak{A})$ is an increasing function of N.
3. If a new partition $\mathfrak{B}$ is formed by subdividing one of the sets of $\mathfrak{A}$, then $H(\mathfrak{B}) \geq H(\mathfrak{A})$.

† C. E. Shannon and W. Weaver, *The Mathematical Theory of Communication*, University of Illinois Press, 1949.

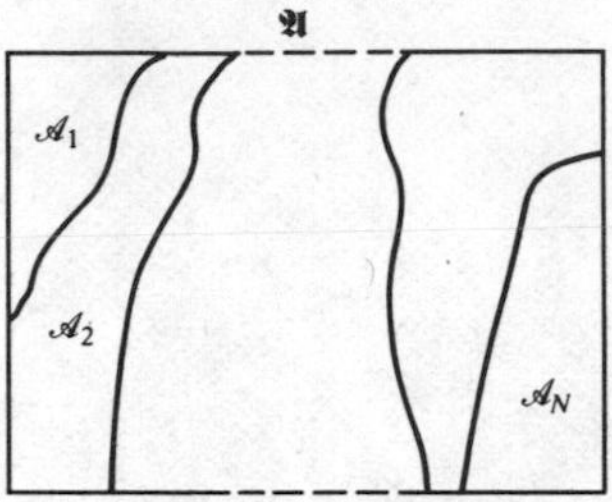

Figure 15-1

It can be shown that the sum

$$H(\mathfrak{A}) = -p_1 \log p_1 - \cdots - p_N \log p_N \tag{15-1}$$†

satisfies these postulates and it is unique within a constant factor. The proof of this assertion is not difficult but we choose not to reproduce it. We propose, instead, to introduce (15-1) as the *definition* of entropy and to develop axiomatically all its properties within the framework of probability. It is true that the introduction of entropy in terms of postulates establishes a link between the sum in (15-1) and our heuristic understanding of uncertainty. However, for our purposes, this is only incidental. In the last analysis, the justification of the concept must ultimately rely on the usefulness of the resulting theory.

The applications of entropy can be divided into two categories. The first deals with problems involving the determination of unknown distributions (Sec. 15-4). The available information is in the form of known expected values or other statistical functionals, and the solution is based on the principle of maximum entropy: We determine the unknown distributions so as to *maximize* the entropy $H(\mathfrak{A})$ of some partition $\mathfrak{A}$ subject to the given constraints (statistical mechanics). In the second category (coding theory), we are given $H(\mathfrak{A})$ (source entropy) and we wish to construct various random variables (code lengths) so as to *minimize* their expected values (Sec. 15-5). The solution involves the construction of optimum mappings (codes) of the random variables under consideration, into the given probability space.

Uncertainty and information In the heuristic interpretation of entropy, the number $H(\mathfrak{A})$ is a measure of our uncertainty about the events $\mathscr{A}_i$ of the partition $\mathfrak{A}$ prior to the performance of the underlying experiment. If the experiment is performed and the results concerning $\mathscr{A}_i$ become known, the uncertainty is removed. We can, thus, say that the experiment provides *information* about the events $\mathscr{A}_i$ equal to the *entropy* of their partition. Thus, uncertainty equals information and both are measured by the sum in (15-1).

Example 15-1(*a*) We shall determine the entropy of the partition $\mathfrak{A}$ = [even, odd] in the fair-die experiment. Clearly, $P\{\text{even}\} = P\{\text{odd}\} = 1/2$. Hence

$$H(\mathfrak{A}) = -\tfrac{1}{2}\log\tfrac{1}{2} - \tfrac{1}{2}\log\tfrac{1}{2} = \log 2$$

† We shall use as logarithmic base either the number 2 or the number e. In the first case, the unit of entropy is the *bit*.

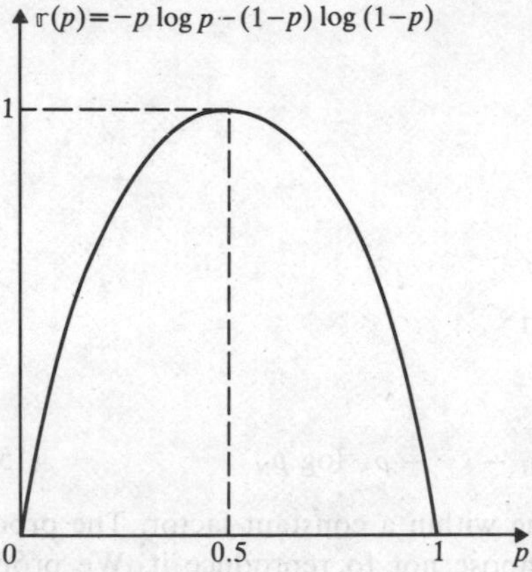

Figure 15-2

(*b*) In the same experiment, $\mathfrak{S}$ is the partition consisting of the elementary events $\{f_i\}$. In this case, $P\{f_i\} = 1/6$, hence

$$H(\mathfrak{S}) = -\tfrac{1}{6} \log \tfrac{1}{6} - \cdots - \tfrac{1}{6} \log \tfrac{1}{6} = \log 6$$

If the die is rolled and we are told which face showed, then we gain information about the partition $\mathfrak{S}$ equal to its entropy log 6. If we are told merely whether "even" or "odd" showed, then we gain information about the partition $\mathfrak{A}$ equal to its entropy log 2. In this case, the information gained about the partition $\mathfrak{S}$ equals again log 2. As we shall see, the difference log 6 − log 2 = log 3 is the uncertainty about $\mathfrak{S}$ assuming $\mathfrak{A}$ (conditional entropy).

Example 15-2 We consider now the coin experiment where $P\{h\} = p$. In this case, the entropy of $\mathfrak{S}$ equals

$$H(\mathfrak{S}) = -p \log p - (1-p) \log (1-p) \equiv \mathbb{r}(p) \qquad (15\text{-}2)$$

The function $\mathbb{r}(p)$ is shown in Fig. 15-2 for $0 \le p \le 1$. This function is symmetrical, convex, even about the point $p = 0.5$, and it reaches its maximum at that point. Furthermore, $\mathbb{r}(0) = \mathbb{r}(1) = 0$.

Historical note The word "entropy" as a scientific concept was first used in thermodynamics (Clausius 1850). Its probabilistic interpretation in the context of statistical mechanics is attributed to Boltzmann (1877). However, the explicit relationship between entropy and probability was recorded several years later (Planck, 1906). Shannon, in his celebrated paper (1948), used the concept to give an economical description of the properties of long sequences of symbols, and applied the results to a number of basic problems in coding theory and data transmission. His remarkable contributions form the basis of modern information theory. Jaynes† (1957) reexamined the method of maximum entropy and applied it to a variety of problems involving the determination of unknown parameters from incomplete data.

† E. T. Jaynes: *Physical Review*, vols. 106–107, 1957.

Maximum entropy and classical definition An important application of entropy is the determination of the probabilities p_i of the events of a partition $\mathfrak{A}$, subject to various constraints, with the method of maximum entropy (MEM). The method states that the unknown p_i's must be so chosen as to maximize the entropy of $\mathfrak{A}$ subject to the given constraints. This topic is considered in Sec. 15-4. In the following we introduce the main idea and we show the equivalence between the MEM and the classical definition of probability (principle of insufficient reason), using as illustration the die experiment.

Example 15-3(*a*) We wish to determine the probabilities p_i of the six faces of a die, having access to no prior information. The MEM states that the p_i's must be such as to maximize the sum

$$H(\mathfrak{S}) = -p_1 \log p_1 - \cdots - p_6 \log p_6$$

Since $p_1 + \cdots + p_6 = 1$, this yields

$$p_1 = \cdots = p_6 = \tfrac{1}{6}$$

in agreement with the classical definition.

(*b*) Suppose now that we are given the following information: A player places a bet of one dollar on "odd" and he wins, on the average, 20 cents per game. We wish again to determine the p_i's using the MEM; however, now we must satisfy the constraints

$$p_1 + p_3 + p_5 = 0.6 \qquad p_2 + p_4 + p_6 = 0.4$$

This is a consequence of the available information because an average gain of 20 cents means that $P\{\text{odd}\} - P\{\text{even}\} = 0.2$. Maximizing $H(\mathfrak{S})$ subject to the above constraints, we obtain

$$p_1 = p_3 = p_5 = 0.2 \qquad p_2 = p_4 = p_6 = 0.133\ldots$$

This agrees again with the classical definition if we apply the principle of insufficient reason to the outcomes of the events {odd} and {even} separately.

Although conceptually the ME principle is equivalent to the principle of insufficient reason, operationally the MEM simplifies the analysis drastically when, as is the case in most applications, the constraints are phrased in terms of probabilities in the space $\mathscr{S}^n$ of repeated trials. In such cases the equivalence still holds, although it is less obvious, but the reasoning is involved and rather forced if we derive the unknown probabilities starting from the classical definition.

The MEM is, thus, a valuable tool in the solution of applied problems. It is used, in fact, even in deterministic problems involving the estimation of unknown parameters from insufficient data. The ME principle is then accepted as a smoothness criterion. We should emphasize, however, that as in the case of the classical definition, the conclusions drawn from the ME principle must be accepted with skepticism particularly when they involve elaborate constraints. This is evident even in the interpretation of the results in Example 15-3: In the absence of prior constraints, we concluded that all p_i's must be equal. This conclusion we accept readily because it is not in conflict with our experience concerning dice. The second conclusion, however, that $p_2 = p_4 = p_6 = 0.133 \ldots$ and $p_1 = p_3 = p_5 = 0.2$ is not as convincing, we would think, even though we have no basis for any other conclusion. In our experience, no crooked dice exhibit such symmetries.

One might argue that this apparent conflict between the MEM and our experience is due to the fact that we did not make total use of our prior knowledge. Had we included among the constraints everything we know about dice, there would be no conflict. This might be true; however, it is not always clear how such constraints can be phrased analytically and, even if they can, how complex the required computations might be.

Typical Sequences and Relative Frequency

Suppose that $\mathfrak{A} = [\mathscr{A}_1, \ldots, \mathscr{A}_N]$ is an N-element partition of an experiment $\mathscr{S}$. In the space $\mathscr{S}^n$ of repeated trials, the elements $\mathscr{A}_i$ of $\mathfrak{A}$ form N^n sequences of the form

$$\{\mathscr{A}_i \text{ occurs } n_i \text{ times in a specific order}\} \tag{15-3}$$

and the probability of each sequence equals

$$p_1^{n_1} \cdots p_i^{n_i} \cdots p_N^{n_N} \tag{15-4}$$

where $p_i = P(\mathscr{A}_i)$. The numbers n_i are arbitrary subject only to the constraint $n_1 + \cdots + n_N = n$. However, according to the relative frequency interpretation of probability, if n is "sufficiently large," then "almost certainly"

$$n_i \simeq np_i \qquad i = 1, \ldots, N \tag{15-5}$$

This is, of course, only a heuristic statement, hence the resulting consequences must be interpreted accordingly. However, as we know, the approximation (15-5) can be given a precise interpretation in the form of the law of large numbers. Following a similar approach, we prove at the end of the section the main consequences [Eq. (15-10)] of (15-5) in the context of entropy.

Guided by (15-5), we shall separate the N^n sequences of the form (15-3) into two groups: (*a*) typical, (*b*) rare. We shall say that a sequence is typical, if $n_i \simeq np_i$. All other sequences will be called *rare*. A typical sequence will be identified with the letter $\mathfrak{t}$:

$$\mathfrak{t} = \{\mathscr{A}_i \text{ occurs } n_i \simeq np_i \text{ times in a specific order}\} \tag{15-6}$$

From the definition it follows that the union of all typical sequences is a set $\mathbb{T}$ in $\mathscr{S}^n$ with

$$P(\mathbb{T}) \simeq 1 \tag{15-7}$$

The complement $\bar{\mathbb{T}}$ of $\mathbb{T}$ is the union of all rare sequences and its probability is negligible for large n

$$P(\bar{\mathbb{T}}) \simeq 0 \tag{15-8}$$

Since

$$n_i \simeq np_i = e^{n \ln p_i}$$

for all typical sequences, (15-4) yields

$$P(\mathfrak{t}) = p_1^{n_1} \cdots p_N^{n_N} \simeq e^{np_1 \ln p_1 + \cdots + np_N \ln p_N}$$

Hence

$$P(\mathfrak{t}) = e^{-nH(\mathfrak{A})} \tag{15-9}$$

where $H(\mathfrak{A})$ is the entropy of the partition $\mathfrak{A}$. Denoting by $n_{\mathbb{T}}$ the number of typical sequences, we conclude from the above that

$$n_{\mathbb{T}} = \frac{P(\mathbb{T})}{P(\mathfrak{t})} \simeq e^{nH(\mathfrak{A})} \tag{15-10}$$

We have thus expressed the number of typical sequences in terms of the entropy of $\mathfrak{A}$. If all the events of $\mathfrak{A}$ are equally likely, then $H(\mathfrak{A}) = \ln N$ and $n_{\mathbb{T}} = N^n$. In all other cases, $H(\mathfrak{A}) < \ln N$ [see (15-38)]. Hence

$$n_{\mathbb{T}} \simeq e^{nH(\mathfrak{A})} \ll N^n \qquad \text{for} \qquad n \gg 1 \tag{15-11}$$

This leads to the important conclusion that, if n is sufficiently large, then *most* sequences are rare even though "almost certainly" none will occur.

Note We should point out that each typical sequence is not more likely than each rare sequence. In fact, the sequence with the largest probability is the rare sequence $\{\mathscr{A}_m$ occurs n times$\}$, where $\mathscr{A}_m$ is the event with the largest probability. As we presently show, the distinction between typical and rare sequences is best expressed in terms of the events

$$\{\mathscr{A}_i \text{ occurs } n_i \text{ times in } any \text{ order}\}$$

As we know [see (3-38)], the probability of these events equals

$$\frac{n!}{k_1! \cdots k_N!} p_1^{k_1} \cdots p_N^{k_N}$$

and it takes significant values only in a small vicinity of the point $(k_1 = n_1 p_1, \ldots, k_N = n_N p_N)$. This follows by repeating the argument leading to (3-17) or, for large n, from the DeMoivre–Laplace approximation (3-39).

Typical Sequences and the Law of Large Numbers

We show next that the preceding results can be reestablished rigorously as consequences of the law of large numbers. For simplicity we consider only two-element partitions and, to be concrete, we assume that $\mathscr{A}$ and $\bar{\mathscr{A}}$ are the events "heads" and "tails," respectively, in the coin experiment. In the space $\mathscr{S}^n$, the probability of the elementary event $\zeta_k = \{k \text{ heads in a specific order}\}$ equals

$$P\{\zeta_k\} = p^k q^{n-k}$$

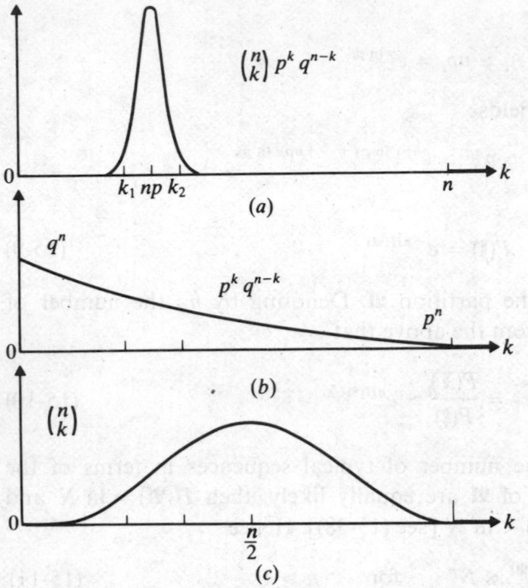

Figure 15-3

and the probability of the event†

$$\mathscr{A}_k = \{k \text{ heads in any order}\}$$

equals

$$P(\mathscr{A}_k) = \binom{n}{k} p^k q^{n-k} \simeq \frac{1}{\sqrt{2\pi npq}} e^{-(k-np)^2/2npq} \tag{15-12}$$

In Fig. 15-3 we plot the probability $P(\mathscr{A}_k)$, the geometric progression $q^n(p/q)^k$, and the binomial coefficients

$$\binom{n}{k} = p^{-k} q^{-(n-k)} P(\mathscr{A}_k) \tag{15-13}$$

as functions of k.

α-typical sequences Given a number α between 0 and 1, we form the number ε such that

$$\alpha = 2\mathbb{G}(\varepsilon\sqrt{n/pq}) - 1 \tag{15-14}$$

where $\mathbb{G}(x)$ is the normal distribution. We shall say that the sequence ζ_k is *α-typical* if k is such that

$$k_1 \le k \le k_2 \qquad \text{where} \qquad k_1 = n(p - \varepsilon) \qquad k_2 = n(p + \varepsilon) \tag{15-15}$$

† The event $\mathscr{A}_k$ is not, of course, an element of the partition $\mathfrak{A} = [\mathscr{A}, \bar{\mathscr{A}}]$.

The union of all α-typical sequences is a set $\mathbb{T}$ consisting of

$$n_{\mathbb{T}} = \sum_{k=k_1}^{k_2} \binom{n}{k} \tag{15-16}$$

elements and its probability equals α [see (3-37)]

$$P(\mathbb{T}) = \sum_{k=k_1}^{k_2} \binom{n}{k} p^k q^{n-k} \simeq 2\mathbb{G}\left(\varepsilon\sqrt{\frac{n}{pq}}\right) - 1 = \alpha \tag{15-17}$$

Fundamental theorem For any $\alpha < 1$, the number $n_{\mathbb{T}}$ of α-typical sequences tends to $e^{nH(\mathfrak{A})}$ in the following sense

$$\frac{\ln n_{\mathbb{T}}}{n} \xrightarrow[n\to\infty]{} H(\mathfrak{A}) \tag{15-18}$$

PROOF If $p = q = 0.5$, then the DeMoivre–Laplace approximation yields

$$\binom{n}{k} \simeq \frac{2^n}{\sqrt{\pi n/2}} e^{-2(k-n/2)^2/n}$$

for k in the $\sqrt{n}$ vicinity of $n/2$. This approximation cannot be used to evaluate the sum in (15-16) for $p \neq 0.5$ because then the center np of the interval (k_1, k_2) is not $n/2$. We shall bound $n_{\mathbb{T}}$ using (15-13) and (15-16). Clearly,

$$n_{\mathbb{T}} = \sum_{k=k_1}^{k_2} p^{-k} q^{k-n} P(\mathscr{A}_k) \tag{15-19}$$

where we assume that $p < q$. As k increases, the term $p^{-k}q^{k-n}$ increases monotonically. Hence

$$q^{-n}\left(\frac{q}{p}\right)^{k_1} \sum_{k=k_1}^{k_2} P(\mathscr{A}_k) < n_{\mathbb{T}} < q^{-n}\left(\frac{q}{p}\right)^{k_2} \sum_{k=k_1}^{k_2} P(\mathscr{A}_k) \tag{15-20}$$

And since [see (15-17)]

$$\sum_{k=k_1}^{k_2} P(\mathscr{A}_k) = P(\mathbb{T}) = \alpha$$

(15-20) yields

$$\frac{\alpha}{q^n}\left(\frac{q}{p}\right)^{k_1} < \sum_{k=k_1}^{k_2} \binom{n}{k} < \frac{\alpha}{q^n}\left(\frac{q}{p}\right)^{k_2} \tag{15-21}$$

Setting $k_1 = np - n\varepsilon$ and $k_2 = np + n\varepsilon$ in the above and using the identity

$$p^{-np} q^{-nq} = e^{-n(p \ln p + q \ln q)} = e^{nH(\mathfrak{A})}$$

we conclude from (15-21) that

$$\alpha e^{nH(\mathfrak{A})}\left(\frac{q}{p}\right)^{-n\varepsilon} < n_{\mathbb{T}} < \alpha e^{nH(\mathfrak{A})}\left(\frac{q}{p}\right)^{n\varepsilon}$$

Hence,

$$nH(\mathfrak{A}) + \ln \alpha - n\varepsilon \log \frac{q}{p} < \ln n_{\mathbb{T}} < nH(\mathfrak{A}) + \ln \alpha + n\varepsilon \ln \frac{q}{p}$$

Dividing by n, we obtain (15-18) because α is constant and, as we see from (15-14), $\varepsilon \to 0$ as $n \to \infty$.

Important conclusion Theorem (15-18) holds for any $\alpha < 1$; it will be assumed, however, that $\alpha \simeq 1$ and the corresponding sequences will be called typical. With this assumption

$$P(\mathbb{T}) = \alpha \simeq 1 \qquad P(\bar{\mathbb{T}}) \simeq 1 - \alpha \simeq 0 \tag{15-22}$$

The probability of an arbitrary event $\mathscr{M}$ equals, therefore, its conditional probability

$$P(\mathscr{M}) = P(\mathscr{M} \mid \mathbb{T})P(\mathbb{T}) + P(\mathscr{M} \mid \bar{\mathbb{T}})P(\bar{\mathbb{T}}) \simeq P(\mathscr{M} \mid \mathbb{T}) \tag{15-23}$$

In other words, in any conclusions concerning probabilities in the space $\mathscr{S}^n$, it suffices to consider the subspace of $\mathscr{S}^n$ consisting of typical sequences only. This is, of course, only approximately true for finite n. It is, however, exact in the limit as $n \to \infty$.

15-2 BASIC CONCEPTS

In this section, we develop deductively the properties of entropy starting with various notations and set operations. At the end of the section, we reexamine the results in terms of the heuristic notion of entropy as a measure of uncertainty, and we conclude with a typical sequence interpretation of the main theorems.

Definitions The notation

$$\mathfrak{A} = [\mathscr{A}_1, \ldots, \mathscr{A}_k] \qquad \text{or simply} \qquad \mathfrak{A} = [\mathscr{A}_i]$$

will mean that $\mathfrak{A}$ is a partition consisting of the events $\mathscr{A}_i$. These events will be called elements† of $\mathfrak{A}$.

I. A partition with only two elements will be called *binary*. Thus

$$\mathfrak{A} = [\mathscr{A}, \bar{\mathscr{A}}]$$

is a binary partition consisting of the event $\mathscr{A}$ and its complement $\bar{\mathscr{A}}$.

II. A partition whose elements are the elementary events $\{\zeta_i\}$ of the space $\mathscr{S}$ will be denoted by $\mathfrak{S}$ and will be called *element partition*.

III. A *refinement* of a partition $\mathfrak{A}$ is a partition $\mathfrak{B}$ such that each element $\mathscr{B}_j$ of $\mathfrak{B}$ is a subset of some element $\mathscr{A}_i$ of $\mathfrak{A}$ (Fig. 15-4). We shall use the notation

† It will be clear from the context whether the word *element* means an event $\mathscr{A}_i$ of a partition $\mathfrak{A}$ or an element ζ_i of the space $\mathscr{S}$.

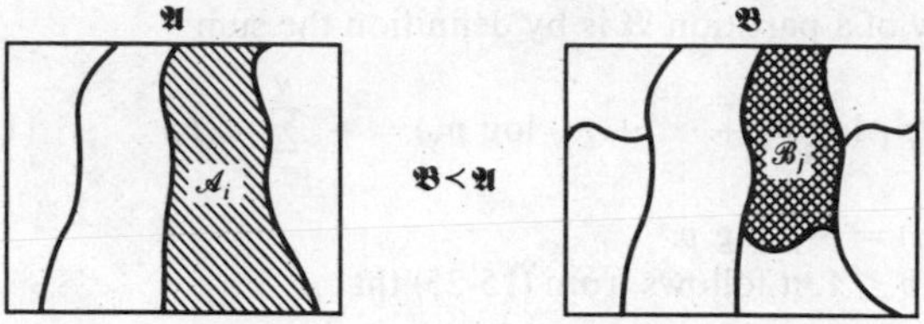

Figure 15-4

$\mathfrak{B} \prec \mathfrak{A}$ to indicate that $\mathfrak{B}$ is a refinement of $\mathfrak{A}$ and we shall say that $\mathfrak{A}$ is larger† than $\mathfrak{B}$. Thus

$$\mathfrak{B} \prec \mathfrak{A} \qquad \text{iff} \qquad \mathscr{B}_j \subset \mathscr{A}_i \tag{15-24}$$

A *common refinement* of two partitions is a refinement of both.

The partition $\mathfrak{D}$ in Fig. 15-5 is a common refinement of the partitions $\mathfrak{A}$ and $\mathfrak{B}$.

IV. The *product*‡ of two partitions $\mathfrak{A} = [\mathscr{A}_i]$ and $\mathfrak{B} = [\mathscr{B}_j]$ is a partition whose elements are all intersections $\mathscr{A}_i\mathscr{B}_j$ of the elements of $\mathfrak{A}$ and $\mathfrak{B}$. This partition will be denoted by

$$\mathfrak{A} \cdot \mathfrak{B}$$

Clearly, $\mathfrak{A} \cdot \mathfrak{B}$ is the largest common refinement of $\mathfrak{A}$ and $\mathfrak{B}$.

Properties From the definition it follows that

$$\mathfrak{S} \prec \mathfrak{A} \qquad \text{for any } \mathfrak{A}$$

$$\mathfrak{A} \cdot \mathfrak{B} = \mathfrak{B} \cdot \mathfrak{A} \qquad \mathfrak{A} \cdot (\mathfrak{B} \cdot \mathfrak{C}) = (\mathfrak{A} \cdot \mathfrak{B}) \cdot \mathfrak{C}$$

If $\quad \mathfrak{A}_1 \prec \mathfrak{A}_2 \prec \mathfrak{A}_3 \quad$ then $\quad \mathfrak{A}_1 \prec \mathfrak{A}_3$

If $\quad \mathfrak{B} \prec \mathfrak{A} \quad$ then $\quad \mathfrak{A} \cdot \mathfrak{B} = \mathfrak{B}$

† The symbol $\prec$ is *not* an ordering of two arbitrary partitions. It has a meaning only if $\mathfrak{B}$ is a refinement of $\mathfrak{A}$.

‡ We should emphasize that partition product is *not* a set operation.

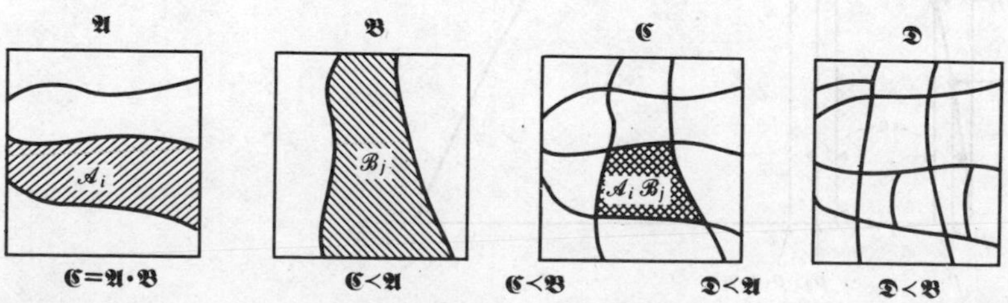

Figure 15-5

Entropy The entropy of a partition $\mathfrak{A}$ is by definition the sum

$$H(\mathfrak{A}) = -(p_1 \log p_1 + \cdots + p_N \log p_N) = -\sum_{i=1}^{N} \varphi(p_i) \tag{15-25}$$

where $p_i = P(\mathscr{A}_i)$ and $\varphi(p) = -p \log p$.

Since $\varphi(p) \geq 0$ for $0 \leq p \leq 1$, it follows from (15-25) that

$$H(\mathfrak{A}) \geq 0 \tag{15-26}$$

where $H(\mathfrak{A}) = 0$ iff one of the p_i's equals 1; all others are then equal to 0.

Binary partitions If $\mathfrak{A} = [\mathscr{A}, \bar{\mathscr{A}}]$ and $P(\mathscr{A}) = p$, then (Fig. 15-2)

$$H(\mathfrak{A}) = -p \log p - (1-p) \log (1-p) = \mathfrak{r}(p) \tag{15-27}$$

Equally likely events If

$$p_1 = p_2 = \cdots = p_N$$

then

$$H(\mathfrak{A}) = -\frac{1}{N} \log \frac{1}{N} \cdots -\frac{1}{N} \log \frac{1}{N} = \log N \tag{15-28}$$

If, in particular, $N = 2^m$, then $H(\mathfrak{A}) = m$.

Inequalities The function $\varphi(p) = -p \log p$ is convex. Therefore (see Fig. 15-6 and Prob. 15-2)

$$\varphi(p_1 + p_2) < \varphi(p_1) + \varphi(p_2) < \varphi(p_1 + \varepsilon) + \varphi(p_2 - \varepsilon) \tag{15-29}$$

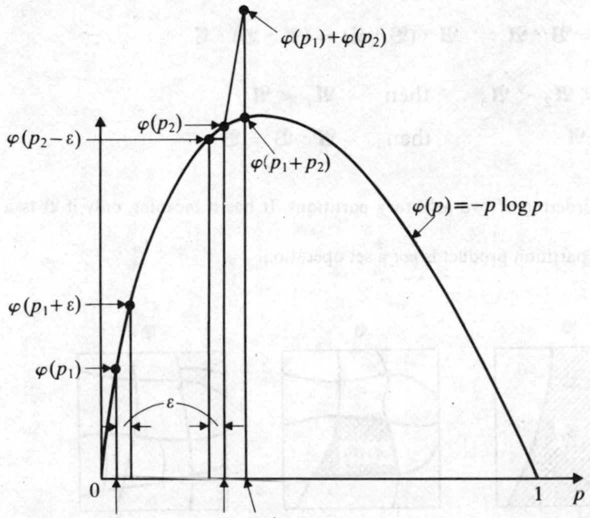

Figure 15-6

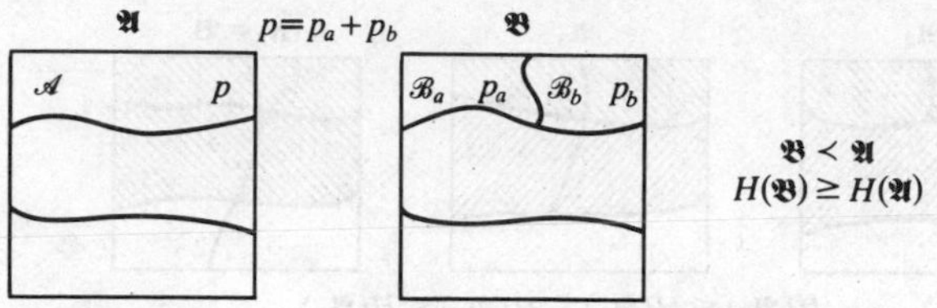

Figure 15-7

where

$$p_1 < p_1 + \varepsilon < p_2 - \varepsilon < p_2 \tag{15-30}$$

This leads to the following properties of entropy:

1. Given a partition $\mathfrak{A} = [\mathscr{A}_1, \mathscr{A}_2, \ldots, \mathscr{A}_N]$ we form the partition $\mathfrak{B} = [\mathscr{B}_a, \mathscr{B}_b, \mathscr{A}_2, \ldots, \mathscr{A}_N]$ obtained by splitting $\mathscr{A}_1$ into the elements $\mathscr{B}_a$ and $\mathscr{B}_b$ as in Fig. 15-7. We maintain that

$$H(\mathfrak{A}) \le H(\mathfrak{B}) \tag{15-31}$$

PROOF Clearly

$$H(\mathfrak{A}) - \varphi(p_a + p_b) = H(\mathfrak{B}) - \varphi(p_a) - \varphi(p_b)$$

because each side equals the contribution to $H(\mathfrak{A})$ and $H(\mathfrak{B})$ respectively due to the common elements of $\mathfrak{A}$ and $\mathfrak{B}$. Hence (15-31) follows from the first inequality in (15-29).

Example 15-4 In the next table we list the probabilities of the events of a partition $\mathfrak{A}$ and of its refinement $\mathfrak{B}$ obtained as above.

$\mathfrak{A}$	$p = 0.4$		0.35	0.25
$\mathfrak{B}$	$p_a = 0.22$	$p_b = 0.18$	0.35	0.25

In this case

$$H(\mathfrak{A}) = -(0.4 \log 0.4 + 0.35 \log 0.35 + 0.25 \log 0.25) = 1.559$$

$$H(\mathfrak{B}) = -(0.22 \log 0.22 + 0.18 \log 0.18 + 0.35 \log 0.35 + 0.25 \log 0.25) = 1.956$$

Thus

$$H(\mathfrak{A}) = 1.559 < 1.956 = H(\mathfrak{B})$$

in agreement with (15-31).

2. If

$$\mathfrak{B} \prec \mathfrak{A} \quad \text{then} \quad H(\mathfrak{B}) \ge H(\mathfrak{A}) \tag{15-32}$$

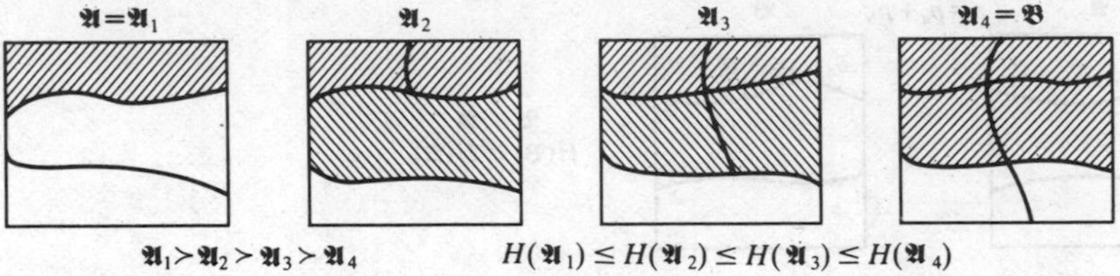

Figure 15-8

PROOF Repeating the construction of Fig. 15-7, we form a chain of refinements

$$\mathfrak{A} = \mathfrak{A}_1 \prec \cdots \mathfrak{A}_{m-1} \prec \mathfrak{A}_m \prec \cdots \prec \mathfrak{A}_n = \mathfrak{B}$$

where $\mathfrak{A}_m$ is obtained by splitting one of the elements of $\mathfrak{A}_{m-1}$ as in Fig. 15-8. From this and (15-31) it follows that

$$H(\mathfrak{A}) = H(\mathfrak{A}_1) \le \cdots \le H(\mathfrak{A}_n) = H(\mathfrak{B})$$

and (15-32) results.

3. For any $\mathfrak{A}$

$$H(\mathfrak{A}) \le H(\mathfrak{S}) \tag{15-33}$$

where $\mathfrak{S}$ is the element partition.

PROOF It follows from (15-31) because $\mathfrak{S}$ is a refinement of $\mathfrak{A}$.

4. For any $\mathfrak{A}$ and $\mathfrak{B}$

$$H(\mathfrak{A} \cdot \mathfrak{B}) \ge H(\mathfrak{A}) \qquad H(\mathfrak{A} \cdot \mathfrak{B}) \ge H(\mathfrak{B}) \tag{15-34}$$

PROOF It follows from (15-31) because $\mathfrak{A} \cdot \mathfrak{B}$ is a refinement of $\mathfrak{A}$ and of $\mathfrak{B}$.

Example 15-5 In the die experiment, the probabilities of the six events $\{f_1\}, \ldots, \{f_6\}$ equal

$$0.1 \qquad 0.1 \qquad 0.15 \qquad 0.2 \qquad 0.2 \qquad 0.25$$

respectively. The probabilities of the events of the partitions

$$\mathfrak{A} = [\text{even, odd}] \qquad \mathfrak{B} = [i \le 3, i > 3]$$

are given by

$$P\{\text{even}\} = 0.55 \qquad P\{\text{odd}\} = 0.45 \qquad P\{i \le 3\} = 0.35 \qquad P\{i > 3\} = 0.65$$

The product $\mathfrak{A} \cdot \mathfrak{B}$ is a partition consisting of the four elements

$$\{f_2\} \qquad \{f_1\, f_3\} \qquad \{f_4\, f_6\} \qquad \{f_5\}$$

with respective probabilities

$$0.1 \qquad 0.25 \qquad 0.45 \qquad 0.2$$

From the above it follows that

$$H(\mathfrak{A}) = 0.993 \qquad H(\mathfrak{B}) = 0.934 \qquad H(\mathfrak{A} \cdot \mathfrak{B}) = 1.815$$

in agreement with (15-34).

5. Suppose that $\mathfrak{A}$ and $\mathfrak{B}$ are two partitions that have the same elements except the first two (Fig. 15-9)

$$\mathfrak{A} = [\mathscr{A}_1, \mathscr{A}_2, \mathscr{A}_3, \ldots, \mathscr{A}_N] \qquad \mathfrak{B} = [\mathscr{B}_1, \mathscr{B}_2, \mathscr{A}_3, \ldots, \mathscr{A}_N]$$

We maintain that if

$$P(\mathscr{A}_1) = p_1 \qquad P(\mathscr{A}_2) = p_2 \qquad P(\mathscr{B}_1) = p_1 + \varepsilon \le p_2 - \varepsilon = P(\mathscr{B}_2)$$

as in (15-30), then

$$H(\mathfrak{A}) \le H(\mathfrak{B}) \tag{15-35}$$

PROOF Clearly

$$H(\mathfrak{A}) - \varphi(p_1) - \varphi(p_2) = H(\mathscr{B}) - \varphi(p_1 + \varepsilon) - \varphi(p_2 + \varepsilon)$$

because each side equals the contribution to $H(\mathfrak{A})$ and $H(\mathfrak{B})$ respectively due to the common elements of $\mathfrak{A}$ and $\mathfrak{B}$. Hence (15-35) follows from the second inequality in (15-29).

Example 15-6 In the next table we list the probabilities of the events of the partitions $\mathfrak{A}$ and $\mathfrak{B}$.

$\mathfrak{A}$	0.1	0.3	0.35	0.25	$p_1 = 0.1$
					$\varepsilon = 0.08$
$\mathfrak{B}$	0.18	0.22	0.35	0.25	$p_2 = 0.3$

In this case

$$H(\mathfrak{A}) = 1.883 \qquad H(\mathfrak{B}) = 1.956$$

in agreement with (15-35).

6. If we equalize the entropies of two elements of a partition, leaving all others unchanged, its entropy increases.

PROOF It follows from the above with $\varepsilon = (p_2 - p_1)/2$.

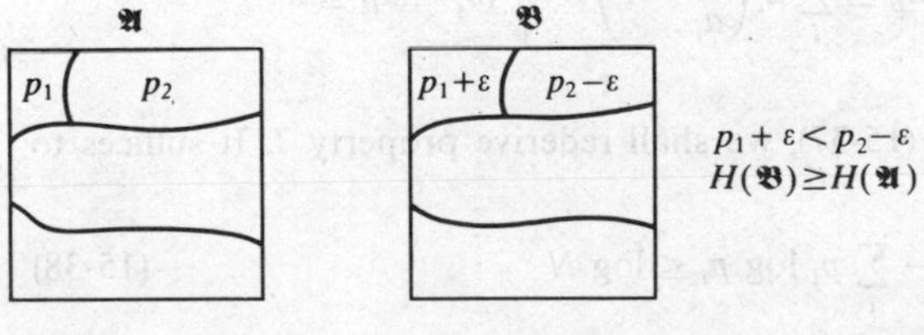

Figure 15-9

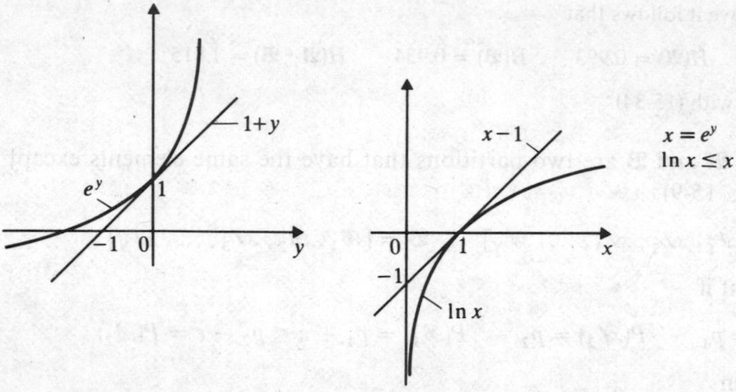

Figure 15-10

7. The entropy of a partition is maximum if all its elements are equally likely as in (15-28).

PROOF Suppose that $\mathfrak{A}$ is a partition such that $H(\mathfrak{A}) = H_m$ is maximum and two of its elements have unequal probabilities. If they are made equal, then (property 6), $H(\mathfrak{A})$ increases. But this is impossible because H_m is maximum by assumption.

A useful inequality If a_i and b_i are N positive numbers such that

$$a_1 + \cdots + a_N = 1 \qquad b_1 + \cdots + b_N \le 1 \tag{15-36}$$

then

$$-\sum_i a_i \log a_i \le -\sum_i a_i \log b_i \tag{15-37}$$

with equality iff $a_i = b_i$.

PROOF From the inequality $e^y \ge 1 + y$ it follows that $\ln x \le x - 1$ (Fig. 15-10). With $x = b_i/a_i$, this yields

$$\ln b_i - \ln a_i = \ln \frac{b_i}{a_i} \le \frac{b_i}{a_i} - 1$$

Multiplying by a_i and adding we obtain

$$\sum_i a_i(\ln b_i - \ln a_i) \le \sum_i a_i\left(\frac{b_i}{a_i} - 1\right) = \sum_i (b_i - a_i) \le 0$$

and (15-37) results.

Maximum entropy Using (15-37), we shall rederive property 7. It suffices to show that

$$-\sum_i p_i \log p_i \le \log N \tag{15-38}$$

PROOF The numbers $a_i = p_i$ and $b_i = 1/N$ satisfy (15-36). Inserting into (15-37), we conclude that

$$-\sum_i p_i \log p_i \le -\sum_i p_i \log \frac{1}{N} = \log N \sum_i p_i = \log N$$

Conditional Entropy and Mutual Information

The conditional entropy of a partition $\mathfrak{A}$ assuming $\mathscr{M}$ is by definition the sum

$$H(\mathfrak{A} \mid \mathscr{M}) = -\sum_{i=1}^{N_{\mathfrak{A}}} P(\mathscr{A}_i \mid \mathscr{M}) \log P(\mathscr{A}_i \mid \mathscr{M}) \tag{15-39}$$

where $\mathscr{M}$ is an event with $P(\mathscr{M}) \ne 0$ and

$$P(\mathscr{A}_i \mid \mathscr{M}) = \frac{P(\mathscr{A}_i \mathscr{M})}{P(\mathscr{M})}$$

As we explain later, $H(\mathfrak{A} \mid \mathscr{M})$ is the uncertainty about $\mathfrak{A}$ in the subsequence of trials in which $\mathscr{M}$ occurs.

Suppose now that $\mathfrak{B}$ is a partition consisting of the $N_{\mathfrak{B}}$ elements $\mathscr{B}_j$. Clearly

$$H(\mathfrak{A} \mid \mathscr{B}_j) = -\sum_{i=1}^{N_{\mathfrak{A}}} P(\mathscr{A}_i \mid \mathscr{B}_j) \log P(\mathscr{A}_i \mid \mathscr{B}_j) \tag{15-40}$$

is the conditional entropy of $\mathfrak{A}$ assuming $\mathscr{B}_j$ defined as in (15-39). The conditional entropy of $\mathfrak{A}$ assuming $\mathfrak{B}$ is the weighted average of $H(\mathfrak{A} \mid \mathscr{B}_j)$

$$H(\mathfrak{A} \mid \mathfrak{B}) = \sum_{j=1}^{N_{\mathfrak{B}}} P(\mathscr{B}_j) H(\mathfrak{A} \mid \mathscr{B}_j) \tag{15-41}$$

This equals the uncertainty about $\mathfrak{A}$ if at each trial we know which of the events $\mathscr{B}_j$ of $\mathfrak{B}$ has occurred.

Example 15-7 We shall determine the conditional entropy $H(\mathfrak{S} \mid \mathfrak{B})$ of the element partition $\mathfrak{S}$ in the fair-die experiment where $\mathfrak{B} = [\text{even, odd}]$.

Clearly $P\{f_i \mid \text{even}\} = \frac{1}{3}$ if i is even and $P\{f_i \mid \text{even}\} = 0$ if i is odd. Similarly, $P\{f_i \mid \text{odd}\} = \frac{1}{3}$ if i is odd and $P\{f_i \mid \text{odd}\} = 0$ if i is even. Hence

$$H(\mathfrak{S} \mid \text{even}) = -(\tfrac{1}{3} \log \tfrac{1}{3} + \tfrac{1}{3} \log \tfrac{1}{3} + \tfrac{1}{3} \log \tfrac{1}{3}) = \log 3 = H(\mathfrak{S} \mid \text{odd})$$

And since $P\{\text{even}\} = P\{\text{odd}\} = 0.5$, we conclude from (15-41) that

$$H(\mathfrak{S} \mid \mathfrak{B}) = 0.5 \log 3 + 0.5 \log 3 = \log 3$$

Thus, in the absence of any information, our uncertainty about $\mathfrak{S}$ equals $H(\mathfrak{S}) = \log 6$. If we know, however, whether at each trial "even" or "odd" showed, then our uncertainty is reduced to $H(\mathfrak{S} \mid \mathfrak{B}) = \log 3$.

Theorem 1
If

$$\mathfrak{B} \prec \mathfrak{A} \qquad \text{then} \qquad H(\mathfrak{A} \mid \mathfrak{B}) = 0 \tag{15-42}$$

PROOF Since $\mathfrak{B}$ is a refinement of $\mathfrak{A}$, each element $\mathscr{B}_j$ of $\mathfrak{B}$ is a subset of some element $\mathscr{A}_k$ of $\mathfrak{A}$ and, therefore, it is disjoint with all other elements of $\mathfrak{A}$. Hence, $\mathscr{A}_i \mathscr{B}_j = \mathscr{B}_j$ if $i = k$ and $\mathscr{A}_i \mathscr{B}_j = 0$ otherwise. This leads to the conclusion that

$$P(\mathscr{A}_i \mid \mathscr{B}_j) = \frac{P(\mathscr{A}_i \mathscr{B}_j)}{P(\mathscr{B}_j)} = \begin{cases} 1 & i = k \\ 0 & i \neq k \end{cases}$$

And since $p \log p = 0$ for $p = 0$ and $p = 1$, we conclude that all terms in (15-40) equal zero, hence

$$H(\mathfrak{A} \mid \mathscr{B}_j) = 0$$

for every j. From this and (15-41) it follows that $H(\mathfrak{A} \mid \mathfrak{B}) = 0$.

Independent partitions Two partitions $\mathfrak{A} = [\mathscr{A}_i]$ and $\mathfrak{B} = [\mathscr{B}_j]$ are called independent if the events $\mathscr{A}_i$ and $\mathscr{B}_j$ are independent for every i and j

$$P(\mathscr{A}_i \mathscr{B}_j) = P(\mathscr{A}_i)P(\mathscr{B}_j) \tag{15-43}$$

Theorem 2 If the partitions $\mathfrak{A}$ and $\mathfrak{B}$ are independent, then

$$H(\mathfrak{A} \mid \mathfrak{B}) = H(\mathfrak{A}) \qquad H(\mathfrak{B} \mid \mathfrak{A}) = H(\mathfrak{B}) \tag{15-44}$$

PROOF Clearly, $P(\mathscr{A}_i \mid \mathscr{B}_j) = P(\mathscr{A}_i)$, hence [see (15-40)]

$$H(\mathfrak{A} \mid \mathscr{B}_j) = -\sum_i P(\mathscr{A}_i) \log P(\mathscr{A}_i) = H(\mathfrak{A})$$

Inserting into (15-41), we obtain

$$H(\mathfrak{A} \mid \mathfrak{B}) = H(\mathfrak{A}) \sum_j P(\mathscr{B}_j) = H(\mathfrak{A})$$

and (15-43) results. We can show similarly that $H(\mathfrak{B} \mid \mathfrak{A}) = H(\mathfrak{B})$.

Theorem 3 For any $\mathfrak{A}$ and $\mathfrak{B}$

$$H(\mathfrak{A} \cdot \mathfrak{B}) \le H(\mathfrak{A}) + H(\mathfrak{B}) \tag{15-45}$$

PROOF As we know [see (2-36)]

$$P(\mathscr{A}_i) = \sum_j P(\mathscr{A}_i \mathscr{B}_j)$$

Hence

$$H(\mathfrak{A}) = -\sum_i P(\mathscr{A}_i) \log P(\mathscr{A}_i) = -\sum_{i,j} P(\mathscr{A}_i \mathscr{B}_j) \log P(\mathscr{A}_i)$$

Writing a similar equation for $H(\mathfrak{B})$ and adding, we obtain

$$H(\mathfrak{A}) + H(\mathfrak{B}) = -\sum_{i,j} P(\mathscr{A}_i \mathscr{B}_j) \log [P(\mathscr{A}_i)P(\mathscr{B}_j)] \tag{15-46}$$

Clearly, $H(\mathfrak{A} \cdot \mathfrak{B})$ is a partition with elements $\mathscr{A}_i \mathscr{B}_j$. Hence

$$H(\mathfrak{A} \cdot \mathfrak{B}) = -\sum_{i,j} P(\mathscr{A}_i \mathscr{B}_j) \log P(\mathscr{A}_i \mathscr{B}_j) \tag{15-47}$$

To prove (15-45), we shall apply (15-37) identifying the numbers a_i and b_i with the numbers $P(\mathscr{A}_i\mathscr{B}_j)$ and $P(\mathscr{A}_i)P(\mathscr{B}_j)$ respectively. We can do so because

$$\sum_{i,j} P(\mathscr{A}_i\mathscr{B}_j) = 1 \qquad \sum_{i,j} P(\mathscr{A}_i)P(\mathscr{B}_j) = 1$$

From (15-37) it follows that the sum in (15-47) cannot exceed the sum in (15-46), hence (15-45) must be true.

Corollary

$$H(\mathfrak{A} \cdot \mathfrak{B}) = H(\mathfrak{A}) + H(\mathfrak{B}) \tag{15-48}$$

iff the partitions $\mathfrak{A}$ and $\mathfrak{B}$ are independent.

PROOF This follows from (15-45) because (15-37) is an equality iff $a_i = b_i$ for every i. Hence (15-45) is an equality iff

$$P(\mathscr{A}_i\mathscr{B}_j) = P(\mathscr{A}_i)P(\mathscr{B}_j)$$

for every i and j.

Theorem 4 For any $\mathfrak{A}$ and $\mathfrak{B}$

$$H(\mathfrak{A} \cdot \mathfrak{B}) = H(\mathfrak{B}) + H(\mathfrak{A} \mid \mathfrak{B}) = H(\mathfrak{A}) + H(\mathfrak{B} \mid \mathfrak{A}) \tag{15-49}$$

PROOF Since

$$P(\mathscr{A}_i\mathscr{B}_j) = P(\mathscr{B}_j)P(\mathscr{A}_i \mid \mathscr{B}_j)$$

we conclude from (15-40) that

$$\begin{aligned} P(\mathscr{B}_j)H(\mathfrak{A} \mid \mathscr{B}_j) &= -\sum_i P(\mathscr{B}_j)P(\mathscr{A}_i \mid \mathscr{B}_j) \log P(\mathscr{A}_i \mid \mathscr{B}_j) \\ &= -\sum_i P(\mathscr{A}_i\mathscr{B}_j)[\log P(\mathscr{A}_i\mathscr{B}_j) - \log P(\mathscr{B}_j)] \\ &= -\sum_i P(\mathscr{A}_i\mathscr{B}_j) \log P(\mathscr{A}_i\mathscr{B}_j) - P(\mathscr{B}_j) \log P(\mathscr{B}_j) \end{aligned}$$

Summing over all j, we obtain

$$\sum_j P(\mathscr{B}_j)H(\mathfrak{A} \mid \mathscr{B}_j) = -\sum_{i,j} P(\mathscr{A}_i\mathscr{B}_j) \log P(\mathscr{A}_i\mathscr{B}_j) + \sum_j P(\mathscr{B}_j) \log P(\mathscr{B}_j)$$

and the first equation in (15-49) follows because the above three sums equal $H(\mathfrak{A} \mid \mathfrak{B})$, $H(\mathfrak{A} \cdot \mathfrak{B})$, and $-H(\mathfrak{B})$, respectively. The second equation follows because $\mathfrak{A} \cdot \mathfrak{B} = \mathfrak{B} \cdot \mathfrak{A}$.

Corollaries The following relationships follow readily from the last two theorems: For any $\mathfrak{A}$ and $\mathfrak{B}$

$$H(\mathfrak{B}) \le H(\mathfrak{A} \cdot \mathfrak{B}) \le H(\mathfrak{A}) + H(\mathfrak{B}) \tag{15-50}$$

$$H(\mathfrak{A} \mid \mathfrak{B}) \le H(\mathfrak{A}) \tag{15-51}$$

$$H(\mathfrak{A}) - H(\mathfrak{A} \mid \mathfrak{B}) = H(\mathfrak{B}) - H(\mathfrak{B} \mid \mathfrak{A}) \tag{15-52}$$

Mutual information The function

$$I(\mathfrak{A}, \mathfrak{B}) = H(\mathfrak{A}) + H(\mathfrak{B}) - H(\mathfrak{A} \cdot \mathfrak{B}) \tag{15-53}$$

is called the mutual information of the partitions $\mathfrak{A}$ and $\mathfrak{B}$. From (15-49) it follows that

$$I(\mathfrak{A}, \mathfrak{B}) = H(\mathfrak{A}) - H(\mathfrak{A} \mid \mathfrak{B}) = H(\mathfrak{B}) - H(\mathfrak{B} \mid \mathfrak{A}) \tag{15-54}$$

Clearly [see (15-51)]

$$I(\mathfrak{A}, \mathfrak{B}) \geq 0 \tag{15-55}$$

As we shall presently see, $I(\mathfrak{A}, \mathfrak{B})$ can be interpreted as the "information about $\mathfrak{A}$ contained in $\mathfrak{B}$" and it equals the "information about $\mathfrak{B}$ contained in $\mathfrak{A}$."

Example 15-8 In the fair-die experiment of Example 15-7

$$H(\mathfrak{S}) = \log 6 \qquad H(\mathfrak{S} \mid \mathfrak{B}) = \log 3 \qquad H(\mathfrak{B}) = \log 2 \qquad H(\mathfrak{B} \mid \mathfrak{S}) = 0$$

Hence

$$I(\mathfrak{S}, \mathfrak{B}) = \log 2$$

Thus, the information about the element partition $\mathfrak{S}$ resulting from the observation of the even-odd partition $\mathfrak{B}$ equals log 2.

Generalizations The preceding results can be readily generalized to an arbitrary number of partitions. We list below several special cases leaving the simple proofs as problems:

(*a*) If

$$\mathfrak{B} \prec \mathfrak{C} \text{ then } H(\mathfrak{A} \mid \mathfrak{B}) \leq H(\mathfrak{A} \mid \mathfrak{C}) \tag{15-56}$$

(*b*) If the partitions $\mathfrak{A}$, $\mathfrak{B}$, and $\mathfrak{C}$ are independent, then

$$H(\mathfrak{A} \cdot \mathfrak{B} \cdot \mathfrak{C}) = H(\mathfrak{A}) + H(\mathfrak{B} \cdot \mathfrak{C}) = H(\mathfrak{A}) + H(\mathfrak{B}) + H(\mathfrak{C}) \tag{15-57}$$

(*c*) *Chain rule* For any $\mathfrak{A}$, $\mathfrak{B}$, and $\mathfrak{C}$

$$H(\mathfrak{B} \cdot \mathfrak{C} \mid \mathfrak{A}) = H(\mathfrak{B} \mid \mathfrak{A}) + H(\mathfrak{C} \mid \mathfrak{A} \cdot \mathfrak{B}) \tag{15-58}$$

$$H(\mathfrak{A} \cdot \mathfrak{B} \cdot \mathfrak{C}) = H(\mathfrak{A}) + H(\mathfrak{B} \cdot \mathfrak{C} \mid \mathfrak{A}) = H(\mathfrak{A}) + H(\mathfrak{B} \mid \mathfrak{A}) + H(\mathfrak{C} \mid \mathfrak{A} \cdot \mathfrak{B}) \tag{15-59}$$

Repeated trials In the space $\mathscr{S}^n$ of repeated trials all outcomes are sequences of the form

$$\zeta_{i_1} \cdots \zeta_{i_k} \cdots \zeta_{i_n} \tag{15-60}$$

where each ζ_{i_k} is an element of $\mathscr{S}$. Consider a partition $\mathfrak{A}$ of $\mathscr{S}$ consisting of N events. At the kth trial, one and only one of these events will occur, namely the even $\mathscr{A}_{j_k}$ that contains the element ζ_{i_k}. The cartesian product

$$\mathscr{A}_{j,k} = \mathscr{S} \times \cdots \mathscr{S} \times \mathscr{A}_{j_k} \times \mathscr{S} \cdots \times \mathscr{S} \qquad \zeta_{i_k} \in \mathscr{A}_{j_k} \tag{15-61}$$

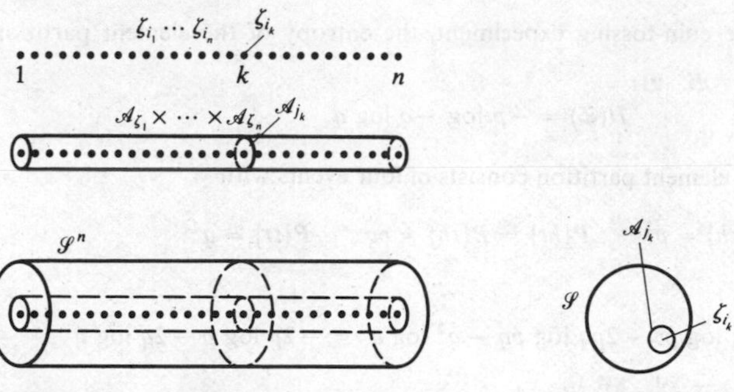

Figure 15-11

is an event in $\mathscr{S}^n$ with probability

$$P(\mathscr{A}_{j,k}) = P(\mathscr{A}_{jk}) \tag{15-62}$$

because it occurs iff the event $\mathscr{A}_{jk}$ occurs at the kth trial. For a specific k, the events $\mathscr{A}_{j,k}$ form an N element partition of the space $\mathscr{S}^n$. This partition will be denoted by $\mathfrak{A}_k$. From (15-62) it follows readily that

$$H(\mathfrak{A}_k) = H(\mathfrak{A}) \tag{15-63}$$

We can define similarly the partition $\mathfrak{B}_k$ of $\mathscr{S}^n$ formed with the elements of another partition $\mathfrak{B}$ of $\mathscr{S}$. Reasoning as in (15-63), we conclude that $H(\mathfrak{B}_k) = H(\mathfrak{B})$ and

$$H(\mathfrak{A}_k \mid \mathfrak{B}_k) = H(\mathfrak{A} \mid \mathfrak{B}) \qquad I(\mathfrak{A}_k, \mathfrak{B}_k) = I(\mathfrak{A}, \mathfrak{B}) \tag{15-64}$$

We next form the product of the n partitions $\mathfrak{A}_k$

$$\mathfrak{A}^n = \mathfrak{A}_1 \cdot \mathfrak{A}_2 \cdots \mathfrak{A}_n \tag{15-65}$$

The elements of this partition are cartesian products of the form

$$\mathscr{A}_{j1} \times \cdots \times \mathscr{A}_{jk} \times \cdots \times \mathscr{A}_{jn} \tag{15-66}$$

If $\mathfrak{A}$ is the element partition of $\mathscr{S}$, then $\mathfrak{A}^n$ is the element partition of $\mathscr{S}^n$. In general, however, the elements of $\mathfrak{A}^n$ are events consisting of a large number of sequences of the form (15-60). If we picture these sequences as wires, then the elements (15-66) of the partition $\mathfrak{A}^n$ can be viewed as cables and their union as a collection of such cables (Fig. 15-11).

From the independence of the trials, it follows that the n partitions $\mathfrak{A}_1, \ldots, \mathfrak{A}_n$ of $\mathscr{S}^n$ are independent. Hence [see (15-57) and (15-63)]

$$H(\mathfrak{A}^n) = H(\mathfrak{A}_1) + \cdots + H(\mathfrak{A}_n) = nH(\mathfrak{A}) \tag{15-67}$$

Defining similarly the partition $\mathfrak{B}^n$, we conclude as in (15-64) that

$$H(\mathfrak{A}^n \mid \mathfrak{B}^n) = nH(\mathfrak{B} \mid \mathfrak{A}) \qquad I(\mathfrak{A}^n, \mathfrak{B}^n) = nI(\mathfrak{A}, \mathfrak{B}) \tag{15-68}$$

Example 15-9 In the coin-tossing experiment, the entropy of the element partition equals

$$H(\mathfrak{S}) = -p \log -q \log q$$

In the space $\mathscr{S}^2$, the element partition consists of four events with

$$P\{hh\} = p^2 \qquad P\{ht\} = P\{th\} = pq \qquad P\{tt\} = q^2$$

Hence

$$H(\mathfrak{S}^2) = -p^2 \log p^2 - 2pq \log pq - q^2 \log q^2 = -2p \log p - 2q \log q$$

Thus

$$H(\mathfrak{S}^2) = 2H(\mathfrak{S})$$

in agreement with (15-67).

Conditional entropy and uncertainty As we have noted, the entropy $H(\mathfrak{A})$ of a partition $\mathfrak{A} = [\mathscr{A}_i]$ gives us a measure of uncertainty about the occurrence of the events $\mathscr{A}_i$ at a given trial. Once the trial is performed and the events $\mathscr{A}_i$ are observed, the uncertainty is removed. We give next a similar interpretation to the conditional entropy $H(\mathfrak{A} \mid \mathscr{M})$ of $\mathfrak{A}$ assuming that the event $\mathscr{M}$ has been observed, and of the conditional entropy $H(\mathfrak{A} \mid \mathfrak{B})$ of $\mathfrak{A}$ assuming that the partition $\mathfrak{B}$ has been observed.†

If in the definition (15-25) of entropy we replace the probabilities $P(\mathscr{A}_i)$ by the conditional probabilities $P(\mathscr{A}_i \mid \mathscr{M})$, we obtain the conditional entropy $H(\mathfrak{A} \mid \mathscr{M})$ of $\mathfrak{A}$ assuming $\mathscr{M}$ [see (15-39)]. The relative frequency interpretation of $P(\mathscr{A}_i \mid \mathscr{M})$ is the same as that of $P(\mathscr{A}_i)$ if we consider not the entire sequence of n trials but only the subsequence of trials in which the event $\mathscr{M}$ occurs. From this it follows that $H(\mathfrak{A} \mid \mathscr{M})$ is the uncertainty about $\mathfrak{A}$ per trial in that subsequence. In other words, if at a given trial we known that $\mathscr{M}$ occurs then our uncertainty about $\mathfrak{A}$ equals $H(\mathfrak{A} \mid \mathscr{M})$; if we know that $\bar{\mathscr{M}}$ occurs, then our uncertainty equals $H(\mathfrak{A} \mid \bar{\mathscr{M}})$. The weighted sum

$$P(\mathscr{M})H(\mathfrak{A} \mid \mathscr{M}) + P(\bar{\mathscr{M}})H(\mathfrak{A} \mid \bar{\mathscr{M}})$$

is the uncertainty about $\mathfrak{A}$ assuming that the binary partition $[\mathscr{M}, \bar{\mathscr{M}}]$ is observed.

Suppose now that at each trial we observe the partition $\mathfrak{B} = [\mathscr{B}_j]$. We maintain that, under this assumption, the uncertainty per trial about $\mathfrak{A}$ equals $H(\mathfrak{A} \mid \mathfrak{B})$. Indeed, in a sequence of n trials, the number of times the event $\mathscr{B}_j$ occurs equals

$$n_j \simeq nP(\mathscr{B}_j)$$

† The expression *a partition* $\mathfrak{B}$ *is observed* will mean that we know which of the events of $\mathfrak{B}$ has occurred.

In this subsequence, the uncertainty about $\mathfrak{A}$ equals $H(\mathfrak{A} \mid \mathscr{B}_j)$ per trial. Hence, the total uncertainty about $\mathfrak{A}$ equals

$$\sum_j n_j H(\mathfrak{A} \mid \mathscr{B}_j) \simeq \sum_j nP(\mathscr{B}_j)H(\mathfrak{A} \mid \mathscr{B}_j) = nH(\mathfrak{A} \mid \mathfrak{B})$$

and the uncertainty per trial equals $H(\mathfrak{A} \mid \mathfrak{B})$.

Thus, the observation of $\mathfrak{B}$ reduces the uncertainty about $\mathfrak{A}$ from $H(\mathfrak{A})$ to $H(\mathfrak{A} \mid \mathfrak{B})$. The difference

$$I(\mathfrak{A}, \mathfrak{B}) = H(\mathfrak{A}) - H(\mathfrak{A} \mid \mathfrak{B})$$

is the reduction of the uncertainty about $\mathfrak{A}$ resulting from the observation of $\mathfrak{B}$. This justifies the statement that the mutual information $I(\mathfrak{A}, \mathfrak{B})$ equals the *information* about $\mathfrak{A}$ contained in $\mathfrak{B}$.

We show next the consistency between the properties of entropy developed earlier and the subjective notion of uncertainty.

1. If $\mathfrak{B}$ is a refinement of $\mathfrak{A}$ and $\mathfrak{B}$ is observed, then we know which of the events of $\mathfrak{A}$ occurred. Hence, $H(\mathfrak{A} \mid \mathfrak{B}) = 0$ in agreement with (15-42).
2. If the partitions $\mathfrak{A}$ and $\mathfrak{B}$ are independent and $\mathfrak{B}$ is observed, no information about $\mathfrak{A}$ is gained. Hence, $H(\mathfrak{A} \mid \mathfrak{B}) = H(\mathfrak{A})$ in agreement with (15-44).
3. If we observe $\mathfrak{B}$, our uncertainty about $\mathfrak{A}$ can only decrease. Hence, $H(\mathfrak{A} \mid \mathfrak{B}) \le H(\mathfrak{A})$ in agreement with (15-51).
4. To observe $\mathfrak{A} \cdot \mathfrak{B}$, we must observe $\mathfrak{A}$ and $\mathfrak{B}$. If only $\mathfrak{B}$ is observed, the information gained equals $H(\mathfrak{B})$. The uncertainty about $\mathfrak{A}$ assuming $\mathfrak{B}$ equals, therefore, the remaining uncertainty $H(\mathfrak{A} \mid \mathfrak{B})$ about $\mathfrak{B}$. Hence, $H(\mathfrak{A} \cdot \mathfrak{B}) - H(\mathfrak{B}) = H(\mathfrak{A} \mid \mathfrak{B})$ in agreement with (15-49).
5. Combining 3 and 4, we conclude that $H(\mathfrak{A} \cdot \mathfrak{B}) - H(\mathfrak{B}) \le H(\mathfrak{A})$ in agreement with (15-45).
6. If $\mathfrak{B}$ is observed, then the information that is gained about $\mathfrak{A}$ equals $H(\mathfrak{A} \mid \mathfrak{B})$. If $\mathfrak{B} \prec \mathfrak{C}$ and $\mathfrak{B}$ is observed, then $\mathfrak{C}$ is known. But knowledge of $\mathfrak{C}$ yields information about $\mathfrak{A}$ equal to $H(\mathfrak{A} \mid \mathfrak{C})$. Hence, if $\mathfrak{B} \prec \mathfrak{C}$, then $I(\mathfrak{A}, \mathfrak{B}) \ge I(\mathfrak{A}, \mathfrak{C})$ or, equivalently, $H(\mathfrak{A} \mid \mathfrak{B}) \le H(\mathfrak{A} \mid \mathfrak{C})$ in agreement with (15-56).

Conditional entropy and typical sequences We give next a typical sequence interpretation of the properties of conditional entropy limiting the discussion to (15-45) and (15-49). The underlying reasoning is used in the proof of the channel capacity theorem (Sec. 15-6).

We denote by $\mathfrak{t}^{\mathfrak{A}}$, $\mathfrak{t}^{\mathfrak{B}}$, and $\mathfrak{t}^{\mathfrak{A} \cdot \mathfrak{B}}$ the typical sequences of the partition $\mathfrak{A}$, $\mathfrak{B}$, and $\mathfrak{A} \cdot \mathfrak{B}$ respectively, and by $\mathbb{T}^{\mathfrak{A}}$, $\mathbb{T}^{\mathfrak{B}}$, and $\mathbb{T}^{\mathfrak{A} \cdot \mathfrak{B}}$ their unions (Fig. 15-12*a*). As we know [see (15-7)]

$$P(\mathbb{T}^{\mathfrak{A}}) \simeq P(\mathbb{T}^{\mathfrak{B}}) \simeq P(\mathbb{T}^{\mathfrak{A} \cdot \mathfrak{B}}) \simeq 1$$

Furthermore, the number of typical sequences in each of the above three sets equals [see (15-10)]

$$n_{\mathbb{T}^{\mathfrak{A}}} \simeq e^{nH(\mathfrak{A})} \qquad n_{\mathbb{T}^{\mathfrak{B}}} \simeq e^{nH(\mathfrak{B})} \qquad n_{\mathbb{T}^{\mathfrak{A} \cdot \mathfrak{B}}} \simeq e^{nH(\mathfrak{A} \cdot \mathfrak{B})} \tag{15-69}$$

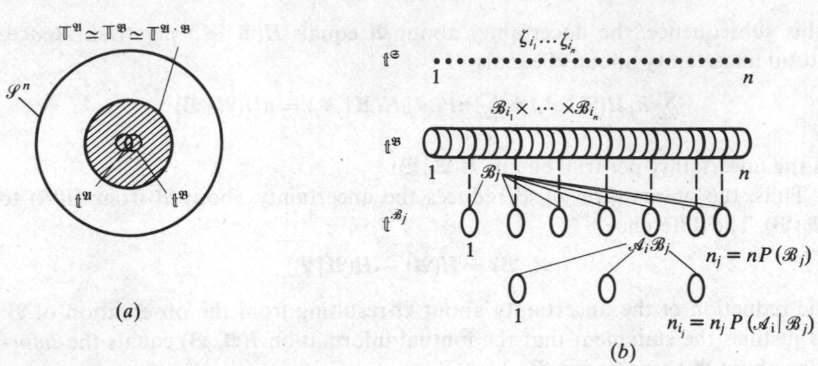

Figure 15-12

I. We maintain that

$$H(\mathfrak{A} \cdot \mathfrak{B}) \leq H(\mathfrak{A}) + H(\mathfrak{B}) \tag{15-70}$$

PROOF Each $t^{\mathfrak{A} \cdot \mathfrak{B}}$ sequence specifies a pair $(t^{\mathfrak{A}}, t^{\mathfrak{B}})$. The total number of such pairs formed with all the elements of $\mathbb{T}^{\mathfrak{A}}$ and $\mathbb{T}^{\mathfrak{B}}$ equals $n_{\mathbb{T}\mathfrak{A}} \cdot n_{\mathbb{T}\mathfrak{B}}$. However, not all such pairs generate $t^{\mathfrak{A} \cdot \mathfrak{B}}$ sequences because, if the partitions $\mathfrak{A}$ and $\mathfrak{B}$ are not independent, then not all pairs can occur. For example, if $\mathscr{A}_i = \mathscr{B}_j$ for some i and j and $\mathscr{A}_i$ occurs at the kth trial, then B_j must also occur at this trial. From the above it follows that

$$n_{\mathbb{T}\mathfrak{A} \cdot \mathfrak{B}} \leq n_{\mathbb{T}\mathfrak{A}} \cdot n_{\mathbb{T}\mathfrak{B}}$$

and (15-70) follows from (15-69).

II. We shall show, finally, that

$$H(\mathfrak{A} \cdot \mathfrak{B}) = H(\mathfrak{B}) + H(\mathfrak{A} \mid \mathfrak{B}) \tag{15-71}$$

PROOF There are $n_{\mathbb{T}\mathfrak{B}}$ sequences in the set $\mathbb{T}^{\mathfrak{B}}$ and $n_{\mathbb{T}\mathfrak{A} \cdot \mathfrak{B}}$ sequences in the set $\mathbb{T}^{\mathfrak{A} \cdot \mathfrak{B}}$. The ratio

$$\frac{n_{\mathbb{T}\mathfrak{A} \cdot \mathfrak{B}}}{n_{\mathbb{T}\mathfrak{B}}} \simeq e^{n[H(\mathfrak{A} \cdot \mathfrak{B}) - H(\mathfrak{B})]}$$

equals, therefore, the number of $t^{\mathfrak{A} \cdot \mathfrak{B}}$ sequences contained in a single $t^{\mathfrak{B}}$ sequence *on the average.* To prove (15-71), we must prove, therefore, that this number equals $e^{nH(\mathfrak{A}\mid\mathfrak{B})}$. We shall prove a stronger statement: The number of $t^{\mathfrak{A} \cdot \mathfrak{B}}$ sequences contained in a *single* $t^{\mathfrak{B}}$ sequence (Fig. 15-13) equals $e^{nH(\mathfrak{A}\mid\mathfrak{B})}$.

As we know [see (1-1)], the number of times the event $\mathscr{B}_j$ occurs in a $t^{\mathfrak{B}}$ sequence "almost certainly" equals

$$n_j \simeq nP(\mathscr{B}_j) \tag{15-72}$$

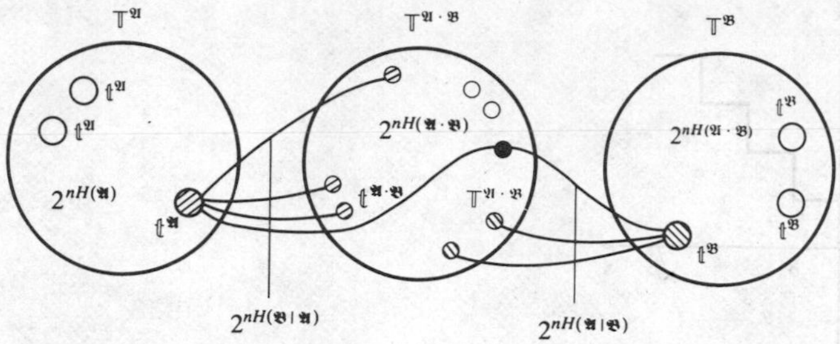

Figure 15-13

We denote by $t^{\mathscr{B}_j}$ a subsequence (Fig. 15-12*b*) of $t^{\mathfrak{B}}$ in which the number of occurrences of $\mathscr{B}_j$ satisfies (15-72). In this subsequence, the relative frequency of the occurrence of an event $\mathscr{A}_i$ equals $P(\mathscr{A}_i \mid \mathscr{B}_j)$ [see (2-32)].

We shall use (15-10) to show that the number of typical sequences formed with the elements $\mathscr{A}_i$ of $\mathfrak{A}$ that are included in a $t^{\mathscr{B}_j}$ sequence equals

$$e^{n_j H(\mathfrak{A}|\mathscr{B}_j)} \simeq e^{nP(\mathscr{B}_j)H(\mathfrak{A}|\mathscr{B}_j)} \tag{15-73}$$

Indeed, this follows from (15-10) if we introduce the following changes: We replace $P(\mathscr{A}_i)$ by $P(\mathscr{A}_i \mid \mathscr{B}_j)$, the length n of the original sequences with the length $n_j \simeq nP(\mathscr{B}_j)$, and the entropy $H(\mathfrak{A})$ of $\mathfrak{A}$ with the conditional entropy $H(\mathfrak{A} \mid \mathscr{B}_j)$.

Returning to the original $t^{\mathfrak{B}}$ sequence, we note that it is formed by combining the $t^{\mathscr{B}_j}$ sequences that are included in $t^{\mathfrak{B}}$. This shows that the total number of $t^{\mathfrak{A}}$ sequences that are included in $t^{\mathfrak{B}}$ equals the product

$$\prod_j e^{nP(\mathscr{B}_j)H(\mathfrak{A}|\mathscr{B}_j)} = e^{nH(\mathfrak{A}|\mathfrak{B})} \tag{15-74}$$

But each $t^{\mathfrak{A}}$ sequence that is included in $t^{\mathfrak{B}}$ is a $t^{\mathfrak{A}\cdot\mathfrak{B}}$ sequence. Hence, the number of $t^{\mathfrak{A}\cdot\mathfrak{B}}$ sequences that is included in $t^{\mathfrak{B}}$ equals $e^{nH(\mathfrak{A}|\mathfrak{B})}$.

15-3 RANDOM VARIABLES AND STOCHASTIC PROCESSES

Entropy is a number assigned to a partition. To define the entropy of an RV we must, therefore, form a suitable partition. This is simple if the RV is of discrete type. However, for continuous-type RVs we can do so only indirectly.

Discrete type Suppose that $\mathbf{x}$ is a RV taking the values x_i with

$$P\{\mathbf{x} = x_i\} = p_i$$

The events $\{\mathbf{x} = x_i\}$ are mutually exclusive and their union is the certain event, hence they form a partition. This partition will be denoted by $\mathfrak{A}_x$ and will be called the partition of $\mathbf{x}$.

Definition The entropy $H(\mathbf{x})$ of a discrete-type RV $\mathbf{x}$ is the entropy $H(\mathfrak{A}_x)$ of its partition $\mathfrak{A}_x$

$$H(\mathbf{x}) = H(\mathfrak{A}_x) = -\sum_i p_i \ln p_i \tag{15-75}$$

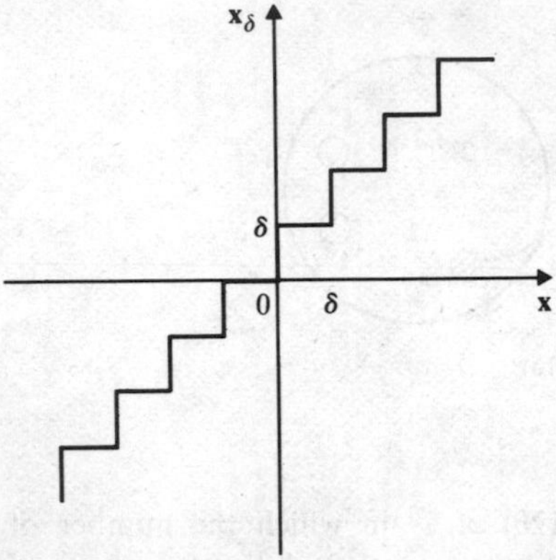

Figure 15-14

Continuous type The entropy of a continuous-type RV cannot be so defined because the events $\{\mathbf{x} = x_i\}$ do not form a partition (they are not countable). To define $H(\mathbf{x})$ we form, first, the discrete-type RV $\mathbf{x}_\delta$ obtained by rounding off $\mathbf{x}$ as in Fig. 15-14

$$\mathbf{x}_\delta = n\delta \qquad \text{if} \qquad n\delta - \delta < \mathbf{x} \le n\delta \tag{15-76}$$

Clearly

$$P\{\mathbf{x}_\delta = n\delta\} = P\{n\delta - \delta < \mathbf{x} \le \delta\} = \int_{n\delta-\delta}^{n\delta} f(x)\,dx = \delta\bar{f}(n\delta)$$

where $\bar{f}(n\delta)$ is a number between the maximum and the minimum of $f(x)$ in the interval $(n\delta - \delta, n\delta)$. Applying (15-75) to the RV $\mathbf{x}_\delta$, we obtain

$$H(\mathbf{x}_\delta) = -\sum_{n=-\infty}^{\infty} \delta\bar{f}(n\delta) \ln [\delta\bar{f}(n\delta)]$$

and since

$$\sum_{n=-\infty}^{\infty} \delta\bar{f}(n\delta) = \int_{-\infty}^{\infty} f(x)\,dx = 1$$

we conclude that

$$H(\mathbf{x}_\delta) = -\ln \delta - \sum_{n=-\infty}^{\infty} \delta\bar{f}(n\delta) \ln \bar{f}(n\delta) \tag{15-77}$$

As $\delta \to 0$, the RV $\mathbf{x}_\delta$ tends to $\mathbf{x}$, however its entropy $H(\mathbf{x}_\delta)$ tends to infinity because $\ln \delta \to \infty$. For this reason, we define the entropy $H(\mathbf{x})$ of $\mathbf{x}$ not as the limit of $H(\mathbf{x}_\delta)$ but as the limit of the sum $H(\mathbf{x}_\delta) + \ln \delta$ as $\delta \to 0$. This yields

$$H(\mathbf{x}_\delta) + \ln \delta \xrightarrow[\delta \to 0]{} -\int_{-\infty}^{\infty} f(x) \ln f(x)\,dx \tag{15-78}$$

Definition The entropy of a continuous-type RV **x** is by definition the integral

$$H(\mathbf{x}) = -\int_{-\infty}^{\infty} f(x) \ln f(x)\, dx \qquad (15\text{-}79)$$

The integration extends only over the region where $f(x) \neq 0$ because $f(x) \ln f(x) = 0$ if $f(x) = 0$.

Example 15-10 If **x** is uniform in the interval $(0, a)$, then

$$H(\mathbf{x}) = -\frac{1}{a}\int_{0}^{a} \ln \frac{1}{a}\, dx = \ln a \qquad (15\text{-}80)$$

Notes 1. The entropy $H(\mathbf{x}_\delta)$ of $\mathbf{x}_\delta$ is a measure of our uncertainty about the RV **x** rounded off to the nearest $n\delta$. If δ is small, the resulting uncertainty is large and it tends to infinity as $\delta \to 0$. This conclusion is based on the assumption that **x** can be *observed perfectly*, i.e., that its various values can be recognized as distinct no matter how close they are. In a physical experiment, however, this assumption is not realistic. Values of **x** that differ slightly cannot always be treated as distinct (noise considerations or round-off errors, for example). The presence of the term $\ln \delta$ in (15-78) is, in a sense, a recognition of this ambiguity.

2. As in the case of arbitrary partitions, the entropy of a discrete-type RV **x** is positive and it is used as a measure of uncertainty about **x**. This is not so, however, for continuous-type RVs. Their entropy can take any value from $-\infty$ to ∞ and it is used to measure only changes in uncertainty. The various properties of partitions apply also to continuous-type RVs if, as is generally the case, they involve only differences of entropies.

Entropy as expected value The integral in (15-79) is the expected value of the RV $\mathbf{y} = -\ln f(\mathbf{x})$ obtained through the transformation $g(x) = -\ln f(x)$

$$H(\mathbf{x}) = E\{-\ln f(\mathbf{x})\} = -\int_{-\infty}^{\infty} f(x) \ln f(x)\, dx \qquad (15\text{-}81)$$

Similarly, the sum in (15-75) can be written as the expected value of the RV $-\ln p(\mathbf{x})$

$$H(\mathbf{x}) = E\{-\ln p(\mathbf{x})\} = -\sum_i p_i \ln p_i \qquad (15\text{-}82)$$

where now $p(x)$ is a function defined only for $x = x_i$ and such that $p(x_i) = p_i$.

Example 5-11 If

$$f(x) = ce^{-cx}U(x) \qquad \text{then} \qquad E\{-\ln f(\mathbf{x})\} = E\{c\mathbf{x} - \ln c\}$$

Since $E\{c\mathbf{x}\} = 1$, this yields

$$H(\mathbf{x}) = 1 - \ln c = \ln \frac{e}{c} \qquad (15\text{-}83)$$

Example 15-12 If

$$f(x) = \frac{1}{\sigma\sqrt{2\pi}} e^{-(x-\eta)^2/2\sigma^2}$$

then

$$E\{-\ln f(\mathbf{x})\} = \ln \sigma\sqrt{2\pi} + E\left\{\frac{(\mathbf{x}-\eta)^2}{2\sigma^2}\right\} = \ln \sigma\sqrt{2\pi} + \frac{\sigma^2}{2\sigma^2}$$

Hence the entropy of a *normal* RV equals

$$H(\mathbf{x}) = \ln \sigma\sqrt{2\pi e} \tag{15-84}$$

Joint entropy Suppose that **x** and **y** are two discrete-type RVs taking the values x_i and y_j respectively with

$$P\{\mathbf{x} = x_i, \mathbf{y} = y_j\} = p_{ij}$$

Their joint entropy, denoted by $H(\mathbf{x}, \mathbf{y})$, is by definition the entropy of the product of their respective partitions. Clearly, the elements of $\mathfrak{A}_x \cdot \mathfrak{A}_y$ are the events $\{\mathbf{x} = x_i, \mathbf{y} = y_j\}$. Hence

$$H(\mathbf{x}, \mathbf{y}) = H(\mathfrak{A}_x \cdot \mathfrak{A}_y) = -\sum_{i,j} p_{ij} \ln p_{ij}$$

The above can be written as expected value

$$H(\mathbf{x}, \mathbf{y}) = E\{-\ln p(\mathbf{x}, \mathbf{y})\}$$

where $p(x, y)$ is a function defined only for $x = x_i$ and $y = y_j$ and it is such that $p(x_i, y_j) = p_{ij}$.

The joint entropy $H(\mathbf{x}, \mathbf{y})$ of two continuous-type RVs **x** and **y** is defined as the limit of the sum

$$H(\mathbf{x}_\delta, \mathbf{y}_\delta) + 2 \ln \delta$$

where $\mathbf{x}_\delta$ and $\mathbf{y}_\delta$ are their staircase approximation. Reasoning as in (15-78), we obtain

$$H(\mathbf{x}, \mathbf{y}) = -\int_{-\infty}^{\infty}\int_{-\infty}^{\infty} f(x, y) \ln f(x, y)\, dx\, dy = E\{-\ln f(\mathbf{x}, \mathbf{y})\} \tag{15-85}$$

Example 15-13 If the RVs **x** and **y** are *jointly normal* as in (6-15), then

$$\ln f(x, y) = \frac{-1}{2(1-r^2)}\left[\frac{(x-\eta_1)^2}{\sigma_1^2} - 2r\frac{(x-\eta_1)(y-\eta_2)}{\sigma_1\sigma_2} + \frac{(y-\eta_2)^2}{\sigma_2^2}\right]$$
$$- \ln 2\pi\sigma_1\sigma_2\sqrt{1-r^2}$$

In this case

$$E\left\{\frac{(\mathbf{x}-\eta_1)^2}{\sigma_1^2} - 2r\frac{(\mathbf{x}-\eta_1)(\mathbf{y}-\eta_2)}{\sigma_1\sigma_2} + \frac{(\mathbf{y}-\eta_2)^2}{\sigma_2^2}\right\} = 1 - 2r^2 + 1$$

Hence

$$E\{-\ln f(\mathbf{x}, \mathbf{y})\} = 1 + \ln 2\pi\sigma_1\sigma_2\sqrt{1 - r^2}$$

From the above and (15-85) it follows that the joint entropy of two jointly normal RVs equals

$$H(\mathbf{x}, \mathbf{y}) = \ln 2\pi e\sqrt{\Delta} \tag{15-86}$$

where

$$\Delta = \mu_{11}\mu_{22} - \mu_{12}^2 \qquad \mu_{11} = \sigma_1^2 \qquad \mu_{22} = \sigma_2^2 \qquad \mu_{12} = r\sigma_1\sigma_2$$

Conditional entropy Consider two discrete-type RVs **x** and **y** taking the values x_i and y_j with

$$P\{\mathbf{x} = x_i \mid \mathbf{y} = y_j\} = \pi_{ji} = p_{ji}/p_j$$

The conditional entropy $H(\mathbf{x} \mid y_j)$ of **x** assuming $\mathbf{y} = y_j$ is by definition the conditional entropy of the partition $\mathfrak{A}_x$ of **x** assuming $\{\mathbf{y} = y_j\}$. From the above and (15-39) it follows that

$$H(\mathbf{x} \mid y_j) = -\sum_i \pi_{ji} \ln \pi_{ji} \tag{15-87}$$

The conditional entropy $H(\mathbf{x} \mid \mathbf{y})$ of **x** assuming **y** is the conditional entropy of $\mathfrak{A}_x$ assuming $\mathfrak{A}_y$. Thus [see (15-41)]

$$H(\mathbf{x} \mid \mathbf{y}) = -\sum_j p_j H(\mathbf{x} \mid y_j) = -\sum_{i,j} p_{ji} \ln \pi_{ji} \tag{15-88}$$

For continuous-type RVs the corresponding concepts are defined similarly

$$H(\mathbf{x} \mid y) = -\int_{-\infty}^{\infty} f(x \mid y) \ln f(x \mid y)\, dx \tag{15-89}$$

$$H(\mathbf{x} \mid \mathbf{y}) = -\int_{-\infty}^{\infty} f(y) H(\mathbf{x} \mid y)\, dy = \int_{-\infty}^{\infty}\int_{-\infty}^{\infty} f(x, y) \ln f(x \mid y)\, dx\, dy \tag{15-90}$$

The above integrals can be written as expected values [see also (7-59)]

$$H(\mathbf{x} \mid y) = E\{-\ln f(\mathbf{x} \mid \mathbf{y}) \mid \mathbf{y} = y\} \tag{15-91}$$

$$H(\mathbf{x} \mid \mathbf{y}) = E\{-\ln f(\mathbf{x} \mid \mathbf{y})\} = E\{E\{-\ln f(\mathbf{x} \mid \mathbf{y}) \mid \mathbf{y}\}\} \tag{15-92}$$

The discrete case leads to similar expressions.

Mutual information Guided by (15-53), we shall call the function

$$I(\mathbf{x}, \mathbf{y}) = H(\mathbf{x}) + H(\mathbf{y}) - H(\mathbf{x}, \mathbf{y}) \tag{15-93}$$

the mutual information of the RVs **x** and **y**.

From (15-81) and (15-85) it follows that $I(\mathbf{x}, \mathbf{y})$ can be written as expected value

$$I(\mathbf{x}, \mathbf{y}) = E\left[\ln \frac{f(\mathbf{x}, \mathbf{y})}{f(\mathbf{x})f(\mathbf{y})}\right] \tag{15-94}$$

Since $f(x, y) = f(x \mid y)f(y)$ it follows from the above and (15-92) that

$$I(\mathbf{x}, \mathbf{y}) = H(\mathbf{x}) - H(\mathbf{x} \mid \mathbf{y}) = H(\mathbf{y}) - H(\mathbf{y} \mid \mathbf{x}) \tag{15-95}$$

Example 15-14 If two RVs $\mathbf{x}$ and $\mathbf{y}$ are *jointly normal* with zero mean, then [see (7-42)] the conditional density $f(x \mid y)$ is normal with mean $r\sigma_x/\sigma_y$ and variance $\sigma_x^2(1 - r^2)$. From this and (15-84) it follows that

$$H(\mathbf{x} \mid y) = E\{-\ln f(\mathbf{x} \mid y)\} = \ln \sigma_x \sqrt{2\pi e(1 - r^2)} \qquad (15\text{-}96)$$

Since this is independent of y, it follows from (15-92) that

$$H(\mathbf{x} \mid \mathbf{y}) = H(\mathbf{x} \mid y) \qquad (15\text{-}97)$$

This yields [see (15-95)]

$$I(\mathbf{x}, \mathbf{y}) = H(\mathbf{x}) - H(\mathbf{x} \mid \mathbf{y}) = -0.5 \ln (1 - r^2) \qquad (15\text{-}98)$$

We note finally that [see (15-86)]

$$H(\mathbf{x} \mid \mathbf{y}) + H(\mathbf{y}) = \ln 2\pi e\sqrt{\Delta} = H(\mathbf{x}, \mathbf{y})$$

SPECIAL CASE Suppose that $\mathbf{y} = \mathbf{x} + \mathbf{n}$ where $\mathbf{n}$ is independent of $\mathbf{x}$ and $E\{\mathbf{n}^2\} = N$. In this case

$$E\{\mathbf{xy}\} = \sigma_x^2 \qquad E\{\mathbf{y}^2\} = \sigma_x^2 + N \qquad r^2 = \frac{\sigma_x^2}{\sigma_x^2 + N}$$

Inserting into (15-98), we obtain

$$I(\mathbf{x}, \mathbf{y}) = 0.5 \ln \left(1 + \frac{\sigma_x^2}{N}\right) \qquad (15\text{-}99)$$

Properties The properties of entropy, developed in Sec. 15-2 for arbitrary partitions, are obviously true for the entropy of discrete-type RVs and can be simply established as appropriate limits for continuous-type RVs. It might be of interest, however, to prove directly theorems (15-45) and (15-49) using the representation of entropy as expected value. The proofs are based on the following version of inequality (15-38): If $\mathbf{x}$ and $\mathbf{y}$ are two RVs with respective densities $a(x)$ and $b(y)$, then

$$E\{\ln a(\mathbf{x})\} \geq E\{\ln b(\mathbf{x})\} \qquad (15\text{-}100)$$

Equality holds iff $a(x) = b(x)$.

PROOF Applying the inequality $\ln z \leq z - 1$ to the function $z = b(x)/a(x)$, we obtain

$$\ln b(x) - \ln a(x) = \ln \frac{b(x)}{a(x)} \leq \frac{b(x)}{a(x)} - 1$$

Multiplying by $a(x)$ and integrating, we obtain

$$\int_{-\infty}^{\infty} a(x)[\ln b(x) - \ln a(x)]\, dx \leq \int_{-\infty}^{\infty} [b(x) - a(x)]\, dx = 0$$

and (15-100) results. The right side is zero because the functions $a(x)$ and $b(x)$ are densities by assumption.

Inequality (15-100) can be readily extended to n-dimensional densities. For example, if $a(x, y)$ and $b(z, w)$ are the joint densities of the RVs $\mathbf{x}$, $\mathbf{y}$ and $\mathbf{z}$, $\mathbf{w}$, respectively, then

$$E\{\ln a(\mathbf{x}, \mathbf{y})\} \geq E\{\ln b(\mathbf{x}, \mathbf{y})\} \tag{15-101}$$

Theorem 1

$$H(\mathbf{x}, \mathbf{y}) \leq H(\mathbf{x}) + H(\mathbf{y}) \tag{15-102}$$

PROOF Suppose that $f_{xy}(x, y)$ is the joint density of the RVs $\mathbf{x}$ and $\mathbf{y}$ and $f_x(x)$ and $f_y(y)$ their marginal densities. Clearly, the product $f_x(z) f_y(w)$ is the joint density of two independent RVs $\mathbf{z}$ and $\mathbf{w}$. Applying (15-101) we conclude that

$$E\{\ln f_{xy}(\mathbf{x}, \mathbf{y})\} \geq E\{\ln [f_x(\mathbf{x}) f_y(\mathbf{y})]\} = E\{\ln f_x(\mathbf{x})\} + E\{\ln f_y(\mathbf{y})\}$$

and (15-102) results.

Theorem 2

$$H(\mathbf{x}, \mathbf{y}) = H(\mathbf{x} \mid \mathbf{y}) + H(\mathbf{y}) = H(\mathbf{y} \mid \mathbf{x}) + H(\mathbf{x}) \tag{15-103}$$

PROOF Inserting the identity $f(x, y) = f(x \mid y) f(y)$ into (15-85), we obtain

$$H(\mathbf{x}, \mathbf{y}) = E\{-\ln f(\mathbf{x}, \mathbf{y})\} = E\{-\ln f(\mathbf{x} \mid \mathbf{y})\} + E\{-\ln f(\mathbf{y})\}$$

and the first equality in (15-103) results. The second follows because $H(\mathbf{x}, \mathbf{y}) = H(\mathbf{y}, \mathbf{x})$.

Corollary Comparing (15-102) with (15-103), we conclude that

$$H(\mathbf{x} \mid \mathbf{y}) \leq H(\mathbf{x}) \tag{15-104}$$

Note If the RV $\mathbf{y}$ is of discrete type, then $H(\mathbf{y} \mid \mathbf{x}) \geq 0$ and (15-103) yields $H(\mathbf{x}) \leq H(\mathbf{x}, \mathbf{y})$. This is not, however, true in general if $\mathbf{y}$ is of continuous type.

Generalizations The preceding results can be readily generalized to an arbitrary number of RVs: Suppose that $\mathbf{x}_1, \ldots, \mathbf{x}_n$ are n RVs with joint density $f(x_1, \ldots, x_n)$. Extending (15-85), we define their *joint entropy* as an expected value

$$H(\mathbf{x}_1, \ldots, \mathbf{x}_n) = E\{-\ln f(\mathbf{x}_1, \ldots, \mathbf{x}_n)\} \tag{15-105}$$

If the RVs $\mathbf{x}_i$ are *independent*, then

$$f(x_1, \ldots, x_n) = f(x_1) \cdots f(x_n)$$

and (15-105) yields

$$H(\mathbf{x}_1, \ldots, \mathbf{x}_n) = H(\mathbf{x}_1) + \cdots + H(\mathbf{x}_n) \tag{15-106}$$

Conditional entropies are defined similarly. For example [see (15-92)]

$$H(\mathbf{x}_n \mid \mathbf{x}_{n-1}, \ldots, \mathbf{x}_1) = E\{-\ln f(\mathbf{x}_n \mid \mathbf{x}_{n-1}, \ldots, \mathbf{x}_1)\} \tag{15-107}$$

Chain rule From the identity [see (8-37)]

$$f(x_1, \ldots, x_n) = f(x_n \mid x_{n-1}, \ldots, x_1) \cdots f(x_2 \mid x_1) f(x_1)$$

and (15-107) it follows that

$$H(\mathbf{x}_1, \ldots, \mathbf{x}_n) = H(\mathbf{x}_n \mid \mathbf{x}_{n-1}, \ldots, \mathbf{x}_1) + \cdots + H(\mathbf{x}_2 \mid \mathbf{x}_1) + H(\mathbf{x}_1) \quad (15\text{-}108)$$

The following relationships are simple extensions of (15-102) and (15-103)

$$H(\mathbf{x}, \mathbf{y} \mid \mathbf{z}) \le H(\mathbf{x} \mid \mathbf{z}) + H(\mathbf{y} \mid \mathbf{z})$$

$$H(\mathbf{x}, \mathbf{y} \mid \mathbf{z}) = H(\mathbf{x} \mid \mathbf{z}) + H(\mathbf{y} \mid \mathbf{x}, \mathbf{z}) \quad (15\text{-}109)$$

$$H(\mathbf{x}_1, \ldots, \mathbf{x}_n) \le H(\mathbf{x}_1) + \cdots + H(\mathbf{x}_n)$$

Example 15-15 If the RVs $\mathbf{x}_i$ are jointly normal with covariance matrix C as in (8-58), then

$$E\{-\ln f(\mathbf{x}_1, \ldots, \mathbf{x}_n)\} = \ln \sqrt{(2\pi)^n \Delta} + \tfrac{1}{2} E\{\mathbf{X} C^{-1} \mathbf{X}^t\} \quad (15\text{-}110)$$

This yields (see Prob. 8-18)

$$H(\mathbf{x}_1, \ldots, \mathbf{x}_n) = \ln \sqrt{(2\pi e)^n \Delta} \quad (15\text{-}111)$$

Transformations of RVs

We shall compare the entropy of the RVs **x** and $\mathbf{y} = g(\mathbf{x})$.

Discrete type If the RV **x** is of discrete type, then

$$H(\mathbf{y}) \le H(\mathbf{x}) \quad (15\text{-}112)$$

with equality iff the transformation $y = g(x)$ has a unique inverse.

Proof Suppose that **x** takes the values x_i with probability p_i and $g(x)$ has a unique inverse. In this case

$$P\{\mathbf{y} = y_i\} = P\{\mathbf{x} = x_i\} = p_i \qquad y_i = g(x_i)$$

hence, $H(\mathbf{y}) = H(\mathbf{x})$. If the transformation is not one-to-one, then $\mathbf{y} = y_i$ for more than one value of **x**. This results in a reduction of $H(\mathbf{x})$ [see (15-31)].

Continuous type If the RV **x** is of continuous type, then

$$H(\mathbf{y}) \le H(\mathbf{x}) + E\{\ln |g'(\mathbf{x})|\} \quad (15\text{-}113)$$

with equality iff the transformation $y = g(x)$ has a unique inverse.

Proof As we know [see (5-5)] if $y = g(x)$ has a unique inverse $x = g^{(-1)}(y)$, then

$$f_y(y) = \frac{f_x(x)}{|g'(x)|} \qquad dy = g'(x)\,dx$$

Hence

$$H(\mathbf{y}) = -\int_{-\infty}^{\infty} f_y(y) \ln f_y(y)\, dy = -\int_{-\infty}^{\infty} f_x(x) \ln \frac{f_x(x)}{|g'(x)|}\, dx$$

$$= -\int_{-\infty}^{\infty} f_x(x) \ln f_x(x)\, dx + \int_{-\infty}^{\infty} f_x(x) \ln |g'(x)|\, dx$$

and (15-113) results.

Several RVs Reasoning as in (15-113), we can similarly show that if

$$\mathbf{y}_i = g_i(\mathbf{x}_1, \ldots, \mathbf{x}_n) \qquad i = 1, \ldots, n$$

are n functions of the RVs $\mathbf{x}_i$, then

$$H(\mathbf{y}_1, \ldots, \mathbf{y}_n) \le H(\mathbf{x}_1, \ldots, \mathbf{x}_n) + E\{\ln |J(\mathbf{x}_1, \ldots, \mathbf{x}_n)|\} \tag{15-114}$$

where $J(x_1, \ldots, x_n)$ is the jacobian of the above transformation [see (8-9)]. Equality holds iff the transformation has a unique inverse.

Linear transformations Suppose that

$$\mathbf{y}_i = a_{i1}\mathbf{x}_1 + \cdots + a_{in}\mathbf{x}_n$$

Denoting by Δ the determinant of the coefficients, we conclude from (15-114) that if $\Delta \neq 0$ then

$$H(\mathbf{y}_1, \ldots, \mathbf{y}_n) = H(\mathbf{x}_1, \ldots, \mathbf{x}_n) + \ln |\Delta| \tag{15-115}$$

because the transformation has a unique inverse and Δ does not depend on $\mathbf{x}_i$.

Stochastic Processes and Entropy Rate

As we know, the statistics of most stochastic processes are determined in terms of the joint density $f(x_1, \ldots, x_m)$ of the RVs $\mathbf{x}(t_1), \ldots, \mathbf{x}(t_m)$. The joint entropy

$$H(\mathbf{x}_1, \ldots, \mathbf{x}_m) = E\{-\ln f(\mathbf{x}_1, \ldots, \mathbf{x}_m)\} \tag{15-116}$$

of these RVs is the *mth order entropy* of the process $\mathbf{x}(t)$. This function equals the uncertainty about the above RVs and it equals the information gained when they are observed.

In general, the uncertainty about the values of $\mathbf{x}(t)$ on the entire t axis or even on a finite interval, no matter how small, is infinite. However, if $\mathbf{x}(t)$ can be expressed in terms of its values on a countable set of points, as is the case for bandlimited processes, then a rate of uncertainty can be introduced. It suffices, therefore, to consider only discrete-time processes.

The mth order entropy of a discrete-time process $\mathbf{x}_n$ is the joint entropy $H(\mathbf{x}_1, \ldots, \mathbf{x}_m)$ of the m RVs

$$\mathbf{x}_n, \mathbf{x}_{n-1}, \ldots, \mathbf{x}_{n-m+1} \tag{15-117}$$

defined as in (15-116). We shall assume throughout that the process $\mathbf{x}_n$ is SSS. In this case, $H(\mathbf{x}_1 \cdots \mathbf{x}_m)$ is the uncertainty about any m consecutive values of the process $\mathbf{x}_n$. The first-order entropy will be denoted by $H(\mathbf{x})$. Thus, $H(\mathbf{x})$ equals the uncertainty about $\mathbf{x}_n$ for a specific n.

Clearly [see (15-109)]

$$H(\mathbf{x}_1, \ldots, \mathbf{x}_m) \le H(\mathbf{x}_1) + \cdots + H(\mathbf{x}_m) = mH(\mathbf{x}) \tag{15-118}$$

Special cases (*a*) If the process $\mathbf{x}_n$ is *strictly white*, i.e., if the RVs $\mathbf{x}_n, \mathbf{x}_{n-1}, \ldots$ are independent, then [see (15-106)]

$$H(\mathbf{x}_1, \ldots, \mathbf{x}_m) = mH(\mathbf{x}) \tag{15-119}$$

(*b*) If the process $\mathbf{x}_n$ is *Markoff*, then [see (12-99)]

$$f(x_1, \ldots, x_m) = f(x_m \mid x_{m-1}) \cdots f(x_2 \mid x_1) f(x_1) \tag{15-120}$$

This yields

$$H(\mathbf{x}_1, \ldots, \mathbf{x}_m) = H(\mathbf{x}_m \mid \mathbf{x}_{m-1}) + \cdots + H(\mathbf{x}_2 \mid \mathbf{x}_1) + H(\mathbf{x}_1) \tag{15-121}$$

From (15-103) and the stationarity of $\mathbf{x}_n$ it follows, therefore, that

$$H(\mathbf{x}_1, \ldots, \mathbf{x}_m) = (m-1)H(\mathbf{x}_1, \mathbf{x}_2) - (m-2)H(\mathbf{x}) \tag{15-122}$$

We have, thus, expressed the mth-order entropy of a Markoff process in terms of its first- and second-order entropies.

Conditional entropy The conditional entropy of order m

$$H(\mathbf{x}_n \mid \mathbf{x}_{n-1}, \ldots, \mathbf{x}_{n-m})$$

of a process $\mathbf{x}_n$ is the uncertainty about its present under the assumption that its m most recent values have been observed. Extending (15-104), we can readily show that

$$H(\mathbf{x}_n \mid \mathbf{x}_{n-1}, \ldots, \mathbf{x}_{n-m}) \le H(\mathbf{x}_n \mid \mathbf{x}_{n-1}, \ldots, \mathbf{x}_{n-m-1}) \tag{15-123}$$

Thus, the above conditional entropy is a decreasing function of m. If, therefore, it is bounded from below, it tends to a limit. This is certainly the case if the RVs $\mathbf{x}_n$ are of discrete type because then all entropies are positive. The limit will be denoted by $H_c(\mathbf{x})$ and will be called the *conditional entropy* of the process $\mathbf{x}_n$

$$H_c(\mathbf{x}) = \lim_{m \to \infty} H(\mathbf{x}_n \mid \mathbf{x}_{n-1}, \ldots, \mathbf{x}_{n-m}) \tag{15-124}$$

The function $H_c(\mathbf{x})$ is a measure of our uncertainty about the present of $\mathbf{x}_n$ under the assumption that its entire past is observed.

Special cases (*a*) If $\mathbf{x}_n$ is *strictly white*, then

$$H_c(\mathbf{x}) = H(\mathbf{x})$$

(*b*) If $\mathbf{x}_n$ is a *Markoff* process, then

$$H(\mathbf{x}_n \mid \mathbf{x}_{n-1}, \ldots, \mathbf{x}_{n-m}) = H(\mathbf{x}_n \mid \mathbf{x}_{n-1})$$

Since $\mathbf{x}_n$ is a stationary process, the above equals $H(\mathbf{x}_2 \mid \mathbf{x}_1)$. Hence

$$H_c(\mathbf{x}) = H(\mathbf{x}_2 \mid \mathbf{x}_1) = H(\mathbf{x}_1, \mathbf{x}_2) - H(\mathbf{x}) \tag{15-125}$$

This shows that if $\mathbf{x}_{n-1}$ is observed, then the past has no effect on the uncertainty of the present.

Entropy rate The ratio $H(\mathbf{x}_1 \cdots \mathbf{x}_m)/m$ is the average uncertainty per sample in a block of m consecutive samples. The limit of this average as $m \to \infty$ will be denoted by $\bar{H}(\mathbf{x})$ and will be called the *entropy rate* of the process $\mathbf{x}_n$

$$\bar{H}(\mathbf{x}) = \lim_{m \to \infty} \frac{1}{m} H(\mathbf{x}_1, \ldots, \mathbf{x}_m) \tag{15-126}$$

If $\mathbf{x}_n$ is *strictly white*, then

$$\bar{H}(\mathbf{x}) = H(\mathbf{x}) = H_c(\mathbf{x})$$

If $\mathbf{x}_n$ is *Markoff*, then [see (15-122)]

$$\bar{H}(\mathbf{x}) = H(\mathbf{x}_1, \mathbf{x}_2) - H(\mathbf{x}) = H_c(\mathbf{x}) \tag{15-127}$$

Thus, in both cases, the limit in (15-126) exists and it equals $H_c(\mathbf{x})$. We show next that this is true in general.

Theorem The entropy rate of a process $\mathbf{x}_n$ equals its conditional entropy

$$\bar{H}(\mathbf{x}) = H_c(\mathbf{x}) \tag{15-128}$$

PROOF This is a consequence of the following simple property of convergent sequences: If

$$a_k \to a \qquad \text{then} \qquad \frac{1}{m} \sum_{k=1}^{m} a_k \to a \tag{15-129}$$

Since $\mathbf{x}_n$ is stationary we conclude, as in (15-108), that

$$H(\mathbf{x}_1, \ldots, \mathbf{x}_m) = H(\mathbf{x}) + \sum_{k=1}^{m} H(\mathbf{x}_n \mid x_{n-1}, \ldots, \mathbf{x}_{n-k})$$

Dividing by m and using (15-129), we obtain (15-128) because

$$H(\mathbf{x}_n \mid \mathbf{x}_{n-1}, \ldots, \mathbf{x}_{n-k})$$

tends to $H_c(\mathbf{x})$ as $k \to \infty$.

Note If $\mathbf{x}_n$ equals the samples $\mathbf{x}(nT)$ of $\mathbf{x}(t)$ then entropy rate is measured in bits per T seconds. If we wish to measure it in bits per second, we must divide by T.

Normal processes We shall show that if $\mathbf{x}_n$ is a normal process with power spectrum $S(\omega)$, then

$$\bar{H}(\mathbf{x}) = \ln \sqrt{2\pi e} + \frac{1}{4\sigma} \int_{-\sigma}^{\sigma} \ln S(\omega)\, d\omega \tag{15-130}$$

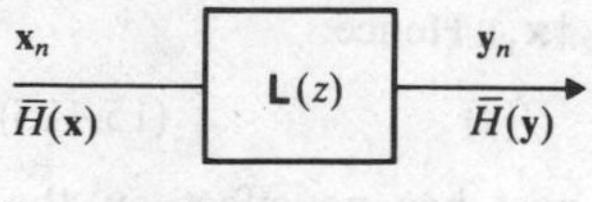

$$\bar{H}(\mathbf{y}) = \bar{H}(\mathbf{x}) + \frac{1}{2\sigma}\int_{-\sigma}^{\sigma} \ln|\mathbf{L}(e^{j\omega T})|\, d\omega$$

Figure 15-15

PROOF As we know, the function $f(x_{m+1} \mid x_m, \ldots, x_1)$ is a one-dimensional normal density with variance Δ_{m+1}/Δ_m [see (13-23) and (13-66)]. Hence

$$H(\mathbf{x}_n \mid \mathbf{x}_{n-1}, \ldots, \mathbf{x}_{n-m}) = \ln \sqrt{\frac{2\pi e\, \Delta_{m+1}}{\Delta_m}} \tag{15-131}$$

as in (15-84). This leads to the conclusion that

$$H_c(\mathbf{x}) = \ln\sqrt{2\pi e} + \frac{1}{2} \lim_{m\to\infty} \ln \frac{\Delta_{m+1}}{\Delta_m} \tag{15-132}$$

and (15-130) follows from (15-128) and Prob. 13-20.

Entropy rate of system response We shall show that the entropy rate $\bar{H}(\mathbf{y})$ of the output $\mathbf{y}_n$ of a linear system $\mathbf{L}(z)$ is given by

$$\bar{H}(\mathbf{y}) = \bar{H}(\mathbf{x}) + \frac{1}{2\sigma}\int_{-\sigma}^{\sigma} \ln|\mathbf{L}(e^{j\omega T})|\, d\omega \tag{15-133}$$

where $\bar{H}(\mathbf{x})$ is the entropy rate of the input $\mathbf{x}_n$ (Fig. 15-15).

Normal processes Suppose, first, that $\mathbf{x}_n$ is a normal process. In this case $\mathbf{y}_n$ is also normal and its entropy rate is given by (15-130) where

$$S(\omega) = S_y(\omega) = S_x(\omega)\,|\mathbf{L}(e^{j\omega T})|^2 \tag{15-134}$$

This yields

$$\bar{H}(\mathbf{y}) = \ln\sqrt{2\pi e} + \frac{1}{4\sigma}\int_{-\sigma}^{\sigma} [\ln S_x(\omega) + \ln|\mathbf{L}(e^{j\omega T})|^2]\, d\omega \tag{15-135}$$

and (15-133) follows.

Arbitrary processes As we know [see (15-115)], if the RVs $\mathbf{y}_1, \ldots, \mathbf{y}_m$ depend linearly on the RVs $\mathbf{x}_1, \ldots, \mathbf{x}_m$, then

$$H(\mathbf{y}_1, \ldots, \mathbf{y}_m) = H(\mathbf{x}_1, \ldots, \mathbf{x}_m) + K_o \tag{15-136}$$

where $K_o = \ln|\Delta|$ is a constant that depends only on the coefficients of the transformation. The process $\mathbf{y}_n$ depends linearly on $\mathbf{x}_n$

$$\mathbf{y}_n = \sum_{k=0}^{\infty} l_k \mathbf{x}_{n+k} \qquad n = -\infty, \ldots, \infty \tag{15-137}$$

where now the transformation matrix is of infinity order. Extending (15-136) to infinitely many variables (see Prob. 15-14) we conclude with (15-126) that

$$\bar{H}(\mathbf{y}) = \bar{H}(\mathbf{x}) + K \tag{15-138}$$

where again K is a constant that depends only on the parameters of the system $\mathbf{L}(z)$. As we have seen, if $\mathbf{x}_n$ is normal, then K equals the integral in (15-133). And since K is independent of $\mathbf{x}_n$, it must equal that integral for any $\mathbf{x}_n$.

15-4 THE MAXIMUM ENTROPY METHOD

The MEM is used to determine various parameters of a probability space subject to given constraints. The resulting problem can be solved, in general, only numerically and it involves the evaluation of the maximum of a function of several variables. In a number of important cases, however, the solution can be found analytically or it can be reduced to a system of algebraic equations. In this section, we consider certain special cases, concentrating on constraints in the form of expected values. The results can be obtained with the familiar variational techniques involving Lagrange multipliers or Euler's equations. For most problems under consideration, however, it suffices to use the following form of (15-100).

If $f(x)$ and $\varphi(x)$ are two arbitrary densities, then

$$-\int_{-\infty}^{\infty} \varphi(x) \ln \varphi(x)\, dx \le -\int_{-\infty}^{\infty} \varphi(x) \ln f(x)\, dx \qquad (15\text{-}139)$$

Example 15-16 In the coin tossing experiment, the probability of heads is often viewed as an RV $\mathbf{p}$. Using the MEM, we shall show that if no prior information about $\mathbf{p}$ is available, then its density $f(p)$ is uniform in the interval (0, 1). In this problem we must maximize $H(\mathbf{p})$ subject to the constraint (dictated by the meaning of $\mathbf{p}$) that $f(p) = 0$ outside the interval (0, 1). The corresponding entropy is, therefore, given by

$$H(\mathbf{p}) = -\int_0^1 f(p) \ln f(p)\, dp$$

and our problem is to find $f(p)$ such as to maximize the above integral.

We maintain that $H(\mathbf{p})$ is maximum if

$$f(p) = 1 \qquad H(\mathbf{p}) = 0$$

Indeed, if $\varphi(p)$ is any other density such that $\varphi(p) = 0$ outside the interval (0, 1), then [see (15-139)]

$$-\int_0^1 \varphi(p) \ln \varphi(p) \le -\int_0^1 \varphi(p) \ln f(p)\, dp = 0$$

Example 15-17 Suppose that $\mathbf{x}$ is an RV vanishing outside the interval $(-\pi, \pi)$. Using the MEM, we shall determine the density $f(x)$ of $\mathbf{x}$ under the assumption that the coefficients c_n of its Fourier series expansion

$$f(x) = \sum_{n=-\infty}^{\infty} c_n e^{jnx} \qquad -\pi \le x \le \pi$$

are known for $|n| \le N$. Our problem now is to maximize the integral

$$H(\mathbf{x}) = -\int_{-\pi}^{\pi} f(x) \ln f(x)\, dx$$

subject to the constraints

$$c_n = \frac{1}{2\pi}\int_{-\pi}^{\pi} f(x) e^{-jnx}\, dx \qquad |n| \le N \tag{15-140}$$

Clearly, $H(\mathbf{x})$ depends on the unknown coefficients c_n and it is maximum iff

$$\frac{\partial H}{\partial c_n} = \frac{\partial H}{\partial f}\frac{\partial f}{\partial c_n} = -\int_{-\pi}^{\pi} [\ln f(x) + 1] e^{jnx}\, dx = 0 \qquad |n| > N$$

This shows that the coefficients γ_n of the Fourier series expansion of the function $\ln f(x) + 1$ in the interval $(-\pi, \pi)$ are zero for $|n| > N$. Hence

$$\ln f(x) + 1 = \sum_{k=-N}^{N} \gamma_k e^{jkx}$$

From the above it follows that

$$f(x) = \exp\left\{-1 + \sum_{k=-N}^{N} \gamma_k e^{jkx}\right\} \qquad -\pi \le x \le \pi \tag{15-141}$$

We have thus shown that the unknown function is given by an exponential involving the parameters γ_k. These parameters can be determined from (15-140). The resulting system is nonlinear and can be solved only numerically.

Constraints as Expected Values

We shall consider now a class of problems involving constraints in the form of expected values. Such problems are common in statistical mechanics. We start with the one-dimension case.

We wish to determine the density $f(x)$ of an RV $\mathbf{x}$ subject to the condition that the expected values η_i of n known functions $g_i(\mathbf{x})$ of $\mathbf{x}$ are given

$$E\{g_i(\mathbf{x})\} = \int_{-\infty}^{\infty} g_i(x) f(x)\, dx = \eta_i \qquad i = 1, \ldots, n \tag{15-142}$$

Using (15-139), we shall show that the MEM leads to the conclusion that $f(x)$ must be an exponential

$$f(x) = A \exp\{-\lambda_1 g_1(x) - \cdots - \lambda_n g_n(x)\} \tag{15-143}$$

where λ_i are n constants determined from (15-142) and A is such as to satisfy the density condition

$$A \int_{-\infty}^{\infty} \exp\{-\lambda_1 g_1(x) - \cdots - \lambda_n g_n(x)\}\, dx = 1 \tag{15-144}$$

PROOF Suppose that $f(x)$ is given by (15-143). In this case

$$\int_{-\infty}^{\infty} f(x) \ln f(x)\, dx = \int_{-\infty}^{\infty} f(x)[\ln A - \lambda_1 g_1(x) - \cdots - \lambda_n g_n(x)]\, dx$$

Hence

$$H(\mathbf{x}) = \lambda_1 \eta_1 + \cdots + \lambda_n \eta_n - \ln A \tag{15-145}$$

To prove (15-143), it suffices therefore to show that, if $\varphi(x)$ is any other density satisfying the constraints (15-142), then its entropy cannot exceed the right side of (15-145). This follows readily from (15-139):

$$\begin{aligned} -\int_{-\infty}^{\infty} \varphi(x) \ln \varphi(x)\, dx &\le -\int_{-\infty}^{\infty} \varphi(x) \ln f(x)\, dx \\ &= \int_{-\infty}^{\infty} \varphi(x)(\lambda_1 g_1(x) + \cdots + \lambda_n g_n(x) - \ln A)\, dx \\ &= \lambda_1 \eta_1 + \cdots + \lambda_n \eta_n - \ln A \end{aligned}$$

We note that, if $f(x) = 0$ outside a certain set R, then $f(x)$ is again given by (15-143) for every x in R and the region of integration in (15-144) is the set R.

Example 15-18 We shall determine $f(x)$ assuming that $\mathbf{x}$ is a positive RV with known mean η. With $g(x) = x$, it follows from (15-143) that

$$f(x) = \begin{cases} Ae^{-\lambda x} & x > 0 \\ 0 & x < 0 \end{cases}$$

We have thus shown that if an RV is positive with specified mean, then its density, obtained with the MEM, is an exponential.

The partition function In certain problems, it is more convenient to express the given constraints in terms of the partition function (Zustandsumme)

$$Z(\lambda_1, \ldots, \lambda_n) = \frac{1}{A} = \int_{-\infty}^{\infty} \exp\{-\lambda_1 g_1(x) - \cdots - \lambda_n g_n(x)\}\, dx \tag{15-146}$$

Indeed, differentiating with respect to λ_i, we obtain

$$-\frac{\partial Z}{\partial \lambda_i} = \int_{-\infty}^{\infty} g_i(x) \exp\left\{-\sum_{k=1}^{n} \lambda_k g_k(x)\right\} dx = Z \int_{-\infty}^{\infty} g_i(x) f(x)\, dx$$

This yields

$$-\frac{1}{Z}\frac{\partial Z}{\partial \lambda_i} = -\frac{\partial}{\partial \lambda_i} \ln Z = \eta_i \qquad i = 1, \ldots, n \tag{15-147}$$

The above is a system of n equations equivalent to (15-142) and can be used to determine the n parameters λ_i.

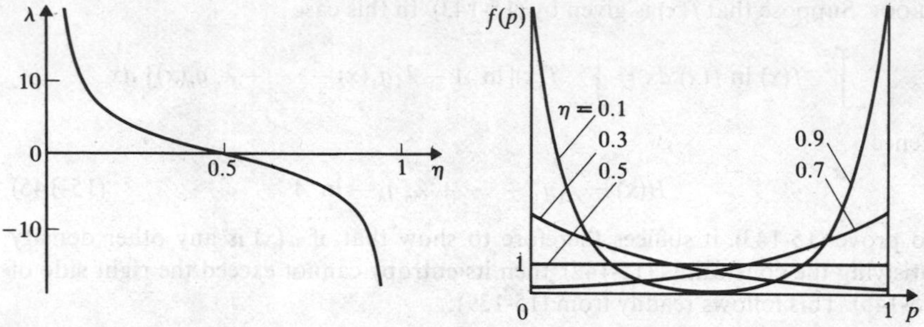

Figure 15-16

Example 15-19 In the coin-tossing experiment of Example 15-16, we assume that **p** is an RV with known mean η. Since $f(p) = 0$ outside the interval (0, 1), (15-143) yields

$$f(p) = \begin{cases} Ae^{-\lambda p} & 0 \le p \le 1 \\ 0 & \text{otherwise} \end{cases} \qquad Z = \int_0^1 e^{-\lambda p}\, dp = \frac{1 - e^{-\lambda}}{\lambda}$$

The constant λ is determined from (15-147)

$$-\frac{1}{Z}\frac{\partial Z}{\partial \lambda} = \frac{1 - e^{-\lambda} - \lambda e^{-\lambda}}{\lambda(1 - e^{-\lambda})} = \eta$$

In Fig. 15-16, we plot λ and $f(p)$ for various values of η. We note that if $\eta = 0.5$, then $\lambda = 0$ and $f(p) = 1$.

Example 15-20 A collection of particles moves in a conservative field whose potential equals $V(x)$. For a specific t, the x component of the position of a particle is an RV **x** with density $f(x)$ independent of t (stationary state). Thus, the probability that the particle is between x and $x + dx$ equals $f(x)\,dx$ and the total energy per unit mass of the ensemble equals

$$I = \int_{-\infty}^{\infty} V(x) f(x)\, dx = E\{V(\mathbf{x})\}$$

We shall find $f(x)$ under the assumption that the function $g(x) = V(x)$ and the mean I of $V(\mathbf{x})$ are given. Inserting into (15-143), we obtain

$$f(x) = \frac{1}{Z} e^{-\lambda V(x)} \tag{15-148}$$

where

$$Z = \int_{-\infty}^{\infty} e^{-\lambda V(x)}\, dx \qquad \frac{1}{Z}\int_{-\infty}^{\infty} V(x) e^{-\lambda V(x)}\, dx = I$$

Special case In a gravitational field, the potential $V(x) = Mgx$ is proportional to the distance x from the ground. Since $f(x) = 0$ for $x < 0$, it follows from (15-148) that

$$f(x) = \frac{Mg}{I} e^{-Mgx/I} U(x)$$

The resulting atmospheric pressure is proportional to $1 - F(x)$.

Example 15-21 We shall find $f(x)$ such that $E\{\mathbf{x}^2\} = m_2$. With $g_1(x) = x^2$, (15-143) yields

$$f(x) = Ae^{-\lambda x^2} \tag{15-149}$$

Thus, if the second moment m_2 of an RV $\mathbf{x}$ is specified, then $\mathbf{x}$ is $N(0, m_2)$. We can show similarly that if the variance σ^2 of $\mathbf{x}$ is specified, then $\mathbf{x}$ is $N(\eta, \sigma^2)$ where η is an arbitrary constant.

SPECIAL CASE We consider again a collection of particles in stationary motion and we denote by $\mathbf{v}_x$ the x component of their velocity. We shall determine the density $f(v_x)$ of $\mathbf{v}_x$ under the constraint that the corresponding average kinetic energy $K_x = E\{M\mathbf{v}_x^2/2\}$ is specified. This is a special case of (15-149) with $m_2 = 2K_x/M$. Hence

$$f(v_x) = \sqrt{\frac{M}{4\pi K_x}}\, e^{-Mv^2/4K_x}$$

Discrete type RVs Suppose that an RV $\mathbf{x}$ takes the values x_k with probability p_k. We shall use the MEM to determine p_k under the assumption that the expected values

$$E\{g_i(\mathbf{x})\} = \sum_k g_i(x_k)p_k = \eta_i \tag{15-150}$$

of the n known functions $g_i(\mathbf{x})$ are given.

Using (15-37), we can show as in (15-143) that the unknown probabilities equal

$$p_k = A \exp\{-\lambda_1 g_1(x_k) - \cdots - \lambda_n g_n(x_k)\} \tag{15-151}$$

where

$$\frac{1}{A} = Z = \sum_k \exp\{-[\lambda_1 g_1(x_k) + \cdots + \lambda_n g_n(x_k)]\} \tag{15-152}$$

The n constants λ_i are determined either from (15-150) or from the equivalent system

$$-\frac{1}{Z}\frac{\partial Z}{\partial \lambda_i} = \eta_i \qquad i = 1, \ldots, n \tag{15-153}$$

Example 15-22 A die is rolled a large number of times and the average number of dots up equals η. Assuming that η is known, we shall determine the probabilities p_k of the six faces f_k using the MEM. For this purpose, we form an RV $\mathbf{x}$ such that $\mathbf{x}(f_k) = k$. Clearly

$$E\{\mathbf{x}\} = p_1 + 2p_2 + \cdots + 6p_6 = \eta$$

With $g(x) = x$, it follows from (15-151) that

$$p_k = \frac{1}{Z}e^{-k\lambda} \qquad Z = w + w^2 + \cdots + w^6$$

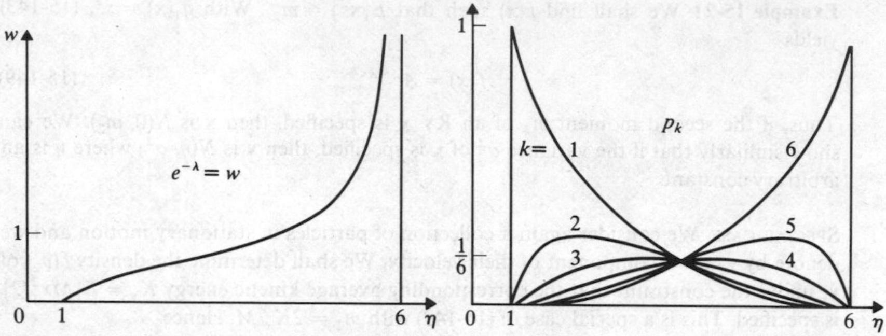

Figure 15-17

where $w = e^{-\lambda}$. Hence

$$p_k = \frac{w^k}{w + w^2 + \cdots + w^6} \qquad \frac{w + 2w^2 + \cdots + 6w^6}{w + w^2 + \cdots + w^6} = \eta$$

as in Fig. 15-17. We note that if $\eta = 3.5$, then $p_k = \frac{1}{6}$.

Joint density The MEM can be used to determine the density $f(X)$ of the random vector $\mathbf{X}$: $[\mathbf{x}_1, \ldots, \mathbf{x}_M]$ subject to the n constraints

$$E\{g_i(\mathbf{X})\} = \eta_i \qquad i = 1, \ldots, n \tag{15-154}$$

Reasoning as in the scalar case, we conclude that

$$f(X) = A \exp\{-\lambda_1 g_1(X) - \cdots - \lambda_n g_n(X)\} \tag{15-155}$$

Second-Order Moments and Normality

We are given the correlation matrix

$$R = E\{\mathbf{X}^t\mathbf{X}\} \tag{15-156}$$

of the random vector $\mathbf{X}$ and we wish to find its density using the MEM. We maintain that $f(X)$ is normal with zero mean as in (8-58)

$$f(X) = \frac{1}{\sqrt{2\pi\Delta}} \exp\{-\tfrac{1}{2}XR^{-1}X^t\} \tag{15-157}$$

PROOF The elements $R_{jk} = E\{\mathbf{x}_j\mathbf{x}_k\}$ of R are the expected values of the M^2 RVs $g_{jk}(\mathbf{X}) = \mathbf{x}_j\mathbf{x}_k$. Changing the subscript i in (15-154) to double subscript, we conclude from (15-155) that

$$f(X) = A \exp\left\{-\sum_{j,k} \lambda_{jk} x_j x_k\right\} \tag{15-158}$$

This shows that $f(X)$ is normal. The M^2 coefficients λ_{jk} can be determined from the M^2 constraints in (15-156). As we know [see (8-58)], these coefficients equal the elements of the matrix $R^{-1}/2$ as in (15-157).

The preceding results are acceptable only if the matrix R is p.d. Otherwise, the function $f(X)$ in (15-157) is not a density. The p.d. condition is, of course, satisfied if the given R is a true correlation matrix. However, even then (15-157) might not be acceptable if only a subset of the elements of R is specified. In such cases, it might be necessary, as we shall presently see, to introduce the unspecified elements of R as auxiliary constraints.

Suppose, first, that we are given only the diagonal elements of R

$$E\{\mathbf{x}_i^2\} = R_{ii} \qquad i = 1, \ldots, M \tag{15-159}$$

Inserting the functions $g_{ii}(x) = x_i^2$ into (15-155), we obtain

$$f(X) = A \exp\{-\lambda_{11}x_1^2 - \cdots - \lambda_{MM}x_m^2\} \tag{15-160}$$

This shows that the RVs $\mathbf{x}_i$ are normal, independent, with zero mean and variance $R_{ii} = 1/2\lambda_{ii}$.

The above solution is acceptable because $R_{ii} > 0$. If, however, we are given $N < M^2$ arbitrary joint moments, then the corresponding quadratic in (15-158) will contain only the terms $x_j x_k$ corresponding to the given moments. The resulting $f(X)$ might not then be a density. To find the ME solution for this case, we proceed as follows: We introduce as constraints the M^2 joint moments R_{jk} where now only N of these moments are given and the other $M^2 - N$ moments are unknown parameters. Applying the MEM, we obtain (15-157). The corresponding entropy equals [see (15-111)].

$$H(\mathbf{x}_1, \ldots, \mathbf{x}_M) - \ln\sqrt{(2\pi e)^m \Delta} \qquad \Delta = |R| \tag{15-161}$$

This entropy depends on the unspecified parameters of R and it is maximum if its determinant Δ is maximum. Thus, the RVs $\mathbf{x}_i$ are again normal with density as in (15-157) where the unspecified parameters of R are such as to maximize Δ.

Note From the above it follows that the determinant Δ of a correlation matrix R is such that

$$\Delta \le R_{11} \cdots R_{MM}$$

with equality iff R is diagonal. Indeed, (15-159) is a restricted moment set, hence the ME solution (15-160) maximizes Δ.

Stochastic processes The MEM can be used to determine the statistics of a stochastic process subject to given constraints. We shall discuss the following case.

Suppose that $\mathbf{x}_n$ is a WSS process with autocorrelation

$$R[m] = E\{\mathbf{x}_{n+m}\mathbf{x}_n\}$$

We wish to find its various densities assuming that $R[m]$ is specified either for some or for all values of m. As we know [see (15-158)] the MEM leads to the conclusion that, in both cases, $\mathbf{x}_n$ must be a normal process with zero mean. This completes the statistical description of $\mathbf{x}_n$ if $R[m]$ is known for all m. If, however, we know $R[m]$ only partially, then we must find its unspecified values. For finite order densities, this involves the maximization of the corresponding entropy with respect to the unknown values of $R[m]$ and it is equivalent to the maximization of the correlation determinant Δ [see (15-161)]. An important special case is the MEM solution to the extrapolation problem considered in Sec. 14-1. We shall reexamine this problem in the context of entropy rate.

We start with the simplest case: Given the average power $E\{\mathbf{x}_n^2\} = R[0]$ of $\mathbf{x}_n$, we wish to find its power spectrum. In this case, the entropy of the RVs

$$\mathbf{x}_n, \ldots, \mathbf{x}_{n+M}$$

is maximum if these RVs are normal and independent for any M [see (15-160)], i.e., if the process $\mathbf{x}_n$ is normal white noise with $R[m] = R[0]\delta[m]$.

Suppose now that we are given the $N + 1$ values (data)

$$R[0], \ldots, R[N]$$

of $R[m]$ and we wish to find the density $f(X)$ of the $M + 1$ RVs $\mathbf{x}_n, \ldots, \mathbf{x}_{n+M}$. If $M \le N$, then the correlation matrix of $\mathbf{X}$ is specified in terms of the data and $f(X)$ is given by (15-157). This is not the case, however, if $M > N$ because then only the center diagonal and the N upper and lower diagonals of the correlation matrix

$$R_{M+1} = \begin{bmatrix} R[0] & \cdots\cdots & R[N] & \cdots\cdots & R[M] \\ \cdots & & & & \text{unknown} \\ R[N] & \cdots\cdots & \text{known} & & \\ & & & & \\ & \text{unknown} & & & \\ R[M] & & \cdots\cdots & & R[0] \end{bmatrix}$$

are known. To complete the specification of R_{M+1}, we maximize the determinant Δ_{M+1} with respect to the unknown values of $R[m]$.

Example 15-23 Given $R[0]$ and $R[1]$, we shall find $R[2]$ using the maximum determinant method. In this case,

$$\Delta = \begin{vmatrix} R[0] & R[1] & R[2] \\ R[1] & R[0] & R[1] \\ R[2] & R[1] & R[0] \end{vmatrix}$$

Hence

$$\frac{\partial \Delta}{\partial R[2]} = -2R[0]R[2] + 2R^2[1] = 0 \qquad R[2] = \frac{R^2[1]}{R[0]}$$

The MEM in spectral estimation We are given again $R[m]$ for $|m| \le N$. The power spectrum

$$S(\omega) = \sum_{m=-\infty}^{\infty} R[m]e^{-jm\omega T}$$

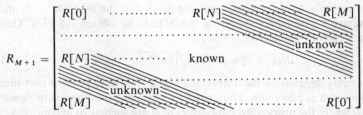

of $\mathbf{x}_n$ involves the values of $R[m]$ for every m. To find its unspecified values, we maximize the correlation determinant Δ_M and examine the form of the resulting $R[m]$ as $M \to \infty$. This is equivalent to the maximization of the *entropy rate* $\bar{H}(\mathbf{x})$ of the process $\mathbf{x}_n$. Using this equivalence, we shall develop a more direct method for determining $S(\omega)$.

As we know, the MEM leads to the conclusion that under the given constraints (second-order moments), the process $\mathbf{x}_n$ must be normal with zero mean. From this and (15-130) it follows that

$$\bar{H}(\mathbf{x}) = \ln \sqrt{2\pi e} + \frac{1}{4\sigma} \int_{-\sigma}^{\sigma} \ln S(\omega)\, d\omega$$

The entropy rate $\bar{H}(\mathbf{x})$ depends on the unspecified values of $R[m]$ and it is maximum if

$$\frac{\partial \bar{H}}{\partial R[m]} = \frac{1}{4\sigma} \int_{-\sigma}^{\sigma} \frac{1}{S(\omega)} e^{-jm\omega T}\, d\omega = 0 \qquad |m| > N \tag{15-162}$$

This shows that the coefficients of the Fourier series expansion of $1/S(\omega)$ are zero for $|m| > N$. Hence

$$\frac{1}{S(\omega)} = \sum_{k=-N}^{N} c_k e^{-jk\omega T}$$

Factoring the resulting $\mathbf{S}(z)$ as in (10-114), we obtain

$$S(\omega) = \frac{1}{|b_0 + b_1 e^{-j\omega T} + \cdots + b_N e^{-jN\omega T}|^2} \tag{15-163}$$

This is the spectrum obtained in Sec. 14-1 [see (14-25)] and it shows that the MEM leads to an AR model. The coefficients b_k can be obtained either from the Yule–Walker equations or from Levinson's algorithm.

Note The MEM has applications also in nonprobabilistic problems involving the determination of unknown parameters from insufficient data. In such cases, probabilistic models are created where the unknown parameters take the form of statistical variables that are determined with the MEM. We should point out, however, that the results obtained are not unique because more than one model can be used in the same problem. In the following, we illustrate this approach using as example the one-dimensional form of an important problem in *crystallography.*

A deterministic application of the MEM We wish to find a nonnegative periodic function $f(x)$ with period 2π

$$0 < f(x) = \sum_{n=-\infty}^{\infty} c_n e^{jnx}$$

having access only to partial information about its Fourier series coefficients

$$c_n = r_n e^{j\varphi_n}$$

The truncation problem We assume that c_n is known only for $|n| \le N$.

SOLUTION 1 We create the following probabilistic model: In the interval $(-\pi, \pi)$, the unknown function $f(x)$ is the density of an RV $\mathbf{x}$ taking values between $-\pi$ and π. We determine $f(x)$ so as to maximize the entropy

$$I = -\int_{-\pi}^{\pi} f(x) \ln f(x)\, dx$$

of $\mathbf{x}$. This yields [see (15-141)]

$$f(x) = \exp\left\{-1 + \sum_{n=-N}^{N} \gamma_n e^{jnx}\right\}$$

The constants γ_n are determined in terms of the given information.

SOLUTION 2 We assume that $f(x)$ is the power spectrum of a stochastic process $\mathbf{x}_n$ and we determine $f(x)$ so as to maximize the entropy rate (we omit incidental constants)

$$I = \int_{-\pi}^{\pi} \ln f(x)\, dx$$

of $\mathbf{x}_n$. In this case, $f(x)$ is given by [see (15-163)]

$$f(x) = \frac{1}{\sum_{n=-N}^{N} d_n e^{jnx}}$$

The constants d_n are again determined in terms of the given information (Levinson's algorithm).

The phase problem We assume that we know only the amplitudes r_n of c_n for $|n| \leq N$.

To solve the problem, we form again the integral I, either as entropy or as entropy rate, and we maximize it with respect to the unknown parameters which are now the coefficients c_n (amplitudes and phases) for $|n| > N$, and the phase φ_n for $|n| \leq N$. An equivalent approach involves the determination of $f(x)$ as in the truncation problem, treating the phases φ_n as parameters, and the maximization of the resulting I with respect to these parameters. In either case, the required computations are not simple.

15-5 CODING

Coding belongs to a class of problems involving the efficient search and identification of an object ζ_i from a set $\mathscr{S}$ of N objects. This topic is extensive and it has many applications. We shall present here merely certain aspects related to entropy and probability, limiting the discussion to binary instantly decodable codes. The underlying ideas can be readily generalized.

Binary coding can be also described in terms of the familiar game of twenty questions: A person selects an object ζ_i from a set $\mathscr{S}$. Another person wants to identify the object by asking "yes" or "no" questions. The purpose of the game is to find ζ_i using the smallest possible number of questions.

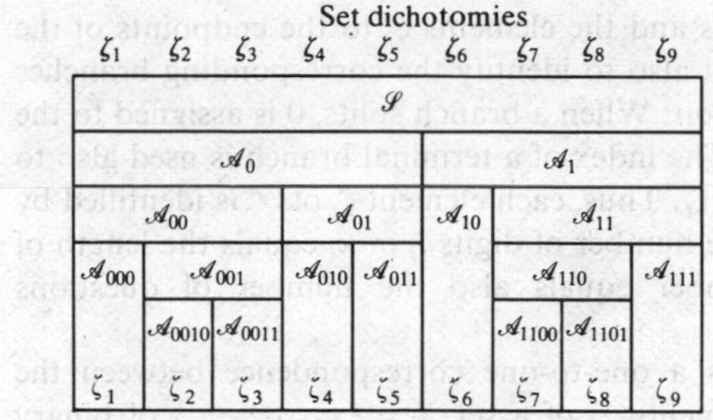

Set dichotomies

ζ_1	ζ_2	ζ_3	ζ_4	ζ_5	ζ_6	ζ_7	ζ_8	ζ_9
				$\mathscr{S}$				
		$\mathscr{A}_0$				$\mathscr{A}_1$		
	$\mathscr{A}_{00}$		$\mathscr{A}_{01}$	$\mathscr{A}_{10}$		$\mathscr{A}_{11}$		
$\mathscr{A}_{000}$	$\mathscr{A}_{001}$		$\mathscr{A}_{010}$	$\mathscr{A}_{011}$		$\mathscr{A}_{110}$		$\mathscr{A}_{111}$
	$\mathscr{A}_{0010}$	$\mathscr{A}_{0011}$				$\mathscr{A}_{1100}$	$\mathscr{A}_{1101}$	
ζ_1	ζ_2	ζ_3	ζ_4	ζ_5	ζ_6	ζ_7	ζ_8	ζ_9

Binary code

x_1	000	0010	0011	010	011	10	1100	1101	111
l_i	3	4	4	3	3	2	4	4	3

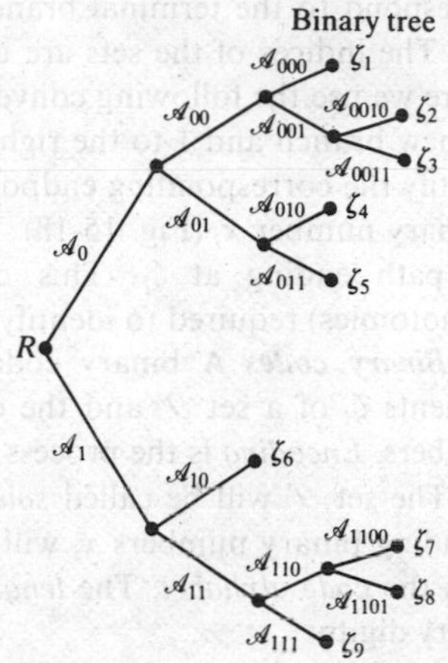

Figure 15-18

The various search techniques can be described in three equivalent forms: (a) as chains of dichotomies of the set $\mathscr{S}$; (b) in the form of a binary tree; (c) as binary codes (Fig. 15-18). We start with an explanation of these approaches, ignoring for the moment optimality considerations. The criteria for selecting the "best" search method will be developed later.

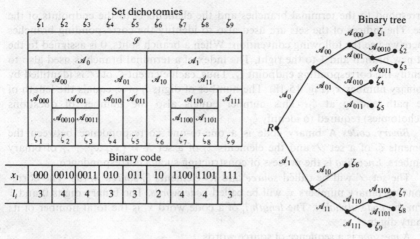

Set dichotomies We subdivide the set $\mathscr{S}$ into two nonempty sets $\mathscr{A}_0$ and $\mathscr{A}_1$ (first generation sets). We subdivide each of the sets $\mathscr{A}_0$ and $\mathscr{A}_1$ into two nonempty sets $\mathscr{A}_{00}$, $\mathscr{A}_{01}$ and $\mathscr{A}_{10}$, $\mathscr{A}_{11}$ (second generation sets). We continue with such dichotomies until the final sets consist of a single element each.

The indices of the sets of each generation are binary numbers formed by attaching 0 or 1 to the indices of the preceding generation sets.

In Fig. 15-18, we illustrate the above with a set consisting of nine elements. We shall use the chain of sets so formed to identify the element ζ_7 by a sequence of appropriate questions (set dichotomies): Is it in $\mathscr{A}_0$? No. Is it in $\mathscr{A}_{10}$? No. Is it in $\mathscr{A}_{110}$? Yes. Is it in $\mathscr{A}_{1100}$? Yes. Hence, the unknown element is ζ_7 because $\mathscr{A}_{1100} = \{\zeta_7\}$.

Binary trees A tree is a simply connected graph consisting of line segments called *branches*. In a binary tree, each branch splits into *two* other branches or it terminates. The points of termination are the *endpoints* of the tree and the starting point R is its *root* (Fig. 15-18). A *path* is a part of the tree from R to an endpoint. The two branches closest to the root are the first-generation branches. They split into two branches each, forming the second generation. Since each branch splits into two or it terminates, the number of branches in each generation is always *even*. The *length* of a path is the total number of its branches.

There is a one-to-one correspondence between set dichotomies and trees. The kth generation sets correspond to the kth generation branches and each set dichotomy to the splitting of the corresponding branch. The terminal sets $\{\zeta_i\}$

correspond to the terminal branches and the elements ζ_i to the endpoints of the tree. The indices of the sets are used also to identify the corresponding branches where we use the following convention: When a branch splits, 0 is assigned to the left new branch and 1 to the right. The index of a terminal branch is used also to identify the corresponding endpoint ζ_i. Thus, each element ζ_i of $\mathscr{S}$ is identified by a binary number x_i (Fig. 15-18). The number of digits l_i of x_i equals the length of the path ending at ζ_i. This number equals also the number of questions (dichotomies) required to identify ζ_i.

Binary codes A binary code is a one-to-one correspondence between the elements ζ_i of a set $\mathscr{S}$ and the elements x_i of a set $X = \{x_1, x_2, \ldots\}$ of binary numbers. *Encoding* is the process of constructing such a correspondence.

The set $\mathscr{S}$ will be called *source* and its elements ζ_i *source words*. The corresponding binary numbers x_i will be called *code words*. The binary digits 0 and 1 form the *code alphabet*. The *length* l_i of a code word x_i is the total number of its binary digits.

A *message* is a sequence of source words

$$\zeta_{i_1} \cdots \zeta_{i_k} \cdots \zeta_{i_n} \qquad \zeta_{i_k} \in \mathscr{S} \tag{15-164}$$

The sequence of the corresponding code words

$$x_{i_1} \cdots x_{i_k} \cdots x_{i_n} \tag{15-165}$$

is a *coded message*.

The indices of the terminal elements of a tree, or equivalently, of a chain of set dichotomies, specify a code. Codes can, of course, be formed in other ways, however, other codes will not be considered here. The term "code" will mean a *binary code* specified by a tree as above.

In Fig. 15-18, we show the code words x_i of a source $\mathscr{S}$ consisting of $N = 9$ elements, and the corresponding word lengths l_i.

Theorem If a source $\mathscr{S}$ has N words and the lengths of the corresponding code words equal l_i, then

$$\sum_{i=1}^{N} \frac{1}{2^{l_i}} = 1 \tag{15-166}$$

Proof The last generation branches of the tree are terminal and they form pairs. The two branches of one such pair are the ends of two paths of length l_r (Fig. 15-19). If they are removed, the tree *contracts* into a tree with $N - 1$ endpoints. In this operation, the two paths are replaced with one path of length $l_r - 1$ and the two terms 2^{-l_r} in (15-166) are replaced with the term $2^{-(l_r - 1)}$. Since

$$2^{-l_r} + 2^{-l_r} = 2^{-(l_r - 1)} \tag{15-167}$$

the sum does not change. Thus, the binary length sum in (15-166) is invariant to a contraction. Repeating the process until we are left with only two first-generation branches, we obtain (15-166) because $2^{-1} + 2^{-1} = 1$.

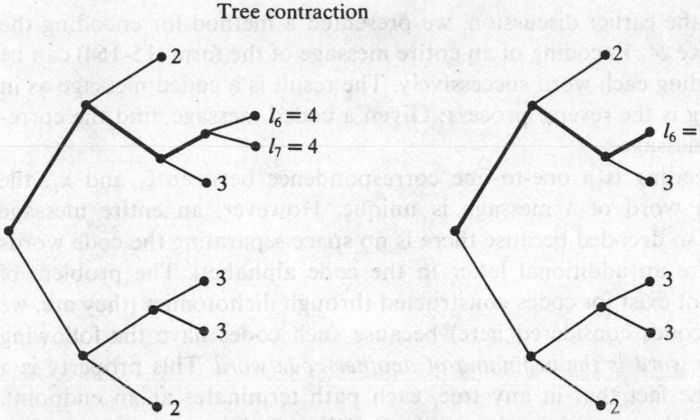

$$\frac{1}{2^4}+\frac{1}{2^4}+\sum_{i=1}^{5}\frac{1}{2^{l_i}} \qquad \frac{1}{2^3}+\sum_{i=1}^{5}\frac{1}{2^{l_i}}$$

Figure 15-19

Converse theorem Given N integers l_i satisfying (15-166), we can construct a code with lengths l_i.

PROOF It suffices to construct a binary tree with path lengths l_i. From (15-166) it follows that if l_r is the largest of the integers l_i, then the number n of lengths that equal l_r is even. Using $n = 2m$ segments, we form the rth (last) generation branches of our tree. If each of the m pairs of integers l_r is replaced by a single integer $l_r - 1$ and all others are not changed, the resulting set of numbers will satisfy (15-166) [see (15-167)]. We can, therefore, continue this process until we are left with only two terms. These terms yield the two first-generation branches. The above is illustrated in Fig. 15-20 for $N = 8$.

Tree construction

l	1	2	3	4	5	6	7	8
l_i	2	2	3	3	4	4	4	4
	2	2	3	3	3		3	
	2	2	2		2			
	1				1			
x_i	11	10	011	010	0011	0010	0001	0000

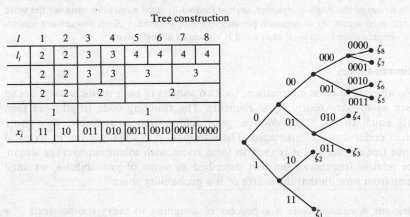

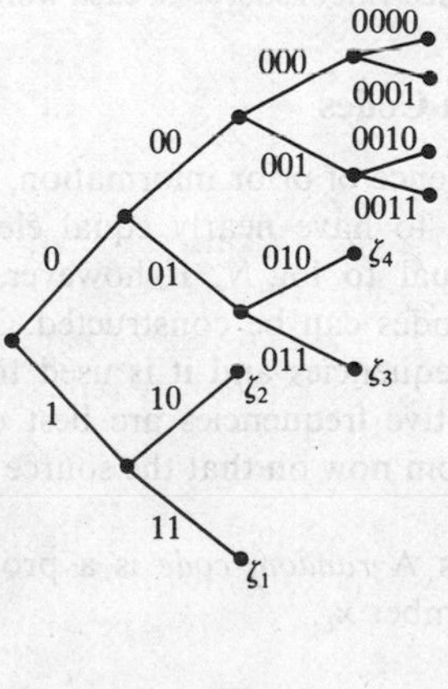

Figure 15-20

Decoding In the earlier discussion, we presented a method for encoding the words ζ_i of a source $\mathscr{S}$. Encoding of an entire message of the form (15-164) can be obtained by encoding each word successively. The result is a coded message as in (15-165). Decoding is the reverse process: Given a coded message, find the corresponding source message.

Since word coding is a one-to-one correspondence between ζ_i and x_i, the decoding of each word of a message is unique. However, an entire message cannot always be so decoded because there is no space separating the code words (this would require an additional letter in the code alphabet). The problem of separation does not exist for codes constructed through dichotomies (they are, we repeat, the only codes considered here) because such codes have the following property: *no code word is the beginning of another code word.* This property is a consequence of the fact that in any tree, each path terminates at an endpoint, therefore, it cannot be part of another path. Codes with this property are called "instantaneous" because they are instantly decodable, i.e., if we start from the beginning of a message, we can identify in real time the end of each word without any reference to the future.

Example 15-24 We wish to decode the message

$$1011010000101000101111000000010$$

formed with the code shown in Fig. 15-18. Starting from the beginning, we identify the code words by underlying them with the help of the table of Fig. 15-18

$$\underline{10}\ \underline{1101}\ \underline{000}\ \underline{010}\ \underline{10}\ \underline{0010}\ \underline{111}\ \underline{1100}\ \underline{000}\ \underline{0010}$$

The corresponding source message is the sequence

$$\zeta_6\zeta_8\zeta_1\zeta_4\zeta_6\zeta_2\zeta_9\zeta_7\zeta_1\zeta_2$$

Note We have identified each source word with a single symbol ζ_i. It is possible, however, that ζ_i might be a grouping of other symbols. For example, the source $\mathscr{S}$ might consist of: all the *letters* of the English alphabet, certain frequently used words (for instance the word *the*) and even a number of common phrases like *happy birthday*. Such sources are equivalent to single-symbol sources if each word is viewed as a single element.

Optimum Codes

In the absence of prior information, the two subsets of each set dichotomy are so chosen as to have nearly equal elements. The resulting code lengths are then nearly equal to log N. If however, prior information is available, then more efficient codes can be constructed. The information is usually given in terms of relative frequencies and it is used to form codes with minimum average length. Since relative frequencies are best described in terms of probabilities, we shall assume from now on that the source $\mathscr{S}$ is a probability space.

Definitions A *random code* is a process of assigning to every source word ζ_i a binary number x_i.

Since ζ_i is an element of the probability space $\mathscr{S}$, a random code defines an RV **x** such that

$$\mathbf{x}(\zeta_i) = x_i$$

The *length* of a random code is an RV **L** such that

$$\mathbf{L}(\zeta_i) = l_i \tag{15-168}$$

where l_i is the length of the code word x_i assigned to the element ζ_i.

The expected value of **L** is denoted by L and it is called the *average length* of the random code **x**. Thus

$$L = E\{\mathbf{L}\} = \sum_i p_i l_i \tag{15-169}$$

where $p_i = P\{\mathbf{x} = x_i\} = P\{\zeta_i\}$.

Optimum code An *optimum code* is a code whose average length does not exceed the average length of any other code. A basic objective of coding theory is the determination of such a code. Optimum codes have the following properties:

1. Suppose that ζ_a and ζ_b are two elements of $\mathscr{S}$ such that

$$p_a = P\{\zeta_a\} \qquad p_b = P\{\zeta_b\} \qquad l(\zeta_a) = l_a \qquad l(\zeta_b) = l_b$$

We maintain that if the code is optimum and

$$p_a > p_b \qquad \text{then} \qquad l_a \le l_b \tag{15-170}$$

PROOF Suppose that $l_a > l_b$. Interchanging the codes assigned to the elements ζ_a and ζ_b, we obtain a new code with average length

$$L_1 = L - (p_a l_a + p_b l_b) + (p_a l_b + p_b l_a) = L - (p_a - p_b)(l_a - l_b)$$

And since $(p_a - p_b)(l_a - l_b) > 0$, we conclude that $L_1 < L$. This, however, is impossible because L is the optimum code length, hence, $l_a \le l_b$.

Repeated application of (15-170) leads to the conclusion that if

$$p_1 \ge p_2 \ge \cdots \ge p_N \qquad \text{then} \qquad l_1 \le l_2 \le \cdots \le l_N \tag{15-171}$$

2. The elements (source words) with the two smallest probabilities p_{N-1} and p_N are in the last generation of the tree, i.e., their code lengths are l_{N-1} and l_N.

PROOF This is a consequence of (15-171) and the fact that the number of branches in each generation is even.

The following basic theorem shows the relationship between the entropy

$$H(\mathfrak{S}) = -\sum_{i=1}^{N} p_i \log p_i$$

of the source word partition $\mathfrak{S}$ and the average length L of an arbitrary random code **x**.

Theorem

$$H(\mathfrak{S}) \leq L \tag{15-172}$$

PROOF As we see from (15-166), if l_i are the lengths of the code words of $\mathbf{x}$ and $q_i = 1/2^{l_i}$, then the sum of the q_i's equals 1. With $a_i = p_i$ and $b_i = q_i$ it follows, therefore, from (15-37) that

$$-\sum_i p_i \log p_i \leq -\sum_i p_i \log q_i = \sum_i p_i l_i = L \tag{15-173}$$

and (15-172) results.

In general, $H(\mathfrak{S}) < L$. We maintain, however, that $H(\mathfrak{S}) = L$ iff the probabilities p_i are binary decimals, i.e., iff $p_i = 1/2^{n_i}$.

PROOF If $H(\mathfrak{S}) = L$, then (15-173) is an equality, hence $p_i = q_i = 1/2^{l_i}$ [see (15-37)] and our assertion is true because the lengths l_i are integers.

Conversely, if $p_i = 1/2^{n_i}$ and n_i are integers, then we can construct a code with lengths $l_i = n_i$ because the sum of the p_i's equals 1. The length L of this code equals $H(\mathfrak{S})$. In other words, if all p_i's are binary decimals, then the code with lengths $l_i = n_i$ is optimum.

Shannon, Fano, and Huffman Codes

The preceding theorem gives us a low bound for the average code length L but it does not say how close we can come to this bound. At the end of the section we show that, if we encode not each word but an entire message, then we can construct codes with average length per word less than $H(\mathfrak{S}) + \varepsilon$ for any $\varepsilon > 0$.

In the following, we present three well-known codes including the optimum code (Huffman). The description of these codes is clarified in Example 15-25.

The Shannon code As we noted, if all probabilities p_i are binary decimals, then the code with lengths $l_i = -\log p_i$ is optimum. Guided by this, we shall construct a code for all other cases.

Each p_i specifies an integer n_i such that

$$\frac{1}{2^{n_i}} \leq p_i < \frac{1}{2^{n_i - 1}} \tag{15-174}$$

where $p_i > 1/2^{n_i}$ for at least one p_i (assumption). With n_m the largest of the integers n_i, it follows from the above that

$$\sum_{i=1}^{N} \frac{1}{2^{n_i}} \leq 1 - \frac{1}{2^{n_m}} \tag{15-175}$$

because the left side is a binary integer smaller than one. If, therefore, n_m is changed to $n_m - 1$, the resulting value of the sum in (15-175) will not exceed one.

We continue the process of reducing the largest integer by one until we reach a set of integers l_i such that

$$\sum_{i=1}^{N} \frac{1}{2^{l_i}} = 1 \qquad l_i \leq n_i \tag{15-176}$$

With this set of integers we construct a code and we denote by L^a its average length. Thus

$$L^a = \sum_{i=1}^{N} p_i l_i \leq \sum_{i=1}^{N} p_i n_i$$

We maintain that

$$H(\mathfrak{S}) \leq L^a < H(\mathfrak{S}) + 1 \tag{15-177}$$

PROOF From (15-174) it follows that $n_i < -\log p_i + 1$. Multiplying by p_i and adding, we obtain

$$\sum_{i=1}^{N} p_i n_i < \sum_{i=1}^{N} p_i(-\log p_i + 1) = H(\mathfrak{S}) + 1$$

and (15-177) results [see (15-172)].

The Fano code We shall describe this code in terms of set dichotomies based on the following rule of subdivision. We number the probabilities p_i in descending order

$$p_1 \geq p_2 \geq \cdots \geq p_N \tag{15-178}$$

and we select the sets $\mathscr{A}_0$ and $\mathscr{A}_1$ of the first generation so as to have equal or nearly equal probabilities. To do so, we determine k such that

$$p_1 + \cdots + p_k \leq 0.5 \leq p_{k+1} + \cdots + p_N$$

and we set $\mathscr{A}_0$ equal to $\{\zeta_0, \ldots, \zeta_k\}$ or to $\{\zeta_0, \ldots, \zeta_{k+1}\}$. The same rule is used in all subsequent subdivisions. As we see in Example 15-25, the length L^b of the resulting code is close to the Shannon code length L^a.

We note that, since there is an ambiguity in the choice of the subsets in each dichotomy, the Fano code is not unique.

The Huffman code We denote by $\mathbf{x}_N^0$ the optimum N-element code and by L_N^0 its average length. We shall determine $\mathbf{x}_N^0$ using the following operation: We arrange the probabilities p_i of the elements ζ_i of $\mathscr{S}$ in descending order as in (15-178) and we number the corresponding elements ζ_i accordingly. We then replace the last two elements ζ_{N-1} and ζ_N with a new element and we assign to this element the probability $p_{N-1} + p_N$. A new source results with $N - 1$ elements. This operation will be called *Huffman contraction.*

In the table of Example 15-25, the new element is identified by a box in which the replaced elements are shown.

Rearranging the probabilities of the new source in descending order, we repeat the above operation until we reach a set with only two elements.

To each element ζ_i of the source $\mathscr{S}$ we shall assign a code word x_i starting from the last digit: We assign the numbers 0 and 1 respectively to the last digits of the code words of the elements ζ_{N-1} and ζ_N. At each subsequent contraction, we assign the numbers 0 and 1 to the *left* of the partially completed code words of all elements that are included in the last two boxes.

The code so formed (Huffman) will be denoted by $\mathbf{x}_N^c$ and its average length by L_N^c. We shall show that this code is optimal.

PROOF The proof of the optimality is based on the following observation. We can readily see that the last two code words x_{N-1} and x_N have the same length l_r. In Example 15-25,

$$N = 9 \qquad x_8 = 00000 \qquad x_9 = 00001 \qquad l_r = 5$$

If we replace these two words with a single word consisting of their common part, we obtain the Huffman code $\mathbf{x}_{N-1}^c$ for the set of $N-1$ elements and the code length of the new element equals l_{r-1}. This leads to the conclusion that

$$L_N^c - (p_{N-1} + p_N)l_r = L_{N-1}^c - (p_{N-1} + p_N)(l_r - 1)$$

Hence

$$L_N^c = L_{N-1}^c + p_{N-1} + p_N \tag{15-179}$$

In the example

$$L_9^c = \sum_{i=1}^{7} p_i l_i + 5p_8 + 5p_9 \qquad L_8^c = \sum_{i=1}^{7} p_i l_i + 4(p_8 + p_9)$$

Induction The Huffman code is optimum for $N = 2$ because there is only one code with two words. We assume that it is optimum for every source with $k \le N - 1$ elements and we shall show that it is optimum for $k = N$. Suppose that there is an N-element source $\mathscr{S}$ for which this is not true, i.e., suppose that

$$L_N^0 < L_N^c \tag{15-180}$$

As we know, the two elements ζ_{N-1} and ζ_N with the smallest probabilities are in the last generation branches of the optimum code tree. If they are removed, the contracted tree specifies a new code with length L_{N-1}. Reasoning as in (15-179) we conclude with (15-180) that

$$L_{N-1} + p_{N-1} + p_N = L_N^0 < L_N^c = L_{N-1}^c + p_{N-1} + p_N$$

hence, $L_{N-1} < L_{N-1}^c$. But this is impossible because the Huffman code of order $N-1$ is optimum by assumption.

Example 15-25 We shall describe the above codes using as source a set $\mathscr{S}$ with 9 elements. Their probabilities are shown in the table below:

i	1	2	3	4	5	6	7	8	9
p_i	0.22	0.19	0.15	0.12	0.08	0.07	0.07	0.06	0.04

The resulting entropy equals

$$H(\mathfrak{S}) = -\sum_{i=1}^{9} p_i \log p_i = 2.703$$

Arbitrary code We form a code using a chain of dichotomies chosen arbitrarily as in Fig. 15-19. In the table below we show the code words and their lengths.

i	1	2	3	4	5	6	7	8	9	
x_i	000	0010	0011	010	011	10	1100	1101	111	$L = \sum_{i=1}^{9} p_i l_i = 3.40$
l_i	3	4	4	3	3	2	4	4	3	

Shannon code In the table below we show the integers n_i determined from (15-174) and the required reductions until the final lengths l_i are reached. The corresponding code tree is shown in Fig. 15-20.

p_i	0.22	0.19	0.15	0.12	0.08	0.07	0.07	0.06	0.04	
	$\frac{1}{2^3} \le p_i < \frac{1}{2^2}$			$\frac{1}{2^4} \le p_i < \frac{1}{2^3}$				$\frac{1}{2^5} \le p_i < \frac{1}{2^4}$		$\sum_{i=1}^{N} \frac{1}{2^{n_i}}$
n_i	3	3	3	4	4	4	4	5	5	12/16
	3	3	3	3	3	4	4	4	4	14/16
l_i	3	3	3	3	3	3	3	4	4	1
x_i	000	001	010	011	100	101	110	1110	1111	$L^a = 3.1$

Fano code In the table below we show the subsets obtained with the Fano dichotomies, and their probabilities. The last generation sets are the elements ζ_i of $\mathscr{S}$; their probabilities are shown on the first row of the table. The dichotomies start with

$$\mathscr{A}_0 = \{\zeta_1, \zeta_2, \zeta_3\} \qquad P(\mathscr{A}_0) = 0.22 + 0.19 + 0.15 = 0.56$$

p_i	0.22	0.19	0.15	0.12	0.08	0.07	0.07	0.06	0.04	
	$\mathscr{A}_0$		0.56		$\mathscr{A}_1$			0.46		
	$\mathscr{A}_{00}$	$\mathscr{A}_{01}$	0.34	$\mathscr{A}_{10}$	0.20		$\mathscr{A}_{11}$		0.24	
		$\mathscr{A}_{010}$	$\mathscr{A}_{011}$	$\mathscr{A}_{100}$	$\mathscr{A}_{101}$	$\mathscr{A}_{110}$	0.14	$\mathscr{A}_{111}$	0.10	
						$\mathscr{A}_{1100}$	$\mathscr{A}_{1101}$	$\mathscr{A}_{1110}$	$\mathscr{A}_{1111}$	
	ζ_1	ζ_2	ζ_3	ζ_4	ζ_5	ζ_6	ζ_7	ζ_8	ζ_9	
x_i	00	010	011	100	101	1100	1101	1110	1111	
l_i	2	3	3	3	3	4	4	4	4	$L^b = 3.02$

Optimum code In the table below we show the original set $\mathscr{S}_9$ consisting of 9 elements and the sets obtained with each Huffman contraction. The elements ζ_i are identified by their indices and the combined elements by boxes. Each box contains all elements ζ_i of the original source involved in each contraction, and the evolution of their code words x_i starting with the last digit. The rows below each $\mathscr{S}_i$ line show the probabilities of the various elements of $\mathscr{S}_i$. For example, the number 0.10 in the line below $\mathscr{S}_7$ is the probability of the box (element of $\mathscr{S}_7$) that contains the elements ζ_8 and ζ_9.

The column at the extreme right shows the sum of the two smallest probabilities of the elements in $\mathscr{S}_i$. This number is used to form the row $\mathscr{S}_{i+1}$ by reordering the elements of $\mathscr{S}_i$.

Evolution of Huffman code

$\mathscr{S}_9$	1	2	3	4	5	6	7	8	9	
$p_{i,9}$	0.22	0.19	0.15	0.12	0.08	0.07	0.07	0.06	0.04	0.10
$\mathscr{S}_8$	1	2	3	4	8	9	5	6	7	
					0	1				
$p_{i,8}$	0.22	0.19	0.15	0.12	0.10		0.08	0.07	0.07	0.14
$\mathscr{S}_7$	1	2	3	6	7	4	8	9	5	
				0	1		0	1		
$p_{i,7}$	0.22	0.19	0.15	0.14		0.12	0.10		0.08	0.18
$\mathscr{S}_6$	1	2	8	9	5	3	6	7	4	
			00	01	1		0	1		
$p_{i,6}$	0.22	0.19		0.18			0.14		0.12	0.26
$\mathscr{S}_5$	6	7	4	1	2	8	9	5	3	
	00	01	1			00	01	1		
$p_{i,5}$		0.26		0.22	0.19		0.18		0.15	0.33
$\mathscr{S}_4$	8	9	5	3	6	7	4	1	2	
	000	001	01	1	100	01	1			
$p_{i,4}$		0.33				0.26		0.22	0.19	0.41
$\mathscr{S}_3$	1	2	8	9	5	3	6	7	4	
	0	1	000	001	01	1	00	01	1	
$p_{i,3}$	0.41			0.33				0.26		0.59
$\mathscr{S}_2$	8	9	5	3	6	7	4	1	2	
	0000	0001	001	01	100	101	11	0	1	
$p_{i,2}$				0.59				0.41		1
$\mathscr{S}_1$	8	9	5	3	6	7	4	1	2	
	00000	00001	0001	001	0100	0101	011	10	11	

The completed code words x_i taken from the last line of the table and their code lengths l_i are listed below.

	1	2	3	4	5	6	7	8	9
x_i	10	11	001	011	0001	0100	0101	00000	00001
l_i	2	2	3	3	4	4	4	5	5

$L^0 = 3.01$

The Shannon Coding Theorem

In the earlier discussion, we considered only codes of the elements ζ_i of a set $\mathscr{S}$ and we showed that the optimum code is between $H(\mathfrak{S})$ and $H(\mathfrak{S}) + 1$

$$H(\mathfrak{S}) \le L^0 \le H(\mathfrak{S}) + 1 \qquad (15\text{-}181)$$

This follows from (15-172) and (15-177). We show next that if we encode not merely single words but entire messages, then the code length per word can be reduced to less than $H(\mathfrak{S}) + \varepsilon$ for any $\varepsilon > 0$.

A message of length n is any element of the product space $\mathscr{S}^n$. The number of such messages is N^n and a code of the space $\mathscr{S}^n$ is a correspondent between its elements and a set of N^n binary numbers. This correspondence defines the RV $\mathbf{x}_n$ (random code) on the space $\mathscr{S}^n$ and the lengths of the code words form another RV $\mathbf{L}_n$ (random code length). The expected value L_n of $\mathbf{L}_n$ is the average code length. From the definition it follows that L_n is the average number of digits required to encode the elements of $\mathscr{S}^n$. The ratio

$$\bar{L} = \frac{L_n}{n} \qquad (15\text{-}182)$$

is the *average code length* per word. The term "word" means, of course, an element of $\mathscr{S}$.

We shall assume that $\mathscr{S}^n$ is the space of n *independent* trials.

Theorem We can construct a code of the space $\mathscr{S}^n$ such that

$$H(\mathfrak{S}) \le \bar{L} \le H(\mathfrak{S}) + \frac{1}{n} \qquad (15\text{-}183)$$

PROOF We shall give two proofs. The first is a direct consequence of (15-181). The second is based on the concept of typical sequences.

1. Applying the earlier results to the source $\mathscr{S}^n$, we construct a code L_n such that

$$H(\mathfrak{S}^n) \le L_n < H(\mathfrak{S}^n) + 1 \qquad (15\text{-}184)$$

This yields (15-183) because $L_n = n\bar{L}$ and $H(\mathfrak{S}^n) = nH(\mathfrak{S})$ [see (15-67)].

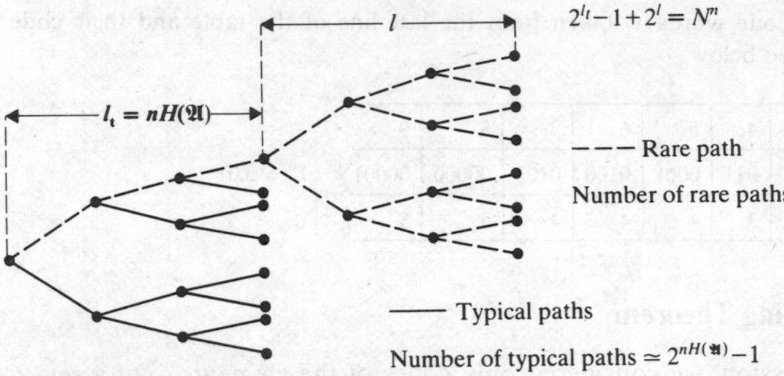

Figure 15-21

2. As we know the space $\mathscr{S}^n$ can be divided into two sets: the set $\mathbb{T}$ of all typical sequences and the set $\bar{\mathbb{T}}$ of all rare sequences. To prove (15-183), we construct a code tree consisting of $2^{nH(\mathfrak{S})} - 1$ short paths of length $l_t = nH(\mathfrak{S})$ and 2^l paths of length $l_t + l$. The short paths are used as the code words of the typical sequencies and the long paths for the long sequences (Fig. 15-21). Since $P(\mathbb{T}) \simeq 1$ and $P(\bar{\mathbb{T}}) \simeq 0$, we conclude that the average length of the resulting code equals

$$L_n = l_t P(\mathbb{T}) + (l + l_t)P(\bar{\mathbb{T}}) \simeq l_t = nH(\mathfrak{S})$$

Thus, $\bar{L} \simeq H(\mathfrak{S})$ and (15-183) results.

We note that (15-184) holds even if the trials are not independent. In this case, the theorem is true if $H(\mathfrak{S})$ is replaced by $H(\mathfrak{S}^n)/n$.

15-6 CHANNEL CAPACITY

We wish to transmit a message from point A to point B by means of a communications channel (a telephone cable, for example). The message to be transmitted is a stationary process $\mathbf{x}_n$ generating at the receiving end another process $\mathbf{y}_n$. The output $\mathbf{y}_n$ depends not only on the input $\mathbf{x}_n$ but also on the nature of the channel. Our objective is to determine the maximum rate of information that can be transmitted through the channel. To simplify the discussion, we make the following assumptions:

1. The channel is *binary*, i.e., the input $\mathbf{x}_n$ and the output $\mathbf{y}_n$ take only the values 0 and 1.
2. The channel is *memoryless*, i.e., the present value of $\mathbf{y}_n$ depend only on the present value of $\mathbf{x}_n$.
3. The input $\mathbf{x}_n$ is *strictly white noise*.

From 2 and 3 it follows that $\mathbf{y}_n$ is also white noise.

4. The messages are transmitted at the rate of *one word per second*.

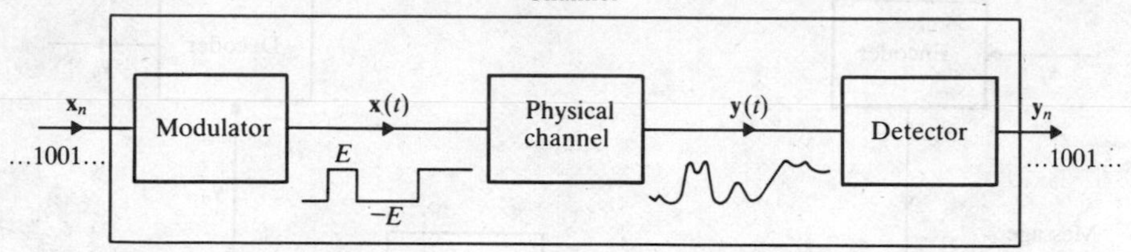

Figure 15-22

This is a mere normalization stating that the duration T of each transmitted state equals one second.

Example 15-26 In Fig. 15-22 we show a simple realization of a channel as a system with input $\mathbf{x}_n$ and output $\mathbf{y}_n$. The input to the physical channel is a time signal $\mathbf{x}(t)$ taking the values E and $-E$ (binary transmission). These values correspond to the two states 1 and 0 of $\mathbf{x}_n$. The received signal $\mathbf{y}(t)$ is a distorted version of $\mathbf{x}(t)$ contaminated possibly by noise. The system output $\mathbf{y}_n$ is obtained by some decision rule (detector) translating the time signal $\mathbf{y}(t)$ into a discrete-time signal consisting of 0s and 1s.

Noiseless Channel

We shall say that a channel is noiseless† if there is a one-to-one correspondence between the input $\mathbf{x}_n$ and the output $\mathbf{y}_n$. For a binary channel this means that if $\mathbf{x}_n = 0$, then $\mathbf{y}_n = 0$; if $\mathbf{x}_n = 1$, then $\mathbf{y}_n = 1$.

In a given channel, the uncertainty per transmitted word equals the entropy rate $\bar{H}(\mathbf{x}) = H(\mathbf{x})$ of the input $\mathbf{x}_n$. If the channel is noiseless, then the observed output $\mathbf{y}_n$ determines $\mathbf{x}_n$ uniquely, hence, it removes this uncertainty. Thus, the rate of transmitted information equals $H(\mathbf{x})$.

Definition of channel capacity The maximum value of $H(\mathbf{x})$, as $\mathbf{x}$ ranges over all possible inputs, is denoted by C and is called the *channel capacity*

$$C = \max_{\mathbf{x}_n} H(\mathbf{x}) \tag{15-185}$$

It appears that C does not depend on the channel but that is not so because the channel determines the number of the input states. If it is binary, then $\mathbf{x}_n$ has two possible states, hence

$$H(\mathbf{x}) = -p \log p - (1 - p) \log (1 - p) = r(p) \tag{15-186}$$

where $r(p)$ is the function of Fig. 15-2. Since $r(p)$ is maximum for $p = 0.5$ and $r(0.5) = 1$, we conclude that the capacity of a binary noiseless channel equals 1 bit/s.

† This definition does not lead to any conclusion about the actual presence of noise in the channel.

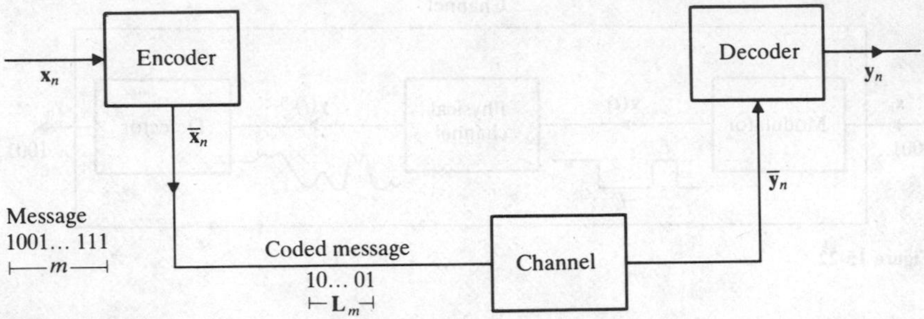

Figure 15-23

Similarly, if the channel accepts N input states, then its capacity equals $\log N$ bits/s.

Rate of information transmission We repeat: The channel transmits messages at the rate of 1 word/s. It transmits information at the rate of $H(\mathbf{x})$ bits/s. This rate depends on the source and it is maximum if the two states of the source are equally likely.

Theorem The maximum rate of 1 bit/s can be reached even if the input $\mathbf{x}_n$ is arbitrary provided that it is properly encoded prior to transmission.

PROOF 1 An m-word message is a binary number with m digits. There are 2^m such messages forming the space $\mathscr{S}_x^m$ and every realization of the input $\mathbf{x}_n$ is a sequence of such messages. We encode optimally the space $\mathscr{S}_x^m$ into a set of binary numbers $\bar{\mathbf{x}}_n$ using the techniques of the last section (Fig. 15-23). The number of digits (code length) of each $\bar{\mathbf{x}}_n$ is an RV $\mathbf{L}_m$ with mean $L_m = E\{\mathbf{L}_m\}$. As we know

$$mH(\mathbf{x}) \le L_m < mH(\mathbf{x}) + 1 \tag{15-187}$$

Hence, $\bar{L}_m \simeq H(\mathbf{x})$ for large m. A code word $\bar{\mathbf{x}}_n$ requires $\mathbf{L}_m$ seconds to be transmitted because it consists of $\mathbf{L}_m$ binary digits. Hence, the average time required to transmit the m-word messages of $\mathbf{x}_n$ in code form equals $L_m \simeq mH(\mathbf{x})$ seconds. And since the information contained in each message equals $mH(\mathbf{x})$ bits, we conclude that the average rate of information transmission equals $mH(\mathbf{x})/mH(\mathbf{x}) = 1$ bit/s.

PROOF 2 We have 2^m messages of length m. In a direct transmission (not encoded), each message requires the same transmission time: m seconds. However, of all these messages, only $2^{mH(\mathbf{x})}$ are likely to occur (typical sequences). To reduce the time of transmission, we encode all typical sequences into words of length $l_t \simeq mH(\mathbf{x})$ as in Fig. 15-21. The rare sequences require longer codes; however, the probability of their occurrence is negligible. Hence, the average time of transmission of each message is reduced from m seconds to $mH(\mathbf{x})$ seconds.

Noisy Channel

Due to a variety of factors, a physical channel establishes not a functional but a statistical relationship between the input $\mathbf{x}_n$ and the output $\mathbf{y}_n$. For a binary channel, this relationship is completely specified in terms of the probabilities

$$P\{\mathbf{x}_n = 0\} = p \qquad P\{\mathbf{x}_n = 1\} = q$$

of the two states of the input, and the conditional probabilities

$$P\{\mathbf{y}_n = j \mid \mathbf{x}_n = i\} = \pi_{ij} \qquad i, j = 0, 1 \tag{15-188}$$

The probabilities of the output states are given by

$$P\{\mathbf{y}_n = 0\} = \pi_{00}\, p + \pi_{10}\, q \qquad P\{\mathbf{y}_n = 1\} = \pi_{01} p + \pi_{11} q \tag{15-189}$$

Definition A noisy channel is a random system establishing a statistical relationship between the input $\mathbf{x}_n$ and the output $\mathbf{y}_n$.

For a memoryless channel, this relationship is completely specified in terms of the *channel matrix* Π whose elements π_{ij} are the conditional probabilities between the input states and the output states. For a binary channel

$$\Pi = \begin{bmatrix} \pi_{00} & \pi_{01} \\ \pi_{10} & \pi_{11} \end{bmatrix} \quad \text{where} \quad \begin{matrix} \pi_{00} + \pi_{01} = 1 \\ \pi_{10} + \pi_{11} = 1 \end{matrix} \tag{15-190}$$

The channel is called *symmetrical* if $\pi_{10} = \pi_{01} = \beta$. In a symmetrical channel, $\pi_{00} = \pi_{11} = 1 - \beta$ and

$$\Pi = \begin{bmatrix} 1-\beta & \beta \\ \beta & 1-\beta \end{bmatrix} \tag{15-191}$$

Example 15-27 To give some idea of the nature of the channel matrix, we show in Fig. 15-24 a simple version of a symmetrical channel. The input $\mathbf{x}(t)$ is a time signal as in Example 15-26, and the resulting output $\mathbf{y}(t)$ is the sum

$$\mathbf{y}(t) = \mathbf{x}(t) + \mathbf{v}_n \qquad nT \le t < nT + T \tag{15-192}$$

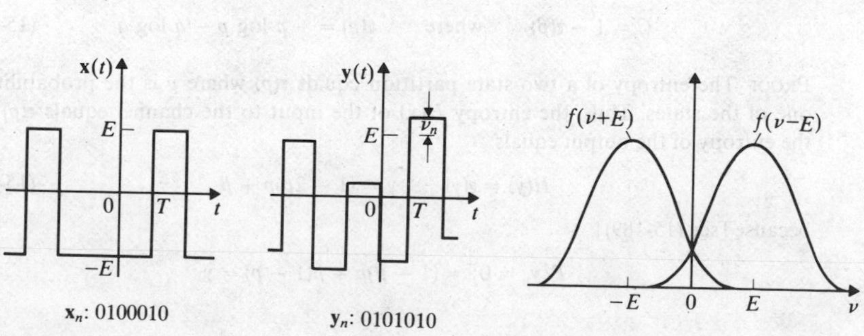

Figure 15-24

where $\mathbf{v}_n$ is a sequence of independent RVs with density the even function $f(v)$. The output states are determined as follows

$$\mathbf{y}_n = \begin{cases} 1 & \text{if} \quad \mathbf{y}(t) \geq 0 \\ 0 & \text{if} \quad \mathbf{y}(t) < 0 \end{cases}$$

From this we conclude that the channel is symmetrical and

$$\beta = P\{\mathbf{y}_n = 1 \mid \mathbf{x}_n = 0\} = \int_0^{\infty} f(v + E)\, dv = P\{\mathbf{v} > E\}$$

Channel capacity Prior to transmission, the uncertainty about the input $\mathbf{x}_n$ equals $H(\mathbf{x})$ per word. In a noiseless channel, the observed output $\mathbf{y}_n$ reduces the uncertainty to zero. This is not so, however, for a noisy channel because $\mathbf{y}_n$ does not determine $\mathbf{x}_n$ uniquely. Knowledge of $\mathbf{y}_n$ reduces the uncertainty about $\mathbf{x}_n$ from $H(\mathbf{x})$ to $H(\mathbf{x} \mid \mathbf{y})$ and the difference

$$I(\mathbf{x}, \mathbf{y}) = H(\mathbf{x}) - H(\mathbf{x} \mid \mathbf{y}) \tag{15-193}$$

is the *rate of information transmission*.†

If the channel is noiseless, then $H(\mathbf{x} \mid \mathbf{y}) = 0$, hence $I(\mathbf{x}, \mathbf{y}) = H(\mathbf{x})$. If the output $\mathbf{y}_n$ is independent of the input, then $H(\mathbf{x} \mid \mathbf{y}) = H(\mathbf{x})$, hence $I(\mathbf{x}, \mathbf{y}) = 0$. In other words, such a channel is useless (it does not transmit any information).

Definition The function $I(\mathbf{x}, \mathbf{y})$ depends on the matrix Π and on the input $\mathbf{x}_n$. The capacity C of a noisy channel is the maximum value of $I(\mathbf{x}, \mathbf{y})$ as $\mathbf{x}_n$ ranges over all possible inputs.

$$C = \max_{\mathbf{x}_n} I(\mathbf{x}, \mathbf{y}) \tag{15-194}$$

This is consistent with (15-185) because, for noiseless channels, $I(\mathbf{x}, \mathbf{y}) = H(\mathbf{x})$.

Example 15-28 We shall show that the capacity of *binary symmetrical channel* with channel matrix as in (15-191) (Fig. 15-25) equals

$$C = 1 - \mathfrak{r}(\beta) \qquad \text{where} \qquad \mathfrak{r}(p) = -p \log p - q \log q \tag{15-195}$$

PROOF The entropy of a two-state partition equals $\mathfrak{r}(p)$ where p is the probability of one of the states. Thus, the entropy $H(\mathbf{x})$ of the input to the channel equals $\mathfrak{r}(p)$ and the entropy of the output equals

$$H(\mathbf{y}) = \mathfrak{r}(\gamma) \qquad \gamma = (1 - 2\beta)p + \beta \tag{15-196}$$

because [see (15-189)]

$$P\{\mathbf{y}_n = 0\} = (1 - \beta)p + \beta(1 - p) = \gamma$$

† The conditional entropy $H(\mathbf{x} \mid \mathbf{y})$ is Shannon's *equivocation*.

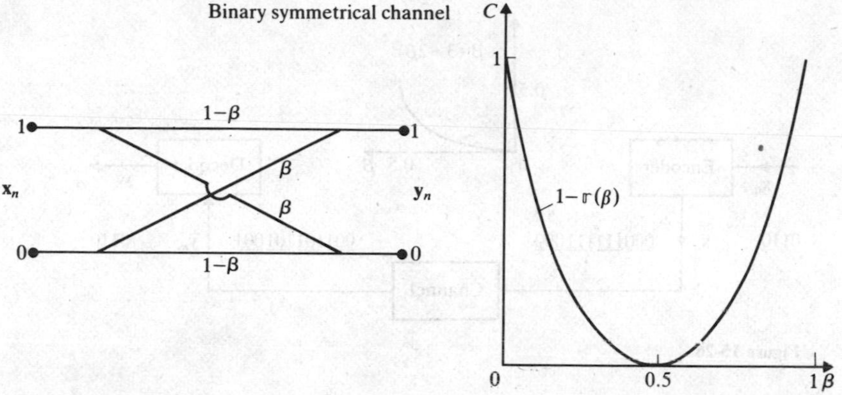

Figure 15-25

The above holds also for conditional entropies. Thus, since

$$P\{\mathbf{y}_n = 0 \mid \mathbf{x}_n = 0\} = P\{\mathbf{y}_n = 1 \mid \mathbf{x}_n = 1\} = 1 - \beta$$

we conclude that

$$H(\mathbf{y} \mid \mathbf{x}_n = 0) = H(\mathbf{y} \mid \mathbf{x}_n = 1) = \mathfrak{r}(1 - \beta)$$

Inserting into (15-41) and using the fact that $\mathfrak{r}(\beta) = \mathfrak{r}(1 - \beta)$, we obtain

$$H(\mathbf{x} \mid \mathbf{y}) = H(\mathbf{y} \mid \mathbf{x}) = p\mathfrak{r}(\beta) + q\mathfrak{r}(\beta) = \mathfrak{r}(\beta).$$

From the above it follows that $I(\mathbf{x}, \mathbf{y}) = \mathfrak{r}(\gamma) - \mathfrak{r}(\beta)$. This yields (15-195) because $\mathfrak{r}(\beta)$ does not depend on p and $\mathfrak{r}(\gamma)$ is maximum if $\gamma = 0.5$.

Redundant and random codes Consider a set $\mathscr{A}$ (source) with N elements and a set $\mathscr{B}$ (code) with M elements where $N < M$. A redundant code is a one-to-one correspondence between the elements of $\mathscr{A}$ and the elements of a subset $\mathscr{B}_1$ of $\mathscr{B}$.

The subset $\mathscr{B}_1$ consists of N elements that can be selected in many ways. If the elements of $\mathscr{B}_1$ are chosen at random from the M elements of $\mathscr{B}$, the resulting code is called *random*. From the definition it follows that the probability that a specific element of $\mathscr{B}$ is in the randomly selected set $\mathscr{B}_1$ equals N/M.

In the next example we show that redundant encoding can be used to reduce the probability of error in transmission.

Example 15-29 In a symmetrical channel, the probability of error equals β. To reduce this error, we encode the input set $\mathscr{A} = \{0, 1\}$ into the subset $\mathscr{B}_1 = \{000, 111\}$ of the set $\mathscr{B}$ of all three-digit binary numbers. In the earlier notation, $N = 2$ and $M = 8$.

The input $\mathbf{x}_n$ is thus encoded into a signal $\bar{\mathbf{x}}_n$ consisting of triplets of 0s and 1s yielding as output a signal $\bar{\mathbf{y}}_n$ (Fig. 15-26). The decoding scheme is the *majority rule*: If the received triplet consists of at least two 0s, then $\mathbf{y}_n = 0$, otherwise $\mathbf{y}_n = 1$.

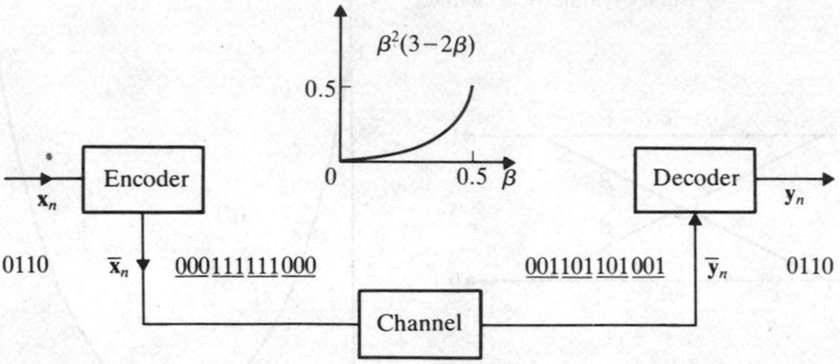

Figure 15-26

It can be readily seen that (Prob. 15-19) the probability that a transmitted word will be detected incorrectly equals $\beta^2(3 - 2\beta)$. This is less than β if $\beta < 0.5$. However, the rate of transmission is also reduced from 1 word per second to 1 word per three seconds.

It appears from the above that reduction of the probability of error by redundant encoding must result in transmission rates that tend to zero as the error tends to zero. This however, is not so. As the following remarkable theorem shows, it is possible to achieve arbitrarily small error probabilities while maintaining the rate of information transmission close to the channel capacity.

The Channel Capacity Theorem

Information can be transmitted through a noisy channel at a rate nearly equal to the channel capacity C with negligible probability of error.

PROOF *Preliminary remarks* From the definition of channel capacity it follows that the maximum of $H(\mathbf{x})$ is at least equal to C because

$$H(\mathbf{x}) = I(\mathbf{x}, \mathbf{y}) + H(\mathbf{x} \mid \mathbf{y}) \geq I(\mathbf{x}, \mathbf{y}) \tag{15-197}$$

This shows that we can find a source with entropy rate as close to C as we want. We shall show that if $\mathbf{x}_n$ is a source with entropy rate

$$H(\mathbf{x}) < C \tag{15-198}$$

then it can be transmitted at the rate of 1 word per second with probability of error less than α for any $\alpha > 0$. This will prove the theorem because the information per word equals $H(\mathbf{x})$.

As in the noiseless case, the proof is based on proper encoding of the space $\mathscr{S}_x^m$ consisting of all possible segments of $\mathbf{x}_n$ of length m. However, as the following remarks show, the objectives are different.

Noiseless channel The code set consists of two groups of binary numbers (Fig. 15-27a). The first group has 2^{m_1} elements of length $m_1 = mH(\mathbf{x})$ and it is used to encode the 2^{m_1} typical sequences of the input space $\mathscr{S}_x^m$. The second group is used to encode the rare sequences of $\mathscr{S}_x^m$. Since the set of all rare sequences has negligible probability, the average length of the code equals m_1.

Thus, in the noiseless case, the purpose of coding is reduction of the time of transmission of m-word messages from m seconds to m_1 seconds. This results in an increase of the rate of information transmission from $mH(\mathbf{x})$ bits per m seconds to $mH(\mathbf{x})$ bits per $m_1 = mH(\mathbf{x})$ seconds.

Noisy channel Reasoning as in (15-197) we conclude that, given $\varepsilon > 0$, we can find a process $\mathbf{z}_n$ such that

$$H(\mathbf{z}) - H(\mathbf{z} \mid \mathbf{y}) > C - \varepsilon \tag{15-199}$$

Choosing $\varepsilon < C - H(\mathbf{x})$, we obtain

$$H(\mathbf{z}) > H(\mathbf{x}) + H(\mathbf{z} \mid \mathbf{y}) \geq H(\mathbf{x}) \tag{15-200}$$

because $H(\mathbf{z} \mid \mathbf{y}) > 0$.

All sequences of $\mathbf{z}_n$ of length m form a space $\mathscr{S}_z^m$ consisting of 2^m elements. We can, therefore, encode the input set $\mathscr{S}_x^m$ into the set $\mathscr{S}_z^m$. The resulting code is one-to-one (Fig. 15-27*b*). The code can, however, be viewed as redundant if we consider only the mapping of the subset $\mathbb{T}(\mathbf{x}_n)$ of all typical sequences of $\mathscr{S}_x^m$ into the subset $\mathbb{T}(\mathbf{z}_m)$ of all typical sequences of $\mathscr{S}_z^m$. Indeed, $\mathbb{T}(\mathbf{x}_n)$ has $N = 2^{mH(\mathbf{x})}$ elements and $\mathbb{T}(\mathbf{z}_n)$ has $M = 2^{mH(\mathbf{z})}$ elements where

$$N = 2^{mH(\mathbf{x})} \ll 2^{mH(\mathbf{z})} = M \tag{15-201}$$

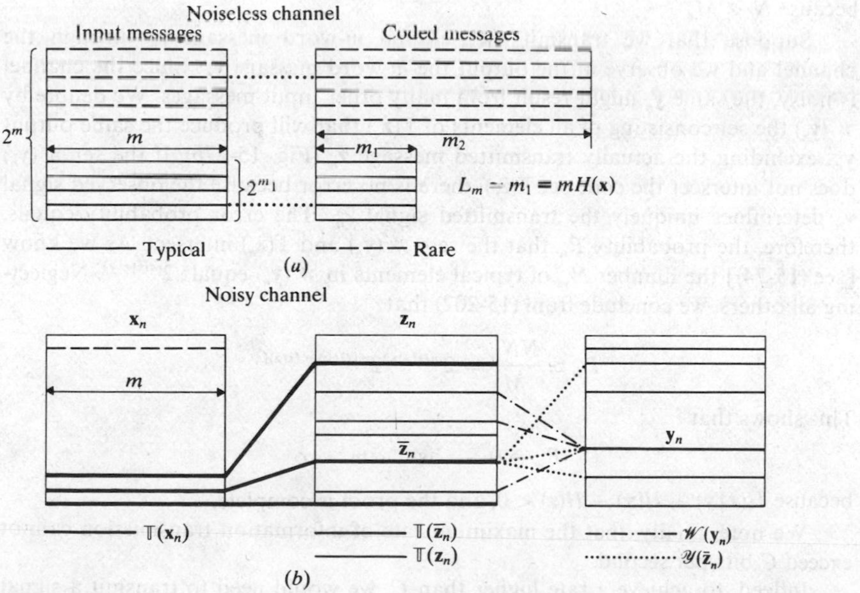

Figure 15-27

because $H(\mathbf{x}) < H(\mathbf{z})$ and $m \gg 1$. We denote by $\bar{\mathbf{z}}_n$ the code word of a typical $\mathbf{x}_n$ message and by $\mathbb{T}(\bar{\mathbf{z}}_n)$ the set of all such code words. Clearly, $\mathbb{T}(\bar{\mathbf{z}}_n)$ is a subset of the set $\mathbb{T}(\mathbf{z}_n)$ consisting of $N \ll M$ elements.

The purpose of the coding is to select the set $\mathbb{T}(\bar{\mathbf{z}}_n)$ such that its elements are at a "large distance" from each other in the following sense: Since the channel is noisy, the output due to a specific element $\bar{\mathbf{z}}_n$ is not unique. We denote by $\mathscr{Y}(\bar{\mathbf{z}}_n)$ the set of all output sequences due to this element, and we attempt to design the code such that the probability of the intersection of the output sets $\mathscr{Y}(\bar{\mathbf{z}}_n)$ as $\bar{\mathbf{z}}_n$ ranges over every element of the set $\mathbb{T}(\bar{\mathbf{z}}_n)$ is negligible. This will ensure the unique determination of $\bar{\mathbf{z}}_n$ in terms of the observed output $\mathbf{y}_n$.

Random code To complete the proof, we shall show that among all N-element subsets of the set $\mathbb{T}(\mathbf{z}_n)$ there exists at least one that meets our requirements. In fact, we shall prove a stronger statement: If we select *at random* N elements $\bar{\mathbf{z}}_n$ from the M elements of $\mathbb{T}(\mathbf{z}_n)$ and use the resulting set $\mathbb{T}(\bar{\mathbf{z}}_n)$ to encode the set $\mathbb{T}(\mathbf{x}_n)$ then, almost certainly, the probability of error in transmission will be negligible.

We note that, once the *code set* $\mathbb{T}(\bar{\mathbf{z}}_n)$ has been selected, the probability that an element of $\mathbb{T}(\mathbf{z}_n)$ is in $\mathbb{T}(\bar{\mathbf{z}}_n)$ equals N/M. From this it follows that, if $\mathscr{W}$ is a randomly selected subset of $\mathbb{T}(\mathbf{z}_n)$ consisting of N_w elements, then the probability P_w that it will intersect the set $\mathbb{T}(\bar{\mathbf{z}}_n)$ equals

$$P_w = 1 - \left(1 - \frac{N}{M}\right)^{N_w} \simeq \frac{NN_w}{M} \tag{15-202}$$

because $N \ll M$.

Suppose that we transmit the selected m-word message $\mathbf{z}_n$ through the channel and we observe at the output the m-word message $\mathbf{y}_n$. Since the channel is noisy, the same $\mathbf{y}_n$ might result from many other input messages. We denote by $\mathscr{W}(\mathbf{y}_n)$ the set consisting of all elements of $\mathbb{T}(\mathbf{z}_n)$ that will produce the same output $\mathbf{y}_n$, excluding the actually transmitted message $\mathbf{z}_n$ (Fig. 15-27*b*). If the set $\mathscr{W}(\mathbf{y}_n)$ does not intersect the code set $\mathbb{T}(\bar{\mathbf{z}}_n)$, there is no error because the observed signal $\mathbf{y}_n$ determines uniquely the transmitted signal $\bar{\mathbf{z}}_n$. The error probability equals, therefore, the probability P_w that the sets $\mathscr{W}(\mathbf{y}_n)$ and $\mathbb{T}(\bar{\mathbf{z}}_n)$ intersect. As we know [see (15-74)] the number N_w of typical elements in $\mathscr{W}(\mathbf{y}_n)$ equals $2^{mH(\mathbf{z}|\mathbf{y})}$. Neglecting all others, we conclude from (15-202) that

$$P_w \simeq \frac{NN_w}{M} = 2^{mH(\mathbf{z}|\mathbf{y})}2^{m[H(\mathbf{x}) - H(\mathbf{z})]}$$

This shows that

$$P_w \to 0 \qquad \text{as} \qquad m \to \infty$$

because $H(\mathbf{z}\,|\,\mathbf{y}) + H(\mathbf{x}) - H(\mathbf{z}) < 0$, and the proof is complete.

We note, finally, that the maximum rate of information transmission cannot exceed C bits per second.

Indeed, to achieve a rate higher than C, we would need to transmit a signal $\mathbf{z}_n$ such that $H(\mathbf{z}) - H(\mathbf{z}\,|\,\mathbf{y}) > C$. This, however, is impossible [see (15-194)].

PROBLEMS

15-1 Show that $H(\mathfrak{A} \cdot \mathfrak{B} \mid \mathfrak{B}) = H(\mathfrak{A} \mid \mathfrak{B})$

15-2 Show that if $\varphi(p) = -p \log p$ and $p_1 < p_1 + \varepsilon < p_2 - \varepsilon < p_2$, then

$$\varphi(p_1 + p_2) < \varphi(p_1) + \varphi(p_2) < \varphi(p_1 + \varepsilon) + \varphi(p_2 - \varepsilon)$$

15-3 In Fig. P15-3*a*, we give a schematic representation of the identities

$$H(\mathfrak{A} \cdot \mathfrak{B}) = H(\mathfrak{A}) + H(\mathfrak{B} \mid \mathfrak{A}) = H(\mathfrak{A}) + H(\mathfrak{B}) - I(\mathfrak{A} \cdot \mathfrak{B})$$

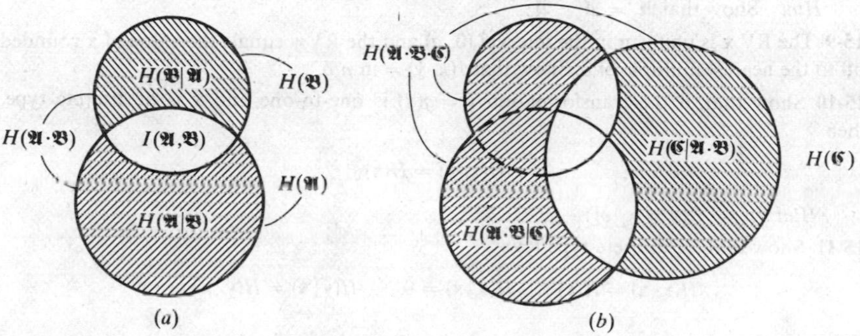

Figure P15-3

where each quantity equals the area of the corresponding region. Extending formally this representation to three partitions (Fig. P15-3*b*), we obtain the identities

$$H(\mathfrak{A} \cdot \mathfrak{B} \cdot \mathfrak{C}) = H(\mathfrak{A}) + H(\mathfrak{B} \cdot \mathfrak{C} \mid \mathfrak{A}) = H(\mathfrak{A} \cdot \mathfrak{B}) + H(\mathfrak{C} \mid \mathfrak{A} \cdot \mathfrak{B})$$

$$H(\mathfrak{A} \cdot \mathfrak{B} \cdot \mathfrak{C}) = H(\mathfrak{A}) + H(\mathfrak{B} \mid \mathfrak{A}) + H(\mathfrak{C} \mid \mathfrak{A} \cdot \mathfrak{B})$$

$$H(\mathfrak{B} \cdot \mathfrak{C} \mid \mathfrak{A}) = H(\mathfrak{B} \mid \mathfrak{A}) + H(\mathfrak{C} \mid \mathfrak{A} \cdot \mathfrak{B})$$

Show that these identities are correct.

15-4 Show that

$$I(\mathfrak{A} \cdot \mathfrak{B}, \mathfrak{C}) + I(\mathfrak{A}, \mathfrak{B}) = I(\mathfrak{A} \cdot \mathfrak{C}, \mathfrak{B}) + I(\mathfrak{A}, \mathfrak{C})$$

and identify each quantity in the representation of Fig. P15-3*b*.

15-5 The conditional mutual information of two partitions $\mathfrak{A}$ and $\mathfrak{B}$ assuming $\mathfrak{C}$ is by definition

$$I(\mathfrak{A}, \mathfrak{B} \mid \mathfrak{C}) = H(\mathfrak{A} \mid \mathfrak{C}) + H(\mathfrak{B} \mid \mathfrak{C}) - H(\mathfrak{A} \cdot \mathfrak{B} \mid \mathfrak{C})$$

(*a*) Show that

$$I(\mathfrak{A}, \mathfrak{B} \mid \mathfrak{C}) = I(\mathfrak{A}, \mathfrak{B} \cdot \mathfrak{C}) - I(\mathfrak{A}, \mathfrak{C}) \qquad \text{(i)}$$

and identify each quantity in the representation of Fig. P15-3*b*.

(*b*) From (i) it follows that $I(\mathfrak{A}, \mathfrak{B} \cdot \mathfrak{C}) \geq I(\mathfrak{A}, \mathfrak{C})$. Interpret this inequality in terms of the subjective notion of mutual information.

15-6 In an experiment $\mathscr{S}$, the entropy of the binary partition $\mathfrak{A} = [\mathscr{A}, \bar{\mathscr{A}}]$ equal $\mathfrak{r}(p)$ where $p = P(\mathscr{A})$. Show that in the experiment $\mathscr{S}^3 = \mathscr{S} \times \mathscr{S} \times \mathscr{S}$, the entropy of the eight-element partition $\mathfrak{A}^3 = \mathfrak{A} \cdot \mathfrak{A} \cdot \mathfrak{A}$ equals $3\mathfrak{r}(p)$ as in (15-67).

15-7 Show that

$$H(\mathbf{x} + a) = H(\mathbf{x}) \qquad H(\mathbf{x} + \mathbf{y} \mid \mathbf{x}) = H(\mathbf{y} \mid \mathbf{x})$$

In the above, $H(\mathbf{x} + a)$ is the entropy of the RV $\mathbf{x} + a$ and $H(\mathbf{x} + \mathbf{y} \mid \mathbf{x})$ is the conditional entropy of the RV $\mathbf{x} + \mathbf{y}$.

15-8 The RVs $\mathbf{x}$, $\mathbf{y}$ are of discrete type and independent. Show that if $\mathbf{z} = \mathbf{x} + \mathbf{y}$ and the line $x + y = z_i$ contains no more than one mass point, then

$$H(\mathbf{z} \mid \mathbf{x}) = H(\mathbf{y}) \leq H(\mathbf{z})$$

Hint: Show that $\mathfrak{A}_z = \mathfrak{A}_x \cdot \mathfrak{A}_y$.

15-9 The RV $\mathbf{x}$ is uniform in the interval $(0, a)$ and the RV $\mathbf{y}$ equals the value of $\mathbf{x}$ rounded off to the nearest multiple of δ. Show that $I(\mathbf{x}, \mathbf{y}) = \ln a/\delta$.

15-10 Show that, if the transformation $\mathbf{y} = g(\mathbf{x})$ is one-to-one and $\mathbf{x}$ is of discrete type, then

$$H(\mathbf{x}, \mathbf{y}) = H(\mathbf{x})$$

Hint: $p_{ij} = P\{\mathbf{x} = x_i\}\delta[i - j]$.

15-11 Show that for discrete-type RVs

$$H(\mathbf{x}, \mathbf{x}) = H(\mathbf{x}) \qquad H(\mathbf{x} \mid \mathbf{x}) = 0 \qquad H(\mathbf{y} \mid \mathbf{x}) = H(\mathbf{y}, \mathbf{x} \mid \mathbf{x})$$

$$H(\mathbf{y} \mid \mathbf{x}_1, \ldots, \mathbf{x}_n) = H\left(\mathbf{y}, \sum_{k=1}^{n} a_k \mathbf{x}_k \mid \mathbf{x}_1, \ldots, \mathbf{x}_n\right)$$

For continuous-type RVs, the relevant densities are singular. The above holds, however, if we set $H(\mathbf{x}, \mathbf{x}) = H(\mathbf{x})$ and use theorem (15-103) and its extensions to several variables to define recursively all conditional entropies.

15-12 The process $\mathbf{x}_n$ is normal white noise with $E\{\mathbf{x}_n^2\} = 5$, and

$$\mathbf{y}_n = \sum_{k=0}^{\infty} 2^{-k} \mathbf{x}_{n-k}$$

(*a*) Find the mutual information of the RVs $\mathbf{x}_n$ and $\mathbf{y}_n$.
(*b*) Find the entropy rate of the process $\mathbf{y}_n$.

15-13 The RVs $\mathbf{x}_n$ are independent and each is uniform in the interval (4, 6). Find the entropy rate of the process

$$\mathbf{y}_n = 5 \sum_{k=0}^{\infty} 2^{-k} \mathbf{x}_{n-k}$$

15-14 Show that the entropy rate of the output $\mathbf{y}_n$ of a minimum-phase system $\mathbf{L}(z)$ equals

$$\bar{H}(\mathbf{y}) = \bar{H}(\mathbf{x}) + K$$

where K is a constant independent of the input $\mathbf{x}_n$.

Hint: Assume first that $l_n = 0$ for $n > N$. Apply (15-115) to the $m + 1$ by $m + 1$ transformation

$$\mathbf{y}_{n-k} = \mathbf{x}_{n-k} l_0 + \cdots + \mathbf{x}_{n-k-N} l_N \qquad 0 \leq k \leq m - N$$

$$\mathbf{w}_{n-k} = \mathbf{x}_{n-k} l_0 + \cdots + \mathbf{x}_{n-m} l_{m-k} \qquad m - N \leq k \leq m$$

and show that

$$\bar{H}(\mathbf{y}) = \lim_{m \to \infty} \frac{1}{m+1} H(\mathbf{y}_n, \ldots, \mathbf{y}_{n-m+N}, \mathbf{w}_{n-m+N-1}, \ldots, \mathbf{w}_{n-m})$$

15-15(*a*) Show that the processes $\mathbf{x}[n]$ and $\mathbf{x}[-n]$ have the same entropy rate.

(*b*) Show that if $h[n]$ is the delta response of an all-pass system $\mathbf{H}(z)$ and

$$\mathbf{y}[n] = \mathbf{x}[n] * h[n] \qquad \text{then} \qquad \mathbf{y}[-n] = \mathbf{x}[-n] * h[n]$$

(see Fig. 13-13).

(*c*) Using the above, show that the entropy rate of the output of an all-pass system equals the entropy rate of the input. This agrees with (15-133) because $|\mathbf{H}(e^{j\omega T})| = 1$.

15-16 In the coin experiment, the probability of "heads" is an RV $\mathbf{p}$ with $E\{\mathbf{p}\} = 0.6$. Using the MEM, find its density $f(p)$.

15-17 (The Brandeis dice problem.†) In a die experiment, the average number of dots up equals 4.5. Using the MEM, find $p_i = P\{f_i\}$.

15-18 Using the MEM, find the joint density $f(x_1, x_2, x_3)$ of the RVs $\mathbf{x}_1, \mathbf{x}_2, \mathbf{x}_3$ if

$$E\{\mathbf{x}_1^2\} - E\{\mathbf{x}_2^2\} = E\{\mathbf{x}_3^2\} = 4 \qquad E\{\mathbf{x}_1\mathbf{x}_2\} = E\{\mathbf{x}_1\mathbf{x}_3\} = 1$$

15-19 A source has seven elements with probabilities

0.3 0.2 0.15 0.15 0.1 0.06 0.04

respectively. Construct a Shannon, a Fano, and a Huffman code and find their average code lengths.

15-20 Show that in the redundant coding of Example 15-29, the probability of error equals $\beta^2(3 - 2\beta)$.

Hint: $P\{\mathbf{y}_n = 1 \mid \mathbf{x}_n = 0\} = \beta^3 + 3\beta^2(1 - \beta)$.

15-21 Find the channel capacity of a symmetrical binary channel if the received information is always wrong.

† E. T. Jaynes, Brandeis lectures, 1962.

BIBLIOGRAPHY

Roman and arabic numbers in parentheses indicate relevant book parts and chapters respectively.

Abramson, N. M.: *Information Theory and Coding*, McGraw-Hill Book Company, New York, 1963. **(15)**

Antoniou, A.: *Digital Filters: Analysis and Design*, McGraw-Hill Book Company, New York, 1979. **(9, 10)**

Ash, R.: *Information Theory*, Interscience Publishers, New York, 1965. **(15)**

Bharucha-Reid, A. T.: *Elements of the Theory of Markov Processes and their Applications*, McGraw-Hill Book Company, New York, 1960. **(12)**

Blackman, R. B., and J. W. Tukey: *The Measurement of Power Spectra*, Dover Publications, Inc., New York, 1959. **(14)**

Blanc-Lapierre, A., and R. Fortet: Théorie des Fonctions Aléatoires, Masson et Cie, Paris, 1953. **(II, III)**

Childers, D. G. ed.: *Modern Spectrum Analysis*, John Wiley and Sons, New York, 1978. **(14)**

Cooper, R. B.: *Introduction to Queueing Theory*, North Holland Publishing Company, New York, 1981. **(12)**

Cramér, H.: *Mathematical Methods of Statistics*, Princeton University Press, Princeton, N.J., 1946. **(I)**

Davenport, W. B., Jr., and W. L. Root: *An Introduction to the Theory of Random Signals and Noise*, McGraw-Hill Book Company, New York, 1958. **(II)**

Doob, J. L.: *Stochastic Processes*, John Wiley and Sons, New York, 1953. **(II, III)**

Feinstein, A.: *Foundations of Information Theory*, McGraw-Hill Book Company, New York, 1958. **(15)**

Feller, W.: *An Introduction to Probability Theory and its Applications*, John Wiley and Sons, New York, vol. I, 1957, vol. II, 1967. **(I)**

Franks, L. E.: *Signal Theory*, Prentice-Hall, Inc., Englewood Cliffs, N.J., 1979. **(II)**

Helstrom, C. W.: *Statistical Theory of Signal Detection*, 2d ed., Pergamon Press, New York, 1968. (**II**)

Jenkins, G. M. and D. G. Watts: *Spectral Analysis and its Applications*, Holden-Day, Inc., Publisher, San Francisco, 1968. (**14**)

Kleinrock, L: *Queueing Systems*, 2 vols., John Wiley and Sons, New York, 1975–1976. (**12**)

Laning, J. H. and R. H. Battin: *Random Processes in Automatic Control*, McGraw-Hill Book Company, New York, 1956. (**II**)

Lebedev, V. L.: "Random Processes in Electrical and Mechanical Systems," NSF and NASA Technical Translations (**9, 10**)

Nahi, N. E.: *Estimation Theory and Applications*, John Wiley and Sons, New York, 1969. (**13**)

Oppenheim, A. V. and R. W. Schafer: Digital Signal Processing, Prentice-Hall, Inc., Englewood Cliffs, N.J., 1975. (**10**)

Papoulis, A.: *The Fourier Integral and its Applications*, McGraw-Hill Book Company, New York, 1962. (**10**)

——— *Signal Analysis*, McGraw-Hill Book Company, New York, 1977. (**10, 13**)

——— *Circuits and Systems: A Modern Approach*, Holt Rinehart and Winston, New York, 1980. (**10**)

——— *Systems and Transforms with Applications in Optics*, Krieger Publishing Company, Malabar, FL, 1981. (**9**)

Parzen, E.: *Modern Probability Theory and its Applications*, John Wiley and Sons, New York, 1960. (**I**)

Proakis, J.: *Introduction to Digital Communications*, McGraw-Hill Book Company, New York, 1983. (**II**)

Schwartz, M.: *Computer-Communication Network Design and Analysis*, Prentice-Hall, Englewood Cliffs, N.J., 1977. (**12**)

Schwartz, M., and L. Shaw: *Signal Processing*, McGraw-Hill Book Company, New York, 1975. (**10, 13**)

Wainstein, L. A., and V. D. Zubakov: *Extraction of Signals from Noise*, translated from Russian, Prentice-Hall, Inc., Englewood Cliffs, N.J., 1962 (**13**)

Wiener, N.: *Extrapolation, Interpolation, and Smoothing of Stationary Time Series*, MIT Press, 1949. (**13**)

Woodward, P.: *Probability and Information Theory with Applications to Radar*, Pergamon Press, New York, 1953. (**II**)

Yaglom, A. M.: *Stationary Random Functions*, translated from Russian, Prentice-Hall, Englewood Cliffs, N.J., 1962. (**II**)

INDEX